HANDBOOK OF STATISTICS
VOLUME 24

Handbook of Statistics

VOLUME 24

General Editor
C.R. Rao

ELSEVIER
NORTH-HOLLAND

AMSTERDAM • BOSTON • HEIDELBERG • LONDON • NEW YORK • OXFORD
PARIS • SAN DIEGO • SAN FRANCISCO • SINGAPORE • SYDNEY • TOKYO

Data Mining and Data Visualization

Edited by

C.R. Rao
Center for Multivariate Analysis
Department of Statistics, The Pennsylvania State University
University Park, PA, USA

E.J. Wegman
Center for Computational Statistics
George Mason University
Fairfax, VA, USA

J.L. Solka
Naval Surface Warfare Center, DD
Dahlgren, VA, USA

ELSEVIER
NORTH-HOLLAND

2005

AMSTERDAM • BOSTON • HEIDELBERG • LONDON • NEW YORK • OXFORD
PARIS • SAN DIEGO • SAN FRANCISCO • SINGAPORE • SYDNEY • TOKYO

ELSEVIER B.V.	ELSEVIER Inc.	ELSEVIER Ltd	ELSEVIER Ltd
Radarweg 29	525 B Street, Suite 1900	The Boulevard, Langford Lane	84 Theobalds Road
P.O. Box 211, 1000 AE Amsterdam	San Diego, CA 92101-4495	Kidlington, Oxford OX5 1GB	London WC1X 8RR
The Netherlands	USA	UK	UK

© 2005 Elsevier B.V. All rights reserved

This work is protected under copyright by Elsevier B.V., and the following terms and conditions apply to its use:

Photocopying
Single photocopies of single chapters may be made for personal use as allowed by national copyright laws. Permission of the Publisher and payment of a fee is required for all other photocopying, including multiple or systematic copying, copying for advertising or promotional purposes, resale, and all forms of document delivery. Special rates are available for educational institutions that wish to make photocopies for non-profit educational classroom use.

Permissions may be sought directly from Elsevier's Rights Department in Oxford, UK: phone (+44) 1865 843830, fax (+44) 1865 853333, e-mail: permissions@elsevier.com. Requests may also be completed on-line via the Elsevier homepage (http://www.elsevier.com/locate/permissions).

In the USA, users may clear permissions and make payments through the Copyright Clearance Center, Inc., 222 Rosewood Drive, Danvers, MA 01923, USA; phone: (+1) (978) 7508400; fax: (+1) (978) 7504744, and in the UK through the Copyright Licensing Agency Rapid Clearance Service (CLARCS), 90 Tottenham Court Road, London W1P 0LP, UK; phone: (+44) 20 7631 5555; fax: (+44) 20 7631 5500. Other countries may have a local reprographic rights agency for payments.

Derivative Works
Tables of contents may be reproduced for internal circulation, but permission of the Publisher is required for external resale or distribution of such material. Permission of the Publisher is required for all other derivative works, including compilations and translations.

Electronic Storage or Usage
Permission of the Publisher is required to store or use electronically any material contained in this work, including any chapter or part of a chapter.

Except as outlined above, no part of this work may be reproduced, stored in a retrieval system or transmitted in any form or by any means, electronic, mechanical, photocopying, recording or otherwise, without prior written permission of the Publisher.
Address permissions requests to: Elsevier's Rights Department, at the fax and e-mail addresses noted above.

Notice
No responsibility is assumed by the Publisher for any injury and/or damage to persons or property as a matter of products liability, negligence or otherwise, or from any use or operation of any methods, products, instructions or ideas contained in the material herein. Because of rapid advances in the medical sciences, in particular, independent verification of diagnoses and drug dosages should be made.

First edition 2005

Library of Congress Cataloging in Publication Data
A catalog record from the Library of Congress.

British Library Cataloguing in Publication Data
A catalogue record is available from the British Library.

ISBN: 0-444-51141-5
ISSN: 0169-7161

⊗ The paper used in this publication meets the requirements of ANSI/NISO Z39.48-1992 (Permanence of Paper).
Printed in The Netherlands.

Preface

It has long been a philosophical theme that statisticians ought to be data centric as opposed to methodology centric. Throughout the history of the statistical discipline, the most innovative methodological advances have come when brilliant individuals have wrestled with new data structures. Inferential statistics, linear models, sequential analysis, nonparametric statistics, robust statistical methods, and exploratory data analysis have all come about by a focus on a puzzling new data structure. The computer revolution has brought forth a myriad of new data structures for researchers to contend with including massive datasets, high-dimensional datasets, opportunistically collected datasets, image data, text data, genomic and proteomic data, and a host of other data challenges that could not be dealt with without modern computing resources.

This volume presents a collection of chapters that focus on data; in our words, it is data-centric. Data mining and data visualization are both attempts to handle nonstandard statistical data, that is, data, which do not satisfy traditional assumptions of independence, stationarity, identically distribution, or parametric formulations. We believe it is desirable for statisticians to embrace such data and bring innovative perspectives to these emerging data types.

This volume is conceptually divided into three sections. The first focuses on aspects of data mining, the second on statistical and related analytical methods applicable to data mining, and the third on data visualization methods appropriate to data mining. In Chapter 1, Wegman and Solka present an overview of data mining including both statistical and computer science-based perspectives. We call attention to their description of the emerging field of massive streaming datasets. Kaufman and Michalski approach data mining from a machine learning perspective and emphasize computational intelligence and knowledge mining. Marchette describes exciting methods for mining computer security data with the important application to cybersecurity. Martinez turns our attention to mining of text data and some approaches to feature extraction from text data. Solka et al. also focuses on text mining applying these methods to cross corpus discovery. They describe methods and software for discovery subtle, but significant associations between two corpora covering disparate fields. Finally Duric et al. round out the data mining methods with a discussion of information hiding known as steganography.

The second section, on statistical methods and related methods applicable to data mining begins with Rao's description of methods applicable to dimension reduction and graphical representation. Hand presents an overview of methods of statistical pattern recognition, while Scott and Sain present an update of Scott's seminal 1992 book on multivariate density estimation. Hubert et al., in turn, describe the difficult problem of analytically determining multivariate outliers and their impact on robustness. Sutton describes recent developments in classification and regression trees, especially the concepts of bagging and boosting. Marchette et al. describe some new computationally effective classification tools, and, finally Said gives an overview of genetic algorithms.

The final section on data visualization begins with a description of rotations (grand-tour methods) for high-dimensional visualization by Buja et al. This is followed by Carr's description of templates and software for showing statistical summaries, perhaps the most novel of current approaches to visual data presentation. Wilhelm describes in depth a framework for interactive statistical graphics. Finally, Chen describes a computer scientist's approach to data visualization coupled with virtual reality.

The editors sincerely hope that this combination of philosophical approaches and technical descriptions stimulates, and perhaps even irritates, our readers to encourage them to think deeply and with innovation about these emerging data structures and develop even better approaches to enrich our discipline.

C.R. Rao
E.J. Wegman
J.L. Solka

Table of contents

Preface   v

Contributors   xiii

Ch. 1.   Statistical Data Mining   1
         Edward J. Wegman and Jeffrey L. Solka

1. Introduction   1
2. Computational complexity   2
3. The computer science roots of data mining   9
4. Data preparation   14
5. Databases   19
6. Statistical methods for data mining   21
7. Visual data mining   29
8. Streaming data   37
9. A final word   44
   Acknowledgements   44
   References   44

Ch. 2.   From Data Mining to Knowledge Mining   47
         Kenneth A. Kaufman and Ryszard S. Michalski

1. Introduction   47
2. Knowledge generation operators   49
3. Strong patterns vs. complete and consistent rules   60
4. Ruleset visualization via concept association graphs   62
5. Integration of knowledge generation operators   66
6. Summary   69
   Acknowledgements   70
   References   71

Ch. 3.   Mining Computer Security Data   77
         David J. Marchette

1. Introduction   77
2. Basic TCP/IP   78

3. The threat   84
4. Network monitoring   92
5. TCP sessions   97
6. Signatures versus anomalies   101
7. User profiling   102
8. Program profiling   104
9. Conclusions   107
   References   107

Ch. 4. **Data Mining of Text Files**   109
         Angel R. Martinez

1. Introduction and background   109
2. Natural language processing at the word and sentence level   110
3. Approaches beyond the word and sentence level   119
4. Summary   129
   References   130

Ch. 5. **Text Data Mining with Minimal Spanning Trees**   133
         Jeffrey L. Solka, Avory C. Bryant and Edward J. Wegman

   Introduction   133
1. Approach   133
2. Results   140
3. Conclusions   167
   Acknowledgements   168
   References   169

Ch. 6. **Information Hiding: Steganography and Steganalysis**   171
         Zoran Duric, Michael Jacobs and Sushil Jajodia

1. Introduction   171
2. Image formats   172
3. Steganography   174
4. Steganalysis   179
5. Relationship of steganography to watermarking   181
6. Literature survey   184
7. Conclusions   186
   References   186

Ch. 7. **Canonical Variate Analysis and Related Methods for Reduction of Dimensionality and Graphical Representation**   189
         C. Radhakrishna Rao

1. Introduction   189
2. Canonical coordinates   190
3. Principal component analysis   197

4. Two-way contingency tables (correspondence analysis) 201
  5. Discussion 209
     References 210

Ch. 8. Pattern Recognition 213
        David J. Hand

  1. Background 213
  2. Basics 214
  3. Practical classification rules 216
  4. Other issues 226
  5. Further reading 227
     References 227

Ch. 9. Multidimensional Density Estimation 229
        David W. Scott and Stephan R. Sain

  1. Introduction 229
  2. Classical density estimators 230
  3. Kernel estimators 239
  4. Mixture density estimation 248
  5. Visualization of densities 252
  6. Discussion 258
     References 258

Ch. 10. Multivariate Outlier Detection and Robustness 263
         Mia Hubert, Peter J. Rousseeuw and Stefan Van Aelst

  1. Introduction 263
  2. Multivariate location and scatter 264
  3. Multiple regression 272
  4. Multivariate regression 278
  5. Classification 282
  6. Principal component analysis 283
  7. Principal component regression 292
  8. Partial Least Squares Regression 296
  9. Some other multivariate frameworks 297
 10. Availability 297
     Acknowledgements 300
     References 300

Ch. 11. Classification and Regression Trees, Bagging, and Boosting 303
         Clifton D. Sutton

  1. Introduction 303
  2. Using CART to create a classification tree 306
  3. Using CART to create a regression tree 315
  4. Other issues pertaining to CART 316

5. Bagging 317
6. Boosting 323
   References 327

Ch. 12. Fast Algorithms for Classification Using Class Cover Catch Digraphs 331
         David J. Marchette, Edward J. Wegman and Carey E. Priebe

1. Introduction 331
2. Class cover catch digraphs 332
3. CCCD for classification 334
4. Cluster catch digraph 338
5. Fast algorithms 340
6. Further enhancements 343
7. Streaming data 344
8. Examples using the fast algorithms 346
9. Sloan Digital Sky Survey 351
10. Text processing 355
11. Discussion 357
    Acknowledgements 357
    References 358

Ch. 13. On Genetic Algorithms and their Applications 359
         Yasmin H. Said

1. Introduction 359
2. History 360
3. Genetic algorithms 361
4. Generalized penalty methods 372
5. Mathematical underpinnings 378
6. Techniques for attaining optimization 381
7. Closing remarks 386
   Acknowledgements 387
   References 387

Ch. 14. Computational Methods for High-Dimensional Rotations in Data Visualization 391
         Andreas Buja, Dianne Cook, Daniel Asimov and Catherine Hurley

1. Introduction 391
2. Tools for constructing plane and frame interpolations: orthonormal frames and planar rotations 397
3. Interpolating paths of planes 403
4. Interpolating paths of frames 406
5. Conclusions 411
   References 412

Ch. 15. Some Recent Graphics Templates and Software for Showing Statistical Summaries 415
         Daniel B. Carr

1. Introduction 415
2. Background for quantitative graphics design 417

3. The template for linked micromap (LM) plots 420
4. Dynamically conditioned choropleth maps 426
5. Self-similar coordinates plots 431
6. Closing remarks 434
   Acknowledgements 435
   References 435

Ch. 16. Interactive Statistical Graphics: the Paradigm of Linked Views 437
Adalbert Wilhelm

1. Graphics, statistics and the computer 437
2. The interactive paradigm 444
3. Data displays 446
4. Direct object manipulation 467
5. Selection 469
6. Interaction at the frame level 475
7. Interaction at the type level 476
8. Interactions at the model level 483
9. Interaction at sample population level 486
10. Indirect object manipulation 487
11. Internal linking structures 488
12. Querying 494
13. External linking structure 496
14. Linking frames 498
15. Linking types 499
16. Linking models 499
17. Linking sample populations 503
18. Visualization of linked highlighting 505
19. Visualization of grouping 508
20. Linking interrogation 508
21. Bi-directional linking in the trace plot 509
22. Multivariate graphical data analysis using linked low-dimensional views 509
23. Conditional probabilities 511
24. Detecting outliers 518
25. Clustering and classification 519
26. Geometric structure 520
27. Relationships 522
28. Conclusion 532
    Future work 533
    References 534

Ch. 17. Data Visualization and Virtual Reality 539
Jim X. Chen

1. Introduction 539
2. Computer graphics 539
3. Graphics software tools 543
4. Data visualization 547
5. Virtual reality 556

   6. Some examples of visualization using VR 560
      References 561

Colour Figures 565

Subject Index 609

Contents of Previous Volumes 619

Contributors

Asimov, Daniel, *Department of Mathematics, University of California, Berkeley, CA 94720 USA*; *e-mail: asimov@msri.org* (Ch. 14).

Bryant, Avory C., *Code B10, Naval Surface Warfare Center, DD, Dahlgren, VA 22448 USA*; *e-mail: bryantac@nswc.navy.mil* (Ch. 5).

Buja, Andreas, *The Wharton School, University of Pennsylvania, 471 Huntsman Hall, Philadelphia, PA 19104-6302 USA*; *e-mail: buja@wharton.upenn.edu* (Ch. 14).

Carr, Daniel B., *Department of Applied and Engineering Statistics, MS 4A7, George Mason University, 4400 University Drive, Fairfax, VA 22030-4444 USA*; *e-mail: dcarr@gmu.edu* (Ch. 15).

Chen, Jim X., *Department of Computer Science, MS 4A5, George Mason University, 4400 University Drive, Fairfax, VA 22030-4444 USA*; *e-mail: jchen@cs.gmu.edu* (Ch. 17).

Cook, Dianne, *Department of Statistics, Iowa State University, Ames, IA 50011 USA*; *e-mail: dicook@iastate.edu* (Ch. 14).

Duric, Zoran, *Center for Secure Information Systems, MS 4A4, George Mason University, 4400 University Drive, Fairfax, VA 22030-4444 USA*; *e-mail: zduric@cs.gmu.edu* (Ch. 6).

Hand, David J., *Department of Mathematics, The Huxley Building, Imperial College London, 180 Queen's Gate, London SW7 2BZ, UK*; *e-mail: d.j.hand@ic.ac.uk* (Ch. 8).

Hubert, Mia, *Department of Mathematics, Katholieke Universiteit Leuven, W. de Croylaan 54, B-3001 Leuven, Belgium*; *e-mail: mai.hubert@wis.kuleuven.ac.be* (Ch. 10).

Hurley, Catherine, *Mathematics Department, National University of Ireland, Maynooth Co., Kildare, Ireland*; *e-mail: churley@maths.may.ie* (Ch. 14).

Jacobs, Michael, *Center for Secure Information Systems, MS 4A4, George Mason University, 4400 University Drive, Fairfax, VA 22030-4444 USA*; *e-mail: mjacobs1@gmu.edu* (Ch. 6).

Jajodia, Sushil, *Center for Secure Information Systems, MS 4A4, George Mason University, 4400 University Drive, Fairfax, VA 22030-4444 USA*; *e-mail: jajodia@ise.gmu.edu* (Ch. 6).

Kaufman, Kenneth A., *School of Computational Sciences, George Mason University, 4400 University Drive, Fairfax, VA 22030-4444 USA*; (Ch. 2).

Marchette, David J., *Code B10, Naval Surface Warfare Center, DD, Dahlgren, VA 22448 USA*; *e-mail: marchettedj@nswc.navy.mil* (Chs. 3, 12).

Martinez, Angel R., *Aegis Metrics Coordinator, NAVSEA Dahlgren – N20P, 17320 Dahlgren Rd., Dahlgren, VA 22448-5100 USA*; e-mail: martinezar@nswc.navy.mil (Ch. 4).

Michalski, Ryszard, *School of Computational Sciences, George Mason University, 4400 University Drive, Fairfax, VA 22030-4444 USA*; e-mail: richard.michalski@gmail.com (Ch. 2).

Priebe, Carey E., *Department of Applied Mathematics and Statistics, Whitehead Hall Room 201, Whiting School of Engineering, Johns Hopkins University, Baltimore, MD 21218-2682 USA*; e-mail: cep@jhu.edu (Ch. 12).

Rao, C.R., *326 Thomas Building, Pennsylvania State University, University Park, PA 16802 USA*; e-mail: crr1@psu.edu (Ch. 7).

Rousseeuw, Peter J., *Department of Mathematics and Computer Science, University of Antwerp, Middleheimlaan 1, B-2020 Antwerpen, Belgium*; e-mail: peter.rousseeuw@ua.ac.be (Ch. 10).

Said, Yasmin, *School of Computational Sciences, George Mason University, 4400 University Drive, Fairfax, VA 22030-4444 USA*; e-mail: ysaid99@hotmail.com, ysaid@gmu.edu (Ch. 13).

Sain, Stephan R., *Department of Mathematics, University of Colorado at Denver, PO Box 173364, Denver, CO 80217-3364 USA*; e-mail: ssain@math.cudenver.edu (Ch. 9).

Scott, David W., *Department of Statistics, MS-138, Rice University, PO Box 1892, Houston, TX 77251-1892 USA*; e-mail: scottdw@rice.edu (Ch. 9).

Solka, Jeffrey L., *Code B10, Naval Surface Warfare Center, DD, Dahlgren, VA 22448 USA*; e-mail: solkajl@nasw.navy.mil (Chs. 1, 5).

Sutton, Clifton D., *Department of Applied and Engineering Statistics, MS 4A7, George Mason University, 4400 University Drive, Fairfax, VA 22030-4444 USA*; e-mail: csutton@gmu.edu (Ch. 11).

Van Aelst, Stefan, *Department of Applied Mathematics and Computer Science, Ghent University, Krijslaan 281 S9, B-9000 Ghent, Belgium*; e-mail: stefan.vanaelst@ughent.be (Ch. 10).

Wegman, Edward J., *Center for Computational Statistics, MS 4A7, George Mason University, 4400 University Drive, Fairfax, VA 22030-4444 USA*; e-mail: ewegman@gmu.edu, ewegman@galaxy.gmu.edu (Chs. 1, 5, 12).

Wilhelm, Adalbert, *School of Humanities and Social Sciences, International University Bremen, PO Box 750 561, D-28725 Bremen, Germany*; e-mail: a.wilhelm@iu-bremen.de (Ch. 16).

Statistical Data Mining

Edward J. Wegman and Jeffrey L. Solka

Abstract

This paper provides an overview of data mining methodologies. We have been careful during our exposition to focus on many of the recent breakthroughs that might not be known to the majority of the community. We have also attempted to provide a brief overview of some of the techniques that we feel are particularly beneficial to the data mining process. Our exposition runs the gambit from algorithmic complexity considerations, to data preparation, databases, pattern recognition, clustering, and the relationship between statistical pattern recognition and artificial neural systems.

Keywords: dimensionality reduction; algorithmic complexity; statistical pattern recognition; clustering; artificial neural systems

1. Introduction

The phrases 'data mining', and, in particular, 'statistical data mining', have been at once a pariah for statisticians and also a darling. For many classically trained statisticians, data mining has meant the abandonment of the probabilistic roots of statistical analysis. Indeed, this is exactly the case simply because the datasets to which data mining techniques are typically applied are opportunistically acquired and were meant originally for some other purpose, for example, administrative records or inventory control. These datasets are typically not collected according to widely accepted random sampling schemes and hence inferences to general situations from specific datasets are not valid in the usual statistical sense. Nonetheless, data mining techniques have proven their value in the marketplace. On the other hand, there has been considerable interest in the statistics community in recent years for approaches to analyzing this new data paradigm.

The landmark paper of Tukey (1962), entitled "The future of data analysis," and later in the book, *Exploratory Data Analysis*, John Tukey (1977) sets forth a new paradigm for statistical analysis. In contrast to what has come to be called confirmatory analysis in which a statistical model is assumed and inference is made on the parameters of that model, exploratory data analysis (EDA) is predicated on the fact that we do not necessarily know that model assumptions actually hold for data under investigation. Because the data may not conform to the assumptions of the confirmatory analysis, inferences made with invalid model assumptions are subject to (potentially gross) errors. The idea then is to explore the data to verify that the model assumptions actually hold for the data in hand. It is a very short leap of logic to use exploratory techniques to discover unanticipated structure in the data. With the rise of powerful personal computing, this more aggressive form of EDA has come into vogue. EDA is no longer used to simply verify underlying model assumptions, but also to uncover unanticipated structure in the data.

Within the last decade, computer scientists operating in the framework of databases and information systems have similarly come to the conclusion that a more powerful form of data analysis could be used to exploit data residing in databases. That work has been formulated as knowledge discovery in databases (KDD) and data mining. A landmark book in this area is (Fayyad et al., 1996). The convergence of EDA from the statistical community and KDD from the computer science community has given rise to a rich if somewhat tense collaboration widely recognized as data mining.

There are many definitions of data mining. The one we prefer was given in (Wegman, 2003). Data mining is an extension of exploratory data analysis and has basically the same goals, the discovery of unknown and unanticipated structure in the data. The chief distinction lies in the size and dimensionality of the data sets involved. Data mining, in general, deals with much more massive data sets for which highly interactive analysis is not fully feasible.

The implication of the distinction mentioned above is the sheer size and dimensionality of data sets. Because scalability to massive datasets is one of the cornerstones of the data mining activity, it is worthwhile for us to begin our discussion of statistical data mining with a discussion of the taxonomy of data set sizes and their implications for scalability and algorithmic complexity.

2. Computational complexity

2.1. Order of magnitude considerations

In Table 1 we present Peter Huber's taxonomy of data set complexity (Huber, 1992, 1994). This was expanded by Wegman (1995). This taxonomy provides a feel for the magnitudes of the datasets that one is often interested in analyzing. We present a descriptor for the dataset, the size of the dataset in bytes, and the appropriate storage mode for such a dataset. Originally, the statistics community was often concerned with datasets that resided in the Small to Medium classes. Numerous application areas such as computer security and text processing have helped push the community into analyzing datasets that reside in the Large or Huge classes. Ultimately we will all be interested in the analysis of datasets in the Massive or Supermassive class. Such monstrous datasets will necessitate the development of recursive methodologies and will require the use of those approaches that are usually associated with streaming data.

Table 1
Huber–Wegman taxonomy of data set sizes

Descriptor	Data set size in bytes	Storage mode
Tiny	10^2	Piece of paper
Small	10^4	A few pieces of paper
Medium	10^6	A floppy disk
Large	10^8	Hard disk
Huge	10^{10}	Multiple hard disks, e.g. RAID storage
Massive	10^{12}	Disk farms/tape storage silos
Supermassive	10^{15}	Distributed data centers

Table 2
Algorithmic complexity

Complexity	Algorithm
$O(r)$	Plot a scatterplot
$O(n)$	Calculate means, variances, kernel density estimates
$O(n \log(n))$	Calculate fast Fourier transforms
$O(nc)$	Calculate singular value decomposition of an rc matrix; solve a multiple linear regression
$O(nr), O(n^{3/2})$	Solve a clustering algorithm with $r \propto \text{sqrt}(n)$
$O(n^2)$	Solve a clustering algorithm with c fixed and small so that $r \propto n$

Let us consider a data matrix consisting of r rows and c columns. One can calculate the total number of entries n as $n = rc$. In the case of higher-dimensional data, we write $d = c$ and refer to the data as d-dimensional. There are numerous types of operations or algorithms that we would like to utilize as part of the data mining process. We would like to be able to plot our data as a scatterplot, to compute summary statistics for our data such as means and variances, to perform probability density estimation on our data set using the standard kernel density estimator approach, we might wish to apply the fast Fourier transform to our dataset, we would like to be able to obtain the singular value decomposition of a multi-dimensional data set in order to ascertain appropriate linear subspaces that capture the nature of the data, we may wish to perform a multiple linear regression on a multi-dimensional dataset, we may wish to applying a clustering methodology with r proportional to sqrt(n), or, finally, we might wish to solve a clustering algorithm with c fixed and small so that r is proportional to n. This list of algorithms/data analysis techniques is not exhaustive but the list does represent many of the tasks that one may need to do as part of the data mining process. In Table 2 we examine algorithmic complexity as a function of these various statistical/data mining algorithms.

Now it is interesting to match the computational requirements of these various algorithms against the various different dataset sizes that we discussed previously. This allows one to ascertain the necessary computational resources in order to have a hope of applying each of the particular algorithms to the various datasets. Table 3 details the number of operations (within a constant multiplicative factor) for algorithms of various computational complexities and various data set sizes.

Table 3
Number of operations for algorithms of various computational complexities and various data set sizes

n	$n^{1/2}$	n	$n \log(n)$	$n^{3/2}$	n^2
Tiny	10	10^2	2×10^2	10^3	10^4
Small	10^2	10^4	4×10^4	10^6	10^8
Medium	10^3	10^6	6×10^6	10^9	10^{12}
Large	10^4	10^8	8×10^8	10^{12}	10^{16}
Huge	10^5	10^{10}	10^{11}	10^{15}	10^{20}
Massive	10^6	10^{12}	1.2×10^{13}	10^{18}	10^{24}
Supermassive	$10^{7.5}$	10^{15}	1.5×10^{16}	$10^{22.5}$	10^{30}

2.2. Feasibility limits due to CPU performance

Next we would like to take these computational performance figures of merit and examine the possibility of executing them on some current hardware. It is difficult to know exactly which machine should be included in the list of current hardware since the machine performance capabilities are constantly changing as the push for faster CPU speeds continues. We will use 4 machines for our computational feasibility analysis. The first machine will be a 1 GHz Pentium IV (PIV) machine with a sustainable performance of 1 gigaflop. The second machine will consist of a hypothetical flat neighborhood network of 12 1.4 GHz Athlons with a sustainable performance of 24 gigaflops. The third machine will be a hypothetical flat neighborhood network of 64 0.7 GHz Athlons with a sustained performance of 64 gigaflops. The fourth and final machine will be a hypothetical massively distributed grid type architecture with a sustained processing speed of 1 teraflop. This list of machines runs the gambit from a relatively low end system to a state of the art high performance "super computer".

In Table 4 we present execution times for a hypothetical 1 GHz PIV machine with a sustained performance level of 1 gigaflop when applied to the various algorithm/dataset combinations. In Table 5 we present execution times for a hypothetical Flat Neighborhood Network of 12 1.4 GHz Athlons with a sustained performance of 24 gigaflops when applied to the various algorithm/dataset combinations. In Table 6 we present execution times for a hypothetical Flat Neighborhood Network of 64 0.7 GHz Athlons with a sustained performance of 64 gigaflops. And finally in Table 7 we present mini-

Table 4
Execution speed of the various algorithm/dataset combinations on a Pentium IV 1 GHz machine with 1 gigaflop performance assumed

n	$n^{1/2}$	n	$n \log(n)$	$n^{3/2}$	n^2
Tiny	10^{-8} s	10^{-7} s	2×10^{-7} s	10^{-6} s	10^{-5} s
Small	10^{-7} s	10^{-5} s	4×10^{-5} s	0.001 s	0.1 s
Medium	10^{-6} s	0.001 s	0.006 s	1.002 s	16.74 min
Large	10^{-5} s	0.1 s	0.78 s	16.74 min	115.7 days
Huge	10^{-4} s	10.02 s	1.668 min	11.57 days	3170 years

Table 5
Execution speed of the various algorithm/dataset combinations on a Flat Neighborhood Network of 12 1.4 GHz Athlons with a 24 gigaflop performance assumed

n	$n^{1/2}$	n	$n\log(n)$	$n^{3/2}$	n^2
Tiny	4.2×10^{-10} s	4.2×10^{-9} s	8.3×10^{-9} s	4.2×10^{-8} s	4.2×10^{-7} s
Small	4.2×10^{-9} s	4.2×10^{-7} s	1.7×10^{-6} s	4.2×10^{-5} s	4.2×10^{-3} s
Medium	4.2×10^{-8} s	4.2×10^{-5} s	2.5×10^{-4} s	4.2×10^{-2} s	42 s
Large	4.2×10^{-7} s	4.2×10^{-3} s	0.03 s	42 s	4.86 days
Huge	4.2×10^{-6} s	0.42 s	4.2 s	11.67 h	133.22 years

Table 6
Execution speed for the various algorithm/dataset combinations on a Flat Neighborhood Network of 64 700 MHz Athlons with a 64 gigaflop performance assumed

n	$n^{1/2}$	n	$n\log(n)$	$n^{3/2}$	n^2
Tiny	1.6×10^{-10} s	1.6×10^{-9} s	3.1×10^{-9} s	1.6×10^{-8} s	1.6×10^{-7} s
Small	1.6×10^{-9} s	1.6×10^{-7} s	6.3×10^{-7} s	1.6×10^{-5} s	41.6×10^{-3} s
Medium	1.6×10^{-8} s	1.6×10^{-5} s	9.4×10^{-5} s	1.6×10^{-2} s	16 s
Large	1.6×10^{-7} s	1.6×10^{-3} s	0.01 s	16 s	1.85 days
Huge	1.6×10^{-6} s	0.16 s	1.6 s	4.34 h	49.54 years

Table 7
Execution speed for the various algorithm/dataset combinations on a massively distributed grid type architecture with a 1 teraflop performance assumed

n	$n^{1/2}$	n	$n\log(n)$	$n^{3/2}$	n^2
Tiny	1×10^{-11} s	1×10^{-10} s	2×10^{-10} s	1×10^{-9} s	1×10^{-8} s
Small	1×10^{-10} s	1×10^{-8} s	4×10^{-8} s	1×10^{-6} s	1×10^{-4} s
Medium	1×10^{-9} s	1×10^{-6} s	6×10^{-6} s	1×10^{-3} s	1 s
Large	1×10^{-8} s	1×10^{-4} s	0.0008 s	1 s	2.78 h
Huge	1×10^{-7} s	0.01 s	0.1 s	16.67 min	3.17 years

mum execution times for a massively distributed grid type architecture with a sustained processing speed of 1 teraflop.

By way of comparison and to give a sense of scale, as of June, 2004, a NEC computer in the Earth Simulator Center in Japan achieved a speed of 35.860 teraflops in 2002 and has a theoretical maximum speed of 40.960 teraflops. The record in the United States is held Lawrence Livermore National Laboratory with a California Digital Corporation computer that achieved 19.940 teraflops in 2004 with a theoretical peak performance 22.038 teraflops. These records of course are highly dependent on specific computational tasks and highly optimized code and do not represent performance that would be achieved with ordinary algorithms and ordinary code. See http://www.top500.org/sublist/ for details.

Tables 4–7 give computation times as a function of the overall the data set size and the algorithmic complexity at least to a reasonable order of magnitude for a variety of computer configurations. This is essentially an updated version of (Wegman, 1995). What is perhaps of more interest for scalability issues is a discussion of what is com-

Table 8
Types of computers for interactive feasibility with a response time less than one second

n	$n^{1/2}$	n	$n \log(n)$	$n^{3/2}$	n^2
Tiny	PC	PC	PC	PC	PC
Small	PC	PC	PC	PC	PC
Medium	PC	PC	PC	FNN24	TFC
Large	PC	PC	PC	TFC	–
Huge	PC	FNN24	TFC	–	–

Table 9
Types of computers for computational feasibility with a response time less than one week

n	$n^{1/2}$	n	$n \log(n)$	$n^{3/2}$	n^2
Tiny	PC	PC	PC	PC	PC
Small	PC	PC	PC	PC	PC
Medium	PC	PC	PC	PC	PC
Large	PC	PC	PC	PC	FNN24
Huge	PC	PC	PC	FNN24	–

putationally feasible and what is feasible from an interactive computation perspective. We consider a procedure to be computationally feasible if the execution time is less than one week. Of course, this would imply that there is essentially no human interaction with the computation process once it has begun. A procedure is thought to be feasible in an interactive mode if the computation time is less than one second. In Table 8 we present needed computational resources for a response time less than 1 s, i.e. the resources needed for interactive feasibility. In Table 9 we present needed computational resources for a response time less than one week, that is resources needed for computational feasibility.

It is clear from Table 8 that interactive feasibility is possible for order $O(n)$ algorithms on relatively simple computational resources, but that for order $O(n^2)$ algorithms, feasibility begins to disappear already for medium datasets, i.e. a million bytes. Note that although polynomial-time algorithms are often regarded as feasible, the combination of dataset size and algorithmic complexity severely limit practical interactive feasibility. Algorithms such as a minimum volume ellipsoid for determining multivariate outliers are exponentially complex and are certainly out of the range of feasibility except possibly for tiny datasets.

Table 9 suggests that even for settings in which we can be satisfied with computational feasibility, by the time we consider order $O(n^2)$ algorithms and huge (10^{10}) datasets, we are already out of the realm of computational feasibility. Even with the World's fastest supercomputers mentioned above, they are capable of only scalar multiples of our hypothetical teraflop computers, and of course, it is highly unlikely that dedicated access for a full week would be give to such an algorithm, e.g. clustering. The message is clear. In order to be scalable to massive or larger datasets, data mining algorithms must be of order no more than $O(n \log(n))$ and preferably of order $O(n)$.

That having been said, although data mining is often justified on the basis of dealing with huge, massive or larger datasets, it is often the case that exemplars of data mining methodology are applied to medium or smaller datasets, i.e. falling in the range of more traditional EDA applications.

2.3. Feasibility limits due to file transfer performance

File transfer speed is one of the key issues in the feasibility of dealing with large datasets. In (Wegman, 1995), network transfer was thought to be a major factor blocking the dealing with massive datasets. With the innovation of gigabit networks and the National LambdaRail (see http://www.nlr.net/) the major infrastructure backbone for dealing with transport of massive datasets seems to be emerging. In Table 10 we present transfer rates associated with several of our datasets for various types of devices. The National LambdaRail technology promises a transfer rate of 10 gigabits per second which translates into a massive dataset moving over the backbone in 13.33 min. This is certainly feasible. Similarly, in a local computer with a 4 GHz clock, data can move from cache memory to CPU in 4.17 min. The difficult lies not in the two ends of the data transfer chain, but in the intermediate steps. Read times from a hard drive have improved considerably but read times even for the fastest disk arrays are still on the order of hours for massive datasets, and with local Ethernet operating at 100 megabits per second transfer of a massive dataset would take almost a whole day. This assumes there is no contention for resources on the fast Ethernet, which is not normally the case.

The conclusion is that with the most esoteric technology in place dealing with huge and massive datasets may become feasible in the next few years. However, with current end-to-end technology, datasets on the order of 10^8, i.e. what we have called large datasets are probably the upper limit that can be reasonably dealt with. It is notable that while current storage capabilities (hard drives) have improved according to Moore's Law at a rate faster than CPU improvements, their data transfer speed has not improved nearly as dramatically.

It is also worth noting that in Table 10, we assume a cache transfer according to a 4 GHz clock. However, typically data transfer from and to peripheral devices such as hard drives and NIC cards is often at a dramatically slower clock speed. For example, high-speed USB connections operate at 480 megabits per second although standard full speed USB connections operate at 12 megabits per second. This is in contrast to a theoretical LambdaRail transfer rate 10 gigabits per second. So clearly the bottlenecks

Table 10
Transfer rates for a variety of data transfer regimes

n	Standard Ethernet 10 Mb/s	Fast Ethernet 100 Mb/s	Cache transfer at 4 GHz	National LambdaRail at 10 Gb/s
	1.25×10^6 B/s	1.25×10^7 B/s	4×10^9 B/s	1.25×10^9 B/s
Tiny	8×10^{-5} s	8×10^{-6} s	2.5×10^{-8} s	8×10^{-8} s
Small	8×10^{-3} s	8×10^{-4} s	2.5×10^{-6} s	8×10^{-6} s
Medium	8×10^{-1} s	8×10^{-2} s	2.5×10^{-4} s	8×10^{-4} s
Large	1.3 min	8 s	2.5×10^{-2} s	8×10^{-2} s
Huge	2.22 h	13.33 min	2.50 s	8 s
Massive	9.26 days	22.22 h	4.17 min	13.33 min

2.4. Feasibility limits due to visual resolution

Perhaps the most interesting and common methodology for exploratory data analysis is based on graphical methods and data visualization. One immediately comes to the issue of how scalable are graphical methods. This question leads naturally to the issue of how acute is human vision. Consider the following thought experiment. Suppose by some immensely clever technique, we could map a multi-dimensional observation into a single pixel. The question of how much data could we see becomes how many pixels can we see?

The answer to this question of course depends on the ability of the eye to resolve small objects and the ability of the display device to display small objects. It is most convenient to think in terms of angular resolution. It is a matter of simple trigonometry to calculate the angle that a display device subtends based on the size of the display device and the distance of the viewer from the display devices. Several hypothetical scenarios are given in Table 11, i.e. watching a 19 inch monitor at 24 inches, watching a 25 inch television at 12 feet, watching a 15 foot home theater screen at 20 feet and perhaps most optimistic, being in an immersive (virtual reality) environment. In the latter scenario, the field of view is approximately 140°. One can test this empirically by spreading one's arms straight out parallel to the plane of the body and then slowly bringing them forward until they appear on one's peripheral vision.

Of course, once the view angle is determined, it need simply be divided by the (horizontal) angular resolution of the human eye to obtain the number of pixels across that can be seen. Experts differ on the angular resolution: Valyus (1962) claims 5 seconds of arc, Wegman (1995) puts the angular resolution at 3.6 minutes of arc, while Maar (1982) puts it at 4.38 minutes of arc. The Valyus figure seems to be based on the minimal size of an object that can be seen at all with the naked eye. Too small an object will not yield enough photons to stimulate the retinal cells. Whereas the Wegman and Maar figures are based on minimal angular resolution that can distinguish two adjacent features. This is more realistic for describing how many pixels can be resolved. Maar asserts that two objects must be separated by approximately nine foveal cones to be

Table 11
Resolvable number of pixels across screen for several viewing scenarios

	19 inch monitor at 24 inches	25 inch TV at 12 feet	15 foot screen at 20 feet	Immersion
Angle	39.005°	9.922°	41.112°	140°
5 seconds of arc resolution (Valyus)	28 084	7144	29 601	100 800
1 minute of arc resolution	2340	595	2467	8400
3.6 minutes of arc resolution (Wegman)	650	165	685	2333
4.38 minutes of arc resolution (Maar 1)	534	136	563	1918
0.486 minutes of arc/foveal cone (Maar 2)	4815	1225	5076	17 284

resolvable, thus $0.486 \times 9 = 4.38$. Because the angular separation between two foveal cones is approximately one minute of arc (2×0.486), we include this angle in Table 11.

The standard aspect ratio for computer monitors and standard NTSC television is 4 : 3, width to height. If we take the Wegman angular resolution in an immersive setting, i.e. 2333 pixels horizontal resolution, then the vertical resolution would be approximately 1750 pixels for a total of 4.08×10^6 resolvable pixels. Notice that taking the high definition TV aspect ratio of 16 : 9 would actually yield fewer resolvable pixels. Even if we took the most optimistic resolution of one minute of arc (implying each pixel falls on a single foveal cone) in an immersive setting, the horizontal resolution would be 8400 pixels and the vertical resolution would be 6300 pixels yielding the total number of resolvable pixels at 5.29×10^7 resolvable pixels. Thus as far as using graphical data mining techniques, it would seem that there is an insurmountable upper bound around 10^6 to 10^7 data points, i.e. somewhere between medium to large datasets. Interestingly enough, this coincides with the approximate number of cones in the retina. According to Osterberg (1935), there are approximately 6.4 million cones in the retina and somewhere around 110 to 120 million rods.

3. The computer science roots of data mining

As we have pointed out before, data mining in some sense flows from a conjunction of both computer science and statistical frameworks. The development of relational database theory and the structured query language (SQL) among information systems specialists allowed for logical queries to databases and the subsequent exploitation of knowledge in the databases. However, being limited to logical queries (and, or, not) was a handicap and the desire to exploit more numerical and even statistically oriented queries led to the early development of data mining. Simultaneously, the early exploitation of supercomputers for physical system modeling using partial differential equations had run its course and by the early 1990s, supercomputer manufacturers were looking for additional marketing outlets. The exploitation of large scale commercial databases was a natural application of supercomputer technology. So there was both an academic pull and a commercial push to develop data mining in the context of computer science. In later sections, we describe relational databases and SQL.

As discussed above, data mining is often defined in terms of approaches that can deal with large to massive data set sizes. An important implication of this definition is that analysis almost by definition has to be automated so that interactive approaches and approaches that exploit very complex algorithms are prohibited in a data mining framework.

3.1. Knowledge discovery in databases and data mining

Knowledge discovery in databases consists of identifying those patterns or models that meet the goals of the knowledge discovery process. So a knowledge discovery engine needs the ability to measure the validity of a discovered pattern, the utility of the pattern, the simplicity or complexity of the pattern and the novelty of the pattern. These "metrics" help to identify the degree of "interestingness" of a discovered pattern. Data mining itself can be defined as a step in the knowledge discovery process consisting of particular algorithms (methods) that under some acceptable objective, produces a particular enumeration of patterns (models) over the data. The knowledge discovery process can be defined as the process of using data mining methods (algorithms) to extract (identify) what is deemed knowledge according to the specifications of measures and thresholds, using a database along with any necessary preprocessing or transformations.

The steps in the data mining process are usually described as follows. First, an understanding of the application domain must be obtained including relevant prior domain knowledge, problem objectives, success criteria, current solutions, inventory resources, constants, terminology cost and benefits. The next step focuses on the creation of a target dataset. This step might involve an initial dataset collection, producing an adequate description of the data, verifying the data quality, and focusing on a subset of possible measured variables. Following this is data cleaning and preprocessing. In this step the data is denoised, outliers are removed from the data, missing field are handled, time sequence information is obtained, known trends are removed from the data, and any needed data integration is performed. Next data reduction and projection is performed. This step consists of feature subset selection, feature construction, feature discretization, and feature aggregation. Finally, one must identify the purpose of the data mining effort such as classification, segmentation, deviation detection, or link analysis. The appropriate data mining approaches must be identified and used to extract patterns or models from the data. These models then need to be interpreted and evaluated and finally the discovered data must be consolidated.

Figure 1 portrays an in-depth flow chart for the data mining process. It might be surprising to note that the vast majority of effort associated with this process is focused

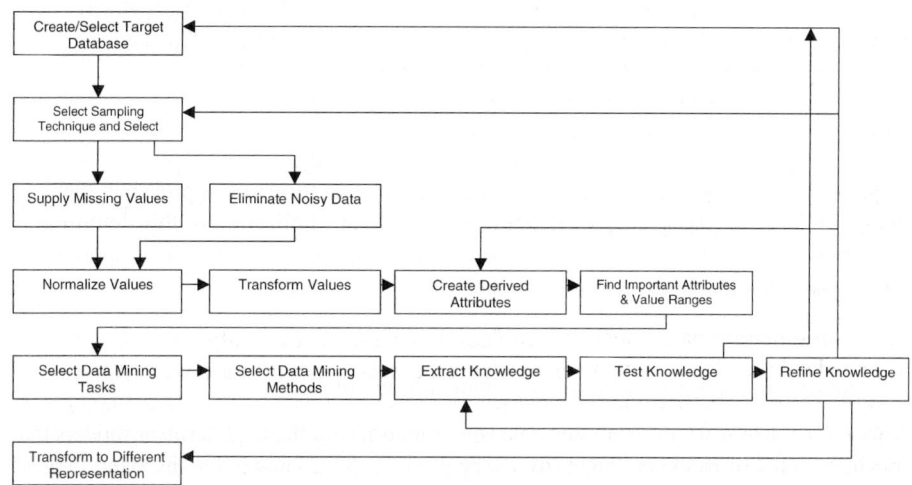

Fig. 1. An in-depth data mining flowchart.

on the data preparation portion of the process. The amount of effort associated with the data collection, cleaning, standardization, etc., can be somewhat daunting and actually far outweigh those steps associated with the rest of the data analysis.

Our current electronic age has allowed for the easy production of copious amounts of data that can be subject to analysis. Each time that an individual sets out on a trip he/she generates a paper trail of credit card receipts, hotel and flight reservation information, and cell phone call logs. If one also includes website utilization preferences, then one can begin to create a unique "electronic fingerprint" for the individual. Once the information is collected then it must be stored in some sort of data repository. Historical precedence indicates that data was initially stored and manipulated as flat files without any sort of associated database structure. There are some data analysis purists who might insist that this is still the best strategy. Currently data mining trends have focused on the use of relational databases as a convenient storage facility. These databases or data warehouses provide the data analyst with a ready supply of real material for use in knowledge discovery. The discovery of interesting patterns within the data provides evidence as to the utility of the collected data. In addition, a well thought-out data warehouse could provide a convenient framework for integration of the knowledge discovery process into the organization.

The application areas of interest to the data mining community have been driven by the business community. Market-based analysis which is an example of tool-based machine learning has been one of the prominent applications of interest. In this case, one can analyze either the customers, in order to discern what a customer purchases in order to provide insight into psychological motivations for purchasing strategies or insight into the products actually purchased. Product-based analysis is usually referred to as market basket analysis. This type of analysis gives insight into the merchandise by revealing those products that tend to be purchased together and those that are most amenable to purchase.

Some of the applications of these types of market-based analysis include focused mailing in direct/email marketing, fraud detection, warranty claims analysis, department store floor/shelf layout, catalog design, segmentation based on transaction patterns, and performance comparison between stores. Some of the questions that might be pertinent to floor/shelf layout include the following. Where should detergents be placed in the store in order to maximize their sales? Are window cleaning products purchased when detergents and orange juice are bought together? Is soda purchased with bananas? Does the brand of the soda make a difference? How are the demographics of the neighborhood affecting what customers are buying?

3.2. Association rules

The general market-based analysis problem may be mathematically stated as follows. Given a database of transactions in which each transaction contains a set of items, find all rules $X \rightarrow Y$ that correlate the presence of one set of items X with another set of items Y. One example of such a rule is when a customer buys bread and butter, they buy milk 85% of the time. Mined association rules can run the gambit from trivial, customers who purchase large appliances are likely to purchase maintenance agreements, to useful, on Friday afternoons convenience store customers often purchase diapers and beer

Table 12
Co-occurrence of products

	OJ	Window cleaner	Milk	Soda	Detergent
OJ	4	1	1	2	1
Window cleaner	1	2	1	1	0
Milk	1	1	1	0	0
Soda	2	1	0	3	1
Detergent	1	0	0	1	2

together, to the inexplicable, when a new super store opens, one of the most commonly sold items is light bulbs.

The creation of association rules often proceeds from the analysis of grocery point-of-sale transactions. Table 12 provides a hypothetical co-occurrence of products matrix.

A cursory examination of the co-occurrence table suggests some simple patterns that might be resident in the data. First we note that orange juice and soda are more likely to be purchases together than any other two items. Next we note that detergent is never purchased with window cleaner or milk. Finally we note that milk is never purchased with soda or detergent. These simple observations are examples of associations and may suggest a formal rule like: if a customer purchases soda, then the customer does not purchase milk.

In the data, two of the five transactions include both soda and orange juice. These two transactions *support* the rule. The support for the rule is two out of five or 40%. In general, of course, data subject to data mining algorithms is usually collected for some other administrative purpose other than market research. Consequently, these data are not considered as a random sample and probability statements and confirmatory inference procedures in the usual statistical sense may not be associated with such data.

This caveat aside, it is useful to understand what might be the probabilistic underpinnings of the association rules. The support of a product corresponds to the unconditional probability, $P(A)$, that a product is purchased. The support for a pair of products corresponds to the unconditional probability, $P(A \cap B)$, that both occur simultaneously.

Because two of the three transactions that contain soda also contain orange juice, there is 67% confidence in the rule 'If soda, then orange juice.' The confidence corresponds to the conditional probability $P(A \mid B) = P(A \cap B)/P(B)$.

Typically, the data miner would require a rule to have some minimum user-specified confidence. Rule 1 & 2 → 3 has a 90% confidence if when a customer bought 1 and 2, in 90% of the cases, the customer also bought 3. A rule must have some minimum user-specified support. By this we mean that the rule 1 & 2 → 3 should hold in some minimum percentage of transactions to have value.

Consider a simple transaction in Table 13. The simple rule 1 → 3 has a minimum support of 50% or 2 transactions and a minimum confidence of 50%. Table 14 provides a simple frequency count for each of the items. The rule 1 → 3 has a support of 50% and a confidence given by Support({1, 3})/Support({1}) = 66%.

Table 13
Simple transaction table

Transaction ID number	Items
1	{1, 2, 3}
2	{1, 3}
3	{1, 4}
4	{2, 5, 6}

Table 14
Simple item frequency table

Frequency of item set	Support
{1}	75%
{2}	50%
{3}	50%
{4}	25%

With this sort of tool in hand one can proceed forward with the identification of interesting associations between various products. For example, one might search for all of the rules that contain "Diet coke" as a result. The results of this analysis might help the store better boost the sales of Diet coke. Alternatively one might wish to find all rules that have "Yogurt" in the condition. These rules may help determining what products may be impacted if the store discontinues selling "Yogurt". As another example one might wish to find all rules that have "Brats" in the condition and "mustard" in the result. These rules may help in determining the additional items that have to be sold together to make it highly likely that mustard will also be sold. Sometimes one may wish qualify their analysis to identify the top k rules. For example, one might wish to find the top k rules that contain the word "Yogurt" in the result. Figure 2 presents a tree-based representation of item associations from the specific level to the general level.

Association rules may take various forms. They can be quantitative in nature. A good example of a quantitative rule is "Age[35,40] and Married[Yes] $\rightarrow$ NumCars[2]." Association rules may involve constraints like "Find all association rules where the prices of items are > 100 dollars." Association rules may vary temporarily. For example, we may have an association rule "Diaper $\rightarrow$ Beer (1% support, 80% confidence)." This

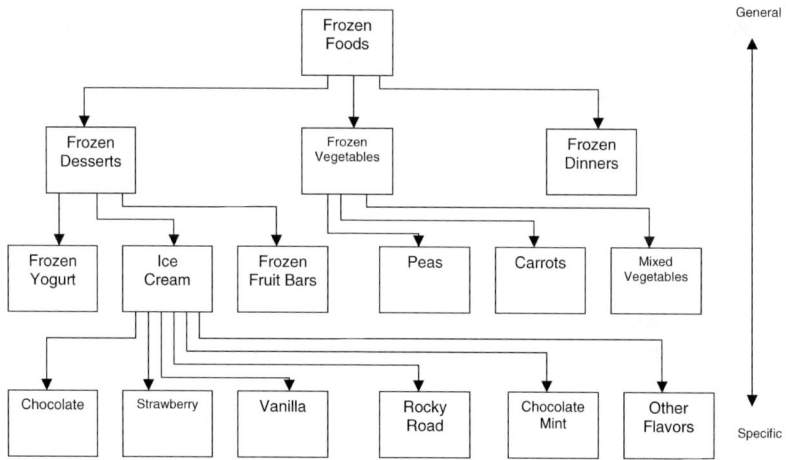

Fig. 2. Item association granularities from general to specific.

rule can be modified to accommodate a different time slot such as "Diaper → Beer (20% support) 7:00–9:00 pm weekdays." They can be generalized association rules consisting of hierarchies over items (UPC codes). For example, even though the rule "clothes → footwear" may hold even if "clothes → shoes" does not.

Association rules may be represented as Bayesian networks. These provide for the efficient representation of a probability distribution as a directed acyclic graph where the nodes represent attributes of interest, the edges direct causal influence between the nodes and conditional probabilities for nodes are given all possible.

One may actually be interested in optimization of association rules. For example, given a rule $(I < A < u)$ and X → Y, find values for I and u such that it has a support greater than certain threshold and maximizes a support confidence or gain. For example, suppose we have a rule "ChkBal$[I\ u]$ → DvDPlayer." We might then ascertain that choosing $I = \$30\,000$ and $u = \$50\,000$ optimizes the support confidence or gain of this rule.

Some of the strengths of market basket analysis are that it produces easy to understand results, it supports undirected data mining, it works on variable length data, and rules are relatively easy to compute. It should be noted that if there are n items under consideration and rules of the form "$A \to B$" are considered, there will be $\binom{n}{2}$ possible association rules. Similarly, if rules of the form "$A\,\&\,B \to C$" are considered, there will be $\binom{n}{3}$ possible association rules. Clearly if all possible association rules are considered, the number grows exponentially with n. Some of the other weaknesses of market basket analysis are that it is difficult to determine the optimal number of items, it discounts rare items, it is limited on the support that it provides.

The other computer science area that has contributed significantly to the roots of data mining is the area of text classification. Text classification has become a particularly important topic given the plethora of readily available information from the World Wide Web. Several other chapters in this volume discuss text mining as another aspect of data mining.

A discussion of the historical roots of the data mining methodologies within the computer science community would not be complete without touching upon terminology. One usually starts the data analysis process with a set of n observations in p space. What is usually referred to as dimensions in the mathematical community are known as variables in statistics or attributes in computer science. Observations within the mathematics community, i.e., one row of the data matrix, are referred to as cases in the statistics community and records in the computer science community. Unsupervised learning in computer science is known as clustering in statistics. Supervised learning in computer science is known as classification or discriminant analysis in the statistics community. It is worthwhile to note that statistical pattern recognition usually refers to both clustering and classification.

4. Data preparation

Much time, effort, and money is usually associated with the actual collection of data prior to data mining analysis. We will assume for the discussions below that the data has already been collected and that we will merely have to obtain the data from its storage facility prior to conducting our analysis. With the data in hand, the first step of the analysis procedure is data preparation. Our experiences seem to indicate that the data preparation phase may require up to 60% of the effort associated with a given project. In fact, data preparation can often determine the success or failure of our analytical efforts.

Some of the issues associated with data preparation include data cleaning/data quality assurance, identification of appropriate data type (continuous or categorical), handling of missing data points, identifying and dealing with outliers, dimensionality reduction, standardization, quantization, and potentially subsampling. We will briefly examine each of these issues in turn. First we note that data might be of such a quality that it does contain statistically significant patterns or relationships. Even if there are meaningful patterns in the data, these patterns might be inconsistent with results obtained using other data sets. Data might also have been collected in a biased manner or since in many cases the data is based on human respondents, the data may be of uneven quality. We finally note that one has to be careful that the discovered patterns are not too specific or too general for the application at hand.

Even when the researcher is presented with meaningful information, one still often must remove noise from the dataset. This noise can take the form of faulty instrument/sensor readings, transmission errors, data entry errors, technology limitations, or naming conventions misused. In many cases, numerous variables are stored in the database that may have nothing whatsoever to do with the particular task at hand. In this situation, one must be willing to identify those variables that are germane to the current analysis while ignoring or destroying the other confounding information.

Some of the other issues associated with data preparation include duplicate data removal, missing value imputation (manually or statistical), identification and removal of data inconsistencies, identification and refreshment of stale or untimely data, and the creation of a unique record or case id. In many cases, the human has a role in interactive procedures that accomplish these goals.

Next we consider the distinction between continuous and categorical data. Most statistical theory and many graphics tools have been developed for continuous data. However, most of the data that is of particular interest to the data mining community is categorical. Those data miners that have their roots in the computer science community often take a set of continuous data and transform the data into categorical data such as low, medium, and high. We will not focus on the analysis of categorical data here but the reader is referred to Agresti (2002) for a thorough treatment of categorical data analysis.

4.1. Missing values and outliers

One way to check for missing values in the case of categorical or continuous data is with a missing values plot. An example missing data plot is provided in Figure 3. Each observation is plotted as a vertical line. Missing values are plotted in black. The data was artificially generated. The missing data plot actually is a special case of the color

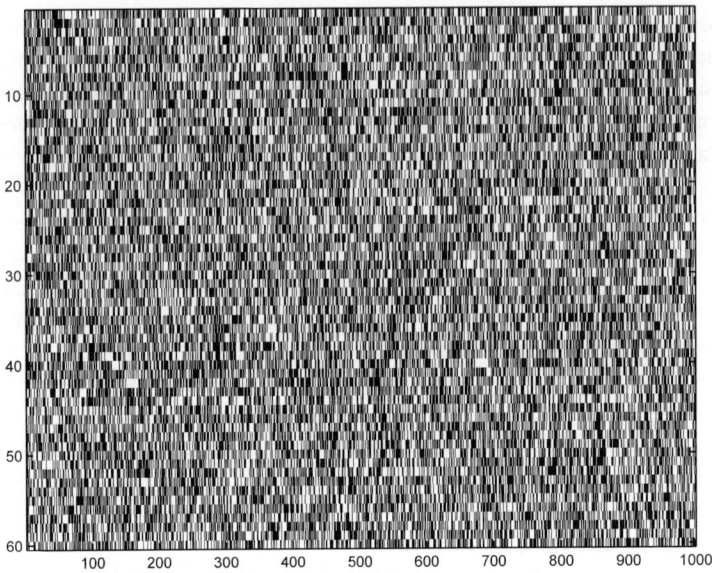

Fig. 3. Missing data plot for an artificial dataset. Each observation is plotted as a vertical bar. Missing values are plotted in black.

histogram. The color histogram first appeared in the paper on the use of parallel coordinates for statistical data analysis (Wegman, 1990). This data analysis technique was subsequently rediscovered by Minnotte and West (1998). They coined the phrase "data image" for this "new" visualization technique in their paper.

Another important issue in data preparation is the removal/identification of outliers. Outliers, while easy to detect in low dimensions, $d = 1, 2$, or 3, their identification in high-dimensional spaces may be more tenuous. In fact, high-dimensional outliers may not actually manifest their presence in low-dimensional projections. For example, one could imagine points uniformly distributed on a hyper-dimensional sphere of large radius with a single outlier point at the center of the sphere. Minimum volume ellipsoid (MVE) (see Poston et al., 1997) has been previously proposed as a methodology for outlier detection but these methods are exponentially computationally complex. Methods based on the use of the Fisher information matrix and convex hull peeling are more feasible but still too complex for massive datasets.

There are even visualization strategies for the identification of outliers in high-dimensional spaces. Marchette and Solka (2003) have proposed a method based on an examination of the color histogram of the interpoint distance matrix. This method provides the capability to identify outliers in extremely high-dimensional spaces for moderately sized, fewer than 10 000 observations, datasets. In Figure 4 we present a plot of the interpoint data image for a collection of 100 observations uniformly distributed along the surface of a 5-dimensional hypersphere along with a single outlier point positioned at the center of the hypersphere. We have notated the outlier in the data

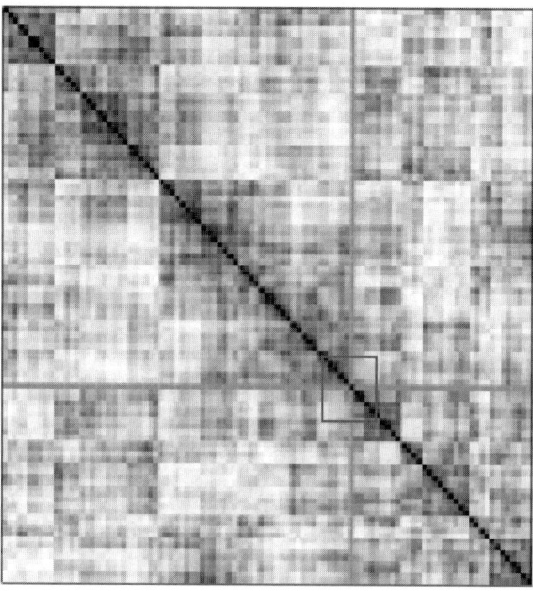

Fig. 4. Interpoint distance data image for a set of 100 observations sampled uniformly from the center of a 5-dimensional hypersphere with an outlier placed at the center of the hypersphere. We have indicated the location of the outlier by a rectangle placed at the center of the data image.

image via a square. The presence of the outlier is indicated in the data image via the characteristic cross like structure.

4.2. Quantization

Once the missing values and outliers have been removed from the dataset, one may be faced with the task of subsampling from the dataset in the case when there are too many observations to be processed. Sampling from a dataset can be particularly expensive in the case where the dataset is stored in some sort of relational databases management system. So we may be faced with squishing the dataset, reducing the number of cases, or squashing the dataset, reducing the number of variables or dimensions associated with the data. One normally thinks of subsampling as the standard way of squishing a dataset but quantization or binning is an equally viable approach. Quantization has a rich history of success in the computer science community in application areas including signal and image processing.

Quantization does possess some useful statistical properties. For example, given that $E[W \mid Q = y_j]$ is the mean of the observations on the jth bin $= y_j$ then the quantizer can be shown to be self-consistent, i.e. $E[W \mid Q] = Q$. The reader is referred to (Khumbah and Wegman, 2003) for a full development of the statistical properties associated with the quantization process. See also (Braverman, 2002).

A perhaps more interesting topic is geometry-based tessellation. One needs space-filling tessellations made up of congruent tiles that are as spherical as possible in order

Fig. 5. Tessellation of 3-soace by truncated octahedra.

to perform efficient space-filling tessellation. In one dimension, one is relegated to a straight line segment and in two dimensions one may choose equilateral triangles, squares, or hexagons. The situation becomes more interesting in three dimensions. In this case, one can use a tetrahedron, a cube, a hexagonal prism, a rhombic dodecahedron, or a truncated octahedron. We present a visualization of three-space tessellated by truncated octahedra in Figure 5.

A brief analysis of the computational complexity of the geometry-based quantization is in order. First we note that the use of up to 10^6 bins is both computationally and visually feasible. The index of x_i in one dimension is given by $j = \text{fixed}[k*(xi-a)/(b-a)]$ for data in the range $[a, b]$, k bins and one-dimensional data. The computational complexity of this method is $4n + 1 = O(n)$ and the memory requirements drop to $3k$, the location of the bin plus the number of items in the bin plus a representer of the bin.

In two dimensions, each hexagon is indexed by 3 parameters. The computational complexity is 3 times the one-dimensional complexity, i.e. $12n + 3 = O(n)$. The complexity for the square is two times the one-dimensional complexity and the storage complexity is still $3k$.

In three dimensions, there are 3 pairs of square sides and 4 pairs of hexagonal sides on a truncated octahedron. The computational complexity of the process is still $O(n)$, $28n + 7$. The computational complexity for the cube is $12n + 3$ and the storage complexity is still $3k$.

In summary, we present the following guidelines. First, optimality in terms of minimizing distortion is obtained using the roundest polytope in d-dimensions. Second, the complexity is always $O(n)$ with an associated storage complexity of $3k$. Third, the number of tiles grows exponentially with dimension, another manifestation of the so-called curse of dimensionality. Fourth, for ease of implementation always use a hypercube or d-dimensional simplex. The hypercube approach is known as a datacube in computer science literature and is closely related to multivariate histograms in the statistical literature. Fifth, this sort of geometric approach to binning is applicable up to around 4- or 5-dimensional space. Sixth, adaptive tilings may improve the rate at which the number of tiles grows, but probably destroys the spherical structure associate with the data. This property relegates its use to large n but makes its use problematic in large d.

5. Databases

5.1. SQL

Knowledge discovery and data mining have many of their roots in database technology. Relational databases and structured query language (SQL) have a 25+ year history. However, the boolean relations (and, or, not) commonly used in relational databases and SQL are inadequate for fully exploring data.

Relational databases and SQL are typically not well-known to the practicing statistician. SQL, pronounced "ess-que-el" is used to communicate with a database according to certain American National Standards Institute (ANSI) standards. SQL statements can be used to store records in a database, access these records, and retrieve the records. Common SQL commands such as "Select", "Insert", "Update", "Delete", "Create", and "Drop" can be used to accomplish most of the tasks that one needs to do with a database. Some common relational database management systems that use SQL include Oracle, Sybase, Microsoft SQL Server, Access, Ingres, and the public license servers MySQL and MSQL.

A relational database system contains one or more objects called tables. These tables store the information in the database. Tables are uniquely identified by their names and are comprised of rows and columns. The columns in the table contain the column name, the data type and any other attribute for the columns. We, statisticians, would refer to the columns as the variable identifiers. Rows contain the records of the database. Statisticians would refer to the rows as cases. An example database table is given in Table 15.

The *select* SQL command is one of the standard ways that data is extracted from a table. The format of the command is: *select* "column1"[,"column2", etc.] *from* "tablename" [*where* "condition"]; . The arguments given in the square brackets are optional. One can select as many column names as they like or use "*" to choose all of the columns. The optional *where* clause is used to indicate which data values or rows should be returned. The operators that are typically used with *where* include = (equal), > (greater than), < (less than), >= (greater than or equal to), <= (less than or equal to), <> (not equal to), and LIKE. The LIKE operator is a pattern matching operator that does support the use of wild-card characters through the use of %. The % can appear at the start or end of a string. Some of the other handy SQL operators include *create table* (to create a new table), *insert* (to insert a row into a table), *update* (to update or change

Table 15
Example relational database

City	State	High	Low
Phoenix	Arizona	105	90
Tucson	Arizona	101	92
Flagstaff	Arizona	88	69
San Diego	California	77	60
Albuquerque	New Mexico	80	72

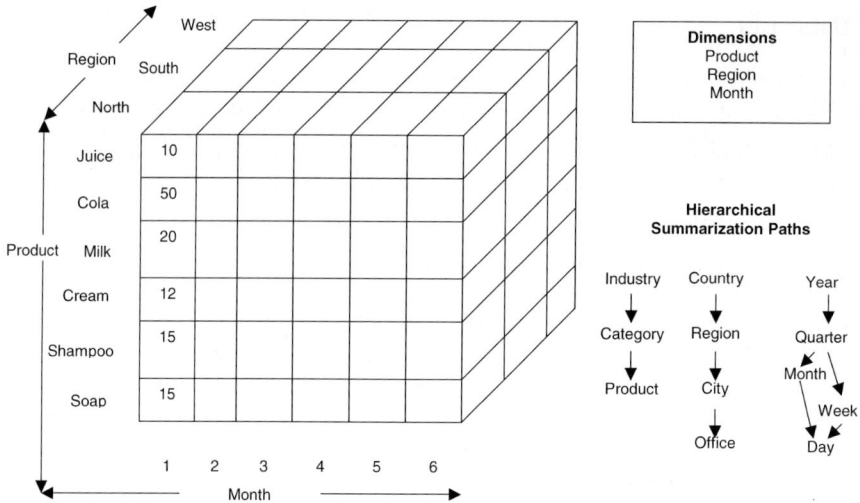

Fig. 6. A datacube with various hierarchical summarization rules illustrated.

those records that match a specified criteria). The *delete* operator is used to delete those designated rows or records from a table.

5.2. Data cubes and OLAP

Computer scientists tend to deal with relational databases accessing them through SQL. Statisticians tend to deal with flat, text files that are space, tab, or comma delimited. Relational databases have more structure, data protection, and flexibility but these incur a large computational overhead. Relational databases are not typically suited for massive dataset analysis except as a means to extract a flat file.

Data cubes and online analytical processing (OLAP) are ideas that have grown out of the database industry. These tools/ideas are most often perceived as a response to business management. These ideas are often applied to a data warehouse. A data warehouse is a central data repository wherein several local databases are assembled.

A data cube is quite simply a multi-dimensional array of data. Each dimension of a data cube is a set of sets representing domain content such as time or geography. The dimensions scaled categorically are such as region of country, state, quarter of year, week of quarter. The cells of the cube represent aggregated measures (usually counts) of variables. An example data cube is illustrated in Figure 6.

Exploration of a data cube involves the operations of *drill down*, *drill up*, and *drill through*. *Drill down* involved splitting an aggregation into subsets, e.g. splitting region of country into states. *Drill up* involves consolidation, i.e. aggregating subsets along a dimension. *Drill through* involves subsets of crossing of sets, i.e. the user might investigate statistics within a state subsetted by time.

OLAP and MOLAP, multi-dimensional OLAP, are techniques that are typically performed on data cubes in an online fashion. Operations are usually limited to simple measures like counts, means, proportions and standard deviations. ROLAP refers to relational OLAP using extended SQL.

In summary, we note that the relational database technology is fairly compute intensive. Because of this, commercial database technology is challenged by the analysis of datasets above about 10^8 observations. This computational limitation applies to many of the algorithms developed by computer scientists for data mining.

6. Statistical methods for data mining

The hunt for structure in numerical data has had a long history within the statistical community. Examples of methodologies that may be used for data exploration in a data mining scenario include correlation and regression, discriminant analysis, classification, clustering, outlier detection, classification and regression trees, correspondence analysis, multivariate nonparametric density estimation for hunting bump and ridges, nonparametric regression, statistical pattern recognition, categorical data analysis, time-series methods for trend and periodicity, and artificial neural networks. In this volume, we have a number of detailed discussions on these techniques including nonparametric density estimation (Scott and Sain, 2005), multivariate outlier detection (Hubert et al., 2005), classification and regression trees (Sutton, 2005), pattern recognition (Hand, 2005), classification (Marchette et al., 2005), and correspondence analysis (Rao, 2005).

To a large extent, these methods have been developed and treated historically as confirmatory analysis techniques with emphasis on their statistical optimality and asymptotic properties. In the context of data mining, where data are often not collected according to accepted statistical sampling procedures, of course, traditional interpretations as probability models with emphasis on statistical properties are inappropriate. However, most of these methodologies can be reinterpreted as exploratory tools. For example, while a nonparametric density estimator is frequently interpreted to be an estimator that asymptotically converges to the true underlying density, as a data mining tool, we can simply think of the density estimator as a smoother which helps us hunt for bumps and other features in the data, where we have no reason to assume that there is a unique underlying density.

No discussion on statistical methods would be complete without referring to data visualization as an exploratory tool. This area also is treated by several authors in this volume, including Carr (2005), Buja et al. (2005), Chen (2005), and Wilhelm (2005). We shall discuss visual data mining briefly in a later section of this paper, reserving this section for some brief reviews of some analytical tools which are not covered so thoroughly in other parts of this volume and which are perhaps a bit less common in the usual statistical literature. Of further interest is the emerging attention in approaches to streaming data. This topic will be discussed separately in yet another section of this chapter.

6.1. Density estimation

As noted earlier in Section 2.1, kernel smoothers are in general $O(n)$ algorithms and thus should in principle be easily adapted to service in a data mining framework. A straightforward adaptation to the kernel density smoother in high dimensions is given by

$$f(\mathbf{x}) = \frac{1}{nh_1h_2\cdots h_d}\sum_{i=1}^{n}K\left(\frac{\mathbf{x}-\mathbf{x}_i}{\mathbf{h}}\right), \quad \mathbf{x}\in\mathcal{R}^d,$$

where n is the sample size, d is the dimension, K is the smoothing kernel, h_j is the smoothing parameter in the jth dimension, $\mathbf{x}_i$ is the ith observation, $\mathbf{x}$ is the point at which the density is estimated, and $\mathbf{h}=(h_1,h_2,\ldots,h_d)$. In general, kernel densities must be computed on a grid since there is no direct closed form.

The so-called curse of dimensionality dramatically affects the requirement for computation on a grid. Consider, for example, a one-dimensional problem in which we have a grid with k points. In d dimensions, in order to maintain the same density on a side, one would require k^d grid points implying that the computational complexity explodes exponentially in the dimension of the data. Moreover, suppose that we have in one dimension 10 cells with an average of 10 observations per cell or 100 observations. In order to maintain the same average cell density in d dimensions, we would required a 10^{d+1} observations. Thus, not only does the computational complexity increase exponentially with dimension, the data requirements also increase exponentially with dimension.

One approach which circumvents this requirement is the one which exploits a finite mixture of normal densities:

$$f(\mathbf{x};\boldsymbol{\psi}) = \sum_{i=1}^{N}\pi_i\phi(\mathbf{x};\mathbf{x}_i,\theta_i)$$

where $\boldsymbol{\psi} = (\theta_1,\ldots,\theta_N,\pi_1,\ldots,\pi_N)$, ϕ is usually taken to be a normal density, N is the number of mixing terms ($N \ll n$), if ϕ is normal, then $\boldsymbol{\theta}_i = (\boldsymbol{\mu}_i, \boldsymbol{\Sigma}_i)$, where $\boldsymbol{\mu}_i$ and $\boldsymbol{\Sigma}_i$ are respectively the mean vector and covariance matrix. The parameters are re-estimated using the EM algorithm. Specifically, we have:

$$\tau_{ij} = \frac{\pi_i\phi(\mathbf{x};\mathbf{x}_j,\boldsymbol{\theta}_i)}{\sum_{i=1}^{N}\pi_i\phi(\mathbf{x};\mathbf{x}_j,\boldsymbol{\theta}_i)}, \qquad \pi_i = \frac{1}{n}\sum_{j=1}^{n}\tau_{ij},$$

$$\boldsymbol{\mu}_i = \frac{1}{n\pi_i}\sum_{j=1}^{n}\tau_{ij}\mathbf{x}_j, \qquad \boldsymbol{\Sigma}_i = \frac{1}{n\pi_i}\sum_{j=1}^{n}\tau_{ij}(\mathbf{x}_j-\boldsymbol{\mu}_i)(\mathbf{x}_j-\boldsymbol{\mu}_i)^{\dagger},$$

where τ_{ij} is the estimated posterior probability that $\mathbf{x}_j$ belongs to component i, π_i is the estimated mixing coefficient, $\boldsymbol{\mu}_i$ and $\boldsymbol{\Sigma}_i$ are the estimated mean vector and covariance matrix, respectively. The EM is applied until convergence is obtained. A visualization of this is given in Figure 7. Mixture estimates are L_1 consistent only if the mixture density is correct and if the number of mixture terms is correct.

Adaptive mixtures are used in another, semiparametric, recursive method discussed by Priebe and Marchette (1993) and Priebe (1994) and provide an alternate formulation avoiding issues of the number of terms being correct. The recursive update equations become:

$$\tau_{i,n+1} = \frac{\pi_{i,n}\phi(\mathbf{x};\mathbf{x}_{n+1},\boldsymbol{\theta}_n)}{\sum_{i=1}^{N}\pi_{i,n}\phi(\mathbf{x};\mathbf{x}_{n+1},\boldsymbol{\theta}_n)}, \qquad \pi_{i,n+1} = \pi_{i,n} + \frac{1}{n}(\tau_{i,n+1}-\pi_{i,n}),$$

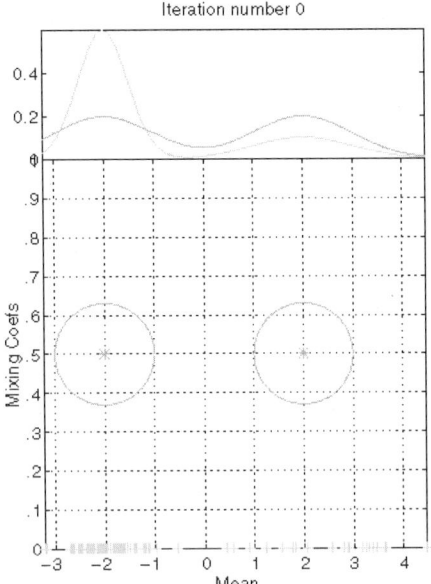

Fig. 7. Visualization of normal mixture model parameters in a one-dimensional setting. For a color reproduction of this figure see the color figures section, page 565.

$$\mu_{i,n+1} = \mu_{i,n} + \frac{\tau_{i,n+1}}{n\pi_{i,n}}(x_{n+1} - \mu_{i,n}),$$

$$\Sigma_{i,n+1} = \Sigma_{i,n} + \frac{\tau_{i,n+1}}{n\pi_{i,n}}\big[(x_{n+1} - \mu_{i,n})(x_{n+1} - \mu_{i,n})^\dagger\big].$$

In addition, it is necessary to have a create rule. The basic idea is that if a new observation is too far from the previously established components (usually in terms of Mahalanobis distance), then the new observation will not be adequately represented by the current mixing terms and a new mixing term should be added. We create a new component centered at x_t with a nominal starting covariance matrix. The basic form of the update/create rule is as follows:

$$\theta_{t+1} = \theta_t + \big[1 - \mathcal{P}_t(x_{t+1}, \theta_t)\big]\mathcal{U}_t(x_{t+1}, \theta_t) + \mathcal{P}_t(x_{t+1}, \theta_t)\mathcal{C}_t(x_{t+1}, \theta_t)$$

where $\mathcal{U}_t$ is the previously given update rule, $\mathcal{C}_t$ is the create rule, and $\mathcal{P}_t$ is the decision rule taking on values either 0 or 1. Figure 8 presents a three-dimensional illustration. This procedure is nonparametric because of the adaptive character. It is generally L_1 consistent under relatively mild conditions, but almost always too many terms are created and pruning is needed.

Solka et al. (1995) proposed the visualization methods for mixture models shown in Figures 7 and 8. Solka et al. (1998) and Ahn and Wegman (1998) proposed alternative effective methods for eliminating redundant mixture terms. We note finally that mixture models are valuable from the perspective that they naturally suggest clusters centered at the mixture means. However, while the model complexity may be reduced,

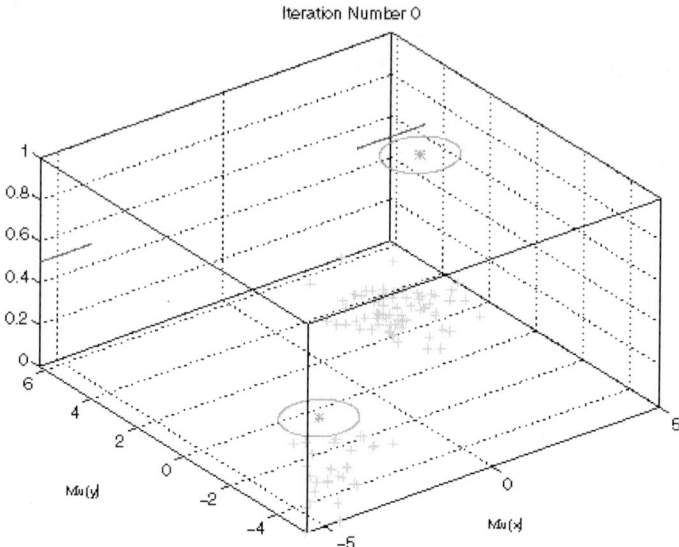

Fig. 8. Visualization of adaptive normal mixture parameters in a two-dimensional setting.

the computational complexity may be increased dramatically. Nonetheless, once established, a mixture model is computationally simple because it does not have to be estimated on a grid.

6.2. Cluster analysis

There are many methods for cluster analysis. In this section, we give a brief overview of classical distance-based cluster analysis. It should be remarked at the outset that traditional distance-based clustering of a dataset of size n requires distance between points to be calculated. Because there are $\binom{n}{2}$ point pairs, the computational complexity of just computing the distance is $O(n^2)$. The implication is immediately that distance-based clustering is not suitable for truly massive datasets.

The basic clustering problem is, then, given a collection of n objects each of which is described by a set of d characteristics, derive a useful division into a number of classes. Both the number of classes and the properties of the classes are to be determined. We are interested in doing this for several reasons, including organizing the data, determining the internal structure of the dataset, predication, and discovery of causes. In general, because we do not know a priori the number of clusters, the utility of a particular clustering scheme can only be measured by the quality. In general, there is no absolute measure of quality.

Suppose that our data is of the form x_{ik}, $i = 1, \ldots, n$ and $k = 1, \ldots, d$. Here the i indexes the number of observations and k indexes the set of characteristics. Distances (or dissimilarities) can be measured in multiple ways. Some frequently used metrics are the following:

Euclidean distance: $$d(x_i., x_j.) = \sqrt{\sum_{k=1}^{d}(x_{ik} - x_{jk})^2};$$

City block metric: $$d(x_i., x_j.) = \sum_{k=1}^{d} |x_{ij} - x_{jk}|;$$

Canberra metric: $$d(x_i., x_j.) = \sum_{i=1}^{d} \frac{|x_{ik} - x_{jk}|}{(x_{ik} - x_{jk})};$$

Angular separation metric: $$d(x_i., x_j.) = \frac{\sum_{k=1}^{d} x_{ik} x_{jk}}{\sqrt{\sum_{k=1}^{d} x_{ik}^2 \sum_{k=1}^{d} x_{jk}^2}}.$$

6.2.1. Hierarchical clustering

The agglomerative clustering algorithm begins with clusters $C_1, C_2, \ldots, C_n$, each cluster initially with a single point. Find the nearest pair C_i and C_j, merge C_i and C_j into C_{ij}, delete C_i and C_j, and decrement cluster count by one. If the number of clusters is greater than one, repeat the previous set until there is only one cluster remaining.

In single linkage (nearest neighbor) clustering, the distance is defined as that of the closet pair of individuals where we consider one individual from each group. We could also form complete linkage (furthest neighbor) clustering where the distance between groups is defined as the most distant pair, one individual from each cluster. Consider the following example with initial distance matrix:

	1	2	3	4	5
1	0.0				
2	2.0	0.0			
3	6.0	5.0	0.0		
4	10.0	9.0	4.0	0.0	
5	9.0	8.0	5.0	3.0	0.0

The distance 2.0 is the smallest distance, so that we join individuals 1 and 2 in the first round of agglomeration. This yields the following distance matrix based on single linkage:

	(1, 2)	3	4	5
(1, 2)	0.0			
3	5.0	0.0		
4	9.0	4.0	0.0	
5	8.0	5.0	3.0	0.0

Now the distance 3.0 is the smallest, so that individuals 4 and 5 are joined in the second round of agglomeration. This yields the following single linkage distance matrix:

	(1, 2)	3	(4, 5)
(1, 2)	0.0		
3	5.0	0.0	
(4, 5)	8.0	4.0	0.0

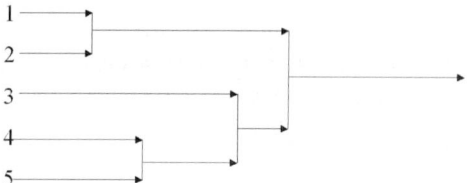

Fig. 9. Dendrogram resulting from single linkage agglomerative clustering algorithm.

In this matrix, the smallest distance is now 4.0, so that 3 is joined into (4, 5). This yields the penultimate distance matrix of

	(1, 2)	(3, 4, 5)
(1, 2)	0.0	
(3, 4, 5)	5.0	0.0

Of course, the last agglomerative step is to join (1, 2) to (3, 4, 5) and to yield the single cluster (1, 2, 3, 4, 5). This hierarchical cluster yields a dendrogram given in Figure 9. In this example, the complete linkage clustering yields the same cluster sequence and the same dendrogram. The intermediate distance matrices are different however.

An alternate approach could be to use group-average clustering. The distance between clusters is the average of the distance between all pairs of individuals between the two groups. We also note that there are methods for divisive clustering, i.e. beginning with every individual in one large cluster and recursively separating into smaller, multiple clusters, until every individual is in a singleton cluster.

6.2.2. The number of groups problem

A significant question is how do we decide on the number of groups. One approach is to maximize or minimize some criteria. Suppose g is the number of groups and n_i is the number of items in the ith group. Consider

$$T = \frac{1}{n} \sum_{i=1}^{g} \sum_{j=1}^{n_i} (x_{ij} - \bar{x})(x_{ij} - \bar{x})^{\dagger},$$

$$W = \frac{1}{n-g} \sum_{i=1}^{g} \sum_{j=1}^{n_i} (x_{ij} - \bar{x}_j)(x_{ij} - \bar{x}_i)^{\dagger},$$

$$B = \sum_{i=1}^{g} n_i (x_{ij} - \bar{x})(x_{ij} - \bar{x})^{\dagger},$$

$$T = W + B.$$

Some optimization strategies are:

(1) minimize *trace(W)*,
(2) maximize *det(T)/det(W)*,
(3) minimize *det(W)*,
(4) maximize *trace(BW⁻¹)*.

It should be noted that all approaches of this discussion have important computational complexity requirements. The number of partitions of n individuals into g groups $N(n, g)$ can be quite daunting. For example,

$$N(15, 3) = 2\,375\,101,$$
$$N(20, 4) = 45\,232\,115\,901,$$
$$N(25, 8) = 690\,223\,721\,118\,368\,580,$$
$$N(100, 5) = 10^{68}.$$

We note that many references to clustering algorithms exist. Two important ones are (Everitt et al., 2001, Hartigan, 1975).

Other approaches to clustering involve minimal spanning trees (see, for example, Solka et al., 2005, this volume), clustering based on mixture densities mentioned in the previous section, and clustering based on Voronoi tessellations with centroids determined by estimating modes (Sikali, 2004).

6.3. Artificial neural networks

6.3.1. The biological basis

Artificial neural networks have been inspired by mathematicians attempting to create an inference network modeled on the human brain. The brain contains approximately 10^{10} neurons that are highly interconnected with an average of several thousand interconnects per neuron. The neuron is composed of three major parts, the cell body, an input structure called dendrites, and an output structure called the axon. The axon of a given cell is connected to the dendrites of other cells. The cells communicate by electrochemical signals. When a given cell is activated, it fires the electrochemical signal along its axon which is transferred to all of the cells with dendrites connected to the cell's axon. The receiving cells may or may not fire depending on whether a receiving cell has received enough total stimulus to exceed its firing threshold.

There is actually a gap called the synapse with the neurotransmitters prepared to transmit the signal across the synapse. The strength of a signal that a neuron receives depends on the efficiency of the synapses. The neurotransmitters are released from the neuron at presynaptic nerve terminal on the axon and cross the synapse to dendrites of the next neuron. The dendrite of the next neuron may have a receptor site for the neurotransmitter. The activation of the receptor site may be either inhibitory or excitatory, which may lower or raise the possibility of the next neuron firing. Learning is thought to take place when the efficiency of the synaptic connection is increased. This process is called neuronal plasticity.

The brain is capable of extremely complex tasks based on these simple units that transmit binary signals (fire, don't fire). Artificial neural networks in principle work the same way except that the overall complexity in terms of processing units and interconnects does not realistically approximate the scale of the brain.

6.3.2. Functioning of an artificial neural network

An artificial neuron has a number of receptors that receive inputs either from data or from the outputs of other neurons. Each input comes by way of a connection with

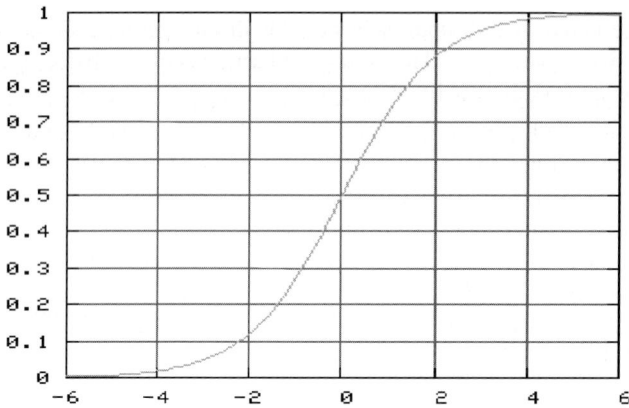

Fig. 10. The sigmoid curve.

a weight (or strength) in analogy to the synaptic efficiency of the biological neuron. Each artificial neuron has a threshold value from which the weighted sum of the inputs minus the threshold is formed. The artificial neuron is not binary as is the biological neuron. Instead, the weighted sum of inputs minus threshold is passed through a transfer function (also called an activation function), which produces the output of the neuron. Although it is possible to use a step-activation function in order to produce a binary output, step activation functions are rarely used. Generally speaking, the most used activation function is a translation of the sigmoid curve given by

$$f(t) = \frac{1}{1 + e^{-t}}.$$

See Figure 10. The sigmoid curve is a special case of the logistic curve.

The most common form of an artificial neural network is called a feed-forward network. The first layer of neurons receives input from the data. The first layer is connected to one or more additional layers, called hidden layers, and the last hidden layer is connected to the output layer. The signal is fed forward through the network from the input layer, through the hidden layer to the units in the output layer. This network is stable because it has no feedback. A network is recurrent if it has a feedback from later neurons to earlier neurons. Recurrent networks can be unstable and, although interesting from a research perspective, are rarely useful in solving real problems.

Neural networks are essentially nonlinear nonparametric regression tools. If the functional form of the relationship between input variables and output variables was know, it would be modeled directly. Neural networks learn the relationship between input and output through training, which determines the weights of the neurons in the network. In a supervised learning scenario, a set of inputs and outputs is assembled and the network is trained (establishes neuron weights and thresholds) to minimize the error of its predictions on the training dataset. The best known example of a training method is back propagation. After the network is trained, it models the unknown function and can be used to make predictions on input values for which the output is not known.

From this description it is clear that artificial neural networks operate on numerical data. Although more difficult, categorical data can be modeled by representing the data by numeric values. The number of neurons required for the artificial neural network is related to the complexity of the unknown function that is modeled as well as to the variability of the additive noise. The number of the required training cases increases nonlinearly as the number of connections in the network increases. A rule-of-thumb is that the number of cases should be approximately ten times the number of connections. Obviously, the more variables involved, the more neurons are required. Thus, variables should be chosen carefully. Missing values can be accommodated if necessary. If there is enough data, observations with missing values should be discarded. Outliers can cause problems and should be discarded.

6.3.3. Back propagation

The error generated by a particular configuration (geometry, neural weights, thresholds) of the network is determined by processing all of the training cases through the network and comparing the network's actual output with the known target outputs of the training cases. The differences are combined by an error function, often a sum of squared errors or a cross entropy function, which is used for maximum likelihood classification. The error surfaces of neural networks are often complicated by such features as local minima, saddle points, and flat or nearly flat spots. Starting at a random point on the error surface, the training algorithm attempts to find a global minimum. In back propagation, the gradient of the error surface is calculated and a steepest descent trajectory is followed. Generally, one takes small steps along the steepest descent trajectory making smaller steps as the minimum is approached. The size of the steps is crucial in the sense that a step size that is too big may alternate between sides of the local minimum well and never converge, or may jump from one local minimum well to another. Step sizes that are too small may converge properly, but require too many iterations. Usually the step size is chosen proportional to the slope of the gradient vector and to a constant known as the learning rate. The learning rate is essentially a fudge factor. It is application dependent and is chosen experimentally.

7. Visual data mining

As much as we earlier protested in Section 2.4 that data mining of massive datasets is unlikely to be successful, still the upper bound of approximately 10^6 allows for visually mining relatively large datasets. Indeed, we have pursued this fairly aggressively. See Wegman (2003). Of course, there are many situations where it is useful to attempt to apply data mining techniques to more modest datasets. In this section, we would like to give some perspective on our views of visual data mining. As pointed out earlier, this volume contains a number of thoughtful chapters on data visualization and the reader is strongly encouraged to examine the variety of perspectives that these chapters represent.

7.1. The four stages of data graphics

We note that data visualization has a long history in the statistics profession. Michael Friendly's website (http://www.math.yorku.ca/SCS/friendly.html) has a wealth of interesting material on the history of data graphics. Anyone interested in this topic should browse Michael's website. As a preface to our view of data graphics, we offer the following perspective. In general, we think that statisticians tend to be methodology centric rather than data centric. By this we mean that statisticians tend to identify themselves by the tools they employ (i.e. their methodology) rather than by the problem they are trying to solve (i.e. the data with which they are presented). In general, we believe that new data types generate new data analysis tools.

The most common form of data graphics is the static graphic. Tufte (1983, 1990, 1997) provides a long significant discussion on the art of static graphics. Wilkinson (1999), on the other hand, presents an elegant discussion on the mathematics of static graphics. In addition to the usual line plots and bar charts, static graphics includes a host of plot devices like scatterplot matrices, parallel coordinate plots, trellis plots, and density plots. Static graphics may include color and anaglyph stereoscopic plots. The second, third, and fourth stages of data graphics have come to being with the computer age. Interactive graphics, the second stage, involves manipulation of a graph data object but not necessarily the underlying data from which the graph data object was creates. Think of the data residing on a server and the graph data object residing on a client (computer). Interactive graphics is possible with relatively large datasets where the resulting graph data object is relatively much smaller. Interactive graphics devices include brushing, 3D stereoscopic plots, rocking and rotation, cropping and cutting, and linked views.

Dynamic graphics involves graphics devices in which the underlying data can be interacted with. There is perhaps a subtle distinction between dynamic graphics and interactive graphics. However, the requirement for dynamic graphics that the underlying data must be engaged to complete the graphic device implies that the scale of data must be considerably smaller than in the case with interactive graphics. Such techniques as grand tour, dynamic smoothing of plots, conditioned chloropleth maps and pixel tours are examples of dynamic graphics.

The final category of graphics is what we have begun to call evolutionary graphics. In this setting, the underlying datasets are no longer static but evolving either by being dynamically adjusted or because the data are streaming. Examples in which the data are dynamically adjusted include data set mapping (Wegman and King, 1990) and iterative denoising (Priebe et al., 2004). Examples of streaming data applications include waterfall plots, transient geographic mapping, and skyline plots. Evolutionary graphics with respect to streaming data is discussed more thoroughly in Section 8.

7.2. Graphics constructs for visual data mining

We take as a given in the visual data mining scenario that we will have relatively high-dimensional data and relatively large datasets. We can routinely handle up to 30 dimensions and 500 000 to 1 million observations. The graphics constructs we routinely use include scatterplot matrices, parallel coordinate plots, 3D stereoscopic plots, density plots, grand tour on all plot devices, linked views, saturation brushing, and pruning and cropping. We have particularly exploited the combination of parallel coordinate plots, k-dimensional grand tours ($1 \leqslant k \leqslant d$, where d is the dimension of the data), and saturation brushing. We briefly describe these below.

Parallel coordinates are a multi-dimensional data plotting device. Ordinary cartesian coordinates fail as a plot device after three dimensions because we live in a world with 3 orthogonal spatial dimensions. The basic idea of parallel coordinates is to give up the orthogonality requirement and draw the axes as parallel to each other. A d-dimensional point is plotted by locating its appropriate component on each of the corresponding parallel axes and interpolating with a straight line between axes. Thus a multi-dimensional point is uniquely represented by a broken line. Much of the elegance of this graph device is due to a projective geometry duality, which allows for interpretation of structure in the parallel coordinate plot. Details are available in (Wegman, 1990, 2003).

The grand tour is a method for animating a static plot by forming a generalized rotation in a k-subspace of the d-space where the data live. The basic idea is that we wish to examine the data from different views uncovering features that may not be visible in a static plot, literally taking a grand tour of the data, i.e. seeing the data from all perspectives. We might add that the grand tour is especially effective at uncovering outliers and clusters, tasks that are difficult analytically because of the computational complexity of the algorithms involved. A discussion of the mathematics underpinning both parallel coordinates and the grand tour can be found in (Wegman and Solka, 2002).

The final concept is saturation brushing. Ordinary brushing involves brushing (or painting) a subset of the data with a color. Using multiple colors in this mode allows for designation of clusters within the dataset or other subsets, for example, negative and positive values of a given variable by means of color. Ordinary brushing does not distinguish between a pixel that represents one point and a pixel that represents 10 000 points. The idea of saturation brushing is to desaturate the color so that it is nearly black and brush with the desaturated color. Then, using the so-called α-channel found on modern graphics cards, add up the color components. Heavy overplotting is represented by a fully saturated pixel whereas a single observation or a small amount of overplotting will remain nearly black. Thus saturation brushing is an effective way of seeing the structure of large datasets.

Combining these methods leads to several strategies for interactive data analysis. The BRUSH-TOUR strategy is a recursive method for uncovering cluster structure. The basic idea is to brush all visible clusters with distinct colors. If the parallel axes are drawn horizontally, then any gap in any horizontal slice separates two clusters. (Some authors draw the parallel coordinates vertically, so in this case any gap in any vertical slice separates two clusters.) Once all visible clusters are marked, initiate the grand tour until more gaps appear. Stop the tour and brush the new clusters. Repeat until no unmarked clusters appear. An example of the use of the BRUSH-TOUR strategy may be found in (Wilhelm et al., 1999).

A second strategy is the TOUR-PRUNE strategy, which is useful for forming tree structures. An example the use of a TOUR-PRUNE is to recursively build a decision tree based on demographic data. In the case illustrated in (Wegman, 2003), we considered a profit variable and demographic data for a number of customers of a bank. The profit variable was binarized by brushing the profit data with red for customers that lost money for the bank and green for customers that made a profit for the bank. The profit variable was taken out of the grand tour and the tour allowed to run on the remaining demographic variables until either a strongly red or strongly green region was found. A strongly red region indicated a combination of demographic variables that represented customers who lost money whereas strongly green region indicated a combination of demographic variables that represented customers who made profits for the bank. By recursively touring and pruning, a decision tree can be built from combinations of demographic variables for the purpose of avoiding unnecessary risks for the bank.

7.3. Example 1 – PRIM 7 data

Friedman and Tukey (1974) introduced the concept of projection pursuit and used this methods to explore the PRIM 7 data. This dataset has become something of a challenge dataset for statisticians seeking to uncover multi-dimensional structure. In addition to Friedman and Tukey, Carr et al. (1986) and Carr and Nicholson (1988) found linear features, Scott (1992, p. 213) reported on a triangular structure found by his student Rod Jee in an unpublished thesis (Jee, 1985, 1987), and Cook et al. (1995) found the linear features hanging off the vertices of the Jee–Scott triangle.

The PRIM 7 data is taken from a high-energy particle-physics scattering experiment. A beam of positively charged pi-mesons with an energy of 16 BeV is collided with a stationary target of protons contained in hydrogen nuclei. In such an experiment, quarks can be exchanged between the pi-meson and proton, with overall conservation of energy and momentum. The data consists of 500 examples. For this experiment, seven independent variables are sufficient to fully characterize the reaction products. A detailed description of the physics of the reaction is given in (Friedman and Tukey, 1974, p. 887). The seven-dimensional structure of this data has been investigated over the years.

The initial configuration of the data is given in a 7-dimensional scatterplot matrix in Figure 13. The BRUSH-TOUR strategy was used to identify substructures of the data. A semifinal brushed view of the data is given in Figure 11. Of particular interest is the view illustrating three triangular features given in Figure 12, which is a view of the data after a GRAND-TOUR rotation. The coloring in Figure 12 is the same as in Figure 11. Of course, two-dimensional features such as a triangle will often collapse into a one-dimensional linear feature or a zero-dimensional point feature in many of the two-dimensional projections. The fundamental question from the exploration of this data is what is the underlying geometric structure. The presence of triangular features suggests that the data form a simplex. It is our conjecture that the data actually form a truncated six-dimensional simplex. Based on this conjecture, we constructed simulated data. The initial configuration of the simulated data is shown in Figure 14, which can be compared directly with the real data in Figure 13.

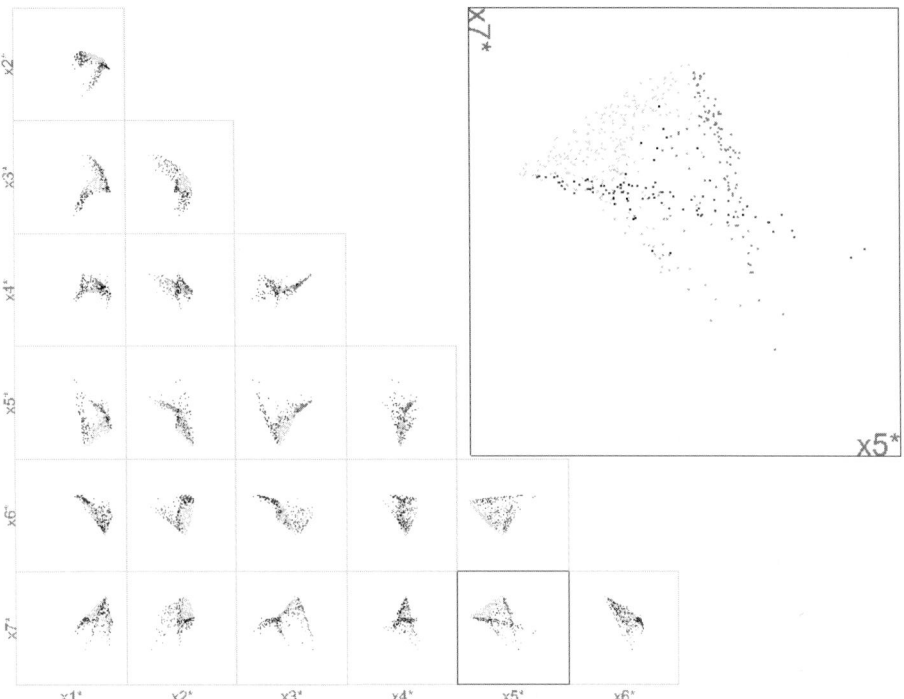

Fig. 11. Scatterplot matrix of the PRIM 7 data after GRAND-TOUR rotation with features highlighted in different colors. For a color reproduction of this figure see the color figures section, page 566.

Fig. 12. A scatterplot of the PRIM 7 data after GRAND-TOUR illustrating the three triangular features in the data, again highlighted in different colors. For a color reproduction of this figure see the color figures section, page 566.

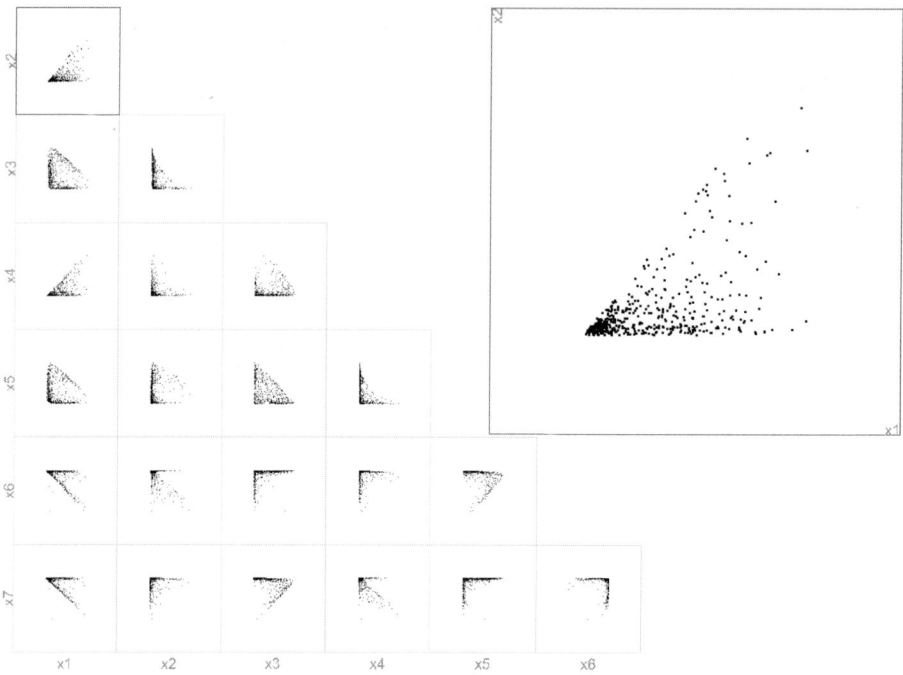

Fig. 13. A scatterplot matrix of the initial unrotated configuration of the PRIM 7 data. After considerable exploration, we conjecture a truncated 6-dimensional simplex based on the multiple triangular features.

7.4. Example 2 – iterative denoising with hyperspectral data

The hyperspectral imagery data consists of 14 478 observations selected from higher-resolution images in 126 dimensions (1 824 228 numbers) arising from six known classes and one unknown class. The six known classes were determined by ground truth to represent pixels coming from runway, water, swamp, grass, scrub, and pine. The goal is to assign a class to the data from the unknown class. The approach here is to use a modified TOUR-PRUNE strategy. We first begin by reducing dimension. We do this by forming the first three principal components. Figure 15 is a scatterplot of the first two principal components. Three components clearly stand apart from the rest. They are respectively water (blue and green) and runway (red). Once these are pruned away, we recompute the principal components based on the now reduced (denoised) dataset.

Figure 16 is the scatterplot for the remaining data. The prominent blue points are the swamp and the prominent cyan points correspond to grass. These are pruned away leaving only scrub, pines and unknown. Again the first three principal components are recomputed, and the resulting (toured) image is displayed in Figure 17. Notice that the blue is scrub, which overlaps somewhat with the other two classes. Removing the scrub and one final recomputation of the principal components, yields the (toured) image in Figure 18. Notice that the red (unknown) and green (pines) are thoroughly intermingled, which suggests that the unknowns are pines. In fact, we do know that they are oak trees,

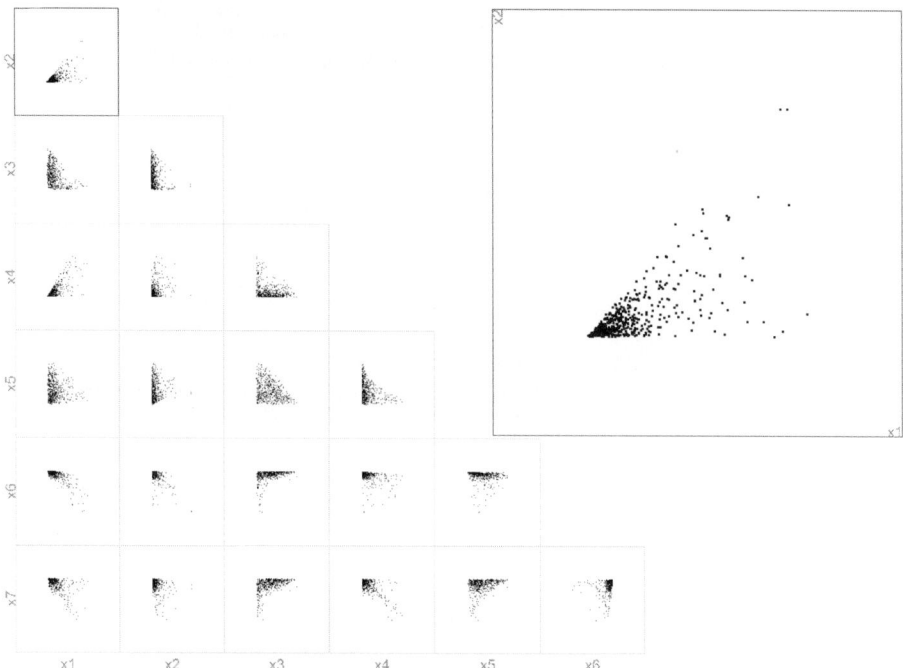

Fig. 14. A scatterplot matrix of the initial unrotated configuration of our simulated PRIM 7 data based on our conjectured truncated 6-dimensional simplex.

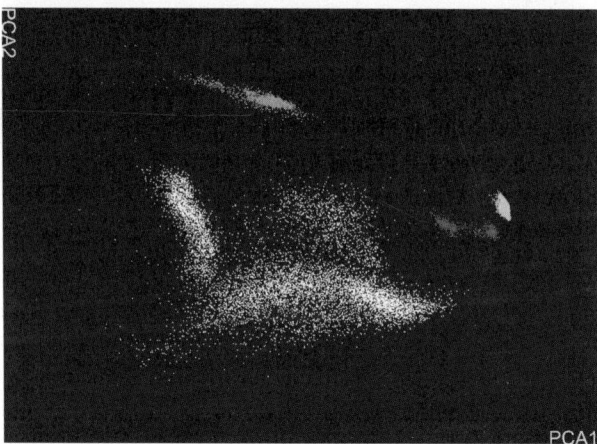

Fig. 15. The first two principal components of the hyperspectral imagery. There are 7 classes of pixels including runway, water, swamp, grass, scrub, pine, and unknown (really oaks). The water and runway are isolated in this figure. For a color reproduction of this figure see the color figures section, page 567.

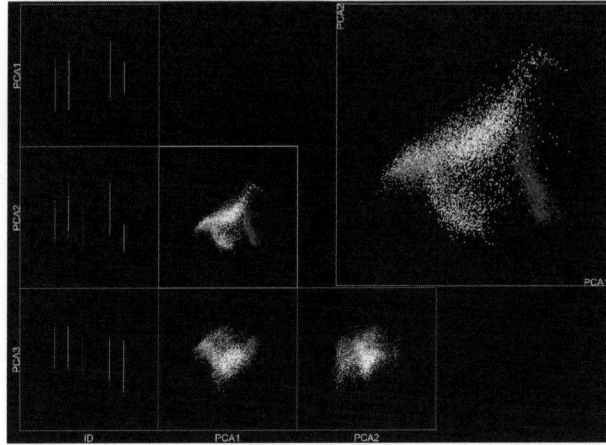

Fig. 16. The recomputed principal component after denoising by removing water and runway pixels. The swamp and grass pixels are colored respectively by cyan and blue. These may be removed and the principal components once again computed. For a color reproduction of this figure see the color figures section, page 567.

Fig. 17. The penultimate denoised image. The scrub pixels are shown in blue. They are removed and one final computation of the principal components is completed. For a color reproduction of this figure see the color figures section, page 568.

so that pines and oaks have similar spectra. This also explains why scrub is closer to the pines and oaks than say the grass or other categories. One last image (Figure 19) is of interest. In this image, we took ten principal components and brushed them with distinct colors for the seven distinct classes. This rotation shows that there is additional structure in the data that is not encoded by the seven class labels. The source of this structure is unknown by the experts who provided this hyperspectral data. Finally, we note in closing this section that the results in Examples 1 and 2 have not previously been published.

Fig. 18. The final denoised image show heavy overlap between the pine and unknown (oak) pixels. Based on this analysis, we classify the unknowns as being closest to pines. In fact, both are trees and are actually intermingled when the hyperspectral imagery was ground truthed. For a color reproduction of this figure see the color figures section, page 568.

Fig. 19. The same hyperspectral data but based on 10 principal components instead of just 3. The plot of PC 5 versus PC 6 shows that there is additional structure in this dataset not captured by the seven classes originally conjectured. For a color reproduction of this figure see the color figures section, page 569.

8. Streaming data

Statisticians have flirted with the concept of streaming data in the past. Statistical process control considers data to be streaming, but at a comparatively low rate because of the limits with which physical manufacturing can take place. Sequential analysis on the other hand take a more abstract perspective and assumes that the data is unending, but very highly structured and seeks to make decisions about the underlying structure as quickly as possible. Data acquisition techniques based on electronic and high-speed computer resources have changed the picture considerably. Data acquired may not be well structured in the sense that the underlying probabilistic structure may not exist and, if it does, is likely to be highly nonstationary. Examples of such data streams abound including Internet traffic data, point of sales inventory data, telephone traffic billing data, weather data, military and civilian intelligence data, NASA's Earth observing system satellite instrument data, high-energy physics particle collider data, and large-scale simulation data. It is our contention that massive streaming data represent a fundamentally new data paradigm and consequently require fundamentally new tools.

In December, 2002 and again in July of 2004, the Committee on Applied and Theoretical Statistics of the US National Academy of Science held workshops on streaming data. Many of the papers presented at the December workshop were fleshed out and published in a special issue of the December, 2003 issue of *Journal of Computational and Graphical Statistics*, vol. 12, no. 4. The reader is directed to this interesting issue.

Our own experience with streaming data has focused on Internet packet header data gathered at both the Naval Surface Warfare Center and at George Mason University. These data are highly non-stationary with time-of-day effects, day-of-week effects, and seasonal affects as well. At current data rates, our estimate in the year 2004 is that at George Mason University we could collect 24 terabytes of streaming Internet traffic header data. Clearly, this volume of streaming is not easily stored so new formulations of data analysis and visualization must be formulated. In the subsequent sections we give some suggestions.

8.1. Recursive analytic formulations

Much of the discussion in this and the next sections is based on (Wegman and Marchette, Marchette and Wegman, 2003, 2004, and Kafadar and Wegman, 2004). Each of these articles, but particularly the first, describe the structure of TCP/IP packet headers. The graphics illustrations we give later are based on Internet traffic data. In essence, the variables involved are source IP (SIP), destination IP (DIP), source port, destination port, length of session, number of packets, and number of bytes. We will not address specific details in this chapter because we are interested in general principles.

Because the data are streaming at a high rate, algorithmic issues must embrace two concepts. First, no data is permanently stored. The implication is that algorithms must operate on the data items and then discard. If each datum is processed and then discarded, we have a purely recursive algorithm. If a small amount of data for a limited time period are stored as in a moving window, we have a block recursion. Second, the algorithms must be relatively computationally simple in order to keep up with the data rates. With these two principles in mind, we can describe some pertinent algorithms.

8.1.1. Counts, moments and densities

Suppose we first agree that X_i, $i = 1, 2, \ldots$, represents the incoming data stream. Clearly, the count of the number of items can be accumulated recursively. In addition, the traditional $\overline{X}_n$ can be computed recursively by

$$\overline{X}_n = \frac{n-1}{n}\overline{X}_{n-1} + \frac{X_n}{n}.$$

Also clear is that moments of all orders can be computed recursively by

$$\sum_{i=1}^{n} X_i^k = \sum_{i=1}^{n-1} X_i^k + X_n^k.$$

A recursive form of the kernel density estimator was formulated by Wolverton and Wagner (1969) and independently by Yamato (1971):

$$f_n^*(x) = \frac{n-1}{n} f_{n-1}^*(x) + \frac{1}{nh_n} K\left(\frac{x - X_n}{h_n}\right)$$

where K is the smoothing kernel and h_n is the usual bandwidth parameter. Wegman and Davies (1979) proposed an additional recursive formulation and showed strong consistency and asymptotic convergence rates for

$$f_n^\dagger(x) = \frac{n-1}{n} \left(\frac{h_{n-1}}{h_n}\right)^{1/2} f_{n-1}^\dagger(x) + \frac{1}{nh_n} K\left(\frac{x - X_n}{h_n}\right),$$

where the interpretation of K and h_n is as above. Finally, we note that the adaptive mixtures described in Section 6.1 is also a recursive formulation.

The difficulty with all of these procedures is that they do not discount old data. In fact, both $1/n$ and $1/(nh_n)$ converge to 0 so that new data rather than old data is discounted. The exponential smoother has been traditionally used to discount older data. The general formulation is

$$Y_t = \sum_{i=0}^{\infty} (1-\theta)\theta^i X_{t-1}^k, \quad 0 < \theta < 1.$$

This may be reformulated recursively as

$$Y_t = \theta Y_{t-1} + (1-\theta) X_t^k.$$

It is straightforward to verify if $E[X_t^k] = E[X^k]$ is independent of t, then $E[Y_t] = E[X_t^k]$.

θ is the parameter that controls the rate of discounting. Small values, i.e. close to zero, discount older data rapidly while values close to one discount older data more slowly. The recursive density formulation can also be reformulated as an exponential smoother:

$$f_n(x) = \theta f_{n-1}(x) + \frac{1-\theta}{h_n} K\left(\frac{x - X_n}{h_n}\right).$$

Of course, discounting older data may be done simply by keeping a moving window of data, thus totally discarding data of a certain age. This is the so-called block recursion form. An additional approach is to use the geometric quantization as described in Section 4.2.

8.2. Evolutionary graphics

8.2.1. Waterfall diagrams and transient geographic mapping

Waterfall diagrams and transient geographic mapping are examples of evolutionary graphics. The idea of waterfall diagrams is to record data for a small epoch of time. In Figure 20, we study the source port as a function of time. In this particular example, we record all source ports for a short epoch. This is essentially binary data. If a source port was observed, it is record as a black pixel. If a source port is not recorded, the corresponding pixel is left white. At the end of the short epoch, the top line of the waterfall diagram is recorded and the next epoch begins. As the second epoch ends, the first line is pushed down in the diagram and the results of the second epoch are recorded on the top line. This procedure repeats until, say, 1000 lines are recorded. The bottom of the diagram represents the oldest epoch, the top the newest. As new data comes in, the oldest epoch is dropped of the bottom and the most recent is appended to the top. The diagonal streaks in Figure 20 correspond to increments in the source port, which is characteristic of the operating system of the particular computer. Such a diagram can readily make inferences about the operating system and detect potential intruders and unauthorized activity.

Transient geographic mapping is much harder to illustrate, but comparatively easy to understand. Most users belong to a class B network. Class A are typically reserved for very large providers of backbone services such as AT&T, Sprint, MCI and the like. Thus the first two octets can typically be identified with a corporate including ISPs, university

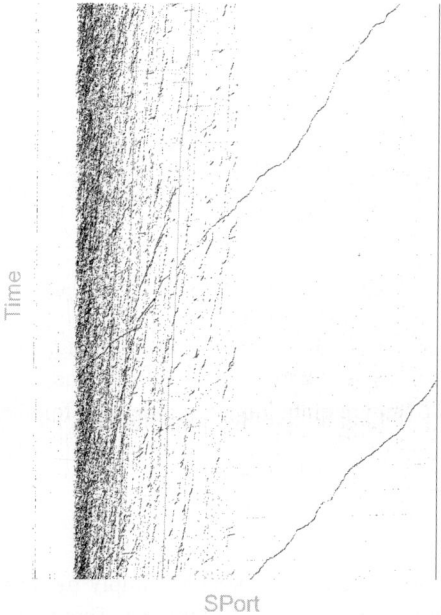

Fig. 20. Waterfall diagram of source port as a function of time. The diagonal lines indicate distinct operating systems with different slopes characteristic of different operating systems.

or government user, whether national or international. International corporations may use the same first two octets in widely geographically distributed regions, but to a large extent the first two octets can be reasonably geographically localized. We suggest two types of transient geographic mapping. The idea is to identify the first two octets with a geographic location, usually the headquarters of the class B network owners. Two forms of transient displays are desirable. First an unthresholded display for which every packet from a source IP lights up the source geographic point with a fairly rapid decay. Thus sessions for which many packets are being sent from a source will have a persistent bright display, with less persistent displays for sources sending fewer packets. This type of display is useful in a benign situation for gathering ground truth average traffic. However, in a denial of service attack, this would be useful for rapidly identifying the sources of the attack.

A second suggestion is to threshold the high-frequency traffic and plot only low-frequency packets with a long persistence. Characteristically, intruders tend to try to attack systems stealthily so that probing packets are sent infrequently so as not to arouse suspicion. Thus making infrequent packets from a particular pair of octets may be quite useful in identifying would-be intruders.

Figure 21 is another variation on an evolutionary graphic. Here a barplot records the number of sessions operating in successive 30-s non-overlapping intervals during one

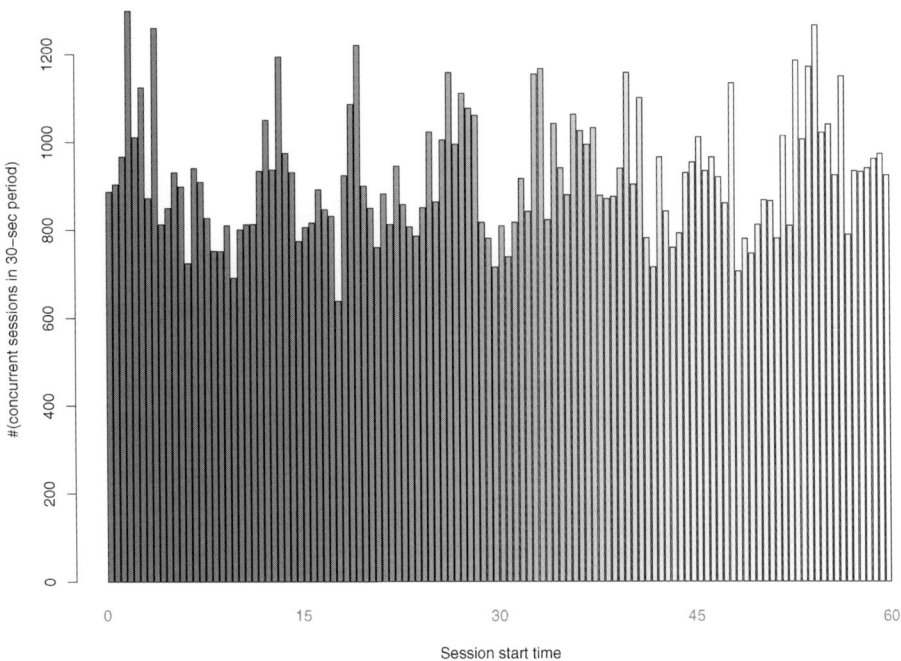

Fig. 21. Number of concurrent sessions plotted against start time. This is also an illustration of an evolutionary graphic where oldest data falls off on the left-hand side and newest data is plotted on the right-hand side.

hour. As a new 30-s epoch begins, the oldest data on the left is discarded and data for the newest 30-s increment is added on the right.

8.2.2. Block-recursive plots and conditional plots

In some circumstances, a desirable plot may not explicitly involve time as a variable. In these cases, we may not wish to discard the data immediately after its first use, but save it for a short period of time. This is explicitly what we mean by block recursion. Figure 22 represents a plot of $\log(1 + \sqrt{\text{number of bytes}})$ versus $\log(1 + \sqrt{\text{session duration}})$. In the case of Internet traffic data, most sessions are very short and involve a small number of bytes, with a smaller number of outliers. The log-sqrt transform rescales these small values so as to make the structure more visible. In Figure 22, there are 135 605 records covering one hour duration. 136 000 records qualifies as a small dataset according to the taxonomy in given in Section 2.1. Here the idea would be to wait for a small epoch, say 30 s to a minute, to accumulate new data, discard the oldest 30 s to one minute worth of data, and redraw the plot. The graphic, while not explicitly showing time, would be evolving as new data comes in. Anomalies, such as the horizontal, vertical, or diagonal linear features could potentially be investigated for their source potentially signifying unwanted intrusions.

The plot in Figure 22 is perhaps too cluttered to view anomalies directly. A second suggestion we make is to consider conditional plots. In Figure 23, we illustrate the same

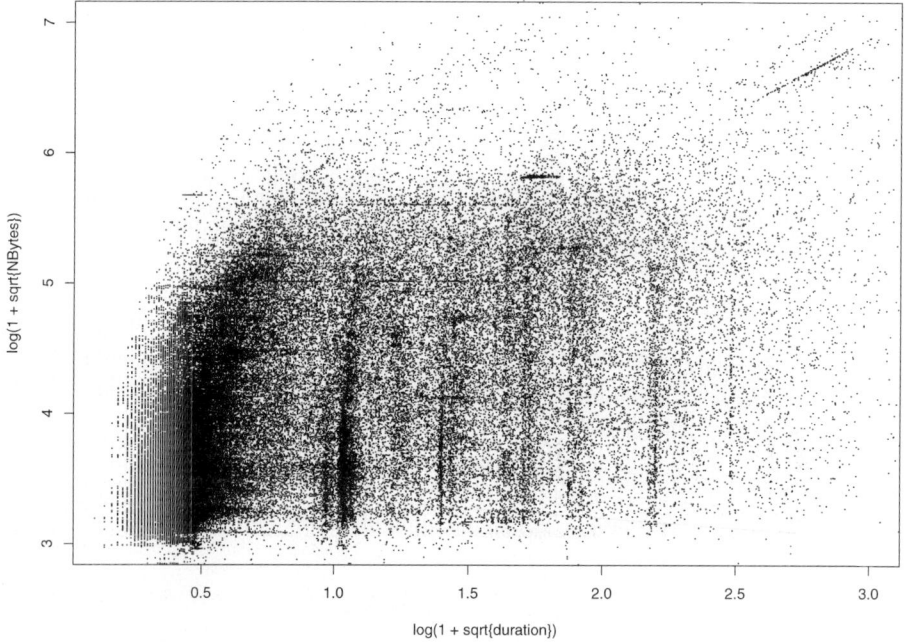

Fig. 22. The number of bytes versus the duration rescaled by a $\log(1 + \sqrt{(\cdot)})$ transformation. This is a block recursive evolutionary graphic.

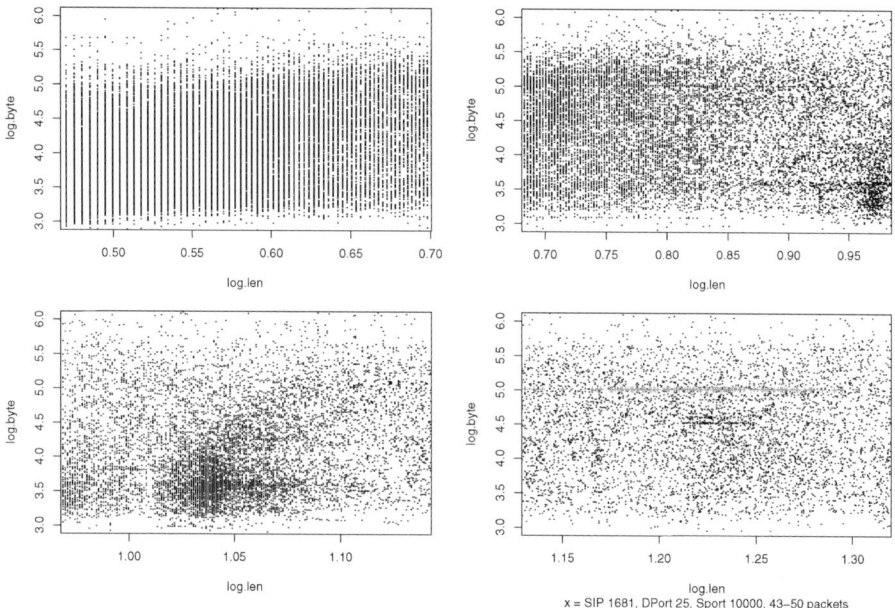

Fig. 23. Conditional block recursive plots of number of bytes versus duration. The conditioning here is on different scales of the duration. The duration is coded as len (for length) in these plots.

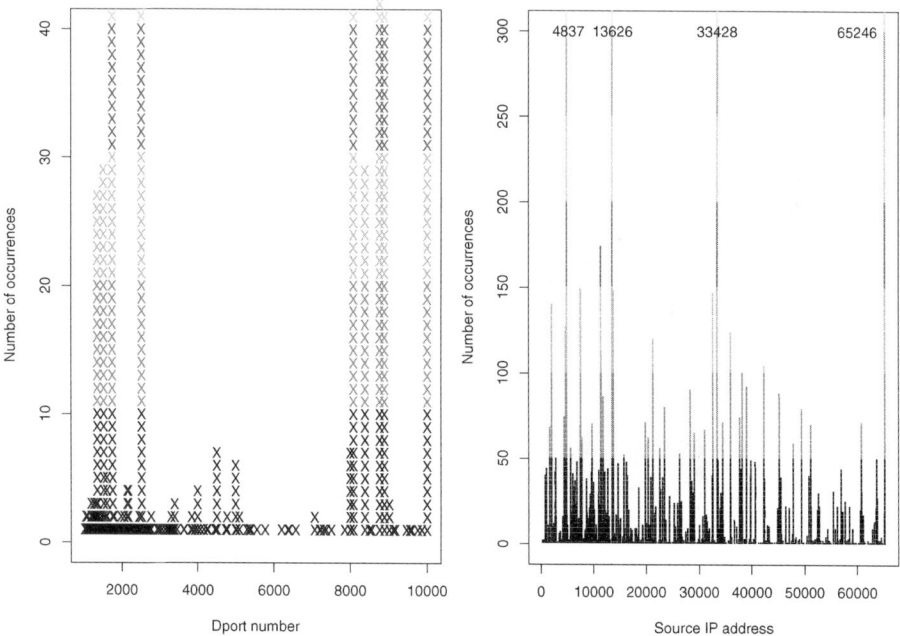

Fig. 24. Skyline plots of destination port and source IP address. This is also a snapshot of a block-recursive evolutionary graphic.

data but conditioned on 4 separate ranges of the independent variable. We could also condition on source ports, destination ports, or source IP address. The latter conditional plots would potentially be tied to specific users and anomalous behaviors at a source IP would suggest that the computer has been hijacked by an unauthorized user.

Finally we ask you to consider Figure 24. This figure is a pair of skyline plots of destination port and source IP address. Again these plots do not explicitly involve time. However, as with earlier block recursion plots, we discard older data and add new data. The height of the bars represents the level of activity of a particular destination port and source IP. As a destination port or source IP becomes more or less active the height of the corresponding bar will increase or decrease, again giving us a visual indication of possibly anomalous activity for a particular destination port or source IP.

9. A final word

In this chapter, we have tried to paint an overview of statistical data mining methods with a broad brush. There are, no doubt, favorite methods and areas of some researchers and users that we have omitted. The whole of this volume is intended to give insight into a vast field in which there are many players. This chapter and this volume reflect personal perspectives, which we hope are found to be interesting and informative.

Acknowledgements

The work of E.J.W. was supported by the Defense Advanced Research Projects Agency via Agreement 8905-48174 with The Johns Hopkins University. This contract was administered by the Air Force Office of Scientific Research. The work of JLS was supported by the Office of Naval Research under "In-House Laboratory Independent Research." Figures 21 through 24 were prepared by Professor Karen Kafadar who spent time visiting E.J.W. During her visit, she was support by a Critical Infrastructure Protection Fellows Program funded at George Mason University by the Air Force Office of Scientific Research. Much of this chapter summarizes work done with a vast array of collaborators of both of us and we gratefully acknowledge their contributions in the form of ideas and inspiration.

References

Agresti, A. (2002). *Categorical Data Analysis*. Wiley, New York.
Ahn, S., Wegman, E.J. (1998). A penalty function method for simplifying adaptive mixtures density estimates. *Comput. Sci. Statist.* **30**, 134–143.
Braverman, A. (2002). Compressing massive geophysical datasets using vector quantization. *J. Comput. Graph. Statist.* **11** (1), 44–62.
Buja, A., Cook, D., Asimov, D., Hurley, C. (2005). Computational methods for high-dimensional rotations in data visualization. In: Rao, C.R., Wegman, E.J., Solka, J.L. (Eds.), *Data Mining and Data Visualization*, *Handbook of Statistics*, vol. 24. Elsevier, Amsterdam. This volume.

Carr, D.B. (2005). Some recent graphics templates and software for showing statistical summaries. In: Rao, C.R., Wegman, E.J., Solka, J.L. (Eds.), *Data Mining and Data Visualization, Handbook of Statistics*, vol. 24. Elsevier, Amsterdam. This volume.
Carr, D.B., Nicholson, W.L. (1988). EXPLOR4: A program for exploring four-dimensional data. In: Cleveland, W.S., McGill, M.E. (Eds.), *Dynamic Graphics for Statistics*. Wadsworth, Belmont, CA, pp. 309–329.
Carr, D.B., Nicholson, W.L., Littlefield, R.J., Hall, D.L. (1986). Interactive color display methods for multivariate data. In: Wegman, E., DePriest, D. (Eds.), *Statistical Image Processing and Graphics*. Dekker, New York, pp. 215–250.
Chen, J.X. (2005). Data visualization and virtual reality. In: Rao, C.R., Wegman, E.J., Solka, J.L. (Eds.), *Data Mining and Data Visualization, Handbook of Statistics*, vol. 24. Elsevier, Amsterdam. This volume.
Cook, D., Buja, A., Cabrera, J., Hurley, C. (1995). Grand tour and projection pursuit. *J. Comput. Graph. Statist.* **4** (3), 155–172.
Everitt, B.S., Landau, S., Leese, M. (2001). *Cluster Analysis*, fourth ed. Oxford University Press, Oxford.
Fayyad, U.M., Piatetsky-Shapiro, G., Smyth, P., Uthurusamy, R. (1996). *Advances in Knowledge Discovery and Data Mining*. AAAI Press/MIT Press, Cambridge, MA.
Friedman, J.H., Tukey, J.W. (1974). A projection pursuit algorithm for exploratory data analysis. *IEEE Trans. Comput.* **C-23** (9), 881–889.
Hand, D.J. (2005). Pattern recognition. In: Rao, C.R., Wegman, E.J., Solka, J.L. (Eds.), *Data Mining and Data Visualization, Handbook of Statistics*, vol. 24. Elsevier, Amsterdam. This volume.
Hartigan, J.A. (1975). *Clustering Algorithms*. Wiley, New York.
Huber, P.J. (1992). Issues in computational data analysis. In: Dodge, Y., Whittaker, J. (Eds.), Computational Statistics, vol. 2, Physica, Heidelberg.
Huber, P.J. (1994). Huge data sets. In: Dutter, R., Grossmann, W. (Eds.), Compstat 1994: Proceedings, Physica, Heidelberg.
Hubert, M., Rousseeuw, P.J., Van Aelst, S. (2005). Multivariate outlier detection and robustness. In: Rao, C.R., Wegman, E.J., Solka, J.L. (Eds.), *Data Mining and Data Visualization, Handbook of Statistics*, vol. 24. Elsevier, Amsterdam. This volume.
Jee, J.R. (1985). A Study of Projection Pursuit Methods. PhD thesis. Department of Mathematical Sciences, Rice University.
Jee, J.R. (1987). Exploratory projection pursuit using nonparametric density estimation. In: *Proceedings of the Statistical Computing Section*. American Statistical Association, Alexandria, VA, pp. 335–339.
Kafadar, K., Wegman, E.J. (2004). Graphical displays of Internet traffic data. In: Compstat 2004: Proceedings, Physica, Heidelberg, pp. 287–301.
Khumbah, N.-A., Wegman, E.J. (2003). Data compression by geometric quantization. In: Akritas, M.G., Politis, D.N. (Eds.), *Recent Advances and Trends in Nonparametric Statistics*. Elsevier (North-Holland), pp. 35–48.
Maar, D. (1982). *Vision*. Freeman, New York.
Marchette, D.J., Solka, J.L. (2003). Using data images for outlier detection. *Comput. Statist. Data Anal.* **43** (4), 541–552.
Marchette, D.J., Wegman, E.J. (2004). Statistical analysis of network data for cybersecurity. *Chance* **17** (1), 8–18.
Marchette, D.J., Wegman, E.J., Priebe, C.E. (2005). Fast algorithms for classification using class cover catch digraphs. In: Rao, C.R., Wegman, E.J., Solka, J.L. (Eds.), *Data Mining and Data Visualization, Handbook of Statistics*, vol. 24. Elsevier, Amsterdam. This volume.
Minnotte, M., West, W. (1998). The data image: A tool for exploring high dimensional data sets. In: *Proceedings of Section on Statistical Graphics*. American Statistical Association, Alexandria, VA.
Osterberg, G. (1935). Topography of the layer of rods and cones in the human retina. *Acta Ophthal.* (suppl. 6), 1–103.
Poston, W.L., Wegman, E.J., Priebe, C.E., Solka, J.L. (1997). A deterministic method for robust estimation of multivariate location and shape. *J. Comput. Graph. Statist.* **6** (3), 300–313.
Priebe, C.E. (1994). Adaptive mixtures. *J. Amer. Statist. Assoc.* **98** (427), 796–806.
Priebe, C.E., Marchette, D.J. (1993). Adaptive mixture density estimation. *Pattern Recogn.* **26** (5), 771–785.
Priebe, C.E., Marchette, D.J., Park, Y., Wegman, E.J., Solka, J.L., Socolinsky, D.A., Karakos, D., Church, K.W., Guglielmi, R., Coifman, R.R., Lin, D., Healy, D.M., Jacobs, M.Q., Tsao, A. (2004). Iterative denoising for cross-corpora discovery. In: Compstat 2004: Proceedings, Physica, Heidelberg, pp. 381–392.

Rao, C.R. (2005). Canonical variate analysis and related methods for reduction of dimensionality and graphical representation. In: Rao, C.R., Wegman, E.J., Solka, J.L. (Eds.), *Data Mining and Data Visualization*, *Handbook of Statistics*, vol. 24. Elsevier, Amsterdam. This volume.

Scott, D.W. (1992). *Multivariate Density Estimation: Theory, Practice, and Visualization*. Wiley, New York.

Scott, D.W., Sain, S.R. (2005). Multi-dimensional density estimation. In: Rao, C.R., Wegman, E.J., Solka, J.L. (Eds.), *Data Mining and Data Visualization*, *Handbook of Statistics*, vol. 24. Elsevier, Amsterdam. This volume.

Sikali, E., 2004. Clustering Massive Datasets. PhD dissertation. School of Information Technology and Engineering, George Mason University.

Solka, J.L., Wegman, E.J., Poston, W.L. (1995). A new visualization technique to study the time evolution of finite and adaptive mixture estimators. *J. Comput. Graph. Statist.* **4** (3), 180–198.

Solka, J.L., Wegman, E.J., Priebe, C.E., Poston, W.L., Rogers, G.W. (1998). A method to determine the structure of an unknown mixture using the Akaike information criterion and the bootstrap. *Statist. Comput.* **8**, 177–188.

Solka, J.L., Bryant, A.C., Wegman, E.J. (2005). Text data mining with minimal spanning trees. In: Rao, C.R., Wegman, E.J., Solka, J.L. (Eds.), *Data Mining and Data Visualization*, *Handbook of Statistics*, vol. 24. Elsevier, Amsterdam. This volume.

Sutton, C.D. (2005). Classification and regression trees, bagging, and boosting. In: Rao, C.R., Wegman, E.J., Solka, J.L. (Eds.), *Data Mining and Data Visualization*, *Handbook of Statistics*, vol. 24. Elsevier, Amsterdam. This volume.

Tufte, E. (1983). *The Visual Display of Quantitative Information*. Graphics Press, Cheshire, CT.

Tufte, E. (1990). *Envisioning Information*. Graphics Press, Cheshire, CT.

Tufte, E. (1997). *Visual Explanations: Images and Quantities, Evidence and Narrative*. Graphics Press, Cheshire, CT.

Tukey, J.W. (1962). The future of data analysis. *Ann. Math. Statist.* **33**, 1–67.

Tukey, J.W. (1977). *Exploratory Data Analysis*. Addison–Wesley, Reading, MA.

Valyus, N.A. (1962). *Stereoscopy*. Focal Press, New York.

Wegman, E.J. (1990). Hyperdimensional data analysis using parallel coordinates. *J. Amer. Statist. Assoc.* **85**, 664–675.

Wegman, E.J. (1995). Huge data sets and the frontiers of computational feasibility. *J. Comput. Graph. Statist.* **4** (4), 281–295.

Wegman, E.J. (2003). Visual data mining. *Statist. Med.* **22**, 1383–1397 + 10 color plates.

Wegman, E.J., Davies, H.I. (1979). Remarks on some recursive estimators of a probability density. *Ann. Statist.* **7**, 316–327.

Wegman, E.J., King, R.D. (1990). A parallel implementation of data set mapping. In: *Proceedings of the Fourth Conference on Hypercubes, Concurrent Computers, and Applications*, pp. 1197–1200.

Wegman, E.J., Marchette, D.J. (2003). On some techniques for streaming data: A case study of Internet packet headers. *J. Comput. Graph. Statist.* **12** (4), 893–914.

Wegman, E.J., Solka, J.L. (2002). On some mathematics for visualizing high dimensional data. *Sanhkya Ser. A* **64** (2), 429–452.

Wilhelm, A. (2005). Interactive statistical graphics: The paradigm of linked views. In: Rao, C.R., Wegman, E.J., Solka, J.L. (Eds.), *Data Mining and Data Visualization*, *Handbook of Statistics*, vol. 24. Elsevier, Amsterdam. This volume.

Wilhelm, A., Wegman, E.J., Symanzik, J. (1999). Visual clustering and classification: The Oronsay particle size data set revisited. *Comput. Statist.* **14** (1), 109–146.

Wilkinson, L. (1999). *The Grammar of Graphics*. Springer-Verlag, New York.

Wolverton, C.T., Wagner, T.J. (1969). Asymptotically optimal discriminant functions for pattern classification. *IEEE Trans. Inform. Theory* **IT-15**, 258–265.

Yamato, H. (1971). Sequential estimation of a continuous probability density function and the mode. *Bull. Math. Statist.* **14**, 1–12.

From Data Mining to Knowledge Mining

Kenneth A. Kaufman and Ryszard S. Michalski

Abstract

In view of the tremendous production of computer data worldwide, there is a strong need for new powerful tools that can automatically generate useful knowledge from a variety of data, and present it in human-oriented forms. In efforts to satisfy this need, researchers have been exploring ideas and methods developed in machine learning, statistical data analysis, data mining, text mining, data visualization, pattern recognition, etc. The first part of this chapter is a compendium of ideas on the applicability of symbolic machine learning and logical data analysis methods toward this goal. The second part outlines a multistrategy methodology for an emerging research direction, called *knowledge mining*, by which we mean the derivation of high-level concepts and descriptions from data through symbolic reasoning involving both data *and* relevant background knowledge. The effective use of background as well as previously created knowledge in reasoning about new data makes it possible for the knowledge mining system to derive useful new knowledge not only from large amounts of data, but also from limited and weakly relevant data.

1. Introduction

We are witnessing the extraordinary expansion of computer accessible data about all kinds of human activities. The availability of these large volumes of data and our limited capabilities to process them effectively creates a strong need for new methodologies for extracting useful, task-oriented knowledge from them. There is also a need for methodologies for deriving plausible knowledge from small and indirectly relevant data, as in many practical areas, only such data may be available, e.g., fraud detection, terrorism prevention, computer intrusion detection, early cancer diagnosis, etc. This chapter addresses issues and methods concerned with developing a new research direction, called *knowledge mining*, which aims at solving both types of problems.

Current tools for analyzing data and extracting from it useful patterns and regularities primarily use conventional statistical methods, such as regression analysis, numerical taxonomy, multidimensional scaling, and more recent data mining techniques, such as classification and regression trees, association rules, and Bayesian nets (e.g., Daniel and

Wood, 1980; Tukey, 1986; Pearl, 1988, 2000; Morgenthaler and Tukey, 1989; Diday, 1989; Sharma, 1996; Neapolitan, 2003). While useful for many applications, these techniques have inherent limitations.

For example, a statistical analysis can determine distributions, covariances and correlations among variables in data, but is not able to characterize these dependencies at an abstract, conceptual level as humans can, and produce a causal explanation why these dependencies exist. While a statistical data analysis can determine the central tendencies and variance of given factors, it cannot produce a qualitative description of the regularities, nor can it determine a dependence on factors not explicitly provided in the data.

Similarly, a numerical taxonomy technique can create a classification of entities, and specify a numerical similarity among the entities assembled into the same or different categories, but it cannot alone build qualitative descriptions of the classes created and present a conceptual justification for including entities into a given category. Attributes and methods that are used to measure the similarity must be specified by a data analyst in advance. Popular classification (or decision) trees or forests can represent a relationship between the input and output variables, but their representation power is very modest. These methods may thus produce a very complex tree even for a conceptually simple relationship. Similarly, association rules, which are popular in data mining, have a limited representation power. Bayesian nets are very attractive for many applications, but typically rely on human input as to their structure, and can automatically determine only relatively simple interrelationships among attributes or concepts.

The above methods typically create patterns that use only attributes that are present in the data. They do not by themselves draw upon background domain knowledge in order to automatically generate additional relevant attributes, nor do they determine attributes' changing relevance to different data analysis problems. In cases where the goal is to address such tasks as those listed above, a data analysis system has to be equipped with a substantial amount of background knowledge, and be able to conduct symbolic reasoning involving that knowledge and the input data.

In efforts to satisfy the growing need for new data analysis tools that can overcome the above limitations, researchers have turned to ideas and methods developed in symbolic machine learning. The field of machine learning is a natural source of ideas for this purpose, because the essence of research in this field is to develop computational models for acquiring knowledge from facts and background knowledge.

The above and related efforts led to the emergence of a research area concerned with logical data analysis, and the development of methods for data mining and knowledge discovery (e.g., Lbov, 1981; Michalski et al., 1982, 1992; Zhuravlev and Gurevitch, 1989; Zagoruiko, 1991; Van Mechelen et al., 1993; Fayyad et al., 1996a, 1996b; Evangelos and Han, 1996; Brachman et al., 1996; Michalski and Kaufman, 1998; Han and Kamber, 2001; Hand et al., 2001; Alexe et al., 2003).

A natural step in this progression appears to be the development of systems that closely integrate databases with inductive learning and data mining capabilities (e.g., Michalski et al., 1992; Khabaza and Shearer, 1995; Han et al., 1996; Imielinski et al., 1996). Such systems would be able to, for example, automatically call upon a decision rule generator, regression analysis, conceptual clusterer or attribute generation operator, depending on the state of data analysis.

To achieve such a capability, a database needs to be integrated with a knowledge base and a wide range of data analysis, inductive learning, and data management methods. We call such systems *inductive databases*, and consider them to be a technical basis for implementing knowledge mining (Michalski and Kaufman, 1998; Kaufman and Michalski, 2003). An inductive database can answer not only those queries for which answers are stored in its memory, but also those that require the synthesis of *plausible knowledge*, generated by inductive inference from facts in the database and prior knowledge in the knowledge base.[1]

It should be mentioned that our meaning of the term "inductive database" is somewhat different from that of Imielinski and Mannila (1996) and De Raedt et al. (2002), by which they mean a database that stores both original data and induced hypotheses. Because from any nontrivial dataset a very large number of inductive hypotheses can be generated, and it is not known a priori which of them may be the most useful for the future tasks of interest, it is better in our view to equip a database with inductive inference capabilities, rather than store only their amassed results. These capabilities should be tightly integrated with the query language, so that they are transparent to the user, who can apply them without having to invoke separate data analysis or inductive learning programs.

This chapter discusses selected ideas and methods for logical (or conceptual) data analysis, initiated by Michalski et al. (1982), which provide a basis for the development of a knowledge mining methodology and its implementation in an inductive database system. The chapter is an update and extension of our earlier work presented in Michalski and Kaufman (1998).

While this chapter presents many issues generally and makes many references to research done by others, its main focus is on the ideas and methods developed by the authors initially at the University of Illinois at Urbana–Champaign, and more recently at the George Mason University Machine Learning and Inference Laboratory. While the methods are presented in the context of analyzing numeric and symbolic data, they can also be applied to text, speech or image mining (e.g., Bloedorn et al., 1996; Umann, 1997; Cavalcanti et al., 1997; Michalski et al., 1998).

2. Knowledge generation operators

This section discusses classes of symbolic learning methods that can serve as *knowledge generation operators* in logical data analysis and knowledge mining.

2.1. Discovering rules and patterns via AQ learning

An important class of tools for knowledge discovery in databases stems from concept learning methods developed in machine learning. Given collections of examples

---

[1] It may be interesting to note that R.S. Michalski first introduced and taught this concept in his course on deductive and inductive databases in 1973 at the University of Illinois at Urbana–Champaign.

User 1 ⇐ session_time_new_window ⩽ 3 hours (368, 8340)
         & #characters_in_protected_words = 0 (110, 919)
         & #processes_in_current_window ⩽ 7 (203, 4240)
         & #windows_opened ⩽ 16 (369, 7813): 65, 0

User 1 ⇐ process_name = explorer (93, 875)
         & #characters_in_protected_words = 9..24 (161, 2999)
         & #processes_in_current_window ⩽ 7 (203, 4240)
         & #windows_opened ⩽ 16 (369, 7813)
         & #protected_words_in_window_title = 1 (109, 3304): 31, 0

Fig. 1. Two rules in User 1's profile learned by AQ21.

of different concepts (or decision classes), the concept learning program hypothesizes a general description of each class. Some inductive methods use a fixed criterion for choosing the description from a large number of possible hypotheses, while others allow the user to define a criterion that reflects the problem at hand. A concept description can be in the form of a set of decision rules, a decision tree, a semantic net, etc. A decision rule can also take on many different forms.

In the AQ learning methodology, which we will discuss here, the general form of a decision (or classification) rule is:

$$\text{CONSEQUENT} \Leftarrow \text{PREMISE} \;|\_\; \text{EXCEPTION} \qquad (1)$$

where CONSEQUENT is a statement indicating a decision, a class, or a concept name to be assigned to an entity (an object or situation) that satisfies PREMISE, provided it does not satisfy EXCEPTION; PREMISE is a logical expression (e.g., a product of logical conditions or a disjunction of such products), EXCEPTION (optional) defines conditions under which the rule does not apply; and ⇐ denotes implication.

If PREMISE is a disjunctive description (a disjunction of products), then rule (1) can be transformed into several rules with the same CONSEQUENT, in which PREMISE is a conjunctive description (a single product of conditions). For example, Figure 1 shows two rules representing a disjunctive description of a computer user profile learned by the AQ21 learning program in a study on learning patterns in computer user behavior (Michalski et al., 2005). AQ21 is the most recent member of the family of AQ inductive learning programs (Wojtusiak, 2004). The rules characterize two different patterns of the behavior of User 1.

The first rule says that User 1 is indicated if the current window was opened less than 3 hours into the session, there are no protected words[2] in the window title, the number of active processes in the current window does not exceed seven, and the total number of windows opened during the session does not exceed 16.

The second rule says that User 1 is also indicated if the name of the active process is explorer, there are 9 to 24 characters in the protected words in the window title, the number of active processes in the current window does not exceed seven, the total number of windows opened during the session does not exceed 16, and there is only one protected word in the window title.

---

[2] In order to protect user privacy, during preprocessing, all words in the window title that were not names of system programs or functions were replaced with randomly selected numbers. The words that were not affected by this sanitization were called "protected."

The pairs of numbers in parentheses after each condition in the rules indicate the number of positive and negative training events, respectively, that support (are covered by) the condition. For example, the first condition in the first rule covers 368 training events of User 1's behavior, and 8340 training events of behavior of the other users. The pair of numbers at the end of each rule indicates the total number of positive and negative training events covered by the rule. For example, the first rule covers 65 events from the training data of User 1, but no events in the training data of other users. Because both rules cover no negative events, they are said to be *consistent* with the training data.

The rules in Figure 1 are examples of simple *attributional rules*, which are decision rules expressed in *attributional calculus*, a logic system used for representing knowledge in AQ learning (Michalski, 2004). A set of attributional rules with the same consequent (indicating the same decision) is called a *ruleset*. A collection of rulesets whose consequents span all values of the output (decision) variable is called a *ruleset family*, or a *classifier*. Rules, rulesets and classifiers are examples of *attributional descriptions*, as they involve only attributes in characterizing entities.

In contrast to attributional descriptions, *relational* (aka *structural*) descriptions employ not only attributes but also multi-argument predicates representing relationships among components of the entities. Such descriptions are produced, for example, by the INDUCE inductive learning programs (Larson, 1977; Bentrup et al., 1987), and by inductive logic programs (e.g., Muggleton, 1992). Constructing structural descriptions requires a more complex description language that includes multi-argument predicates, for example, PROLOG, or Annotated Predicate Calculus (Michalski, 1983; Bratko et al., 1997).

For database exploration, attributional descriptions appear to be the most important and the easiest to implement, because most databases characterize entities in terms of attributes. As one can see, they are also easy to interpret and understand. A simple and popular form of attributional description is a decision or classification tree. In such a tree, nodes correspond to attributes, branches stemming from the nodes correspond to attribute values, and leaves correspond to individual classes (e.g., Quinlan, 1986). A decision tree can be transformed into a set of simple attributional rules (a ruleset family) by traversing all paths from the root to individual leaves. Such rules can often be simplified by detecting superfluous conditions in them, but such a process can be computationally very costly (e.g., Quinlan, 1993). The opposite process of transforming a ruleset family into a decision tree is simple, but may introduce superfluous conditions because a tree representation is less expressive than a rule representation (Imam and Michalski, 1993).

The attributional calculus distinguishes between many different types of attributes, such as nominal, rank, cyclic, structured, interval, ratio, absolute, set-valued, and compound (Michalski, 2004). By distinguishing so many attribute types, attributional calculus caters to the needs of knowledge mining. By taking into consideration different attribute types, a learning system can be more effective in generating inductive generalizations, because different generalization rules apply to different attribute types (Michalski, 1983). Thus, a specification of such attribute types constitutes a form of background knowledge used in knowledge mining.

The aforementioned AQ21 is a multipurpose learning system for generalizing cases into rules and detecting patterns in data. It is an updated version of AQ19, which was used as a major module of the INLEN system developed for testing initial ideas and methods for knowledge mining (see Section 5). The input to AQ21 consists of a set of training examples representing different concepts (classes, decisions, predictions, etc.), parameters defining the type of description to be learned and how it should be learned, a multi-criterion measure of description quality, and background knowledge, which includes a specification of domains and types of attributes, hierarchical structures defining structured domains, and arithmetic and logical rules suggesting ways to improve the representation space and/or define constraints on it. The measure of description quality (or preference criterion) may refer to computational simplicity of the description, its generality level, the cost of measuring attributes in the description, or an estimate of its predictive ability.

Many symbolic learning programs learn rules that are *consistent* and *complete* with regard to the input data. This means that they completely and correctly classify every distinct training example. Others select rules according to a description quality criterion that does not necessarily give maximal weight to a description's consistency. The AQ21 learning program, depending on the setting of its parameters, can generate either complete and consistent descriptions, or strong patterns that can be partially inconsistent and incomplete.

2.2. Types of problems in learning from examples

Descriptions generated from examples by symbolic learning programs may take two forms, depending on whether the learning goal is to describe members of a particular concept (a group of entities), or contrast them against other groups. Descriptions that enumerate the common properties of the entities in each group are called *characteristic descriptions*. Descriptions that specify differences between groups are called *discriminant descriptions*.

Some methods for concept learning assume that examples do not have errors, that all attributes have a specified value in them, that all examples are located in the same database, and that concepts to be learned have a precise ("crisp") description that does not change over time. In many situations one or more of these assumptions may not hold. This leads to a variety of more complex machine learning and data mining problems and methods for solving them:

- *Learning from noisy data*, i.e., learning from examples that contain a certain amount of errors or noise (e.g., Quinlan, 1990; Michalski and Kaufman, 2001). These problems are particularly important for data and knowledge mining because databases frequently contain some amount of noise.
- *Learning from incomplete data*, i.e., learning from examples in which the values of some attributes are unknown (e.g., Dontas, 1988; Lakshminarayan et al., 1996).
- *Learning from distributed data*, i.e., learning from spatially distributed collections of data that must be considered together if the patterns within them are to be exposed (e.g., Ribeiro et al., 1995).

- *Learning drifting or evolving concepts*, i.e., learning concepts that are not stable but changing over time, randomly or in a certain general direction. For example, the "area of interest" of a computer user is usually an evolving concept (e.g., Widmer and Kubat, 1996).
- *Learning concepts from data arriving over time*, i.e., incremental learning in which currently held hypotheses characterizing concepts may need to be updated to account for the new data (e.g., Maloof and Michalski, 2004).
- *Learning from biased data*, i.e., learning from a data set that does not reflect the actual distribution of events (e.g., Feelders, 1996).
- *Learning flexible concepts*, i.e., concepts that inherently lack precise definition and whose meaning is context-dependent; approaches concerned with this topic include *fuzzy sets* (e.g., Zadeh, 1965; Dubois et al., 1993), *two-tiered concept representations* (e.g., Michalski, 1990; Bergadano et al., 1992), and *rough sets* (e.g., Pawlak, 1991; Slowinski, 1992; Ziarko, 1994).
- *Learning concepts at different levels of generality*, i.e., learning descriptions that involve concepts from different levels of generalization hierarchies (e.g., Kaufman and Michalski, 1996). An example of such problem is learning the concept of a liver disease versus the concept of liver cancer.
- *Integrating qualitative and quantitative discovery*, i.e., determining sets of equations that fit a given set of data points, and qualitative conditions for the application of these equations (e.g., Falkenhainer and Michalski, 1990).
- *Qualitative prediction*, i.e., discovering patterns in sequences or processes and using these patterns to qualitatively predict the possible continuation of the given sequences or processes (e.g., Davis, 1981; Michalski et al., 1985, 1986; Dieterich and Michalski, 1986).

Each of these problems is relevant to the derivation of useful knowledge from a collection of data and knowledge. Therefore, it can be asserted that methods for solving these problems developed in the area of machine learning are directly relevant to logical data analysis and knowledge mining.

2.3. Clustering of entities into conceptually meaningful categories

Another class of machine learning methods relevant to knowledge mining concerns the problem of building a conceptual classification of a given set of entities. The problem is similar to that considered in traditional cluster analysis, but is defined in a different way, which allows for knowledge-based constructions. Given a set of attributional descriptions of entities, a description language for characterizing classes of such entities, and a classification quality criterion, the problem is to partition entities into classes that have a simple and meaningful description in the given description language and maximize the classification quality criterion. Thus, a conceptual clustering method seeks not only a classification structure of entities, but also an understandable description of the proposed classes (clusters). An important, distinguishing aspect of conceptual clustering is that, unlike in similarity-based cluster analysis, the properties of these class descriptions are taken into consideration in the process of determining the partition of entities into clusters (e.g., Michalski and Stepp, 1983).

To clarify the difference between conceptual clustering and conventional clustering, notice that a conventional clustering method typically determines clusters on the basis of a similarity measure that is a function solely of the properties (attribute values) of the entities being compared, and not of any other factors:

$$\text{Similarity}(A, B) = f\big(\text{properties}(A), \text{properties}(B)\big) \tag{2}$$

where A and B are entities being compared.

In contrast, a conceptual clustering program creates clusters based on *conceptual cohesiveness*, which is a function of not only properties of the entities, but also of two other factors: the *description language* L, which the system uses for describing the classes of entities, and of the *environment* E, which is the set of neighboring examples:

$$\text{Conceptual cohesiveness}(A, B) = f\big(\text{properties}(A), \text{properties}(B), L, E\big). \tag{3}$$

Thus, two objects may be *similar*, i.e., close according to some distance measure, while having a low conceptual cohesiveness, or *vice versa*. An example of the first situation is shown in Figure 2. The points (black dots) A and B are "close" to each other; in fact, they are closer than any other pair of points in the figure. They would therefore be placed into the same cluster by any technique based solely upon the distances between the points. However, these points have small conceptual cohesiveness, because they can be viewed as belonging to configurations representing different concepts.

A conceptual clustering method, if equipped with an appropriate description language, would cluster the points in Figure 2 into a circle and a horizontal line, as people normally would. A classification quality criterion used in conceptual clustering may involve a variety of factors, such as the *fit* of a cluster description to the data (called sparseness), the *simplicity* of the description, and other properties of the entities or the concepts that describe them (Michalski and Stepp, 1983; Stepp and Michalski, 1986). Ideas on employing conceptual clustering for structuring text databases and creating concept lattices for discovering dependencies in data are described by Carpineto and Romano (1995a, 1995b). The concepts created through the clustering are linked in lattice structures that can be traversed to represent generalization and specialization relationships.

Recent advances in traditional, similarity-based clustering have attempted to go beyond the limitations described above as well. Subspace clustering (e.g., Agrawal et al., 1998, 1999; Wang et al., 2004) alters the description language and the environment before creating groups by projecting the event space on a subspace that produces better clusters. Various approaches to manifold learning (Tenenbaum et al., 2000;

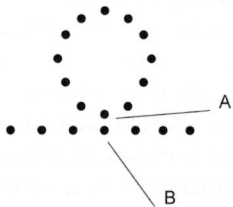

Fig. 2. An illustration of the difference between closeness and conceptual cohesiveness.

Saul and Roweis, 2003; Belkin and Niyogi, 2004) also attempt to reduce the dimensionality of the problem; they differ from subspace clustering in that their focus is preserving the relationships among the datapoints, rather than compacting them.

2.4. Automated improvement of the search space: constructive induction

Most methods for learning from examples assume that the attributes used for describing examples are sufficiently relevant to the learning problem at hand. This assumption does not always hold in practice. Some attributes used in the examples may not be directly relevant (e.g., if the target concept is based on density, and we only have mass and volume attributes), and others may be irrelevant or *nonessential* (e.g., if the target concept is predicting a student's performance in a class, and an attribute indicates the student's height). An important characteristic of logical analysis methods is that they can relatively easily determine irrelevant or nonessential attributes.

An attribute is *nonessential* if there is a complete and consistent description of the concepts to be learned that does not use this attribute. Thus, a nonessential attribute may be either irrelevant or relevant, but will by definition be dispensable. There may exist a set of attributes, each one by itself nonessential, yet some member of the set must be present in order to generate a complete and consistent attributional description. Inductive learning programs such as the rule-learning program AQ21, or the decision tree-learning C4.5, can cope relatively easily with a large number of nonessential attributes in their input data.

If there are very many nonessential attributes in the input data descriptions, the complexity of the learning process may significantly increase, along with the execution time, and the risk for generating spurious rather than relevant knowledge may increase. Such a situation calls for a method that can efficiently determine the most relevant attributes for the given problem from among all those given initially. Only the most relevant attributes should be used in the learning process.

Determining the most relevant attributes is therefore a useful data exploration operator. Such an operator can also be useful for the data analyst on its own merit, as it may be important to know which attributes are most discriminatory for a given learning task. By removing less relevant attributes, the representation space is reduced, and the problem becomes simpler. Thus, such a process is a form of improving the representation space. Some methods for finding the most relevant attributes are described by Zagoruiko (1972), Baim (1982), Fayyad and Irani (1992), Caruana and Freitag (1994), Langley (1994).

In applications in which the attributes originally given may only be weakly or indirectly relevant to the problem at hand, there is a need for generating new, more relevant attributes that may be functions of the original attributes. These functions may be simple, e.g., a product or sum of a set of the original attributes, or very complex, e.g., a Boolean attribute based on the presence or absence of a straight line or circle in an image (Bongard, 1970). Finally, in some situations, it will be desirable to abstract some attributes, that is, to group some attribute values into units, and thus reduce the attribute's range of possible values. A quantization of continuous attributes is a common example of such an operation (e.g., Kerber, 1992).

All the above operations – removing less relevant attributes, adding more relevant attributes, and abstracting attributes – are different means of improving the original representation space for learning. A learning process that consists of two (intertwined) phases, one concerned with the construction of the "best" representation space, and the second concerned with generating the "best" hypothesis in the found space is called *constructive induction* (Michalski, 1978, 1983; Bloedorn et al., 1993; Wnek and Michalski, 1994; Bloedorn and Michalski, 1998). An example of a constructive induction program is AQ17 described in Bloedorn et al. (1993), which performs all three types of operators for improving the original representation space. In AQ17, the process of generating new attributes is performed through the combination of existing attributes using mathematical and/or logical operators, and then selecting the "best" combinations.

2.5. Reducing the amount of data: selecting representative examples

When a database is very large, determining general patterns or rules characterizing different concepts may be very time-consuming. To make the process more efficient, it may be useful to extract from the database the most representative or important cases (examples) of given classes or concepts. Even a random extraction should not be costly in the case of very large datasets, as the selected set will likely be quite representative. Most methods of heuristic selection of examples attempt to select those that are either most typical or most extreme (assuming that there is not too much noise in the data). A method for determining the most representative examples, called *"outstanding representatives"*, is described by Michalski and Larson (1978).

2.6. Integrating qualitative and quantitative methods of numerical discovery

In a database that contains numerical and symbolic attributes, a useful discovery could be an equation binding numerical attributes. A standard statistical technique for this purpose is regression analysis. This technique requires that the general form of the equation is provided to the system, as in multivariate linear regression. The application of machine learning to quantitative discovery produced another approach to this problem that does not require a specification of the form of the equation.

For instance, from a table of planetary data including planets' names, planet's masses, their densities, distances from the sun, periods of rotation, lengths of local years, and the number of moons, a quantitative discovery system would derive Kepler's Law, which states that the cube of the planet's distance from the sun is proportional to the square of the length of its year. The attributes such as the planet's name and the number of moons would be ignored.

Research on quantitative discovery was pioneered by the BACON system (Langley et al., 1983), and then followed by many other systems, such as COPER (Kokar, 1986), FAHRENHEIT (Zytkow, 1987), and ABACUS (Falkenhainer and Michalski, 1990). Similar problems have been explored independently by Zagoruiko (1972) in Russia under the name of empirical prediction.

Some equations may not apply directly to data, because of an inappropriate value of a constant, or different equations may apply under different qualitative conditions. For example, in applying Stoke's Law to determine the velocity of a falling ball, if the ball is falling through a vacuum, its velocity depends on the length of time it has been falling and on the gravitational force being exerted upon it. A ball falling through some sort of fluid will reach a terminal velocity dependent on the radius and mass of the ball and the viscosity of the fluid.

The program ABACUS (Greene, 1988; Falkenhainer and Michalski, 1990; Michael, 1991) is able to determine quantitative laws under different qualitative conditions. It does so by partitioning the data into subsets, each of which adheres to a different equation determined by a quantitative discovery module. The qualitative discovery module can then determine conditions/rules that characterize each of these example sets. For example, given a table containing data on how fast different balls fall through different media, ABACUS can discover these patterns based on the medium of descent:

if Medium = vacuum then $v = 9.8175t$,

if Medium = glycerol then $vr = .9556m$,

if Medium = castor oil then $vr = .7336m$.

2.7. Predicting processes qualitatively

Most programs for learning rules from examples determine them from examples of various classes of objects. An example of a concept represents that concept regardless of its relationship to other examples. Contrast that with a sequence prediction problem, in which a positive example of a concept is directly dependent on the position of the example in the sequence.

For example, Figure 3 shows a sequence of nine figures. One may ask what object plausibly follows in the tenth position. To answer such a question, one needs to search for a pattern in the sequence, and then use the pattern to predict a plausible sequence continuation. In *qualitative prediction*, the problem is not to predict a specific value of a variable (as in time series analysis), but rather to *qualitatively* characterize a plausible subsequent object, that is, to describe plausible properties of that future object.

In the example in Figure 3, one may observe that the sequence consists of circles with parallel shading and squares with dark shapes inside. The figures may be rotated in different orientations at 45-degree intervals. But is there a consistent pattern?

To determine such a pattern, one can employ different *descriptive models*, and instantiate the models to fit the particular sequence. The instantiated model that best fits the data is then used for prediction. Such a method was initially developed by Dieterrich and Michalski (1986), and then generalized in SPARC/G system to handle arbitrary sequences of entities described by attributes (Michalski et al., 1986). The method employs three descriptive models – periodic, decomposition, and DNF.

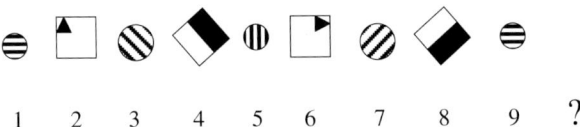

Fig. 3. An example of a qualitative prediction problem.

The *periodic model* is used to detect repeating patterns in a sequence. For example, Figure 3 depicts a recurring pattern that alternates round and square objects. In general, there can also be periodic subsequences within the periodic sequences. In the figure, the round objects form a subsequence in which individual objects rotate leftward by 45 degrees and alternate between small and large. The square objects have a subsequence alternating between those with a triangle in the corner and those half-filled. Each subsequence is rotating clockwise.

The second model, the *decomposition model*, is used to characterize a sequence by decision rules in the following general form: "If one or more of the previous elements of the sequence have a given set of characteristics, then the next element will have the following characteristics ..." One such rule that applies to the sequence in Figure 3 would state that if an element in the sequence has a triangular component, then the next element in the sequence will have a diagonally shaded component; otherwise it will have no diagonal shading.

The third model, the DNF (disjunctive normal form) or "catch-all" model, tries to capture general properties characterizing the whole sequence. For example, for the sequence in Figure 3, it could instantiate to a statement such as "all elements in the sequence are round or square, their interiors are either shaded, or contain a dark rectangle or triangle, etc.

Given the problem in Figure 3, SPARC/G would find the following pattern based on the periodic model:

$$\text{Period}([\text{shape} = \text{circle}]$$
$$\&\ [\text{shading} = \text{parallel}][\text{orientation}(i+1) = \text{orientation}(i) + 45],$$
$$[\text{shape} = \text{square}]\ \&\ [\text{orientation}(i+1) = \text{orientation}(i) + 45]).$$

The pattern can be paraphrased: there are two phases in a repeating period (their descriptions are separated by a comma). The first phase involves a circular figure, and the second phase a square figure. The circular figure is shaded and rotates to the right and the square figure also rotates to the right by 45 degrees in relation to its predecessor. Based on this pattern, a plausible next figure in the sequence would be a square figure rotated clockwise 45 degrees in relation to the previous square figure. This rule does not specify the contents of that square.

The qualitative prediction capabilities described above can be useful for conceptual exploration of temporal databases in many application domains, such as agriculture, medicine, robotics, economic forecasting, computer intrusion detection, etc.

2.8. Knowledge improvement via incremental learning

One of the very important aspects of the application of machine learning to logical data analysis is the existence of methods for incremental learning that can improve data generalizations when new data become available. This is analogous to Bayesian learning, but it is not the posterior probability of a description that being improved, but rather the description itself.

Incremental learning can take three forms, depending on how much data from which the prior knowledge was generated is available. Zero-memory learning, in which none of the earlier data is retained, is more economical, while full-memory incremental learning, in which all earlier training examples are retained, is likely to result in more accurate descriptions, provided that issues of concept drift can be accounted for. Partial-memory incremental learning is an attempt to strike a balance between these two extremes, by selecting for retention only the cases most likely to be of use later on.

The zero-memory algorithm is straightforward. New data that contradicts prior hypotheses is integrated into the prior hypotheses through specialization operators that reshape the hypotheses into ones consistent with the new data. The best of these modified hypotheses are then input to the learner along with the new data points (e.g., Michalski and Larson, 1983).

In the full-memory incremental learning, the new data that contradict the previous description are first filtered by removing any examples which were identical to earlier training examples. The prior hypotheses are then appropriately specialized or generalized to account for new data, while preserving consistency and completeness of the description with regard to all past examples (e.g., Reinke and Michalski, 1988).

The partial memory method utilizes a selection of prior data for retention. Some partial memory systems select examples that are near the perceived boundaries of the concepts, either based on the incoming datastream (e.g., Kibler and Aha, 1987), or on induced rules (e.g., Maloof and Michalski, 2000, 2004). Others retain a seed positive example (e.g., Elio and Watanabe, 1991), or maintain only negative examples of a concept so as to define a boundary (e.g., Iba et al., 1988).

2.9. Summarizing the logical data analysis approach

To help the reader develop a rough sense of what is different and new in the above, let us consider operations typically performed by traditional multivariate data analysis methods. These include computing mean-corrected or standardized variables, variances, standard deviations, covariances and correlations among attributes; principal component analysis; factor analysis; cluster analysis; regression analysis; multivariate analysis of variance; and discriminant analysis. All these methods can be viewed as primarily oriented toward numerical characterizations of data.

In contrast, the logical data analysis approach, described above, focuses on developing symbolic logic-style descriptions of data, which may characterize data qualitatively, differentiate among classes, create a "conceptual" classification of data, qualitatively predict sequences, etc. These techniques are particularly well-suited for developing descriptions and seeking patterns in data that involve nominal (categorical), rank, and structured attributes (with hierarchically-ordered domains), although they can handle all types of attributes.

Another important distinction between the two approaches to data analysis is that purely statistical methods are particularly useful for globally characterizing a set of objects, but not so for determining a description for predicting class membership of individual objects (with some exceptions, e.g., classification trees). A statistical operator may determine, for example, that the average lifespan of a certain type of automobile is 7.3 years, but it may not provide conditions indicating the lifespan of an automobile with particular characteristics, nor the ability to recognize the type of a specific automobile from its description. A symbolic machine learning approach is particularly useful for such tasks. It may create a description such as "if the front height of a vehicle is between 5 and 6 feet, body color is silver or gray, and the driver's seat is 2 to 3 feet above the ground, then the vehicle is likely to be a minivan of brand X." Such descriptions are particularly suitable for classifying future, not yet observed entities based on their properties.

The knowledge mining methodology aims at integrating a wide range of strategies and operators for data exploration based on both machine learning research and statistical methods. The reason for such a multistrategy approach is that a data analyst may be interested in many different types of information about the data, requiring different exploratory strategies and different operators.

3. Strong patterns vs. complete and consistent rules

In its early stages of development, machine learning was oriented primarily toward methods that produce consistent and complete descriptions of the training data, that is, descriptions that explain ("cover") all positive training examples of the target concepts, and none of the negative examples. In practical applications, however, data frequently contain some errors; therefore, a complete and consistent description will likely overfit the data, producing incorrect micro-patterns. Also, in practice, one may be more interested in determining a simple but not completely correct pattern than a complex but a correct one.

There have been several methods developed to determine such patterns using the symbolic learning approach. One method is through postprocessing of learned descriptions using ruleset optimization (e.g., Bergadano et al., 1992). The well-known decision tree pruning is a simple form of the same idea (e.g., Quinlan, 1993). In this method, an initially learned complete and consistent description is simplified by removing statistically insignificant components (subtree pruning in decision tree learning, or rule truncation in AQ learning), or optimizing some of its components (rule optimization in AQ learning).

Another method is to optimize descriptions during the rule generation process. Such a method employs a rule quality criterion, defined by the user, that specifies a tradeoff between completeness and consistency of a rule. At each stage of rule learning, candidate hypotheses are overgeneralized (introducing inconsistency, but increasing rule coverage), and then evaluated using the rule quality criterion. Whichever variant of the original hypothesis scores best is retained as input to the next iteration of rule learning. In this way, negative examples are ignored if the creation of a strong pattern requires it.

Such a method was implemented in AQ learning, as an additional option to rule truncation (Michalski and Kaufman, 2001). The method uses a rule quality measure $Q(w)$, where w is a user-specified weight parameter controlling the relative importance of rule coverage in relation to rule consistency gain.

Specifically, given a training dataset consisting of P positive examples of a concept and N negative examples (examples of other concepts), and given a rule R that covers p positive examples and n negative examples, the rule's coverage (relative support) is defined as

$$\text{cov}(R) = p/P. \tag{4}$$

The consistency of rule R is defined as the fraction of covered examples that are positive (correctly classified), or

$$\text{cons}(R) = p/(p+n). \tag{5}$$

However, without taking into account the distribution of training examples, a rule's consistency alone does not provide a strong indication of the predictive utility of the rule. Thus, we instead apply *consistency gain* (cgain), which measures the rule's improvement in performance over the expected performance of blindly guessing the positive class. A normalization factor ensures that this measure will be zero when the rule performs no better than such a blind guess, and 1 when the rule achieves 100% consistency.

$$\text{cgain}(R) = \bigl(p/(p+n)\bigr) - \bigl(P/(P+N)\bigr)\bigl((P+N)/N\bigr). \tag{6}$$

The $Q(w)$ formula then combines the coverage and consistency gain terms through multiplication (so that Q will be 1 when both terms are 1, and Q will be 0 when either term is 0), and accordingly, the weight w is computed as an exponent in the equation. Specifically:

$$Q(w) = \text{cov}(R)^w \text{cgain}(R)^{1-w}. \tag{7}$$

It should be noted that both cov(R) and cgain(R) are functions of the rule's positive and negative support. Other programs typically also use various functions of rule's positive and negative support in evaluating the descriptions they generate.

Table 1 presents examples of how different methods can choose differently from the same set of candidate rules (Kaufman and Michalski, 1999). In the table, three separate data sets are assumed, each with 1000 training examples. In data set A, 200 of the training examples are in the positive class; in data set B, 500 training examples are, and in data set C, 800 training examples are. For each data set, seven rules are hypothesized, each covering different numbers of positive and negative examples. The table shows how each set of seven rules would be ranked by information gain, by the programs PROMISE (Baim, 1982), CN2 (Clark and Niblett, 1989), and RIPPER (Cohen, 1995), and by $Q(w)$ for $w = 0, 0.25, 0.5, 0.75$, and 1. In each column, "1" indicates the rule determined by the method to be the best (the highest rank), and "7" indicates the worst.

This is not meant to suggest that any of these ranking methods are superior or inferior to any other. Rather, it serves to indicate how by changing the rule quality criterion, one can often alter which rules will be selected, and demonstrates the flexibility of the $Q(w)$ measure to emulate several different rule quality criteria through adjustment of its weight. This research thus shows that by controlling the w parameter in the AQ learning program, one can obtain rulesets representing different tradeoffs between consistency and completeness, and approximate behavior of different learning programs.

Fürnkranz and Flach (2003) have studied the behavior of different rule quality measures, and present a means for showing graphically how these measures can be intuitively visualized and compared.

Table 1
How different methods rank different rules

Data set	Pos	Neg	Ranks								
			Inf. gain	PROMISE	CN2	RIPPER	Q(0)	Q(0.25)	Q(0.5)	Q(0.75)	Q(1)
A	50	5	7	7	4	7	4	7	7	7	6
	50	0	6	6	1	6	1	6	6	6	6
200	200	5	1	1	2	1	2	1	1	1	1
pos	150	10	2	2	3	2	3	2	2	2	2
	150	30	3	3	6	3	6	3	3	3	2
800	100	15	5	5	5	5	5	4	4	5	5
neg	120	25	4	4	7	4	7	5	5	4	4
B	50	5	7	7	3	7	3	7	7	7	7
	250	25	6	5	3	5	3	5	5	5	5
500	500	50	1	1	3	1	3	1	1	1	1
pos	500	150	2	3	7	3	7	6	4	2	1
	200	5	5	6	1	6	1	4	6	6	6
500	400	35	3	2	2	2	2	2	2	3	3
neg	400	55	4	4	6	4	6	3	3	4	3
C	50	5	7	–	3	7	3	6	6	6	7
	250	25	5	–	3	5	3	2	5	4	5
800	500	50	1	–	3	1	3	3	1	1	1
pos	500	150	6	–	7	3	7	7	7	7	1
	200	5	3	–	1	6	1	1	3	5	6
200	400	35	2	–	2	2	2	2	2	2	3
neg	400	55	4	–	6	4	6	5	4	3	3

4. Ruleset visualization via concept association graphs

When working with symbolic knowledge as described above, it is desirable for a data analyst to be able to visualize the results of the learning process. The purpose of such visualization operators is to relate visually the input data to the rules that have been learned from them, to see which datapoints would corroborate or contradict these rules, to identify possible errors, etc. To this end, programs are needed that are specialized toward the visualization of data and attributional knowledge. Two such approaches are the *diagrammatic visualization* method implemented in the KV program (Zhang, 1997), and the *concept association graph* (e.g., Michalski and Kaufman, 1997; Kaufman and Michalski, 2000). The latter approach is particularly oriented toward problems of data and knowledge mining, due to a lack of complications arising from scaling up to many large attribute domains.

Concept association graphs were developed as a tool for visualizing attributional rulesets, or more generally the relationships between consequents and premises (Michalski and Kaufman, 1997). Attributes, rules and their relationships are displayed in a graphical form using nodes and links. There are three different types of nodes: input nodes, output nodes and rule nodes.

Input nodes are nodes that represent components of the premise of a rule, while output nodes represent the consequent of a rule. Rule nodes represent the relationships between one or more input attributes and one or more output attributes. All of the conditions in the premise of a rule are linked to its rule-node, which is then linked to the output node(s). Input and output nodes appear as ovals in a concept association graph, and rule nodes appear as rectangles.

There are two types of links, presented as continuous links and dotted links. The dependency between input and output nodes is represented with continuous links of different thickness. The thicker the link, the stronger the relationship. The thickness of the link can be computed using many different methods, which may take into consideration for example, the completeness (what percentage of positive examples of the consequent class are covered by the condition) or the consistency (what percentage of the examples covered by the condition are of the target class). A third method combines completeness and consistency using the $Q(w)$ measure (Section 3). These links are labeled by annotations that specify the values of the attribute represented by the associated input node that satisfy the condition represented by the link. This can be done either through a specification of attribute values (as seen, for example, in Figure 4) or, more simply, through one of four symbols that characterize those values (as seen, for example, in Figure 5).

The abstraction to four symbols can be used in the case of linear or binary attributes. The symbol '+' indicates a positive relationship between the attribute (or higher values of it) and the rule. The symbol '−' indicates a negative relationship; a rank attribute should have a low value, or a binary attribute should be false. Linear (rank, interval or ratio) attributes can also be characterized by the symbols '∧' and '∨', which indicate respectively that the attribute should have central or extreme values in the given condition.

Dotted links in concept association graphs are used to display generalization (is-a) relationship between nodes. For example, in the mushroom domain (see Figure 4), a dotted link shows that the output node [class = poisonous] is an instantiation of the classes-of-mushrooms node. Dotted links are optional, and are used primarily when output nodes can also serve as input nodes for another diagrammed rule.

The major advantage of a concept association graphs is that it can visualize multivariate relationships (rules) with a graphical indication of the strength of individual condition in these rules. The visualization method using concept association graphs has been implemented in program CAG1, which reads in a set of attributional rules learned by AQ-type learning program, and then displays a concept association graph. The program allows the user to modify the graph.

To illustrate different forms of concept association graphs we will use attributional rules learned from the "mushroom dataset" obtained from the data repository at the University of California at Irvine, and rules learned from a medical database representing patients with histories of different diseases.

The mushroom dataset contains the examples of more than 8000 different species of mushrooms, classified as edible or poisonous. Each mushroom is described in terms of 23 attributes, of which 22 are discrete input attributes (nominal or ordinal), and one is an output attribute, with the domain {edible, poisonous}. There were 3916 examples of poisonous mushrooms, and 4208 examples of edible mushrooms. The attributional rules learned from these examples are:

[class = poisonous]
    ⇐ [odor = creosote or fishy or foul or musty or pungent or spicy: 3796, 0]:
        $p = 3796, n = 0$
    ⇐ [cap_color ≠ cinnamon: 3904, 4176]
        & [gill_spacing = close or distant 3804, 3008]
        & [stalk_root ≠ equal: 3660, 2624]
        & [stalk_surface_above_ring = fibrous or silky or smooth: 3908, 4192]
        & [ring_type ≠ flaring: 3916, 4160]
        & [spore_print_color = green or purple or chocolate or yellow or white: 3468, 672]
        & [habitat ≠ waste: 916, 4016]: $p = 3440, n = 24$

[class = edible]
    ⇐ [odor = almond or anise or none: 4208, 120]: $p = 4208, n = 120$

Thus, there are two rules for poisonous mushrooms, and one rule for edible mushrooms. The pairs of numbers after ":" in each condition in each rule denote the number of positive and negative examples covered by this condition, respectively. Parameters p and n after each rule denote the total number of positive and negative examples covered by each rule, respectively. Thus, the first rule for the poisonous mushrooms covers 3796 examples of poisonous mushrooms and zero examples of edible mushrooms.

Given these rules, and numbers of positive and negative examples associated with each condition in each rule, CAG1 generated a concept association graph presented in Figure 4. The thickness of the lines connecting conditions with the class is proportional

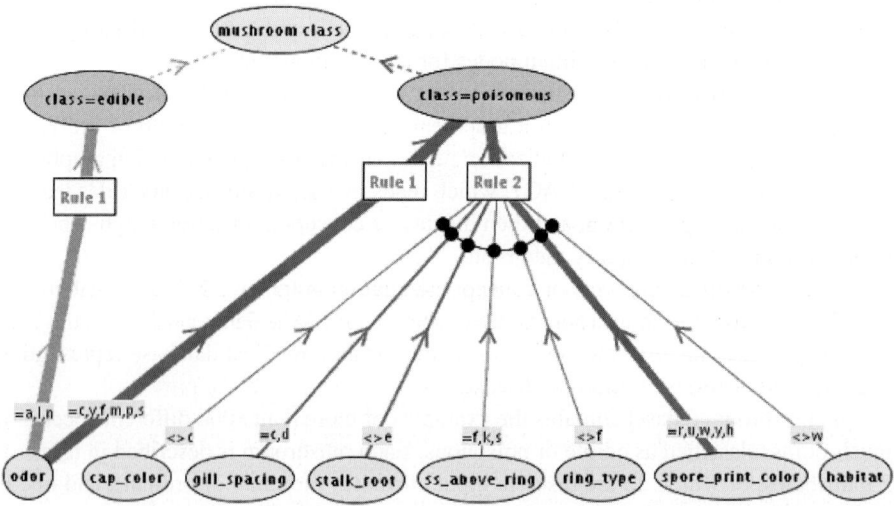

Fig. 4. A concept association graph representing rules for distinguishing edible from poisonous mushrooms.

to the consistency of the conditions, measured by $(p/p + n)$. Thus, in Rule 2 for the poisonous class, the link corresponding to the condition [spore_print_color = green or purple or chocolate or yellow or white] is much thicker than the others, as the consistency of this condition is approximately 84%, while the next strongest conditions, [stalk_root ≠ equal] and [gill_spacing = close or distant], have consistencies of 58 and 56%, respectively. The links associated with them are somewhat thicker than the links representing the rule's remaining conditions, which have consistencies around 50%.

CAG1 was also used to visualize attributional rules learned from data collected by the American Cancer Society on lifestyles and diseases of nonsmoking men, aged 50–65. The data consist of over 73 000 records describing them in terms of 32 attributes; 25 are Booleans indicating the occurrence or nonoccurrence of various classes of disease, and the other 7 describe elements of their lifestyles. Six of the seven are discrete attributes, with 2–7 linearly ordered values, and the seventh, representing how long the respondent had lived in the same neighborhood, is numeric. Among the discovered patterns were:

```
[Arthritis = present]
  ⇐ [High_Blood_Pressure = present] (432, 1765)
    & [Education < grad school] (940, 4529)
    & [Rotundity > very_low] (1070, 5578)
    & [Years_in_Neighborhood >= 1] (1109, 5910): 325, 1156
[Colon_Polyps = present]
  ⇐ [Prostate_Disease = present] (34: 967)
    & [Sleep = 5, 9] (16, 515)
    & [Years_in_Neighborhood >= 8] (33, 1477)
    & [Rotundity = average] (58, 2693)
    & [Education < college degree] (83, 4146): 5, 0
[Diverticulosis = present]
  ⇐ [stroke = absent] (257, 7037)
    & [Arthritis = present] (70, 1033)
    & [Rotundity >= average] (170, 4202)
    & [Education >= some college] (176, 4412)
    & [Sleep = 7..9] (205, 5743)
    & [Years_in_Neighborhood > 10] (134, 3846): 24, 115
[Stomach_Ulcer = present]
  ⇐ [Arthritis = present] (107, 1041)
    & [Education <= college degree] (305, 5276)
    & [Exercise >= medium] (298, 5606): 79, 668
[Asthma = present]
  ⇐ [Hay_Fever = present] (170, 787): 170, 187
```

The first rule, for example, states that occurrence of arthritis is associated with high blood pressure, education below graduate school, rotundity (a relation of patient's weight to height) above very low, and that patients moved into their current neighborhood at least a year ago.

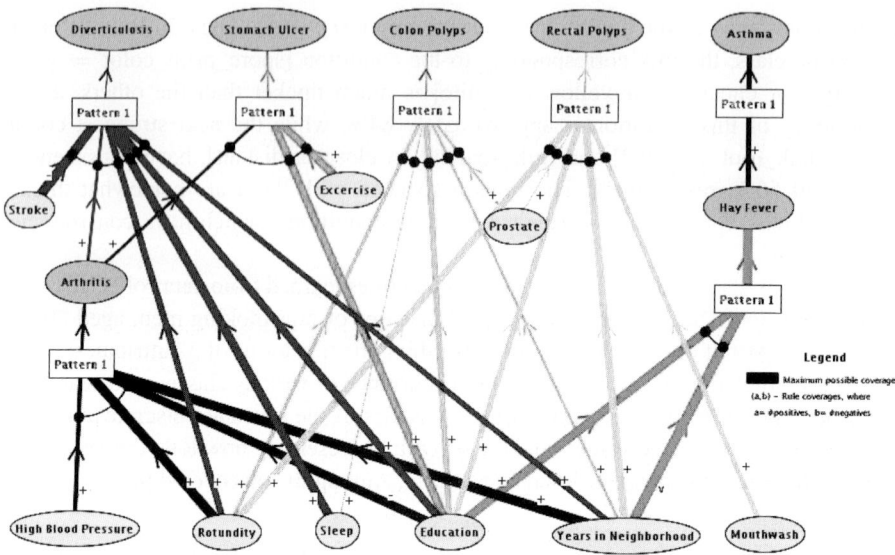

Fig. 5. A concept association graph representing discovered multivariate relationships between diseases and lifestyles. Link thicknesses represent relative support.

When the above rules and several others were input to CAG1, it resulted in the generated graph shown in Figure 5. In this CAG, links are color-coded according to which rule they apply to for ease of viewing. The concepts are present at a higher level of abstraction than in Figure 4; as discussed above. The relationship between nodes is represented with symbols rather than with an exact "relation reference". That is, instead of the list of values for attributes shown on the input links of Figure 4, the links are instead annotated with the four symbols +, −, ∧ and ∨.

In these graphs, the output nodes are also used directly as input nodes, without linking to an intermediate node to signify their values. This is possible because the output attributes are all binary, and the value true (or in this case "present") is understood.

In the graph in Figure 5, link thicknesses are based on completeness (support). For example, two of the links comprising the Stomach Ulcer rule in Figure 5 are noticeably thicker than the third, because the conditions involving Education and Exercise had approximately three times the support of the arthritis condition.

5. Integration of knowledge generation operators

To make the data exploration operations described above easily available to a data analyst, and applicable in sequences in which the output from one operation is an input to another one, programs performing these operations are best integrated into one system. This idea underlay the INLEN system (Michalski et al., 1992; Michalski and Kaufman, 1997), and its successor, VINLEN (Kaufman and Michalski, 2003), which is currently under development. The INLEN system integrates machine learning programs, simple

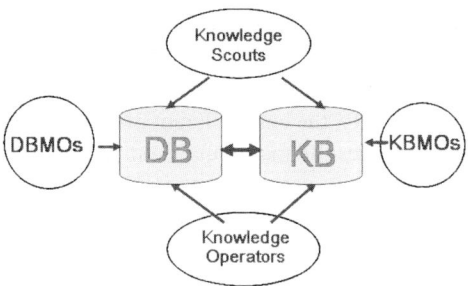

Fig. 6. A general schema of the INLEN inductive database system.

statistical data analysis tools, data tables, a knowledge base, inference procedures, and various supporting programs under a unified architecture and graphical interface. The knowledge base is used for storing, updating and applying rules that may be employed for assisting data exploration, and for reporting results from it.

The general architecture of INLEN is shown in Figure 6. The system consists of *knowledge systems*, which maintain the data and knowledge relevant to a given application domains. Each knowledge system is associated with a database (DB) and a knowledge base (KB), both of which can be accessed by a set of operators. The operators are divided into three classes:

- *DBMOs*: Data Management Operators, which operate on the database. These are conventional data management operators that are used for creating, modifying and displaying relational tables.
- *KBMOs*: Knowledge Management Operators, which operate on the knowledge base. These operators play a similar role to the DBMOs, but apply to the rules and other structures in the knowledge base.
- *KGOs*: Knowledge Generation Operators, which operate on both the data and knowledge bases. These operators perform symbolic and numerical data exploration tasks. They are based both on various machine learning and inference programs and on conventional data exploration techniques.

The execution of a KGO usually requires some background knowledge, and is guided by control parameters (if some parameters are not specified, default values are used). The background knowledge may contain some general knowledge, previously discovered knowledge, and knowledge specifically relevant to a given application domain, such as a specification of the value sets and types of attributes, the constraints and relationships among attributes, initial rules hypothesized by an expert, etc. The KGOs can be classified into groups, based on the type of operation they perform, each of which includes a number of specific operators that are instantiated by a combination of parameters. For example one group consists of operators for learning decision rules, another for selecting attributes, another for applying knowledge, and so forth.

One INLEN operator extends beyond the traditional learning and discovery operators, and thus merits further discussion. Specifically, the *Scout Access* operator is used to build *knowledge scouts* – scripts to serve as intelligent agents performing discovery tasks in the inductive database (Michalski and Kaufman, 2000).

To explore the idea of a knowledge scout further, consider the task of data exploration and knowledge mining. Typically, the entire plan of discovery can not be determined in its entirety beforehand. Some results may require no action, others may require some action to be taken, and occasionally, some may warrant a new and completely unplanned course of action. Yet it is time-consuming and subject to errors for an analyst to stand over every stage of the discovery process and respond appropriately to the output from each of those steps. Thus, the idea of a knowledge scout is that of a mechanism that can encapsulate the user's knowledge of how to react to different contingencies.

For instance, an experiment based on 1993 World Factbook data found the following rule describing 25 of the 55 countries with low ($< 1\%$) population growth:

PopGrRate $< 1\%$ if:	pos	neg
1. BirthRate is 10..20 or 50..60	46	20
2. FertRate is 1..2 or > 7	32	17
3. Religion is Protestant or Roman_Catholic or Eastern_Orth or Shinto or Bulgarian_Orth or Russian_Orth or Romanian_Orth or Greek_Orth	38	32
4. NetMigRate $\leqslant +10$	54	123

The first and strongest condition is surprising. Birth rates ranged in the data from 10 to 60, and while the low birth rate is intuitive, the very high one is not. Looking at the 25 countries that satisfy the rule, 24 of them had birth rates less than 20. Only one country, Malawi, had a birth rate above 50. Such a counterintuitive result could instigate a series of experiments to determine the cause of such behavior. In fact, subsequent investigation of Malawi compared to the rest of the countries quickly turned up an explanation: an outward net migration rate that dwarfs those of all the other countries.

Thus, a goal of knowledge scout would be to be able to specify in a script anomalies to be detected and what should be done in response to them (either logging them for a human, or calling upon new discovery operators). A knowledge scout needs the means to specify its plan of action, specifically, a language rich enough to specify the operators available to it and the means to select and execute actions. For instance, M-SQL extends the SQL data query language by adding to it the ability to query for certain types of rules and to invoke an association rule generating operator (Imielinski et al., 1996). Thus, it has access to conventional database operators plus a data mining operator.

Built for the INLEN environment which contained a variety of learning and discovery operators, each of which offered a range of parameter settings, KGL was designed as a language for specifying detailed plans in such an environment (Michalski and Kaufman, 2000). By combining the means for specifying the steps to be taken and the means for specifying the control structure, KGL code, such as the set of instructions shown in Figure 7, could be provided to the program. In that figure, the program is asked to examine the PEOPLE data table extracted from the CIA's World Factbook, and to take action based on the rules it finds. Comments (bracketed and in italic) have been added to explain each line.

The presented script learns rules for each possible decision attribute, then executes three tasks. First it counts the number of rules whose consequent predicts a country's

```
open PEOPLE                                    {Select PEOPLE database}
do CHAR(decision=all, pfile=people1.lrn)       {Characterize concepts
                                                representing single values
                                                of all attributes, using
                                                parameters specified in file
                                                people1.lrn}
strongPGrules1 = #rules(PGR, compl >= 60)      {Count rules for Population}
strongPGrules2 = #rules(PGR, supp >= 25)       {Growth Rate that satisfy}
strongPGrules3 = #rules(PGR,                   {three different conditions}
    num_conds(cons >= 50 and supp > 10) > 2)   {for threshold of strength}
print "Number of strong PGR rules:
    Type 1 = ", strongPGrules1, ",
    Type 2 = ", strongPGrules2, ",
    Type 3 = ", strongPGrules3
if #conditions(Fert) > 150                     {Is Fert ruleset too}
  begin                                        {complex?}
  do SELECT(attributes, decision=Fert,
      thresh=4, out=PEOPLE2, criterion=max)    {If so, find "thresh" best}
  do CHAR(pfile=people1.lrn, decision=Fert)    {independent attributes,}
  end                                          {then recharacterize}
for i = 1 to 6
  begin                                        {For each value of i, 1-6,}
  print "Number of LE conditions with p/n      {count & display number of}
      ratio of at least", i, ":1 =",           {Life Expectancy conditions}
      #conditions(LE, cons >= i/(i+1))         {with consistency ≥ i/(i+1)}
  end
```

Fig. 7. KGL code defining a knowledge scout for exploring a World Factbook data table.

population growth rate that are strong according to three criteria: high relative support, high absolute support, and containing at least two conditions that have both absolute support and standalone consistency above the given thresholds. Then it tests the ruleset that determines a country's likely fertility rate for complexity (based on the total number of conditions in the ruleset); if it is too complex, the four most relevant attributes to the task are chosen, and the learning is repeated on this streamlined dataset. Finally, it reports on the number of conditions in the life expectancy rule base that have confidence levels above different thresholds.

In summary, an inductive database integrates a database with a set of operators for performing various types of operations on the database, on the knowledge base, or on the data and knowledge bases combined.

6. Summary

The main thesis of this chapter is that modern methods for symbolic machine learning have a direct and important application to logical data analysis and the development of a new research direction, called knowledge mining. Knowledge mining has been characterized as a derivation of human-like knowledge from data and prior knowledge. It was indicated that a knowledge mining system can be implemented using inductive database technology that deeply integrates a database, a knowledge base, and operators for data and knowledge management and knowledge generation.

Among knowledge generation operators are operators for inductive learning of attributional rules or trees characterizing the relationship between designated output and input attributes, for creating conceptual hierarchies (conceptual clustering) from data, for selecting most relevant attributes, and for visualizing data and rules learned from the data. The learned rules represent high-level knowledge that can be of great value to a data analyst, and used for human decision-making or for automated classification. Other important operators include construction of equations along with logical preconditions for their application, determination of symbolic descriptions of temporal sequences of multi-attribute events, automated generation of new, more relevant attributes, and selection of representative examples.

The underlying theme of all these methods is the ability to generate knowledge that is easily understood and articulated. Visualization techniques such as concept association graphs facilitate the presentation of broad concepts to the user.

The knowledge generation capability was illustrated by presenting results from several application domains. In analyzing demographic data, the knowledge mining approach helped to discover the anomalous Malawi's population changes. In analyzing medical data, it showed in understandable terms relationships between the occurrences of diseases and the presence or absence of other diseases as well as factors in individuals' lifestyles. In a computer intrusion detection domain, it created symbolic user models from process table data characterizing users' activities (Michalski et al., 2005).

In contrast to many data mining approaches, the methodology presented can utilize various types of background knowledge regarding the domain of discourse. This background knowledge may include, for example, a specification of the domain and the type of the attributes, the known relationships among attributes, prior concept descriptions, and other high-level knowledge. An important aspect of the methodology is its ability to take advantage of this knowledge.

We presented KGL as a language for developing knowledge scouts. KGL was designed for an environment in which the data was not stored in a full-scale DBMS. VINLEN, which works in conjunction with an SQL-accessible relational database, requires a knowledge scout language that is tailored to such an environment. Thus, one topic of ongoing research is the development of a *Knowledge Query Language* – a language of SQL-like form that extends its capabilities into a knowledge mining paradigm.

Acknowledgements

The authors thank their past and current collaborators who provided research feedback, comments, and assistance at different stages of the development of methods and computer programs described in this chapter. We are grateful to the students and researchers who contributed in a constructive and helpful way in this research over the years, in particular, to Jerzy Bala, Eric Bloedorn, Michal Draminski, Mark Maloof, Jarek Pietrzykowski, Jim Ribeiro, Bartlomiej Sniezynski, Janusz Wnek, Janusz Wojtusiak, and Qi Zhang.

This research was conducted in the Machine Learning and Inference Laboratory of George Mason University. Support for the Laboratory's research related to the presented results has been provided in part by the National Science Foundation under Grants No. DMI-9496192, IRI-9020266, IIS-9906858 and IIS-0097476; in part by the UMBC/LUCITE #32 grant; in part by the Office of Naval Research under Grant No. N00014-91-J-1351; in part by the Defense Advanced Research Projects Agency under Grant No. N00014-91-J-1854 administered by the Office of Naval Research; and in part by the Defense Advanced Research Projects Agency under Grants No. F49620-92-J-0549 and F49620-95-1-0462 administered by the Air Force Office of Scientific Research.

The findings and opinions expressed here are those of the authors, and do not necessarily reflect those of the above sponsoring organizations.

References

Aggarwal, C.C., Procopiuc, C., Wolf, J., Yu, P.S., Park, J.S. (1999). Fast algorithms for projected clustering. In: *Proceedings of the 1999 ACM SIGMOD International Conference on Management of Data*. ACM Press, pp. 61–72.

Aggarwal, R., Gehrke, J., Gunopulos, D., Raghavan, P. (1998). Automatic subspace clustering of high dimensional data for data mining applications. In: *Proceedings of the 1998 ACM SIGMOD International Conference on Management of Data*. ACM Press, pp. 94–105.

Alexe, S., Blackstone, E., Hammer, P. (2003). Coronary risk prediction by logical analysis of data. *Ann. Oper. Res.* **119**, 15–42.

Baim, P.W. (1982). The PROMISE method for selecting most relevant attributes for inductive learning systems. Report No. UIUCDCS-F-82-898. Department of Computer Science, University of Illinois, Urbana.

Belkin, M., Niyogi, P. (2004). Semi-supervised learning on Riemannian manifolds. *Machine Learning* **56**, 209–239.

Bentrup, J.A., Mehler, G.J., Riedesel, J.D. (1987). INDUCE 4: a program for incrementally learning structural descriptions from examples. Report of the Intelligent Systems Group ISG 87-2. UIUCDCS-F-87-958. Department of Computer Science, University of Illinois, Urbana.

Bergadano, F., Matwin, S., Michalski, R.S., Zhang, J. (1992). Learning two-tiered descriptions of flexible concepts: the POSEIDON system. *Machine Learning* **8**, 5–43.

Bloedorn, E., Michalski, R.S. (1998). Data-driven constructive induction. In: *Special Issue on Feature, Transformation and Subset Selection, IEEE Intell. Syst.* (March/April), 30–37.

Bloedorn, E., Mani, I., MacMillan, T.R. (1996). Machine learning of user profiles: representational issues. In: *Proceedings of the Thirteenth National Conference on Artificial Intelligence*. AAAI-96, Portland, OR, AAAI Press.

Bloedorn, E., Wnek, J., Michalski, R.S. (1993). Multistrategy constructive induction. In: *Proceedings of the Second International Workshop on Multistrategy Learning*. Harpers Ferry, WV, George Mason University Press, Fairfax, VA, pp. 188–203.

Bongard, N. (1970). *Pattern Recognition*. Spartan Books, New York. A translation from Russian.

Brachman, R.J., Khabaza, T., Kloesgen, W., Piatetsky-Shapiro, G., Simoudis, E. (1996). Mining business databases. *Comm. ACM* **39** (11), 42–48.

Bratko, I., Muggleton, S., Karalic, A. (1997). Applications of inductive logic programming. In: Michalski, R.S., Bratko, I., Kubat, M. (Eds.), *Machine Learning and Data Mining: Methods and Applications*. Wiley, London.

Carpineto, C., Romano, G. (1995a). Some results on lattice-based discovery in databases. In: *Workshop on Statistics, Machine Learning and Knowledge Discovery in Databases*. Heraklion, Greece, pp. 216–221.

Carpineto, C., Romano, G. (1995b). Automatic construction of navigable concept networks characterizing text databases. In: Gori, M., Soda, G. (Eds.), *Topics in Artificial Intelligence*. In: *Lecture Notes in Artificial Intelligence*, vol. 992. Springer-Verlag, pp. 67–78.

Caruana, R., Freitag, D. (1994). Greedy attribute selection. In: *Proceedings of the Eleventh International Conference on Machine Learning.* Morgan Kaufmann, pp. 28–36.

Cavalcanti, R.B., Guadagnin, R., Cavalcanti, C.G.B., Mattos, S.P., Estuqui, V.R. (1997). A contribution to improve biological analyses of water through automatic image recognition. *Pattern Recognition Image Anal.* **7** (1), 18–23.

Clark, P., Niblett, T. (1989). The CN2 induction algorithm. *Machine Learning* **3**, 261–283.

Cohen, W. (1995). Fast effective rule induction. In: *Proceedings of the 12th International Conference on Machine Learning.* Morgan Kaufmann.

Daniel, C., Wood, F.S. (1980). *Fitting Equations to Data.* Wiley, New York.

Davis, J. (1981). CONVART: a program for constructive induction on time-dependent data. MS Thesis. Department of Computer Science, University of Illinois, Urbana.

De Raedt, L., Jaeger, M., Lee, S.D., Mannila, H. (2002). A theory of inductive query answering. In: *Proceedings of the 2002 IEEE International Conference on Data Mining.* ICDM'02, Maebashi, Japan, IEEE Press, pp. 123–130.

Diday, E. (Ed.) (1989). *Proceedings of the Conference on Data Analysis, Learning Symbolic and Numeric Knowledge.* Nova Science, Antibes.

Dieterrich, T., Michalski, R.S. (1986). Learning to predict sequences. In: Michalski, R.S., Carbonell, J.G., Mitchell, T.M. (Eds.), *Machine Learning: An Artificial Intelligence Approach, vol. 2.* Morgan Kaufmann, pp. 63–106.

Dontas, K., (1988). APPLAUSE: an implementation of the Collins–Michalski theory of plausible reasoning. MS Thesis. Computer Science Department, The University of Tennessee, Knoxville, TN.

Dubois, D., Prade, H., Yager, R.R. (Eds.) (1993). *Readings in Fuzzy Sets and Intelligent Systems.* Morgan Kaufmann.

Elio, R., Watanabe, L. (1991). An incremental deductive strategy for controlling constructive induction in learning from examples. *Machine Learning* **7**, 7–44.

Evangelos, S., Han, J. (Eds.) (1996). *Proceedings of the Second International Conference on Knowledge Discovery and Data Mining.* Portland, OR, AAAI Press.

Falkenhainer, B.C., Michalski, R.S. (1990). Integrating quantitative and qualitative discovery in the ABACUS system. In: Kodratoff, Y., Michalski, R.S. (Eds.), *Machine Learning: An Artificial Intelligence Approach, vol. III.* Morgan Kaufmann, San Mateo, CA, pp. 153–190.

Fayyad, U.M., Irani, K.B. (1992). The attribute selection problem in decision tree generation. In: *Proceedings of the Tenth National Conference on Artificial Intelligence.* San Jose, CA, AAAI Press, pp. 104–110.

Fayyad, U.M., Haussler, D., Stolorz, P. (1996a). Mining scientific data. *Comm. ACM* **39** (11), 51–57.

Fayyad, U.M., Piatetsky-Shapiro, G., Smyth, P., Uhturusamy, R. (Eds) (1996b). *Advances in Knowledge Discovery and Data Mining.* AAAI Press, San Mateo, CA.

Feelders, A. (1996). Learning from biased data using mixture models. In: *Proceedings of the Second International Conference on Knowledge Discovery and Data Mining.* Portland, OR, AAAI Press, pp. 102–107.

Fürnkranz, J., Flach, P. (2003). An analysis of rule evaluation metrics. In: *Proceedings of the Twentieth International Conference on Machine Learning.* ICML 2003, Washington, DC, AAAI Press, pp. 202–209.

Greene, G. (1988). The Abacus.2 system for quantitative discovery: using dependencies to discover non-linear terms. *Reports of the Machine Learning and Inference Laboratory* MLI 88-4. Machine Learning and Inference Laboratory, George Mason University, Fairfax, VA.

Han, J., Kamber, M. (2001). *Data Mining: Concepts and Techniques.* Morgan Kaufmann, San Francisco, CA.

Han, J., Fu, Y., Wang, W., Chiang, J., Gong, W., Koperski, K., Li, D., Lu, Y., Rajan, A., Stefanovic, N., Xia, B., Zaiane, O.R. (1996). DBMiner: a system for mining knowledge in large relational databases. In: *Proceedings of the Second International Conference on Data Mining and Knowledge Discovery.* KDD'96, Portland, OR, AAAI Press, pp. 250–255.

Hand, D., Mannila, H., Smyth, P. (2001). *Principles of Data Mining.* MIT Press, Cambridge, MA.

Iba, W., Woogulis, J., Langley, P. (1988). Trading simplicity and coverage in incremental concept learning. In: *Proceedings of the Fifth International Conference on Machine Learning.* Morgan Kaufmann, San Francisco, CA, pp. 73–79.

Imam, I.F., Michalski, R.S. (1993). Should decision trees be learned from examples or from decision rules? In: *Proceedings of the Seventh International Symposium on Methodologies for Intelligent Systems.* ISMIS-93, Trondheim, Norway, Springer-Verlag.

Imielinski, T., Mannila, H. (1996). A database perspective on knowledge discovery. *Comm. ACM* **39**, 58–64.
Imielinski, T., Virmani, A., Abdulghani, A. (1996). DataMine: application programming interface and query language for database mining. In: *Proceedings of the Second International Conference on Knowledge Discovery and Data Mining*. AAAI Press, pp. 256–261.
Kaufman, K.A., Michalski, R.S. (1996). A method for reasoning with structured and continuous attributes in the INLEN-2 multistrategy knowledge discovery system. In: *Proceedings of the Second International Conference on Knowledge Discovery and Data Mining*. Portland, OR, AAAI Press, pp. 232–237.
Kaufman, K., Michalski, R.S. (1999). Learning from inconsistent and noisy data: the AQ18 approach. In: *Proceedings of the Eleventh International Symposium on Methodologies for Intelligent Systems*. Warsaw, Poland, Springer-Verlag, pp. 411–419.
Kaufman, K., Michalski, R.S. (2000). A knowledge scout for discovering medical patterns: methodology and system SCAMP. In: *Proceedings of the Fourth International Conference on Flexible Query Answering Systems*. FQAS'2000, Warsaw, Poland, Springer-Verlag, pp. 485–496.
Kaufman, K., Michalski, R.S. (2003). The development of the inductive database system VINLEN: a review of current research. In: *International Intelligent Information Processing and Web Mining Conference*. Zakopane, Poland, Springer-Verlag.
Kerber, R. (1992). Chimerge: discretization for numeric attributes. In: *Proceedings of the Tenth National Conference on Artificial Intelligence*. AAAI-92, San Jose, CA, AAAI Press, pp. 123–128.
Khabaza, T., Shearer, C. (1995). Data mining with clementine. In: *Colloquium on Knowledge Discovery in Databases*. The Institution of Electrical Engineers.
Kibler, D., Aha, D. (1987). Learning representative exemplars of concepts: a case study. In: *Proceedings of the Fourth International Conference on Machine Learning*. Morgan Kaufmann, San Francisco, CA, pp. 24–30.
Kokar, M.M. (1986). Coper: a methodology for learning invariant functional descriptions. In: Michalski, R.S., Mitchell, T.M., Carbonell, J.G. (Eds.), *Machine Learning: A Guide to Current Research*. Kluwer Academic, Boston, MA.
Lakshminarayan, K., Harp, S.A., Goldman, R., Samad, T. (1996). Imputation of missing data using machine learning techniques. In: *Proceedings of the Second International Conference on Knowledge Discovery and Data Mining*. Portland, OR, AAAI Press, pp. 140–145.
Langley, P. (1994). Selection of relevant features in machine learning. In: *AAAI Fall Symposium on Relevance*, pp. 140–144.
Langley, P., Bradshaw, G.L., Simon, H.A. (1983). Rediscovering chemistry with the BACON system. In: Michalski, R.S., Carbonell, J.G., Mitchell, T.M. (Eds.), *Machine Learning: An Artificial Intelligence Approach*. Morgan Kaufmann, San Mateo, CA, pp. 307–329.
Larson, J.B. (1977). INDUCE-1: an interactive inductive inference program in VL21 logic system. Report No. 876. Department of Computer Science, University of Illinois, Urbana.
Lbov, G.S. (1981). *Metody Obrabotki Raznotipnych Ezperimentalnych Dannych* (Methods for Analysis of Multitype Experimental Data). Akademia Nauk USSR, Sibirskoje Otdelenie, Institut Matematiki, Izdatelstwo Nauka, Novosibirsk.
Maloof, M., Michalski, R.S. (2000). Selecting examples for partial memory learning. *Machine Learning* **41**, 27–52.
Maloof, M., Michalski, R.S. (2004). Incremental learning with partial instance memory. *Artificial Intelligence* **154**, 95–126.
Michael, J. (1991). Validation, verification and experimentation with Abacus2. *Reports of the Machine Learning and Inference Laboratory* MLI 91-8. Machine Learning and Inference Laboratory, George Mason University, Fairfax, VA.
Michalski, R.S. (1978). A planar geometrical model for representing multi-dimensional discrete spaces and multiple-valued logic functions. ISG Report No. 897. Department of Computer Science, University of Illinois, Urbana.
Michalski, R.S. (1983). A theory and methodology of inductive learning. *Artificial Intelligence* **20**, 111–161.
Michalski, R.S. (1990). Learning flexible concepts: fundamental ideas and a method based on two-tiered representation. In: Kodratoff, Y., Michalski, R.S. (Eds.), *Machine Learning: An Artificial Intelligence Approach, vol. III*. Morgan Kaufmann, San Mateo, CA, pp. 63–102.

Michalski, R.S. (2004). Attributional calculus: a logic and representation language for natural induction. *Reports of the Machine Learning and Inference Laboratory* MLI 04-2. George Mason University, Fairfax, VA.

Michalski, R.S., Kaufman, K.A. (1997). Multistrategy data exploration using the INLEN system: recent advances. In: *Sixth Symposium on Intelligent Information Systems*, IIS'97, Zakopane, Poland, IPI PAN.

Michalski, R.S., Kaufman, K.A. (1998). Data mining and knowledge discovery: a review of issues and a multistrategy approach. In: Michalski, R.S., Bratko, I., Kubat, M. (Eds.), *Machine Learning and Data Mining: Methods and Applications*. Wiley, London, pp. 71–112.

Michalski, R.S., Kaufman, K. (2000). Building knowledge scouts using KGL metalanguage. *Fund. Inform.* **40**, 433–447.

Michalski, R.S., Kaufman, K. (2001). Learning patterns in noisy data: the AQ approach. In: Paliouras, G., Karkaletsis, V., Spyropoulos, C. (Eds.), *Machine Learning and its Applications*. Springer-Verlag, pp. 22–38.

Michalski, R.S., Larson, J.B. (1978). Selection of most representative training examples and incremental generation of VL1 hypotheses: the underlying methodology and the description of programs ESEL and AQ11. Report No. 867. Department of Computer Science, University of Illinois, Urbana.

Michalski, R.S., Larson, J. (1983). Incremental generation of VL1 hypotheses: the underlying methodology and the description of program AQ11. *Reports of the Intelligent Systems Group, ISG* 83-5, UIUCDCS-F-83-905. Department of Computer Science, University of Illinois, Urbana.

Michalski, R.S., Stepp, R. (1983). Learning from observation: conceptual clustering. In: Michalski, R.S., Carbonell, J.G., Mitchell, T.M. (Eds.), *Machine Learning: An Artificial Intelligence Approach*. Tioga Publishing, Palo Alto, CA.

Michalski, R.S., Baskin, A.B., Spackman, K.A. (1982). A logic-based approach to conceptual database analysis. In: *Sixth Annual Symposium on Computer Applications in Medical Care*. SCAMC-6, George Washington University, Medical Center, Washington, DC, IEEE Press, pp. 792–796.

Michalski, R.S., Ko, H., Chen, K. (1985). SPARC/E(V. 2), an eleusis rule generator and game player. *Reports of the Intelligent Systems Group, ISG* 85-11, UIUCDCS-F-85-941. Department of Computer Science, University of Illinois, Urbana.

Michalski, R.S., Ko, H., Chen, K. (1986). Qualitative prediction: a method and a program SPARC/G. In: Guetler, C. (Ed.), *Expert Systems*. Academic Press, London.

Michalski, R.S., Kerschberg, L., Kaufman, K., Ribeiro, J. (1992). Mining for knowledge in databases: the INLEN architecture, initial implementation and first results. *J. Intelligent Inform. Systems: Integr. AI Database Technol.* **1**, 85–113.

Michalski, R.S., Rosenfeld, A., Duric, Z., Maloof, M., Zhang, Q. (1998). Application of machine learning in computer vision. In: Michalski, R.S., Bratko, I., Kubat, M. (Eds.), *Machine Learning and Data Mining: Methods and Applications*. Wiley, London.

Michalski, R.S., Kaufman, K., Pietrzykowski, J., Wojtusiak, J., Sniezynski, B. (2005). Learning user behavior and understanding style: a natural induction approach. *Reports of the Machine Learning and Inference Laboratory*. George Mason University, Fairfax, VA. In press.

Morgenthaler, S., Tukey, J.W. (1989). The next future of data analysis. In: Diday, E. (Ed.), *Proceedings of the Conference on Data Analysis, Learning Symbolic and Numeric Knowledge*. Nova Science, Antibes.

Muggleton, S. (Ed.) (1992). *Inductive Logic Programming*. Morgan Kaufmann.

Neapolitan, R.E. (2003). *Learning Bayesian Networks*. Prentice Hall.

Pawlak, Z. (1991). *Rough Sets: Theoretical Aspects of Reasoning about Data*. Kluwer Academic, Dordrecht.

Pearl, J. (1988). *Probabilistic Reasoning in Intelligent Systems: Networks of Plausible Inference*. Morgan Kaufmann, San Mateo, CA.

Pearl, J. (2000). *Causality: Models, Reasoning and Inference*. Cambridge University Press.

Quinlan, J.R. (1986). Induction of decision trees. *Machine Learning* **1**, 81–106.

Quinlan, J.R. (1990). Probabilistic decision trees. In: Kodratoff, Y., Michalski, R.S. (Eds.), *Machine Learning: An Artificial Intelligence Approach, vol. III*. Morgan Kaufmann, San Mateo, CA, pp. 140–152.

Quinlan, J.R. (1993). *C4.5: Programs for Machine Learning*. Morgan Kaufmann, Los Altos, CA.

Reinke, R.E., Michalski, R.S. (1988). Incremental learning of concept descriptions: a method and experimental results. *Machine Intelligence* **11**, 263–288.

Ribeiro, J.S., Kaufman, K.A., Kerschberg, L. (1995). Knowledge discovery from multiple databases. In: *Proceedings of the First International Conference on Knowledge Discovery and Data Mining*. Montreal, PQ, AAAI Press, pp. 240–245.

Saul, L.K., Roweis, S.T. (2003). Think globally, fit locally: unsupervised learning of low dimensional manifolds. *J. Machine Learning Res.* **4**, 119–155.

Sharma, S. (1996). *Applied Multivariate Techniques*. Wiley, London.

Slowinski, R. (Ed.) (1992). *Intelligent Decision Support: Handbook of Applications and Advances of the Rough Sets Theory*. Kluwer Academic, Dordrecht.

Stepp, R., Michalski, R.S. (1986). Conceptual clustering: inventing goal-oriented classifications of structured objects. In: Michalski, R.S., Carbonell, J., Mitchell, T.M. (Eds.), *Machine Learning: An Artificial Intelligence Approach, vol. II*. Morgan Kaufmann.

Tenenbaum, J.B., de Silva, V., Langford, J.C. (2000). A global geometric framework for nonlinear dimensionality reduction. *Science* **290**, 2319–2323.

Tukey, J.W. (1986). In: Jones, L.V. (Ed.), *The Collected Works of John W. Tukey, vol. V. Philosophy and Principles of Data Analysis: 1965–1986*. Wadsworth & Brooks/Cole, Monterey, CA.

Umann, E. (1997). Phons in spoken speech: a contribution to the computer analysis of spoken texts. *Pattern Recognition Image Anal.* **7** (1), 138–144.

Van Mechelen, I., Hampton, J., Michalski, R.S., Theuns, P. (Eds.) (1993). *Categories and Concepts: Theoretical Views and Inductive Data Analysis*. Academic Press, London.

Wang, H., Chu, F., Fan, W., Yu, P.S., Pei, J. (2004). A fast algorithm for subspace clustering by pattern similarity. In: *Proceedings of the 16th International Conference on Scientific and Statistical Database Management*. SSDBM'04, Santorini Island, Greece, IEEE Press.

Widmer, G., Kubat, M. (1996). Learning in the presence of concept drift and hidden concepts. *Machine Learning* **23**, 69–101.

Wnek, J., Michalski, R.S. (1994). Hypothesis-driven constructive induction in AQ17-HCI: a method and experiments. *Machine Learning* **14**, 139–168.

Wojtusiak, J. (2004). AQ21 user's guide. *Reports of the Machine Learning and Inference Laboratory* MLI 04-3. George Mason University, Fairfax, VA.

Zadeh, L. (1965). Fuzzy sets. *Inform. and Control* **8**, 338–353.

Zagoruiko, N.G. (1972). *Recognition Methods and their Application*. Sovietskoe Radio, Moscow. In Russian.

Zagoruiko, N.G. (1991). *Ekspertnyie Sistemy I Analiz Dannych (Expert Systems and Data Analysis)*. Wychislitelnyje Sistemy, vol. 144. Akademia Nauk USSR, Sibirskoje Otdelenie, Institut Matematiki, Novosibirsk.

Zhang, Q. (1997). Knowledge visualizer: a software system for visualizing data, patterns and their relationships. *Reports of the Machine Learning and Inference Laboratory* MLI 97-14. George Mason University, Fairfax, VA.

Zhuravlev, Y.I., Gurevitch, I.B. (1989). Pattern recognition and image recognition. In: Zhuravlev, Y.I. (Ed.), *Pattern Recognition, Classification, Forecasting: Mathematical Techniques and their Application. Issue 2*. Nauka, Moscow, pp. 5–72. In Russian.

Ziarko, W.P. (Ed.) (1994). *Rough Sets, Fuzzy Sets and Knowledge Discovery*. Springer-Verlag, Berlin.

Zytkow, J.M. (1987). Combining many searches in the FAHRENHEIT discovery system. In: *Proceedings of the Fourth International Workshop on Machine Learning*. Irvine, CA, Morgan Kaufmann, pp. 281–287.

Mining Computer Security Data

David J. Marchette

1. Introduction

Any system connected to the Internet has the potential to be compromised, in which data may be lost, or the use of the system by authorized users may be reduced or eliminated. While this has always been true, it has become much more common in recent years for attacks against computers to be publicized, and several famous ones (for example, the Melissa and "I Love You" viruses; the Slammer, Nimda and Code Red worms) have raised the public's awareness of the problem. This has made the protection of computer systems and networks an important and recognized problem. This chapter will discuss some of the techniques that data mining and statistical pattern recognition can bring to the problem.

Computer security data comes in three basic forms. There is network data, which consists of measurements on packets as they traverse the network; there are log files on each computer that store information about unusual events or errors; and there are data derived from monitoring the activities of programs or users. A multilevel computer security system would include all of these types of data, in addition to taking certain precautions to ensure that the system is as secure as possible. I will look at some of these data with an eye to demonstrating how data mining and statistical pattern recognition can aid in the understanding of these data and the detection of intruders.

The purpose of this chapter is not to develop specific intrusion detection systems. Rather, I am interested in exploratory data analysis applied to computer security data. I will illustrate various types of information that can be obtained by mining these data, and discuss how they apply to the overall problems of computer security.

One of the important problems in computer security is the detection of an attack on a system or network. Thus one of the reasons I look at these data is to determine methods for detecting attacks. Another purpose is to detect misconfigurations or vulnerabilities that might lead to a successful attack. Detecting these provides the security analyst with the opportunity to close the holes before an attack is mounted, thus making the system more secure.

A final purpose is the analysis of the system or network behavior and context to understand it, and provide metrics to assess its performance. Several researchers have taken this perspective, including (Leland et al., 1994; Feldmann et al., 1997, 1998, 1999;

Willinger et al., 1997; and Park and Willinger, 2000). There has also been work showing that this kind of analysis can be of use in understanding attacks. See, for example, (Ogielski and Cowie, 2002) and http://www.renesys.com/projects/bgp_instability/.

Properly secure systems are supposed to ensure that no unauthorized person can access the computer, or that data cannot be damaged or made inaccessible, either inadvertently or maliciously. As with all things, there is no such thing as a perfectly secure system (except possibly one which contains no data and is turned off). Thus, another important use for data mining tools is the determination of the severity of a successful attack, or the estimation of the severity of a potential attack.

In this chapter I will first give a brief overview of the basics of networking, for the purpose of understanding the data that can be collected from the network. This discussion, Section 2, will provide the necessary background to understand the sections which deal with network monitoring. I give some examples of typical attack scenarios in Section 3, providing some indications of the kind of data that must be collected to detect the attacks. Section 4 describes some of the statistical properties of network data and gives an introduction to network based intrusion detection. Section 5 describes the analysis of session data, which looks at aggregates of the network data corresponding to the actual communications sessions. I consider the distinction between signature based techniques and the detection of anomalous behavior in Section 6. Sections 7 and 8 consider attacks at the machine level, where I look at profiling users and programs in an attempt to detect masquerading users and compromised programs.

Several books are available that describe the basic ideas in intrusion detection. These include (Amoroso, 1999; Anon., 1997; Bace, 2000; Escamilla, 1998; Marchette, 2001; Northcutt et al., 2001; and Proctor, 2001).

2. Basic TCP/IP

Computer networks require specific protocols in place to allow communication between computers. These protocols specify how communication can take place and insure that the computers, and the applications running on them, are able to decode the information on the network.

I will describe the basics of networking at the minimal level necessary to understand the data. I focus primarily on the TCP (Transmission Control Protocol), which is used for a significant portion of Internet traffic. The other protocols are important, and I ignore them only in the interest of space. Interested readers are encouraged to explore the references.

2.1. Overview of networking

The Internet uses what is referred to as TCP/IP, which is a suite of network protocols, all under the umbrella of the IP (Internet Protocol). A network protocol is a set of rules for the coordination of communications across a network. These rules are set up to ensure the proper functioning of the network, and to see that the information is properly delivered, or, if delivery is impossible, indicate the proper procedure for handling the error conditions.

Messages are broken into discrete units, called *packets*, which are individually routed from their source to their intended destination. Each packet contains information about the protocol they are under, along with the information necessary to deliver the packet, and to properly interpret it once it is received, as well as any data that is sent.

In order to accomplish all this, networking is implemented in a layered manner. For our purposes we can think of this as a four layer process (some authors divide this up differently, but this is adequate for this discussion). The layers are depicted in Figure 1. The hardware, or link layer, is tasked with physically placing packets onto the network, and taking them off at their destination. The IP layer handles the basic address and routing information. This is where the source and destination of the packet are set and interpreted. The protocol layer implements the various protocols. Finally, the application layer is where the different applications are implemented, such as email and web. Different applications can implement their own protocols, in order to facilitate the information flow that they require.

The main lesson for the data miner here is that these different layers provide different types of information, and provide the attacker with different methods of attack. Further, the packets themselves reflect this layering. Each layer prepends a header onto the packet, as indicated in Figure 2. Thus, one can process the packet by sequentially pulling off the headers in the order of IP, protocol, application. (In this chapter I will

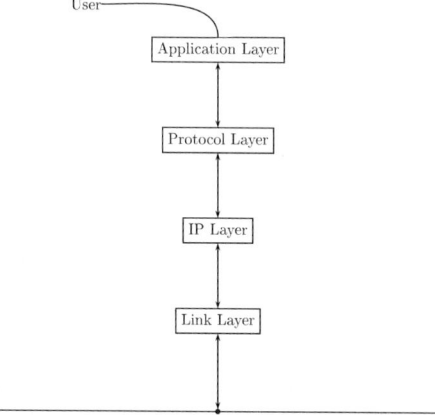

Fig. 1. An illustration of network layering. The user interacts with the application layer, which determines the protocol to use and sends the information on to the protocol layer, where it is implemented. The protocol layer then passes it down to the IP layer, where the source and destination of the packet are set, and finally the link layer puts the packet onto the network.

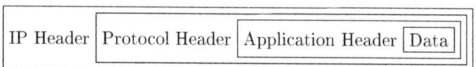

Fig. 2. Data encapsulation. Each layer places a header on the front of the packet with layer-specific information. The link layer then places network-specific information as a header (and possibly trailer) on the packet (not shown).

Version	Length	Type of Service	Total Length	
Identification			Flags	Fragment Offset
Time to Live		Protocol	Header Checksum	
Source IP Address				
Destination IP Address				
Options (if any)				
Data				

Fig. 3. The IP header. Each row corresponds to a 32-bit word. All fields (with the exceptions of Flags and Fragment offset) are either 4-, 8-, 16-, or 32-bit numbers.

only be concerned with TCP/IP packets, and will ignore the network-specific (usually Ethernet) headers that are attached to the packet.) Figure 2 illustrates this encapsulation.

The IP layer is the constant in TCP/IP. All packets (essentially) on the Internet are IP packets. This discussion is relevant to IP version 4, which represents the current universally implemented version. Eventually, this will be replaced by IP version 6 (with a corresponding change in the headers) and protocols. The IP header consists of a variable number of bytes (not to exceed 60), and is illustrated in Figure 3. The fields are:

Version. The version number (currently 4) is 4 bits.
Length. The length is a 4-bit number indicating the length of the IP header in bytes.
Type of service. Also called ToS, this is an 8-bit field whose value indicates the routing priority for the packet.
Total length. This is a 16-bit integer holding the total length of the packet in bytes, from the IP header through the data.
Identification. A 16-bit number guaranteed to be unique to this packet (at least for the time the packet is likely to be alive on the network).
Flags. Three 1-bit flags indicating the fragmentation status of the packet. Fragmentation will be discussed briefly below.
Fragment offset. If the packet is fragmented, this indicates the offset (in bytes) for this fragment.
Time to live. Abbreviated TTL, this 8-bit number is decremented at each router, and the packet is dropped if the value ever reaches 0. Thus, this implements a finite lifespan for packets on the Internet.
Protocol. An 8-bit value indicating the protocol to be found in the protocol header.
Header checksum. A checksum to ensure the packet header is not corrupted.
Source IP address. The IP address of the source machine. IP addresses are 32-bit numbers, often represented in "dot" notation, such as: 10.10.165.23, where each number is between 0 and 255.
Destination IP address. The IP address of the destination machine.
Options. Various options are available, such as specifying the route that must be taken, or recording the route that was taken. These are rarely used.

Fragmentation occurs when a router receives a packet that is too large to be forwarded. Different routers enforce different maximum packet sizes, in order to better match the packets to the underlying network. If a packet is too big to forward, the router checks one of the flags in the IP header (called the DF (Do not Fragment) flag). If this is set, the packet is dropped and an error message is sent back (using ICMP, the Internet Control Message Protocol). If this flag is not set, the packet is broken up into pieces of the correct size, and these are sent along. Each one (except the last) will have the MF (More Fragments coming) flag set, and they will each have an offset value allowing the original packet to be reassembled at the destination.

The main protocols of the Internet are TCP (Transmission Control Protocol), UDP (User Datagram Protocol) and ICMP (Internet Control Message Protocol). These protocols are used for the vast majority of communications on the Internet, particularly those generated by users sending email, browsing web pages, and downloading files. Since most of the data that I will be considering in this chapter is TCP, I will focus on the TCP protocol. Interested readers can check (Marchette, 2001; Northcutt et al., 2001) for more information about these protocols as they relate to intrusion detection, and (Stevens, 1994) for the definitive reference on the protocols themselves.

The IP protocol does not implement any guarantees of delivery of the packets. In order to ensure that packets are not lost in transit, a protocol was needed to ensure reliable communications. The transmission control protocol (TCP) was thus designed to implement reliable communications. The idea is to initiate a communications session, with acknowledgments of receipt of packets, and a mechanism to resend packets that are lost or corrupted along the way. In order to implement this, TCP makes use of two basic ideas. The first is the concept of ports, which are used to define the communication session. These can be thought of as logical connectors between the two computers, linking the applications at each end and providing a communication link between the computers. The second is the use of flags and sequence numbers to control the connection, and specify the packets and packet order in a unique way.

The TCP header is shown in Figure 4. The fields are:

Source port. A 16-bit number giving an initiating port number to the session.

Destination port. A 16-bit number giving a destination port number to the session. The destination port usually corresponds to a unique application, such as email, web, ftp, etc.

Sequence number. A 32-bit number that sequences the packets going from source to destination.

Source Port			Destination Port	
Sequence Number					
Acknowledgment Number					
Length	Reserved	Flags	Window Size	
Checksum			Urgent Pointer	
Options (if any)					
Data					

Fig. 4. The TCP header.

Acknowledgment number. A 32-bit number containing the sequence number for the packet(s) being acknowledged. This number is actually the sequence number for the next packet the host expects to see.

Length. A 4-bit word indicating the length of the header in bytes.

Reserved. Not used.

Flags. 6 1-bit flags controlling the session. The flags are URG (urgent), ACK (acknowledgment), PSH (push), RST (reset) SYN (synchronize), and FIN (finish). These will be discussed in more detail below.

Window size. A 16-bit number used by applications to control the amount of data they receive.

Checksum. A checksum to ensure the header is not corrupted.

Urgent pointer. Used for emergency control, such as when a user tries to stop a transfer.

Options. Various options are available to control the TCP session.

The session is initiated with a three-way handshake. This is illustrated in Figure 5. First the initiating machine sends a SYN packet (a packet in which the only flag set is the SYN (or synchronize) flag). This packet also contains the initial sequence number for the session. The destination machine, assuming it agrees to communicate, responds with a SYN/ACK packet. It sends a sequence number of its own, and an acknowledgment number corresponding to the number expected to be seen next. Finally, the initiating computer acknowledges (with an ACK packet) receipt of the SYN/ACK, and the session is established. In Figure 5, only one set of sequence/acknowledgment numbers is shown.

A note about terminology is in order. As seen above, one often refers to a packet with only the SYN flag set as a SYN packet. Similarly, a packet with only the SYN and

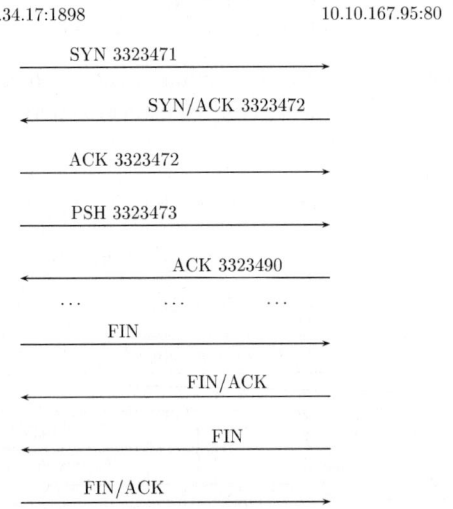

Fig. 5. A "typical" web (port 80) session. This is a session between 10.10.34.17 on port 1898 and 10.10.167.95 on port 80 (http). The handshake is followed by data transfered back and forth, until the session is closed in each direction. The sequence numbers are shown along with the flags above the arrows.

ACK flags set is referred to as a SYN/ACK packet (or simply a SYN/ACK), while, for example, a packet with the flags SYN, ACK, PSH flags set would be referred to as a SYN/ACK/PSH (or sometimes a SAP) packet. This shorthand will be used throughout, without further comment.

It is not necessary to acknowledge every packet individually. Often one will see several packets acknowledged at once, reducing the amount of back-and-forth transmission and hence the load on the network. ACK packets may also contain data going the other way, to improve the efficiency of the session.

Several of the values in the header are of interest, both from the perspective of intrusion detection and from simply a data analysis perspective. Consider the TTL values depicted in Figure 6. Note the clearly multi-modal aspect of these data. This is typical of these values. The reason for it is the habit of operating systems to select a fixed value for their time to live values, usually a power of two (or 255, the maximum value possible). Thus, the values that are observed are these numbers decremented by the number of routers through which the packet has traveled.

This allows one to estimate the distance the packet has traveled. It can also be used, in conjunction with other packet attributes, to determine the operating system of the machine that sent the packet. This is called *fingerprinting* the operating system, and when it is done on received packets such as these (rather than as responses to carefully crafted packets) it is referred to as *passive fingerprinting*.

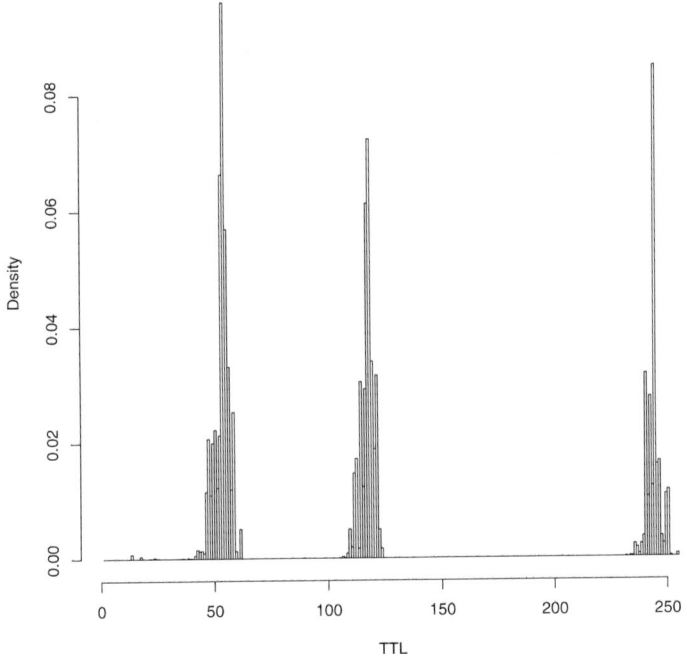

Fig. 6. A histogram of time-to-live values for incoming packets to a network on a Wednesday morning, 11 am. These data represent 2 800 302 packets.

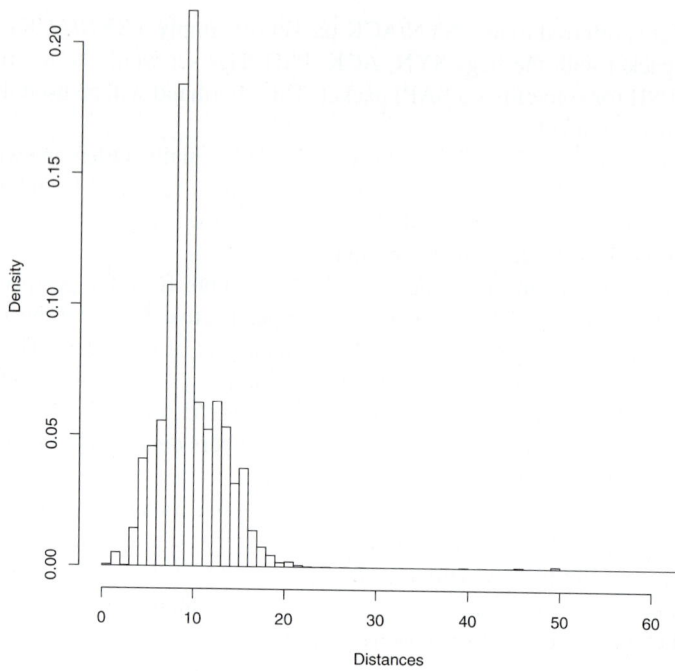

Fig. 7. A histogram of an estimate of the distance traveled for incoming packets to a network on a Wednesday morning, 11 am. These data represent 2 800 302 packets.

Figure 7 shows the distances traveled by the packets, as measured by a first-cut estimate of the original TTL value. In this figure, the TTL is subtracted from the closest, without exceeding, of 64, 128, or 255. The slight skewness is a result of two problems with this analysis. First, not all operating systems choose these values (some pick 30, 32, 60, etc.). Second, there are times when the TTL value is deliberately set to a particular value. This happens with traceroute, a utility that maps the route between two hosts by sending packets with sequentially increasing TTL values, to see which routers respond when the TTL decrements to zero. That said, the plot does show that to first order the vast majority of machines are within 20 hops (routers) of this network. This estimation shows a slightly lower value than reported in (Biagioni et al., 2000), where an average value of 20 is given. It should be noted, however, that Biagioni et al. (2000) are reporting average distance of all hosts, while I am concerned with the average distance of hosts that visit this site, and so the lower value in this study is as would be expected.

3. The threat

Attacks against networks and computers come in three basic flavors. There are denial of service attacks, which attempt to shut down computers or services or otherwise make resources unavailable. Attacks that gain access to an account on a computer, or access/change information on a computer (such as defacing a web site or illegally obtaining source code) are another common grouping of attacks. Finally, there are malicious codes such as viruses, worms and Trojan programs. Each type of attack requires a different approach to detection and analysis, and often entirely different data must be collected.

The threat is described in (Belger et al., 2002), as well as other sites on the web, such as http://www.nipc.gov (Belger et al., 2002) report nearly 40 attacks per week per company in the last quarter of 2001. In addition to attacks, there are a number of reconnaissance activities, such as scanning for hosts or applications, fingerprinting of operating systems, etc. Most attacks aim at specific known vulnerabilities of a particular operating system or software, although many, particularly denial of service attacks, simply use the properties of Internet communication to provide the desired effect.

3.1. Probes and scans

Prior to mounting an attack, an attacker may want to gather information about the victim host or network. This reconnaissance comes in three general categories: network scans, port scans, and operating system fingerprinting.

Network (or IP) scans search the target network for active machines. This is accomplished by sending packets to a list of IP addresses and seeing which addresses respond. For example, if one sends SYN packets to every possible IP address on a network, those machines that respond exist, and as an added benefit one knows whether they are running the application associated with the destination port of the packet. Alternatively, if one sends RST packets, one only receives a response if the packet is destined for an IP address that does not exist. In this case the final router will respond with a "host unreachable" message, indicating that the IP address does not correspond to an existing machine. This so-called inverse scan provides a listing of the machines on the network by considering those addresses that did not generate a response (Green et al., 1999).

Application (or port) scans search for known applications on a host or network. For example, an attacker who knows of a vulnerability in ftp might want to determine if an IP address is running an ftp server. To do this, a packet is sent to port 21, and the response indicates whether the service is running or not. Another attacker may just want to know what services are running, and so sends a sequence of packets to a large list (or all) of the ports.

Operating systems are constrained in how they respond to certain legitimate packets. This is required for the proper functioning of the protocols. However, it is possible to send "illegal" packets, and operating systems are free to respond (or not) to these packets in a variety of ways. For example, there is no legitimate reason for a packet to have both the SYN and FIN flags set (simultaneously initiating and closing a session). An operating system may respond to such a packet with a RST, a FIN/ACK, or simply ignore the packet. By sending a collection of carefully crafted packets and considering the response, the attacker can, with high probability, learn the operating system that the victim machine is running. It is often possible to determine even the version number of the operating system with these techniques. This can be important information for the attacker, since different operating systems (and versions) are vulnerable to different attacks.

3.2. Denial of service attacks

The basic idea of most denial of service attacks is to flood a computer with bogus requests, or otherwise cause it to devote resources to the attack at the expense of the legitimate users of the system. A classic in this genre is the SYN flood. The attacker sends SYN packets requesting a connection, but never completes the handshake. One way to do this is to set the source IP address to a nonexistent address (this process of changing the source address is called "spoofing" the address). For each SYN packet, the victim computer allocates a session and waits a certain amount of time before "timing out" and releasing the session. If enough of these "bogus" SYN packets are sent, all the available sessions are devoted to processing the attack, and no legitimate users can connect to the machine.

A related attack is to send packets that are out of sequence, or errors, forcing the victim computer to spend time handling the errors. For example, if a SYN/ACK packet is sent without having received an initiating SYN packet, the destination computer generates and sends a RST (reset) packet. If the attacker can arrange to have millions of SYN/ACK packets sent, the victim computer will spend all its resources handling these errors, thus denying service to legitimate users. One way to arrange this, is through a distributed denial of service tool, such as trinoo or TFN2k. These tools compromise a set of computers and use these to send thousands of packets to the victim. Each packet is spoofed to have a random IP address, so the attacking machines cannot be identified. See (Marchette, 2001; Chen, 2001; and Northcutt et al., 2001) for descriptions of some distributed denial of service attacks.

The result of such an attack is a number of reset (or other) packets appearing at random sites around the Internet, with no obvious session or initiating packets to explain them. See Figure 8. This is used by Moore et al. (2001) to estimate the number of denial of service attacks during three one-week periods, by counting how many unsolicited packets are seen addressed to one of the 2^{24} possible IP addresses they monitored. A similar experiment is depicted in Figure 9. In this experiment, a reset packet was determined to be "unsolicited" if no packets between the source and destination machines had been seen within a time period (14 minutes in this experiment) prior to the reset packet. These unsolicited packets are referred to as "backscatter" from the denial of service attack. See also (Marchette, 2002).

As can be seen from the figure, there is a certain level of background noise, the result of the fact that the criterion for determining a packet to be unsolicited is not perfect. These may also be the result of slow network scans, which send only a few packets per hour to avoid detection. In addition, there are a number of large spikes, indicating a large number of reset packets. These are indicative of an attack of some kind, either an attack against the monitored network, or the result of backscatter from a denial of service attack as described above.

Figure 10 shows similar data for the month of September 2001. In this all unsolicited TCP packets were recorded, resulting in a total of 1 102 799 packets. These have been

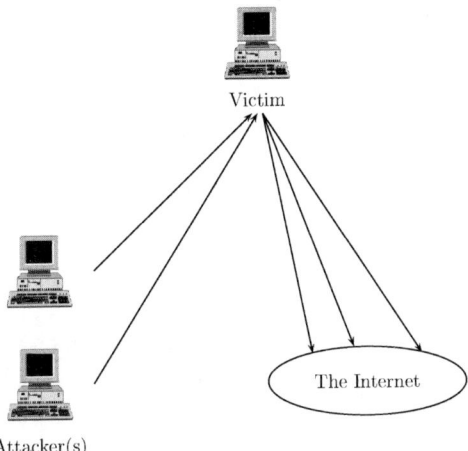

Fig. 8. Backscatter from a denial of service attack. Packets are sent to the victim from one or more attackers. These packets have spoofed IP addresses, which cause the victim's response to be sent to random addresses within the Internet.

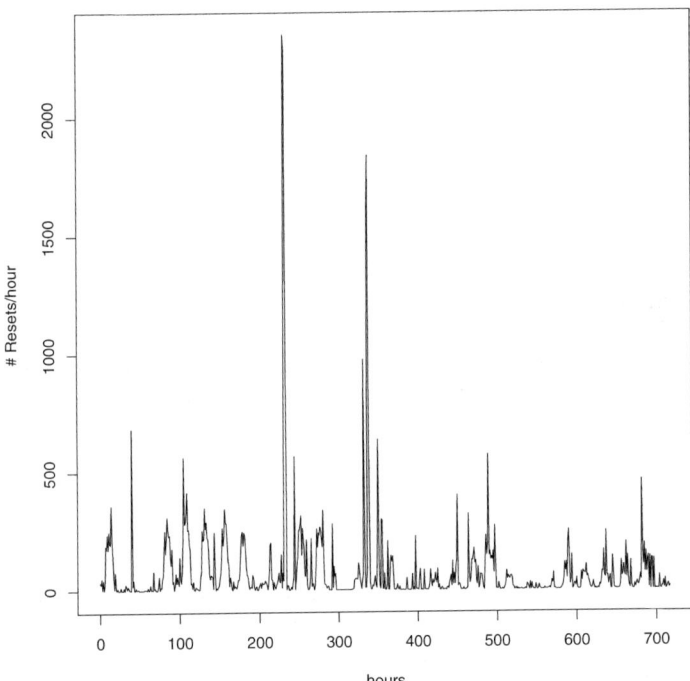

Fig. 9. The number of unsolicited resets seen addressed to a network of 2^{16} possible IP addresses, tabulated on an hourly basis. These data were taken during the month of June, 2001.

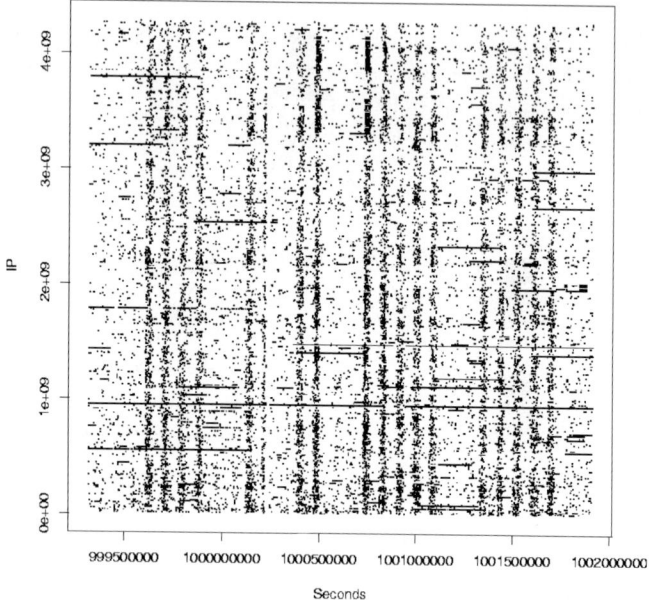

Fig. 10. Unsolicited TCP packets seen at a network of 2^{16} possible IP addresses. The horizontal axis corresponds to time (seconds since 1970) while the vertical axis corresponds to a arbitrary number associated with each source IP address.

depicted in the figure by placing a dot at the time of receipt of a packet (the x-axis) with a y value corresponding to an arbitrary assignment of a unique number to each source IP address. What is clear in this picture is that the weekdays show up very well. There are two interesting weeks here. The first week contained Labor Day, and so only four week days show up. The second week has a small number of packets for Tuesday (September 11) and fewer Wednesday (September 12).

This weekday banding may be an artifact of errors in the selection of "unsolicited" packets. In this experiment, a packet was considered unsolicited if there had been no legitimate session between the IPs for a period of time previous to receipt of the packet (14 minutes in this experiment). The sensor used here is not perfect, and dropped packets have been observed. Also, some applications (notably web servers) keep sessions alive much longer than the standard TCP timeout period. So some of these "unsolicited" packets are undoubtedly a result of these effects.

We see that there are many very short sequences of packets, and a few much longer ones. The longer sequences are the results of attacks, either scans of the monitored network or backscatter from denial of service attacks. Some of these consist of over 100 000 packets. Some persist for nearly the full month. Most of these are indeed the result of backscatter from denial of service attacks (for example SYN floods) rather than a scanning attempt against the monitored network.

The basic idea of backscatter analysis is that the packets sent to flood the victim generate responses in the form of packets sent to the source IP addresses on the flooding packets. Since the attack tools spoof the IP addresses (selecting them at random from the set of 2^{32} possible addresses) some of these packets arrive at the monitored network. If one assumes that there are m packets in the attack, and n IP addresses are monitored, then the probability of detecting the attack is

$$P[\text{detect attack}] = 1 - \left(1 - \frac{n}{2^{32}}\right)^m. \tag{1}$$

This results in an expected number of backscatter packets of $nm/2^{32}$. The probability of seeing exactly j packets is

$$P[j \text{ packets}] = \binom{m}{j}\left(\frac{n}{2^{32}}\right)^j\left(1 - \frac{n}{2^{32}}\right)^{m-j}. \tag{2}$$

From this, one can obtain an estimate of the size of the attack, given that j packets have been observed, as

$$\widehat{m} = \left\lfloor \frac{j 2^{32}}{n} \right\rfloor. \tag{3}$$

The assumption that the source IP addresses are randomly selected is critical in this analysis. An attack tool could just as easily select IP addresses sequentially. However, such an attack would be easy to filter.

The assumption of random source IPs gives us a way to eliminate scans (see Section 3.1) from the backscatter packets. If the destination IPs appear random, then the attack is probably backscatter, while if it is patterned (say, sequential) then it is probably a scan. (Recall that since the backscatter is the response to the attack, the original source IP of the attack packets becomes the destination IP of the backscatter packets.) Thus, one can use a goodness of fit test to determine whether the packets from a particular IP correspond to backscatter or an attack on the monitoring network.

More sophisticated analysis can be used to eliminate some "non-backscatter" packets from the data set. For example, if a sequence of PSH/ACK packets are seen, with incrementing sequence numbers, then it is likely that this is the result of an ongoing session, rather than the result of backscatter, since it is unlikely that a machine will acknowledge an unsolicited PSH packet. This can be checked by looking for initiating PSH packets going the other way. Also, a large number of unsolicited packets to a single machine is probably an attack against that machine rather than the result of backscatter. This kind of analysis is critical to cleaning the data to ensure that the packets remaining are the result of backscatter.

Given a data set such as the September 2001 backscatter data depicted in Figure 10, a question arises as to how one determines that packets correspond to backscatter, as opposed to either a direct attack on the monitored network, or normal packets that were improperly denoted backscatter. As mentioned above, one way to determine that the packets are backscatter is to do a randomness test on the destination IP addresses. Unfortunately, attackers are free to modify the attack as they see fit, either to make denial of service attacks seem less random or to make scanning attacks against the monitored network seem more random. Other methods are clearly needed.

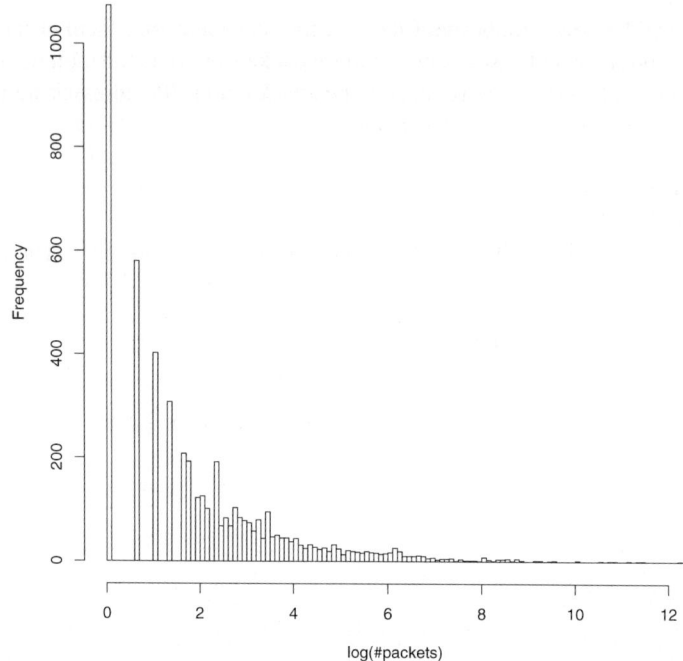

Fig. 11. Histogram of the number of packets per source IP address for the January 2002 backscatter data.

Figure 11 depicts the histogram for the number of packets per source IP. Each source IP corresponds to a victim of a denial of service attack (if the packets do in fact correspond to backscatter). In this figure it is easy to see that over 1000 source IPs had a single packet in the data set. These are almost certainly not the result of backscatter from denial of service attacks. Most likely, these are the result of web servers sending single resets long after the client has stopped using the session.

The source IP with the most packets (199 703 packets) had the statistics tabulated in Table 1. This looks like the backscatter from a SYN flood, where each SYN packet generates a SYN/ACK. In this case, some of these packets are followed, after a short period, by a RST/ACK. The data in the table corresponded to two distinct attacks, one for six days and one for 11 days, with a 20 hour gap in between. The longest gap between packets was an hour and a quarter. Note that this pairing of SYN/ACK and RST/ACK

Table 1
Statistics for the source IP with the most unsolicited packets destined to the monitored network

Number of packets		199 703
Number of destinations		26 078
Flag combinations	AS	170 723
	AR	28 979
	R	1

violates the independence assumptions of the above analysis (Eqs. (1)–(3)), and so the estimates would need to be done (for example) on the SYN/ACK packets alone.

An interesting problem would be to use the estimate $\hat{m}$ to determine whether the attack was distributed across multiple attacking machines, and to estimate the number of attacking machines. This would require some way of determining the average number of packets per machine per attack, and de-interleaving the packets corresponding to the different machines in the attack. Another problem of interest would be to infer the severity of the attack, by both the number of packets and the pattern of the time between packets. Presumably, as the victim machine becomes overloaded, the interval between responses increases, and the number of backscatter packets per time unit also decreases. An abrupt cut-off can be evidence of a cessation of attack, or the success of the attack. Methods for analyzing these data to determine which is the case would be of value.

3.3. Gaining access

For other kinds of attacks, such as an unauthorized login to a computer, other kinds of data must be collected. One could look inside the data of the packets, searching for strings that are indicative of attacks. This requires considerable processing power, however, if it is to be done on all the packets on the network. It also requires the ability to decrypt encrypted sessions, which may not be practical (or desirable). Thus in most cases it makes more sense to do this kind of analysis on the individual hosts. In order to do this, one must investigate the security log files.

Log files are generated describing logins, errors, and providing messages from various utilities as they run. For example, I have used a program called portsentry which monitors the ports on my computer and warns me (through a log file) when a machine attempts to connect to a closed port. Typically, log files are minimally formatted text files, containing a time stamp, the name of the program that generated the entry, and a string containing the log entry. Extracting useful information from a log file is a challenge. An example of how one might use log files to detect intrusions would be to check the log file for login attempts which failed, attempts to log into accounts that are not normally accessed (such as lp or daemon) or logins at unusual times or from unusual hosts.

There are several ways an attacker might gain unauthorized access to a computer. Several of these are discussed in (Marchette, 2001). Some of these rely on subterfuge, such as stealing passwords, and are difficult to detect, since the actual login itself may use the proper name and password. Others rely on bugs in software, or unforeseen vulnerabilities in protocols or applications. Many of these are difficult to detect in log files, and other detection methods are necessary.

One approach to detecting these kinds of attacks is through monitoring the system calls that programs make for unusual patterns (Forrest et al., 1994, 1997; Forrest and Hofmeyr, 2001; Naiman, 2002). The idea is that an attack will result in unusual patterns of execution of the attacked program. By monitoring the system calls and characterizing the "normal" patterns, it is hoped that the attack can be detected through the detection of "abnormal" patterns.

Another approach is to try to recognize that the user typing at the keyboard is not the authorized user. This can take the form of monitoring keystrokes, where, for example, the time between characters in common words is used to construct a "fingerprint" for the user. See, for example, (Bleha et al., 1990; Bleha and Obaidat, 1991; Bleha and Gillespie, 1998; Lin, 1997; Obaidat and Sadoun, 1997; and Robinson et al., 1998). This approach does not work for remote accesses, however. The process of formatting the packets, routing them through the network, then processing the packets eliminates any user-specific timing information that might be used to validate users.

One can also monitor program usage by the user (Schonlau et al., 2001). By noting which commands the user typically uses, and in what order, a profile can be developed to distinguish the authorized user from a masquerader. Unlike keystroke monitoring, this can be used to authenticate remote users. I will discuss these ideas in more detail in Section 7.

4. Network monitoring

In order to detect an attack on a network, the network must be monitored. This means collecting information about the packets transiting the network. This is accomplished via a *network sensor* or *sniffer*, a program that collects packets or packet headers. These are then processed by an intrusion detection system.

In this section I will look only at TCP packets. In all the examples I will consider data taken from a sensor outside the firewall of a class B network (at most 2^{16} IP addresses). Because the sensor sits outside the firewall, it sees all attacks, even those that are denied by the firewall. It also sees all legitimate traffic, going into and out of the protected network.

There are a number of attacks that can be detected by a network monitor. I discussed denial of service attacks in Section 3.2. Most of the attacks discussed there can be detected by looking for unusual or improperly formatted packets or large numbers of packets to a particular host. One can also detect scans on the network, where the attacker sends packets to a large number of hosts or ports.

In addition, there are a number of activities that are easily determined to be attacks. For example, an attempt to access a print server from outside the network is unlikely to be a legitimate activity. Thus, the security officer can list a number of disallowed activities, such as telnet, remote procedure calls, etc., that the network monitor can watch for and flag as suspicious.

One task for the data miner would be to determine the set of allowed activities from the data. A simple example of this would be to keep track of the number of connections to each port on each machine, and define "normal" activity as connections to those machine/port combinations that are commonly accessed. More sophisticated rules could be developed by maintaining statistics about these "normal" sessions, and use these to detect sessions that deviate from these statistics. For example, a session with an unusual amount of data transfer might be the result of a Trojan program installed on that port.

Network monitoring can also be used for passive fingerprinting, determining the operating system of a machine by observing the packets that it sends. One reason to do this would be to determine information about the attacker. Different attack tools have been written for different operating systems, and knowing the operating system of the attacker can thus provide information about the attack, and about the attacker's sophistication.

Network data is quite complex, and tools are needed to analyze packet data. Some exploratory data analysis can provide interesting insight into the data.

I examined TTL in Section 2.1, Figures 6 and 7. Now let me consider the source port of the packets. When a connection is initiated, the initiating machine chooses an arbitrary number for the source port of the connection. This port is usually greater than 1024, and is generally selected sequentially. Figure 12 shows the six most active machines during an hour starting at 6 am in January, 2002. These data represent 25 047 SYN packets. Note that the sequential nature of these data is obvious in this plot. However, it is also clear that the data lie in distinct bands. For example, the red and green IP addresses are always in a band below 10 000, while the blue, magenta and black all take their source ports from above 30 000 (at least in this plot). The fact that the red, green, and cyan IPs repeat leads us to infer that they are constrained to a subrange of the possible port range.

There are three possible reasons for this banding. One is that it is driven by the application. That is, it could be the case that ftp uses one range, HTTP (web) another, etc., and that each IP was requesting a single service. An investigation shows this not to be the case. Another possibility is that the choice of source port is driven by the operating system. This is a reasonable hypothesis. To test this (on admittedly a tiny sample), I de-

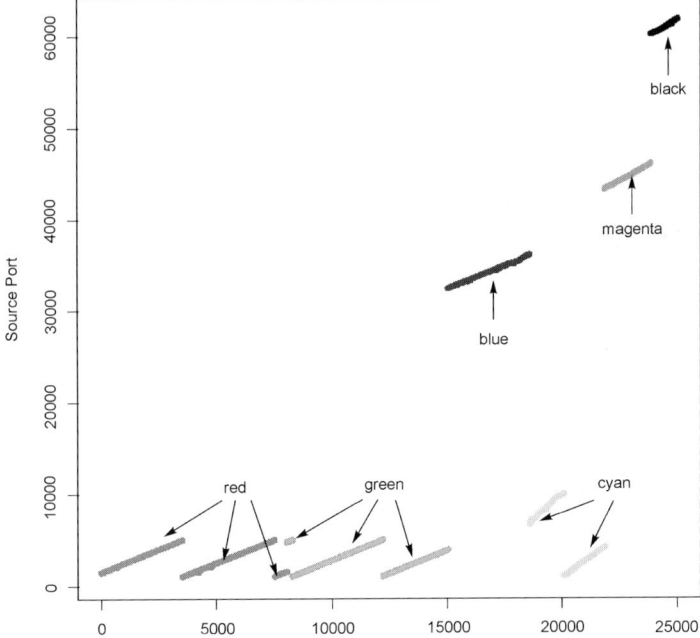

Fig. 12. Source ports for the 6 most active IP addresses at a site during the hour of 6 am, January 23, 2002. The observations are plotted in the order in which they were received. The colors correspond to distinct source IP addresses.

termined, to the best of my ability, the operating systems of the machines in the figure. The red, green and black are Windows 98 machines, the blue is a Windows 2000, and the others are unknown. This gives some (imperfect) support to the hypothesis. However, the fact that the black IP address is in the 60 000 s shows that it is not perfect, and I have not adequately characterized the operating system with this small sample. (Note that traditional statistical analysis would generally not consider a sample of size 25 000 to be "small".)

There is another reason that the source ports chosen by a particular machine may appear to fall in a band, as indicated by the red, green and cyan machines in Figure 12. It is possible that this is a result of the machines being turned off at night. Thus, the only source ports observed are the ones that had been reached prior to the end of the day, with the port range reinitialized the following morning. A larger data set is needed to determine if this is the case.

A further study is depicted in Figure 13. This represents data collected over a one month period (January 1, 2002 through February 4, 2002) on a network consisting of 3983 active machines. Table 2 shows the different operating systems in the data. The operating systems were determined from the accreditation database, which is supposed to be updated every time a new machine is installed or an operating system changes. This database is not completely accurate, as can be seen by the entries "dos", indicating a Microsoft operating system, and "mac", indicating an Apple Macintosh operating system. There is no data on how often the database is updated to reflect the installation of new versions of operating systems, and so the version numbers indicated may not be reliable.

Figure 13 depicts a data image of the source port data. The rows correspond to operating systems, indicated by the letters, in the order of Table 2, from bottom to top. The

Table 2
Operating systems of 3983 machines used to investigate the correlation between source port and operating system

OS	#M	#P	OS	#M	#P
dos	69	258 989	macos-8.6	9	77 182
irix-6.1	1	137	macos-9	26	294 806
irix-6.2	3	131	solaris-2.4	1	35 861
irix-6.5	38	52 878	solaris-2.5.1	10	182 298
linux-6.0	2	79	solaris-2.6	19	148 749
linux-6.1	5	15 771	solaris-2.7	9	11 284
linux-6.2	20	83 836	solaris-7	16	11 953
linux-7.0	7	66 685	solaris-8	21	6 163
linux-7.1	8	19 472	windows-2000	896	6 359 055
linux-7.2	15	39 177	windows-3.1	6	35 596
mac	13	28 436	windows-95	414	1 931 628
macos-7.5.5	1	441	windows-98	1530	10 135 511
macos-8	13	38 287	windows-me	15	90 140
macos-8.1	8	22 283	windows-nt-4.0	792	7 319 801
macos-8.5	8	86 245	windows-xp	8	41 811

The data were collected by monitoring the outgoing SYN packets from a network for one month. The second column shows the number of machines (#M), while the third shows to the number of packets (#P).

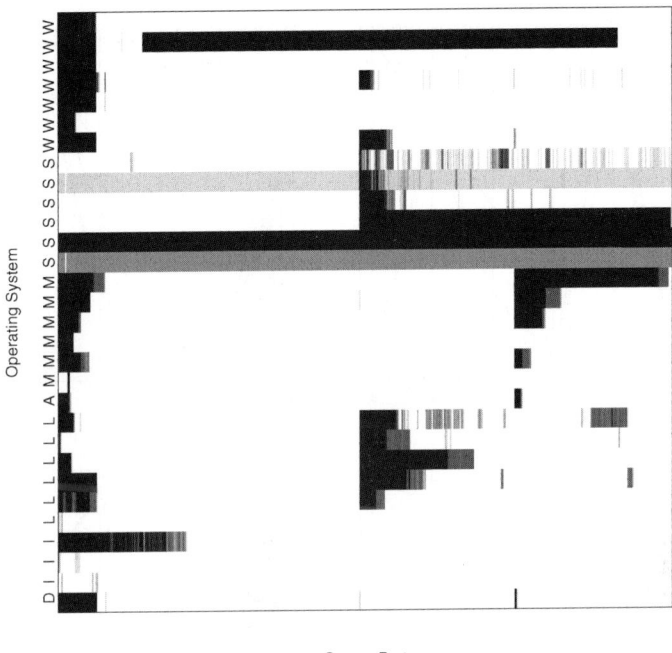

Fig. 13. Source port versus operating system for the data from Table 2. Gray values indicate the number of packets for that combination, with black corresponding to greater than 255. The operating systems are indicated by the first letter of their name, and are (from the bottom): dos, Irix, Linux, Apple (generic mac), MacOS, Solaris, Windows.

columns correspond to source ports, from 1024 to 65 535. The number of times a particular port was used by a machine of the corresponding operating system is indicated by a gray value, with black indicating that the combination was detected more than 255 times.

While the source port is clearly not perfectly correlated with operating system, it is clear that the different operating systems do make slightly different choices of ranges for source ports. Thus, source ports, in conjunction with other attributes, have a potential for allowing some measure of passive operating system fingerprinting. There may still be an effect due to machines being turned off before cycling through their full port range. This remains to be investigated.

Another study is presented in Figure 14. In this, six features were collected from each packet: source port, window size, max segment size, time-to-live, header length and whether the don't-fragment flag (DF) was set. These have been scaled and plotted in a parallel coordinates plot for each operating system class. In this study, an operating system class consists of all observed versions of a particular operating system, such as Irix, Linux, MacOS, Solaris, or Windows. These plots represent 3901 individual machines, where those denoted "dos" and "mac" have been removed. As can be seen by

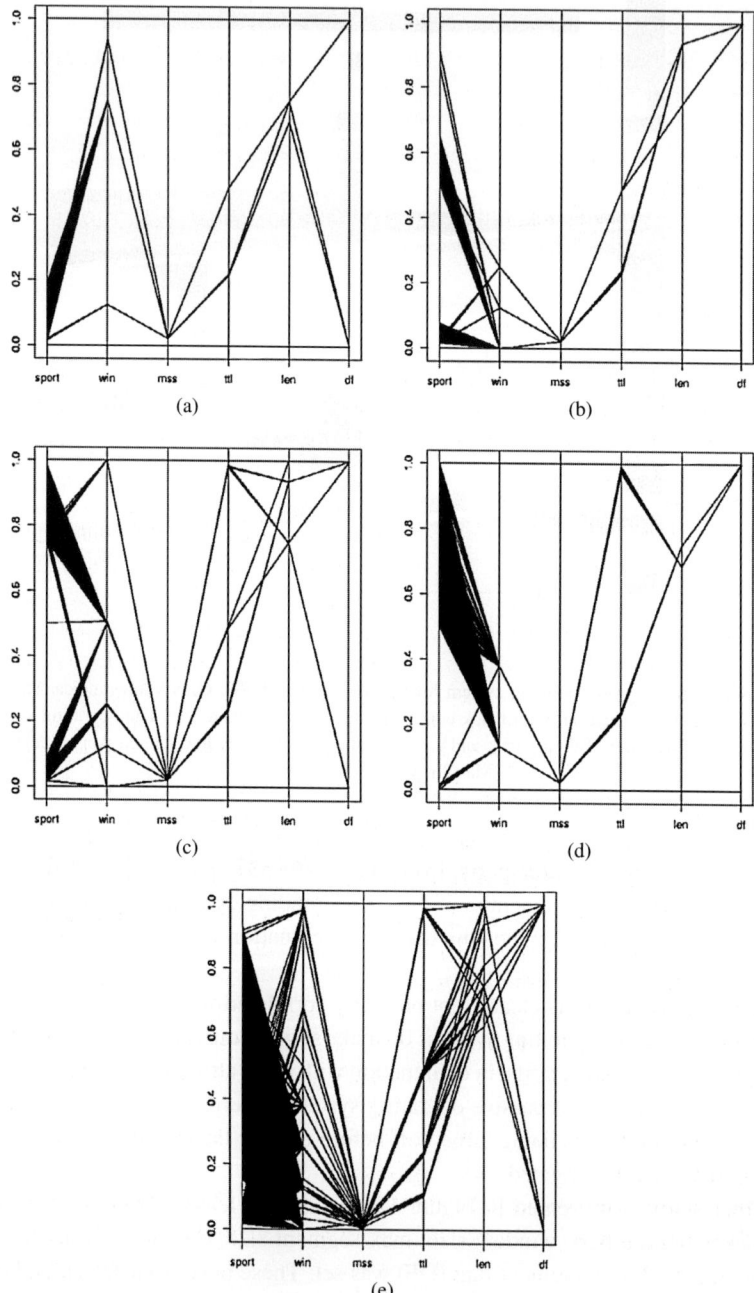

Fig. 14. Parallel coordinates plots of packet header values for the operating systems: (a) Irix, (b) Linux, (c) MacOS, (d) Solaris, and (e) Windows. The features are: source port, window size, max segment size, time-to-live, header length, and DF.

the plots, it is possible, in many cases, to distinguish the operating system class using these features, at least for these 5 operating system classes.

Knowing the operating system of a victim can be very useful for an attacker. By the same token, knowing the operating system of the attacker can be useful for the defender. This provides information as to the likely attack scripts that may be used. Whether or not common attack scripts are used in the attack provides information about the likely sophistication of the attacker. This information can be useful in mounting a defense or in cleaning up from the attack.

5. TCP sessions

Considering individual packets allows for the detection of some kinds of attacks, such as those which rely on malformed packets (active operating system fingerprinting and some denial of service attacks), network and port scans, and attempts to access forbidden services. However, many attacks do not fall into these categories. Some of these may be detected by considering TCP sessions instead of the individual packets.

It is obvious that there should be a correlation between the number of packets in a session and the number of data bytes sent during the session. What may not be obvious, is that this correlation is dependent to some degree on the application that is running on the session. Consider Figure 15. The top figure shows a scatter plot of number of packets against number of data bytes. While there is an obvious correlation, there is clearly a mixture of different processes represented in this picture. To investigate this, the bottom picture shows the same plot for the applications sendmail (email on port 25) and telnet (login on port 23). As can be seen by the fitted lines, the two application have distinctly different character. The reason for this is that telnet is an interactive session. The user types commands, which are generally sent one character per packet, and the machine sends back the results of the commands (the text that would have been displayed on the screen). This kind of interaction is relatively low bandwidth, and thus there are generally many small packets per session. Email tends to push the whole message through as fast as it can, and so tends to have larger packets. A session such as ftp, with large data transfers, will tend to have even larger packets, and also a larger number of packets per session than email.

One application of this observation is the detection of tunneling. Most networks have restrictions on the applications that are allowed in from the Internet. Telnet is an example of an application that is usually blocked at the firewall, due in part to the danger of having passwords stolen, since telnet is unencrypted. However, many networks allow some applications, such as web, through their firewall.

Consider a user who wishes to be able to telnet in to his work computer from home. Although telnet is blocked at the firewall, web is allowed through. The user sends packets that appear to be destined for a web server, but instead are interpreted as telnet sessions once they arrive at the destination machine.

I have heard that it is possible to accomplish this through fragmented packets, but have not been able to verify this. The idea would be as follows. The user sends the first fragment (which contains the header) exactly as though it was destined for a web

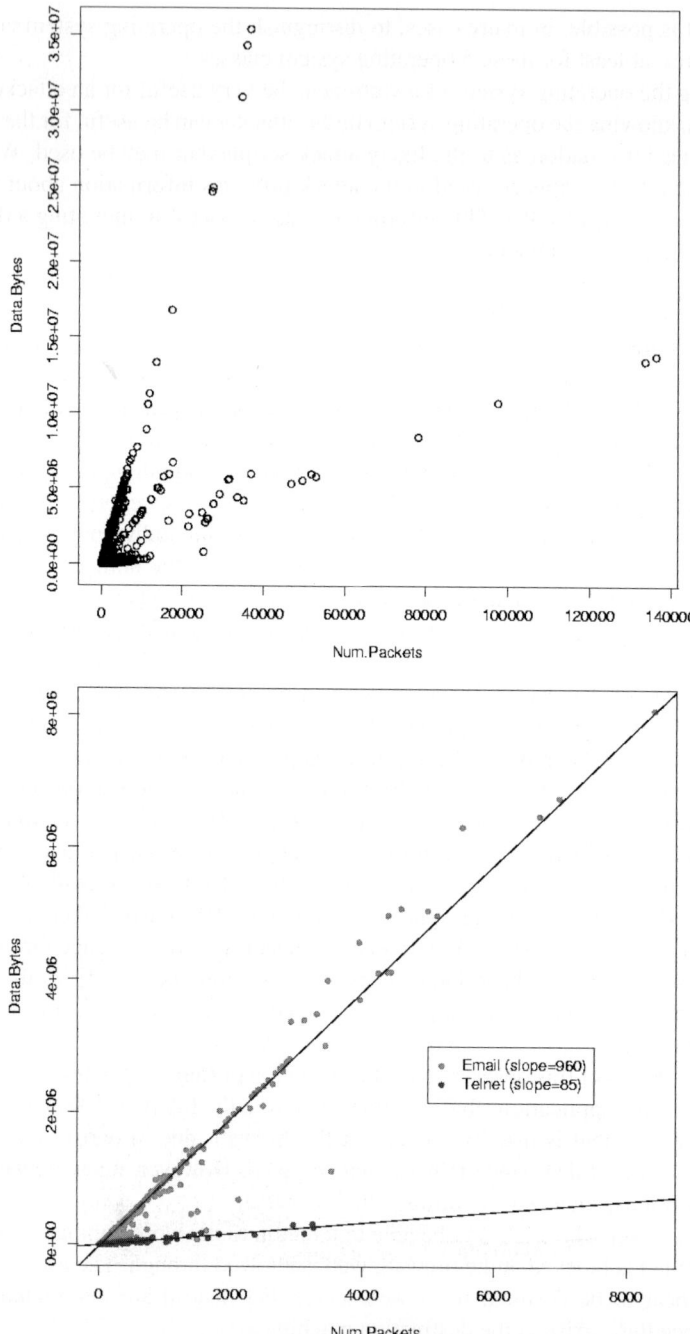

Fig. 15. Number of packets versus number of data bytes per session. The top figure shows the results for all sessions within a fixed time period, while the bottom shows only those corresponding to email (port 25) and telnet (port 23).

server (destination port 80). The fragment offset on the first fragment is 0. The next few packets are sent fragmenting the TCP header, offset to overlap the first packet's TCP header. This header is that of a telnet packet. Thus, the original (web) header gets overwritten by subsequent packets, and when it gets sent to the TCP layer it looks like a telnet packet. As noted, this attack is speculative, as I have not seen such an attack implemented.

There are other methods for tunneling applications past firewalls. Some Trojan programs implement tunneling, and some legitimate programs, such as ssh, allow tunneling of other applications (like X-windows). In this latter case, the session is encrypted, and there is no simple way to tell if tunneling is taking place. However, by characterizing the sessions (for example, by considering the ratio of packets to bytes in a session, as was done in Figure 15) one may be able to detect an attempt to tunnel an application past the firewall. This kind of analysis can also be used to detect Trojan programs, by characterizing the "normal" sessions, and looking for "abnormal" sessions.

Another illustration of this kind of analysis is given in Figure 16. Here I have displayed a parallel coordinates plot showing the total number of packets, the number of incoming packets (P–>), outgoing packets (<–P), packets containing data (DP), data bytes (DB), incoming and outgoing data bytes (DB–> and <–DB, respectively), and duration for a number of web sessions. Sessions are color coded according to the number of packets, in order to see the relationships of these different types of sessions. In this

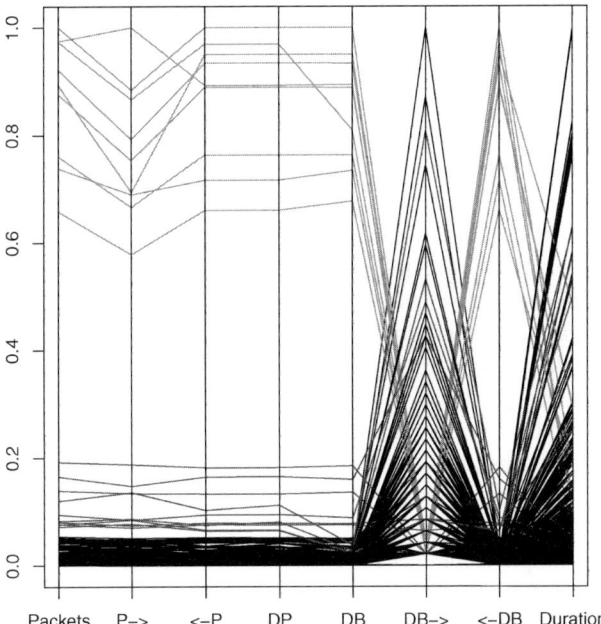

Fig. 16. Parallel coordinates plot of web sessions. The axes are the total number of packets, the number of packets in each direction (–> corresponds to incoming, <– to outgoing), the number of data packets and data bytes (total and in each direction) and the duration of the session.

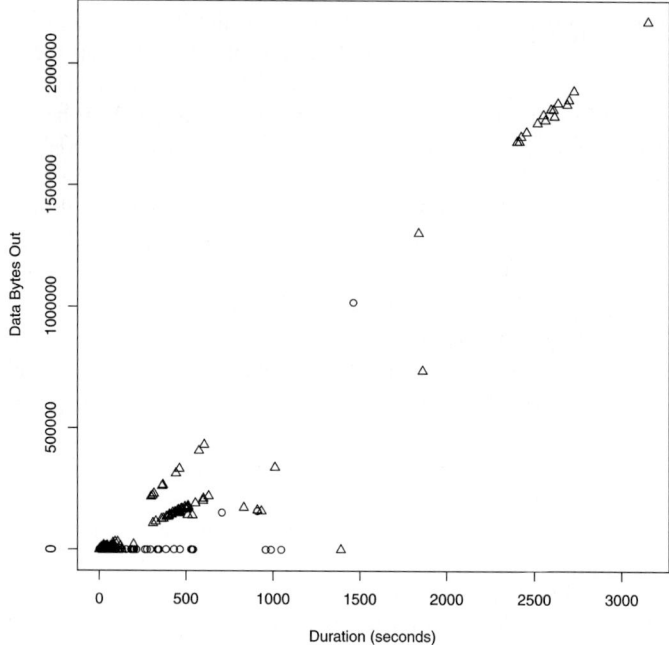

Fig. 17. Number of data packets versus session duration for a set of secure shell (ssh) sessions. Day sessions are shown as circles, night sesssions as triangles.

plot, the axes have been scaled individually, due to the large difference in the ranges of the variables.

As can be seen, the sessions with a large number of packets tend to be the ones with a large amount of outgoing data. These probably correspond to the transfer of images, or other large data files. These sessions also have no incoming data, indicating further that these are data transfers from the server to the client.

Another example of session analysis is depicted in Figure 17. In this plot, the number of data packets is plotted against duration for a set of secure shell (ssh: port 22) sessions. On this machine, the majority of daytime ssh sessions are interactive, with little or no data transfered. Each night, a data set is downloaded from another server (corresponding to packets collected by a sensor during the previous day). The day and night sessions are coded by distinct symbols in the figure. As can be seen, there is a strong correlation of large transfers with long durations at night, while most day time activity corresponds to relatively few data packets.

This is a very simple technique for detecting the transfer of data, which may correspond to a tunneled application if the application does not normally have large data transfers. More sophisticated analysis is required for more complicated situations. For example, I do not know whether this kind of analysis can be used to distinguish between normal data transfers (scp, the secure copy utility of ssh) or, for example, a tunneled X session in ssh. This is an area for future investigation.

6. Signatures versus anomalies

Many intrusion detection algorithms rely on a set of signatures of attacks. For example, a UDP packet with source port 7 (echo) and destination port 19 (chargen) is an attack. The source and destination machines are set to be two machines within the victim's network. This sets up a denial of service in which the destination machine generates characters, which are sent to the source machine, which echoes the packets back, generating more characters. The resultant load on the machines reduces their abilities to service legitimate requests. Since there is no legitimate reason to generate such a packet, any such can be characterized as an attack.

A problem with signature-based systems is that they require knowledge about the attacks, which, in turn, means that new attacks may go undetected. Thus, an alternative approach is to detect "abnormal" activity. This requires a model of "normal" and a method for determining when an activity is sufficiently far from normal to raise an alarm.

There are many ways to characterize the data on a network. For the detection of attacks against TCP, one might choose to determine which services are normally served by the different machines on one's network, and flag attempts to access rare or unused services. One may also want to characterize typical times of usage, number of bytes transfered, connecting machines, and similar attributes of sessions.

This kind of analysis can also be used to cluster the machines on a network for the purposes of determining the natural groupings of the various servers and workstations on the network according to how they are actually used. Figure 18 depicts a clustering of machines by activity. In this experiment machines were characterized according to the proportion of sessions to each port within a fixed time period. This results in a vector of "activity levels" for each machine, which can be compared to cluster the machines into natural groups. In the figure, the machines are labeled according to their most active port. All ports numbered above 6020 were collapsed into a single bin at 6020.

The clustering technique used here is a hierarchical clustering algorithm. It is an agglomerative technique, which successively combines closest groups until all points

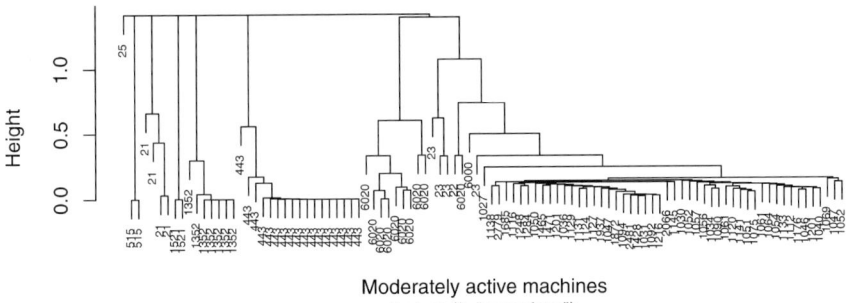

Fig. 18. Clusters for activity levels of machines with a moderate amount of activity. The machines are labeled according to the port with the largest amount of activity.

are contained in a single group. This produces a dendrogram displaying the order in which the points were grouped into clusters as the algorithm proceeds. See (Everitt, 1993) or (Duda et al., 2000) for more details on this and other clustering techniques. See also (Marchette, 1999) for a similar study.

As can be seen in the figure, machines tend to be well characterized according to their most used application, at least for the more common applications. There is some variability among groups, indicating that these do have other applications running, and so the problem of characterizing activity must take this into account. For example, the machines running a secure web server (port 443) are very well grouped, except for three machines, which are slightly different from the pack. Analysis of these three machines may provide information on how their usage differs from the others. This information may be used to provide a more carefully tailored model of "normal" activity for these machines.

7. User profiling

An alternative to profiling network or machine activity to detect attacks would be to profile users. This could be used to detect masqueraders who have gained access to a legitimate user's account.

Many methods have been proposed for profiling users. At one extreme the user could be required to provide fingerprints, retinal scans, or even DNA to verify that they are who they claim to be. If one believes James Bond movies, all of these can be easily defeated by a sufficiently sophisticated attacker.

Several researchers have looked at characterizing keystroke timings as a method of user authentication. See, for example, (Bleha et al., 1990; Bleha and Obaidat, 1991; Bleha and Gillespie, 1998; Lin, 1997; Obaidat and Sadoun, 1997; and Robinson et al., 1998). These typically measure the times between keystrokes for a word or phrase, and characterize the patterns associated to each individual. One application of this is password authentication. The theory is that since a password or passphrase is typed often by the corresponding user, the user develops a characteristic pattern of typing that can be used to detect masqueraders. This approach has had mixed results, as the references show. While some measure of detection is indeed possible, the false alarms tend to be rather high. For user authentication, where the timings are used in addition to the password to provide additional authentication, one may be willing to accept false alarm rates of a percent or two (provided the user is allowed to try again). If the system were to monitor typing during the full session, one may be unwilling to accept false alarms that might cause the system to lock legitimate users out more than once every few months.

An example of keystroke timing data is depicted in Table 3. In this, three users were timed while they typed the nonsense phrase: home mark dart start hello dash fast task past mask. The goal was to distinguish between these three users. This is a confusion matrix for a 5-nearest neighbor classifier. As can be seen, user D is most often confused with the others. For example, 7% of the time the classifier incorrectly labeled a passage

Table 3
11-nearest neighbor classifier performance for keystroke timing data

	D	J	T
D	95	3	2
J	7	88	6
T	4	2	94

Three users, D, J and T were timed on a 10-word phrase, and the confusion matrix is presented below.

Table 4
11-nearest neighbor classifier performance for keystroke timing data

	D	J	T
D	100	0	0
J	5	93	2
T	7	0	93

Three users, D, J and T were timed on a 10-word phrase. In this example, each of the first 4 words was classified independently, then the results multiplied to obtain a joint classification.

typed by J as one typed by D. Similarly, user J is the most often misclassified as one of the other users.

This study treated the entire sequence as a single phrase. If one considers the first four words, and classifies each individually, multiplying the resulting probabilities to obtain the joint classification, one obtains the results reported in Table 4. This shows that the overall performance can easily be brought into the mid 90% range.

This example is misleading in that the words chosen were short, and the phrase does not lend itself well to specialization by different users. It is a reasonable assumption that longer words, and phrases that are meaningful to the user, would be more useful for this kind of a task. However, this does indicate that there is a measure of difference in the typing pattern of users.

Keystroke timings cannot be used to authenticate remote users, since the timings are destroyed by the transmission of the packets across the network. An alternative technique is to characterize the commands that are used. The statistics on the commands used, command sequences, and other information about how the user typically interacts with the computer, such as the use of the mouse, windows, etc., can be used to detect masqueraders.

Schonlau et al. (2001) provide a description of such an experiment. In this, the authors monitored a collection of 50 users, recording the commands they used for several months. These command sequences were then used to characterize the users, with an aim at detecting when an attacker has gained access to the user's account. Several modeling techniques were proposed and evaluated. Again, the reported performance was not as good as one might wish. With false alarm rates around 1%, the algorithms investigated detected between 30 and 70% of the masqueraders. At a 1% false alarm rate, some of us would be raising (false) alarms several times a week, which would tend to get rather annoying.

The above observations show that the problem of user profiling is a difficult one. Some ideas for improving the models are to include other information such as:

- misspellings and mistypings;
- use of aliases;
- files and directories accessed;
- command line arguments;
- use of the mouse;
- use of accelerators such as command completion and shell histories.

Another issue is that users are not static, their behavior changes over time due to changes in workload, interests, knowledge, and other factors. This is more than a simple drift in the distribution of whatever measures are taken, but instead can be rather dramatic changes. Methods for identifying, tracking and assessing these changes would be important to a system for user profiling.

This discussion illustrates some of the difficulties of user profiling, and some of the interesting sources of data that can be used for this purpose. The models used for these applications tend to be fairly complex, and are generally not thought to properly model the underlying stochastic nature of the process, but rather to simply provide a good functional fit to the discriminant region.

8. Program profiling

Program profiling is analogous to the user profiling described in (Schonlau et al., 2001). Instead of tracking the commands that a user makes, one tracks the system calls that a program makes. The idea is that if a program is attacked, for example via a buffer overflow attack, the pattern of system calls will change, giving an unusual pattern which may be detected as an anomaly. This approach has been developed in (Forrest et al., 1994, 1997; Forrest and Hofmeyr, 2001), and recently analyzed by Naiman (2002). As with user command profiling, program profiling relies on characterizing sequences of system calls and looking for deviations from normal sequences.

Naiman (2002) analyzed data taken from an httpd (web) server over a period of several months. In order to determine the variability of the sequence call patterns, he analyzed the patterns found at the various levels of the process tree. As with many modern programs, httpd forks processes to accomplish various tasks, and by keeping track of the parent and child processes a process tree is built from the root (httpd) node. Naiman found that to a very large degree, all the variation in the sequence calls was found at process level 2 of the tree, and the tree never went beyond a process level of 5.

The lengths of the level 2 processes are depicted in Figure 19. The bimodal and skewed nature of these data are readily apparent in these plots. In the top a kernel estimator (Silverman, 1986) is overlaid on the histogram, while in the bottom a filtered kernel estimator, using two bandwidths, is shown (Marchette et al., 1996).

Figure 20 shows the number of system calls as a function of position in the process. As can be seen the variability jumps dramatically after a position of about 300 in the process.

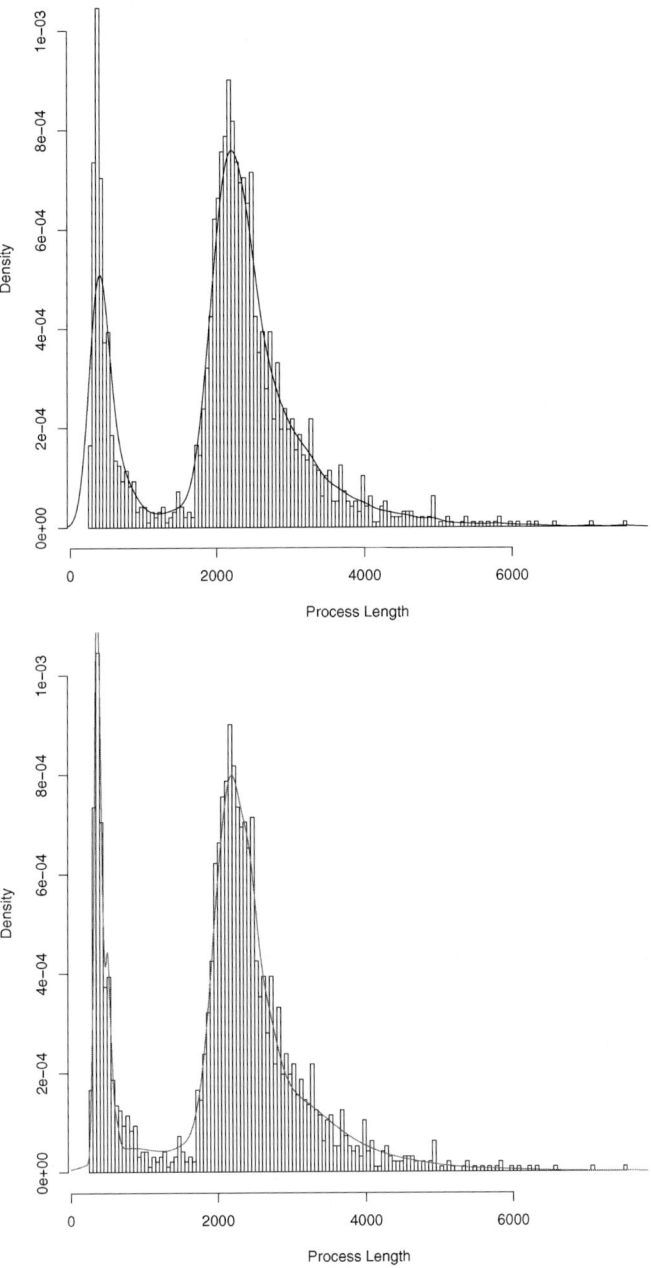

Fig. 19. Histogram of the lengths of processes at level 2 for the httpd data with a kernel estimator (top) and filtered kernel estimator (bottom) overlaid on the plot.

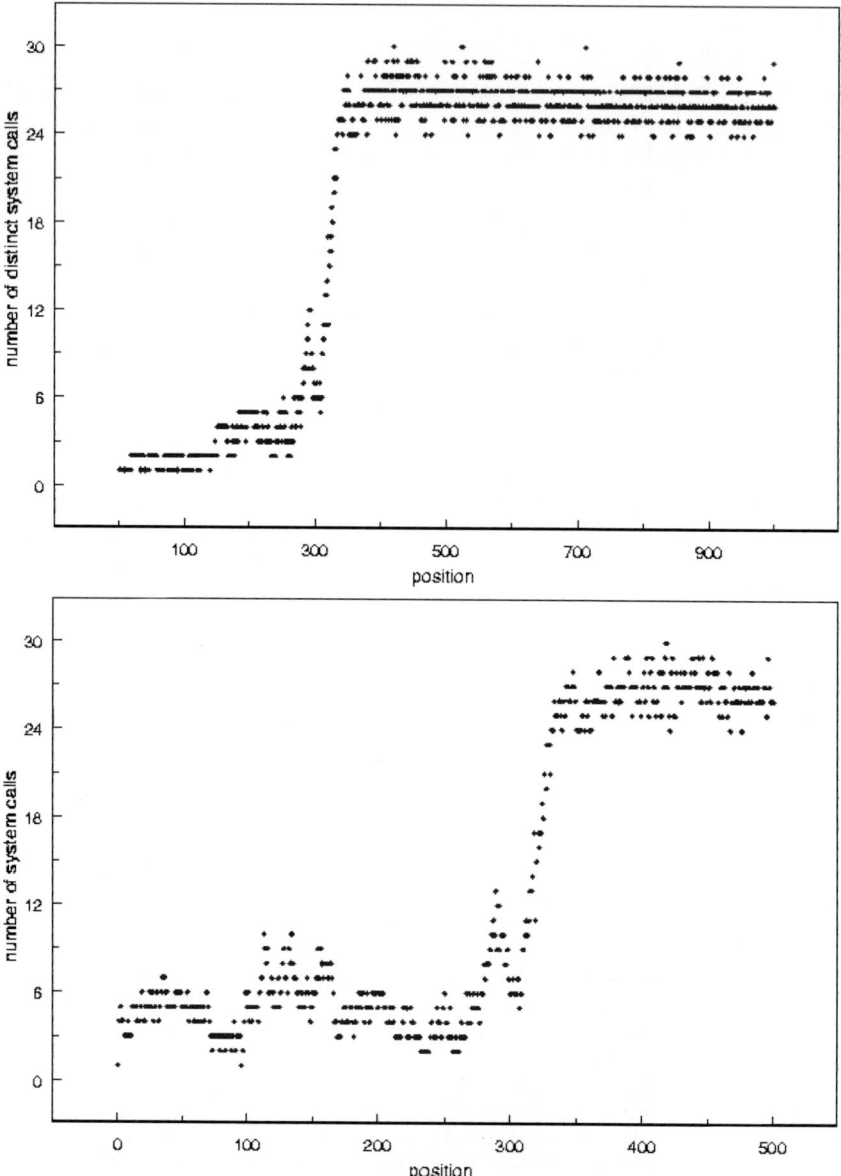

Fig. 20. Number of system calls as a function of position within the process. Figures courtesy of Dan Naiman.

This analysis shows that the problem of program profiling, at least for this version of httpd, requires two basic approaches. Outside of the level 2 processes, a simple signature is adequate, since the processes are very deterministic. Level 2 processes may lend themselves to statistical methods, such as the n-gram approach of Forrest, or hidden Markov models, and this is an area of current research.

9. Conclusions

Perfect computer security is a chimera. There will always be bugs, undocumented features, and design flaws that can be exploited. The result is that there will always be a need for intrusion detection systems, and these systems will require the expertise of statisticians and data miners.

Intrusion detection takes many forms. Network intrusion detection systems try to detect attacks by analyzing the packet streams for attack signatures or unusual behavior. Host based systems look at logs, program or user activity, file integrity, or other measures of system health and normal operation. These diverse data sets provide a wealth of information for data miners to investigate. They also provide a number of challenges in both the modeling and interpretation of the data.

A topic not covered in this chapter, but which is of paramount importance, is that of fusing the information from multiple intrusion detection systems. Since no one system, whether network based, host based, or a hybrid, can do a perfect job of detecting all intrusions, methods need to be developed to combine the alerts from multiple sensors. This has the potential to dramatically reduce false alarms, which in turn may allow the sensitivity of the individual systems to be increased, thus improving detection. While some work has been done in this area, much still needs to be done. See, for example, (Ye et al., 1998; Bass, 2000).

More information on visualization and processing of network data can be found in (Wegman and Marchette, 2003), which discusses the issues of streaming data. Marchette and Wegman (2004) provide an overview of some of the issues discussed in this chapter.

References

Amoroso, E. (1999). *Intrusion Detection: An Introduction to Internet Surveillance, Correlation, Trace Back, Traps, and Response*. Intrusion.net Books, Sparta, NJ.

Anonymous (1997). *Maximum Security*. Sams.net Publishing, Indianapolis, IN.

Bace, G.B. (2000). *Intrusion Detection*. MacMillan, Indianapolis, IN.

Bass, T. (2000). Intrusion detection systems and multisensor data fusion. *Commun. ACM* **43**, 99–105.

Belger, T., Yoran, E., Higgins, M., Dunphy, B., Odom, J. (2002). Riptech Internet security threat report. Available on the web from Riptech, Inc. at http://www.securitystats.com/reports/Riptech-Internet_Security_Threat_Report_vII.20020708.pdf.

Biagioni, E., Hinely, P., Liu, C., Wang, X. (2000). Internet size measurements. Available on the web at http://www2.ics.hawaii.edu/~esb/prof/proj/imes/index.html.

Bleha, S., Gillespie, D. (1998). Computer user identification using the mean and the median as features. In: *IEEE International Conference on Systems, Man, and Cybernetics*. IEEE Press, pp. 4379–4381.

Bleha, S.A., Obaidat, M.S. (1991). Dimensionality reduction and feature extraction applications in identifying computer users. *IEEE Trans. Syst. Man Cybernetics* **21** (2), 452–456.

Bleha, S., Slivinsky, C., Hussien, B. (1990). Computer-access security systems using keystroke dynamics. *IEEE Trans. Pattern Anal. Machine Intelligence* **12** (12), 1217–1222.

Chen, E.Y. (2001). AEGIS: An active-network-powered defense mechanism against DDoS attacks. In: Marshall, I.W., Nettles, S., Wakamiya, N. (Eds.), *Active Networks: IFIP-TC6 Third International Working Conference*. In: *Lecture Notes in Comput. Sci.*, vol. 2207. Springer-Verlag, pp. 1–15.

Duda, R.O., Hart, P.E., Stork, D.G. (2000). *Pattern Classification*. Wiley, New York.

Escamilla, T. (1998). *Intrusion Detection: Network Security Beyond the Firewall*. Wiley, New York.

Everitt, B.S. (1993). *Cluster Analysis*, third ed. Wiley, New York.

Feldmann, A., Gilbert, A.C., Willinger, W., Kurtz, T.G. (1997). Looking behind and beyond self-similarity: On scaling phenomena in measured WAN traffic. In: *Proceedings of the 35th Annual Allerton Conference on Communication, Control and Computing*, pp. 269–280.

Feldmann, A., Gilbert, A.C., Willinger, W., Kurtz, T.G. (1998). The changing nature of network traffic: Scaling phenomena. *ACM SIGCOMM Comput. Commun. Rev.* **28**, 5–29.

Feldmann, A., Gilbert, A.C., Huang, P., Willinger, W. (1999). Dynamics of IP traffic: A study of the role of variability and the impact of control. In: *Proceedings of the ACM/SIGCOMM'99*. ACM Press.

Forrest, S., Hofmeyr, S.A. (2001). Immunology as information processing. In: Segel, L.A., Cohen, I. (Eds.), *Design Principles for the Immune System and Other Distributed Autonomous Systems*. In: *Santa Fe Inst. Stud. Sci. Complexity Lecture Notes*. Oxford University Press, Oxford, UK. Also available at http://www.cs.unm.edu/~forrest/ism_papers.htm.

Forrest, S., Perelson, A.S., Allen, L., Cherukuri, R. (1994). Self–nonself discrimination in a computer. In: *1994 IEEE Symposium on Research in Security and Privacy*. IEEE Press. Also available at http://www.cs.unm.edu/~forrest/isa_papers.htm.

Forrest, S., Hofmeyr, S.A., Somayaji, A. (1997). Computer immunology. *Commun. ACM* **40**, 88–96.

Green, J., Marchette, D., Northcutt, S., Ralph, B. (1999). Analysis techniques for detecting coordinated attacks and probes. In: *USENIX Workshop on Intrusion Detection and Network Monitoring*, ID '99, Proceedings, pp. 1–9.

Leland, W.E., Taqqu, M.S., Willinger, W., Wilson, D.V. (1994). On the self-similar nature of Ethernet traffic (extended version). *IEEE/ACM Trans. Network.* **2** (1), 1–15.

Lin, D.-T. (1997). Computer-access authentication with neural network based keystroke identity verification. In: *International Conference on Neural Networks*. IEEE Press, pp. 174–178.

Marchette, D.J. (1999). A statistical method for profiling network traffic. In: *USENIX Workshop on Intrusion Detection and Network Monitoring*, ID '99, Proceedings, pp. 119–128.

Marchette, D.J. (2001). *Computer Intrusion Detection and Network Monitoring: A Statistical Viewpoint*. Springer-Verlag, New York.

Marchette, D.J. (2002). A study of denial of service attacks on the internet. In: *Proceedings of the Eighth Annual Army Conference on Applied Statistics*, pp. 41–60.

Marchette, D.J., Priebe, C.E., Rogers, G.W., Solka, J.L. (1996). Filtered kernel density estimation. *Comput. Statist.* **11** (2), 95–112.

Marchette, D.J., Wegman, E.J. (2004). Statistical analysis of network data for cybersecurity. *Chance* **17**, 9–19.

Moore, D., Voelker, G.M., Savage, S. (2001). Inferring Internet denial-of-service activity. Available on the web at http://www.usenix.org/publications/library/proceedings/sec01/moore.html.

Naiman, D.Q. (2002). Statistical anomaly detection via httpd data analysis. *Comput. Statist. Data Anal.*.

Northcutt, S., Novak, J., McLaclan, D. (2001). *Network Intrusion Detection. An Analyst's Handbook*. New Riders, Indianapolis, IN.

Ogielski, A.T., Cowie, J.H. (2002). Spatio–temporal analysis of Internet routing: Discovery of global routing instabilities due to worm attacks and other events. *Comput. Sci. Statist.* **34**, 440 (Abstract).

Obaidat, M.S., Sadoun, B. (1997). Verification of computer users using keystroke dynamics. *IEEE Trans. Syst., Man Cybernetics* **27** (2), 261–269.

Park, K., Willinger, W. (Eds.) (2000). *Self-Similar Network Traffic and Performance Evaluation*. Wiley, New York.

Proctor, P.E. (2001). *The Practical Intrusion Detection Handbook*. Prentice Hall, Englewood Cliffs, NJ.

Robinson, J.A., Liang, V.M., Chambers, J.A.M., MacKenzie, C.L. (1998). Computer user verification using login string keystroke dynamics. *IEEE Trans. Syst. Man Cybernetics* **28** (2), 236–241.

Schonlau, M., DuMouchel, W., Ju, W.-H., Karr, A.F., Theus, M., Vardi, Y. (2001). Computer intrusion: Detecting masquerades. *Statist. Sci.* **16**, 58–74.

Silverman, B.W. (1986). *Density Estimation for Statistics and Data Analysis*. Chapman and Hall, New York.

Stevens, W.R. (1994). *TCP/IP Illustrated, Volume 1: The Protocols*. Addison–Wesley, Reading, MA.

Wegman, E.J., Marchette, D.J. (2003). On some techniques for streaming data: A case study of Internet packet headers. *J. Comput. Graph. Statist.* **12** (4), 893–914.

Willinger, W., Taqqu, M.S., Sherman, R., Wilson, D.V. (1997). Self-similarity through high-variablility: statistical analysis of Ethernet LAN traffic and the source level. *IEEE/ACM Trans. Network.* **5**, 71–86.

Ye, N., Giordano, J., Feldman, J., Zhong, Q. (1998). Information fusion techniques for network intrusion detection. In: *IEEE Information Technology Conference*. IEEE Press, pp. 117–120.

Data Mining of Text Files

Angel R. Martinez

Abstract

The goal of this chapter is to present textual data mining from a broad perspective, in addition to discussing several methods in computational statistics that can be applied to this area. I begin by discussing natural language processing at the word and sentence level, since a textual data mining system that seeks to discover knowledge requires methods that will capture and represent the semantic content of the text units. This section includes descriptions of hidden Markov models, probabilistic context-free grammars, and various supervised and unsupervised methods for word sense disambiguation. Next, I look at approaches beyond the word and sentence level, such as vector space models for information retrieval, latent semantic indexing, and a new approach based on a bigram proximity matrix. I conclude with a brief description of self-organizing maps.

Keywords: Information retrieval; Knowledge discovery; Text understanding; Computational linguistics; Probabilistic context-free grammars; Mutual information; Word sense disambiguation; Natural language processing; Model-based clustering; Bigram proximity matrix; Latent semantic indexing

1. Introduction and background

In this chapter I will paint a picture in broad strokes of the field of textual data mining (TDM) and some of the methods of computational statistics being applied to it.

The methods I discuss are not specific to TDM but to a larger set of areas of study and application (sometimes not clearly differentiated), areas like text retrieval (TR), information retrieval (IR), information extraction (IE), knowledge discovery in databases (KDD), text understanding, and data mining (DM). And although various disciplines, like computational linguistics, database technology, machine learning, pattern recognition, artificial intelligence, and visualization may claim ownership to some of these areas, at the core, many of the methods used are statistical methods.

TDM, as is the case of its close relative DM, goes beyond the mere retrieval of factual quanta. The goal is the discovery or derivation of new information, or the finding of trends and patterns (Hearst, 1999). As is also the case with DM, the objects of study of

TDM are very large sets of 'data'. In TDM work, text bases, or corpora, containing hundreds of thousands or millions of documents is the norm. Working with these massive 'data sets' requires techniques beyond classical statistics (Hand et al., 2001), so some of the methods and approaches discussed in this chapter fall in the area of computational statistics.

TDM work seems to have two loci: a "search-centric" (Hearst, 1999) one as in TR and a "knowledge-centric" one as in text understanding. Our discussion will follow a course keeping in sight these two loci, but organized into two main sections more or less defined by the text units to which computational methods are applied. Thus, the first part will address natural language processing at the word and sentence level, while the second part will look at the larger picture of whole documents and collections of these. In TDM, both views are tightly linked. The word and sentence levels provide the finer semantic features from which new information can be derived, while the document and corpus levels provide the framework from which relevant text will be extracted, classified and clustered.

2. Natural language processing at the word and sentence level

A TDM system that places emphasis on knowledge discovery requires methods to help capture and represent the semantic content of the text units at some useful level. What this useful level is has been debated for the past 20 years (Cowie and Lehnert, 1996). Although the subject of the debate is not of concern here, the underlying premise is: syntactic analysis is essential for information extraction, and thus, for TDM.

In order to delve into the meaning of text units, the process of parsing the constituent sentences takes place first. Parsing groups sentence components into grammatical structures, which in turn helps to determine meaning. However, parsing of natural language from an unbounded domain has proven difficult for practical applications. The main reason for this is the richness of natural languages; complex grammar rules, as well as word ambiguity, contribute to this richness. Setting this larger problem of parsing language from unbounded domains aside for the moment, successful and computationally efficient limited focus parsing has been attained through the use of hidden Markov models (HMMs) and probabilistic context-free grammars (PCFGs). I consider HMMs and PCFGs in the next subsection.

2.1. Hidden Markov models

Parsing a language's utterances requires that the words constituting the language have been tagged with their syntactic categories. This tagging process is called *part-of-speech tagging* or POST. Charniak's (1996) discussion of POST cannot be improved, and I make extensive reference of it here.

The assignment of syntactic tags, e.g., N for noun, V for verb, VP for verb phrase, etc., to the words of a corpus can be automated after the tagger has been trained with manually tagged corpora. This can be accomplished using hidden Markov models (HMMs). HMMs are a generalization of Markov chains. However, unlike Markov

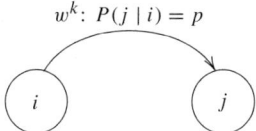

Fig. 1. Sample transition in an HMM.

chains, a state may have several transitions out of it, all with the same symbol. More formally, an HMM is a four-tuple $\langle s^1, S, W, E \rangle$, where S is the set of states, $s^1 \in S$ is the initial state of the model, W is a set of output symbols, and E is a set of transitions.

Transitions are four-tuples $\langle s^i, s^j, w^k, p \rangle$, where $s^i \in S$ is the state where the transition originates, $s^j \in S$ is the state where the transition ends, $w^k \in W$ is an output symbol accepted or generated by the model (depending on whether the model is an acceptor or a generator, respectively), and p is the probability of taking that transition.

The transition in Figure 1, from state i to state j, will take place with probability p and will accept/produce w^k. Shorthand for this is

$$s^i \xrightarrow{w^k} s^j. \tag{1}$$

State i can be the starting state for several transitions that have the same accepting/generating symbol, but go to different ending states. In these cases, it is impossible to know the sequence of states transitioned by the HMM just by looking at the string that has been accepted/generated; the sequence is hidden. Generalizing what has been stated so far and making use of the Markov assumption, we have

$$P(w_n, s_{n+1} \mid w_{1,n-1}, s_{1,n}) = P(w_n, s_{n+1} \mid s_n) = P\left(s^i \xrightarrow{w^k} s^j\right). \tag{2}$$

Extending Eq. (2), we can state that the probability of a sequence $w_{1,n}$ is

$$P(w_{1,n}) = \sum_{S_{1,n+1}} P(w_{1,n}, s_{1,n+1}). \tag{3}$$

That is, the probability of all possible paths through the HMM that can produce the sequence.

The summation indices $S_{1,n+1}$ in Eq. (3) vary over all possible state sequences, with the assumption that there are transitions from all states to every other state, with all possible outputs, and with some transition probabilities equal to zero. Using Eq. (3) as a starting point, we can then compute the probabilities of sequences using HMM in the following way (Charniak, 1996):

$$P(w_{1,n}) = \sum_{S_{1,n+1}} P(w_{1,n}, s_{1,n+1})$$
$$= \sum_{S_{1,n+1}} P(s_1) P(w_1, s_2 \mid s_1) \cdots P(w_n, s_{n+1} \mid w_{1,n-1}, s_{1,n})$$

$$= \sum_{s_{1,n+1}} P(w_1, s_2 \mid s_1) P(w_2, s_3 \mid w_1, s_2) \cdots P(w_n, s_{n+1} \mid s_n)$$

$$= \sum s_{1,n+1} \prod_{i=1}^{n} P(w_i, s_{i+1} \mid s_i) = \sum_{s_{1,n+1}} \prod_{i=1}^{n} P\left(s_i \xrightarrow{w^k} s_{i+1}\right).$$

Applying HMMs to the task of tagging words with their part-of-speech classes has the very desirable benefit of providing the means for automating the training process. As seen above, HMMs are comprised of three components: outputs, transitions, and states. In the application of HMMs to part-of-speech tagging, this needs to be changed by replacing the states with the part-of-speech tags. This is assuming, of course, that there is some connection between the states and the parts-of-speech. Under this assumption, Eq. (3) becomes

$$P(w_{1,n}) = \sum_{t_{1,n+1}} P(w_{1,n}, t_{1,n+1}), \qquad (4)$$

where $t_{1,n+1}$ is the sequence of $n+1$ parts of speech or tags. That is n tags for n words, plus a pointless prediction for the non-existent word w_{n+1}.

Eq. (4) indicates that using a tagged corpus, finding all occurrences where $w_{1,n}$ is associated with some specific $t_{1,n+1}$, and repeating the process of finding other values where the same words appear with other sequences of part-of-speech tags, we begin to cover all possible assignments of parts of speech to those words. This process provides an assignment of probability of the words for any part of speech. More specifically, part-of-speech tagging is finding the $t_{1,n}$ that maximizes $P(t_{1,n} \mid w_{1,n})$.

We assumed above the possibility of substituting states with tags. There are several ways by which this can be accomplished. The easiest is to relate each state in the HMM to the part of speech of the word produced next. The resulting model equation is then

$$P(w_{1,n}) = \sum_{t_{1,n+1}} P(w_{1,n}, t_{1,n+1}) = \sum_{t_{1,n+1}} \prod_{i=1}^{n} P(w_i \mid t_i) P(t_{i+1} \mid t_i). \qquad (5)$$

A more complete coverage of the subject can be found in (Manning and Schütze, 2000).

2.2. Probabilistic context-free grammars

As mentioned above, when pursuing the extraction of semantic context, sentence components are grouped into syntactic structures by the process called parsing. Natural language parsing is usually done using context-free grammars (CFGs).

CFGs are part of the hierarchy of languages developed by Noam Chomsky in the mid-fifties (Chomsky, 1956, 1959). This hierarchy of languages is part of the larger theory of generative grammars that revolutionized the world of linguistics. Generative grammars opened the door to the automatic parsing of natural language. However, the recursive nature of the grammar rules and unguided rule selection during parsing has proven to be computationally very costly. The following example demonstrates this on a small scale. The example also introduces the concept of probabilistic context-free grammars (PCFGs) (Charniak, 1996; Manning and Schütze, 2000), a probabilistic approach to address the computational problem of CFGs, as well as making possible the automatic learning of grammar rules from tagged corpora.

Let us define a grammar G as follows:

$$G = (\{S, NP, N, V\}, \{Fido, runs\}, \mathfrak{S}, F), \qquad (6)$$

where S stands for sentence, NP represents a noun phrase, N is a noun, and V denotes a verb. {Fido runs} is the set of "terminals", i.e., words in the language. The character $\mathfrak{S}$ is a starting symbol, and the set of grammar rules or productions F is defined as follows:

1. $\mathfrak{S} \to S$
2. $S \to NP\ V$
3. $S \to NP$
4. $NP \to N$
5. $NP \to N\ N$
6. $N \to Fido$
7. $N \to runs$
8. $V \to runs$

If a parser were to produce the sentence *Fido runs* using the grammar G (or equivalently, check if *Fido runs* is a sentence in the language defined by G), it would check the production rules in the manner shown in Examples 1 and 2. The corresponding tree structures illustrated in Figures 2 and 3 are called parse trees. They are used to indicate the possible derivations or parses of the sentence.

EXAMPLE 1.

1. $\mathfrak{S} \to S$
2. $S \to NP\ V$
3. $NP \to N$
4. $S \to N\ V$
5. $N \to Fido$
6. $S \to Fido\ V$
7. $V \to runs$
8. $S \to Fido\ runs$

The sentence *Fido runs* belongs to the language. However, the machine would not have stopped there; other parses are possible. The machine would also produce, for example,

EXAMPLE 2.

1. $\mathfrak{S} \to S$
2. $S \to NP$
3. $NP \to N\ N$
4. $S \to N\ N$
5. $N \to runs$
6. $S \to runs\ N$
7. $N \to Fido$
8. $S \to runs\ Fido$

This sentence belongs to the language, but does not make sense.

Fig. 2. Parse tree for the parsing in Example 1.   Fig. 3. Parse tree for the parsing in Example 2.

It is clear from the examples above that the two major problems of generative grammars in general and CFG in particular are: (1) a deterministic approach results in a combinatorial explosion of possible paths, and (2) an inherent inability to correctly choose rules in the presence of ambiguity. The example above is a simple one. CFGs used in NLP (natural language processing) are large and complex, exacerbating the problems by many orders of magnitude. Some in the AI–NLP (artificial intelligence–NLP) community realized that no progress would take place unless solutions were sought outside the bounds of the traditional computational linguistics and AI–NLP approaches. Methods from the area of probability and statistics were borrowed with fruitful results, one being probabilistic context-free grammars (PCFG).

The advantages of PCFG over the traditional CFG are obvious once we look at the complete definition of one. For example, our sample G grammar above is now transformed into G′, a PCFG:

1. $\mathfrak{S} \to$ S   1.0
2. S $\to$ NP V   0.7
3. S $\to$ NP   0.3
4. NP $\to$ N   0.8
5. NP $\to$ N N   0.2
6. N $\to$ Fido   0.6
7. N $\to$ runs   0.4
8. V $\to$ runs   1.0

The probabilities (shown to the right) assigned to each of the productions are based on tagged documents in a corpus. These probabilities remove some of the ambiguity regarding parse path selection and reduce dramatically the number of unnecessary parses. For example, the correct parsing (Example 1) used productions 1, 2, 4, 6, and 8, yielding a joint probability of $1.0 \times 0.7 \times 0.8 \times 0.6 \times 1.0 = 0.336$. While Example 2 productions (1, 5, 6, and 7) have a joint probability of 0.048. What may be the most important benefit of this approach is that it opens the door to *statistical grammar learning*, or *grammar induction* as it is also called, by adjusting production probabilities as the parser is exposed to new documents in the corpus.

Stating the above more formally: a PCF grammar G′ is an ordered four-tuple $(V_N, V_T, \mathfrak{S}, F)$, where V_N is a set of non-terminal symbols $\{N^1, \ldots, N^\nu\}$, V_T is a set of terminal symbols $\{t^1, \ldots, t^\omega\}$, $\mathfrak{S}$ is the starting symbol and F is a set of production rules, each of which is of the form $N^i \to \zeta^j$, where ζ^j is a string of terminals and non-terminals. Each rule has probability $P(N^i \to \zeta^j)$. The probabilities for the rules that expand the same non-terminal must add to one. The probability of a sentence in PCFG is the sum of the probabilities of all possible parses for the sentences (Charniak, 1996). If $p_{1,n}$ represents the parsing of all possible trees from syntactic symbol (tag) 1 to n, and $t_{1,n}$ represents the sequence of terminal symbols (words) parsed on each tree, then the probability of a sentence in a PCFG is given by

$$P(t_{1,n}) = \sum_{p_{1,n}} P(t_{1,n}, p_{1,n}) = \sum_{p_{1,n}} P(p_{1,n}) P(t_{1,n} \mid p_{1,n}). \tag{7}$$

Although some improvement over the previous state of affairs was attained with PCFGs, the power to disambiguate is not that great. All PCFGs can do is to prefer those constructions more common over the less common ones. Syntactic disambiguation involving less common constructions often fails because there is little difference between their parsing probabilities; deciding information is needed from some other source. As Charniak (1996) points out, the approach can be supplemented by paying special attention to prepositional phrases using the same probabilistic techniques already applied, as well as by considering the frequency of tags occurring in relation to specific words.

2.3. Word sense disambiguation

When the textual data mining approach depends on a finer granularity of language, i.e., the one provided by individual words, the problem of *polysemy* arises. Polysemy, that is, when a word has more than one meaning or sense, is usually approached using one of two ways: supervised disambiguation or unsupervised disambiguation. Both approaches make use of the immediate context of the ambiguous word. However, one approach considers the context as a 'bag of words'; that is, it disregards syntactic information. The other approach makes use of syntactic information to help in the disambiguation process.

2.3.1. Supervised disambiguation

Representative of the former approach is that of Gale et al. (1992), where Bayes classification is used to examine a context window of the words of the ambiguous word w. The underlying assumption is that each context word contributes some information toward the determination of the correct sense of w.

Let us consider the set $s_1, s_2, \ldots, s_M$ of possible senses of word w, and the set $c_1, c_2, \ldots, c_K$ of possible contexts of w. Bayes decision rule is then given by:

$$\text{Decide for sense } s_i \text{ if } P(s_i \mid c) > P(s_k \mid c) \quad \text{for } s_i \neq s_k. \tag{8}$$

Seldom is the value of $P(s_i \mid c)$ known; however, through Bayes' rule

$$P(s_i \mid c) = \frac{P(c \mid s_i)}{P(c)} P(s_i), \tag{9}$$

where $P(s_i)$ is the prior probability of sense s_i.

The equation is simplified by eliminating $P(c)$, which is constant for all senses, and using logs of probabilities. Thus, we would assign sense s^* to word w where

$$s^* = \arg\max_{s_i} \left[\log P(c \mid s_i) + \log P(s_i) \right]. \tag{10}$$

The implicit assumption of Gale et al. is that the features used to classify are conditionally independent. This is not true, of course. The words v_j in the context of w follow the rules of syntax, making word collocations dependent of word types. This simplifying assumption, however, can be "effective despite its shortcomings" (Manning and Schütze, 2000).

The decision rule can then be stated as follows:

$$s^* = \arg\max_{s_i} \left[\log P(s_i) + \sum_{V_j \in C} \log P(v_j \mid s_i) \right], \tag{11}$$

with $P(v_j \mid s_i)$ and $P(s_i)$ being computed using maximum-likelihood estimation from the training corpus (words are labeled with their senses):

$$P(v_j \mid s_i) = \frac{C(v_j, s_i)}{C(s_i)}, \tag{12}$$

$$P(s_i) = \frac{C(s_i)}{C(w)}. \tag{13}$$

Here $C(v_j, s_i)$ is the number of occurrences of v_j in the context of sense s_i in the training corpus, $C(s_i)$ is the number of occurrences of s_i in the training corpus, and $C(w)$ is the total number of occurrences of the ambiguous word w.

Whereas Gale et al.'s approach assumes that each word in the context of w sheds some light on w's sense, the approach considered next looks for the contextual feature to make that determination. Additionally, while their approach treats w's context as a "bag of words", and thus assumes conditional independence, the information-theoretic approach now considered makes use of collocation features of the language *and* no assumption of conditional independence.

A good example of these collocation restrictions is how verbs impose selectional bounds on those objects (nouns) that effect the actions. For example, *a dog eats*, or *a door opens*, *a man thinks*, etc. This collocation expectation can be exploited using the concept of mutual information.

Mutual information is defined as

$$I(X; Y) = \sum_{x \in X} \sum_{y \in Y} p(x, y) \log \frac{p(x, y)}{p(x) p(y)}. \tag{14}$$

More specific to our verb example, we have

$$I(v; c) = \sum_{v \in V} \sum_{c \in N} p(v, c) \log \frac{p(v, c)}{p(v) p(c)}, \tag{15}$$

where v is a verb and c is the class of nouns, $I(v; c)$ is the probability of v taking any noun in the class c as its direct object divided by the product of the probability of v and the probability of any member of c being the direct object of any verb (Charniak, 1996).

In the example above, a labeled corpus showing all possible direct objects that would go with each verb was used. Brown et al. (1991) use a bilingual dictionary as the labeled corpus.

2.3.2. Unsupervised disambiguation

The presence of tagged corpora to be used as training sets greatly facilitates the process of disambiguation. However, there are instances where such an advantage is not available. For example, in information retrieval, any field of knowledge is a legitimate target, preempting the use of domain-specific dictionaries and other tagged corpora. In this situation, disambiguation as the idea of selecting a sense s^* for word w is not possible. Sense discrimination is then the next best thing.

Sense discrimination can be done in a fully unsupervised manner via the EM algorithm. Following Manning and Schütze (2000), I present here the algorithm they used.

After the random initialization of $P(v_j \mid s_k)$, the $P(c_j \mid s_k)$ is computed for each context c_i of w; that is, the probability that context c_i 'was generated by' sense s_k. The preliminary grouping of contexts obtained serves as a training set with which $P(v_j \mid s_k)$ is re-estimated to maximize the likelihood of the data given the model.

The algorithm from Manning and Schütze (2000) follows. (In this notation, K is the number of desired senses, $c_i, i = 1, \ldots, I$, are the contexts of the ambiguous word, and $v_j, j = 1, \ldots, J$, are the words used as features to disambiguate.)

EM ALGORITHM.

1. Randomly initialize the parameters $P(v_j \mid s_k)$ and $P(s_k)$ of the model μ.
2. Compute the log of the likelihood of the corpus C given the model:

$$l(C \mid \mu) = \log \prod_{i=1}^{I} \sum_{k=1}^{K} P(c_i \mid s_k) P(s_k) = \sum_{i=1}^{I} \log \sum_{k=1}^{K} P(c_i \mid s_k) P(s_k).$$

3. *E-Step*: Estimate the posterior probability that s_k generated c_i as follows

$$h_{ik} = \frac{P(c_i \mid s_k)}{\sum_{k=1}^{K} P(c_i \mid s_k)},$$

where we make the naive Bayes assumption:

$$P(c_i \mid s_k) = \prod_{v_j \in C} P(v_j \mid s_k).$$

4. *M-Step*: Update the parameters $P(v_j \mid s_k)$ and $P(s_k)$ using maximum likelihood estimation

$$P(v_j \mid s_k) = \frac{\sum_{(c_i: v_j \in c_i)} h_{ik}}{Z_j},$$

where the sum is over all contexts in which v_j occurs and Z_j is a normalizing constant given by

$$Z_j = \sum_{k=1}^{K} \sum_{(c_i: v_j \in c_i)} h_{ik}.$$

Also update the probabilities of the senses using

$$P(s_k) = \frac{\sum_{i=1}^{I} h_{ik}}{\sum_{k=1}^{K} \sum_{i=1}^{I} h_{ik}}.$$

5. Repeat from step 2 until the likelihood (computed in step 2) is not increasing significantly.

Using the model μ thus obtained, the Bayes decision rule becomes:

Choose sense s^* if: $\quad s^* = \arg\max \left[\log P(s_k) + \sum_{v_j \in C} \log P(v_j \mid s_k) \right].$

An interesting approach to the discovery of word senses is the one proposed by Pantel and Lin (2002). Most approaches will not distinguish between the multiple senses of polysemous words. For example, the word *plant* will be grouped with *shrub*, as well as with *factory*. However, Pantel's and Lin's method, while using the same immediate context of the word, distinguishes between the *organic* sense and the *manufacturing* sense of the word.

In the same manner as with the other methods, the method by Pantel and Lin starts with the assumption that words found in similar contexts will have similar senses. However, in variance to those other methods, theirs separates the features in the context that are most similar from those that are 'marginally' similar. This is accomplished through their Clustering by Committee (CBC) algorithm.

The CBC is a clustering algorithm whose cluster centroids are constructed by averaging the feature vectors of a subset of the cluster measures. This subset acts as a committee determining which other members belong to the cluster. The key to also finding and separating less frequent senses is the removal of the common features between a candidate vector and the vectors of a cluster centroid. The features left behind point to secondary senses.

More specifically, each word is represented by a feature vector, where each feature corresponds to a context in which the word occurs. The value of the feature is the measure of mutual information (Manning and Schütze, 2000) between the word and the feature. Using Pantel's and Lin's nomenclature:

$$\mathrm{mi}_{w,c} = \log \frac{\frac{F_c(w)}{N}}{\frac{\sum_i F_i(w)}{N} \times \frac{\sum_j F_c(j)}{N}}, \quad (16)$$

where c is the context and $F_c(w)$ is the frequency count of word w in context c; $N = \sum_i \sum_j F_i(j)$ is the total frequency counts for all words and their contexts.

In order to compensate for the bias of the mutual information measure found in frequent words or features, a discounting factor is used:

$$\frac{F_c(w)}{F_c(w)+1} \times \frac{\min\{\sum_i F_i(w), \sum_j F_c(j)\}}{\min\{\sum_i F_i(w), \sum_j F_c(j)\}+1}. \quad (17)$$

A similarity measure is then computed for two words w_i and w_j using the cosine coefficient (Salton and McGill, 1983) of their mutual information vectors:

$$\mathrm{sim}(w_i, w_j) = \frac{\sum_i \mathrm{mi}_{w_i c} \times \mathrm{mi}_{w_j c}}{\sqrt{\sum_i \mathrm{mi}_{w_i c}^2 \times \sum_j \mathrm{mi}_{w_j c}^2}}. \quad (18)$$

After all pairwise similarity measures are computed, the top k similar elements to each element are determined. Using the top k elements, a collection of tight clusters is constructed. The elements of each of these clusters form a committee. The algorithm tries to form as many committees as possible under the constraint that each new committee is not very similar to the other committees. Each element is essentially assigned to its most similar clusters. This similarity is computed against the constraint of the committee members. This is similar to k means clustering, however the number of clusters is not fixed and the centroids do not change.

3. Approaches beyond the word and sentence level

In a TDM application, before any consideration can be given to the contribution of words in their immediate context to the problem of information mining, we need first to be able to collect all relevant documents.

3.1. Information retrieval

Document retrieval, more often called text retrieval (TR), is sometimes embedded in the concept of information retrieval (IR) (Lewis and Sparck Jones, 1996). However, the emphasis of IR seems to be more on the retrieval of useful information than the retrieval of documents containing some key words in common with the user's query. That is, information about a subject is the goal of IR (Baeza-Yates and Ribeiro-Neto, 1999).

A useful taxonomy of IR models is found in (Baeza-Yates and Ribeiro-Neto, 1999). Three classical models are presented: Boolean, vector and probabilistic. The Boolean model looks for matches of an index set of terms in documents and queries. In the vector model, the queries and documents are represented as vectors in d-dimensional space, where d is the size of the working vocabulary. In the probabilistic model, both queries and documents are modeled via probabilistic reparameterizations.

Basic approaches supporting these classic IR models have matured and in some instances transmuted into different manifestations of the original technologies. In this section, I will discuss some computational methods used in the most popular of the classic approaches, as well as in recent popular methods.

3.1.1. Vector space model
The vector space model, introduced by Salton et al. (1975), seems the most widely used method for IR work. It is the preferred method for many search engines. For this reason, I will focus on this model and some of its most recent variations. (For a more complete consideration of IR models, see Baeza-Yates and Ribeiro-Neto, 1999.)

3.1.1.1. Generic implementation. The vector space model is used to encode terms and documents in a text collection. Each document in the collection is represented by a vector whose elements are values associated with the terms in the document. These values are weighted so as to represent the importance of the terms in the semantics of the document.

A collection of documents comprised of n documents is represented by an $m \times n$ matrix where m is the number of terms used to index the documents. The column space of this term-by-document matrix determines the semantics of the collection.

Preliminary processing is performed by eliminating words that occur too frequently and thus have little or no discriminating use. These are usually called *stop* or *noise* words. In many applications, an additional preprocessing step is taken by reducing all words to their stems or roots.

Let $m \times n$ matrix $\hat{\mathbf{A}}$'s entries $\hat{a}_{ij}$ represent the frequencies of terms i in documents j. A $1 \times m$ query vector $\mathbf{q}$ is created by giving a value of 1 to the $\mathbf{q}$-vector element corresponding to the words indexed by the rows of matrix $\hat{\mathbf{A}}$.

Using matrix $\hat{\mathbf{A}}$ and query vector $\mathbf{q}$, compute the cosines of the angles formed by this query vector and each column (document) vector of $\hat{\mathbf{A}}$, using

$$\cos \theta_j = \frac{\sum_{i=1}^{m} a_{ij} q_i}{\sqrt{\sum_{i=1}^{m} a_{ij}^2} \sqrt{\sum_{i=1}^{m} q_i^2}}. \tag{19}$$

If both the query vector and the columns of $\hat{\mathbf{A}}$ have been normalized, the computation is just the inner product.

A simple example will illustrate. Let us assume that we have the following set of terms and documents:

t_1: industry d_1: vacation travel plans
t_2: car d_2: your vacation destination
t_3: travel d_3: do your travel by car or plane
t_4: vacation d_4: the car industry
t_5: destination d_5: car manufacture industry
t_6: manufacture
t_7: plane

Matrix $\hat{\mathbf{A}}$ would be:

$$\hat{\mathbf{A}} = \begin{bmatrix} 0 & 0 & 0 & 1 & 1 \\ 0 & 0 & 1 & 1 & 1 \\ 1 & 0 & 1 & 0 & 0 \\ 1 & 1 & 0 & 0 & 0 \\ 0 & 1 & 0 & 0 & 0 \\ 0 & 0 & 0 & 0 & 1 \\ 0 & 0 & 1 & 0 & 0 \end{bmatrix}.$$

If our interest is to find documents related to vacation travel our $\mathbf{q}$-vector would be

$$\mathbf{q} = (0 \ \ 0 \ \ 1 \ \ 1 \ \ 0 \ \ 0 \ \ 0)^T,$$

where the superscript T denotes the transpose. When elements in $\hat{\mathbf{A}}$ are scaled so that the Euclidean norm for each column equals one, we obtain matrix $\mathbf{A}$:

$$\mathbf{A} = \begin{bmatrix} 0 & 0 & 0 & 0.7071 & 0.5774 \\ 0 & 0 & 0.5774 & 0.7071 & 0.5774 \\ 0.7071 & 0 & 0.5774 & 0 & 0 \\ 0.7071 & 0.7071 & 0 & 0 & 0 \\ 0 & 0.7071 & 0 & 0 & 0 \\ 0 & 0 & 0 & 0 & 0.5774 \\ 0 & 0 & 0.5774 & 0 & 0 \end{bmatrix}.$$

Computing the cosines for the angles between the query vector and each of the document vectors of $\mathbf{A}$, we get:

$\cos \theta$ for $d_1 = 1.00$,
$\cos \theta$ for $d_2 = 0.50$,

$\cos\theta$ for $d_3 = 0.41$,
$\cos\theta$ for $d_4 = 0.00$,
$\cos\theta$ for $d_5 = 0.00$.

According to the cosine values, documents 1, 2, and 3 would be retrieved. However, document 3 is not relevant to our vacation query. If a threshold such as $|\cos\theta_j| \geq 0.5$ is set, then document 3 would not be retrieved.

It is obvious, by the previous example, that term frequencies by themselves are not sufficient in providing adequate results. One enhancement to the method described is the use of term weights.

3.1.1.2. Using term weights. One would like to retrieve relevant information while leaving behind that which is irrelevant. In order to retrieve relevant documents for a given query, one would use those words that appear in many documents in the collection, i.e., those words with high frequency. However, this broad netting of documents needs to be balanced by narrowing the precision by using those infrequent or more specific terms that will match to the most relevant documents.

The elements a_{ij} of the term-by-document matrix **A** are transformed by applying two types of weights: local and global. Different implementations may apply to one set or both. That is, matrix element $a_{ij} = l_{ij} g_i d_j$ is weighted using local (l_{ij}) and or global (g_i) weights plus document normalization (d_j) or not (Berry and Browne, 1999).

Some of the formulas for local and global weights found in the literature are:

Local weights:
  Binary (Salton and Buckley, 1988): $I(f_{ij})$
  Logarithmic (Harman, 1992): $\log(1 + f_{ij})$
  Term frequency (Salton and Buckley, 1988): f_{ij}
Global weights:
  Entropy (Dumais, 1991): $1 + \sum_j (p_{ij} \log(p_{ij}))/\log n$
  Inverse document frequency
    (Salton and Buckley, 1988; Dumais, 1991): $\log(n/\sum_j I(f_{ij}))$
  Probabilistic inverse
    (Salton and Buckley, 1988; Harman, 1992): $\log((n - \sum_j I(f_{ij}))/\sum_j I(f_{ij}))$
Normalization:
  cosine (Salton and Buckley, 1988): $(\sum_i (g_i l_{ij})^2)^{-1/2}$

In the above, we have

$$I(x) = \begin{cases} 1 & \text{if } x > 0, \\ 0 & \text{if } x = 0, \end{cases} \quad \text{and} \quad l_{ij} = g_i = d_j = 1$$

if the type of weight is not applied.

Text retrieval approaches based on weighted word frequencies, as described above, have a fundamental deficiency: they do not account for polysemy nor synonymy. That is, queries are fashioned using terms not necessarily indexed in the corpora or indexed in a context that affixes a different sense to the term. A text retrieval method devised to counter this problem is Latent Semantic Indexing (LSI) (Deerwester et al., 1990).

3.1.2. Latent Semantic Indexing (LSI)

LSI approximates the original d-dimensional term space by the first k principal component directions in this space, using the term-by-document matrix to estimate the directions. Precision of retrieval is increased by this method compared to the vector-space approaches described above (Hand et al., 2001).

Using singular value decomposition (SVD) to decompose the term-by-document matrix $\mathbf{A}$, the latent semantic model takes the form of $\mathbf{A} \approx \tilde{\mathbf{A}} = \mathbf{TSD}^T$, where $\tilde{\mathbf{A}}$ is approximately equal to $\mathbf{A}$ and is of rank k (with k less than the number of documents). The dimensionality reduction in $\tilde{\mathbf{A}}$ is obtained by selecting the highest singular values of $\mathbf{S}$ and entering zeros in the remaining positions in the diagonal. Zeros are also entered in the corresponding positions in $\mathbf{T}$ and $\mathbf{D}^T$. The rows of the reduced matrices of singular vectors represent terms and documents as points in a k-dimensional space. The inner products between points are then used to compare the similarity of corresponding objects.

Three comparisons are of interest: comparing two terms, comparing two documents, and comparing a term and a document. These comparisons can be stated respectively as: (1) how semantically similar are two terms; (2) how semantically similar are two documents; and (3) how associated are term i and document j (Deerwester et al., 1990). Comparison between two terms can be attained by the term-to-term inner products contained in the matrix $\tilde{\mathbf{A}}\tilde{\mathbf{A}}^T$. Comparison between two documents can be attained by the document-to-document inner products, contained in the matrix $\tilde{\mathbf{A}}^T\tilde{\mathbf{A}}$. Comparison between a term and a document is provided by the value of the cell $\tilde{a}_{ij}$ of $\tilde{\mathbf{A}}$.

In contrast to other conceptual indexing schemes, LSI provides a more sophisticated tool for semantic investigations. The SVD representation replaces individual terms with derived orthogonal factor values. These factor values may be thought of as artificial concepts that represent extracted common meaning components of many different words and documents. Each term or document is then characterized by a vector of weights indicating its strength of association with each other or these underlying concepts. That is, the meaning of a particular term, query, or document can be expressed by k factor values, or equivalently, the location of the vector in the k-space defined by the factors (Deerwester et al., 1990). LSI has been applied to semantic-related measures and indicators like textual coherence, discourse segmentation and semantic distances in text (Foltz et al., 1998).

3.2. Other approaches

An interesting comment from Hand et al. (2001) pertaining to document classification is that classification should not be thought of as 1 : 1, but 1 : m, where m is the number of topics or classes. Recent research points to this fact as discussed below.

The methods discussed so far represent documents by a list of terms lifted out of their contexts. Information is lost with the loss of context (Hand et al., 2001). An attempt to preserve contextual information is found in the approach described next.

3.2.1. The bigram proximity matrix

The bigram proximity matrix (BPM) is a non-symmetric matrix that captures the number of word co-occurrences in a moving 2-word window (Martinez and Wegman, 2002,

2003; Martinez, 2002). It is a square matrix whose column and row headings are the alphabetically ordered entries of the lexicon. Matrix element b_{ij} is the number of times word i appears immediately before word j in the unit of text. The size of the BPM is determined by the size of the lexicon created by listing alphabetically the unique occurrences of the words in the text. It is asserted that the BPM representation of the semantic content preserves enough unique features to be semantically separable from BPMs of other thematically unrelated collections.

The rows in the BPM represent the first word in the pair, and the second word is shown in the column. For example, the BPM for the sentence or text stream,

The wise young man sought his father in the crowd.

is shown in Table 1. We see that the matrix element located in the third row (*his*) and the fifth column (*father*) has a value of one. This means that the pair of words '*his father*' occurs once in this unit of text. It should be noted that in most cases, depending on the size of the lexicon and the size of the text stream, the BPM will be very sparse.

3.2.1.1. Measures of semantic similarity. Documents represented as BPMs based on a common lexicon are examined for semantic closeness by means of similarity measures or distances. When similarity measures are used, these are transformed to distances using the following transformation: Let $C = \{c_{ij}\}$ be the similarity matrix and assume it is positive definite, then

$$d_{ij} = \sqrt{c_{ii} - 2c_{ij} + c_{jj}},$$

where i and j represent two different documents (du Toit et al., 1986).

Two types of similarity (distance) measures were used: binary and probabilistic. Examples of the binary measures used are:

3.2.1.2. Matching coefficient

$$S_{\mathbf{X},\mathbf{Y}} = \mathbf{X} \text{ AND } \mathbf{Y}, \qquad (20)$$

Table 1
Example of bigram proximity matrix

	·	crowd	his	in	father	man	sought	the	wise	young
crowd	1									
his					1					
in								1		
father				1						
man							1			
sought			1							
the		1							1	
wise										1
young						1				

Note: Zeros in empty boxes are removed for clarity.

where **X** and **Y** are different BPMs whose elements have been converted to binary values, and AND is the logical operator (Everitt et al., 2001).

3.2.1.3. Jaccard coefficient

$$S_{\mathbf{X},\mathbf{Y}} = \frac{\mathbf{X} \text{ AND } \mathbf{Y}}{\mathbf{X} \text{ OR } \mathbf{Y}}, \qquad (21)$$

where OR is the logical operator (Jaccard, 1901).

3.2.1.4. Ochiai measure (also called cosine)

$$S_{\mathbf{X},\mathbf{Y}} = \frac{\mathbf{X} \text{ AND } \mathbf{Y}}{\sqrt{|\mathbf{X}||\mathbf{Y}|}}, \qquad (22)$$

where $|\mathbf{X}|$ represents the total number of non-zero elements in matrix **X** (Ochiai, 1957).

Examples of probabilistic measures are:

3.2.1.5. L_1 distance

$$S_{p,q} = \sum |p_{ij} - q_{ij}|, \qquad (23)$$

where p and q represent probability distributions, and p_{ij} and q_{ij} are the corresponding elements of the qth (pth) BPM after the original elements are divided by the total number of non-zero entries in the matrix. The L_1 distance is symmetric and well defined for arbitrary p and q. It can be interpreted as the expected proportion of events that are going to be different between distributions p and q (Manning and Schütze, 2000).

3.2.1.6. Information radius measure (IRad)

$$S_{p,q} = D\left(p \,\Big\|\, \frac{p+q}{2}\right) + D\left(q \,\Big\|\, \frac{p+q}{2}\right), \qquad (24)$$

where

$$D(p \,\|\, q) = \sum_{i,j} p_{ij} \log \frac{p_{ij}}{q_{ij}}. \qquad (25)$$

IRad is based on the Kullback–Leibler measure; however, it does not suffer from two of its problems: lack of symmetry and possible infinite results (Manning and Schütze, 2000). One interpretation of the IRad measure is how much information is lost if we describe two proximity matrices with their average similarity measure (it is sometimes called "total divergence to the average") (Dagan et al., 1999).

3.2.1.7. Document classification via supervised learning. As with other methods, stop words were excluded and the remaining words were reduced to their roots. Using a subset of the Topic Detection and Tracking (TDT) Pilot Corpus (Linguistic Data Consortium, Philadelphia, PA) and k nearest neighbor classifier (kNN) (Everitt et al., 2001), 503 documents were classified into 16 topics, shown in Table 2.

Table 2
List of 16 topics

Topic number	Topic description	Topic number	Topic description
4	Cessna on the White House	15	Kobe, Japan Quake
5	Clinic Murders (Salvi)	16	Lost in Iraq
6	Comet into Jupiter	17	NYC Subway Bombing
8	Death of Kim Jong II's Father	18	Oklahoma City Bombing
9	DNA in OJ Trial	21	Serbians Down F-16
11	Hall's Copter in N. Korea	22	Serbs Violate Bihac
12	Flooding Humble, TX	24	US Air 427 Crash
13	Justice-to-be Breyer	25	WTC Bombing Trial

Table 3
Classification results

	$k=1$	$k=3$	$k=5$	$k=7$	$k=10$
Matching coefficient	0.9483	0.9443	0.9742	0.9722	0.9761
Jaccard	0.9841	0.9801	0.9801	0.9781	0.9881
Ochiai	0.9901	0.9841	0.9861	0.9861	0.9881
L_1 norm	0.9881	0.9801	0.9821	0.9801	0.9781
IRad	0.9920	0.9881	0.9841	0.9821	0.9781

The method of cross-validation was used to evaluate the performance of the classifier. A correct classification ratio was computed for each similarity measure-k value pair as shown in Table 3.

Full dimensionality of the space (7, 146^2) was used in the tests reported above. This was reduced to only 2 through 6 dimensions using the non-linear dimensionality reduction method Isometric Feature Mapping (Isomap) (Tenenbaum, 2000), and experiments were repeated with essentially the same results.

3.2.1.8. Document classification via model-based clustering. Unsupervised learning experiments were conducted in the reduced space produced through Isomap. The method chosen for the unsupervised learning experiments is called model-based clustering (Banfield and Raftery, 1993; Fraley and Raftery, 1998). This method is based on finite mixtures where the output model is a weighted sum of c multivariate normals:

$$f(\mathbf{x}) = \sum_{k=1}^{c} p_k \phi(\mathbf{x}; \mu_k, \Sigma_k). \tag{26}$$

In general, with model-based clustering the idea is to generate estimates based on Eq. (26), where constraints are imposed on the covariance matrices. The best estimate and model (i.e., number of components, parameter estimates, and form of the covariance matrices) is chosen based on the model that yields the highest value of the Bayesian Information Criterion (BIC), given by

$$\text{BIC} = 2L_M(\mathbf{x}; \widehat{\theta}) - m_M \log n, \tag{27}$$

where m_M is the number of parameters in model M and L_M is the log likelihood (Banfield and Raftery, 1993).

A foundational assumption of model-based clustering is that the data are generated by a mixture of probability distributions in which each component represents a different group or cluster. That is, in the general case, given observations $\mathbf{x}_1, \ldots, \mathbf{x}_n$, let $f(\mathbf{x}_i \mid \theta_k)$ be the density of an observation $\mathbf{x}_i$ from the kth component, where θ_k is a vector containing the parameters for the component. One way to model the composite of the clusters is via the mixture likelihood which maximizes

$$L(\theta_1, \ldots, \theta_c; \tau_1, \ldots, \tau_c \mid \mathbf{x}) = \prod_{i=1}^{n} \sum_{k=1}^{c} \tau_k f_k(x_i \mid \theta_k), \tag{28}$$

where c is the number of components in the mixture, and τ_k is the probability that an observation belongs to the kth component.

In model-based clustering, the multivariate normal is used as the density for $f(\mathbf{x}_i \mid \theta_k)$, with θ_k consisting of a vector of means μ_k and a covariance matrix Σ_k. Key to this method is the fact that the covariance matrix determines important geometric characteristics of the clusters. Banfield and Raftery (1993) developed a model-based clustering framework by parameterizing the covariance matrix in terms of eigenvalue decomposition, as follows

$$\Sigma_k = \lambda_k \mathbf{D}_k \mathbf{A}_k \mathbf{D}_k^T, \tag{29}$$

where $\mathbf{D}_k$ is the orthogonal matrix of eigenvectors, $\mathbf{A}_k$ is a diagonal matrix whose elements are proportional to the eigenvalues of Σ_k, and λ_k is a scalar. By means of this decomposition of the covariance matrix Σ_k, geometric characteristics of the distributions can be imposed and a suitable model can be generated. Orientation, volume and shape of the clusters can be specified by using the models. See Fraley and Raftery (1998) or Celeux and Govaert (1995) for a description of the various models that can be obtained by parameterizing Eq. (29). The determination of the component parameters for each of the models is done via the Expectation–Maximization algorithm (EM), which is described in (Dempster et al., 1977).

Results from the unsupervised learning experiments (using the same TDT subset) were reviewed using the visualization tool ReClus (Martinez, 2002). ReClus takes the output from the model-based clustering procedure and draws one large rectangle. This rectangle is subdivided into n smaller rectangles, where n is the number of clusters chosen according to the highest BIC value. The area of each smaller rectangle is proportional to the number of cases (documents in this context) in the cluster. Inside each rectangle and for each case assigned to that cluster, the class number is printed, or optionally, the case number is printed. Each number is color-coded to denote the degree of certainty about the particular case belonging to the cluster. A threshold is set to print in black bold type when the certainty is 0.8 or above.

ReClus, thus, provides a quick visual way to examine the results from model-based clustering. Although, judging between two results entails a degree of subjectivity, this is a problem only where results are close. Additionally, ReClus provides information to guide the examination of confounding factors in the clustering process.

The results from one of the experiments are shown in Figure 4, where the 503 TDT documents have been clustered. Each number is an instance of a document, and the number itself is the true topic label.

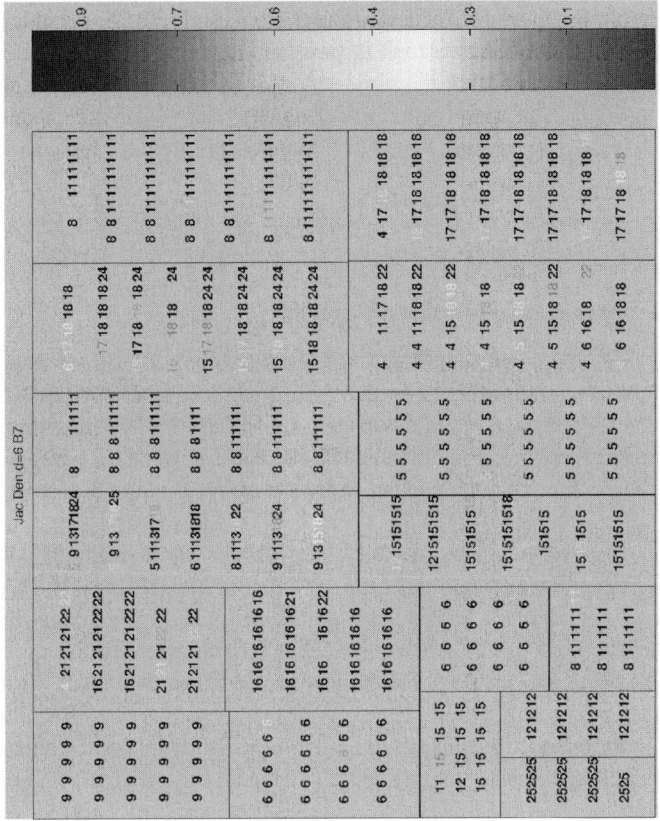

Fig. 4. ReClus plot showing output from model-based clustering.

Figure 4 shows 17 clusters where there should be 16. However, two clusters have the same topic label #6 (on the left). A careful reading of the documents belonging to topic #6 showed the existence of two sub-topics. Topic #6 was comprised of newscasts on the crash of the comet on the surface of Jupiter. One subset of the newscasts discussed the impending event, and the other subset discussed the occurrence and the aftermath of the event. This appears to be the reason for the two #6 clusters in ReClus. Other instances of what seems to be the detection of latent topics can be pointed out, as well as the clustering together of related information found in different topics. However, this is beyond the scope of this chapter. The interested reader is referred to Martinez (2002).

3.2.2. Towards knowledge discovery

The ultimate goal of TDM is the discovery of new information. Key to this goal is the ability to uncover the relationships between the entities or subjects discussed in the text. Lin and Pantel (2001) proposed a way of automatically discovering inference rules implied by entity-relationship content.

Their method, Discovery of Inference Rules from Text (DIRT), extends the assumption that words in same contexts tend to have similar meanings. That is, if two paths in dependency trees tend to link the same sets of words, then their meanings are likely to be similar. Given that these paths model binary relationships, an inference rule is generated for each pair of paths. The idea is to discover inference rules like

$$\text{X is author of Y} \approx \text{X wrote Y}$$

or

$$\text{X solved Y} \approx \text{X found solution to Y.}$$

Dependency trees are a representation of the relationships between words in a sentence. Specifically, binary relationships between a lead word and its modifiers. In one work by Lin and Pantel (2001), a broad coverage English parser was used. An example of a dependency tree is shown in Figure 5. Selectional constraints on the paths were imposed to minimize the computational load, as well as to concentrate on the most meaningful paths.

Given that the paths are the objects of interest, the words filling the slots (i.e., the end points of dependency relations) become the contexts to the paths. In preparation to compute path similarity, a triples database is constructed by counting the number of paths and the slot fillers for these in a corpus. For a path p connecting words w_1 and w_2, the triples (p, SlotX, w_1) and (p, SlotY, w_2) are added to the count. (SlotX, w_1) and (SlotY, w_2) are features of path p.

The similarity measure used is based on the information theoretic concept of mutual information applied in this case to a feature and a path. In this context mutual information is given by

$$\text{mi}(p, \text{SlotY}, w) = \log\left(\frac{|p, \text{Slot}, w| \times |*, \text{Slot}, *|}{|p, \text{Slot}, *| \times |*, \text{Slot}, w|}\right), \qquad (30)$$

where $|p, \text{SlotX}, w|$ denotes the frequency count of the triple (p, SlotX, w) and $|p, \text{Slot},*|$ denotes the summation $\sum_w |p, \text{Slot}, w|$.

Similarity between a pair of slots $\text{Slot}_1 = (p_1, S)$ and $\text{Slot}_2 = (p_2, S)$ is defined as

$$\text{sim}(\text{Slot}_1, \text{Slot}_2) = \frac{\sum_{w \in T(p_1,S) \cap T(p_2,S)} \text{mi}(p_1, S, w) + \text{mi}(p_2, S, w)}{\sum_{w \in T(p_1,S)} \text{mi}(p_1, S, w) + \sum_{w \in T(p_2,S)} \text{mi}(p_2, S, w)}, \qquad (31)$$

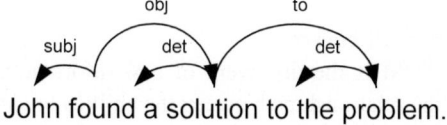

Fig. 5. Example of dependency tree.

where p_1 and p_2 are paths, S is a slot, $T(p_i, S)$ is the set of words that fill in the slot S of path p_i.

The similarity of a pair of paths p_1 and p_2 is determined as the geometric average of the similarities of their SlotX and SlotY:

$$\text{sim}(p_1, p_2) = \sqrt{\text{sim}(\text{SlotX}_1, \text{SlotX}_2) \times \text{sim}(\text{SlotY}_1, \text{SlotY}_2)}, \qquad (32)$$

where SlotX_i and SlotY_i are path i's SlotX and SlotY slots.

3.2.3. WEBSOM

This chapter would not be complete without the mention of the application of self-organizing maps (SOM) (Kohonen, 1982) to the organization of a large collection of documents.

SOM is an unsupervised learning method based on neural network concepts that produces a 'map' or graph of input data. The SOM map is a similarity graph not unlike multidimensional scaling 2D graphs. Input data are approximated by a finite set of models. The models are associated with nodes (neurons) arranged in a 2D grid. The models are the result of a learning process that organizes them on the 2D grid along their mutual similarity.

As part of the WEBSOM2 system, the SOM algorithm is applied to weighted document vectors whose dimensionality has been reduced using a method called random mapping (Kaski, 1998). Followed by the initialization of the models (nodes) of smallest SOM, the iterative learning process begins and continues until the desired size of the map is reached. Via a suitable interface, the final map is offered to the user for exploration (Kohonen et al., 2000).

4. Summary

In this chapter, I did not intend to describe any one system that would perform TDM from beginning to end. Rather, I presented key technological areas and the methods and approaches employed therein, whose collective contributions are in line with TDM goals.

I examined foundational issues and methods like text parsing using probabilistic context-free grammars (PCFGs), the need to tag words with their syntactic categories, and the use of hidden Markov models (HMMs) for this purpose. I considered the problem of polysemy and presented an innovative approach by Pantel and Lin, Clustering by Committee (CBC), to disambiguate word senses. These methods, and others covered in the chapter, are essential to the part of TDM where information and knowledge are distilled from the text. However, before these finely focused approaches are used, relevant documents must be retrieved from corpora.

In the area of text retrieval, I covered the most popular paradigm: the vector space representation of documents. An interesting variant of the vector space model, Latent Semantic Indexing (LSI) was discussed. The power of LSI is that it goes beyond mere key-word matching to what could be called 'conceptual matching' or 'sense matching.'

LSI and vector-space approaches pull the words out of their sentence context with resulting loss of information. The bigram proximity matrix (BPM), a recent research approach, endeavors to keep more contextual information by representing text as matrices of word pairs collected and organized by running a two-word window through the text. Results from unsupervised and supervised learning experiments show the benefits of capturing words' contextual information.

As TDM matures, we should see more attention being paid to approaches that efficiently capture and represent knowledge (Pantel's and Lin's DIRT algorithm is an interesting step in that direction) and to efficient computational methods for the summarization, storage and retrieval of knowledge bearing structures.

References

Baeza-Yates, R., Ribeiro-Neto, B. (1999). *Modern Information Retrieval*. ACM Press, New York.
Banfield, J.D., Raftery, A.E. (1993). Model-based Gaussian and non-Gaussian clustering. *Biometrics* **49**, 803–821.
Berry, M.W., Browne, M. (1999). *Understanding Search Engines: Mathematical Modeling and Text Retrieval*. SIAM, Philadelphia, PA.
Brown, P.F., Della Pietra, S.A., Della Pietra, V.J., Mercer, R.L. (1991). Word-sense disambiguation using statistical methods. In: *Proceedings of the 29th Annual Meeting of the Association for Computational Linguistics*. Assoc. Comput. Linguistics, pp. 264–270.
Celeux, G., Govaert, G. (1995). Gaussian parsimonious clustering models. *Pattern Recognition* **28**, 781–793.
Charniak, E. (1996). *Statistical Language Learning*. MIT Press, Cambridge, MA.
Chomsky, N. (1956). Three models for the description of languages. *IRE Trans. Inform. Theory* **2**, 113–124.
Chomsky, N. (1959). On certain formal preoperation of grammar. *Inform. and Control* **2**, 137–167.
Cowie, J., Lehnert, W. (1996). Information extraction. *Comm. ACM* **39**, 80–91.
Dagan, I., Lee, L., Pereira, F. (1999). Similarity-based models of word cooccurrence probabilities. *Machine Learning* **34**, 43–69.
Deerwester, S., Dumais, S.T., Furnas, G.W., Landauer, T.K. (1990). Indexing by latent semantic analysis. *J. Amer. Soc. Inform. Sci.* **41**, 391–407.
Dempster, A.P., Laird, N.M., Rubin, D.B. (1977). Maximum likelihood from incomplete data via the EM algorithm. *J. R. Stat. Soc. B* **39**, 1–38.
Dumais, S.T. (1991). Improving the retrieval of information from external sources. *Behavior Res. Methods Instruments Comput.* **23**, 229–236.
du Toit, S.H.C., Steyn, A.G.W., Stumpf, R.H. (1986). *Graphical Exploratory Analysis*. Springer-Verlag, New York.
Everitt, B.S., Landau, S., Leese, M. (2001). *Cluster Analysis*. Arnold, London.
Foltz, P.W., Kintsch, W., Landauer, T.K. (1998). The measurement of textual coherence with Latent Semantic Analysis. *Discourse Process.* **25**, 285–307.
Fraley, C., Raftery, A.E. (1998). How many clusters? Which clustering method? – Answers via model-based cluster analysis. *Comput. J.* **41**, 578–588.
Gale, W.A., Church, K.W., Yarowsky, D. (1992). A method for disambiguating word senses in large corpus. *Comput. Human.* **25**, 415–439.
Hand, D., Mannila, H., Smyth, P. (2001). *Principles of Data Mining*. MIT Press, Cambridge, MA.
Harman, D. (1992). Ranking algorithms. In: Frakes, W.B., Baeza-Yates, R. (Eds.), *Information Retrieval: Data Structures & Algorithms*. Prentice Hall, Englewood Cliffs, NJ, pp. 363–392.
Hearst, M.A. (1999). Untangling text data mining. In: *Proceedings of ACL'99: The 37th Annual Meeting of the Association for Computational Linguistics*. University of Maryland, Assoc. Comput. Linguistics.
Jaccard, P. (1901). Étude comparative de la distribution Horale dans une portion des Alpes et des Jura. *Bull. Soc. Vaudoise Sci. Nat.* **37**, 547–549.

Kaski, S. (1998). Dimensionality reduction by random mapping: fast similarity computation for clustering. In: *Proceedings IJCNN'98, International Joint Conference on Neural Networks, vol. 1*. IEEE, pp. 413–418.

Kohonen, T. (1982). Self-organized formation of topologically correct feature maps. *Biol. Cybern.* **43**, 59–69.

Kohonen, T., Kaski, S., Lagus, K., Salogärui, J., Honkela, J., Paatero, V., Saarela, A. (2000). Self organization of a massive document collection. *IEEE Trans. Neural Netw.* **11**, 574–585.

Lewis, D.D., Sparck Jones, K. (1996). Natural language processing for information retrieval. *Comm. ACM* **29**, 99–101.

Lin, D., Pantel, P. (2001). DIRT – discovery of inference rules from text. In: *Proceedings of the ACM SIGKDD Conference on Knowledge Discovery and Data Mining*. ACM Press, pp. 323–328.

Pantel, P., Lin, D. (2002). Discovering word senses from text. In: *Proceedings of the ACM SIGKDD Conference on Knowledge Discovery and Data Mining*. ACM Press, pp. 613–619.

Manning, Ch.D., Schütze, H. (2000). *Foundations of Statistical Natural Language Processing*. MIT Press, Cambridge, MA.

Martinez, A.R. (2002). A Framework for the representation of semantics. *PhD Dissertation*. George Mason University.

Martinez, A.R., Wegman, E.J. (2002). A text stream transformation for semantic-based clustering. *Comput. Sci. Statist.* **34**, 184–203.

Martinez, A.R., Wegman, E.J. (2003). Encoding of text to preserve meaning. In: *Proceedings of the Eighth Annual US Army Conference on Applied Statistics*. US Army Research Office, pp. 27–39.

Ochiai, A. (1957). Zoographic studies on the soleoid fishes found in Japan and its neighboring regions. *Bull. Japan Soc. Sci. Fish.* **22**, 526–530.

Salton, G., Buckley, C. (1988). Term weighting approaches in automatic text retrieval. *Inform. Process. Management* **24**, 513–523.

Salton, G., McGill, M.J. (1983). *Introduction to Modern Information Retrieval*. McGraw-Hill, New York.

Salton, G., Wong, A., Yang, L.S. (1975). A vector space model for information retrieval. *J. Amer. Soc. Inform. Sci.* **18**, 613–620.

Tenenbaum, J.B., Vin deSilva, Langford, J.C. (2000). A global geometric framework for nonlinear dimensionality reduction. *Science* **290**, 2319–2323.

Text Data Mining with Minimal Spanning Trees

Jeffrey L. Solka, Avory C. Bryant and Edward J. Wegman

Introduction

This paper details some of our recent work in the application of graph theoretic techniques to facilitate the data mining process. We are particularly interested in the case where one is presented with a set of n observations in p-dimensional space along with a set of class labels. These class labels can be provided by a human operator or may be provided based on some other automated classification strategy. In either case, we are interested in employing these strategies to facilitate two goals. The first goal is the discovery of unusual and subtle relationships that may exist between the observations in one class and the observations in another class. The second goal is that we are interested in being able to cluster the set of observations and ultimately to weigh the clustering results against the provided class labels. In order to illustrate the success of our formulated approaches, we will provide results based on mining two small sets of text documents. We do wish to emphasize the graph theoretic approaches and visualization frameworks are not necessarily predicated on the data type being text data, but are equally appropriate for other data types. In fact, one must merely be able to compute some sort of interpoint distance measure between the observations. This interpoint distance measure does, of course, define a complete graph on the set of vertices (observations) and we can proceed forward from there with our analysis.

1. Approach

As mentioned in the introduction, we will be using the text data mining problem as a representative application of our methodologies. We must first decide upon a way to capture the semantic content of the documents, extract features from the documents, and then we must be able to compute interpoint distance measures on these features. Once we have the interpoint distance matrix in hand, we can proceed forward with our analysis.

We have chosen to use the bigram proximity matrix of Martinez (2002) as a means for capturing the semantic content of the documents. The bigram proximity matrix (BPM) encodes the document as an n-words by n-words matrix where an entry in the ij

position of this matrix indicates that the ith word in the multi-corpus word list is proximate to the jth word in the multi-corpus word list in one of the sentences in the current document of interest. It is important to point out that stopper or noise words have been removed from the document prior to the computation of the BPM. The BPM entries that were studied by Martinez were either 0 or 1 to indicate the absence or presence of a particular word pair relationship or else a count of the number of times that a particular word pair appears in the document. The important point to note is that the research of Martinez strongly supported the claim that the BPM adequately captures the semantic content of documents using either coding scheme. He tested the semantic capture capability of the BPM and Trigram Proximity Matrix (TPM) via supervised learning testing, unsupervised learning testing, and three different hypothesis tests.

Given the set of BPMs that describe our document set, we need a means of measuring the distances between the documents. Martinez compared over ten different ways of measuring distances between the BPMs. We have chosen to use the Ochiai similarity measure

$$S(X, Y) = \frac{|X \text{ and } Y|}{\sqrt{|X||Y|}}. \tag{1}$$

This similarity is then converted into a distance using

$$d(X, Y) = \sqrt{2 - 2S(X, Y)}. \tag{2}$$

We have chosen this similarity/distance measure in part because of the favorable results obtained by Martinez using them.

At this point in time, we have an interpoint distance matrix where hopefully articles that are close semantically will have a small value for that row column pair in the matrix and articles that are far apart semantically will have a high value for that row column pair in the matrix. We also have a previously obtained categorization of the articles. So how do we explore/exploit this interpoint distance matrix in order to facilitate the discovery of interesting relationships and clusters? Let us first consider the problem of interesting relationship (serendipity) discovery. The simplest thing that one could do is to find the closest articles between each pair of categories. We initially tried this and although the results were interesting we found them less than fully satisfying. Alternative exploration strategies are suggested by considering the discriminant boundaries that separates the articles from the two categories. One might, for example, consider looking for the articles that are along the boundary that separates the two categories. Taking a step up from this "first-order approximation", one could look for articles that have the same "relationship" to the discriminant boundary. Finally one might wish to provide the user with the capability to actually be able to select a local "focus of attention", to make a more in-depth study of the local geometry based on this focus of attention.

So what is required is a means to ascertain the relationships of the observations to the discriminant boundary. The key to understanding this relationship is to realize that if we view the articles as vertices in a graph, the interpoint distance matrix defines a complete graph on this set of vertices. Exploration of the complete graph is impossible so we need a subgraph of the complete graph that captures the information in the complete graph and in particular the relationship of the observations to the discriminant boundary. One natural candidate to capture this information is the minimal spanning tree (MST). The MST is loosely defined as a tree that traverses all of the nodes in the complete graph with a minimal sum of edge weights.

We (Solka and Johannsen, 2002) had previously used methods based on the MST, as computed on the interpoint distance matrix, to characterize discriminant surface complexity. This led us to believe that this would be an appropriate tool to facilitate the exploration of cross corpora (class) relationships in the document set. Our previous efforts had suggested that the number of edges that travel between observations of different classes is a good surrogate measure for classification complexity. This would suggest that one would have a convenient means to provide a human operator with the capability to explore the cross corpora relationships by providing him with a tool that facilitated an efficient exploration of the MST as computed on the two class set of documents along with the associated cross-corpora edges.

The only missing piece in the puzzle is a way to provide a user with a "drill down" capability so that they can focus their attention on a set of nodes rather than the full MST. We chose to accomplish this by allowing the user to choose a subset of observations (nodes) and a depth exploration parameter k. This parameter specifies how far, in terms of number of edges, we wish to travel from our original node to include in our subsequent analysis. For example, the user might select nodes 14 and 2 for his focus of attention and selected a search parameter of $k = 3$. The user selected nodes are colored red while the nodes within 3 links of these are colored blue. In the next stage of the analysis, the user is presented with the complete subgraph. This is rather busy of course and it is necessary for the user to have the capability to remove some of the edges, based on edge strength, before they proceed forward with their analysis. We will have more to say about this when we discuss our software implementation of the user environment. Figure 1 presents a "cartoon" of how the selected nodes interact with the depth search parameter.

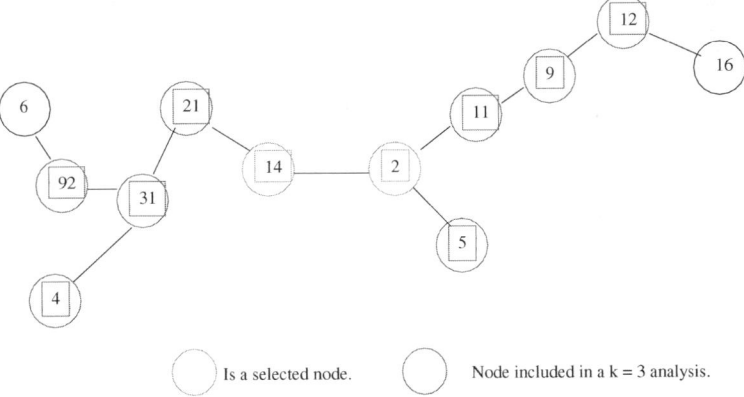

Fig. 1. User specification of drill down focus region. Nodes 14 and 2 have been selected. For a color reproduction of this figure see the color figures section, page 569.

We were also interested in providing the user with the capability to explore clusters of observations (documents). Previous analysis in the literature had indicated that one can obtain all the possible single-linkage hierarchical clusterings of a set of observations merely by sequentially deleting the edges from the MST of the complete graph starting with the largest edge. There had been previous work appearing in the literature on the application of MSTs to geospatial data clustering (Guo et al., 2002, 2003) and gene expression analysis (Xu et al., 2001, 2002). We choose to utilize an MST as computed on the full set of observations, all classes, as a starting point for our cluster analysis. We proceed forward cutting edges in the MST in order to obtain a clustering at a particular number of clusters c. We choose which edge to cut based on the edge strength divided by the mean of the associated edges of path length k adjacent to the chosen edge.

Let us now turn our attention to our software implementation of these algorithmic schemes. The BPM and similarity matrix calculations were implemented in C#. The MST calculations were implemented based on Kruskal's algorithm in JAVA. The visualization environment was implemented in JAVA and the graph layout was accomplished using the general public license package TouchGraph (available at http://www.touchgraph.com). TouchGraph supports zooming, rotation, hyperbolic manipulation, and graph dragging.

We originally implemented the software system with automated serendipity exploration in mind. Figure 2 presents the opening screen of our automated serendipity system. This screen presents the user with a webpage that provides links to all of the "choose 2" corpora within the text dataset. The user can easily navigate in order to study a particular comparison by first clicking the appropriate corpora link at the top

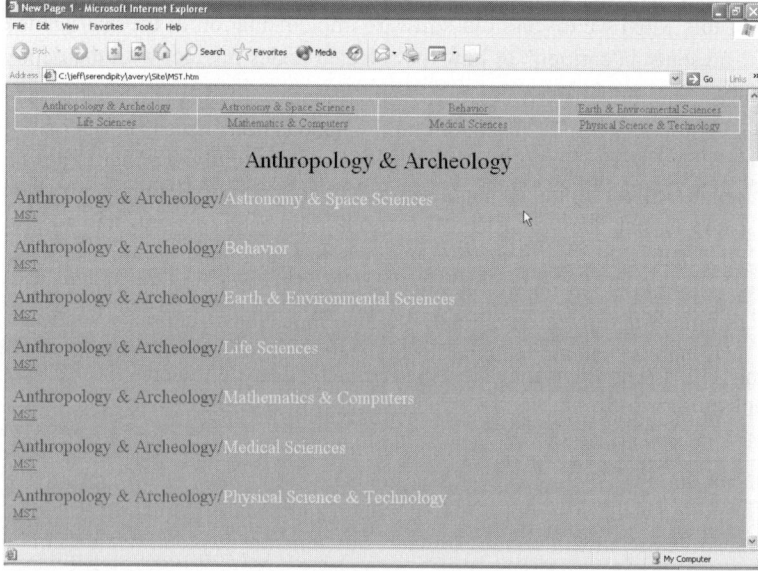

Fig. 2. Opening screen of the automated serendipity search engine. For a color reproduction of this figure see the color figures section, page 570.

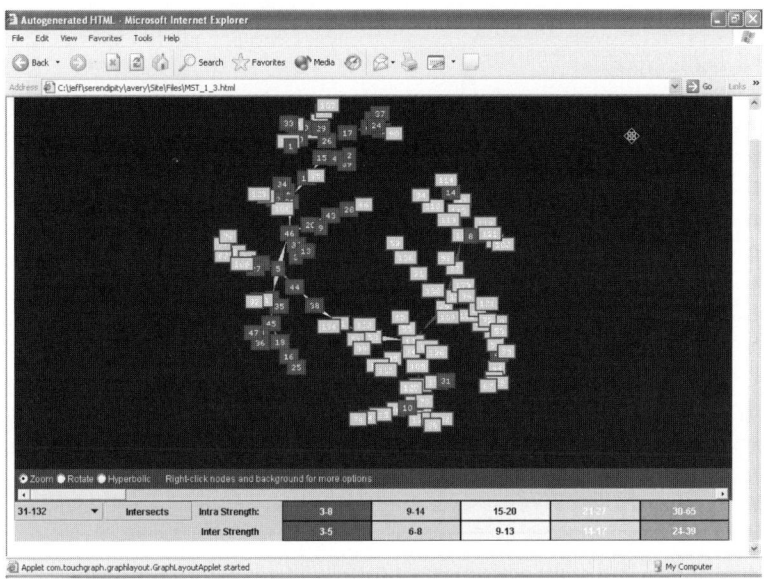

Fig. 3. An example of the MST layout screen. For a color reproduction of this figure see the color figures section, page 570.

of the screen and then subsequently clicking on the corpora that the user would like to compare with the currently focused corpora to. Alternatively, the user may simply scroll down to the point in the webpage that contains the association of interest.

Once a user chooses two categories to explore, they are presented with a rendering of the MST based on a spring-based layout model. Figure 3 presents one example of this screen. We draw the users attention to the fact that we have provided an edge weight legend at the bottom of the plot. We have used a blue, green, yellow, orange, and red color scheme that is spread proportionally among the edge weight values. We have chosen to "reuse" the color map. We use it once to represent the intra-class edges and once to represent the inter-class edges. It is usually easily discernible to the user which situation that they are dealing with. We have also provided the user with a drop down menu on the lower left that has the edges sorted on inter-class strength from the weakest at the top of the list to the strongest at bottom of the list. In this way, the user can rapidly explore the serendipitous-type relationships. Experience also led us to change the symbols of the nodes as they are explored. In this way, the user can keep track of his exploration of the graph.

Once a user has selected a particular edge to display, they are presented with a comparison file that highlights the word pairs from the BPM that the two articles have in common and also enumerates them; please see Figure 4.

The software system that provides a point picking capability and the software system that allows clustering are extensions of the basic framework that has been previously discussed. We think it prudent to say a few words about each of these in turn. First we will discuss the automated serendipity system that allows point picking. Figure 5 shows

138   J.L. Solka et al.

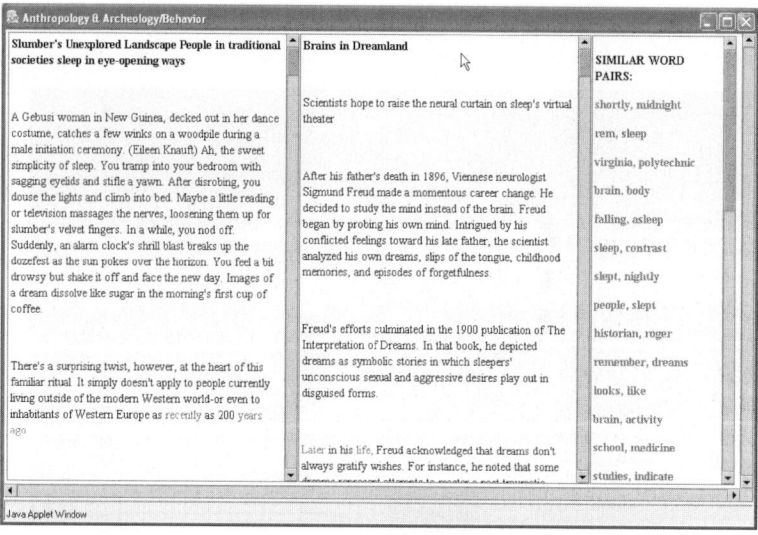

Fig. 4. A sample comparison file. For a color reproduction of this figure see the color figures section, page 571.

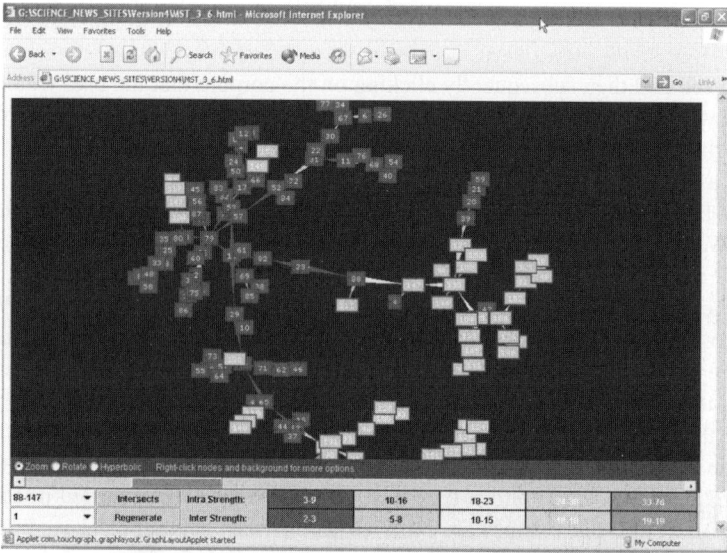

Fig. 5. An example screen for the automated serendipity system with included point picking. For a color reproduction of this figure see the color figures section, page 571.

a sample screen from this version of the software. The main difference is that the user is provided with the capability to select two or more nodes for a focus of attention along with a regeneration parameter k and a button to pick to regenerate the complete graph with the vastly limited focus of attention.

Text data mining with minimal spanning trees   139

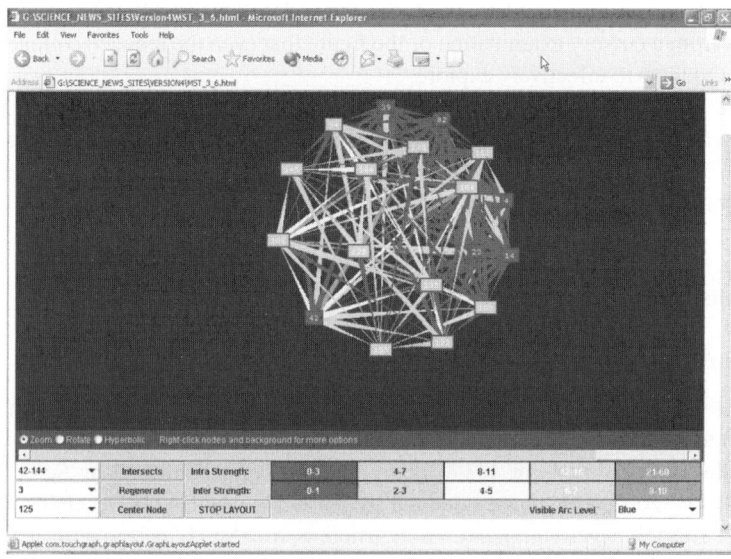

Fig. 6. An example screen for the automated serendipity system with point picking where the user has chosen to limit the focus of attention based on $k = 3$. For a color reproduction of this figure see the color figures section, page 572.

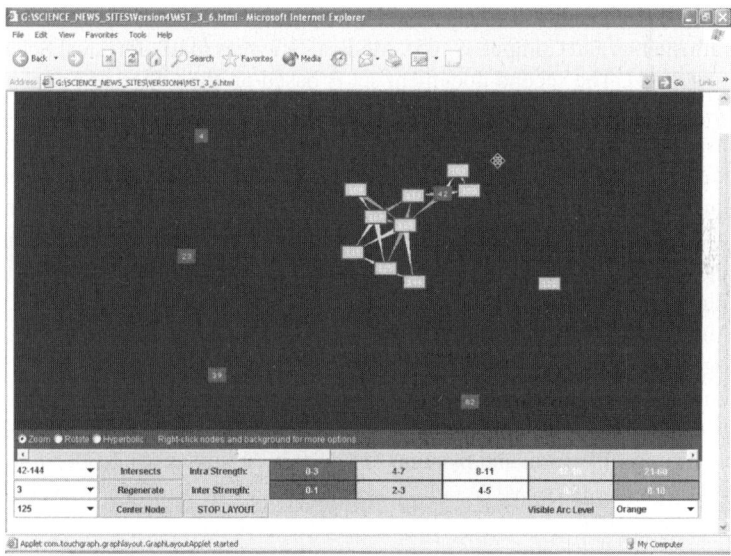

Fig. 7. An example screen for the automated serendipity system with included point picking with point picking where the user has chosen to limit the focus of attention based on $k = 3$ after the user has deleted the edges with a strength less than orange. For a color reproduction of this figure see the color figures section, page 572.

Figure 6 shows the complete graph for the reduced focus of attention. It is quite apparent from a cursory examination of the figure that even the reduced complete graph is much too busy to effectively explore.

Figure 7 shows the complete graph for the reduced focus of attention where the user has chosen only to show the edge weights with a strength of orange or greater. Things are more manageable now. We will relegate the discussions of the clustering version of the software till the results section.

2. Results

2.1. Datasets

Now that we have briefly outlined the visualization framework, let us examine its application to two different text multi-corpora document sets. The first document set that will be discussed consists of 1117 documents that were published by Science News (SN) in 1994–2002. They were obtained from the SN website on December 19, 2002 using the LINUX wget command. Each article is roughly 1/2–1 page in length. The corpus html/xml code was subsequently parsed into straight text. The corpus was read through and categorized into 8 categories by one of us, A.B. The 8 categories were based on the fact that a very small subset of the articles had already been placed in these categories on the Science News website. The categories and their respective article counts are as follows: Anthropology and Archeology (48), Astronomy and Space Sciences (124), Behavior (88), Earth and Environmental Sciences (164), Life Sciences (174), Mathematics and Computers (65), Medical Sciences (310), Physical Sciences and Technology (144). The second document set is based on the Office of Naval Research (ONR) In-house Laboratory Independent Research (ILIR) Program. This program is the primary way that the Office of Naval Research funds the Naval Warfare Centers to conduct fundamental research. This set of documents was provided to us and consists of 343 documents each of which is about 1/2 page in length. Avory Bryant and Jeff Solka jointly read through these documents and classified them into previously defined ONR Technology Focus Areas. The number of documents in each of the ONR Focus Areas are as follows: Advanced Naval Materials (82), Air Platforms and Systems (23), Electronics (0), Expeditionary/USMC (0), Human Performance/Factors (49), Information Technology and Operations (18), Manufacturing Technologies (21), Medical Science and Technology (19), Naval and Joint Experimentation (0), Naval Research Enterprise Programs (0), Operational Environments (27), RF Sensing, Surveillance, and Countermeasures (27), Sea Platform and Systems (38), Strike Weapons (0), Undersea Weapons (0), USW-ASW (5), USW-MIW (17), and Visible and IR Sensing, Surveillance and Countermeasures (17).

2.2. Feature extraction

Both sets of documents were processed to remove standard stopper words prior to the extraction of the bigram proximity matrices. The remaining words were not stemmed in that the current version of the software does not readily support stemming. The BPMs were then extracted for each document. The Ochiai similarity measure and associated distance was then computed on each document pair using the equations described above.

2.3. Automated serendipity extraction on the Science News data set with no user driven focus of attention

Let us begin our analysis of these document sets by first considering the Science News corpus. We will first analyze this set of data using the automated serendipity software system that does not support point picking. First we choose Mathematics and Computer Sciences as compared to Physical Sciences and Technology. Given the strong overlap in these two discipline areas, we would expect that there might be strongly associated articles across the discipline areas. Figure 8 shows the MST for these two categories. We have selected the edge that represents the strongest association between the two categories. Right clicking on this link and choosing the view comparison file leads us to Figure 9. There are numerous associations between these two articles. Some of these are highly relevant such as (allow, electrons), (electronic, devices), (circuit, components), (atomic, scale), (biological, molecules), and some may be either identified as spurious or as encoding "meta associations" that may or may not be relevant such as (santa, barbara), (packard, laboratories), (palo, alto), (hewlett, packard) and others. There are even associations that are clearly "useless" such as (research, team) and (research, group). The "human identified" connection between the two articles is that fact that the first article from the Mathematics and Computer Science area is about computer chip design

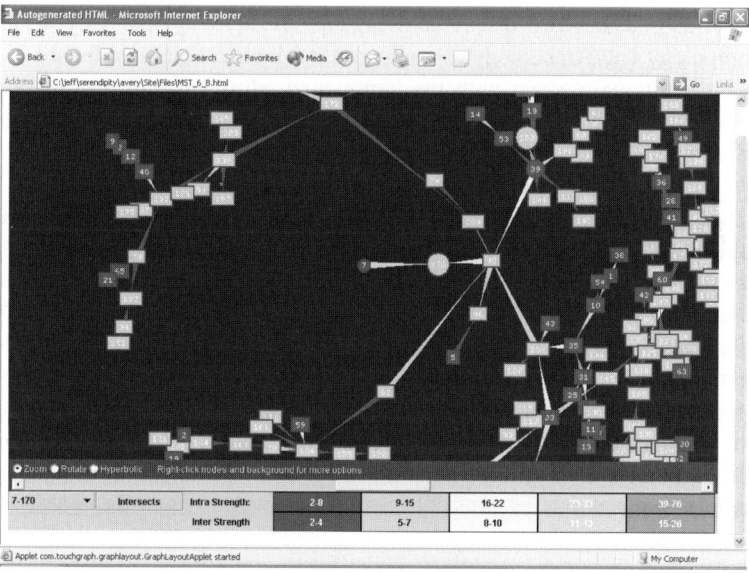

Fig. 8. The MST for the Mathematics and Computer Sciences category as compared to the Physical Sciences and Technology category. The edge indicating the strongest association has been highlighted. For a color reproduction of this figure see the color figures section, page 573.

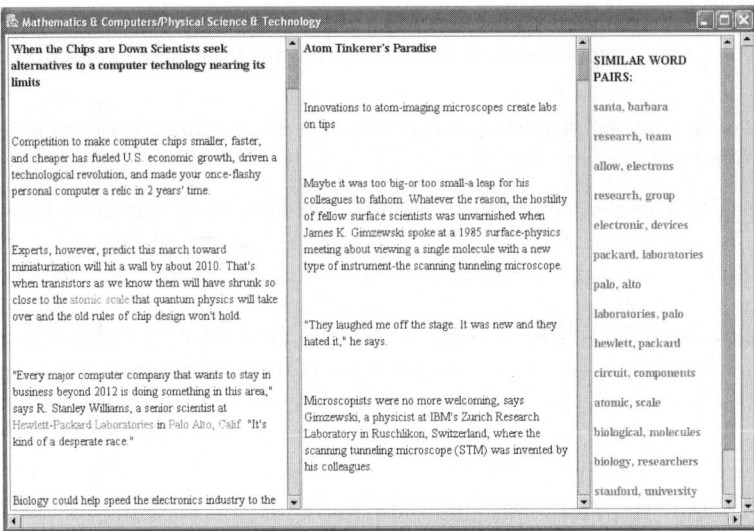

Fig. 9. The BPM comparison file for the strongest association between the Mathematics and Computer Sciences category and the Physical Sciences and Technology category. For a color reproduction of this figure see the color figures section, page 573.

while the second article from the Physical Sciences and Technology arena is about materials science design at the nano level.

Next let us consider two discipline areas that are more conceptually distant such as Anthropology and Archeology as compared to Medical Sciences. Figure 10 shows the MST for this case along with the highlighted edge that indicates the strongest association between these two discipline areas. Right clicking on this link and choosing the view comparison file leads us to Figure 11.

Once again the methodology has identified two articles with a large overlap in their BPMs and hence a large similarity in semantic content. A few of the very important word associations include (major, depression), (family, community), (mental, health), and (mental, disorders). Once again their are also several meta-level associations identified such as (delhi, india) and (anthropologist, veena). There are also some association that are spurious such as (nearly, years). A cursory examination of the two articles does indicate that the association is relevant. The first article from the Anthropology and Archeology category is about depression in people that are living in violent societies while the second article from the Medical Sciences category is about depression in individuals that suffer from disease such as AIDS that of the lead to many in society stigmatizing them.

Let us consider two more discipline pairs before moving on to the ILIR data. The first article pair comes from the Earth and Environmental Sciences and Physical Sciences and Technology categories. We have already provided enough examples of the MST pictures to forgo the display of these in the following example. There are some interesting things to say about the location of the associated articles in the MST, but

Text data mining with minimal spanning trees 143

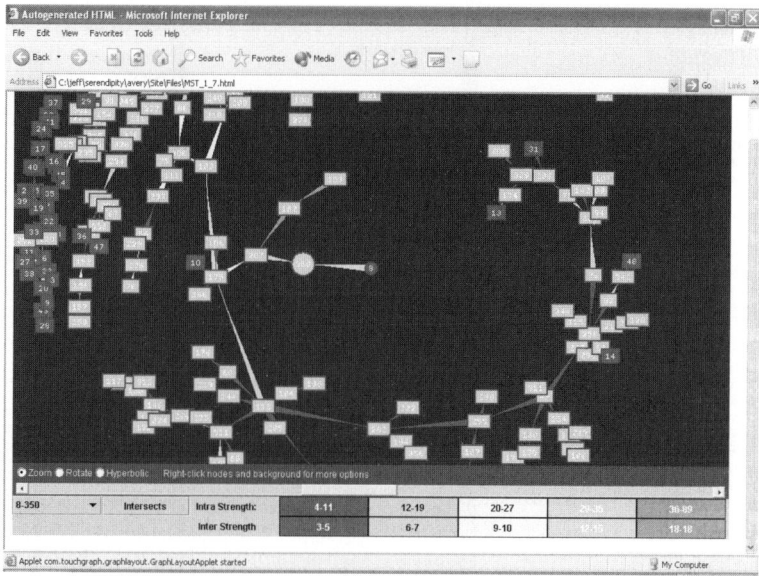

Fig. 10. The MST and highlighted strongest edge between the Anthropology and Archeology discipline area and the Medical Sciences discipline area. For a color reproduction of this figure see the color figures section, page 574.

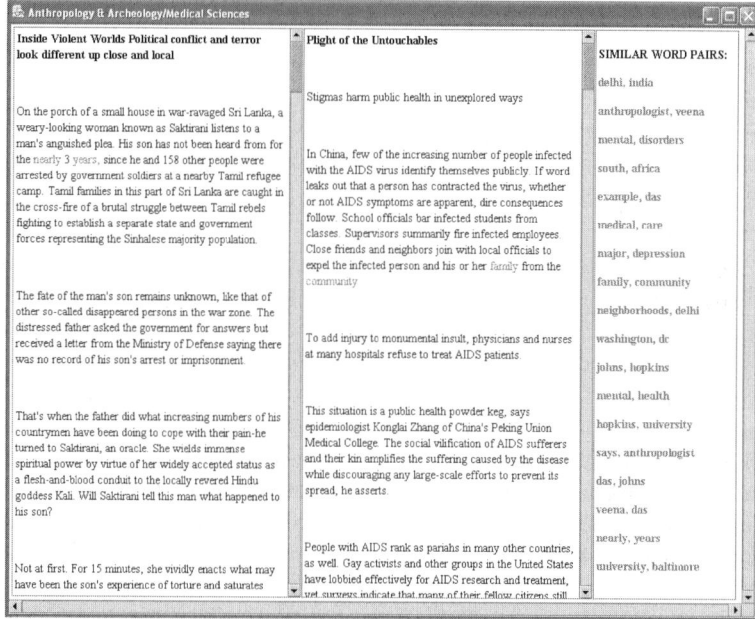

Fig. 11. The BPM comparison file for the strongest association between the Anthropology and Archeology category and the Medical Sciences category. For a color reproduction of this figure see the color figures section, page 574.

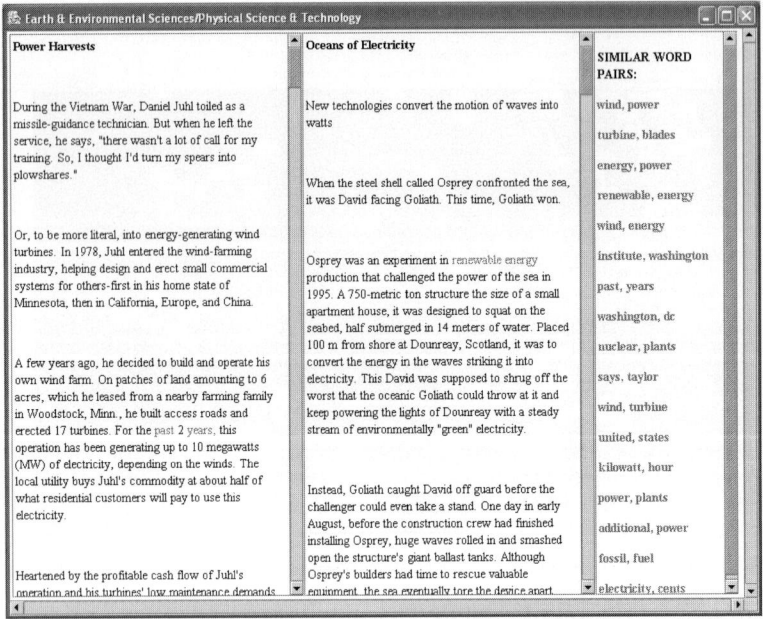

Fig. 12. The BPM comparison file for the strongest association between the Earth and Environmental Sciences category and the Physical Sciences and Technology category. For a color reproduction of this figure see the color figures section, page 575.

this type of analysis will not be performed in this chapter. We proceed exactly to the two strongest associated articles between these two categories, see Figure 12. The first article from the Earth and Environmental Sciences category, entitled "Power Harvests", is about wind generated power. The second article, entitled "Oceans of Electricity", is about new methodologies for generation of power from ocean waves. These two articles are highly associated including numerous word pairs. It seems very reasonable to assume that those researchers that are involved in work on wind generated power would be interested in new technologies that may have been developed for ocean generated power. Ocean generated power research might include work in power conversion and battery storage. Both of these would be relevant to wind power research.

The last discipline area combination that we wish to consider is Anthropology and Archeology and Mathematics and Computer science. This might be considered the most interesting combination so far in that one would not expect that there would necessarily be a natural association between these areas. Once again we will proceed directly to the comparison file between two articles as identified from a cursory examination of the MST. We point out the fact that this is not the strongest association, but merely an interesting one, see Figure 13. The Anthropology and Archeology article is entitled "An artist's timely riddles" and is about the artist Marcel Duchamp, while the Mathematics and Computer Science article is entitled "Visions of infinity tiling a hyperbolic floor inspires both mathematics and art" and is about the tilings of the Poincaré disk

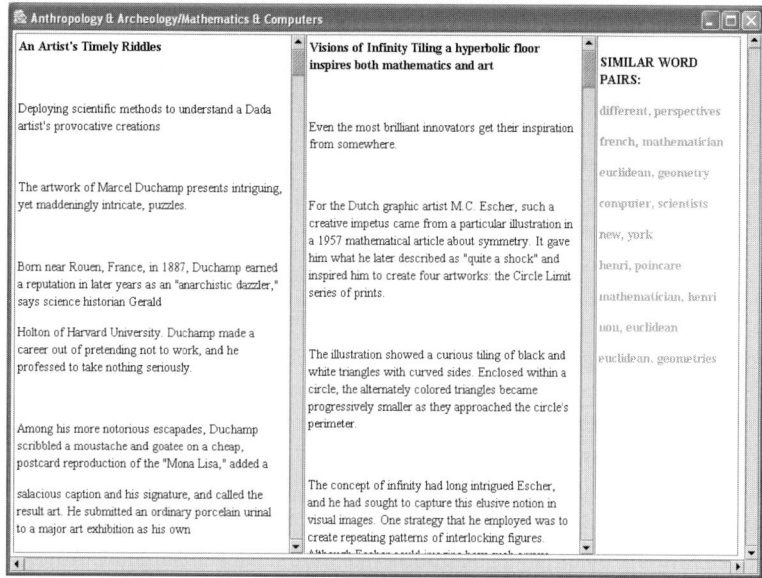

Fig. 13. The BPM comparison file for an interesting association between the Anthropology and Archeology and Mathematics and Computer Sciences categories. For a color reproduction of this figure see the color figures section, page 575.

by M.C. Escher. The number of BPM commonalities between these two articles is not that large, 9, but there are highly relevant associations among the list of associations including (different, perspectives), (euclidean, geometry) and (non, euclidean). It seems that Duchamp was also involved in explorations of non-Euclidean geometries as was Escher. The work by Escher is very well known but the authors had no idea about the Duchamp work.

2.4. Automated serendipity extraction on the ONR ILIR data set with no user driven focus of attention

Next we turn our attention to an analysis of the ONR ILIR data using the same tool. The application of the tool to the ILIR articles was compounded by two facts. First, the ILIR articles are much shorter in length than the Science News articles. Second, the ILIR articles are a somewhat more orthogonal set of articles than the Science News articles. There are many more articles that come from the same type of discipline within the Science News dataset. This last fact may be in part due to the nature of the ILIR dataset and to the fact that there is just a shorter set of articles.

The first pair of articles that we will consider come from the Air Platform and Systems and the Sea Platform and Systems category. The two articles are presented in Figure 14. These two articles are about computational fluid dynamics as applied to the two discipline area. The list of commonalities in the BPMs is short but contains several meaning associations such as (navier, stokes), (high, reynolds), (reynolds, averaged),

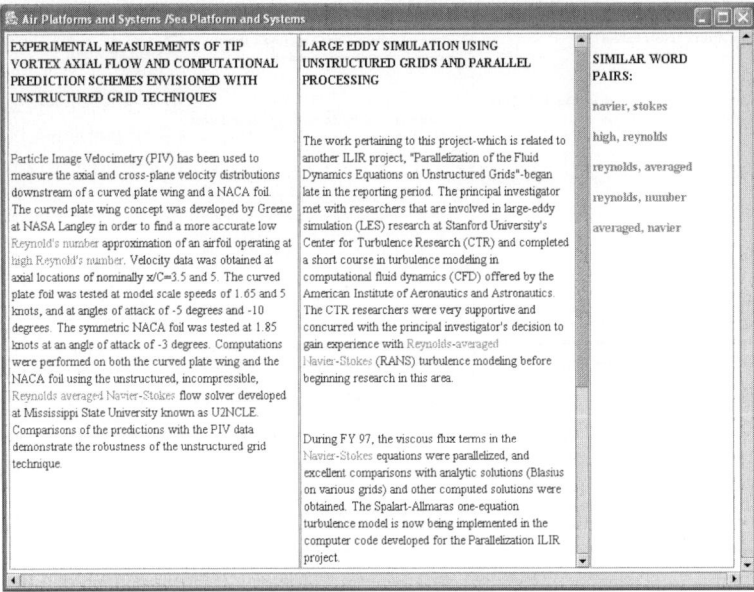

Fig. 14. The BPM comparison file for a computational fluid dynamics association between the Air Platform and Systems and the Sea Platform and Systems classes. For a color reproduction of this figure see the color figures section, page 576.

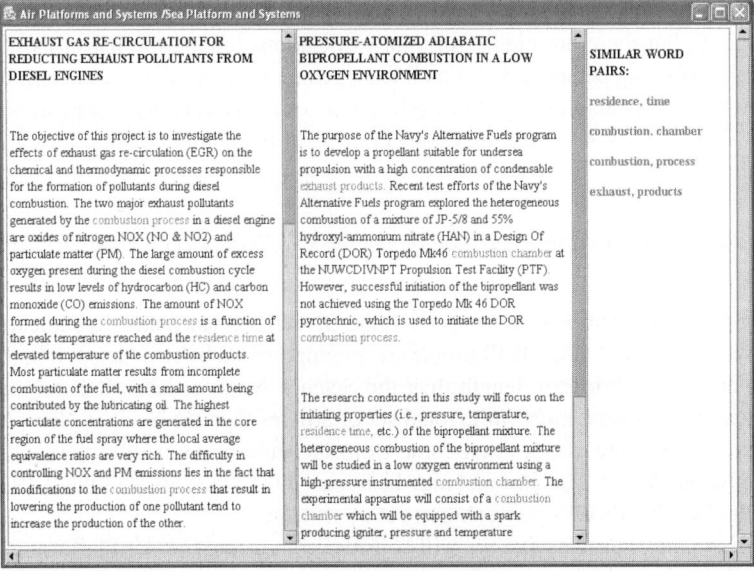

Fig. 15. The BPM comparison file for a combustion physics association between the Air Platform and Systems and the Sea Platform and Systems classes. For a color reproduction of this figure see the color figures section, page 576.

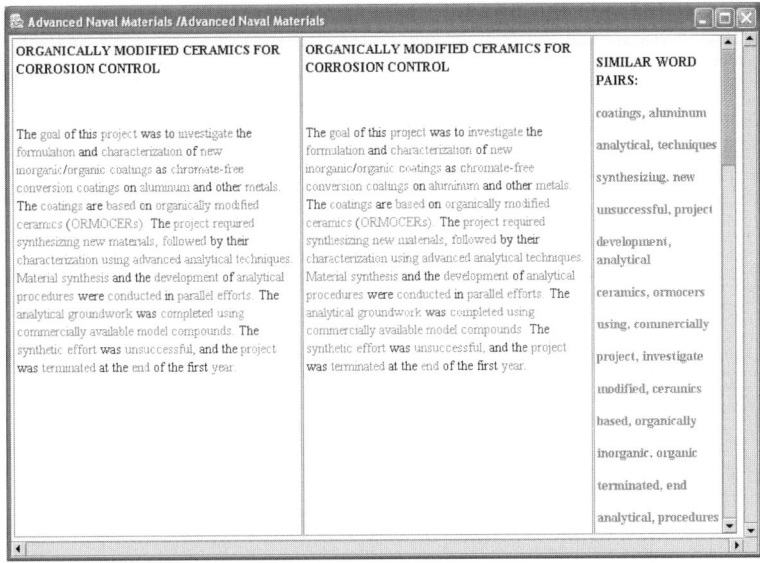

Fig. 16. An apparent duplication entry in the Advanced Naval Material category. For a color reproduction of this figure see the color figures section, page 577.

and (reynolds, number). The two projects were executed at different labs, NSWCDD and NAVSTO, during different years, FY01 and FY99/00. Clearly there would be benefits between collaborative discussions between the two investigators or the later investigator could benefit from some of the insights obtained by the earlier group.

The next two articles also come from the Air Platform and Systems and the Sea Platform and Systems arena. These two articles are both about combustion physics, see Figure 15. One of the projects was executed at NSWCDD in FY99 and the other project was executed at NUWC in FY01. Each project was conducted by a different investigator.

Another capability that our proposed automated serendipity approach possesses is anomaly identification, consider Figure 16. A project has clearly been entered into the database twice.

The next example illustrates another not quite so blatant anomaly, see Figure 17. These two abstracts come from two different projects executed by the same investigator. One project was executed during FY99 and the other was executed during FY01. The two abstracts seem to contain a section that is virtually identical but one might not have guessed this based on the titles of the two abstracts.

The next ILIR example that we will consider is perhaps one of the most curious ones that we uncovered in the database, see Figure 18. One of these articles was entitled "Direct adaptive, gradient descent and genetic algorithm techniques for fuzzy controllers" and another was entitled "Mission scenario classification using parameter space concept learning." A quick read of the titles of these two efforts would not lead one to believe that there would be any association between them but these two articles (abstracts) are

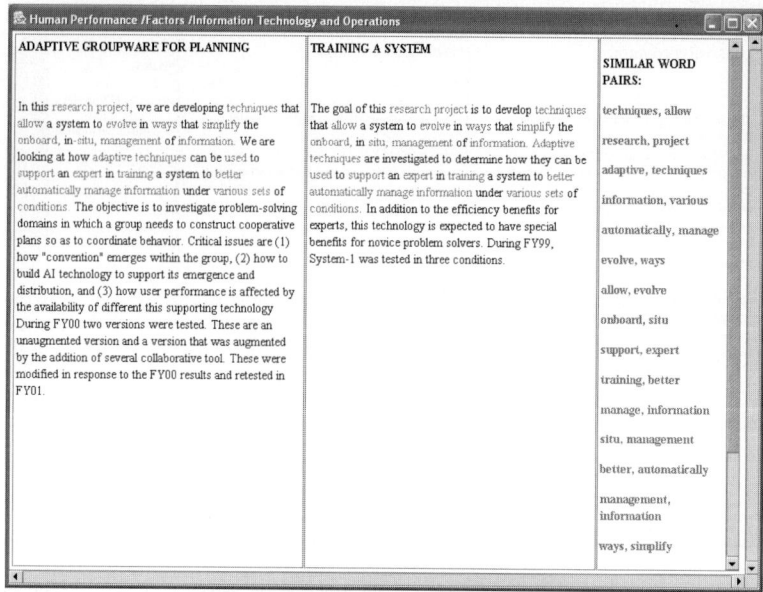

Fig. 17. An unusual association across the Human Performance Factors and the Information Technology and Operation focus areas. For a color reproduction of this figure see the color figures section, page 577.

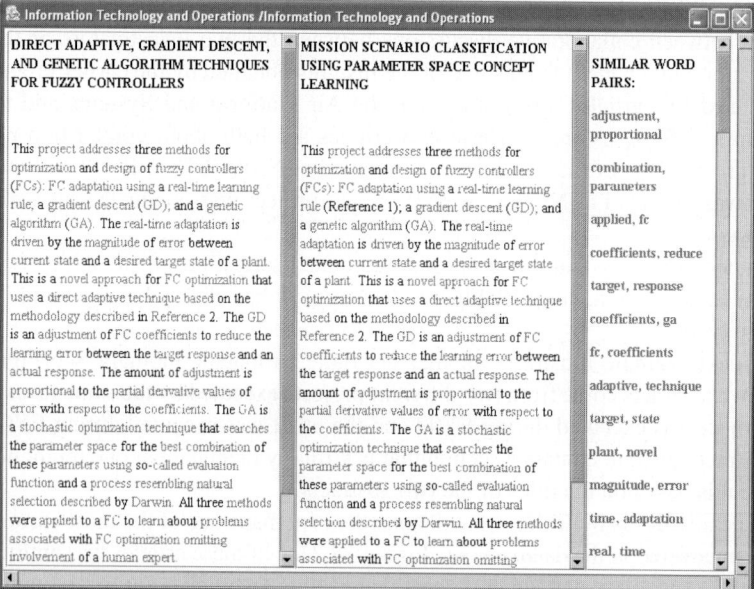

Fig. 18. A highly unusual association between two articles in the Information Technology and Operations focus areas. For a color reproduction of this figure see the color figures section, page 578.

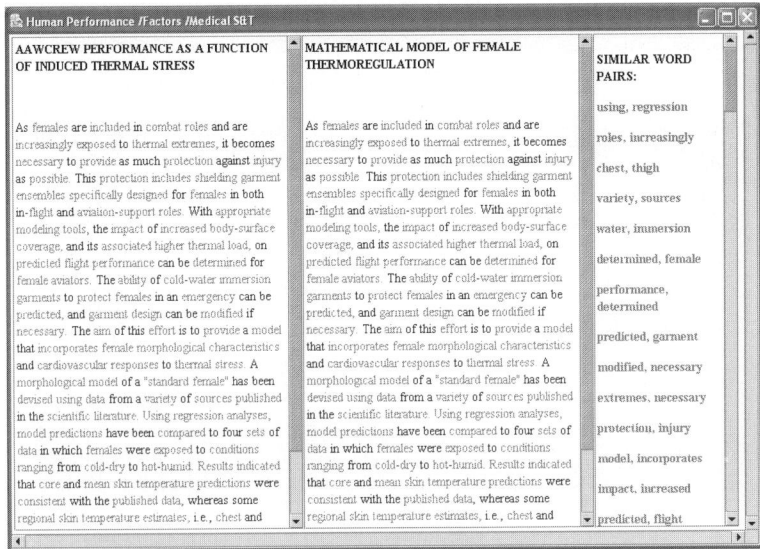

Fig. 19. A highly unusual association between two articles in the Human Performance Factors and the Medical Sciences and Technology focus areas. For a color reproduction of this figure see the color figures section, page 578.

virtually identical and perhaps even more curious was the fact that these two abstracts reflect two projects that were being executed by the same investigator during the same years, FY99/00, at the same laboratory, NUWC. There is clearly something unusual going on here that, while not nefarious, is certainly curious.

We close this subsection with another cross focus area curiosity, see Figure 19. These two articles are entitled "AAWCREW performance as a function of induced thermal stress" (the Human Performance Factors article), and "Mathematical model of female thermoregulation" (the Medical Science and Technology article). Both of these articles seem to be about effects of temperature changes on female AAW crew members. Both of the efforts are being funded by the same lab, NAVSTO, in the same funding cycle (FY99/FY00), and being conducted by the same researcher. The two articles are virtually identical. So one would surmise that this represents a two effort thrust that is being funded in the same discipline area. So we have effectively identified an artifact of the funding strategy of the organization.

2.5. Automated serendipity extraction on the Science News data set with user driven focus of attention

Now we will consider how the insights obtained using the automated serendipity system are changed by allowing the user to focus their attention on a subset of the full set of nodes within the MST, to specify a depth around these nodes to explore and finally to examine the complete graph surrounding these nodes for interesting serendipitous types of relationships. We will illustrate this methodology via an analysis of the Science News dataset. Let us consider the Earth and Environmental Sciences category

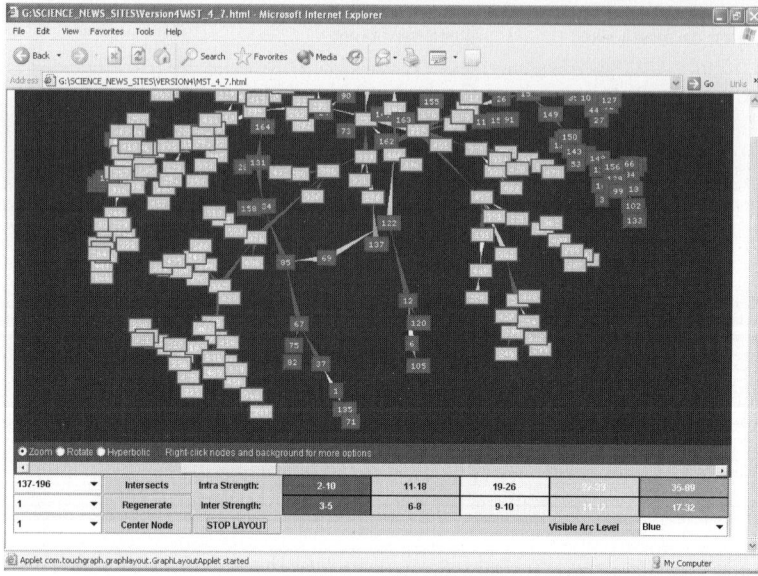

Fig. 20. The user has selected nodes 196 and 137 as the "focus of attention". For a color reproduction of this figure see the color figures section, page 579.

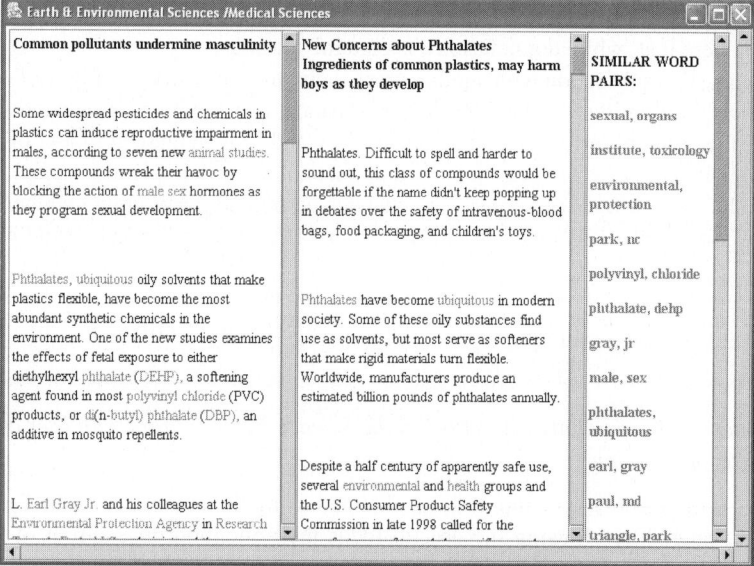

Fig. 21. A comparison for the two articles selected for the "focus of attention". For a color reproduction of this figure see the color figures section, page 579.

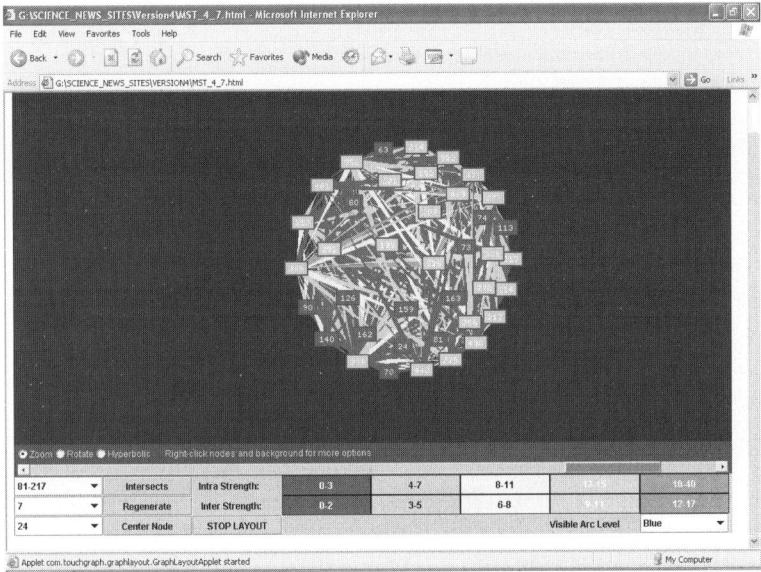

Fig. 22. The complete graph based on those articles within 7 links of articles 196 and 197. For a color reproduction of this figure see the color figures section, page 580.

vs. the Medical Sciences category. First the user must choose the nodes to focus attention on, see Figure 20. These two articles are entitled "Common pollution undermines masculinity" and "New concerns about phthalets ingredients of common plastics, may harm boys as they develop," see Figure 21. The two articles have 32 commonalities in their BPM and the first article details studies concerning the effect of phthalates on animals while the second paper details similar types of studies in humans. After specifying a depth parameter of 7, one obtains the following complete graph, Figure 22. We remind the reader that nodes 196 and 137 have been removed prior to the rendering of the complete graph. The complete graph is rather busy, looking much like a ball of rubber bands. Let us now, using the tool, remove the edges weaker than a strength of orange, see Figure 23. Now some structure is clearly discernible in the graph. There is a central cluster and several singleton groups surrounding this central cluster. Let us take a moment to discuss the structure of the central cluster. There are nodes (70, 217) with 17 associations. Node 70 comes from the Earth and Environmental sciences group and is entitled "Banned pollutant's legacy: lower IQs," while node 21 comes from the Medical Sciences category and is entitled "Lead therapy won't help most kids." Node 70 details the effects of PCBs on IQ while node 217 discusses the effect of lead. Node 70 also shares numerous (12) associations with node 231. The article corresponding to node 231 is entitled "Burned by flame retardants," which comes from the Medical Sciences category. Node 231 discusses PCBs in flame retardants. Node 231 also shares 12 associations with node 90 from the Earth and Environmental Sciences area. The article corresponding to node 231 is entitled "The rise in toxic tides" and discusses various sorts of biologically produced toxins killing land animals (cows) and marine animals

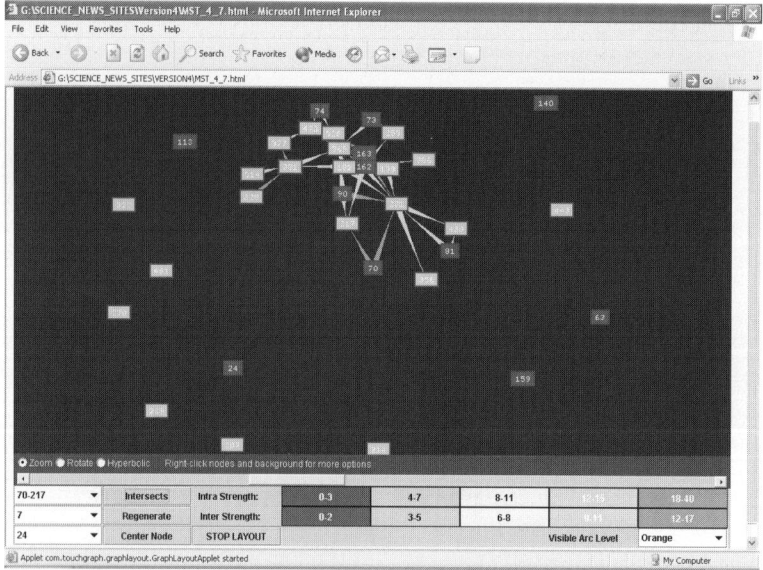

Fig. 23. The complete graph based on those articles within 7 links of articles 196 and 137 with the edges of strength less than orange removed. The graph has been centered on the edge extending between nodes 70 and 217. For a color reproduction of this figure see the color figures section, page 580.

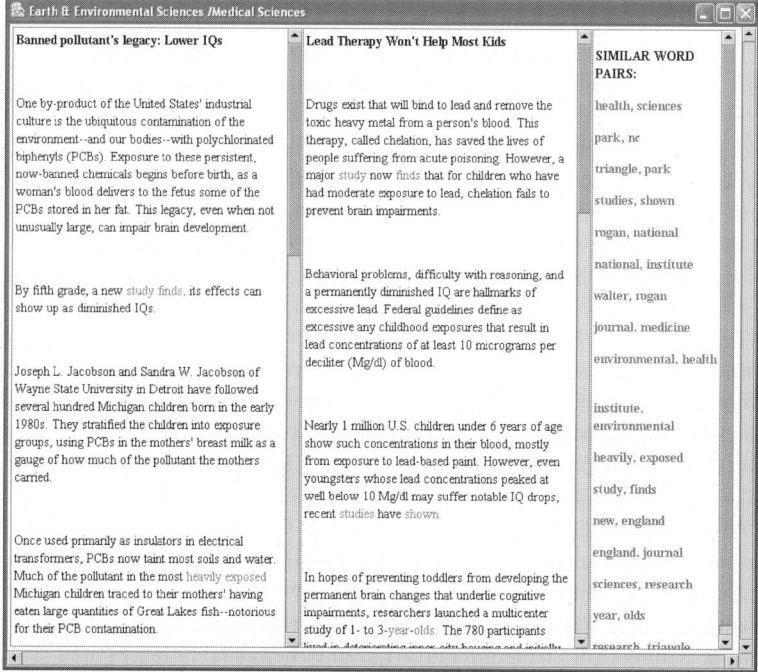

Fig. 24. A closer look at the associations between nodes 70 and 217. For a color reproduction of this figure see the color figures section, page 581.

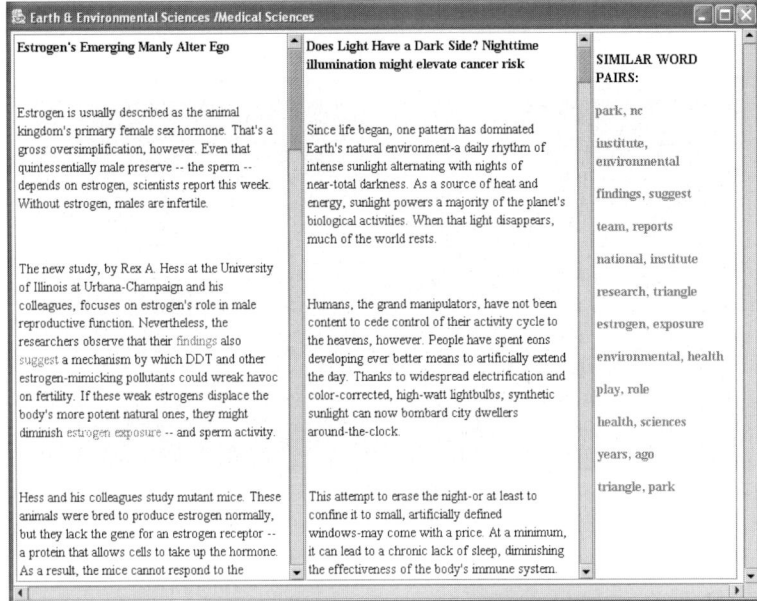

Fig. 25. A closer look at the associations between nodes 162 and 369. For a color reproduction of this figure see the color figures section, page 581.

(fish). Finally, we point out 12 associations between nodes 162 and 369. Node 162 is entitled "Estrogen's emerging manly alter ego" and details DDTs mimicking estrogen, while node 369 is entitled "Does light have a dark side? Nighttime illumination might elevate cancer risk" and discusses the fact that night lights may effect estrogen related maladies such as breast cancer.

Now, after we have briefly outlined some of the interesting associations in this central cluster, let us take a little time to look a bit closer at some of the more interesting associations. First, we will consider the node (70, 217) pair, see Figure 24. There are numerous associations between these two articles, some of which are clearly spurious such as (national, institute) and some of which are highly relevant such as (environmental, health) and (heavily, exposed). There are even meta associations such as a common researcher (Walter, Rogan). The totality of these associations is sufficient in order to make a meaningful connection and, in fact, the authors posit the idea that it might be fruitful for investigators to examine the synergistic effect of PB and PCB exposure.

It is worthwhile to more closely examine one other pair, (162, 369), see Figure 25. Linkages between these articles on topic of environmental health and estrogen exposure were sufficient to lead to a meaningful association. These two might lead one to consider synergistic effects between DDTs lowered estrogen exposure due to DDT substitution and night time illumination or EM fields (increased estrogen exposure).

Let us conclude this discussion by performing a similar analysis on the categories Behavior and Mathematics/Computer Science. Figure 26 illustrates the MST for this combination along with our nodes of interest (88, 147). The comparison file is presented

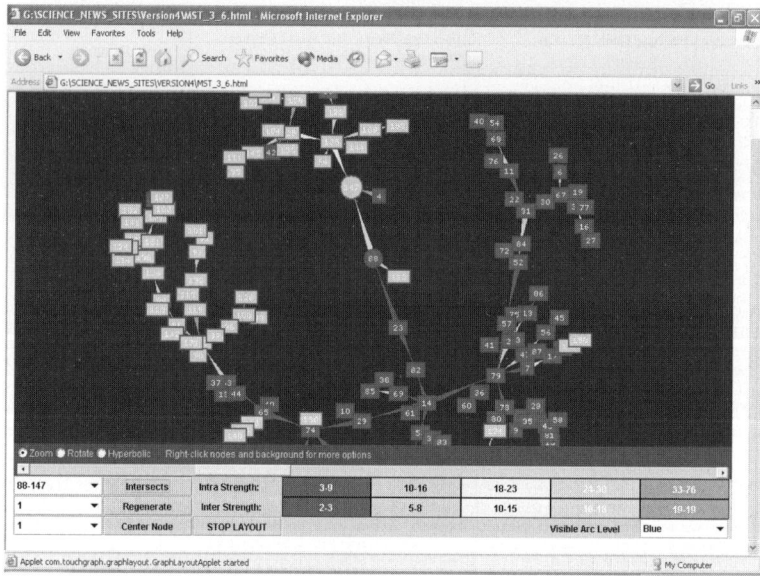

Fig. 26. The MST and nodes 88 and 147 of the Behavior and Computer Science and Mathematics categories. For a color reproduction of this figure see the color figures section, page 582.

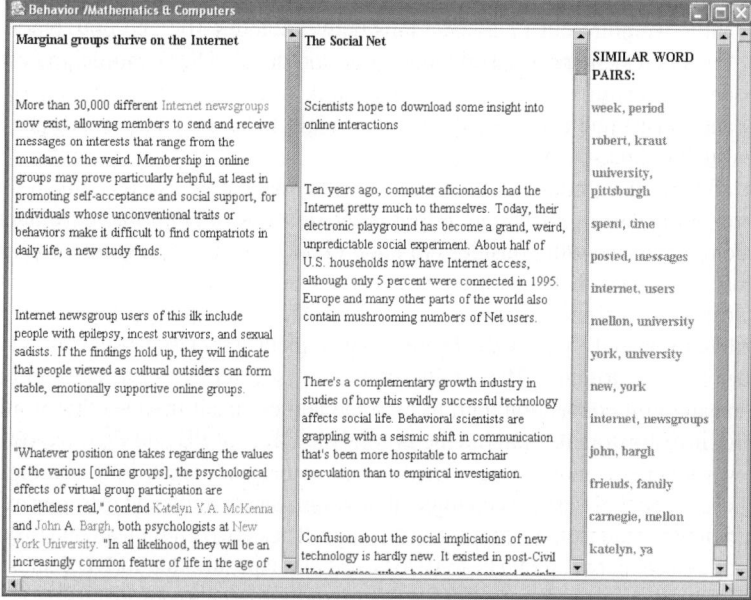

Fig. 27. The BPM comparison file nodes 88 and 147 of the Behavior and Computer Science and Mathematics categories. For a color reproduction of this figure see the color figures section, page 582.

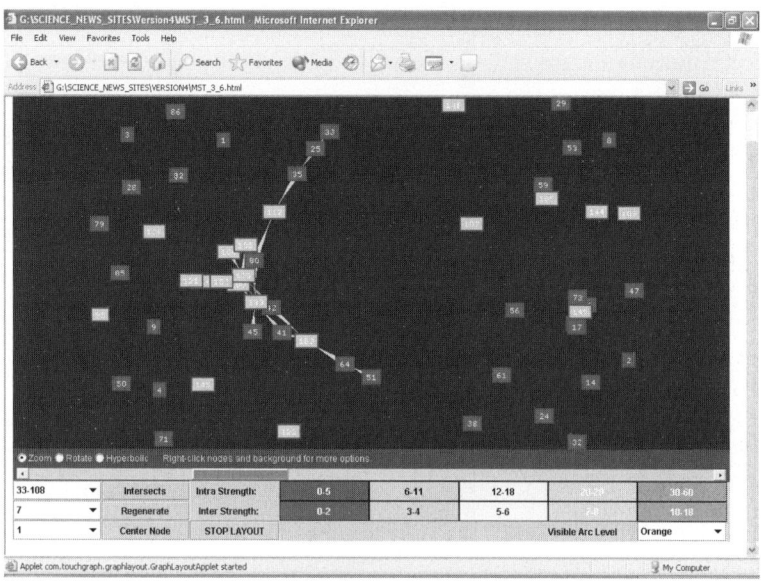

Fig. 28. The articles within 7 links of 88 and 147 with a strength of at least orange. For a color reproduction of this figure see the color figures section, page 583.

in Figure 27. The first article from the Behavior category, entitled "Marginal groups thrive on the internet," is about the existence of marginal groups in cyber space, while the second article from the Computer Science and Mathematics category is entitled "The social net" and is about the application of social network theory to network based interactions. The two articles share many meaningful associations. Choosing these articles as the focus of attention along with a number of links parameter of 7 and then removing the links that are weaker than a strength value of orange yield Figure 28. Let us take a moment to discuss the articles that remain in the central group. Nodes (10, 108) possess 10 associations and are entitled "Simple minds, smart choices," an article from the Behavior category about "bounded rationality", and "The soul of a chess machine," an article from the Mathematics and Computer Science category about computer chess programs. We will have much more to say about this article pair shortly. The next article pair is (80, 133). Article 133 from the Mathematics and Computer science category, entitled "Playing your cards right," is about a computer poker program. The next article pair is (80, 135). Article 135 from the Mathematics and Computer Science category is entitled "Agents of cooperation" and is about intelligent agents. Let us take a closer look at the node (80, 108) associations, see Figure 29.

There are 10 associations between this pair of articles, however some of these are spurious. The set of non-spurious associations is small, but relevant and includes (did, better), (investment, portfolios), (computer, program), (best, possible), and (human, judgment). When we originally formulated our research agenda in the automated serendipity program, we thought of the field of genetic algorithms as one of the areas where an insightful researcher (numerous researchers claim credit in this case) dis-

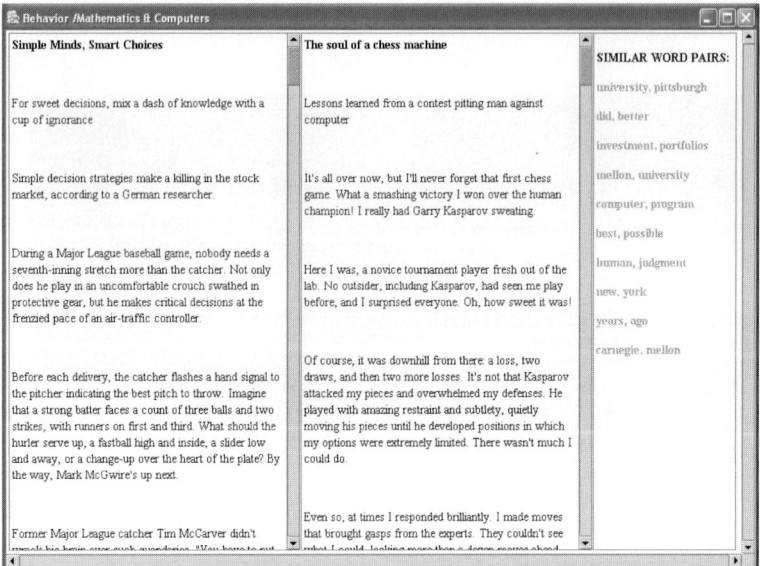

Fig. 29. A closer look at the between-article associations between Behavior article 80 and Mathematics and Computer Science article 108. For a color reproduction of this figure see the color figures section, page 583.

covered the relationship between the fields of evolutionary biology and mathematical optimization. We decided that whatever software that we developed would be "successful" if it could discover such an association. When we first encountered the above association, we felt that perhaps we might have discovered such a "nugget". The first article, the Behavioral one, discusses the technique that humans use when they are faced with an overwhelming optimization problem for which they are presented with limited time to make a decision. The example provided in the article is a catcher in baseball. He must make a decision of the best pitch for the pitcher to throw while worrying about the runners stealing bases. He also has a limited time to make his decision. So one might wonder if the bounded rationality scheme might be relevant to formulating strategies for chess programs. Well one of the authors, J.L.S., has a little experience playing chess and one of his first instructions that his tutor offered him was that "human chess masters don't necessarily think as many moves deep as a computer but they have the ability to rapidly evaluate numerous positions that are several moves deep." So the instructor indicated that human beings employ bounded rationality. It is not much of a stretch to believe that it might be appropriate for a computer to employ bounded rationality in its approach to chess. The potential applicability of the concept of bounded rationality to other areas is supported by a consideration of the comparison files between nodes (80, 133), see Figure 30.

The comparison file between these two includes the word pairs (artificial, intelligence), (real, world), (world, situations), (computer, program), (best, possible), (decision, making), and (probability calculations). Once again we see that the article discussing bounded rationality has been linked to one discussing computer game strategy.

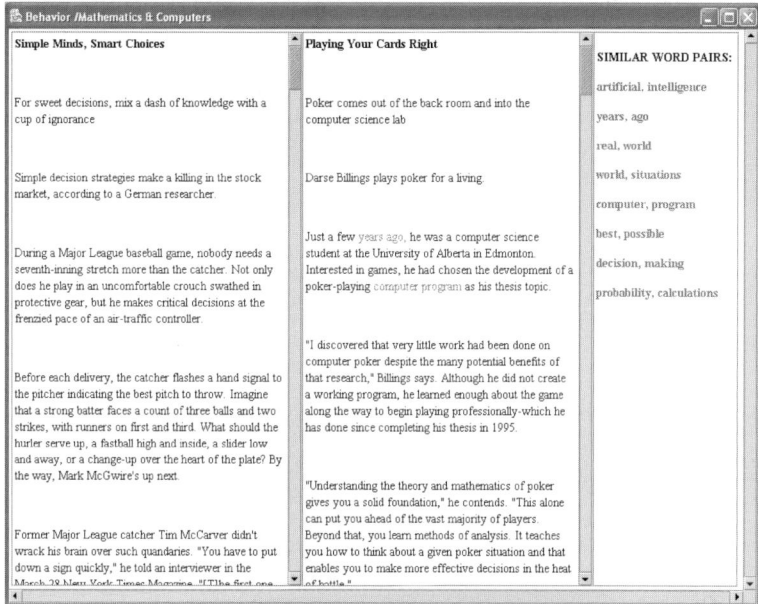

Fig. 30. A closer look at the between article associations between article 80 a Behavior article and article 133 a Mathematics and Computer Science article. For a color reproduction of this figure see the color figures section, page 584.

We may have saved some of the strongest evidence of the relevancy of our discovered association to the last. Dr. Stephen Lubard, the Director of the Office of Naval Research, made a presentation at the Dahlgren Division of the Naval Surface Warfare Center in May of 2004 (Lubard, 2004). During this briefing, he alluded to research underway aimed at applying the concepts of "bounded human rationality" to the Naval Command 21 Knowledge. We consider the fact that the Navy was already looking into the incorporation of bounded rationality into its information theoretic warfare strategies as the strongest endorsement of our auto serendipitous "discovery".

2.6. Clustering results on the ONR ILIR dataset

As indicated above, we postponed our full treatment of the clustering software until now. We have chosen to cut the minimal spanning tree into subtrees at a particular edge based on a criteria determined by the strength of the associated edge divided by the mean strength of the edges connected to the nodes of the edge in question and of a path length of no more than k edges. In terms of implementation details, the user first runs a JAVA program that creates the associated node lists based on a user specification of the maximum number of clusters, the minimum number of points per cluster, and the parameter k. Once this computation is performed, the user is presented with the following opening screen, see Figure 31. The user can select how many clusters to break the data into and these clusters are presented to the user in a sequence of tab panel windows.

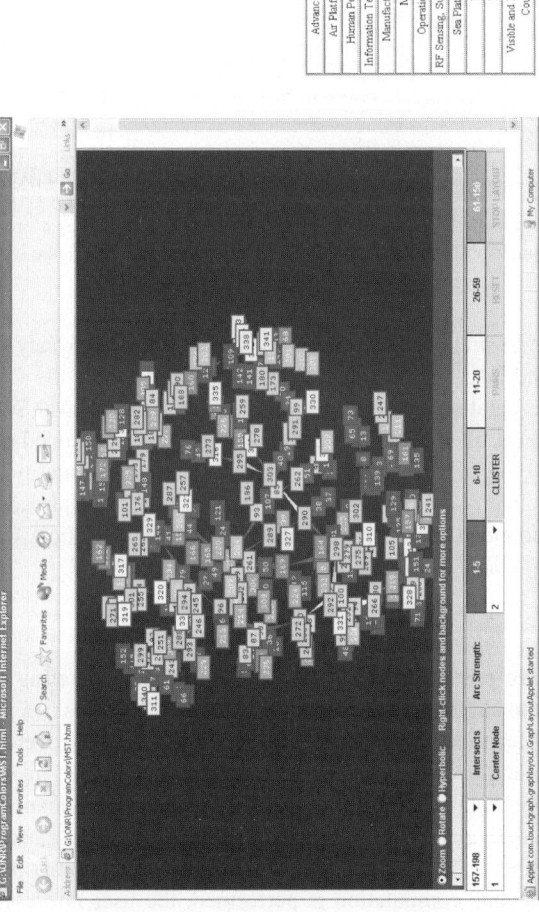

Fig. 31. The opening screen for the clustering program when loaded with the ILIR data and associated color key. For a color reproduction of this figure see the color figures section, page 585.

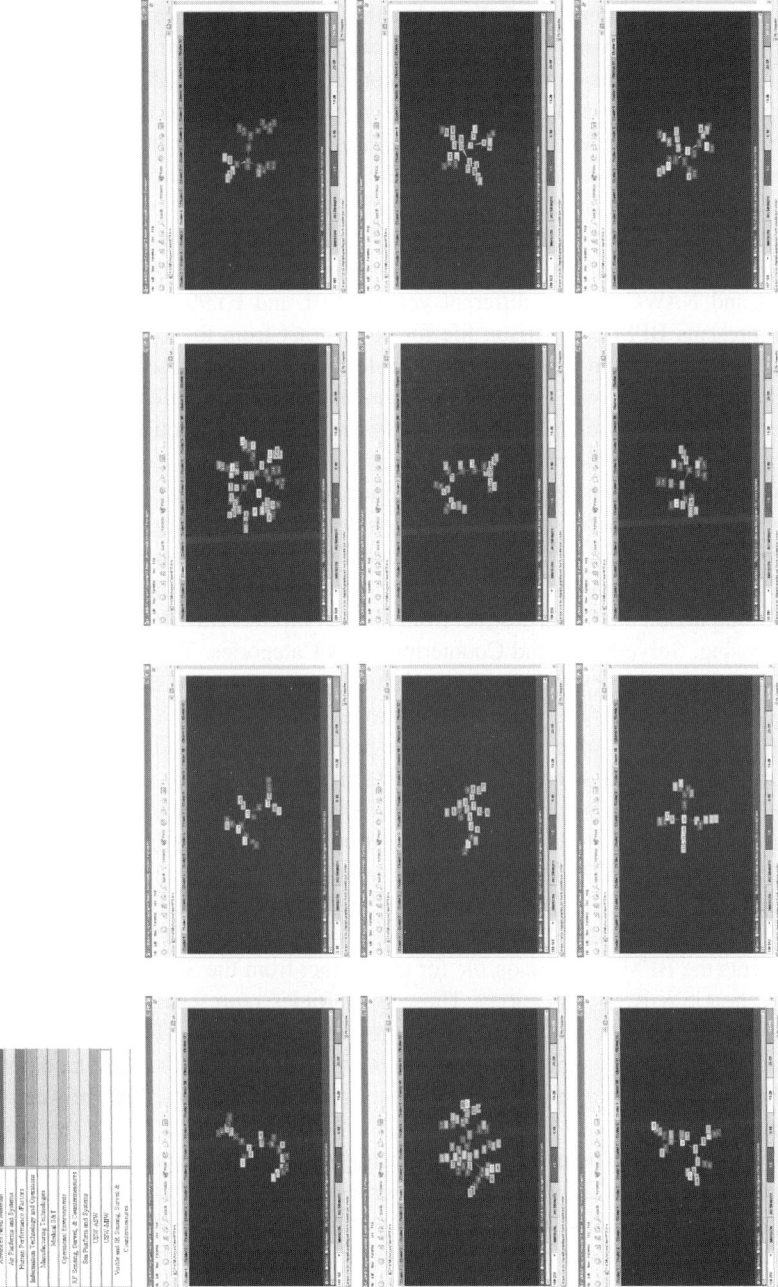

Fig. 32. Overview of the ILIR cluster structure. For a color reproduction of this figure see the color figures section, page 586.

In Figure 32, we present an overview of the ILIR cluster structure along with the associated color map. A quick visual scan of the clusters does not reveal much uniformity in color among the clusters. We believe that this is in part because $S(X, Y) = 0 \Rightarrow D(X, Y) = \sqrt{2}$ for many of node pairs. When we run our algorithm and ask for 8 cuts, these $\sqrt{2}$ numerator terms dominate our decision as to where we wish to make our cuts, and hence we effectively make 8 random cuts on the MST. It turns out that cutting the tree into 8 pieces does effectively help us to make a better exploration of the tree and we can identify a few other anomalies that were not evident in the tree originally.

Let us first explore cluster 2 a little, see Figure 33. The window on the left contains the BPM comparison file for two articles that contain virtually identical abstracts but yet represent projects that were executed by different investigators at different labs, NAVSTO and NAWC, during different years, FY01 and FY99/00. The window on the right presents BPM comparison file for an abstract from the USW-MIW and RF Sensing, Surveillance, and Countermeasures categories. Both of these efforts were conducted during the same year, FY01, at the same lab NUWC, by different investigators. They seem to be closely related technologically and there clearly could have been fruitful interactions between the two investigators.

Let us next explore cluster 3 a little, see Figure 34. The window on the left contains the BPM comparison file for two articles that contain virtually identical abstracts representing work being executed at the same organization, NAVSTO, during the same year, FY99/FY00, with different authors and abstract titles. The window on the right presents the BPM comparison file for an abstract from the Advanced Naval Materials and Visible and IR Sensing, Surveillance and Countermeasures Categories. The two articles seem to be a continuing project for the an investigator. It just so happens that the article title and our categorization of the article changed from FY99/FY00 to FY01.

There are at least two other cluster 3 article pairs that warrant additional discussion, see Figure 35. The window on the left presents an interesting association between an Advanced Naval Materials and Visible and IR Sensing, Surveillance and Countermeasures categories. These projects were executed by different investigators at different labs, NAWV and NAVSTO, during different years, FY99/FY00 and FY01. The two projects are on finite-difference time-domain methods. Once again we can make a strong case for successful collaborations among the researchers. The window on the right presents the BPM comparison file for an abstract from the Sea Platform and Systems and the USW-ASW categories. These two articles represent projects that were executed at different labs, NUWC and SSC, during the same year, FY01, by different investigators. They both seem to be about nonlinear interactions in the acoustic domain. The commonality of topic might once again suggest fruitful interactions between the investigators.

Next we would like to point out some interesting cluster 5 article pairs, see Figure 36. The articles described in the left window describe two efforts at NSWCDD that were executed during FY01. Both of these articles come form the Sea Platform and Systems category. Both of these articles are on the application of Fourier–Kochin theory. One can only hope that the two investigators were in contact with one another during the execution of their respective projects. The article pair detailed in the right window both come from the Operational Environment category. There are many interesting

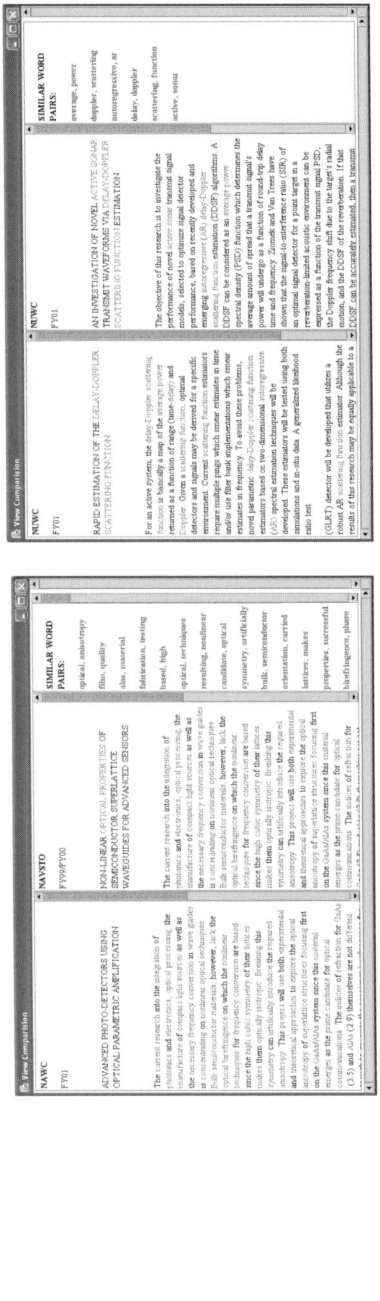

Text data mining with minimal spanning trees 161

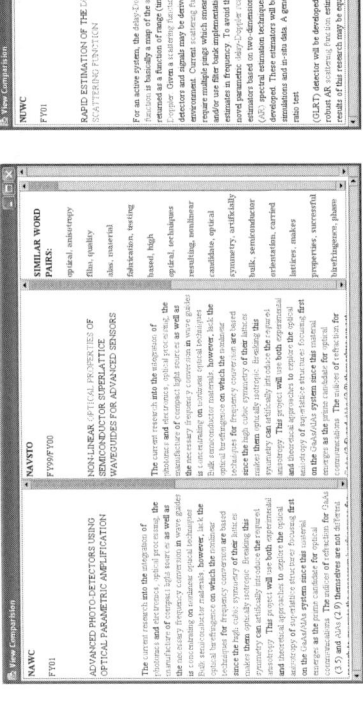

Fig. 33. Two interesting article pairs from cluster 2 of the ILIR dataset. For a color reproduction of this figure see the color figures section, page 587.

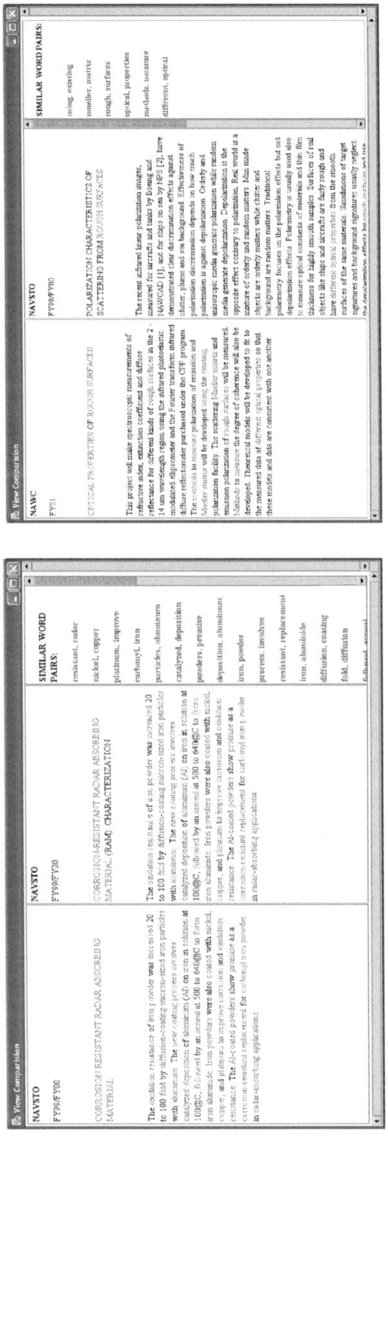

Fig. 34. Two interesting article pairs from cluster 3 of the ILIR dataset. For a color reproduction of this figure see the color figures section, page 587.

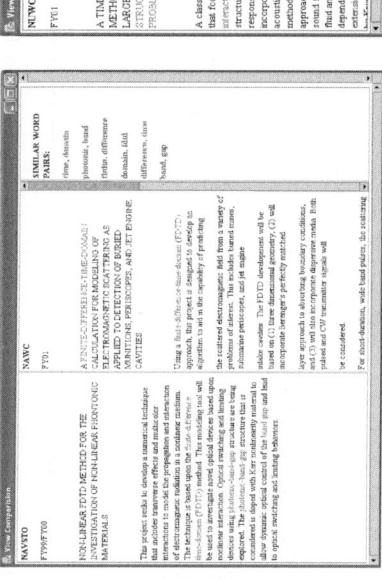

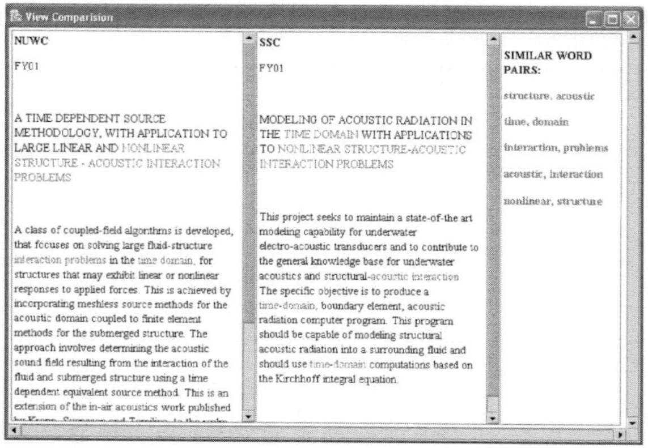

Fig. 35. Two interesting article pairs from cluster 3 of the ILIR dataset. For a color reproduction of this figure see the color figures section, page 588.

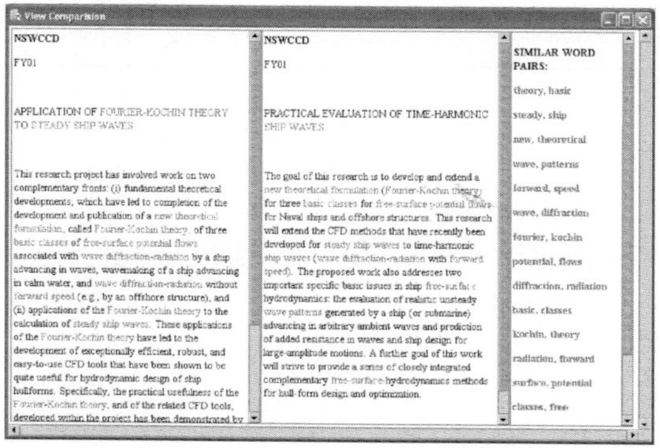

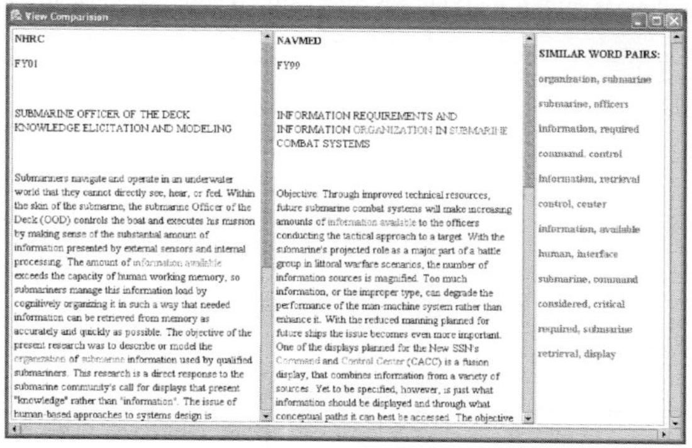

Fig. 36. Some interesting article pairs as revealed in cluster 5 of the ILIR dataset. For a color reproduction of this figure see the color figures section, page 588.

Text data mining with minimal spanning trees 163

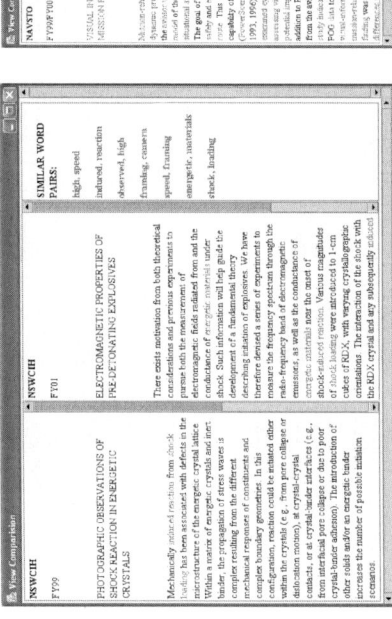

Fig. 37. Some interesting article pairs as revealed in clusters 7, left figure, and 12, right figure, of the ILIR dataset. For a color reproduction of this figure see the color figures section, page 589.

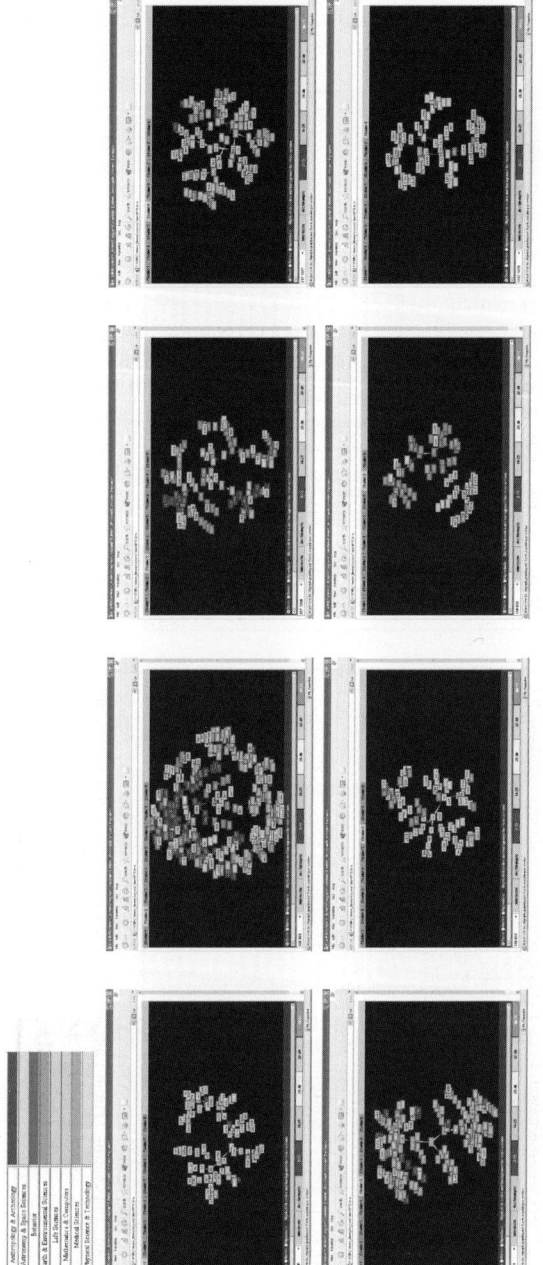

Fig. 38. Overview of the Science News cluster structure. For a color reproduction of this figure see the color figures section, page 590.

associations between these two articles. The first one is entitled "Submarine officer of the deck knowledge elicitation and modeling," while the second article is entitled "Information requirements and information organization in submarine combat systems." The executor of the first project hopefully was aware of the work that had been previously performed by the other investigator.

We conclude this section with one last set of article pairs, see Figure 37. The article pair in the left figure, from cluster 7, represents an interesting association between Advanced Naval Materials and Manufacturing Technologies articles. The word pairs, (high, speed), (induced, reaction), (observed, high), (framing, camera), (speed, framing), (energetic, materials), and (shock, loading), seem to indicate that both articles are studying fundamental properties of explosives. The article pair on the right, from cluster 12, presents a more interesting association. Both of the articles come from the Air Platforms and Systems category. Both of these projects were being executed during the same time frame, FY99/FY00 at the same lab, NAVSTO. It is surprising to see the very large number of associations between the two articles given the different titles. One would not have suspected such a strong association based on their titles, "Visual information requirements for mission rehearsal" and "Dynamic assessment of situation awareness, cognitive."

2.7. Clustering results on the Science News dataset

Next, we turn our attention to the results obtained on the Science News dataset using the clustering program, see Figure 38. There seems to be a little more uniformity of color in the various clusters, as compared to the ILIR dataset. Let us take a moment to explore some of the substructure of some of the clusters. Let us first consider an interesting substructure in cluster 1, see Figure 39. This subcluster consisting of nodes 429, 483, 451, 509, and 543 is focused on animal behavior and sexuality.

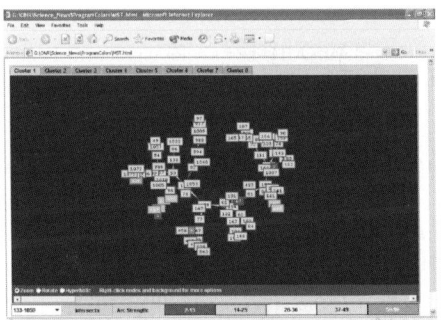
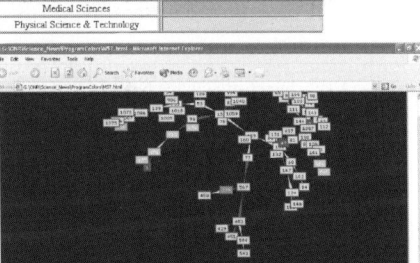

Fig. 39. A subcluster of Science News cluster 1 that is focused on animal behavior and sexuality. For a color reproduction of this figure see the color figures section, page 591.

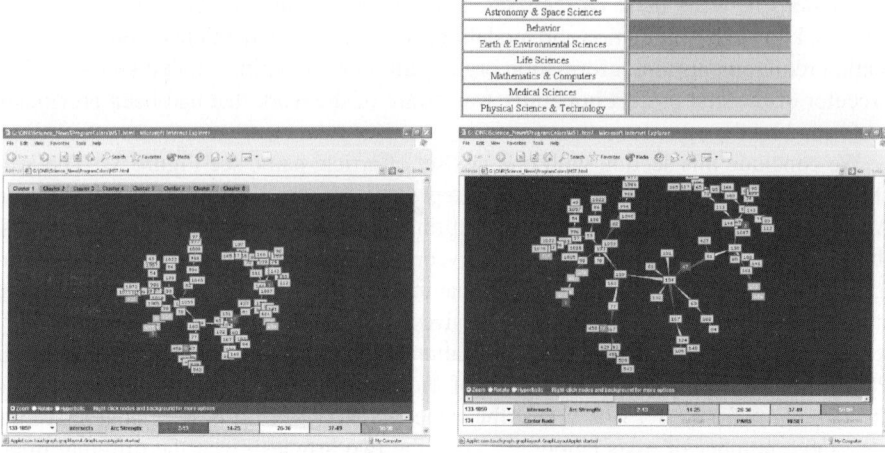

Fig. 40. A subcluster of Science News cluster 1 that is focused on an infrared camera and its applications. For a color reproduction of this figure see the color figures section, page 591.

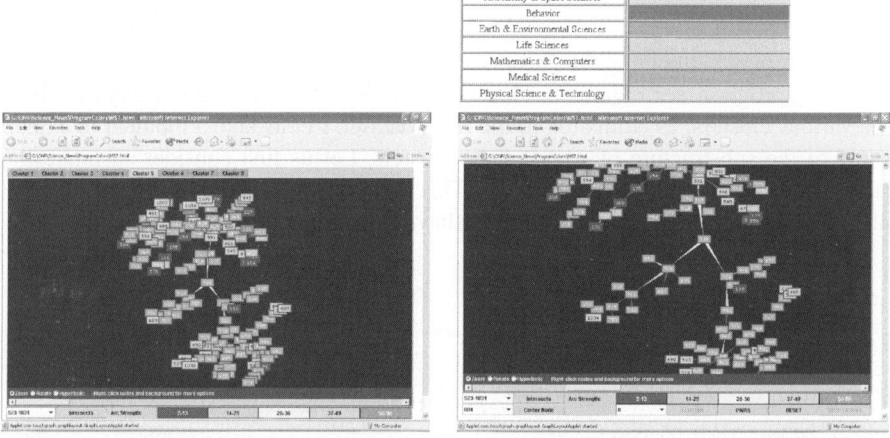

Fig. 41. A subcluster of Science News cluster 5 that is associated with AIDs. For a color reproduction of this figure see the color figures section, page 592.

Let us consider another interesting subcluster of cluster 1, see Figure 40. This subcluster consists of nodes 131, 47, 60, 102, 84, 167, 124, 105, 148, 132, 61. These nodes form the spokes of a wheel about the article 134, "Infrared camera goes the distance." All of these articles are about applications of infrared sensors to various problems in cosmology and other disciplines.

Let us next consider a subcluster of cluster 5, see Figure 41. This subcluster consists of nodes 933, 804, 855, 823, 911, 800, 710, 877, 755, 754, 956, and 1034 of articles about AIDs.

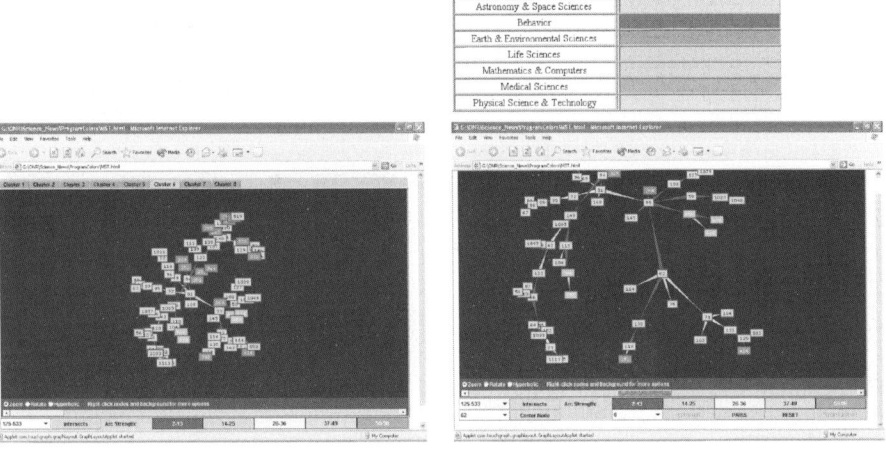

Fig. 42. A subcluster of Science News cluster 6 that is associated with solar activity. For a color reproduction of this figure see the color figures section, page 592.

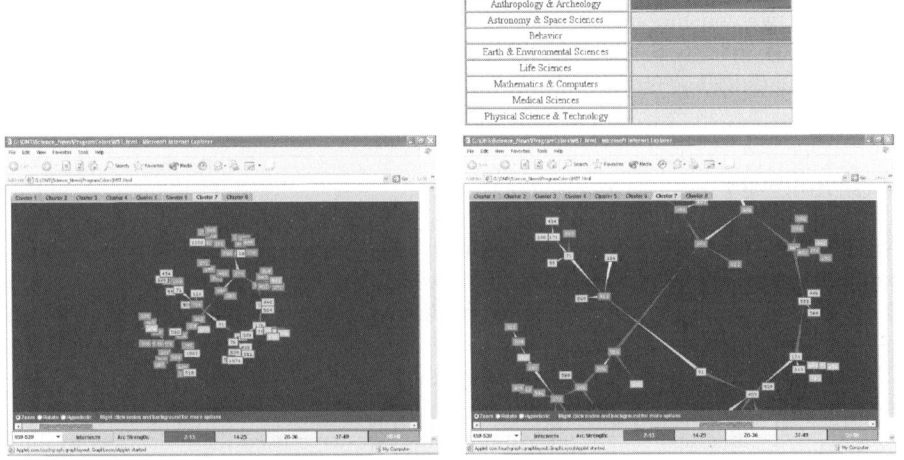

Fig. 43. A subcluster of Science News cluster 7 that is associated with the origins of life. For a color reproduction of this figure see the color figures section, page 593.

Next, we consider a subcluster of cluster 6, see Figure 42. This subcluster consists of nodes 62, 114, 130, 119, 332, 75, 73, 163, 121, 125, 414, 533, and 164 of articles about solar activity.

Next, we consider a subcluster of cluster 7, see Figure 43. This subcluster consists of nodes 314, 508, 55, 71, 136, 171, 454, 269, and 116 of articles on the origins of life. We note that the node placement has been rendered using a slightly different solution to the spring equations than was presented in the cluster overview in Figure 38.

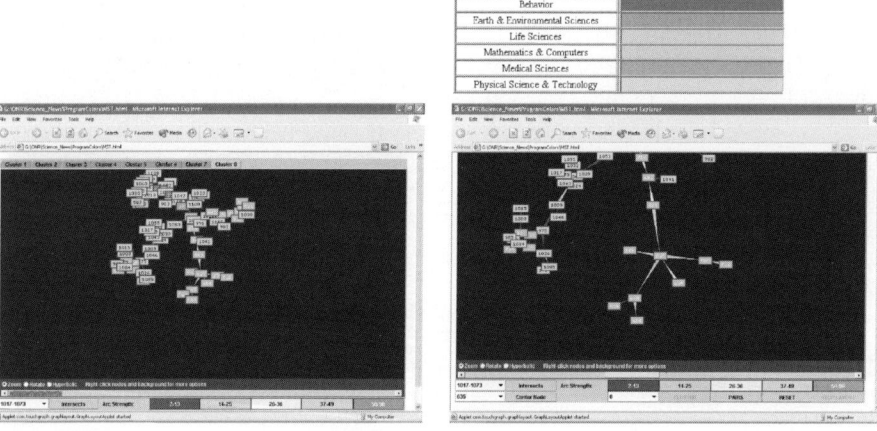

Fig. 44. A subcluster of Science News cluster 8 that is associated with artificial intelligence. For a color reproduction of this figure see the color figures section, page 593.

Finally, we consider a subcluster of cluster 8, see Figure 44. This subcluster consists of nodes 618, 635, 643, 655, 621, 614, 645, 654, and 604 of articles on artificial intelligence (gray nodes).

3. Conclusions

We have presented a new method for effective exploration of a labeled interpoint distance matrix. Our methodology supports the discovery of serendipitous relationships between the various categories encoded within the matrix. The methodology also supports cluster structure exploration.

We have illustrated the application of our developed techniques with a Science News extended abstract dataset and with a short ONR ILIR abstract dataset consisting of roughly 1/2 paragraph length project abstracts. We have used these datasets to illustrate both the automated serendipitous discovery and the cluster exploration. We have discussed many examples using both datasets. We have shown one example, that we feel is particularly relevant, that links the fields of human bounded rationality and computer artificial intelligence for game theory. We have provided evidence to support this particular linkage via current Naval investment strategies in the application of human bounded rationality to shipboard decision making.

This work represents the tip of the iceberg of a new area that is not only of strategic importance to the United States, but is also highly relevant to all who are currently conducting research in any discipline. The identification of effective research strategies is an area that is of paramount strategic importance as the number of possible beneficial connections that exists between various scientific disciplines increases and the amount of funding that can be allocated to these areas decreases.

Acknowledgements

This work was sponsored by the Defense Advanced Research Projects Agency under "Novel Mathematical and Computational Approaches to Exploitation of Massive, Nonphysical Data." Some of the authors (J.L.S. and A.C.B.) also acknowledge the support of the Office of Naval Research (ONR) under "In-House Laboratory Independent Research." The DARPA support of one of the authors (E.J.W.) was obtained via Agreement 8905-48174 with The Johns Hopkins University. His DARPA contract was administered by the AFOSR. The views and conclusions contained in this document are those of the authors and should not be interpreted as representing the official policies, either explicitly or implied, of DARPA, ONR, or the US Government. Approved for a public release, distribution unlimited.

References

Guo, D., Peuquet, D., Gahegan, M. (2002). Opening the black box: Interactive hierarchical clustering for multivariate spatial patterns. In: *Geographic Information Systems 2002*. ACM Press, pp. 131–136.

Guo, D., Peuquet, D., Gahegan, M. (2003). Iceage: Interactive clustering and exploration of large and high-dimensional geodata. *GeoInformatica* **7** (3), 229–253.

Lubard, S. (2004). Briefing to the Naval Surface Warfare Center Dahlgren Division.

Martinez, A.R. (2002). A framework for the representation of semantics. PhD dissertation. Computational Sciences and Informatics, George Mason University.

Solka, J.L., Johannsen, D.A. (2002). Classifier optimization via graph complexity measures. In: *Proc. of the Army Conference on Applied Statistics 2002*, pp. 130–143.

Xu, Y., Olman, V., Xu, D. (2001). Minimum spanning trees for gene expression clustering. *Genome Informatics* **12**, 24–33.

Xu, Y., Olman, V., Xu, D. (2002). Clustering gene expression data using a graph-theoretic approach: an application of minimum spanning trees. *Bioinformatics* **18** (4), 536–545.

Information Hiding: Steganography and Steganalysis

Zoran Duric, Michael Jacobs and Sushil Jajodia

Abstract

The goal of steganography is to insert a message into a carrier signal so that it cannot be detected by unintended recipients. Due to their widespread use and availability of bits that can be changed without perceptible damage of the original signal images, video, and audio are widespread carrier media. Steganalysis attempts to discover hidden signals in suspected carriers or at the least detect which media contain hidden signals. Therefore, an important consideration in steganography is how robust to detection is a particular technique. We review the existing steganography and steganalysis techniques and discuss their limitations and some possible research directions.

Keywords: Information hiding; Steganography; Steganalysis; Watermarking

1. Introduction

Information hiding comprises such diverse topics as steganography, anonymity, covert channels, and copyright marking (Bauer, 1997). Most of these topics have a long history (Kahn, 1996; Petitcolas et al., 1999) and have found their way into everyday life and the popular culture (Johnson et al., 2000). With the advent of the Internet and the widespread use of digital media these topics have become especially important to people interested in either communicating secretly or protecting their digital works from unauthorized copying. Steganography, anonymity, and covert channels all refer to secret communications. Anonymity refers to communicating in a way that protects the identity of the sender and/or the receiver (Petitcolas et al., 1999). Covert channels refer to the means of secretly communicating a small amount of information across tightly monitored channels by exploiting possible loopholes in the communication protocols (Simmons, 1998). Digital steganography refers to modifying a digital object (cover) to encode and conceal a sequence of bits (message) to facilitate covert communication. Digital steganalysis refers to efforts to detect (and possibly prevent) such communication. Copyright marking refers to modifying a digital object so that its owner(s) can be identified. The goal of digital watermarking is to modify a digital object in a way that will both identify the owner and be hard to remove without destroying the object or making it unusable.

Many types of digital objects can be considered (for information hiding) including text, arbitrary files, software, images, audio, and video files (Petitcolas et al., 1999; Katzenbeisser and Petitcolas, 2000; Johnson et al., 2000). In this survey we will focus on digital steganography of image files. There are several reasons for limiting survey of steganography to images only. Images are widespread on the Internet and can be used as carrier objects without raising much suspicion. They come in many formats, although JPEG and GIF are most dominant. (Conversely, audio files are usually in the MPEG3 format and videos are in either MPEG1 or MPEG2 formats.) Finally, most image files are quite large and have a lot of capacity for modification without noticeable damage to the image content. In digital watermarking, we will briefly discuss audio and video watermarking as well since there are many concerns about copyright protection of these object.

Most steganographic methods operate in two steps. First, a cover object is analyzed to determine to what extent it can be modified so that the modifications will not be easily observable. Second, the message bits are inserted into the cover object by replacing the selected cover bits by the message bits to create an altered cover object. In this chapter, cover object is an image in either bitmap or JPEG formats. The perceptually insignificant bits frequently correspond to the *least significant bits* (LSBs) in the image representation: in bitmap images these bits correspond to a subset of the LSBs of the image pixels, in JPEG images they correspond to a subset of LSBs of the JPEG coefficients.

The outline of this chapter is as follows. In Section 2 we describe the basics of image file formats as they relate to this chapter. In Section 3 we describe the principles of the existing image steganography. In Section 4 we describe some existing steganalysis methods. In Section 5 we briefly describe the relationship of steganography and watermarking. In Section 6 we review the existing work in this field. Finally, in Section 7 we make a brief summary of the chapter and discuss some open issues in steganography and steganalysis.

2. Image formats

An image is a two-dimensional array of image points or pixels. In gray level images, each pixel is described by one number corresponding to its brightness. In color images, each pixel is described by three numbers corresponding to the brightnesses of the three primary colors – e.g., red, green, and blue. A typical image file has two parts, the header and the raster data. The header contains "the magic number" identifying the format, the image dimensions, and other format-specific information that describes how the raster data relates to image points or pixels. The raster data is a sequence of numbers that contains specific information about colors and brightnesses of image points.

In a raw image format (e.g., BMP and PGM), the header contains the magic number and image dimensions. The raster data is a sequence of numbers corresponding to either one or three color values at each pixel (e.g., BMP and PGM). Raw (uncompressed) images are quite large and require a lot of storage space. Some space saving can be obtained by compressing the raster data by using loss-less compression (e.g., TIFF format). Frequently, raw images contain much more information than required. For example, in images saved for human viewing both the reduction of the number of possible colors and removal of some image details do not cause perceptible changes. In some formats (e.g., GIF), the image is saved using a reduced color set. Each pixel value is represented by a single number corresponding to an index in the color palette that is stored in the image header. The palette contains the information needed to restore colors of all image pixels. The image file can be made even smaller by using loss-less compression of the raster data – e.g., run-length encoding represents a sequence of several pixels of the same color by just two numbers, the length and the color of the sequence (run).

JPEG image format removes some image details to obtain considerable saving of storage space without much loss of image quality. The savings are based on the fact that humans are more sensitive to changes in lower spatial frequencies than in the higher ones. In addition, it is believed that humans perceive brightness more accurately than chromaticity. JPEG thus uses $YCrCb$ format to save brightness information (Y) in full resolution and chromaticity information (Cr and Cb) in half resolution. At the encoder side each channel is divided into 8×8 blocks and transformed using the two-dimensional discrete cosine transform (DCT). Let $f(i, j), i, j = 0, \ldots, N - 1$, be an $N \times N$ image block in any of the channels and let $F(u, v), u, v = 0, \ldots, N - 1$, be its DCT transform. The relationship between $f(i, j)$ and $F(u, v)$ is given by

$$F(u, v) = \frac{2}{N} C(u) C(v) \sum_{i=0}^{N-1} \sum_{j=0}^{N-1} f(i, j) \cos\left(\frac{\pi u(2i + 1)}{2N}\right)$$
$$\times \cos\left(\frac{\pi v(2j + 1)}{2N}\right), \quad (1)$$

$$f(i, j) = \frac{2}{N} \sum_{u=0}^{N-1} \sum_{v=0}^{N-1} C(u) C(v) F(u, v) \cos\left(\frac{\pi u(2i + 1)}{2N}\right)$$
$$\times \cos\left(\frac{\pi v(2j + 1)}{2N}\right) \quad (2)$$

where $C(u) = 1/\sqrt{2}$ for $u = 0$ and $C(u) = 1$ otherwise. A DCT of an 8×8 block of integers is an 8×8 block of real numbers. The coefficient $F(0, 0)$ is the DC coefficient and all others are called the AC coefficients. JPEG divides the coefficients by values from a quantization table to replace the real number values by integers (see Table 1). It is expected that many coefficients for higher values of $u + v$ become zero and that only a fraction of all coefficients will remain nonzero. The coefficients are reordered into a linear array by placing higher frequency coefficients (higher values of $u + v$) at the end of the array; those coefficients are most likely to be zeroes. Huffman coding is applied to all coefficients from all blocks in the image; zero-valued coefficients are encoded separately using special markers and their count for additional saving. A header

Table 1
Default JPEG quantization table

(u, v)	0	1	2	3	4	5	6	7
0	16	11	10	16	24	40	51	61
1	12	12	14	19	26	58	60	55
2	14	13	16	24	40	57	69	56
3	14	17	22	29	51	87	80	62
4	18	22	37	56	68	109	103	77
5	24	35	55	64	81	104	113	92
6	49	64	78	87	103	121	120	101
7	72	92	95	98	112	100	103	99

The coefficients are divided by their corresponding values and then rounded to the nearest integer.

consists of image type and dimensions, compression parameters, and the quantization table. It is combined with the Huffman encoded coefficients packed as a sequence of bits to form a JPEG encoded image. On the decoder side the integer valued coefficients are restored by Huffman decoding. The quantization is reversed and the inverse DCT (see Eq. (2)) is applied to obtain the image. Huffman coding is loss-less; the losses occur in quantization process.

3. Steganography

A typical steganographic embedding method has two parts. In the first part, a message is prepared and in the second it is embedded in the carrier object – e.g., an image. The preparation can include compression and encryption. It can be assumed that the sender and the intended recipient share a key that is used to encrypt the message. Either compression or the encryption is expected to produce a random-looking sequence of bits. In addition, since the message embedding needs to be reversible a small amount of formatting data such as a password and a message length needs to be included with the message. The formatting information and the message are concatenated and embedded into the cover using the same algorithm. Therefore, it will be assumed here that a message is a random bit sequence and only the message embedding and extraction processes will be considered.

Similar to image compression, message embedding can rely on two different approaches. Loss-less embedding process modifies an image file in a fully reversible way. For example, in a PGM file format the header can include comment lines. Additional comment lines could be added to embed any desired information. Alternatively, the image file could be enlarged by adding extra rows or columns and the information about the image modification could be added to the header. One example would be modifying run length encoding process to embed messages. During the encoding process the method checks all run lengths longer than one pixel. Suppose that a run length of ten pixels is considered and that one bit needs to be embedded. To embed a bit one, the run length is split into two parts whose lengths add to ten, say nine and one; to embed a bit zero the run length is left unmodified. The receivers check all run lengths. Two run lengths of the same color are decoded as a one; otherwise a run length longer than one pixel, preceded and followed by run lengths of different colors, is decoded as a zero. Clearly, this technique relies on obscurity since detecting a file with information embedded by this technique is not hard. In the remainder of this chapter, the focus will be on the lossy embedding techniques.

Lossy steganography modifies the raster data of the original image file to embed a message. Steganography relies on the same basic idea that underlies lossy image compression – i.e., changing image bits that are not perceptually significant usually does not affect image quality. For example, in an uncompressed gray-level image, each pixel is represented by eight bits. A message can be embedded into an image – i.e., cover – by modifying the least significant bits of the pixel values to correspond to the message. An image of size 410×614 has 251 740 pixels (see upper row of Figure 1). If all LSBs are used it is possible to embed 94 402 bytes into this image; this corresponds to a high-quality JPEG image of the same size (see lower row of Figure 1). Insertion into the LSBs of a typical color image cannot be observed by a naked eye. If the message and the cover are uncorrelated, approximately 50% of the cover bits need to be changed to embed a message. If the number of message bits is larger than the number of pixels the message could be embedded using two or more LSBs per pixel. Conversely, if a message is shorter than the number of available cover LSBs it could be embedded into a fraction of the cover image – i.e., some cover pixels do not need to be considered.

3.1. Embedding by modifying carrier bits

Steganographic methods use various approaches to embed messages into covers. First approach identifies the carrier bits – i.e., the bits that will encode a message – and modifies them to encode the message. These carrier bits could be one or more LSBs of selected bytes of raster data – the selection process itself can use a key to select these bytes in pseudo-random order. Also, the raster data can be either raw image bytes (brightnesses and colors), or JPEG coefficients. Embedding is done by modifying the carrier bits suitably to encode the message. The message can be decoded from the carrier bits only – i.e., the receiver identifies the carrier bits and extracts the message using the key and the algorithm. These techniques can be compared using the following criteria (Westfeld, 2001):

1. The *embedding rate* – the number of embedded bits per a carrier bit.
2. The *embedding efficiency* – the expected number of embedded message bits per modified carrier bit.
3. The *change rate* – the average percentage of modified carrier bits.

Several techniques are used to embed messages into carrier bits. Three techniques will be described here:

1. Changing the carrier bits to match the message bits.
2. Using bit parity of bit blocks to encode message bits (Anderson and Petitcolas, 1998).
3. Matrix encoding of message bits into carrier bits (Westfeld, 2001).

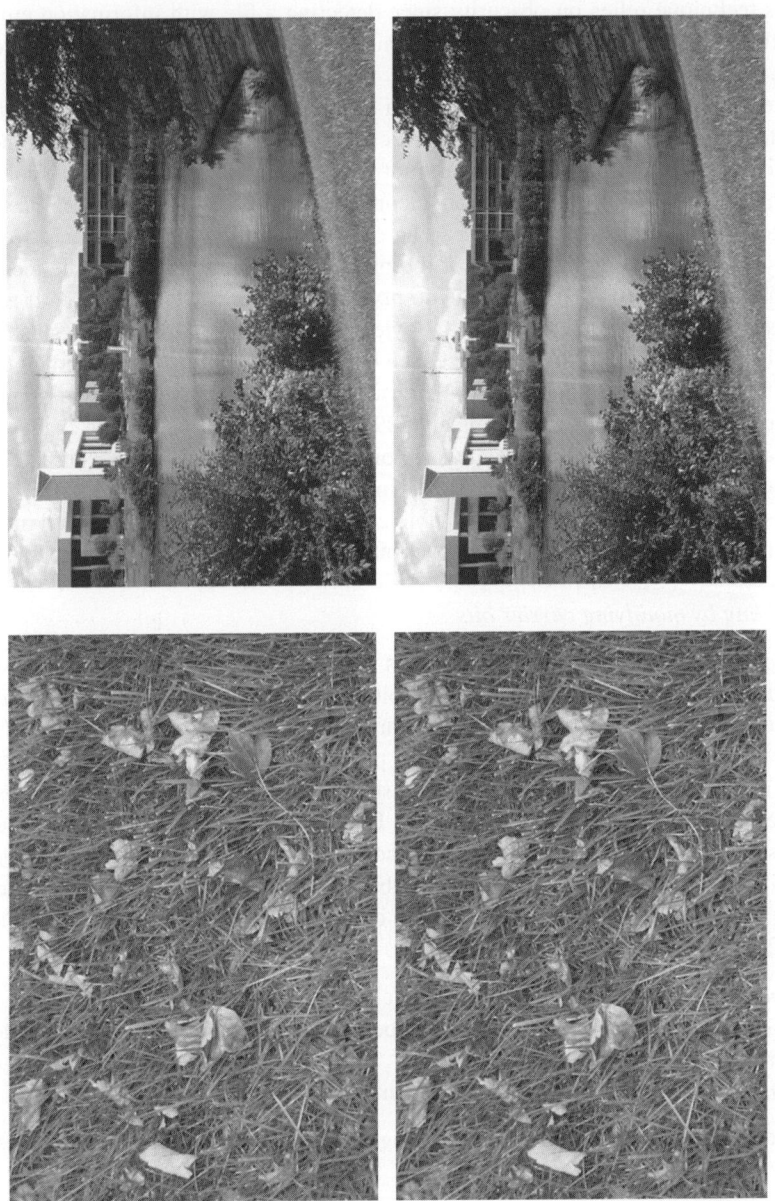

Fig. 1. Uncompressed color images (upper row) were saved as JPEG images at 80% quality (lower row). The JPEG images were inserted into the LSBs of the uncompressed images. Differences between uncompressed original images cannot be observed by a naked eye. For a color reproduction of this figure see the color figures section, page 594.

First, the embedding algorithm can compare the carrier bits and the message bits and change the carrier bits to match the message. Bit modification can be done by either bit flipping – i.e., 0 → 1 or 1 → 0 – or by subtracting 1 from the byte value. For example, let the raster data bytes be

$$01000111 \quad 00111010 \quad 10011000 \quad 10101001,$$

where the identified carrier bits are underlined. Using flipping to embed the message bits 0010 into the carrier bytes produces

$$01000110 \quad 00111010 \quad 10011001 \quad 10101000,$$

where the modified bits are underlined. Using subtraction to embed the same message bits (0010) into the carrier bytes produces

$$01000110 \quad 00111010 \quad 10010111 \quad 10101000,$$

where the modified bits are underlined. Clearly subtraction produces more bit modifications, but the perceptual changes would be about the same as in the case of bit flipping. This technique has been used by various steganographic algorithms to embed messages in raw image data (gray and color images) and JPEG coefficients. Clearly, the embedding rate is 1, the embedding efficiency is 2, since about 50% of carrier bits get modified, and the change rate is 50%.

An embedding method can consider blocks of carrier and message bits at a time to embed a message into a cover. Examples of these techniques include using bit parity and the matrix encoding. The bit parity approach can be used when the number of available carrier bits is at least $n \geq 1$ times larger than the number of message bits. Block of n carrier bits are considered and their parity compared to the corresponding message bits. If the parity matches the message bit nothing is done, otherwise any of the n bits in the current block can be modified to make the parity and the message bit match. In this approach, the embedding rate is $1/n$, the embedding efficiency is 2, and the change rate is $50\%/n$.

The matrix encoding embeds k message bits using n cover bits, where $n = 2^k - 1$. The method embeds a k-bit code word $\mathbf{x}$ into an n-bit cover block $\mathbf{a}$. Let the bits of $\mathbf{x}$ be x_i, $i = 1, \ldots, k$, and let the bits of $\mathbf{a}$ be a_j, $j = 1, \ldots, n$. Let f be defined as *xor* of carrier bit indexes weighted by the bit values, i.e.,

$$f(\mathbf{a}) = \bigoplus_{j=1}^{n} a_j \cdot j$$

and let

$$s = \mathbf{x} \oplus f(\mathbf{a}).$$

A modified cover block $\mathbf{a}'$ is then computed as

$$\mathbf{a}' = \begin{cases} \mathbf{a}, & s = 0 \ (\Leftrightarrow \mathbf{x} = f(\mathbf{a})), \\ a_1 a_2 \ldots \neg a_s \ldots a_n, & s \neq 0. \end{cases}$$

On the decoder side a k-bit message block $\mathbf{x}$ is obtained from an n-bit carrier block $\mathbf{a}$ by computing $\mathbf{x} = f(\mathbf{a})$. As an example let $\mathbf{x} = 101$ and let $\mathbf{a} = 1001101$. Therefore,

$$f(1001101) = 001 \oplus 100 \oplus 101 \oplus 111 = 111$$
$$\rightarrow s = 101 \oplus 111 = 010 \rightarrow \mathbf{a}' = 1101101,$$

i.e., the second bit was flipped to obtain $f(\mathbf{a}') = f(1101101) = 101$. The embedding rate of matrix encoding is $k/n \equiv k/(2^k - 1)$; the embedding efficiency is $k2^k/(2^k - 1)$; and the change rate is $1/(n + 1) \equiv 2^{-k}$ (for any $(n = 2^k - 1)$-bit carrier block there are n matched k-bit code words and one that is mismatched). Note that these numbers can change somewhat when JPEG coefficients are used to embed messages (see Westfeld (2001) for details).

3.2. Embedding using pairs of values

Second approach utilizes perceptually similar *pairs of values* (PoVs) in raster data and modifies them to embed steganographic data. The PoVs are divided into *even* and *odd* elements. Embedding is done by modifying selected raster data to match the message. There are four cases.

1. The raster symbol is an *even* element (s_0) of some PoV (s_0, s_1) and the message bit is 0: leave s_0 unchanged.
2. The raster symbol is an *even* element (s_0) of some PoV (s_0, s_1) and the message bit is 1: replace s_0 by s_1.
3. The raster symbol is an *odd* element (s_1) of some PoV (s_0, s_1) and the message bit is 0: replace s_1 by s_0.
4. The raster symbol is an *odd* element (s_1) of some PoV (s_0, s_1) and the message bit is 1: leave s_1 unchanged.

If the message bits and raster data are uncorrelated, and the proportion of ones and zeros in the message is equal approximately half of the raster data need to be modified to embed a message. On the receiver (decoder) side the raster data are examined. Each raster symbol is interpreted as either even or odd element of some PoV. Even elements are decoded as zeros, odd elements are decoded as ones.

An example of a steganographic technique that uses PoVs to embed messages is *EzStego* (by Romana Machado) used in reduced-color-set images. In a full-color-set (RGB) image, each color is represented by three values corresponding to the *red*, *green*, and *blue* intensities. In a reduced color set, the colors are sorted in lexicographic order; the sorted list of colors is called a *palette*. The palette is stored in the image header and the raster data are formed by replacing the colors by the corresponding indexes in the palette. If the palette has less than 256 colors the three-bytes per pixel full-color image can be represented using just one byte per pixel. To recover actual colors, both the raster data and the palette are needed. Each raster data value is replaced by the corresponding color. Note that colors that are neighbors in the palette, and therefore are assigned indexes that differ by one, can correspond to colors that look very different. For example, it is possible that the palette colors with indexes 0, 100, and 101 correspond

1. $D \leftarrow \{c_0\}$, $C \leftarrow C_{\text{old}} \backslash \{c_0\}$, $c \leftarrow c_0$
2. Find color $d \in C$ that is most distant from c
3. $D \leftarrow \{c_0, d\} \equiv \{d_0, d_1\}$, $C \leftarrow C \backslash \{d\}$
4. *while* $C \neq \emptyset$ *do*
5.   Find color $d \in C$ that is most distant from c
6.   Find two colors $\{d_i, d_{i+1}\} \in D$ so that $\delta\{d_i, d\} + \delta\{d, d_{i+1}\}$ is minimal
7. $D \leftarrow \{d_0, \ldots, d_i, d, d_{i+1}, \ldots\}$, $C \leftarrow C \backslash d$, $c \leftarrow d$
8. *endwhile*

Fig. 2. Sorting a color palette.

to RGB colors $(5, 5, 5)$, $(255, 5, 0)$, and $(10, 10, 10)$, respectively. Thus, flipping a bit and changing a color from, say, 100 to 101 could create a visible artifact in the image.

In EzStego this problem is solved by re-sorting the palette and grouping the new indexes pairwise so that they can be used as PoVs. Let the original palette be $C_{\text{old}} = \{c_i, i = 0, \ldots, n-1\}$, let $I(c_i, C_{\text{old}}) \equiv i$ be index of c_i in C_{old} and let $\delta(a, b)$ be the distance between colors a and b. Sorting is done using the algorithm in Figure 2. Note that this algorithm finds an approximation to the Traveling Salesperson problem in the color palette C_{old}, where colors correspond to cities. The PoVs that the algorithm uses correspond to the indexes of sorted colors in the original palette – i.e., the PoVs are $(I(d_{2k}, C_{\text{old}}), I(d_{2k+1}, C_{\text{old}}))$, $k = 0, \ldots, n/2$, where $I(d_i, C_{\text{old}})$ is the index of color d_i in C_{old} and $D = \{d_0, d_1, \ldots, d_{n-1}\}$ is the sorted palette.

4. Steganalysis

A goal of digital steganalysis is detecting image files that contain hidden information and if possible extracting that information. There are several levels of difficulty in this problem. In the easiest case, both the cover (clean) and the suspect image are given and a goal is to determine if they are different. Simple image processing, including image subtraction, can be used to accomplish this task. In a more difficult case, only the suspect image is known, but both the encoding/embedding and the message decoding/extraction algorithms are known and available. The information is protected by a secret key (password) and the information can be extracted if the key can be guessed (e.g., see Wayner (2002)).

In a more difficult case, given a suspect image, a goal is to determine whether the image contains an embedded message. The result is given as *certainty/probability* that the image contains hidden information. The *steganalyst* – i.e., the agent performing steganalysis – has access to images that have similar statistics as the suspect image. Westfeld and Pfitzmann (1999) developed a simple but effective method for detecting steganographic content in images. The method uses a statistics that is derived from relative frequencies of pairs of values (PoV). It is described here for LSB steganographic insertion in JPEG image files. It was originally developed to detect messages inserted using the JPEG/Jsteg method.

The method begins with a histogram of JPEG coefficients that is created as they are output by the decoder. Let the JPEG coefficients be drawn from the set A. Even though $0 \in A$ and $1 \in A$, JPEG/Jsteg embeds messages by modifying LSBs of all coefficients, except for 0 and 1; the other methods proposed in this paper will also not use 0 and 1 for information hiding. A PoV is the pair of coefficients that have identical binary representations except for their LSBs which are 0 and 1, respectively. For a binary number (bit string) a, we denote the corresponding PoV with a' and a'', where $a' = 2a$ and $a'' = 2a+1$. For example, 118 and 119 are a PoV since their binary representations are (1110110, 1110111), when $a = 59$ or 111011 in binary.

Let $A \setminus \{0, 1\} = \{a'_1, a''_1, a'_2, a''_2, a'_3, a''_3, \ldots, a'_k, a''_k\}$ be organized into k PoV. At any step as the decoder processes the coefficients, a histogram bin for $x \in A$ contains the number of coefficients x output so far. In particular, h'_i is the number of occurrences of a'_i and h''_i is the number of occurrences of a''_i.

In an unaltered image, it could be expected that the frequencies of PoVs significantly differ. In the portion of an image altered by JPEG/Jsteg, these frequencies only depend on the relative frequencies of zeroes and ones in the hidden message. Therefore, if the frequencies of zeroes and ones in the message are approximately equal it could be expected that the PoVs have similar frequencies. The χ^2 statistic with k degrees of freedom is given by

$$\chi_k^2 = \frac{1}{2} \sum_{i=0}^{k} \frac{(h'_i - h''_i)^2}{h'_i + h''_i}. \tag{3}$$

The probability that the image contains steganographic content is estimated by integrating the density function

$$p = 1 - \frac{1}{2^{k/2}\Gamma(k/2)} \int_0^{\chi_k^2} e^{-x/2} x^{k/2} \, dx. \tag{4}$$

This method is illustrated on two images in Figure 3 for two cases. In the first case, a random bit sequence of the same length as the number of available bits (number of coefficients different than 0 and 1) was inserted in each of the images. In the second case, only 50% of the coefficients were used for insertion – i.e., every other coefficient was skipped. Probabilities computed using Eq. (4) are shown in Figure 4. It can be seen that in the full capacity insertion case the probability was almost always very close to 1. However, in the half capacity case (shown by the dashed line and marked by 1/2) the probability becomes 0 after more than 4000 coefficients are used for the computation. When the density of insertion is 1/4 or less probabilities computed using Eq. (4) become 0 everywhere. Based on these and other tests, it can be concluded that the method is quite effective for detecting steganographic content when all of the redundant bits are used. Similar (high detection) results can be obtained for short messages if they are inserted into available bits (coefficients) at the beginning of the image file.

When a shorter message is spread over all available bits this detection method (Westfeld and Pfitzmann, 1999) becomes much less effective. However, it is not true that χ^2 statistics are not useful for detection in this situation. If χ^2 statistic is computed using Eq. (3) it can be seen that insertion of random bit sequences lowers χ^2 considerably. Figure 5 shows χ^2 curves of the original (unaltered) images in Figure 3,

Fig. 3. Two natural images: a falcon (top) and a barley field (bottom). Image sizes are 450 × 292 (falcon) and 480 × 320 (barley). For a color reproduction of this figure see the color figures section, page 595.

as well as χ^2 curves of the altered images with insertion frequencies 1, 1/2, 1/4, and 1/8. Note that bit insertion does not change the number of degrees of freedom for χ^2 statistic. Since any bit insertion generally lowers χ^2 statistic, a new method for detecting steganographic content based on a direct use of the statistic could possibly be designed.

5. Relationship of steganography to watermarking

The goal of digital watermarking is to modify a digital object in a way that will both identify the owner and be hard to remove without destroying the object or making it unusable. In addition, it is desirable that digital watermarks survive normal image processing and manipulation including resizing, printing and scanning, and some compression, e.g., high-quality JPEG (Cox et al., 2001). Note that steganography has different goals – i.e., hide the existence of message – and that information hidden using approaches described in Section 3 would not survive those kinds of processing (Johnson et al., 2000).

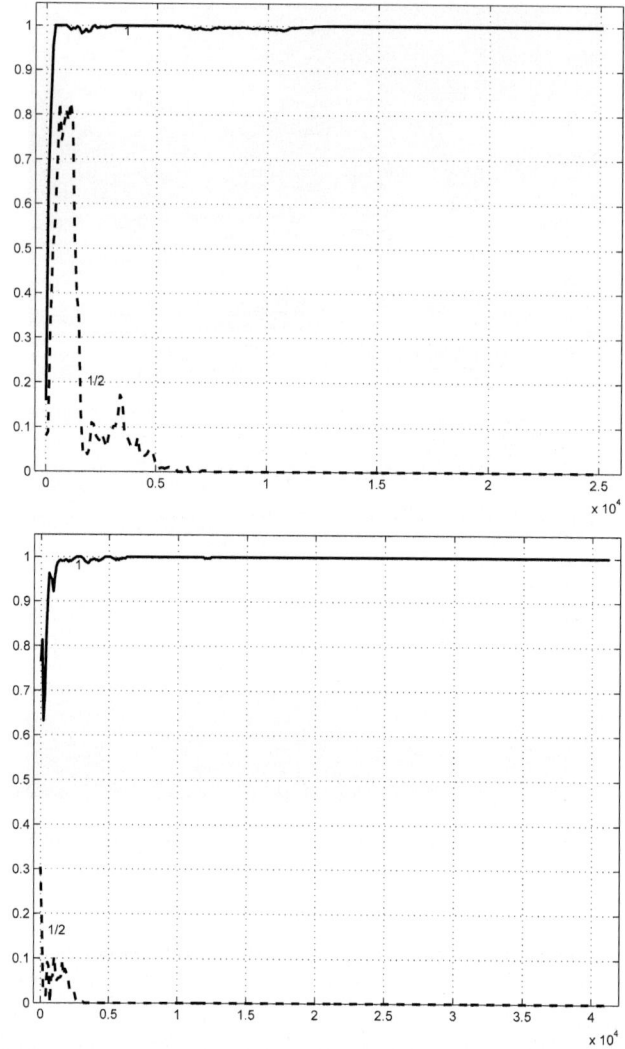

Fig. 4. Probabilities of detection of steganographic content for the images in Figure 3. Top: falcon, bottom: barley field. Graphs show the probabilities computed for full capacity insertion (1: full line) and a half capacity insertion (1/2: dashed line).

Another very important difference is in the amount of information that is transferred in steganography and watermarking. While a goal of steganography is to transfer messages usually many bits long, a digital watermark transfers one bit of information only. Namely, the detector checks an image and reports the degree of certainty that it contains a particular watermark. Nevertheless, a watermarked image is typically more significantly modified by the watermarking algorithm than is desirable in steganography. The main reason is that a key parameter used to determine how strong a watermark can

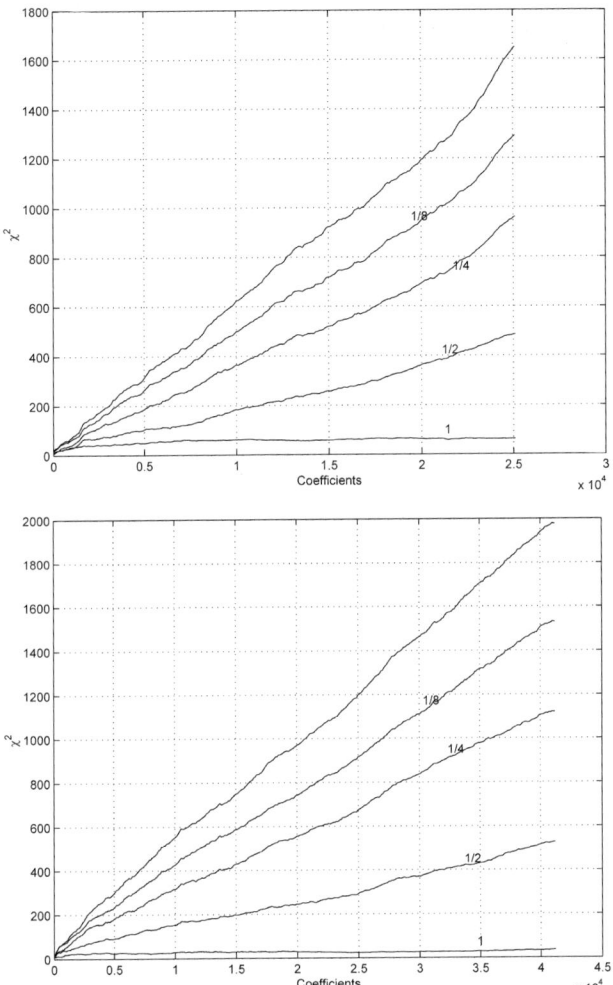

Fig. 5. χ^2 statistic computed for the images in Figure 3. Number of skipped bits is varied. The topmost curves correspond to the original (unaltered) images, the curves marked by 1, 1/2, 1/4, and 1/8 correspond to the relative insertion frequencies, e.g. 1/4 corresponds to changing 1 out of every 4 bits. The "falcon" image (top) has 25 179 coefficients available for embedding and the numbers of altered coefficients in the four cases are: 12 606, 6279, 3118, and 1569. The "barley" image (bottom) has 41 224 coefficients available for embedding and the numbers of altered coefficients in the four cases are: 20 544, 10 256, 5099, and 2545.

be is the *Just Noticeable Difference* (JND) – i.e., a watermark strength has to be such that the difference between a watermarked image and its unwatermarked version is not noticeable by human viewing. To determine the JND, mathematical models of human perceptual system are employed.

In both steganography and watermarking, a key can be used to determine which parts of an image will be used for embedding. However, a major difference is that in watermarking the key defines the watermark. For example, in the Patchwork algorithm (Gruhl and Bender, 1998) n pixel pairs are chosen according to a key K_p. Let the pixel values be (a_i, b_i), $i = 1, \ldots, n$. For randomly chosen pixels, it is expected that

$$\sum_{i=1}^{n}(a_i - b_i) \approx 0.$$

The watermark is embedded by replacing pixel values a_i by $a_i + d$ and b_i by $b_i - d$ in the image data; $d \geq 1$ is chosen to be small. To check if an image contains a watermark embedded using key K_p, the key is used to select n pixel pairs (a_i, b_i), $i = 1, \ldots, n$. A watermark is present if

$$\sum_{i=1}^{n}(a_i - b_i) \approx 2nd.$$

Other watermarking algorithms generate random sequences of bits that are inserted into, say, medium frequency DCT coefficients of an image (Cox et al., 2001). Watermarks are detected by computing cross-correlation between the watermark and the coefficients. Since the watermark is random and uncorrelated with the coefficients the result is significantly different from zero only if the watermark is present.

6. Literature survey

Digital steganography is a relatively new research field (Anderson and Petitcolas, 1998; Johnson and Jajodia, 1998a; Katzenbeisser and Petitcolas, 2000). The term steganalysis was introduced by Johnson and Jajodia (1998b) who discussed various ways that anomalies introduced by some steganographic techniques could be exploited to detect the presence of steganographic data in images. Detailed survey of early algorithms and software for steganography and steganalysis can be found in (Katzenbeisser and Petitcolas, 2000; Johnson et al., 2000; Wayner, 2002).

Maes (1998) demonstrated that artifacts produced by steganographic software can sometimes be discovered by simple histogram analysis of images filtered by a Laplacian operator. The first quantitative technique for steganalysis was designed by Westfeld and Pfitzmann (1999). They exploited the fact that many steganographic techniques change the frequencies of pairs of values (pairs of colors, gray levels, or JPEG coefficients) during a message embedding process. Their method was shown to be effective in detecting messages hidden by several steganographic techniques. This research prompted interest in both improving statistical detection techniques (Dumitrescu et al., 2002; Farid, 2002; Lyu and Farid, 2002; Fridrich et al. 2000, 2001a, 2001b, 2002; Westfeld, 2002) as well as building new steganographic methods that would be difficult to detect by statistical methods (Frantz, 2002; Newman et al., 2002; Provos, 2001; Westfeld, 2001).

Various attempts have been made to make steganograms difficult to detect including reducing their capacity or payload and spreading the message across the whole carrier. Anderson and Petitcolas (1998) suggested using the parity of bit groups to encode zeroes and ones; large groups of pixels could be used to encode a single bit, the bits that need to be changed could be chosen in a way that would make detection hard. Others have suggested matching the cover and the message (Crandall, 1998; Provos, 2001). From a set of possible covers, the one that most closely matches the message would be used. Provos (2001) designed a steganographic method *OutGuess* that spreads a message over a JPEG file; the unused coefficients are adjusted to make the coefficient histogram of the modified file as similar as possible to the histogram of the original non-stego image. This limits the bandwidth of the method since at any frequency at least 50% of all coefficients must remain unchanged for the adjustment step. Provos also experimented with matching the message and the cover as was first suggested by Crandall (1998); however, although he compared his messages with some thousands of images he reported that in the best case the best match between the cover and the message was only slightly better than 50% (chance), which is corroborated by results on string matching (Atallah et al., 1993). Westfeld (2001) designed a steganographic algorithm $F5$ that uses matrix coding to minimize the modifications of the LSBs. His method first discovers the number of available bits and then spreads the message bits over the whole file. To reduce detectability, $F5$ decrements the coefficient value to change bits unlike Jsteg (Upham, 2004) and OutGuess (Provos, 2001) that use bit complement to change bits.

Farid (2002; Lyu and Farid, 2002) successfully used statistical learning techniques to train classifiers to detect images containing hidden messages. Higher-order statistics of the wavelet coefficients of images were used as features; each image was represented by several hundred features. Experiments were performed on 10 000 natural images; several steganographic methods were tested. Training was done using all original images and a small subset of altered images; testing was done on the remaining altered images. It is not clear if this technique could be extended to work on unseen images, i.e. when the known clean images are not available.

Fridrich et al. (2001a) developed a method for detecting changes in uncompressed images previously stored in JPEG format. Very small changes on uncompressed images could be detected using this method. Recently Newman et al. (2002) published a steganographic method that preserved JPEG compatibility. The method stored at most one bit of information per an 8×8 image block. In other work, Fridrich et al. (2000, 2001a, 2001b, 2002) reported several techniques for detecting steganographic content in images. If a message is inserted into the LSBs of an image, some features of the image change in a manner that depends on the message size. A possible shortcoming of these methods is that they depend on empirical observations about the feature values of clean – i.e., unmodified – images. However, the authors have demonstrated very promising results on their data sets. Dumitrescu et al. (2002) present an approach related to Fridrich et al. (2001b) in which they use several features of *sample pairs* computed from a modified image to estimate the percentage of changed LSBs. The method relies on an assumption about the feature values in unmodified images. Frantz (2002) recently proposed using mimicking functions (Wayner, 2002) to ensure that the JPEG coefficient histogram of a modified image is the same as the histogram of the original – i.e., unmodified – image. The author assumed that given two binary sequences of the same length the second sequence could be transformed to have the same proportions of zeroes and ones as the first sequence. The method was not fully implemented since the author assumed that the transformation could be carried out.

7. Conclusions

Survey of methods for steganography and steganalysis was presented. It was shown that most methods used in digital steganography can be divided into embedding by modifying carrier bits and embedding using pairs of values. These methods were described formally and their uses, merits, and requirements were discussed. Various approaches to steganalysis were discussed; chi-square method was described and examples of its application were shown. Relationship between steganography and watermarking was also discussed and it was argued that, although superficially similar, these two fields have many differences and should be treated separately.

There are very few formal results in steganography and steganalysis. It is not known what are practical limits on message size so that detection can be avoided using any detection approach based on image statistics. Practical limits on message size so that detection can be avoided by existing methods of steganalysis have been partially established empirically. It has been proposed (Wayner, 2002) that codes mimicking natural image statistics could be designed and used in image steganography; however, practical algorithms utilizing this idea have not been developed yet.

References

Anderson, R.J., Petitcolas, F. (1998). On the limits of steganography. *IEEE J. Selected Areas Commun.* **16**, 474–481.
Atallah, M.J., Jacquet, P., Szpankowski, W. (1993). A probabilistic analysis of pattern matching problem. *Random Structures Algorithms* **4**, 191–213.
Bauer, F.L. (1997). *Decrypted Secrets – Methods and Maxims of Cryptology.* Springer-Verlag, Berlin, Germany.
Cox, I.J., Miller, M.L., Bloom, J.A. (2001). *Digital Watermarking.* Morgan Kaufmann, San Francisco, CA.
Crandall, R. (1998). Some notes on steganography. Available at http://os.inf.tu-dresden.de/~westfeld/crandall.pdf.
Dumitrescu, S., Wu, X., Wang, Z. (2002). Detection of LSB steganography via sample pair analysis. In: *Lecture Notes in Comput. Sci.*, vol. 2578. Springer-Verlag, Berlin–Heidelberg, pp. 355–372.
Farid, H. (2002). Detecting hidden messages using higher-order statistical models. In: *Proc. Int. Conf. on Image Processing.* Rochester, NY, IEEE.
Frantz, E. (2002). Steganography preserving statistical properties. In: *Lecture Notes in Comput. Sci.*, vol. 2578. Springer-Verlag, Berlin–Heidelberg, pp. 278–294.
Fridrich, J., Du, R., Long, M. (2000). Steganalysis of LSB encoding in color images. In: *Proc. IEEE Int. Conf. on Multimedia and Expo*, vol. 3. IEEE, pp. 1279–1282.
Fridrich, J., Goljan, M., Du, R. (2001a). Steganalysis based on JPEG compatibility. In: *Proc. SPIE Multimedia Systems and Applications IV.* Denver, CO, SPIE.
Fridrich, J., Goljan, M., Du, R. (2001b). Detecting LSB Steganography in color and gray-scale images. In: *IEEE Multimedia Magazine.* IEEE, pp. 22–28.
Fridrich, J., Goljan, M., Hogea, D. (2002). Steganalysis of JPEG images: breaking the F5 algorithm. In: *Lecture Notes in Comput. Sci.*, vol. 2578. Springer-Verlag, Berlin–Heidelberg, pp. 310–323.

Gruhl, D., Bender, W. (1998). Information hiding to foil the casual counterfeiter. In: *Information Hiding Workshop*. In: *Lecture Notes in Comput. Sci.*, vol. 1525. Springer-Verlag, New York.

Johnson, N., Jajodia, S. (1998a). Exploring steganography: seeing the unseen. *IEEE Comput.* **31**, 26–34.

Johnson, N., Jajodia, S. (1998b). Steganalysis of images created using current steganography software. In: *Proc. Int. Workshop on Information Hiding*. In: *Lecture Notes in Comput. Sci.*, vol. 1525. Springer-Verlag, Berlin, pp. 273–289.

Johnson, N., Duric, Z., Jajodia, S. (2000). *Information Hiding: Steganography and Watermarking – Attacks and Countermeasures*. Kluwer Academic, Boston.

Joint Photographic Experts Group. http://www.jpeg.org/public/jpeghomepage.htm.

Kahn, D. (1996). *The Codebreakers – The Story of Secret Writing*. Scribner, New York.

Katzenbeisser, S., Petitcolas, F.A.P. (Eds.) (2000). *Information Hiding: Techniques for Steganography and Digital Watermarking*. Artech House, Norwood, MA.

Lyu, S., Farid, H. (2002). Detecting hidden messages using higher-order statistics and support vector machines. In: *Lecture Notes in Comput. Sci.*, vol. 2578. Springer-Verlag, Berlin–Heidelberg, pp. 34–354.

Maes, M. (1998). Twin peaks: the histogram attack on fixed depth image watermarks. In: *Proc. Int. Workshop on Information Hiding*. In: *Lecture Notes in Comput. Sci.*, vol. 1525. Springer-Verlag, Berlin, pp. 290–305.

Newman, R.E., Moskowitz, I.S., Chang, L.W., Brahmadesam, M.M. (2002). A steganographic embedding undetectable by JPEG compatibility steganalysis. In: *Lecture Notes in Comput. Sci.*, vol. 2578. Springer-Verlag, Berlin–Heidelberg, pp. 258–277.

Petitcolas, F., Anderson, R.J., Kuhn, M. (1999). Information hiding – a survey. *Proc. IEEE* **87**, 1062–1078.

Provos, N. (2001). Defending against statistical steganalysis. In: *Proc. 10th USENIX Security Symposium*. USENIX, pp. 323–325.

Simmons, G.J. (1998). The history of subliminal channels. *IEEE J. Selected Areas Commun.* **16**, 452–462.

Upham, D. (2004). Jsteg V4. Available at http://www.funet.fi/pub/crypt/steganography.

Wayner, P. (2002). *Disappearing Cryptography*, 2nd ed. Morgan Kaufmann, San Francisco, CA.

Westfeld, A. (2001). F5 – a steganographic algorithm: high capacity despite better steganalysis. In: *Proc. Information Hiding Workshop*. In: *Lecture Notes in Comput. Sci.*, vol. 2137. Springer-Verlag, Berlin, pp. 289–302.

Westfeld, A. (2002). Detecting low embedding rates. In: *Proc. Information Hiding Workshop*. In: *Lecture Notes in Comput. Sci.*, vol. 2578. Springer-Verlag, Berlin–Heidelberg, pp. 324–339.

Westfeld, A., Pfitzmann, A. (1999). Attacks on stenographic systems. In: *Proc. Information Hiding Workshop*. In: *Lecture Notes in Comput. Sci.*, vol. 1768. Springer-Verlag, New York, pp. 61–75.

Canonical Variate Analysis and Related Methods for Reduction of Dimensionality and Graphical Representation

C. Radhakrishna Rao

1. Introduction

The problem we consider may be stated as follows. Let $X = (X_1 : \cdots : X_m)$ be $p \times m$ data matrix where the i-th p-vector X_i represents p measurements made on the i-th population or individual, referred to generally as the i-th profile. The measurements may be on a ratio scale giving the dimensions of a specified set of characteristics on a profile, or frequencies (as in two-way contingency tables) or compositions of different elements (as major oxide compositions of rocks in geology). The profiles can be represented as points in a p-dimensional space R^p. The object is to find the best possible representation of the profiles as points in a lower-dimensional Euclidean space to enable a visual study of the *configuration* of profiles.

In some situations, the available data may be in the form of an $m \times m$ dissimilarity matrix (δ_{ij}) where δ_{ij} is a chosen measure of dissimilarity between the i-th and j-th profiles. The problem is to represent the profiles as points in an Euclidean space of 2 or 3 dimensions.

The answer depends on how the configuration of the profiles is defined and what criterion is used to minimize the difference between the configurations in the original space R^p (vector space of p dimensions) and the reduced space E^k (Euclidean space of k dimensions).

It may be of interest in some situations to reverse the roles of profiles and variables, and represent the variables also in the same manner in a low-dimensional Euclidean space. When both profiles and variables are represented in the same graph, it is usually called a *biplot*. There are various ways of representing the variables depending on the type of information we are seeking. It may be to study the interrelationships between variables through their variances and covariances or to examine to what extent each of the original measurements is represented in the profiles in the reduced space. A discussion of the different types of biplots is given in Section 3.2 of this paper.

2. Canonical coordinates

The concept of canonical coordinates (variates) was introduced in an early paper (Rao, 1948) for graphical representation of taxonomical units characterized by multiple measurements for a visual examination of their interrelationships. This is perhaps the first attempt to give a general theory for reduction of high-dimensional data for graphical representation, which included the principal component analysis as a special case (Pearson, 1901; Hotelling, 1933; Rao, 1964). Since then, graphical representation of multivariate data for visual examination of clusters, outliers and other structures in the data has been an active field of research. Some of the developments are biplots (Gabriel, 1971; Grower and Hand, 1996), multidimensional scaling (Carroll, 1972; Kruskal and Wish, 1978; Davison, 1983; Eslava-Gomez and Marriott, 1993), correspondence analysis (Benzécri, 1973; Greenacre, 1984; Rao, 1995), Chernoff faces (Chernoff, 1973), parallel coordinates (Wegman, 1990), Andrews plots (Andrews, 1972), and parallel profiles (Mahalanobis et al., 1949).

2.1. Mahalanobis space

The m profiles, $X_1, \ldots, X_m$, can be represented as points in R^p endowed with an inner product and norm

$$(x, y) = x'My, \quad x, y \in R^p,$$
$$\|x\| = (x, x)^{1/2} = (x'Mx)^{1/2}, \quad x \in R^p, \tag{2.1}$$

where M is a $p \times p$ positive definite matrix. The distance between the points represented by x and $y \in R^p$ is

$$\|x - y\| = \left[(x - y)'M(x - y)\right]^{1/2} \tag{2.2}$$

known as Mahalanobis distance. When $M = I$, R^p becomes Euclidean space E^p.

The structure of m profiles, $X = (X_1 : \cdots : X_m)$, can be described by an $m \times m$ configuration matrix providing the lengths of profiles and angles between them:

$$C_m = (X - \xi 1')'M(x - \xi 1') \tag{2.3}$$

where $\xi \in R^p$ is some reference vector and 1 represents a vector of appropriate dimensions with all its elements equal to unity. The structure of profiles can also be described by a $p \times p$ dispersion matrix giving the weighted sums of squares and products of the measurements:

$$D_p = (X - \xi 1')W(X - \xi 1')' \tag{2.4}$$

where W is a diagonal matrix of weights $w_1 \geq 0, \ldots, w_m \geq 0, \sum w_i = 1$. The matrices M and W are at our choice depending on the nature of the data matrix X. (The matrix W can be a positive definite matrix, not necessarily diagonal, for theoretical purposes.)

DEFINITION 2.1. The norm of a matrix A, satisfying the usual definition of a norm, is said to be (M, N) invariant if

$$\|GAH\| = \|A\| \tag{2.5}$$

for all G and H such that $G'MG = M$, $H'NH = N$.

The main tool in dimension reductionality is the SVD (Singular Value Decomposition) of a $p \times m$ matrix B in the form

$$B = \lambda_1 U_1 V_1' + \cdots + \lambda_p U_p V_p' \tag{2.6}$$

where $U_1, \ldots, U_p$ are orthonormal p-vectors, i.e., $U_i'U_j = 0$ and $U_i'U_i = 1$ and $V_1, \ldots, V_p$ are orthonormal m-vectors. The non-negative constants $\lambda_1, \ldots, \lambda_p$ are called singular values of B, usually ordered as $\lambda_1 \geqslant \lambda_2 \geqslant \cdots \geqslant \lambda_p \geqslant 0$. Some of them may be zero. See (Rao and Rao, 1998) for details.

Computation of SVD. The SVD of a matrix is obtained by using the subroutine A.13 of Khattree and Naik (1999, p. 309).

The symmetric square root of a non-negative definite matrix A, denoted by $A^{1/2}$ which we need in our computations, is obtained by the subroutine A.12 of Khattree and Naik (1999, p. 309).

2.2. Canonical coordinates

First, we consider the problem of representing the points (profiles), $X = (X_1 : \cdots : X_m)$, in R^p in a subspace S of R^p of k dimensions spanned by k independent columns of a $p \times k$ matrix A, so that the m profiles in S can be written as $X_{(k)} = AB$ where B is a $k \times m$ matrix. We determine A and B such that the dispersion matrices, as defined in (2.4), of X and $X_{(k)}$ are as close as possible. Defining closeness as any (M, M) invariant norm of the differences in dispersion matrices

$$\|(X - \xi 1')W(X - \xi 1')' - ABWB'A'\| \tag{2.7}$$

we minimize (2.7) with respect to A and B. We choose $\xi = (1'W1)^{-1}XW1$ as the weighted average of the profile vectors. Using the results of Rao (1979, 1980) which are generalizations of a theorem of Eckart and Young (1936), we find the optimum choices of A and B as

$$\begin{aligned} A_0 &= (U_{1*} : \cdots : U_{k*}) = (v_{ij0}), \quad p \times k \text{ matrix}, \\ B_*' &= (\lambda_1 V_{1*} : \cdots : \lambda_k V_{k*}) = (p_{ij}), \quad m \times k \text{ matrix}, \end{aligned} \tag{2.8}$$

which can also be expressed as

$$\begin{aligned} A_* &= (\lambda_1 U_{1*} : \cdots : \lambda_k U_{k*}) = (v_{ij}), \quad p \times k \text{ matrix}, \\ B_0' &= (V_{1*} : \cdots : V_{k*}) = (p_{ij0}), \quad m \times k \text{ matrix}, \end{aligned} \tag{2.9}$$

using the first k terms in the SVD:

$$M^{1/2}(X - \xi 1')W^{1/2} = \lambda_1 U_1 V_1' + \cdots + \lambda_k U_k V_k' + \cdots + \lambda_p U_p V_p', \quad (2.10)$$
$$(X - \xi 1') = \lambda_1 U_{1*} V_{1*}' + \cdots + \lambda_p U_{p*} V_{p*}' \quad (2.11)$$

where $V_{i*} = W^{-1/2} V_i$ and $U_{i*} = M^{-1/2} U_i$, $i = 1, 2, \ldots$.

2.3. Graphical display of profiles and variables

2.3.1. Canonical coordinates for profiles

There are various ways of plotting the profiles and variables (measurements). First to obtain a plot of the profiles in E^k, we use the elements of the $m \times k$ matrix (p_{ij}) of (2.8) given in Table 2.1.

The i-th row in the (p_{ij}) matrix are the coordinates of the i-th profile in E^k. The configuration matrix of the profiles in E^k,

$$(p_{ij})(p_{ij})'$$

is a good approximation to the configuration matrix in R^p,

$$(X - \xi 1')' M (X - \xi 1'),$$

and the distances between the profile points in E^k are good approximations to the corresponding Mahalanobis distances in R^p. The exact nature of approximation is given in Rao (1995).

It may be seen that the matrix (p_{ij}) has an alternative representation:

$$(p_{ij}) = (X - \xi 1')' M (U_{1*} : \cdots : U_{k*}) \quad (2.12)$$

where U_{i*} is the i-th eigenvector corresponding to the i-th eigenvalue λ_i^2 of the dispersion matrix

$$D_p = (X - \xi 1')' W (X - \xi 1')'$$

with respect to M^{-1}, i.e.,

$$D_p U_{i*} = \lambda_i^2 M^{-1} U_{i*}, \quad i = 1, 2, \ldots.$$

Table 2.1
Canonical coordinates for profiles

Profile	dim 1	dim 2	...	dim k
1	p_{11}	p_{12}	...	p_{1k}
2	p_{21}	p_{22}	...	p_{2k}
⋮	⋮	⋮	⋱	⋮
m	p_{m1}	p_{m2}	...	p_{mk}

Table 2.2
Canonical coordinates for variables

Variable	dim 1	dim 2	...	dim k
1	v_{11}	v_{12}	...	v_{1k}
2	v_{21}	v_{22}	...	v_{2k}
⋮	⋮	⋮	⋱	⋮
p	v_{p1}	v_{p2}	...	v_{pk}

2.3.2. Canonical coordinates for variables

There are two ways for plotting the variables. One is to choose the matrix (v_{ij}) of Table 2.2 and use the canonical coordinates for the variables in different dimensions.

The dispersion matrix computed from the above coordinates,

$$(v_{ij})(v_{ij})' = \lambda_1^2 U_{1*} U'_{1*} + \cdots + \lambda_k^2 U_{k*} U'_{k*}, \qquad (2.13)$$

is a good approximation to the dispersion matrix, based on all the observations:

$$(X - \xi 1')W(X - \xi 1')' = \lambda_1^2 U_{1*} U'_{1*} + \cdots + \lambda_k^2 U_{k*} U'_{k*} + \cdots + \lambda_r^2 U_{p*} U'_{p*}$$
$$= (b_{ij}). \qquad (2.14)$$

The diagonal term b_{ii} in (b_{ij}) is the total variation in the i-th variable, while the contribution to variation of the i-th variable in dimension j is v_{ij}^2, which in terms of proportion is $f_{ij}^2 = v_{ij}^2/b_{ii}$ and it is seen that

$$\sum_1^p f_{ij}^2/b_{ii} = 1;$$

$$\sum_1^k f_{ij}^2/b_{ii} \qquad (2.15)$$

is the total variation in the i-th measurement explained by the canonical coordinates in E^k. We consider the standardized coordinates for variables given in Table 2.3.

The last column indicates the extent to which each variable is represented in the reduced set of coordinates. We may plot the variables in standardized coordinates in the same plot for the canonical coordinates of the profiles. It is seen that all the variables points lie inside a unit sphere in k dimensions (circumference of a unit circle when $k = 2$) and the variables close to the surface have greater influence on the canonical coordinates.

The matrix of standardized canonical coordinates for the variables (f_{ij}) can be written in the form

$$A_S = \left(\lambda_1 \Delta^{-1} U_{1*} : \cdots : \lambda_k \Delta^{-1} U_{k*} \right) = (v_{ijs}), \quad p \times k \text{ matrix}, \qquad (2.16)$$

where Δ is the diagonal matrix with its elements as the square roots of the diagonal elements in (b_{ij}) defined in (2.14). For biplots intended to study the affinities of profiles

Table 2.3
Standardized canonical coordinates for variables and variation explained in each dimension

Variables	Standardized coordinates			Variance explained			Total
	dim 1	...	dim k	dim 1	...	dim k	
1	f_{11}	...	f_{1k}	f_{11}^2	...	f_{1k}^2	$\sum f_{1i}^2$
⋮	⋮	⋱	⋮	⋮	⋱	⋮	⋮
p	f_{p1}	...	f_{pk}	f_{p1}^2	...	f_{pk}^2	$\sum f_{pi}^2$

and the influence of individual measurements, we use the coordinates given in Table 2.1 for profiles, and the standardized coordinates given in Table 2.3 for variables.

In the literature on biplots (Gabriel, 1971; Grower and Hand, 1996), several ways of plotting the coordinates in the same graph have been mentioned. One can choose the pairs of matrices (p_{ij}) and (v_{ij}), or (p_{ij}) and (v_{ij0}), or (p_{ij0}) and (v_{ij}) for profiles and variables. The method suggested in this paper is the biplot of the coordinates in the rows of matrices (p_{ij}) and (v_{ijs}) defined in (2.8)–(2.11). Not all choices produce meaningful graphs in all situations. See Section 3.2 for further discussion on individual and biplots.

2.4. Loss of information due to dimensionality reduction

In terms of the SVD given in (2.10), the full configuration matrix of the profiles is

$$C_m = W^{-1/2}(\lambda_1^2 V_1 V_1' + \cdots + \lambda_p^2 V_p V_p') W^{-1/2},$$

while that based on the canonical coordinates in E^k is

$$C_k = W^{-1/2}(\lambda_1^2 V_1 V_1' + \cdots + \lambda_k^2 V_k V_k') W^{-1/2},$$

and the deficiency is

$$L_1 = C_m - C_k = W^{-1/2}(\lambda_{k+1}^2 V_{k+1} V_{k+1}' + \cdots + \lambda_p^2 V_p V_p') W^{-1/2}.$$

An overall measure of loss of information is the ratio

$$\frac{\text{trace } W^{1/2} L_1 W^{1/2}}{\text{trace } W^{1/2} C_m W^{1/2}} = 1 - \frac{\lambda_1^2 + \cdots + \lambda_k^2}{\lambda^2} \tag{2.17}$$

where $\lambda^2 = \lambda_1^2 + \cdots + \lambda_p^2$. It is customary to compute

$$\frac{\lambda_1^2}{\lambda^2}, \frac{\lambda_2^2}{\lambda^2}, \ldots$$

expressed as percentages to indicate the relative importance of the successive canonical coordinates and to take a decision on the sufficient number of dimensions needed for graphical representation of data.

It is more important to assess the distortion in inter profile squared distances due to reduction of dimensionality. Using a formula due to Torgerson (1958), the $m \times m$ matrix of squared distances d_{ij}^2, $i, j = 1, \ldots, m$ between the profiles is

$$S_{(p)} = c1' + 1c' - 2C_m = (d_{ij}^2)$$

where c is the vector of diagonal elements of C_m. The corresponding matrix in E^k is

$$S_{(k)} = c_{(k)} 1' + 1' c_{(k)} - 2C_k,$$

so that the deficiency is measured by

$$L_2 = S_{(p)} - S_{(k)} = (d_{ij(k)}^2).$$

An overall measure of deficiency is

$$\frac{\sum\sum w_i w_j d_{ij(k)}^2}{\sum\sum w_i w_j d_{ij}^2} = \frac{\lambda_{k+1}^2 + \cdots + \lambda_p^2}{\lambda^2}$$

which is the same as (2.17).

2.5. An example

An example of the biplot given in Rao (1948, 1995) is as follows. The mean values of 9 anthropometric measurements made on individuals from 17 populations constitute the 9×17 data matrix X. A pooled estimate of within population variance–covariance matrix Σ is computed and M is chosen as Σ^{-1}, which is generally used in computing Mahalanobis distance. The weights $w_1, \ldots, w_m$ to be attached to the profiles are all taken to be equal, although in some situations w_i can be chosen as n_i/n when n_i is the sample size on which the mean values for the i-th profile are computed and n is the total number of individuals measured in all the populations. The plots based on (p_{ij})

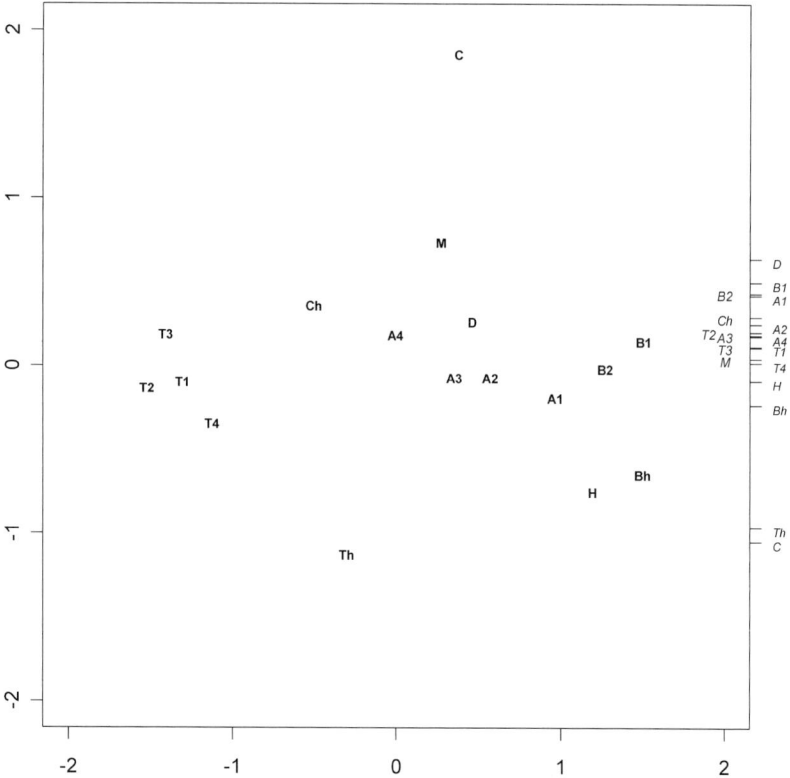

Fig. 1. Configuration of canonical coordinates for caste groups. The coordinates in the third dimension for the caste groups are represented on a separate line.

and (v_{ijs}) are given in Figures 1 and 2. The basic data and the computed values of the canonical coordinates for profiles and standardized canonical coordinates for variables can be found in Rao (1995). It is seen that most of the basic measurements are well represented in the first two canonical coordinates as the points in Figure 2 are close to the circumference of the circle.

A two-dimensional representation may not show the exact relative positions of the populations in the original space and the distortion may be large in particular cases. In general, it will be wise to take into account the differences in the third as well as the higher dimensions while interpreting the distances in the two-dimensional representation. An innovation in Figures 1 and 2 is the representation of the coordinates in the third dimension on a separate line, which shows the additional differences among the groups. (Note that the square of the distance between any two groups in the three-dimensional representation is the square of the distance in the two-dimensional representation plus the square of the difference in the coordinates of the third dimension.)

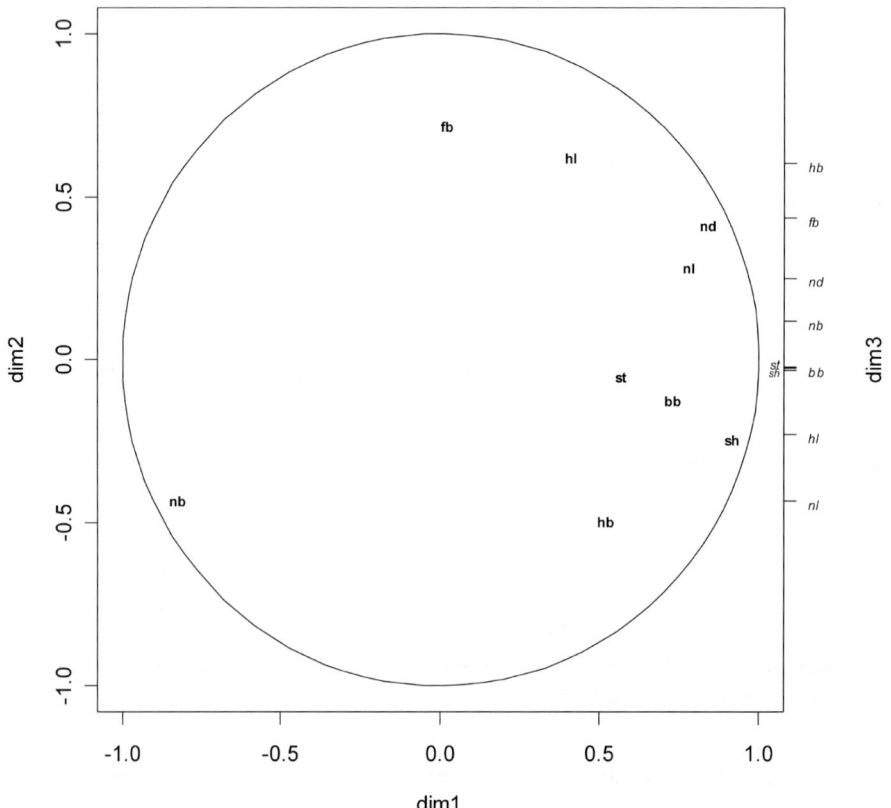

Fig. 2. Configuration of standardized canonical coordinates for variables.

A general recommendation is to consider a two-dimensional plot supplemented by representation of the points in other dimensions on separate lines as parallel coordinates (for details, see Wegman, 1990).

2.5.1. Typical (or eigen) profiles

The p-vectors defined in Table 2.2,

$$\lambda_1 U_{1*}, \ldots, \lambda_k U_{k*}, \ldots, \tag{2.18}$$

considered as hypothetical profiles, were termed as typical profiles in (Rao, 1964), fundamental patterns in (Holter et al., 2000) and eigenprofiles in engineering literature. By their construction, the dispersion matrix (2.13) computed from the first few of them would be a good approximation to the dispersion matrix (2.14) obtained from the whole set of original measurements. A measure of the loss of information in using the first k typical profiles is $(\lambda_{k+1}^2 + \cdots + \lambda_p^2)/(\lambda_1^2 + \cdots + \lambda_p^2)$ expressed as a percentage.

Another way of looking at the typical profiles is to examine to what extent an individual profile can be recovered by fitting a linear combination of the first k typical profiles. The best fit to $X_c = (X - \xi 1')$ by using the first k typical profiles is

$$\widehat{X}_c = \lambda_1 U_{1*} U_{1*}' + \cdots + \lambda_k U_{k*} U_{k*}' \tag{2.19}$$

and the difference $X_c - \widehat{X}_c$ gives the amount of deviation. An overall measure of deficiency in the fit for the i-th profile is d_{ii}, the i-th diagonal element in the matrix

$$(X_c - \widehat{X}_c)' M (X_c - \widehat{X}_c) = (d_{ij}). \tag{2.20}$$

An overall measure of deficiency for all the profiles is

$$\sum_1^m d_{ii} = \lambda_{k+1}^2 + \cdots + \lambda_p^2 \tag{2.21}$$

which is usually expressed as a percentage of $\lambda_1^2 + \cdots + \lambda_p^2$. If any particular value of d_{ii} is large, the corresponding profile may be considered as an outlier.

3. Principal component analysis

3.1. Preprocessing of data

A special case of canonical coordinate analysis with $M = I$ and $W = I$ is known as Principal Component Analysis (PCA). For a general theory of PCA and its applications, reference may be made to Hotelling (1933) and Rao (1964).

However, in practice some preprocessing of the $p \times m$ data matrix $X = (x_{ij})$ is recommended before applying PCA depending on the nature of the data and the object of analysis. Preprocessing consist of the following operations.

(i) Centering the observations in each row:

$$X_{(r)} = X(I - m^{-1}11') = (x_{ij} - \bar{x}_{i.}) \tag{3.1}$$

where x_{ij} is the (i, j)-th element in X and $\bar{x}_{i.}$ is the average of the i-th row.

(ii) Centering the observations in each column:
$$X_{(c)} = (I - p^{-1}11')X = (x_{ij} - \bar{x}_{.j}) \qquad (3.2)$$
where $\bar{x}_{.j}$ is the average of the j-th column.

(iii) Centering the observations in each row and column:
$$X_{(rc)} = (I - p^{-1}11')X(I - m^{-1}11') = (x_{ij} - \bar{x}_{i.} - \bar{x}_{.j} + \bar{x}_{..}) \qquad (3.3)$$
where $\bar{x}_{..}$ is the overall average.

(iv) Standardizing the observation in each row:
$$X_{(rs)} = S_r X_{(r)} \qquad (3.4)$$
where S_r is the $p \times p$ diagonal matrix with its elements as the square roots of the reciprocals of the diagonal elements in $X_{(r)}X'_{(r)}$.

(v) Standardizing the observations in each column:
$$X_{(cs)} = X_{(c)} S_c \qquad (3.5)$$
where S_c is the $m \times m$ diagonal matrix with its elements as the square roots of the reciprocals of the diagonal elements in $X'_{(c)}X_{(c)}$. In some cases, standardization is done without centering.

(vi) Centering and standardizing both rows and columns:

This is done by first standardizing the rows of X yielding a matrix Y_1 (say), then standardizing the columns Y_1 yielding Y_2, then standardizing the rows of Y_2 yielding Y_3 and so on. The operations are repeated till the rows and columns are approximately standardized.

For situations in which the preprocessed data described in (i)–(v) are used in PCA, reference may be made to Pielou (1984) and Jackson (1991), and for (vi) to Holter et al. (2000).

(vii) Transformation of compositional data:

Aitchison and Greenacre (2002) consider an example of compositional data, where the rows represent six-part color compositions (variables) in 22 paintings (profiles). The problem is to study the affinities between profiles. Other examples of compositional data, where the sum of the different compositions expressed as proportions is unity, can be found in Aitchison (1986).

Let $x_{1i}, \ldots, x_{pi}$ be a p-part composition of the i-th profile such that $x_{1i} + \cdots + x_{pi} = 1$ and $x_{ji} > 0$ for all j. Some suggested transformations are:

(a) $\log x_{ji} - \log x_{j'i}$, $j < j'$, leading to $\frac{1}{2}p(p-1)$ variables, and $\qquad (3.6)$

(b) $\log x_{ji} - \frac{1}{p}\sum_j \log x_{ji}$, leading to $(p-1)$ variables. $\qquad (3.7)$

Aitchison and Greenacre (2002) demonstrate biplots after centering rows of the transformed variables. Implicit in such an analysis is the Euclidean distance between the profiles represented by transformed variables as a measure of affinity between profiles. Another transformation which may be considered is

(c) $\sqrt{x_{ji}}$, $j = 1, \ldots, p$, leading to p variables. $\qquad (3.8)$

3.2. Individual and biplots

In the literature on graphical display of profiles and variables based on PCA, several ways of plotting have been described (Gabriel, 1971; Grower and Hand, 1996). A general tendency is to represent the profiles and variables in the same graph, although there may not be always any meaningful interpretation of the relationships between the profile and variable points. First, we describe the individual plots and suggest some useful joint displays. Some identities derivable from the SVD of preprocessed X which are useful in the interpretation of graphs are as follows. From (2.11), we have by choosing $M = I$ and $W = I$:

$$X = U \Lambda V' = U_{(k)} \Lambda_{(k)} V'_{(k)} + R$$

where $U_{(k)}$, $\Lambda_{(k)}$ and $V_{(k)}$ are matrices obtained from U, Λ and V by retaining only the first k columns. We then have:

$$UU' = X(X'X)^{-1}X' \simeq U_{(k)}U'_{(k)}, \tag{3.9}$$

$$VV' = X'(XX')^{-1}X \simeq V_{(k)}V'_{(k)}, \tag{3.10}$$

$$U\Lambda^2 U' = XX' \simeq U_{(k)}\Lambda^2_{(k)}U'_{(k)}, \tag{3.11}$$

$$V\Lambda^2 V = X'X \simeq V_{(k)}\Lambda^2_{(k)}V'_{(k)}. \tag{3.12}$$

In (3.9) and (3.10), generalized inverse (Rao, 1973, pp. 24–27) is used if the ranks of $X'X$ and XX' are deficient.

Let x_i and x_j be the i-th and j-th columns of X representing the p-vectors in full space R^p of the i-th and j-th profiles and y_i and y_j be the i-th and j-th rows of X representing the m-vectors in the full space R^m of measurements on the i-th and j-th variables.

P (profile) plot. Choosing ξ_i, the i-th column k-vector of $\Lambda_{(k)} V'_{(k)}$, we have k coordinates for plotting the i-th profile as a point in the reduced space E^k. Then from (3.12), we find that

$$(\xi_i - \xi_j)'(\xi_i - \xi_j)$$

the square of the distance between the i-th and j-th profiles in E^k is an approximation to

$$(x_i - x_j)'(x_i - x_j)$$

the corresponding distance in the full space R^p.

V (variable) plot. Choosing η_j, the j-th row k-vector of $U_{(k)}\Lambda_{(k)}$, we have k coordinates for plotting the j-th variable as a point in the reduced space E^k. Then $\eta_i \eta'_j$, the scalar product of η_i and η_j in E^k is an approximation to $y_i y'_j$, the corresponding product in R^p.

P_0 *(unweighted profile) plot.* Choosing α_i, the i-th column k-vector in V_k', we have k coordinates for plotting the i-th profile as a point in E^k. Then from (3.10) we find that

$$(\alpha_i - \alpha_j)'(\alpha_i - \alpha_j)$$

the square of the distance between the points for the i-th and j-th profiles in E^k is an approximation to

$$(x_i - x_j)'(XX')^{-1}(x_i - x_j)$$

which is Mahalanobis distance in some sense in the full space R^p.

V_0 *(unweighted variable) plot.* Choosing β_j, the j-th row k-vector of $U_{(k)}$, we have k coordinates for plotting the j-th variable as a point (or an arrow emphasizing the direction) in E^k. From (3.9) we find that $\beta_j \beta_j'$ is an approximation to $y_i(X'X)^{-1}y_j'$, the cross product of vectors in E^p with respect to $(X'X)^{-1}$.

PV_0 *biplot.* When the P-plot and V_0-plot are shown in the same graph we call it the PV_0 biplot in E^k. The projections of the i-th profile point on the vectors of the p variables are approximations to the elements of x_i, the vector of measurements on the i-th profile.

V_{0n} *(unweighted normalized variable) plot.* The coordinates for plotting the point for the j-th variable are the elements of $\beta_j/|\beta_j|$, where $|\beta_j| = \sqrt{\beta_j \beta_j'}$, so that all the points for variables lie on k-dimensional sphere (circumference of a circle when $k = 2$).

P_s *(standardized profile) plot.* The coordinates for plotting the point for the i-th profile are the elements of $\xi_i/|\xi_i|$, where $|\xi_i| = \sqrt{\xi_i' \xi_i}$ so that all the points for profiles lie on the k-dimensional sphere.

V_n *(normalized variable) plot.* The coordinates for plotting the point for the j-th variable are the elements of $\eta_j/|\eta_j|$ where $|\eta_j| = \sqrt{\eta_j \eta_j'}$, so that all the points for the variables lie on a k-dimensional sphere.

$P_0 V_n$ *biplot.* This biplot is known as MDPREF suggested by Chang and Carroll (1968) in the context of data giving the preferences of individuals for a number of objects.

V_s *(standardized variable) plot.* Consider the matrix XX' and let b_{jj} be its j-th diagonal element. Choosing $\zeta_j = b_{jj}^{-1/2} \eta_j$, we have k coordinates for plotting the j-th variable as a point in E^k. The interpretation of a standardized variable is discussed in the earlier Section 2.3.2.

PV_s *biplot.* This is the plot used in Section 2.5. The interpretation of such a plot is discussed in Sections 2.3.2 and 2.5.

In practical applications one has to choose an appropriate individual or biplot.

4. Two-way contingency tables (correspondence analysis)

We consider dichotomous categorical data with s rows and m columns and n_{ij} observations in the (i, j)-th cell. Define

$$N = (n_{ij}), \qquad n_{i.} = \sum_{j=1}^{m} n_{ij}, \qquad n_{.j} = \sum_{i=1}^{s} n_{ij}, \qquad n = \sum_{1}^{s} \sum_{1}^{m} n_{ij},$$

$$R = \text{Diag}(n_{1.}/n, \ldots, n_{s.}/n), \qquad C = \text{Diag}(n_{.1}/n, \ldots, n_{.m}/n),$$

$$P = n^{-1} N C^{-1} = \begin{pmatrix} p_{1|1} & \cdots & p_{1|m} \\ \vdots & \ddots & \vdots \\ p_{s|1} & \cdots & p_{s|m} \end{pmatrix} \quad \text{(column profiles)},$$

$$Q = n^{-1} R^{-1} N = \begin{pmatrix} q_{1|1} & \cdots & q_{m|1} \\ \vdots & \ddots & \vdots \\ q_{1|s} & \cdots & q_{m|s} \end{pmatrix} \quad \text{(row profiles)}, \tag{4.1}$$

$$p = R\mathbf{1}, \qquad q = C\mathbf{1}. \tag{4.2}$$

The problem is to represent the column (row) profiles as points in E^k, $k < s, m$, such that the Euclidean distances between points reflect specified affinities between the corresponding column (row) profiles.

The technique developed for this purpose by Benzécri (1973) is known as correspondence analysis (CA) which can be identified as canonical coordinate analysis. For instance, for representing the column profiles by this method, one chooses

$$X = P, \qquad M = R^{-1}, \qquad W = C \tag{4.3}$$

and applies the analysis described in Section 2.2. Thus one finds the SVD of

$$R^{-1/2}(P - p\mathbf{1}')C^{1/2} = \lambda_1 U_1 V_1' + \cdots + \lambda_s U_s V_s' \tag{4.4}$$

giving the coordinates for the column profiles in E^k:

$$\lambda_1 C^{-1/2} V_1, \quad \lambda_2 C^{-1/2} V_2, \quad \ldots, \quad \lambda_k C^{-1/2} V_k \tag{4.5}$$

where the components of i-th vector are the coordinates of the profiles in the i-th dimension. The standardized canonical coordinates in E^k for the rows, as described in (2.16), obtained from the same SVD as in (4.4) are

$$\lambda_1 \Delta^{-1} R^{1/2} U_1, \quad \lambda_2 \Delta^{-1} R^{1/2} U_2, \quad \ldots, \quad \lambda_k \Delta^{-1} R^{1/2} U_k \tag{4.6}$$

where the components of the i-th vector are the coordinates of the rows in the i-th dimension and Δ is a diagonal matrix with the i-th diagonal element as the square root of the i-th diagonal element in the matrix

$$R^{1/2} \left(\lambda_1^2 U_1 U_1' + \cdots + \lambda_s^2 U_s U_s' \right) R^{1/2} = (P - p\mathbf{1}')C(P - p\mathbf{1}')'. \tag{4.7}$$

Implicit in this analysis is the choice of measure of affinity between the i-th and j-th profiles as

$$d_{ij}^2 = \frac{(p_{1|i} - p_{1|j})^2}{p_1} + \cdots + \frac{(p_{s|i} - p_{s|j})^2}{p_s} \tag{4.8}$$

which is the chi-square distance. The squared Euclidean distance in E^k, the reduced space, between the points representing the i-th and j-th profiles is an approximation to (4.8). Thus the clusters we see in the Euclidean space representation is based on the affinities measured by the chi-square distance (4.8).

The coordinates (4.6) do not represent the row profiles but are useful in interpreting the different dimensions of the column profiles. The coordinates for representing the row profiles in correspondence analysis are obtained as follows.

Note that the expression in (4.4) is

$$R^{-1/2}(P - p1')C^{1/2} = R^{1/2}(Q - 1q')C^{-1/2} = T, \qquad (4.9)$$

so that if we need a representation of the row profiles in E^k, we use the same SVD as in (4.4):

$$R^{1/2}(Q - 1q')C^{-1/2} = \lambda_1 U_1 V_1' + \cdots + \lambda_s U_s V_s', \qquad (4.10)$$

leading to the row (profile) coordinates

$$\left(\lambda_1 R^{-1/2} U_1 : \cdots : \lambda_k R^{-1/2} U_k\right), \qquad (4.11)$$

so that no extra computations are needed if we want a representation of the row profiles also.

The standardized coordinates for the columns (as variables) are

$$\lambda_1 \Delta_1^{-1} C^{1/2} V_1, \quad \ldots, \quad \lambda_k \Delta_1^{-1} C^{1/2} V_k \qquad (4.12)$$

where Δ_1 is the diagonal matrix with its i-th diagonal element as the square root of the i-th diagonal element of $(Q - 1q')'R(Q - 1q')$.

In correspondence analysis, it is customary to plot the coordinates (4.5) for column profiles and the coordinates (4.11) for row profiles in the same graph. Alternative choices for plotting on the same graph are the coordinates (4.5) and (4.6), to study column profiles and the coordinates (4.11) and (4.12) to study row profiles.

An alternative to the chi-square distance which has some advantages is the Hellinger distance (HD) between the i-th and j-th profiles defined by

$$d_{ij}^2 = \left(\sqrt{p_{1|i}} - \sqrt{p_{1|j}}\right)^2 + \cdots + \left(\sqrt{p_{s|i}} - \sqrt{p_{s|j}}\right)^2 \qquad (4.13)$$

which depends only on the i-th and j-th column profiles. In such a case, the Euclidean distance in the reduced space between the i-th and j-th column profiles is an approximation to (4.13). For the derivation of canonical coordinates of the column profiles (considered as population) using HD, we consider

$$X = \begin{pmatrix} \sqrt{p_{1|1}} & \cdots & \sqrt{p_{1|m}} \\ \vdots & \ddots & \vdots \\ \sqrt{s_{1|1}} & \cdots & \sqrt{p_{s|m}} \end{pmatrix},$$

$$M = I, \qquad W = C = \text{Diag}(n_{.1}/n, \ldots, n_{.m}/n),$$

and consider the SVD

$$(X - \xi 1')C^{1/2} = \lambda_1 U_1 V_1' + \cdots + \lambda_s U_s V_s'. \tag{4.14}$$

We may choose $\xi' = (\xi_1, \ldots, \xi_s)$ as

$$\xi_i = \sqrt{p_i} = \sqrt{n_{i.}/n} \quad \text{or} \tag{4.15}$$
$$= n^{-1}\left(n_{.1}\sqrt{p_{i|1}} + \cdots + n_{.m}\sqrt{p_{i|m}}\right) \tag{4.16}$$

where p_i is the i-th element of p defined in (4.2). The canonical coordinates in E^k for the column profiles choosing ξ as in (4.15) or (4.16) are

$$\lambda_1 C^{-1/2} V_1, \quad \lambda_2 C^{-1/2} V_2, \quad \ldots, \quad \lambda_k C^{-1/2} V_k \tag{4.17}$$

where the components of the i-th vector in (4.17) are the coordinates of the m column profiles in the i-th dimension. The standardized coordinates in E^k for the variables, i.e., the row categories, obtained as described in (2.16) from the same SVD as in (4.14) are

$$\lambda_1 \Delta^{-1} U_1, \quad \lambda_2 \Delta^{-1} U_2, \quad \ldots, \quad \lambda_k \Delta^{-1} U_k \tag{4.18}$$

where Δ is a diagonal matrix with the i-th diagonal element as the square root of the i-th diagonal element of

$$\lambda_1^2 U_1 U_1' + \cdots + \lambda_s^2 U_s U_s' = (X - \xi 1')C(X - \xi 1')'. \tag{4.19}$$

The s components of $\lambda_i \Delta^{-1} U_i$ in (4.18) are the coordinates of the s variables in the i-th dimension.

If a representation of the row profiles is also needed, we take $X = \text{sqrt}(Q')$, i.e., the elements of X are the square roots of the elements of Q' where Q is the matrix defined in (4.1) and compute the SVD:

$$(X - \eta 1')R^{1/2} = \mu_1 A_1 B_1' + \cdots + \mu_s A_s B_s', \tag{4.20}$$

leading to the canonical coordinates for row profiles

$$\mu_1 R^{-1/2} B_1, \quad \mu_2 R^{-1/2} B_2, \quad \ldots, \quad \mu_k R^{-1/2} B_k. \tag{4.21}$$

The corresponding standardized coordinates for the columns considered as variables are

$$\mu_1 \Delta_2^{-1} A_1, \quad \mu_2 \Delta_2^{-1} A_2, \quad \ldots, \tag{4.22}$$

where Δ_c is a diagonal matrix with the i-th diagonal element as the square root of the i-th diagonal element of

$$\mu_1^2 A_1 A_1' + \cdots + \mu_s^2 A_s A_s'. \tag{4.23}$$

If we choose ξ as in (4.15), then the matrix in (4.14) is

$$(X - \xi 1')C^{1/2} = \left(\sqrt{\frac{n_{ij}}{n}} - \sqrt{\frac{n_{i.}\,n_{.j}}{n\,n}}\right),$$

which is symmetric in i and j. Then, the same SVD as in (4.14) could be used for computing the canonical coordinates

$$\lambda_1 R^{-1/2} U_1, \quad \lambda_2 R^{-1/2} U_2, \quad \ldots, \quad \lambda_k R^{-1/2} U_k$$

for the row profiles, as in the case of CA.

We consider the data from Greenacre (1993) on 796 scientific researchers classified according to their scientific disciplines (as populations) and funding categories (as variables) as shown in Table 4.1.

The canonical coordinates for the scientific disciplines (considered as profiles) in the first three dimensions and percentage of variance explained by each are given in Table 4.2 for the analyses based on the chi-square distance (correspondence analysis) and the Hellinger distance (alternative). The formula (4.10) is used for the analysis based on chi-square and the formula (4.20) for that based on Hellinger distance.

Table 4.1
Scientific disciplines by research funding categories

Scientific discipline		Funding category					Total
		a	b	c	d	e	
Geology	G	3	19	39	14	10	85
Biochemistry	B_1	1	2	13	1	12	29
Chemistry	C	6	25	49	21	29	130
Zoology	Z	3	15	41	35	26	120
Physics	P	10	22	47	9	26	114
Engineering	E	3	11	25	15	34	88
Microbiology	M_1	1	6	14	5	11	37
Botany	B_2	0	12	34	17	23	86
Statistics	S	2	5	11	4	7	29
Mathematics	M_2	2	11	37	8	20	78
Total		31	128	310	129	198	796

Table 4.2
Canonical coordinates for the scientific disciplines in the first three dimensions

Subjects	Chi-square distance			Hellinger distance		
	dim 1	dim 2	dim 3	dim 1	dim 2	dim 3
G	0.0764	0.3025	−0.0877	−0.0311	0.1674	−0.0482
B_1	0.1798	−0.4549	−0.1517	−0.1293	−0.2421	−0.0776
C	0.0376	0.0733	0.0423	−0.0211	0.0404	0.0282
Z	−0.3273	0.1022	0.0645	0.1388	0.0452	0.0568
P	0.3155	0.0269	0.1086	−0.1653	0.0106	0.0238
E	−0.1174	−0.2917	0.1073	0.0494	−0.1299	0.0829
M_1	0.0127	−0.1096	−0.0414	−0.0049	−0.0525	−0.0084
B_2	−0.1786	−0.0385	−0.1290	0.1514	−0.0365	−0.1080
S	0.1246	0.0141	0.1071	−0.0666	0.0117	0.0525
M_2	0.1067	−0.0613	−0.1756	−0.0503	−0.0375	−0.0780
% var.	47.20	36.66	13.11	45.87	34.10	16.57

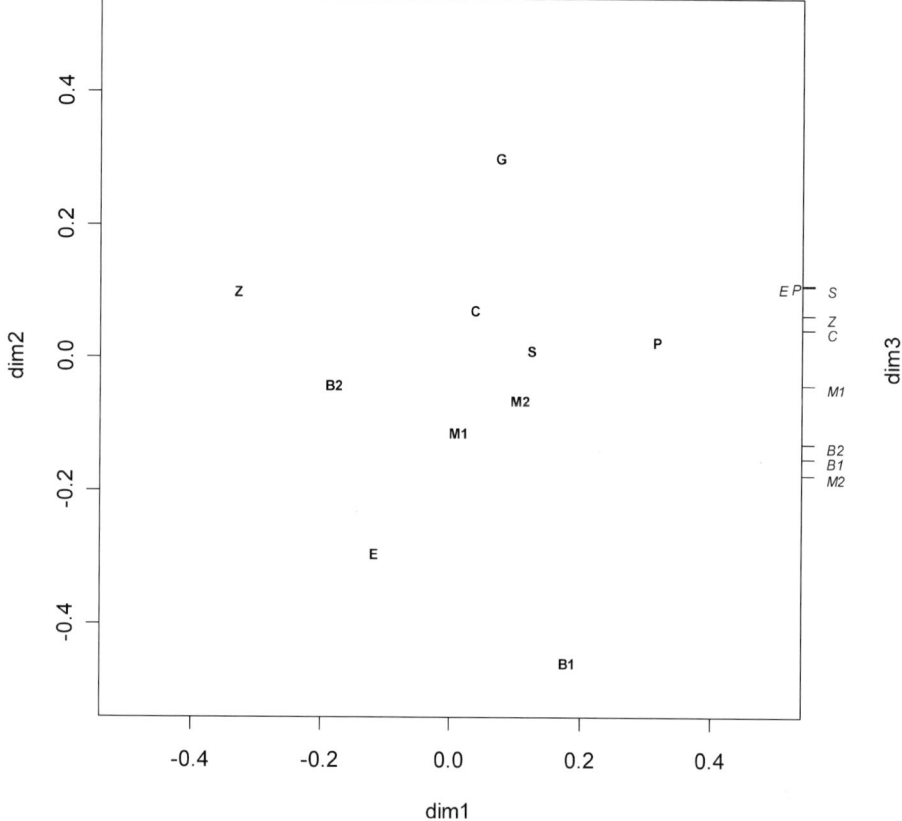

Fig. 3. Configuration of scientific disciplines using chi-square distance (correspondence analysis).

For Hellinger distance analysis, the central point is chosen according to the formula (4.16).

The plots of the scientific disciplines (subjects) using the canonical coordinates based on the chi-square and Hellinger distances are given in Figures 3 and 4, respectively. The coordinates in the third dimension are plotted on a line on the right-hand side of the two-dimensional plot. This will be of help in visualizing the plot in three dimensions and in interpreting the distances in the two-dimensional plot. Thus, although B_2 and E appear to be close to each other in the two-dimensional chart, they are clearly separated in the third dimension. No additional distances in the third dimension are involved in the case of P, C, S, Z and E.

It is of interest to note in this example that the configuration of the scientific disciplines in three dimensions obtained by both the methods are very similar. The percentage of variance explained in each dimension is nearly the same for both the methods.

The standardized canonical coordinates for the funding categories (considered as variables) are computed using the formula (4.12) for the chi-square and the formula (4.21) for the Hellinger distance analysis. These are obtained from the *same SVD* used

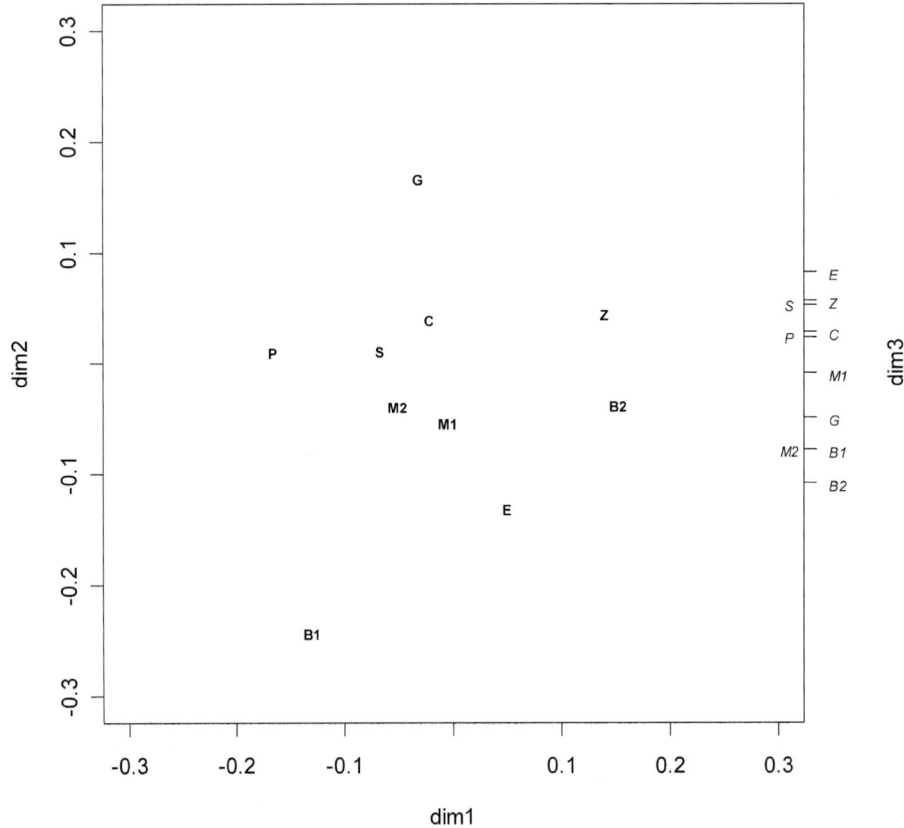

Fig. 4. Configuration of scientific disciplines using Hellinger distance (alternative to correspondence analysis).

Table 4.3
Standardized canonical coordinates for funding categories (variables) in the first three dimensions

Funding category	Chi-square distance				Hellinger distance			
	dim 1	dim 2	dim 3	% var.	dim 1	dim 2	dim 3	% var.
a	0.758	0.114	−0.619	97.1	0.796	0.164	−0.573	98.9
b	0.535	0.728	−0.137	83.5	0.438	0.766	−0.008	77.9
c	0.583	0.352	0.694	94.6	0.501	0.327	0.759	93.4
d	−0.426	0.331	−0.172	99.8	−0.888	0.358	−0.285	99.7
e	−0.108	−0.909	−0.081	99.6	−0.088	−0.978	−0.159	98.9

to compute the canonical coordinates for the scientific disciplines. Table 4.3 gives the standardized canonical coordinates for the funding categories, a, b, c, d, e using the two methods.

The standardized canonical coordinates for the funding categories are plotted in Figures 5 (for chi-square distance) and 6 (for Hellinger distance). It may be noted that

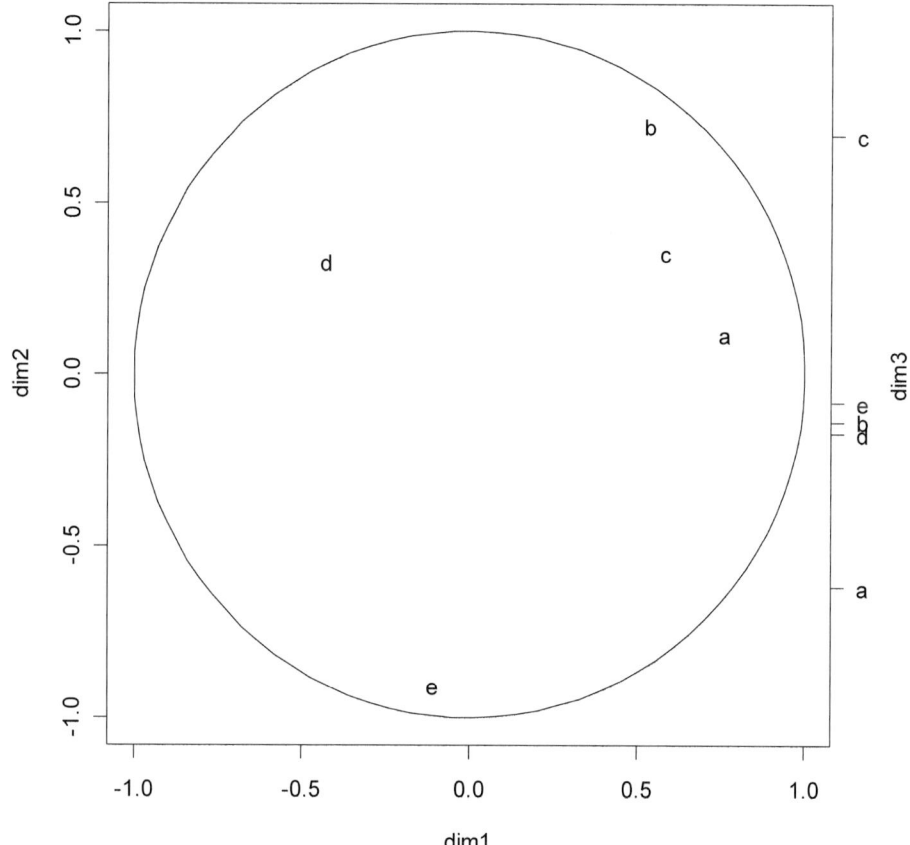

Fig. 5. Configuration of funding categories using standardized canonical coordinates based on chi-square distance.

all the points lie within the unit circle. It is customary to represent the canonical coordinates for the subjects and variables in one chart. We are using separate charts in order to explain the salient features of the configuration of the variables. The following interpretations emerge from the study of Table 4.3 and Figures 5 and 6:

1. The configurations of the funding categories as exhibited by Figures 5 and 6 obtained by using chi-square and Hellinger distances are very similar.
2. Almost all the variation in the funding categories a, d and e is captured in the first canonical coordinates of the scientific disciplines. A large percentage of variation in b and c is explained by the first three coordinates.
3. The first dimension is strongly influenced by a, d, the second dimension by b, e and the third dimension by a, c.

Thus the use of standardized coordinates for variables enables us to interpret the different dimensions in terms of observed variables.

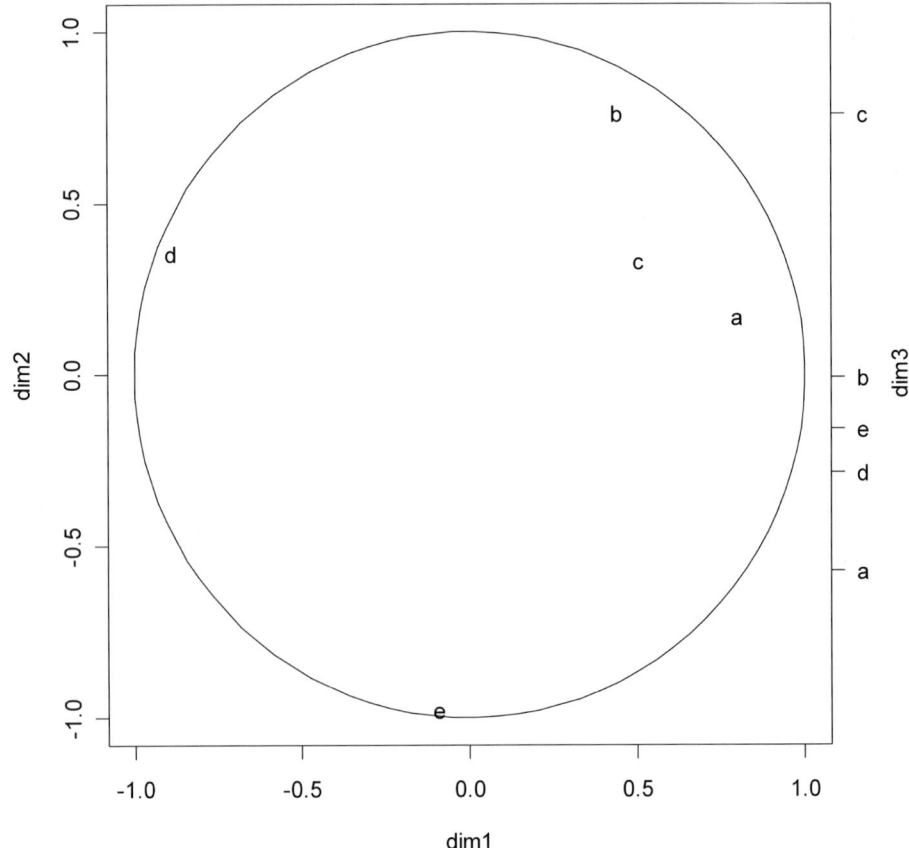

Fig. 6. Configuration of funding categories using standardized canonical coordinates based on Hellinger distance.

NOTE. In computing the canonical coordinates based on Hellinger distance (HD) using the formula (4.14), we chose the relative sample sizes as the weights to be attached to the populations. We could have chosen an alternative set of weights if we wanted distances between a specified subset of populations to be better preserved in the reduced space than the others. In particular, we could have chosen uniform weights for all populations. In fact, such an option could be exercised if the sample sizes of different populations were widely different. Unfortunately, no such options are available in CA.

In the example we considered, there was a perfect match between the plots based on CA and HD. This probably demonstrates that the method of derivation of canonical coordinates is somewhat robust to the choice of the distance measure as well as to the weights. However, the choice of HD provides an insurance against possible distortion due to variations in sample sizes for the populations. See Rao (1995) for an example where the results of chi-square and Hellinger distance analysis could be different.

NOTE. The appropriate software for correspondence analysis and related methods described in Section 4 can be found in Khattree and Naik (2000).

5. Discussion

The general problem considered in the paper is that of visual representation of objects, individuals or populations, generally called profiles, characterized by a set of measurements as points in a low-dimensional Euclidean space for an exploratory data analysis of the affinities between profiles. The *first step* in such a problem is the specification of the basic metric space S in which the profiles with the given set of measurements can be considered as points. The *second* is the development of a methodology for transforming the points in S to an Euclidean space E^k of k dimensions, preferably for $k = 2$ or 3. The choice of the basic space S and the metric (distance between points) in S has to be made on practical considerations relevant to the problem under investigation. The methodology reviewed in the paper, known by different names in statistical literature, is based on the choice of S as a vector space R^p of p dimensions where the metric, the square of the distance between x_1 and $x_2 \in R^p$ is defined as

$$(y_1 - y_2)' M (y_1 - y_2), \qquad (5.1)$$

where y_i may be the p-vector of original measurements or an s-vector, $s \leqslant p$,

$$y_i = \begin{pmatrix} f_1(x_i) \\ \vdots \\ f_s(x_i) \end{pmatrix}, \qquad (5.2)$$

choosing suitable functions $f_1, \ldots, f_s$. These functions constitute what is described as preprocessing data (transforming the original measurements) in Section 3.1 of this paper.

The basic tool used is the singular value decomposition (SVD) of the $s \times m$ matrix Y whose columns are y vectors (defined in (5.2)) corresponding to m profiles. Some of the relevant indices used in interpreting the observed affinities in a low-dimensional representation of the profiles are measures of loss of information in terms of residual singular values of Y as in (2.16) and (2.20), and the proportion of variance in each component of y explained by the reduced coordinates as in (2.15).

In some situations, the basic data may be available in the form of an $m \times m$ matrix with the (i, j)-th element δ_{ij} representing a measure of dissimilarity (or distance) between profiles i and j. The object in such a case is to represent the profiles as points in a low-dimensional Euclidean space such that the corresponding Euclidean distances d_{ij}'s are as close as possible to δ_{ij}'s, or at least approximately satisfy the monotonic constraint

$$\delta_{ij} < \delta_{i'j'} \quad \Rightarrow \quad d_{ij} < d_{i'j'}. \qquad (5.3)$$

The methodology in such a case is known as MDS (multidimensional scaling).

There are two types of MDS, one called metric scaling (Torgerson, 1958; Rao, 1964; Pielou, 1984), where an underlying Euclidean structure of δ_{ij} is assumed and the coordinates for representing profiles in a low-dimensional Euclidean space are obtained by minimizing an aggregated measure of the differences $|\delta_{ij}^2 - d_{ij}^2|$. The second is called non-metric scaling for which an appropriate methodology is developed to satisfy the constraints (5.3). For a description of MDS and its applications, reference may be made to (Kruskal and Wish, 1978; Shepard, 1962). A number of interesting applications of MDS can be found in (Shepard et al., 1972). Applications of MDS to seriation of Plato's works and filiation (linkage) of manuscripts are given in (Kendall, 1971).

We have not used any probability model for the observed measurements in statistical analysis. A technical report by Rios et al. (1994) discusses the same problem as in the present paper, viz., simultaneous representation of profiles and variables, under the assumption of an underlying parametric model. They call the method as IDA (intrinsic data analysis) and is based on the Riemannian structure induced by the Fisher information metric and *Rao distance* based on it (Rao, 1948). Their method is applied to multivariate normal and multinomial distributions. In the multinomial case, the *Rao distance* ρ between two multinomial populations defined by the class probabilities, $p_1, \ldots, p_s$ and $q_1, \ldots, q_s$ is proportional to

$$s = 2\cos^{-1} \sum_{i=1}^{n} \sqrt{p_i q_i},$$

which is a monotone transformation of Hellinger distance.

References

Aitchison, J. (1986). *The Statistical Analysis of Compositional Data*. Chapman and Hall, London.
Aitchison, J., Greenacre, M. (2002). Biplots of compositional data. *J. R. Statist. Soc. Ser. C Appl. Statist.* **51**, 375–392.
Andrews, D.F. (1972). Plots of high dimensional data. *Biometrics* **28**, 125–136.
Benzécri, J.P. (1973). *L'analyse de données*, vols. I, II. Dunod, Paris.
Carroll, J.D. (1972). Individual differences and multidimensional scaling. In: Sheppard, R.N., Romney, A.K., Nerlove, S.B. (Eds.), *Multidimensional Scaling: Theory and Applications in Behavioral Sciences*, vol. 1. Seminar Press, New York.
Chang, J.J., Carroll, J.D. (1968). How to use MDPREF, a computer program for multidimensional analysis of preference data. Bell Telephone Laboratories, Murray Hill, NJ.
Chernoff, H. (1973). The use of faces to represent points in k-dimensional space graphically. *J. Amer. Statist. Assoc.* **68**, 361–368.
Davison, M.L. (1983). *Multidimensional Scaling*. Wiley, New York.
Eckart, C., Young, G. (1936). The approximation of one matrix by another of lower rank. *Psychometrika* **1**, 211–308.
Eslava-Gomez, G., Marriott, F.H.C. (1993). Criteria to represent groups in the plane when the grouping is unknown. *Biometrics* **49**, 1088–1098.
Gabriel, K.R. (1971). The biplot-graphic display of matrices with application to principal component analysis. *Biometrika* **58**, 453–467.
Greenacre, M.J. (1984). *Theory and Applications of Correspondence Analysis*. Academic Press, London.
Greenacre, M.J. (1993). Biplots in correspondence analysis. *J. Appl. Statist.* **20**, 251–269.
Grower, J.C., Hand, D. (1996). *Biplots*. Chapman and Hall, London.

Holter, N.S., Mitra, M., Mariton, A., Cieplak, M., Banavar, J.R., Fedoroff, N.V. (2000). Fundamental patterns underlying gene expression profiles: Simplicity from complexity. *Proc. Natl. Acad. Sci. USA* **97**, 8409–8414.

Hotelling, H. (1933). Analysis of a complex of statistical variables into principal components. *J. Educ. Psychol.* **24**, 417–441, 498–520.

Jackson, J.E. (1991). *A User's Guide to Principal Components*. Wiley, New York.

Hodson, F.R., Kendall, D.G., Tautu, P. (Eds.) (1971). *Mathematics in the Archaeological and Historical Sciences*. Proceedings of the Anglo–Romanian Conference, Mamaia, 1970, Edinburgh University Press.

Khattree, R., Naik, D.N. (1999). *Applied Multivariate Statistics with SAS Software*, second ed. SAS Institute.

Khattree, R., Naik, D.N. (2000). *Multivariate Data Reduction and Discrimination with SAS Software*. SAS Institute.

Kruskal, J.B., Wish, M. (1978). *Multidimensional Scaling*. Sage, Beverly Hills, CA.

Mahalanobis, P.C., Mazumdar, D.N., Rao, C.R. (1949). Anthropometric survey of United Provinces, 1941 – A Statistical Study. *Sankhyā* **9**, 90–324.

Pearson, K. (1901). On lines and planes of closest fit to a system of points in space. *Philos. Mag.* **2**, 557–572.

Pielou, E.C. (1984). *The Interpretation of Ecological Data*. Wiley, New York.

Rao, C.R. (1948). The utilization of multiple measurements in problems of biological classification (with discussion). *J. R. Statist. Soc. B* **10**, 159–193.

Rao, C.R. (1964). The use and interpretation of principal component analysis in applied research. *Sankhyā* **26**, 329–357.

Rao, C.R. (1973). *Linear Statistical Inference and its Applications*, second ed. Wiley, New York.

Rao, C.R. (1979). Separation theorems for singular values of matrices and their applications in multivariate analysis. *J. Multivariate Anal.* **9**, 362–377.

Rao, C.R. (1980). Matrix approximations and reduction of dimensionality in multivariate statistical analysis. In: Krishnaiah, P.R. (Ed.), *Multivariate Analysis V*. North-Holland, Amsterdam, pp. 3–22.

Rao, C.R. (1995). A review of canonical coordinates and an alternative to correspondence analysis using Hellinger distance. *Qüestió* **9**, 23–63.

Rao, C.R., Rao, M.B. (1998). *Matrix Algebra and its Applications to Statistics and Econometrics*. World Scientific, Singapore.

Rios, M., Villarroya, A., Oller, J.M., 1994. Intrinsic data analysis: A method for the simultaneous representation of populations and variables. Mathematics preprint series 160. Universitat de Barcelona.

Shepard, R.N. (1962). The analysis of proximities: Multidimensional scaling with an unknown distance function. *Psychometrika* **27**, 125–139. 219–246.

Shepard, R.N., Romney, A.K., Nerlove, S.B. (Eds.) (1972). *Multidimensional Scaling: Theory and Applications in the Behavioral Sciences, vol. 1*. Seminar Press, New York.

Torgerson, W.S. (1958). *Theory and Methodology of Scaling*. Wiley, New York.

Wegman, E.J. (1990). Hyperdimensional data analysis using parallel coordinates. *J. Amer. Statist. Assoc.* **85**, 664–675.

Pattern Recognition

David J. Hand

1. Background

The aim of statistical pattern recognition is to assign objects into one of a set of classes. In *supervised pattern recognition* or *supervised classification* we are given a sample of objects, for each of which we have observed a vector of descriptive measurements, and for each of which we know its true class, and the aim is to use this sample to construct a rule which will allow us to assign new objects, for which only the vector of measurements is known, to an appropriate class. In *unsupervised pattern recognition*, we are given an unclassified sample of objects and the aim is to divide them into (usually mutually exclusive) classes. This latter exercise also goes under terms such as cluster analysis, segmentation analysis, or partitioning. This article focuses on supervised methods.

The problem of supervised pattern recognition is ubiquitous and as a consequence has been explored in a variety of data analytic disciplines as well as a range of different application areas. Researchers in statistics, machine learning, computational learning theory, data mining, as well as those working primarily in statistical pattern recognition have all made very substantial contributions to the area. Although tackling a common problem, researchers from these distinct areas have brought their own particular interests and emphases to bear, so that there are differences in flavor between the methods which they have developed. This will be illustrated below. The differences have also led to intellectual tensions, which have subsequently resulted in increased understanding and deeper theory. This theory has its counterparts in other areas of statistics and data analysis, so that statistical pattern recognition might be described as encapsulating, in microcosm, the main problems of wider statistical modelling and data analysis.

The areas of application of statistical pattern recognition are too many to list. They include areas such as medical diagnosis and prognosis, genomics and bioinformatics, speech recognition, character recognition, biometrics (areas such as face, fingerprint, iris, and retinal recognition), credit scoring, stock market analysis, vehicle classification, target recognition, fault and anomaly detection, fraud detection, and a huge variety of other areas. Some of the applications are exotic, while others have already surreptitiously entered our lives (a consequence of the ubiquity of computers being increasingly embedded in our everyday environment). It will be clear from this illustrative list that the vectors of descriptive measurements will depend very much on the application domain. In medical diagnosis these measurements will include symptoms (which may be continuous, binary, categorical, or even merely ordinal or nominal variables), while in speech recognition they will be descriptions of the waveform of the sound, and in credit scoring they will be characteristics of the person seeking a loan. The diversity of literatures which have explored the problem mean that different names have been used for these descriptive measurements, including *variables*, *features*, *characteristics* and *covariates*.

2. Basics

We begin by describing how we could tackle supervised classification if we had complete knowledge about the various distributions involved, rather than merely a sample from these distributions. Later sections are concerned with how these basic ideas must be modified to allow for the fact that we do have only a sample of data.

Suppose we have N classes, ω_i, $i = 1, \ldots, N$, and suppose that their relative sizes are $P(\omega_i)$, $i = 1, \ldots, N$. These relative sizes are often called *priors* in the statistical pattern recognition literature, reflecting the fact that, before obtaining any other knowledge about an object (such as vectors of descriptive characteristics), they give the probabilities that the object will come from each class. On this basis, in the absence of further information, we would minimise the proportion of new objects misclassified if we assigned an object to class i whenever

$$P(\omega_i) > P(\omega_j) \quad \text{for all } j \neq i. \tag{1}$$

Ties can be broken arbitrarily. When we do have further information on an object, given by the vector of descriptive variables, $\mathbf{x}$, we can do better if we condition on $\mathbf{x}$, and use the rule: assign an object with measurement vector $\mathbf{x}$ to class i if

$$P(\omega_i \mid \mathbf{x}) > P(\omega_j \mid \mathbf{x}) \quad \text{for all } j \neq i. \tag{2}$$

As will be seen, many methods are based on the direct estimation of the $P(\omega_i \mid \mathbf{x})$ probabilities. This leads to the *diagnostic paradigm*, because of its focus on the (conditional) distinction between the classes. Contrasted with this is the *sampling paradigm*, which estimates the posterior probabilities in (2) indirectly from estimates of the class conditional distributions via Bayes theorem:

$$P(\omega_i \mid \mathbf{x}) = \frac{P(\mathbf{x} \mid \omega_i) P(\omega_i)}{P(\mathbf{x})}. \tag{3}$$

Note that this approach will require separate estimates of the class priors to be provided.

Since the rule given in (2) will lead to minimising the proportion of objects with measurement vector $\mathbf{x}$ which are misclassified, for all $\mathbf{x}$, it will also (by integrating over the distribution of $\mathbf{x}$) lead to minimising the overall misclassification rate (also called *error rate*).

The simple rule in (2) (or its equivalent form via (3)) is all very well, but we may not want simply to minimise the proportion of new objects which are misclassified. Often, certain types of misclassification are regarded as more severe than others. For example, misclassifying someone with an easily curable but otherwise fatal disease as healthy will generally be far more serious than the reverse. To cater for this, we need to take the relative *costs* of the different kinds of misclassification into account. So, let c_{ji} be the cost of assigning an object to class ω_i when it comes from class ω_j. Then the overall expected cost (also called the *risk*) is

$$r = \sum_{i=1}^{N} \int_{\Omega_i} \sum_{j=1}^{N} c_{ji} P(\omega_j \mid \mathbf{x}) P(\mathbf{x}) \, d\mathbf{x} \tag{4}$$

where the classification rule assigns points in region Ω_i to class ω_i. The risk (4) is minimised by choosing the Ω_i such that $\mathbf{x} \in \Omega_i$ if

$$\sum_{j=1}^{N} c_{ji} P(\omega_j \mid \mathbf{x}) \leq \sum_{j=1}^{N} c_{jk} P(\omega_j \mid \mathbf{x}), \quad k = 1, \ldots, N. \tag{5}$$

Note that, if $c_{rs} = 0$ for correct classifications (that is, when $r = s$) and $c_{rs} = 1$ for incorrect classifications (when $r \neq s$), this reduces to the rule given in (2).

The two class case is especially important, partly because this is the most common special case and partly because multiple class cases can often be reduced to multiple two class cases. In this case, the rule implied by (5) becomes: assign an object to class 1 if

$$\frac{P(\omega_1 \mid \mathbf{x})}{P(\omega_2 \mid \mathbf{x})} > \frac{c_{22} - c_{21}}{c_{11} - c_{12}} \tag{6}$$

(assuming correct classifications cost less than incorrect classifications) and this is further specialised, when $c_{11} = c_{22} = 0$, to: assign an object to class 1 if

$$\frac{P(\omega_1 \mid \mathbf{x})}{P(\omega_2 \mid \mathbf{x})} > \frac{c_{21}}{c_{12}}. \tag{7}$$

Note that if the sampling approach is adopted, (7) can be expressed as

$$\frac{P(\mathbf{x} \mid \omega_1)}{P(\mathbf{x} \mid \omega_2)} > \frac{c_{21} P(\omega_2)}{c_{12} P(\omega_1)}. \tag{8}$$

A key feature of (8) is that the prior probability for class i and the costs of misclassifying points from class i appear only when multiplied together. Thus changing costs has the same effect as changing class sizes.

Risk, and its specialisation to misclassification rate, is not the only criterion which has been used. Others include the Neyman–Pearson criterion, which minimises one type of misclassification subject to a fixed level of the other, and the minimax criterion, which seeks to minimise the largest type of misclassification. Yet other measures appear in the context of assessing the performance of classification rules.

At this point it is convenient to introduce some further terminology. The *decision surface* is the surface separating regions of the $\mathbf{x}$ space in which new objects are assigned to different classes. For example, in the two class case derived from rule (2) this is the set of $\mathbf{x}$ for which $P(\omega_1 \mid \mathbf{x}) = P(\omega_2 \mid \mathbf{x})$. In the two class case, a *discriminant function* is any function which is monotonically related to the ratio $P(\omega_1 \mid \mathbf{x})/P(\omega_2 \mid \mathbf{x})$ – such a function allows one to *discriminate* between the classes. To use a discriminant function to make a classification, we have to compare it with a *classification threshold*. For example, in rule (7), the classification threshold is the ratio c_{21}/c_{12}.

3. Practical classification rules

Up to this point we have assumed that all the required probability distributions were known. Unfortunately, in practice we have to estimate these and any other necessary functions from a sample of data. These data will consist of the measurement vectors and known true classes for a sample of objects. The usual considerations apply about the representativeness of this sample: if it is a distorted sample, then any classification rule built using it has the potential to be misleading. Because the classification rule will be built from this sample, it is called the *design set* (in statistics) or the *training set* (in machine learning). And it is because the true classes are known, as if they were provided by a 'supervisor', that the methods are known as 'supervised' pattern recognition. The performance of classification rules is often evaluated on a separate independent data set, and this is called a *test set*. Sometimes, choice between competing classifiers is made on the basis of yet a third data set, called a *validation set*.

In the vast majority of supervised pattern recognition problems, the support regions of the class distributions overlap. This means that there will not exist any decision surface which will give an error rate of zero. The minimal possible error rate is called the *Bayes error rate*. In a few rare problems the classes are perfectly separable: at each $\mathbf{x}$ points from only one class arise, so that the Bayes error is zero. Since such problems are rare, generally occurring in artificial situation where the classes are defined symbolically by humans, we will not discuss them. This has little practical consequence, though it has led to interesting theoretical developments in computational learning theory.

Our aim is to use the design data to build a model which can then be used to predict the likely class value from a new $\mathbf{x}$ vector. At first glance, a sensible aim seems to be to construct our model (whether it is based on the $P(\omega_i \mid \mathbf{x})$, the $P(\mathbf{x} \mid \omega_i)$, or a discriminant function), so that it maximises classification accuracy on the design set: to set out deliberately to perform poorly on the design set would be perverse. However, our aim is not simply to classify the design set points (we already know their classes). Rather, our aim is to generalise to new points from the same distribution. Since the design data are merely a sample from this distribution, a model which has too much fidelity in fitting them will be unlikely to generalise to new points very well. A classic illustration of this is a perfect interpolation between points which have arisen from a true regression model of the form $y = \alpha + \beta x + \varepsilon$. The perfect interpolation will be a wobbly line which may depart substantially from the true straight line. This means that our model should not be too flexible: too highly parameterised a model will risk fitting aspects of the data which may change substantially between different random design sets (called *overfitting*). Rather, we want to capture that aspect which is essentially common to different design sets. Put another way, the variance of a flexible, highly parameterised model is large: its prediction at any particular value of **x** may change dramatically from design set to design set.

In contrast to this, and working in the opposite direction, our model should not be too simple. Once again, a classic illustration of this is fitting a quadratic regression 'truth' with a straight line. Although the fitted straight line is likely to vary little between different design sets, it is possible that it will be substantially biased for some values of the predictor variable (i.e., the prediction from the linear model will differ substantially from the quadratic truth).

These two aspects, the high variance arising from the over-complex model, and the high bias arising from the over-simple model, work in opposite directions, and the aim of modelling (in general, not merely in pattern recognition) is to attain some acceptable or optimal compromise. The overfitting arising from too complex a model appears in various disguises, and will be discussed further below. Whether or not a model is 'too complex' will also depend on the size of the data set. A model with 100 free parameters will tend to overfit a sample of just 200 points, but will not overfit a sample of 100 000 points.

3.1. Linear discriminant analysis

The earliest method to have been formally developed appears to be (Fisher's 1936) linear discriminant analysis (LDA). For the two class case, he considered linear combinations of the components of **x**, **w**'**x**. He showed that the linear combination for which the difference between the means relative to the pooled within class variance is largest is given by

$$\mathbf{w} \propto \mathbf{S}^{-1}(\bar{\mathbf{x}}_1 - \bar{\mathbf{x}}_2). \tag{9}$$

Here **S** is the within group pooled covariance matrix of the **x**, and $\bar{\mathbf{x}}_i$ is the centroid of the ith group. A classification of a new point **x** is made by comparing **w**'**x** (with some constant of proportionality) with a classification threshold. This classification procedure is equivalent to using a hyperplanar decision surface in the original **x** space. Note that this is very much a discriminant function approach, and that no distributional assumptions have been made. However, since the formulation is entirely expressed in terms of first and second order statistics, it will probably come as no surprise to discover that this is also the optimal solution if one assumes that the two classes have multivariate normal distributions (which are defined in terms of their first and second order moments) with common covariance matrices. The equality of the covariance matrices is important, because it means that the quadratic terms, $\mathbf{x}'\mathbf{\Sigma}^{-1}\mathbf{x}$, which arise from the exponents of the normal distributions of the two classes, $-\{(\mathbf{x} - \boldsymbol{\mu}_i)'\mathbf{\Sigma}_i^{-1}(\mathbf{x} - \boldsymbol{\mu}_i)\}/2$, cancel out, leaving only the linear terms. From this perspective, the method clearly belongs to the sampling paradigm. This sampling paradigm perspective leads to immediate generalisation to multiple classes, with each class being estimated by a multivariate normal distribution, $P(\mathbf{x} \mid \omega_i) = \text{MVN}(\boldsymbol{\mu}_i, \mathbf{\Sigma})$, and classifications being made on the basis of which of the estimated $P(\mathbf{x} \mid \omega_i)P(\omega_i)$ is the largest. One often reads that Fisher's LDA method requires the assumption of multivariate normality. As can be seen from the above, this is incorrect. Certainly the method is optimal for such distributions, but it is more generally optimal under any ellipsoidal distributions (with equal covariance matrices). More generally still, however, the method has been proven by many decades of experience to perform well even with data which depart very substantially from this.

Because it was one of the earliest methods to be developed, a solid software base has built up around the method, with a wide variety of diagnostic measures and supplementary devices, such as variable selection, readily available in standard software packages. The method continues to be widely used, especially in areas such as psychology. However, in such applications interest often lies more widely in interpreting the weights ('which variables are most influential in discriminating between the groups?') rather than merely in classification problems.

Although we have described the linear discriminant analysis method as linear in the variables $\mathbf{x} = (x_1, \ldots, x_p)$ which were originally measured, it is easy enough to extend it to include transformed x_j (e.g., squared values, $\exp(x_j)$, etc.) and combinations of the x_j (e.g., products). The resulting decision surface will then be linear in the new variables, and hence nonlinear in the original $(x_1, \ldots, x_p)$. This means a very flexible classifier can result, but two issues must be borne in mind. Firstly, of course, there is the danger of the method being too flexible and overfitting the design data. Secondly, there is the problem of deciding which of the transformed variables to include, since the range of possibilities is unlimited. With too many, overfitting will certainly be a problem.

One extension of linear discriminant analysis is particularly important because it is a natural extension of the basic linear form. This is quadratic discriminant analysis (QDA). If, in the multivariate normal derivation in the two class case, the covariance matrices are not assumed to be equal, then the quadratic terms do not cancel out and a quadratic decision surface results. Clearly a quadratic decision surface is more flexible than a linear one, and so is less susceptible to bias. However, it is also adding many more parameters and so risks overfitting. In fact, early empirical studies (before the overfitting issues were properly understood) on data sets of at most a few hundred data points showed that LDA generally outperformed QDA, for this very reason. The development of theoretical understanding of overfitting issues, and of the bias/variance trade-off, opened up the possibility of other ways of tackling the problem. In particular, one strategy is to develop a complex model, and shrink it towards a simpler model. We will see illustrations of this below, but in the context of LDA, Friedman (1989) proposed shrinking a quadratic model towards a linear model. This is equivalent to shrinking the separate covariance matrix estimates towards a common form. Of course, if something is known about the likely pattern of covariances between the measurements then this can be used to good effect in constraining the number of parameters in $\mathbf{\Sigma}$. For example, if the measurements are repeated measurements of the same variable, taken at different times (e.g., at 1 hour, 2 hours, 3 hours, etc., after ingestion of a medicine) then one might expect measurements which are closer in time to be more highly correlated.

3.2. Logistic discrimination

In some application domains (for example, medicine and credit scoring) linear discriminant analysis has been replaced in popularity by logistic discrimination. This also yields a decision surface which is linear in the components of $\mathbf{x}$, but is based on the diagnostic paradigm. In particular, in the two class case, it estimates the probabilities $P(\omega_1 \mid \mathbf{x})$ from a linear combination $\mathbf{w}'\mathbf{x}$ using a logistic transformation:

$$P(\omega_1 \mid \mathbf{x}) = \frac{\exp(\mathbf{w}'\mathbf{x})}{1 + \exp(\mathbf{w}'\mathbf{x})}. \tag{10}$$

Inverting this transformation yields

$$\ln\left[\frac{P(\omega_1 \mid \mathbf{x})}{P(\omega_2 \mid \mathbf{x})}\right] = \mathbf{w}'\mathbf{x}. \tag{11}$$

This method originates in work of Cox (1966) and Day and Kerridge (1967), but was developed in detail by Anderson (reviewed in Anderson, 1982). It was subsequently extended to handle more than two classes, for example in the form

$$\ln\left[\frac{P(\omega_i \mid \mathbf{x})}{P(\omega_N \mid \mathbf{x})}\right] = \mathbf{w}_i'\mathbf{x}, \quad i = 1, \ldots, N-1, \tag{12}$$

and is now conveniently described as a type of generalised linear model (McCullagh and Nelder, 1989). The parameters are generally estimated by maximum likelihood via an iteratively weighted least squares approach.

Because of the natural log transformation in (11), it turns out that the model in (10) is optimal if the distributions of the two classes are multivariate normal with equal covariance matrices (the equality of these matrices again leads to the terms quadratic in $\mathbf{x}$ cancelling). This leads to an obvious question: since both LDA and logistic discrimination are optimal under the assumption of multivariate normal distributions with equal covariance matrices, in what ways do they differ? The answer is that, since LDA is based on the sampling paradigm, if the data really do arise from ellipsoidal distributions with equal covariance matrices, then this method will use more information, and will lead to more accurate estimates. On the other hand, the logistic approach is also optimal under a wider variety of situations, so that if the data do not arise from ellipsoidal distributions, then this may yield less biased estimates.

3.3. The naive Bayes model

The logistic approach is closely related to a popular and very simple sampling paradigm model (called, variously, the naive Bayes model, the independence Bayes model, or idiot's Bayes) which makes the assumption that the variables are conditionally independent, given each class:

$$P(\mathbf{x} \mid \omega_i) = \prod_{j=1}^{p} P(x_j \mid \omega_i), \quad i = 1, \ldots, N. \tag{13}$$

Clearly this assumption is generally unrealistic, and yet the model has often been found to perform surprisingly well in real applications. Several explanations have been put forward for this, including the fact that, in many studies, the variables have been preselected to remove those highly correlated with variables which have already been selected, that there are fewer parameters to be estimated (the overfitting, bias/variance point mentioned above), and that bias in the probability estimates may not matter in a classification rule, where all that counts is whether a predicted probability is above the classification threshold, not by how much. Such matters are discussed in Hand and Yu (2001). By virtue of the independence assumption, construction of such classification rules is very simple, being restricted to construction of univariate marginal distributions.

The relationship between the logistic method and the naive Bayes model can be seen in the two class case by considering the ratio $P(\omega_1 \mid \mathbf{x})/P(\omega_2 \mid \mathbf{x})$. In general, making no simplifying assumptions at all, we have

$$\frac{P(\omega_1 \mid \mathbf{x})}{P(\omega_2 \mid \mathbf{x})} = \frac{P(\mathbf{x} \mid \omega_1)P(\omega_1)}{P(\mathbf{x} \mid \omega_2)P(\omega_2)}. \tag{14}$$

The naive Bayes model then makes the stringent independence assumption, $P(\mathbf{x} \mid \omega_i) = \prod_j P(x_j \mid \omega_i)$, yielding

$$\frac{P(\omega_1 \mid \mathbf{x})}{P(\omega_2 \mid \mathbf{x})} = \frac{\prod_j P(x_j \mid \omega_1)P(\omega_1)}{\prod_j P(x_j \mid \omega_2)P(\omega_2)}. \tag{15}$$

However, a weaker assumption would be that

$$P(\mathbf{x} \mid \omega_i) = H(\mathbf{x}) \prod_{j=1}^{p} Q(x_j \mid \omega_i), \quad i = 1, 2, \tag{16}$$

where H is an arbitrary function common to the two classes. The H function subsumes the dependencies between the components of $\mathbf{x}$, and its commonality means that these dependencies, while not having to be zero, are the same in the two classes. The common H assumption in (16) means that

$$\frac{P(\omega_1 \mid \mathbf{x})}{P(\omega_2 \mid \mathbf{x})} = \frac{\prod_j Q(x_j \mid \omega_1)P(\omega_1)}{\prod_j Q(x_j \mid \omega_2)P(\omega_2)}, \tag{17}$$

the same structure as in (15), though the Q functions are not (necessarily) the marginal $P(x_j \mid \omega_i)$ probability distributions, so that this model is more flexible. By taking logs of (17) we see that it reduces to a function linear in the $\ln[Q(x_j \mid \omega_1)/Q(x_j \mid \omega_2)]$. That is, the model in (17) is equivalent to the logistic model based on simple univariate transformations of the x_j. Since the Q are not the marginal distributions, estimation is more complicated – the iteratively weighted least squares of logistic discrimination.

3.4. The perceptron

Linear discriminant analysis has its origins in the statistical research community. Another classifier linear in the x_j has its origins in the machine learning and early pattern recognition communities. This is the *perceptron*. The LDA method described above seeks weights $\mathbf{w}$ which maximise a measure of separability between the projections of the data points on $\mathbf{w}$, and finds an explicit solution. In contrast, the perceptron is based on an iterative updating (*error-correcting*) procedure, one version of which cycles through the design set points sequentially, and adjusts estimates of $\mathbf{w}$ whenever a point is misclassified. An alternative procedure classifies all the design set points, and updates $\mathbf{w}$ using all those which have been misclassified, in one pass, and then repeats this. The details are as follows. Our aim is to find a vector **w** which assigns as many as possible of the design set elements to the correct class. For any vector of measurements **x** define $\mathbf{y} = (\mathbf{x}', 1)'$, and define $\mathbf{v} = (\mathbf{w}', -t)'$. Now the classification rule 'assign **x** to class ω_1 if $\mathbf{w}'\mathbf{x} > t$ and to class ω_2 otherwise' becomes rule 'assign **x** to class ω_1 if $\mathbf{v}'\mathbf{y} > 0$ and to class ω_2 otherwise', and we seek a **v** which maximises the number of design set points correctly classified using this rule. The exercise is simplified yet further if, for the design set, we define $\mathbf{z} = \mathbf{y}$ for design set objects in class ω_1 and $\mathbf{z} = -\mathbf{y}$ for design set objects in class ω_2. In fact, with this, we now need to find that vector **v** for which $\mathbf{v}'\mathbf{z} > 0$ for *all* design set objects, if we can, since this would mean that **v** was correctly classifying all design set points. With this as the aim, the perceptron algorithm proceeds as follows. If $\mathbf{v}_t$ is the current estimate of **v**, classify all the design set points using $\mathbf{v}_t$. Let M be the set of those which are misclassified. Then update $\mathbf{v}_t$ to

$$\mathbf{v}_t = \mathbf{v}_t + \rho \sum_{\mathbf{z}_i \in M} \mathbf{z}_i \tag{18}$$

where ρ is the iterative step size. In fact, a little algebra shows that this is the iterative step in a gradient minimisation procedure using the *perceptron criterion function*

$$C_P(\mathbf{v}) = \sum_{\mathbf{z}_k \in M} (-\mathbf{v}'\mathbf{z}). \tag{19}$$

If there does exist a hyperplanar surface which can separate the two design set classes, then this algorithm is guaranteed to find it. There are various extensions to this algorithm, to guarantee convergence when there is no true linear separating surface, such as letting ρ decrease at each step, or letting ρ depend on the size of the errors $\mathbf{v}'\mathbf{z}$. Another variant, important in the linearly separable case, maximises the distance between the decision surface and the closest design set point. This is important because the idea has been taken up in *support vector machines*, about which more later.

A point about perceptron methods worth noting is that they do not require global models to be assumed. In particular, they make no assumption about the forms of the distributions far from the decision surface. If one has made such an assumption (e.g., multivariate normality) and this assumption is incorrect, then it will distort the fit of the decision surface in the region where it matters.

3.5. Tree classifiers

Linear decision surfaces have the attraction of simplicity, at least if one regards a weighted sum as simple. This is one reason that they are very widely used in commercial applications. Another type of classifier, the tree or recursive partitioning classifier, is also simple, though in a very different way.

Given a design set, we might observe that we can classify many of the points correctly merely by assigning all those with x_1 value greater than some threshold t_1 to class 1, and all the others to class 2. We might further observe that, if we then classified all those with $x_1 \leqslant t_1$ which also had $x_2 > t_2$ to class 1 and all those with $x_1 \leqslant t_1$ which also had $x_2 \leqslant t_2$ to class 2, then even better design set classification accuracy resulted. We could repeat this splitting exercise as long as we liked.

The basic idea here is to take a data set and find a single split on a single variable which increases the overall *purity* when the data are split into subsets, where a set is more pure the more it is dominated by objects from a single class. Each such subset is then further split, again leading to increased purity of each of its component subsets. Splitting the design set in this way leads to a partition of the **x** space. This exercise has an obvious graphical representation as a tree graph, with the entire **x** space as the root node, with edges leading to nodes representing the cells into which it is split, and with edges leading from these to nodes representing the cells into which it is split, and so on. In practical implementations, each split is chosen by an extensive search, over all possible splits on all of the variables, to find that one which leads to maximum improvement in overall purity.

Various decisions have to be made in building such a tree. Should each split be binary? Binary trees are the most common, though some have argued for the merits of shallower trees in which each split is into multiple cells. For how long should the process continue? If it is continued for too long, there is a real danger of overfitting. The most popular modern method for tackling this is to grow a large tree (many splits) and prune it back, measuring classification accuracy using cross-validation or some similar approach. How should one measure purity? That is, how should one decide on how effective a split is? Various impurity measures have been proposed, all tapping into the decrease in heterogeneity of classes within each node which results when a split is made.

New points are classified by dropping them through the tree, deciding which branch to follow simply by making the appropriate comparisons with the thresholds at each node. When a leaf node is reached, the new object is classified as belonging to the majority class in that node (appropriately modified by relative costs, if necessary). The simplicity of the sequential process is especially attractive in some disciplines, such as medicine, where it can be written down as a protocol to be followed in making a decision, such as a diagnosis.

3.6. Local nonparametric methods

All of the methods discussed so far have found parametric models for the decision surface (that is, decision surfaces which are described in terms of relatively few parameters). An alternative strategy is to use notions of nonparametric function estimation. Broadly speaking, there are two approaches (though, of course, as with the other methods described in this article, they have been extended in many directions). The first, based on the sampling paradigm, uses *kernel density estimation* to provide estimates of the class conditional distributions $P(\mathbf{x} \mid \omega_i)$, which are then inverted via Bayes theorem to give classifications. The second is the *k-nearest neighbour* method, based on the diagnostic paradigm. At its simplest, this finds the k nearest design set points to the point to be classified, in terms of distance in **x** space, and assigns the new point to the majority class amongst these k. Of these two classes of method, the nearest neighbour approach seems to be the most widely used.

Nearest neighbour methods are attractive for a variety of reasons. They produce a highly flexible decision surface, with the degree of flexibility being controlled by the choice of k. They can very easily be updated, simply by adding new cases to the database (and deleting old ones, if that is felt necessary) – no parameters need to be recalculated. In general, they do not require extensive preprocessing, since no parameters need to be estimated (and are therefore a type of 'lazy learning', in machine learning terminology). Of course, there are downsides. One is that they involve an extensive search through the design set to locate the k nearest neighbours. If classification speed is important, this can be an issue. This problem has been tackled by accelerated search procedures (such as branch and bound) and also by reduced, condensed, and edited nearest neighbour procedures, which preprocess the design set to eliminate points which do not influence the classification. For example, a single isolated point from class ω_1 in a region dense with points from class ω_2 will be irrelevant if k is not too small. And in regions which are very densely populated with points from a single class the classification results would be the same even if some of these points were removed.

Of course, in order to use nearest neighbour methods one has to choose a value for k and a metric through which to define 'near'. There is early theoretical work relating k to things such as design set size and number of variables, but a more modern approach is to use cross-validation. For the two class case, an optimal metric would be distance orthogonal to the contours of constant probability of $P(\omega_1 \mid \mathbf{x})$. Since, of course, we do not know these contours (indeed, it is these probabilities which we are trying to estimate) iterative procedures have been proposed. Since, moreover, we will not know the contours exactly, one should relax the notion of strict distance from the contours. Thus, for example, suppose that the direction orthogonal to the contours at $\mathbf{x}$ is estimated to be $\mathbf{w}$. Then one can define distance between points in the vicinity of $\mathbf{x}$ as

$$d(\mathbf{x}, \mathbf{y}) = \left\{ (\mathbf{x} - \mathbf{y})'(\mathbf{I} + \lambda \mathbf{w}\mathbf{w}')(\mathbf{x} - \mathbf{y}) \right\}^{1/2} \tag{20}$$

where λ is a parameter which can be chosen by fit to the data (e.g., by cross-validation). The larger λ is, the more strictly the metric measures deviation from the estimated contour. More sophisticated variants let the orientation of these contours depend on the position in the $\mathbf{x}$ space.

3.7. Neural networks

Neural networks were originally proposed as models of biological nervous systems, which consist of large numbers of relatively simple processing units connected in a dense network of links. However they have subsequently been intensively developed as tools for pattern recognition and classification. Their earliest manifestation is in the perceptron, outlined above. However, the perceptron fits a *linear* decision surface. Such a decision surface has some strong limitations. In particular, it cannot perfectly solve the 'exclusive-or' problem. For example, if, in a bivariate $\mathbf{x}$ space, the points $(0, 0)$ and $(1, 1)$ belong to one class and the points $(0, 1)$ and $(1, 0)$ to another, then no linear decision surface can perfectly classify them. This problem can only be solved by introducing nonlinear aspects. Neural networks, or multilayer perceptrons, solve this problem by combining several simple weighted sums of the components of $\mathbf{x}$ (i.e., several simple perceptrons), but nonlinearly transforming them before combining them.

The final model has the form

$$f(\mathbf{x}) = \sum_{k=1}^{m} v_k \phi_k(\mathbf{w}'_k \mathbf{x}). \tag{21}$$

Here $\mathbf{w}_k$ is the weight vector for the kth linear combination of the raw variables, ϕ_k is a nonlinear transformation, and $\mathbf{v} = (v_1, \ldots, v_m)$ is a weight vector to produce the final result. The model in (21) has a graphical representation in which the x_j correspond to input nodes, linked by edges weighted by the w_{jk} to *hidden nodes* corresponding to the $\phi_k(\mathbf{w}'_k \mathbf{x})$, which are in turn linked by edges weighted by the v_k to the output node f. In the case of multiple classes, there will be several output functions (and nodes), f_r, defined via different weight vectors.

Note that, without the nonlinear transformations ϕ_k, (21) would reduce to a linear combination of linear combinations, and this is simple a linear combination. That is, the nonlinear transformations are essential: without them, we are back to a simple linear separating surface. Various nonlinear transformations are in common use, with one of the most popular being a logistic transformation. This means that logistic discrimination, outlined above, can be regarded as a particularly simple neural network, with no hidden node (since it has only one $\phi_k(\mathbf{w}'_k \mathbf{x})$, which will then equal f).

The basic model in (21) has been extended in a wide variety of ways, most importantly by including further hidden layers. Thus the f_r may themselves not be directly observed, but (nonlinear transformations of) linear combinations of them may contribute to a weighted sum which is the final output.

The result of all this is a highly flexible (and highly parameterised) model, which has great power to fit arbitrarily complex decision surfaces. Of course, as has been noted above, such power carries its own danger and various smoothing methods have been developed to overcome the risk of overfitting. A further complication arises in fitting such a model, again because of the large number of parameters. In fact, the idea for generalising perceptrons in this way arose in the 1960s, but was not pursued because of the computational difficulties involved in the estimations. It was not until a couple of decades had elapsed, and the power of computers had progressed substantially, that effective fitting algorithms arose. The first of these was a steepest descent algorithm, called *back propagation*, which sought to minimise the discrepancies between the target class labels and the predicted class labels. Since then, however, a wide variety of other algorithms have been developed.

3.8. Support vector machines

If we have three different points, belonging to two classes, lying in a bivariate space, then (apart from cases where the three points lie in a univariate subspace) we can always guarantee being able to separate the classes by a linear surface. This idea generalises. If we have 100 points from two classes lying in a 99-dimensional space (again ignoring pathological special cases) we can guarantee being able to find a hyperplane which will separate the classes. The basic principle behind support vector machines is to take the original design set, defined in terms of (say) a p-dimensional measurement vector $\mathbf{x}$, and increase the dimensionality dramatically by adding combinations and transformations of the original variables, so that perfect separation by a hyperplane is guaranteed in this new high-dimensional space. Once one is in the situation of guaranteed separability by a linear function, one can apply the error correcting ideas of perceptrons. We see from this that the perceptron was the intellectual progenitor of both neural networks and support vector machines, although they are quite different kinds of offspring.

In mathematical terms, the function which arises from this process can be expressed as

$$f(\mathbf{x}) = \sum_{k=1}^{K} v_k \phi_k(\mathbf{x}) \tag{22}$$

where K is the large number of the derived variables $\phi_k(\mathbf{x})$ obtained by transforming the $\mathbf{x}$, and v_k are the weights in the perfectly separating hyperplane. Of course, we immediately see a problem: if K is large (even infinite), then actually estimating the v_k may be difficult.

This is overcome by an elegant mathematical trick, and it was this which opened up such models to practical use. It is possible to show that (22) can be re-written in the equivalent, 'dual' formulation

$$f(\mathbf{x}) = \sum_{i=1}^{N} \alpha_i t_i \boldsymbol{\phi}(\mathbf{x}_i)' \boldsymbol{\phi}(\mathbf{x}) = \sum_{i=1}^{N} \alpha_i t_i K(\mathbf{x}_i, \mathbf{x}) \tag{23}$$

where the t_i are the true classes, α_i are weights to be estimated,

$$\boldsymbol{\phi}(\mathbf{x}) = \big(\phi_1(\mathbf{x}), \ldots, \phi_K(\mathbf{x})\big)', \tag{24}$$

$K(\mathbf{x}_i, \mathbf{x}) = \boldsymbol{\phi}(\mathbf{x}_i)' \boldsymbol{\phi}(\mathbf{x})$ is the *kernel* function (a term inherited from integral operator theory), and N is the design set size (with $\mathbf{x}_i$ being the ith design set point). The key fact here is that we do not need to make the separate transformations, $\phi_k(\mathbf{x})$. All we need is the kernel function in order to be able to evaluate (23), and hence (22). The choice of kernel function is equivalent to a choice of features, and we can derive the prediction $f(\mathbf{x})$ without explicitly evaluating the $\phi_k(\mathbf{x})$.

Models of this kind have also been substantially extended, for example to allow for the case where the classes cannot be perfectly separated, even in the transformed space. Support vector machines are a relatively recent innovation, and although they have shown promise on a number of practical problems, more experience needs to be accumulated.

3.9. Other approaches

Many other types of classifiers have been developed, typically variants of those described above, and many of these are described in the general references listed below. However, a particular advance which is widely applicable should be mentioned. This is the process of combining classifiers into *multiple classifier systems* or *ensemble classifiers*. This can be achieved in various ways (and, indeed, with various theoretical objectives).

As has been seen, flexible classifiers are susceptible to overfitting problems. This can be tackled by various methods, including adding complexity measures to the criterion which is optimised to fit such models (so that they are encouraged not to be too flexible), stopping optimisation algorithms early (a rather ad hoc strategy), or to overfit a model and then smooth it (this is the pruning strategy often adopted with tree classifiers). Another approach is to build many slightly different classification models (for example, on randomly selected subsets of the training data) and to calculate average predictions from these models. This idea of using an *ensemble* of classifiers, called *bagging*, short for *bootstrap aggregating*, is closely related to Bayesian ideas, and can be regarded as an attempt to allow for uncertainty in the parameter estimates.

Another important strategy for combining classifiers to yield an overall improved model is *boosting*. This weights points according to how difficult they are to classify correctly, but has been elegantly described in terms of generalised additive models (Friedman et al., 2000).

4. Other issues

This article has primarily focused on describing some of the most important kinds of classification rule, but there are other issues which are also central in constructing an effective pattern recognition system. Two important ones are those of *feature selection* and *performance assessment*.

The key to an accurate pattern recognition system is that the predictor variables contain sufficient information to separate the classes well. One might think, therefore, that the more variables one had, the better. However, as we have seen, too many variables can lead to overfitting the design data. One is then often faced with having to select from the proposed variables. Choosing a subset also has the merit that then only these need be measured in the future – not important in automated data collection systems, such as microarray data, but vital in social problems, where each variable may correspond to a question in a questionnaire. Feature selection algorithms are generally based on simple measures of separability between the classes, but, of course, the feature selection and classifier construction process are really parts of an integrated whole.

Of fundamental importance in building and choosing a pattern recognition system is being able to estimate its likely future performance accurately. 'Performance', however, can be measured in various ways. Even for the two class case, there are two different kinds of misclassification and, as noted above, these may not be regarded as equally serious. If they are, then error rate is an appropriate measure. But even this is not the end of the story, since estimating likely future error rate on the basis of the available design sample involves subtle issues. For example, simply calculating the proportion of the design set cases which are misclassified by the classification rule is likely to underestimate the future error rate on new objects. Many methods for estimating error rates have been developed. A review of recent developments is given in Schiavo and Hand (2000). If the two types of misclassification costs are not equal, but are still known, then cost weighted error rate can be used. However, all too often the costs are not known, and then a more generally measure of separability between the distributions of the estimates

$\widehat{P}(\omega_1 \mid \mathbf{x})$ for the two classes is required. The most popular of these is the 'area under the ROC curve', equivalent to the test statistic used in the Mann–Whitney–Wilcoxon two sample test.

This article has given a lightning review of generic pattern recognition systems, but particular problems will require their own particular refinements. Sometimes one class is much smaller than the other (e.g., in fraud detection in personal banking, less than 0.1% of transactions will typically be suspect). Often there are problems of missing or distorted data (sometimes even the class labels are incorrect). Modern automated data collection has led to some special problems (e.g., microarray data) in which there may be tens of thousands of variables but only a handful of cases, so that the design data lie in a subspace of the $\mathbf{x}$ space. Many situations are susceptible to *population drift*, in which the distributions $P(\mathbf{x} \mid \omega_i)$ evolve over time (for example, predicting the classes of people in customer relationship management systems, when changes in the economy induces changes in spending behaviour). Some applications place a great emphasis on ease of interpretability of a classifier (e.g., the legal requirement that a bank must be able to explain the reason for rejecting an applicant for a loan). Some applications require very rapid classifications (e.g., target recognition). And so on. Given this, it is not surprising that specialist areas of pattern recognition have developed certain classes of methods to a great level of sophistication (e.g., speech recognition has developed highly sophisticated hidden Markov models).

5. Further reading

There are now many books giving overall reviews of statistical pattern recognition. For example, Ripley (1996) presents a theoretically deep overview of the area. Hastie et al. (2001) is an excellent overall review of the ideas and methods. Webb (2002) gives a highly accessible introduction to the subject. McLachlan (1992) gives a detailed discussion from the statistical perspective (though not including neural networks or support vector machines, which indicates the speed with which the discipline has developed). Hand (1997) reviews the different approaches and discusses performance assessment in detail. Fine (1999) gives a thorough mathematical introduction to feed-forward neural networks. Cristianini and Shawe-Taylor (2000) provides a first class introduction to support vector machines.

References

Anderson, J.A. (1982). Logistic discrimination. In: Krishnaiah, P.R., Kanal, L.N. (Eds.), *Classification, Pattern Recognition and Reduction of Dimensionality*. In: *Handbook of Statistics*, vol. 2. North-Holland, Amsterdam, pp. 169–191.

Cox, D.R. (1966). Some procedures associated with the logistic qualitative response curve. In: David, F. (Ed.), *Research Papers on Statistics: Festschrift for J. Neyman*. Wiley, New York, pp. 55–71.

Cristianini, N., Shawe-Taylor, J. (2000). *An Introduction to Support Vector Machine and Other Kernel-Based Learning Methods*. Cambridge University Press, Cambridge.

Day, N.E., Kerridge, D.F. (1967). A general maximum likelihood discriminant. *Biometrics* **23**, 313–324.

Fine, T.L. (1999). *Feedforward Neural Network Methodology*. Springer-Verlag, New York.
Fisher, R.A. (1936). The use of multiple measurements in taxonomic problems. *Ann. Eugenics* **7**, 179–184.
Friedman, J. (1989). Regularized discriminant analysis. *J. Amer. Statist. Assoc.* **84**, 165–175.
Friedman, J., Hastie, T., Tibshirani, R. (2000). Additive logistic regression: a statistical view of boosting (with discussion). *Ann. Statist.* **28**, 307–337.
Hand, D.J. (1997). *Construction and Assessment of Classification Rules*. Wiley, Chichester.
Hand, D.J., Yu, K. (2001). Idiot's Bayes – not so stupid after all? *Internat. Statist. Rev.* **69**, 385–398.
Hastie, T., Tibshirani, R., Friedman, J. (2001). *The Elements of Statistical Learning Theory*. Springer-Verlag, New York.
McCullagh, P., Nelder, J.A. (1989). *Generalized Linear Models*, 2nd ed. Chapman and Hall, London.
McLachlan, G.J. (1992). *Discriminant Analysis and Statistical Pattern Recognition*. Wiley, New York.
Ripley, B.D. (1996). *Pattern Recognition and Neural Networks*. Cambridge University Press, Cambridge.
Schiavo, R., Hand, D.J. (2000). Ten more years of error rate research. *Internat. Statist. Rev.* **68**, 295–310.
Webb, A. (2002). *Statistical Pattern Recognition*, 2nd ed. Wiley, Chichester.

Multidimensional Density Estimation

David W. Scott[1] and Stephan R. Sain[2]

Abstract

Modern data analysis requires a number of tools to undercover hidden structure. For initial exploration of data, animated scatter diagrams and nonparametric density estimation in many forms and varieties are the techniques of choice. This article focuses on the application of histograms and nonparametric kernel methods to explore data. The details of theory, computation, visualization, and presentation are all described.

Keywords: Averaged shifted histograms; Contours; Cross-validation; Curse of dimensionality; Exploratory data analysis; Frequency polygons; Histograms; Kernel estimators; Mixture models; Visualization

1. Introduction

Statistical practice requires an array of techniques and a willingness to go beyond simple univariate methodologies. Many experimental scientists today are still unaware of the power of multivariate statistical algorithms, preferring the intuition of holding all variables fixed, save one. Likewise, many statisticians prefer the familiarity of parametric statistical techniques and forgo exploratory methodologies. In this chapter, the use of density estimation for data exploration is described in practice with some theoretical justification. Visualization is an important component of data exploration, and examples of density surfaces beyond two dimensions will be described.

We generally consider the analysis of a d-variate random sample $(\mathbf{x}_1, \ldots, \mathbf{x}_n)$ from an unknown density function, $f(\mathbf{x})$, where $\mathbf{x} \in \Re^d$. It is worth reminding ourselves that (theoretically) for the analysis of a random sample, perfect knowledge of the density functions $f(\mathbf{x})$ or $f(\mathbf{x}, y)$ means that anything we may need to know can be computed. In practice, the computation may be quite complicated if the dimension of the data is high, but the greatest challenge comes from not knowing a parametric form for the

---
[1] This research was supported in part by the National Science Foundation grants NSF EIA-9983459 (digital government) and DMS 02-04723 (nonparametric methodology).
[2] This research was supported in part through the Geophysical Statistics Project at the National Center for Atmospheric Research under National Science Foundation grant DMS 9815344.

density $f(\mathbf{x})$. Fisher (1932) referred to this step in data analysis as the problem of specification. Nonparametric methodology provides a consistent approach for approximating in a large class of unknown densities, at a cost of less efficient estimation when the correct parametric form is known. Of course, if an incorrect parametric form is specified, then bias will persist.

2. Classical density estimators

The statistical analysis of continuous data is a surprisingly recent development. While data collection such as a population census can be traced back thousands of years, the idea of grouping and tabulating data into bins to form a modern frequency curve seems to have only arisen in the seventeenth century. For example, John Graunt (1662) created a crude histogram of the age of death during the English plague years using the bills of mortality, which listed the cause and age of death week by week. In Figure 1, we analyze the closing price, $\{x_t\}$, of the Dow Jones Industrial (DJI) average from 2/3/1930 to 2/27/2004. A plot and histogram of the 18 598 daily change ratios, x_{t+1}/x_t, are shown. While the eye is drawn to the days when the ratio represents more than a 5% change, the histogram emphasizes how rare such events are. The eye is generally a poor judge of the frequency or density of points in a scatter diagram or time series plot.

By the nineteenth century, the histogram was in common use, as was its continuous cousin the frequency polygon. (The frequency polygon interpolates the midpoints of a histogram in a piecewise linear fashion.) The first publication to advocate a systematic rule for the construction of a histogram was due to Sturges (1926). Sturges was motivated by the notion of an ideal histogram in the case of normal data. Observe that the simplest example of a discrete density that is approximately normal is a binomial distribution with $p = 1/2$. Imagine a histogram of some binomial data with k bins, labeled $0, 1, \ldots, k-1$ and $p = 1/2$. Now the binomial density is $f(i) = \binom{k-1}{i} 2^{-(k-1)}$. Sturges argued that the binomial coefficient, $\binom{k-1}{i}$, could be taken as the idealized bin count of normal data as the number of bins, k, grows, since the binomial density looks normal as

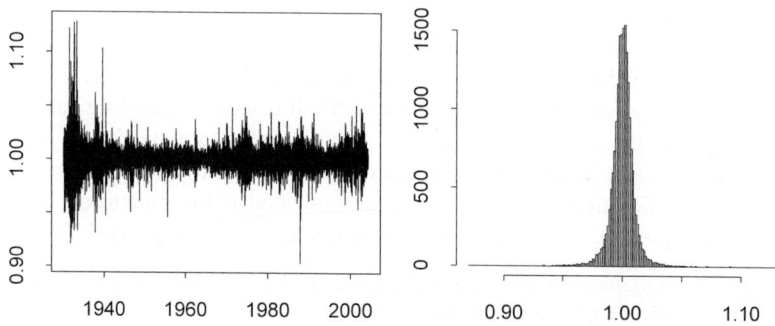

Fig. 1. Plot and histogram of daily change ratio, x_{t+1}/x_t, of the DJI.

k increases. Let v_k denote the bin count of the k-th bin. Then the total bin count is

$$n = \sum_{i=0}^{k-1} v_k = \sum_{i=0}^{k-1} \binom{k-1}{i} = \sum_{i=0}^{k-1}(1+1)^{k-1} = 2^{k-1};$$

hence, given n data points, the appropriate number of bins, k, may be solved as $k = 1 + \log_2(n)$. Let the bin width of a histogram be denoted by h, then Sturges' rule for the bin width of a histogram of a random sample $\{x_1, x_2, \ldots, x_n\}$ may be expressed as

$$h = \frac{x_{(n)} - x_{(1)}}{1 + \log_2(n)}, \tag{1}$$

where $x_{(i)}$ is the i-th order statistic of the sample. Equation (1) is still in common use in most statistical software packages. However, we shall see that this rule-of-thumb is far from optimal in a stochastic setting.

A histogram is a nonparametric estimator because it can successfully approximate almost any density as $n \to \infty$. The only alternative to the histogram before 1900 was a parametric model, $f(x \mid \theta)$, and θ was usually estimated by the method of moments. Pearson (1902) introduced a hybrid density estimator from the family of solutions to the differential equation

$$\frac{d \log f(x)}{dx} = \frac{x - a}{bx^2 + cx + d}. \tag{2}$$

The parameters (a, b, c, d) are estimated by the method of moments. (One parameter serves as a normalization constant, leaving three degrees-of-freedom). The particular "type" or solution of Eq. (2) depends upon the roots of the denominator and special values of the parameters. A number of common distributions satisfy this differential equation, including the normal and t densities. In fact, Gosset made use of this fact to derive the t-distribution (Student, 1908).

The Pearson family of densities is not actually nonparametric since many densities are not in the family. However, the Pearson family is still used in many simulations today and is in the modern spirit of letting the data "speak" rather than imposing a fully parametric assumption.

In the following sections, the properties of the modern histogram and frequency polygon are examined.

2.1. Properties of histograms

In the univariate setting, the frequency histogram plots raw data counts and hence is not actually a density function. We limit our discussion to continuous data and a true density histogram, which is defined over a general mesh on the real line. Let the k-th bin correspond to the interval $B_k = [t_k, t_{k+1})$. Let the number of samples falling into bin B_k be denoted by v_k. Then the density histogram is defined as

$$\hat{f}(x) = \frac{v_k}{n(t_{k+1} - t_k)} \quad \text{for } x \in B_k. \tag{3}$$

In practice, an equally-spaced mesh is often chosen, with $t_{k+1} - t_k = h$ for all k and with bin origin $t_0 = 0$. In this case, the histogram estimator is simply

$$\hat{f}(x) = \frac{v_k}{nh} \quad \text{for } x \in B_k. \tag{4}$$

Now the histogram has only one unknown parameter, namely, the bin width h. Yet as we shall see, the histogram can asymptotically approximate any continuous density and hence earns the label "nonparametric". Some early writers suggested that nonparametric estimators were infinite-dimensional. In fact, the number of parameters has little bearing on whether an estimator is parametric or nonparametric. Rather, the important distinction is the local behavior of the estimator.

2.1.1. Maximum likelihood and histograms

How can maximum likelihood be applied to the density histogram? Let us begin with a general binned estimator of the form

$$\hat{f}(x) = f_k \quad \text{for } x \in B_k = [t_k, t_{k+1}).$$

Then the log-likelihood of the histogram is

$$\sum_{i=1}^{n} \log \hat{f}(x_i) = \sum_k v_k \log f_k \quad \text{subject to} \quad \sum_k (t_{k+1} - t_k) f_k = 1,$$

where we define $f_k = 0$ for bins where $v_k = 0$. The Lagrangian is

$$L(\mathbf{f}, \lambda) = \sum_k v_k \log f_k + \lambda \left[1 - \sum_k (t_{k+1} - t_k) f_k \right].$$

The stationary point for f_ℓ leads to the equation

$$\frac{\partial L(\mathbf{f}, \lambda)}{\partial f_\ell} = \frac{v_\ell}{f_\ell} - \lambda(t_{\ell+1} - t_\ell) = 0.$$

The constraint leads to $\lambda = n$; hence, $\hat{f}_\ell = v_\ell/n(t_{\ell+1} - t_\ell)$ as in Eq. (3). Thus the histogram is in fact a maximum likelihood estimator (MLE) within the class of simple functions (that is, given a pre-determined mesh).

If we extend the MLE to the estimation of h, or indeed of the entire mesh $\{t_k\}$, then we find that the likelihood is unbounded as $h \to 0$ or as $t_{k+1} - t_k \to 0$. Thus Duin (1976) introduced the leave-one-out likelihood. In the context of estimation of h in Eq. (4), the log-likelihood becomes

$$h^* = \arg\max_h \sum_{i=1}^{n} \log \hat{f}_{-i}(x_i \mid h), \tag{5}$$

where the leave-one-out density estimate is defined by

$$\hat{f}_{-i}(x_i) = \frac{v_k - 1}{(n-1)h}, \quad \text{assuming } x_i \in B_k.$$

While good results have been reported in practice, the procedure cannot be consistent for densities with heavy tails. Consider the spacing between the first two order statistics, $x_{(1)}$ and $x_{(2)}$. If $h < x_{(2)} - x_{(1)}$ and the bin count in the first bin is 1, then the likelihood in Eq. (5) will be $-\infty$, since $\hat{f}_{-(1)}(x_{(1)}) = 0$. Since a necessary condition for a histogram to be consistent will be shown to be that $h \to 0$ as $n \to \infty$, the spacings between all adjacent order statistics must vanish as well; however, such is not the case for many densities.

Thus most theoretical work on histograms (and other nonparametric estimators) has focused on distance-based criteria rather than the likelihood criterion. There are four common distance criteria between an estimator $\hat{f}(x)$ and the true but unknown density $g(x)$ (switching notation from $f(x)$) including:

$$\int |\hat{f}(x) - g(x)| \, dx \qquad \text{(integrated absolute error)},$$

$$\int \hat{f}(x) \log[\hat{f}(x)/g(x)] \, dx \qquad \text{(Kullback–Liebler distance)},$$

$$\int [\hat{f}(x)^{1/2} - g(x)^{1/2}]^2 \, dx \qquad \text{(Hellinger distance)},$$

$$\int [\hat{f}(x) - g(x)]^2 \, dx \qquad \text{(integrated squared error)}.$$

The first three are dimensionless, a characteristic which provides many potential benefits in practice. The second is basically the likelihood criterion. The fourth is the most amenable to theoretical investigation and calibration in practice. Integrated squared error (ISE) is also the L_2 distance between the estimator and the true density.

2.1.2. L_2 theory of histograms

We spend a little time outlining the derivation of the ISE for a histogram, since every estimator shares this problem. The derivation is quite straightforward for the histogram, and we will not provide these details for other estimators. We limit our discussion to equally-spaced histograms. Rosenblatt (1956) showed in a general fashion that no nonparametric estimator can be unbiased, and that the rate of convergence of any measure of error cannot achieve the parametric rate of $O(n^{-1})$. Since the histogram estimate, $\hat{f}(x)$, for a fixed x cannot be unbiased, then mean square error (MSE) is a natural criterion pointwise. Globally, MSE can be integrated over x to give the integrated mean square error (IMSE). By Fubini's theorem, IMSE is the same as mean integrated square error (MISE):

$$\text{IMSE} = \int \text{MSE}(x) \, dx = \int E[\hat{f}(x) - g(x)]^2 \, dx$$

$$= E \int [\hat{f}(x) - g(x)]^2 \, dx = \text{MISE}.$$

Finally, since MSE = Var + Bias^2, the IMSE is the sum of the mean integrated variance (IV) and the mean integrated squared bias (ISB).

The bin count, v_k, is a binomial random variable, $B(n, p_k)$, with probability given by the actual bin probability, $p_k = \int_{t_k}^{t_{k+1}} g(x)\,dx$. Hence, for $x \in B_k$, the pointwise variance of the histogram estimate given in Eq. (4) equals

$$\mathrm{Var}\,\hat{f}(x) = \frac{\mathrm{Var}(v_k)}{(nh)^2} = \frac{p_k(1-p_k)}{nh^2}.$$

Since the variance is identical for any $x \in B_k$, the integral of the variance over B_k multiplies this expression by the bin width, h. Therefore,

$$\mathrm{IV} = \sum_k \int_{B_k} \mathrm{Var}(\hat{f}(x))\,dx = \sum_k \frac{p_k(1-p_k)}{nh^2} \times h = \frac{1}{nh} - \sum_k \frac{p_k^2}{nh}, \quad (6)$$

since $\sum_k p_k = \int g(x)\,dx = 1$. Next, by the mean value theorem, $p_k = \int_{B_k} g(x)\,dx = h \cdot g(\xi_k)$ for some $\xi_k \in B_k$; thus, the final sum equals $n^{-1}\sum_k g(\xi_k)^2 h$, or approximately $n^{-1}\int g(x)^2\,dx$. Thus the variance of a histogram pointwise or globally may be controlled by collecting more data (larger n) or having sufficient data in each bin (wider h).

The bias is only a little more difficult to analyze. Clearly,

$$\mathrm{Bias}\,\hat{f}(x) = \mathrm{E}\hat{f}(x) - g(x) = \frac{p_k}{h} - g(x) \quad \text{for } x \in B_k.$$

Again using the fact that $p_k = h \cdot g(\xi_k)$, $\mathrm{Bias}\,\hat{f}(x) = g(\xi_k) - g(x) = O(hg'(x))$, assuming the unknown density has a smooth continuous derivative, since a Taylor's series of $g(\xi_k)$ equals $g(x) + (\xi_k - x)g'(x) + o(h)$ and $|\xi_k - x| < h$ as both ξ_k and x are in the same bin whose width is h.

Thus the squared bias of $\hat{f}(x)$ is of order $h^2 g'(x)^2$, and the integrated squared bias is of order $h^2 \int g'(x)^2\,dx$. In contrast to the manner by which the variance is controlled, the bias is controlled by limiting the size of the bin width, h. In fact, we require that $h \to 0$ as $n \to \infty$. From Eq. (6), $nh \to \infty$ is also necessary.

Combining, we have that the global error of a fixed-bin-width histogram is

$$\mathrm{IMSE} = \mathrm{IV} + \mathrm{ISB} = \frac{1}{nh} + \frac{1}{12}h^2 \int g'(x)^2\,dx + O\left(\frac{1}{n} + h^4\right), \quad (7)$$

where the factor $1/12$ results from a more careful analysis of the difference between $g(\xi_k)$ and $g(x)$ for all x in B_k; see Scott (1979).

The IMSE in Eq. (7) is minimized asymptotically by the choice

$$h^* = \left[\frac{6}{nR(g')}\right]^{1/3} \quad \text{and} \quad \mathrm{IMSE}^* = (9/16)^{1/3} R(g')^{1/3} n^{-2/3}, \quad (8)$$

where the "roughness" of $g(x)$ is summarized by $R(g') \equiv \int g'(x)^2\,dx$. Indeed, the rate of convergence of the IMSE falls short of the parametric rate, $O(n^{-1})$.

2.1.3. Practical histogram rules

We support the idea that histograms constructed with bin widths far from h^* are still of potential value for exploratory purposes. Larger bandwidths allow for a clearer picture of the overall structure. Smaller bandwidths allow for fine structure, which may or may not be real (only a larger sample size will clarify the true structure). Smaller bandwidths may also be useful when the true density is a mixture of components, say, for example, a normal mixture $wN(0, 1) + (1 - w)N(\mu, \sigma^2)$. Obviously there is a different bandwidth appropriate for each component, and h^* represents a compromise between those bandwidths. Using a smaller bandwidth may allow excellent estimation of the narrower component, at the price of making the other component undersmoothed. Unless one uses the more general mesh in Eq. (3), such compromises are inevitable. Data-based algorithms for the general mesh are considerably more difficult than for a single parameter and may in fact perform poorly in practice.

Expressions such as those for the optimal parameters in Eq. (8) may seem of limited utility in practice, since the unknown density $g(x)$ is required. However, a number of useful rules follow almost immediately. For example, for the normal density, $N(\mu, \sigma^2)$, the roughness equals $R(g') = (4\sqrt{\pi}\sigma^3)^{-1}$. Therefore,

$$h^* = \left[\frac{24\sqrt{\pi}\sigma^3}{n}\right]^{1/3} \approx 3.5\sigma n^{-1/3}. \tag{9}$$

Compare this formula to Sturges' rule in Eq. (1). Since the logarithm is a very slowly increasing function, Sturges' bin width is too slow to decrease as the sample size increases, at least with respect to IMSE.

Formulae such as Scott's rule (9), with σ replaced by an estimate $\hat{\sigma}$, are variations of so-called normal reference rules. Almost any other density can be shown to be more complex, resulting in an optimal bin width that is narrower. In fact, we can make this idea quite explicit. Specifically, a calculus of variations argument can be formulated to find the smoothest ("easiest") density that has a given variance, σ^2. Terrell and Scott (1985) showed that this density is given by

$$g_1(x) = \frac{15}{16\sqrt{7}\sigma}\left(1 - \frac{x^2}{7\sigma^2}\right)^2, \quad -\sqrt{7}\sigma < x < \sqrt{7}\sigma,$$

and zero elsewhere. Thus, for all densities with variance σ^2, $R(g') \geq R(g'_1)$. Since $R(g'_1) = 15\sqrt{7}/343\sigma^3$, the optimal bandwidth for density g_1 is

$$h_1^* = h_{OS} = \left[\frac{686\sigma^3}{5\sqrt{7}n}\right]^{1/3} \approx 3.73\sigma n^{-1/3},$$

where OS $\equiv$ oversmoothed. The normal reference bandwidth is only 6% narrower, confirming that the normal density is very close to the "oversmoothed" or "smoothest" density, which is in fact Tukey's biweight function. Since any other density is rougher than g_1, the optimal bandwidth satisfies the inequality $h^* \leq h_{OS}$. Since σ can be estimated quite reliably from the data, we have bounded the search region for h^* to the interval $(0, h_{OS})$. This is a very useful result in practice.

In fact, we can use cross-validation to obtain an estimate of h^* itself. Rudemo (1982) and Bowman (1984) showed that the integrated squared error can be estimated for each choice of h by replacing the second term in the expanded version of the integrated squared error:

$$\text{ISE} = \int \left[\hat{f}(x \mid h) - g(x)\right]^2 dx$$
$$= \int \hat{f}(x \mid h)^2 dx - 2\int \hat{f}(x \mid h)g(x) dx + \int g(x)^2 dx \qquad (10)$$

with the unbiased estimator

$$-\frac{2}{n}\sum_{i=1}^{n} \hat{f}_{-i}(x_i \mid h) \approx -2E\left[\hat{f}(X \mid h)\right] = -2\int \hat{f}(x \mid h)g(x) dx.$$

The final integral in (10), $\int g(x)^2 dx$, is unknown but is constant with respect to h and may be ignored. Thus the least-squares or unbiased cross-validation (UCV) criterion which estimates ISE $- \int g(x)^2 dx$ is

$$\int \hat{f}(x \mid h)^2 dx - \frac{2}{n}\sum_{i=1}^{n} \hat{f}_{-i}(x_i \mid h). \qquad (11)$$

This approach requires the additional assumption that the true density is square integrable for all choices of the smoothing parameter.

There have been a number of further algorithms proposed that focus on improved point estimates of the roughness functional, $R(g')$. However, such approaches are still limited to a single bandwidth. Rudemo (1982) showed how unbiased cross-validation may easily be extended to the variable mesh case. Wegman (1975) demonstrated the strong consistency of a variant of the variable-mesh histogram based upon maintaining a minimum number of samples in each bin. The idea was ultimately the motivation for the random bin width estimators of Hearne and Wegman (1994). Kogure (1987) and Kanazawa (1992) also examine variable mesh histograms. Variable smoothing is discussed further in Section 3.4.

As an example, consider again the ratio of DJI closing prices, $\{x_{t+1}/x_t\}$, since 1930. The oversmoothed bin width is $3.73\hat{\sigma}n^{-1/3} = 0.00137$, which was used to construct the histogram in Figure 1. Sturges' rule suggests $1 + \log_2(18598) = 15.2$ bins. The left frame in Figure 2 uses 16 bins over the sample range. To see if there is structure

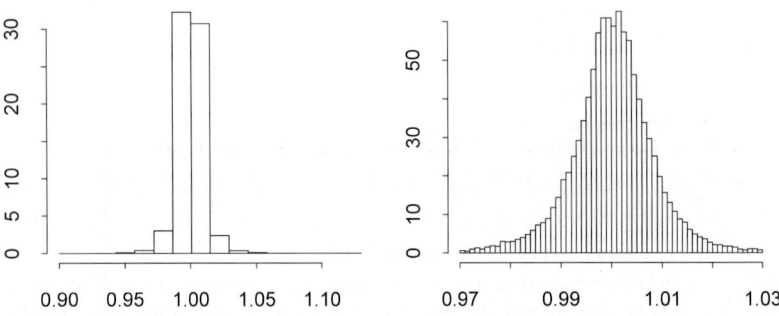

Fig. 2. Histograms of x_{t+1}/x_t of the DJI chosen by Sturges' rule and by eyeball.

that might be revealed if $h < h_{\text{OS}}$, we show a detail in the right frame of Figure 2 with $h = 0.001$, which is about 27% narrower than h_{OS}. The minimizer of the UCV criterion (11) occurs at $h = 0.00092$. Note that strictly speaking, this example does not represent a random sample since the data are a correlated time series; however, the rules of thumb still seem useful in this case. Observe that Sturges' rule grossly oversmooths the data in any case.

2.1.4. Frequency polygons

The use of the piecewise linear frequency polygon (FP) in place of the underlying histogram would seem mainly a graphical advantage. However, Fisher (1932 p. 37) suggested that the advantage was "illusory" and that a frequency polygon might easily be confused with the smooth true density. "The utmost care should always be taken to distinguish" the true curve and the estimated curve, and "in illustrating the latter no attempt should be made to slur over this distinction."

However, Scott (1985a) showed that the smoothness of the frequency polygon reduced not only the bias but also the variance compared to a histogram. A similar analysis of the estimation errors leads to the expression

$$\text{MISE} = \frac{2}{3nh} + \frac{49}{2880} h^4 R(g'') + O\left(\frac{1}{n} + h^6\right),$$

where $R(g'') = \int g''(x)^2 \, dx$. Thus the best choice of the bin width for the underlying histogram is not that given in Eq. (7), but rather

$$h^* = c_0 c_g n^{-1/5} \quad \text{and} \quad \text{MISE}^* = c_1 c_g n^{-4/5},$$

where $c_0 = 1.578$, $c_1 = 0.528$, and $c_g = R(g'')^{-1/5}$. For $N(\mu, \sigma^2)$ data, $h^* = 2.15\sigma n^{-1/5}$, which, for large n, will be much wider than the corresponding histogram formula, $h^* = 3.5\sigma n^{-1/3}$. For example, when $n = 10^5$ with normal data, the optimal FP bin width is 185% wider than that of the histogram; the MISE of the FP is reduced by 81%. The wider bins allow the FP to achieve lower variance. The piecewise linear FP more closely approximates the underlying smooth density than the piecewise constant histogram. (In fact, piecewise quadratic estimates can achieve even closer approximation. However, such estimates often take on negative values and do not offer sufficient improvement in practice to recommend their use.).

In Figure 3, we display the common logarithm of the Canadian lynx data together with a histogram using $h = 0.4$, which is slightly less than the normal reference rule bandwidth $h = 0.47$. (The UCV bandwidth is 0.461.) In Figure 4, we display two shifted versions of the histogram in Figure 3. The theoretical analyses of the MISE for both the histogram and FP indicate that the effect of choosing the bin origin t_0 is relegated to the remainder (low-order) terms. However, the graphical impact is not negligible. The bimodal feature varies greatly among these three histograms. For the FP, the wider bins of its underlying histogram suggest that the choice of t_0 matters more with the FP than with the histogram. We revisit this below in Section 3.1.

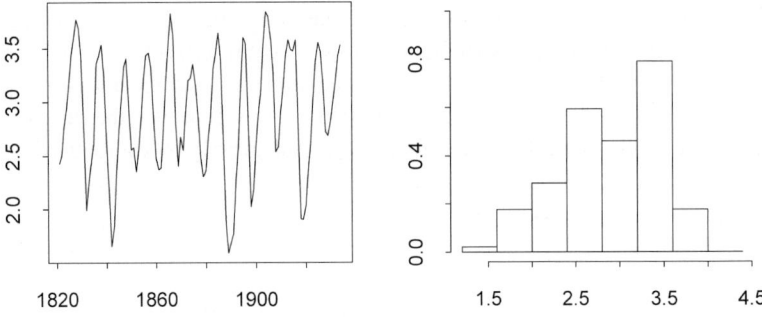

Fig. 3. Canadian lynx data ($\log_{10} x_t$) and its histogram ($h = 0.4$ and $t_0 = 2$).

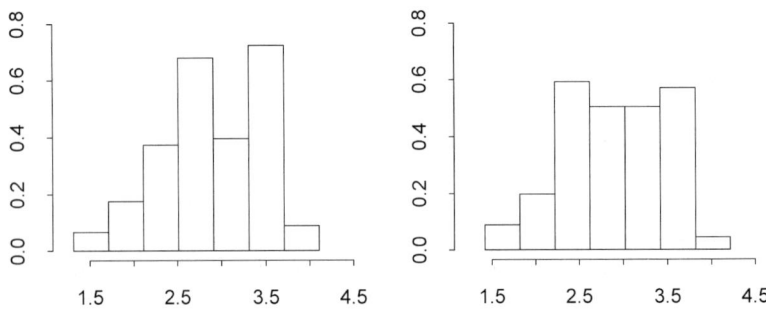

Fig. 4. Two shifted lynx histograms ($t_0 = 2.1$ and $t_0 = 2.2$) with $h = 0.4$.

2.1.5. Multivariate frequency curves

The power of nonparametric curve estimation is in the representation of multivariate relationships. In particular, density estimates in dimensions 3, 4, and even 5 offer great potential for discovery. Examples and visualization techniques are described in Section 5.

Beyond four dimensions, the effects of the so-called curse of dimensionality must be considered. The bias-variance tradeoff is subject to failure since the optimal bin widths must be large, and are generally too wide to avoid substantial bias. Imagine a large sample of 10^6 data uniformly distributed on the unit hypercube in $\Re^{10}$. If each axis is divided into 5 equal bins, then the hypercube has 5^{10} or almost ten million bins. Even such a crude binning leaves 90% of the bins empty. If each axis were divided into only 3 bins, then each bin would still have only 17 points on average. Thus these estimators must be quite biased, even with a truly enormous sample size.

However, the extension of the MISE analyses to the multivariate case is straightforward and serves to better quantify the effects of the curse of dimensionality. For a multivariate histogram with cubical bins of volume h^d, the IV is $O(1/nh^d)$ while the ISB remains of $O(h^2)$. Thus

$$\text{Histogram:} \quad h_d^* = O\left(n^{-1/(d+2)}\right) \quad \text{and} \quad \text{MISE}_d^* = O\left(n^{-2/(d+2)}\right).$$

The IV and ISB for the multivariate frequency polygon are $O(1/nh^d)$ and $O(h^4)$, respectively. Thus the situation is significantly improved (Scott, 1985a):

$$\text{FP:} \quad h_d^* = O\left(n^{-1/(d+4)}\right) \quad \text{and} \quad \text{MISE}_d^* = O\left(n^{-4/(d+4)}\right).$$

Perhaps the most encouraging observation is that the MISE convergence rate of order $n^{-2/5}$ is achieved not only by histograms in $d = 3$ dimensions but also by frequency polygons in $d = 6$ dimensions. Since a number of scientists have successfully used histograms in 3 (and even 4) dimensions, we believe that it is reasonable to expect useful nonparametric estimation in at least six dimensions with frequency polygons and other more smooth estimators. That is more than sufficient for graphical exploratory purposes in dimensions $d \leqslant 5$. Complete nonparametric estimation of a density function in more than six dimensions is rarely required.

3. Kernel estimators

3.1. Averaged shifted histograms

The Canadian lynx example in Figures 3 and 4 indicates that both the bin width and the bin origin play critical roles for data presentation. One approach would be to use unbiased cross-validation to search for the best pair (h, t_0). However, Scott (1985b) suggested that t_0 should be viewed as a nuisance parameter which can be eliminated by averaging several shifted histograms. The details of averaging m shifted histograms are easiest if each is shifted by an amount $\delta = h/m$ from the previous mesh. The averaged shifted histogram (ASH) will be constant over intervals of width δ, so we redefine the bin counts, $\{v_k\}$, to correspond to the mesh $B_k' = [t_0 + k\delta, t_0 + (k+1)\delta)$. Then Scott (1985b) shows that

$$\hat{f}_{\text{ASH}}(x) = \frac{1}{nh} \sum_{i=1-m}^{m-1} \left(1 - \frac{|i|}{m}\right) v_{k+i} \quad \text{for } x \in B_k'. \tag{12}$$

In Figure 5, all shifted histograms have $h = 0.4$. The first two frames show individual histograms with $t_0 = 2.0$ and $t_0 = 2.2$. The ASH with $m = 2$ is shown in the third frame, and so on. Eliminating t_0 shows that the data are clearly bimodal, with a

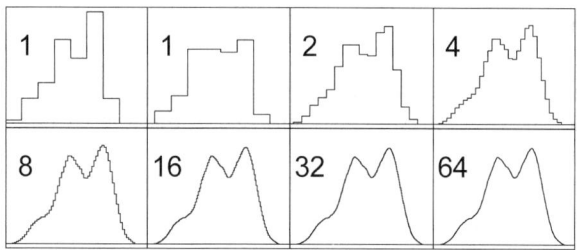

Fig. 5. Original averaged shifted histograms of lynx data.

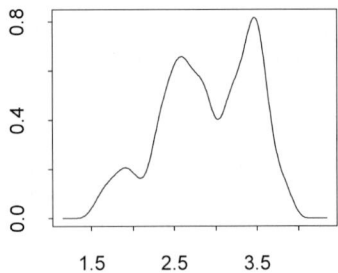
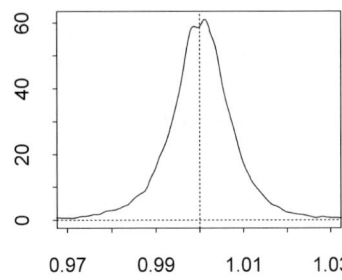

Fig. 6. Averaged shifted histograms of the lynx and DJI data using a smoother weight sequence $w_m(i) \propto (1 - (i/m)^2)_+^5$.

hint of a small bump on the left. The limiting ASH is continuous, which provides visual advantages. Connecting the midpoints of the ASH like a frequency polygon (called the FP-ASH) has some theoretical value, although for $m \geqslant 32$ in Figure 5, the discontinuous nature of the ASH is not visible.

Uniform weights on the shifted histograms are not the only choice. Choose a smooth symmetric probability density, $K(x)$, defined on $[-1, 1]$ that satisfies $K(\pm 1) = 0$. Define the weight function

$$w_m(i) = \frac{m \cdot K(i/m)}{\sum_{j=1-m}^{m-1} K(j/m)} \quad \text{for } i = 1-m, \ldots, m-1. \tag{13}$$

Then the generalized ASH in Eq. (12) is

$$\hat{f}_{\text{ASH}}(x) = \frac{1}{nh} \sum_{i=1-m}^{m-1} w_m(i) v_{k+i} \quad \text{for } x \in B'_k. \tag{14}$$

The weight function for the original ASH in Eq. (12) is the triangle kernel, $K(x) = 1 - |x|$, for $|x| < 1$ and zero elsewhere. Kernels in the shifted Beta family, $K_\ell(x) \propto (1 - x^2)_+^\ell$ are popular in practice. Tukey's biweight kernel corresponds to $\ell = 2$ while the normal kernel is well-approximated for large ℓ. Some examples of ASH estimates with $\ell = 5$ are shown in Figure 6. Notice that the use of a differentiable kernel makes the ASH visually smoother and the small bump on the left now appears clearer. The ASH of the DJI ratio reveals a small bimodal feature suggesting the DJI tries to avoid closing at the same level two days in a row.

3.2. Kernel estimators

As the number of shifts $m \to \infty$ in Eq. (13), the ASH approximates the so-called kernel estimator

$$\hat{f}_K(x) = \frac{1}{nh} \sum_{i=1}^{n} K\left(\frac{x - x_i}{h}\right) = \frac{1}{n} \sum_{i=1}^{n} K_h(x - x_i), \tag{15}$$

where the kernel, K, corresponds to K in Eq. (13) and the scaled kernel function, $K_h(x)$, is defined as $K_h(x) = h^{-1} K(h^{-1} x)$. Thus a kernel estimator is an equal mixture of n

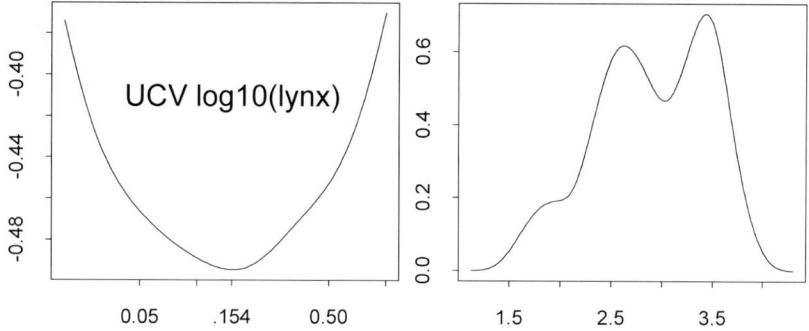

Fig. 7. Unbiased cross-validation and data-based optimal Gaussian kernel estimator.

kernels, centered at the n data points. For large n, the ASH requires much less work, since determining the bin counts is a linear operation, and the smoothing is a discrete convolution on the bin counts. (The kernel estimator may be viewed as a continuous convolution on all n points.) If one wanted to use the normal kernel, then much of the computational efficiency of the ASH would be lost. However, the Fast Fourier Transform can be used in that case; see Silverman (1982) for details. Using the FFT limits the ability to use boundary kernels or to estimate over a subset of the domain.

Choosing a good value for the bandwidth, h, is the most difficult task. The normal reference rule using a normal kernel is $h = 1.06\sigma n^{-1/5}$ for univariate data. More sophisticated plug-in rules have been described by Sheather and Jones (1991). However, we continue to recommend least-squares or unbiased cross-validation algorithms, which are well-studied for kernel estimators; see Rudemo (1982), Bowman (1984), and Sain et al. (1994). For the lynx data transformed by $\log_{10}$, the unbiased cross-validation function in Eq. (11) with the normal kernel suggests using the bandwidth $h = 0.154$; see Figure 7. The corresponding Gaussian kernel estimate is shown in the right frame of this figure. This estimator is slightly less rough than the ASH estimate shown in Figure 6, which was chosen by eye to highlight the small bump/mode near $x = 1.9$. However, at that narrow bandwidth, an extra bump seems to be present near $x = 2.8$. Using a single bandwidth for the entire domain implies such compromises. Locally adaptive smoothing is a possibility and is discussed in Section 3.4.

The unbiased cross-validation of the DJI time series also suggests a wider bandwidth than used for the ASH in Figure 6. The slightly bimodal feature at $x = 1$ disappears. However, care should be exercised when using cross-validation on time series data, since serial correlation is present. Specialized algorithms exist in this situation; see Hart (1984).

The choice of kernel is largely a matter of convenience. The family of scaled Beta densities provides collection of useful polynomial kernels of the form $K(x) \propto (1-x^2)_+^{\ell}$ on the interval $(-1, 1)$. As $\ell \to \infty$, this kernel converges to the normal kernel. The normal kernel has one advantage in practice; namely, as the smoothing parameter h increases, the number of modes is monotone non-increasing (Silverman, 1981). This property led Minnotte and Scott (1993) to propose the "mode tree", which plots the location of modes of a normal kernel estimator as a function of h. Minnotte (1997) proposed a local bootstrap test for the veracity of individual modes by examining the size of modes at critical points in the mode tree. Chaudhuri and Marron (1999) have introduced a graphical tool called SiZer to test the features in a kernel density.

3.3. Multivariate kernel options

The extension of the kernel estimator to vector-valued data, $\mathbf{x} \in \Re^d$, is straightforward for a normal kernel, $K \sim N(0, \Sigma)$:

$$\hat{f}(\mathbf{x}) = \frac{1}{n(2\pi)^{d/2}|\Sigma|^{1/2}} \sum_{i=1}^{n} \exp\left[-\frac{1}{2}(\mathbf{x} - \mathbf{x}_i)' \Sigma^{-1} (\mathbf{x} - \mathbf{x}_i)\right]. \quad (16)$$

It is convenient to separate the "size" of Σ from the "orientation" of Σ. To that end, write $\Sigma = h^2 A$, where $|A| = 1$. Thus, the size of Σ is $|h^2 A| = h^{2d}$. The Gaussian kernel estimate becomes

$$\hat{f}(\mathbf{x}) = \frac{1}{n(2\pi)^{d/2}h^d} \sum_{i=1}^{n} \exp\left[-\frac{1}{2}\left(\frac{\mathbf{x} - \mathbf{x}_i}{h}\right)' A^{-1} \left(\frac{\mathbf{x} - \mathbf{x}_i}{h}\right)\right]. \quad (17)$$

Since A is a symmetric, positive-definite matrix, the symmetric, positive-definite square-root matrix $A^{-1/2}$ exists. Hence, Eq. (17) becomes

$$\hat{f}(\mathbf{x}) = \frac{1}{n(2\pi)^{d/2}h^d} \sum_{i=1}^{n} \exp\left[-\frac{1}{2} \frac{(A^{-1/2}(\mathbf{x} - \mathbf{x}_i))'}{h} \frac{(A^{-1/2}(\mathbf{x} - \mathbf{x}_i))}{h}\right]. \quad (18)$$

This equation proves it is equivalent to rotate the data by the transformation $A^{-1/2}$ and then apply the $N(0, I_d)$ kernel. This transformation is almost into the principal components, except that the final scaling is not applied to make the variances all equal. In this transformed space, the kernel estimate is

$$\hat{f}(\mathbf{x}) = \frac{1}{n(2\pi)^{d/2}h^d} \sum_{i=1}^{n} \exp\left[-\frac{1}{2}\left(\frac{\mathbf{x} - \mathbf{x}_i}{h}\right)'\left(\frac{\mathbf{x} - \mathbf{x}_i}{h}\right)\right]$$

$$= \frac{1}{n} \sum_{i=1}^{n} K_h(\mathbf{x} - \mathbf{x}_i),$$

where $K_h(\mathbf{x}) = h^{-d} K(\mathbf{x}/h) = \prod_{k=1}^{d} \phi(x^{(k)} \mid 0, h)$.

We recommend working with transformed data and using either the normal kernel or, more generally, a product kernel, possibly with different smoothing parameter, h_k, in the k-th direction:

$$\hat{f}(\mathbf{x}) = \frac{1}{n} \sum_{i=1}^{n} \left[\prod_{k=1}^{d} K_{h_k}\left(x^{(k)} - x_i^{(k)}\right)\right]. \quad (19)$$

The *ashn* software (see Section 5) computes an approximation of the multivariate product kernel estimate with kernels selected from the rescaled Beta family.

The multivariate rule-of-thumb for the bandwidth h is surprisingly simple. Assuming a normal product kernel and a true density that is also normal with $\Sigma = I_d$, then to close approximation

$$h^* = n^{-1/(d+4)} \quad \text{or} \quad h_k^* = \hat{\sigma}_k n^{-1/(d+4)}$$

for the general normal product estimator (19). For other choices of kernel, Scott (1992) provides a table of constants by which to multiple h_k^*.

Full cross-validation may be used to estimate h or $(h_1, \ldots, h_d)$ from the data; see Sain et al. (1994). Estimating the shape of the kernel parameters in the matrix A is generally not advisable, as there are too many parameters in high dimensions. We demonstrate below in Section 3.4 that useful estimates of A may be obtained in two dimensions. Wand and Jones (1994) describe multivariate plug-in bandwidth rules, which can be more stable. However, it is important to note that kernel methods cannot handle rank-deficient data. Such degenerate cases can often be detected by computing the principal components and throwing away dimensions where the eigenvalues are essentially zero.

3.4. Locally adaptive estimators

As any practitioner will note, more smoothing is needed to counter the excessive variation in the tails of a distribution where data are scarce while less smoothing is needed near the mode of a distribution to prevent important features from being diminished in the resulting estimate. Several situations have been discussed (e.g. multimodal and multivariate distributions) where the bias-variance trade-off that drives most global bandwidth choices can lead to estimates that lack visual appeal and make feature recognition difficult.

These situations have often motivated the notion of a variable bandwidth function that allows different amounts of smoothing depending on the various characteristics of the data and the density being estimated. Two simplified forms of such estimators have been studied extensively. The first, the *balloon* estimator, varies the bandwidth with the estimation point. The second varies the bandwidth with each estimation point and is referred to as the *sample point* estimator. Jones (1990) gives an excellent comparison of such estimators in the univariate case while Terrell and Scott (1992) and Sain (2002) examined each of the two different formulations in the multivariate setting.

3.4.1. Balloon estimators

The basic form of the balloon estimator is a generalization of Eq. (18):

$$\hat{f}_B(\mathbf{x}) = \frac{1}{n|\mathbf{H}(\mathbf{x})|^{1/2}} \sum_{i=1}^n K\left(\mathbf{H}(\mathbf{x})^{-1/2}(\mathbf{x} - \mathbf{x}_i)\right) = \frac{1}{n} \sum_{i=1}^n K_{\mathbf{H}(\mathbf{x})}(\mathbf{x} - \mathbf{x}_i)$$

where $\mathbf{H}(\mathbf{x})$ is a positive-definite smoothing matrix associated with the estimation point $\mathbf{x}$. Note that $\mathbf{H}$ corresponds to $h^2 A$. At a particular estimation point $\mathbf{x}$, the balloon estimator and the fixed bandwidth are exactly the same. Both place kernels of the same size and orientation at each of the data points and the estimate is constructed by averaging the values of the kernels at $\mathbf{x}$.

Taking K to be a uniform density on the unit sphere with $\mathbf{H}(\mathbf{x}) = h_k(\mathbf{x})\mathbf{I}_d$ and letting $h_k(\mathbf{x})$ the distance from $\mathbf{x}$ to the k-th nearest data point, one has the k-nearest neighbor estimator of Loftsgaarden and Quesenberry (1965). Much has been written about this early balloon estimator that tries to incorporate larger bandwidths in the tails. (Where data are scarce, the distances upon which the bandwidth function is based should be larger.) The estimator is not guaranteed to integrate to one (hence, the estimator is not a density) and the discontinuous nature of the bandwidth function manifests directly into discontinuities in the resulting estimate. Furthermore, the estimator has severe bias problems, particularly in the tails (Mack and Rosenblatt, 1979, and Hall, 1983) although it seems to perform well in higher dimensions (Terrell and Scott, 1992).

In general, the identical construction of the balloon estimator and the fixed bandwidth estimator results in identical pointwise error properties. However, there are certain regions of the underlying density, typically in the tails, where the size and orientation of the kernels can be chosen to yield a higher-order bias (Terrell and Scott, 1992) or even eliminate it completely (Sain, 2001, 2003; Hazelton, 1998; Sain and Scott, 2002; Devroye and Lugosi, 2000).

3.4.2. Sample point estimators

The multivariate sample-point estimator is defined to be

$$\hat{f}_S(\mathbf{x}) = \frac{1}{n} \sum_{i=1}^{n} \frac{1}{|\mathbf{H}(\mathbf{x}_i)|^{1/2}} K\left(\mathbf{H}(\mathbf{x}_i)^{-1/2}(\mathbf{x} - \mathbf{x}_i)\right)$$

$$= \frac{1}{n} \sum_{i=1}^{n} K_{\mathbf{H}(\mathbf{x}_i)}(\mathbf{x} - \mathbf{x}_i), \tag{20}$$

where $\mathbf{H}(\mathbf{x}_i)$ is a positive-definite smoothing matrix associated with the i-th data point, $\mathbf{x}_i$. In contrast to the balloon estimator, this estimator still places a kernel at each data point and the estimator is still constructed by averaging the kernel values at $\mathbf{x}$. However, the size and orientation of each kernel is different and is constant over the entire range of the density to be estimated.

Early efforts with such estimators proposed $\mathbf{H}(\mathbf{x}_i) \propto f(\mathbf{x}_i)^{-\alpha} \mathbf{I}_d$. Breiman et al. (1977) suggested using nearest-neighbor distances which is equivalent to using $\alpha = 1/d$. Abramson (1982) suggested using $\alpha = 1/2$ regardless of the dimension. Pointwise, it can be shown that this parameterization of the bandwidth function can yield a higher-order behavior of the bias (Silverman, 1986; Hall and Marron, 1988; Jones, 1990) and empirical results show promise for smaller sample sizes. However, this higher-order behavior does not hold globally due to bias contributions from the tails (Hall, 1992; McKay, 1993; Terrell and Scott, 1992; Hall et al., 1994; Sain and Scott, 1996) and any gains can be lost as the sample size increases.

Sain and Scott (1996) and Sain (2002) suggest using a binned version of the sample-point estimator in (20). Such an estimator has the form

$$\hat{f}_{Sb}(\mathbf{x}) = \frac{1}{n} \sum_{j=1}^{m} \frac{n_j}{|\mathbf{H}(\mathbf{t}_j)|^{1/2}} K\left(\mathbf{H}(\mathbf{t}_j)^{-1/2}(\mathbf{x} - \mathbf{t}_j)\right) = \frac{1}{n} \sum_{j=1}^{m} n_j K_{\mathbf{H}(\mathbf{t}_j)}(\mathbf{x} - \mathbf{t}_j)$$

where n_j is the number of data points in the j-th bin centered at $\mathbf{t}_j$ and $\mathbf{H}(\mathbf{t}_j)$ is a positive-definite bandwidth matrix associated with the j-th bin. Using such an estimator, the MISE can easily be examined by recognizing that only the n_j are random and follow a multinomial distribution with cell probabilities given by $p_j = \int_{B_j} f(\mathbf{x}) \, d\mathbf{x}$ where B_j denotes the j-th bin. The MISE for a binned estimator with normal kernels is then given by

$$\text{MISE} = \frac{1}{n(2\sqrt{\pi})^d} \sum_j \frac{p_j(1-p_j) + np_j^2}{|\mathbf{H}_j|^{1/2}} + \frac{n-1}{n} \sum_{i \neq j} p_i p_j \phi_{\mathbf{H}_i + \mathbf{H}_j}(\mathbf{t}_i - \mathbf{t}_j)$$

$$- \frac{2}{n} p_j \int \phi_{\mathbf{H}_j}(\mathbf{x} - \mathbf{t}_j) f(\mathbf{x}) \, d\mathbf{x} + R(f)$$

where $\mathbf{H}_j = \mathbf{H}(\mathbf{t}_j)$. Sain and Scott (1996) used this formulation to examine univariate sample-point estimators and showed that while the MISE did not appear to converge at a faster rate, significant gains over fixed bandwidth estimators could be theoretically obtained for a wide variety of densities. Sain (2002) showed similar results in the multivariate setting.

3.4.3. Parameterization of sample-point estimators

Designing practical algorithms that actually achieve some of the gains predicted in theory has been a difficult task and much of the promise depends on how the bandwidth function is parameterized. It seems to be widely held that the sample-point estimator shows more promise, perhaps since the estimator is a bona fide density by construction. However, n positive-definite smoothing matrices must be estimated for the sample-point estimator and it is clear that some sort of dimension reduction must be utilized.

The binning approach outlined in the previous section is one possible approach to reduce the number of smoothing matrices that must be estimated. In addition, further reduction could be had by restricting the form of the smoothing matrices. For example, one could let the kernels be radially symmetric and just vary the size of the kernels, effectively letting $\mathbf{H}(\mathbf{x}_i) = h(\mathbf{x}_i)\mathbf{I}_d$. This leaves just one parameter to be estimated for each bin. A step up is to allow different amounts of smoothing for each dimension using the product kernel form. This would reduce the bandwidth function to $\mathbf{H}(\mathbf{x}_i) = \text{diag}(h_1(\mathbf{x}_i), \ldots, h_d(\mathbf{x}_i))$ where diag indicates a diagonal matrix. Each kernel would be elliptical with the axis of each ellipse aligned with the coordinate axis and d parameters would be estimated for each bin.

In two dimensions, there are three free parameters in the full smoothing matrix. While the product kernel formulation allows for some dimension reduction, many other formulations are possible. For example, Banfield and Raftery (1993) reparameterize covariance matrix for normal components of a mixture as $\mathbf{\Sigma}_k = \lambda_k \mathbf{D}_k \mathbf{A}_k \mathbf{D}'_k$. In this formulation, λ_k controls the volume while the matrix $\mathbf{A}_k$ controls the shape and is a diagonal matrix of the form $\mathbf{A}_k = \text{diag}(1, \alpha_2, \ldots, \alpha_k)$ for $1 \geqslant \alpha_2 \geqslant \cdots \geqslant \alpha_k$. The matrix $\mathbf{D}_k$ is an orthogonal matrix that controls the orientation. The three free parameters for $d = 2$ are then λ_k, α_2, and any one of elements of $\mathbf{D}_k$ (the other elements of $\mathbf{D}_k$ can be obtained from the constraints imposed by orthogonality, i.e. $\mathbf{D}_k \mathbf{D}'_k = \mathbf{I}_d$). Any combination of these terms could be held constant or allowed to vary yielding a great many different parameterizations that could be effective for densities of different shapes.

A comparison of the size and shape of optimal kernels for the three basic forms is given in Figure 8 for a bivariate standard normal density $n = 1000$ and the binned

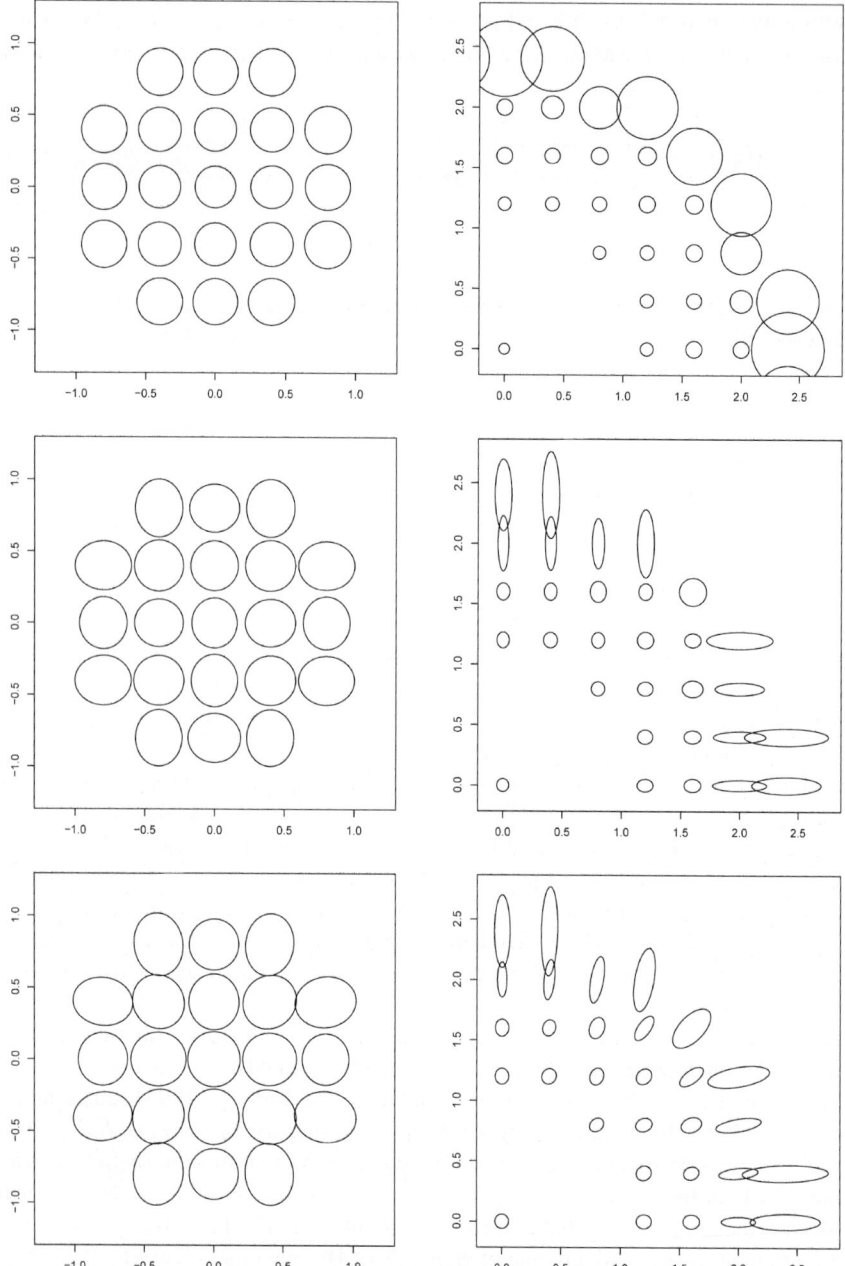

Fig. 8. Ellipses showing relative sizes and shapes of sample point kernels using the binned estimator and a bivariate standard normal density. Left column shows kernels for radially symmetric kernels, middle column shows kernels for diagonal bandwidth matrices, while the right column shows the unrestricted kernels. Top frames show kernels inside the unit circle while the bottom frames shows kernels in the first quadrant and outside the unit circle.

sample-point estimator. A fixed mesh is laid down over the range of the density, bin probabilities are calculated, and the MISE is minimized. Bins in the corners with very low bin probabilities were excluded from the optimization.

Kernels near the mode (inside the unit circle) are nearly circular and are very similar, regardless of the parameterization. As the bins move further out into the tails, the size of the kernels get larger and the product kernels and fully parameterized kernels become more and more elliptical. As expected, the kernels for the product kernel estimator are circular on the diagonals and elliptical on the coordinate axis reflecting the nature of that particular restriction. As in Sain (2002), the MISE for the sample-point estimator with fully parameterized smoothing matrices is the smallest, followed by the product kernel formulation.

3.4.4. Estimating bandwidth matrices

Estimating variable bandwidth matrices from data continues to be a difficult problem. Sain (2002) outlines a cross-validation algorithm based on the binned estimator that involves finding the collection of bandwidth matrices that minimize

$$\text{UCV} = R(\hat{f}) - \frac{2}{n} \sum_{i=1}^{n} \hat{f}_{-i}(\mathbf{x}_i)$$

$$= R(\hat{f}) - \frac{2}{n} \sum_{i=1}^{n} \left[\frac{1}{n-1} \sum_{j=1}^{m} n_{ij}^* K_{\mathbf{H}_j}(\mathbf{x}_i - \mathbf{t}_j) \right]$$

where $n_{ij}^* = n_j - 1$ if $\mathbf{x}_i \in B_j$ or n_j, otherwise; here n_j is the number of data points in the j-th bin, B_j. In practice, a parameterization for the bandwidth matrices is chosen, a mesh is laid down over the data, the bin counts computed, and the UCV criterion minimized.

An example is shown in Figure 9 for Fishers' iris data (sepal length and petal length). A simple parameterization is chosen using radially symmetric smoothing matrices and the left frame shows ellipses representing the estimated kernels. For reference, the ellipse in the upper left-hand corner of the left frame is the cross-validated fixed bandwidth kernel. As expected, kernels in the tails and valleys between the modes are larger

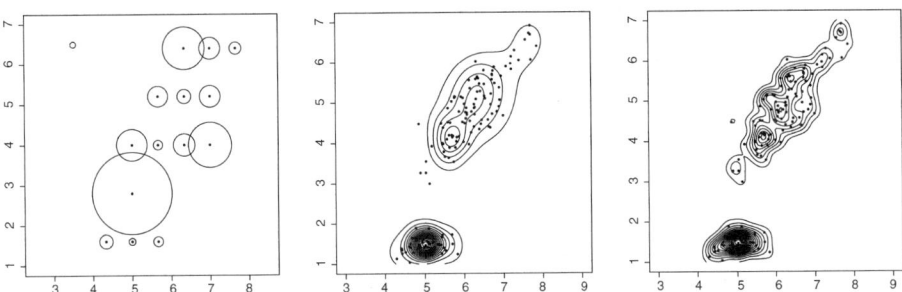

Fig. 9. Example using sepal and petal length for the iris data. Left frame shows ellipses representing the cross-validation estimated kernels with the middle frame the resulting density. The ellipse in the upper left corner of the left frame represents the fixed bandwidth kernel and the resulting estimate is in the right frame.

than near the modes. The variable bandwidth estimate is shown in the middle frame while the fixed bandwidth estimator is shown in the right frame. At first glance, the fixed-bandwidth estimate appears undersmoothed, possibly resulting from UCV's well-known tendency to pick bandwidths smaller than necessary (an improved estimate could possibly be found using, for example, the multivariate plug-in approach of (Wand and Jones, 1994)). However, the estimated bandwidth is clearly focusing on the mode in lower left of the frame (note the similarity between the fixed bandwidth and the variable bandwidths corresponding to this mode). This mode represents one of the species of iris present in the data and has a much smaller scale than the other modes in the data corresponding to the other two species. In contrast, the variable bandwidth estimate, despite being based on just a few bins, is clearly able to adapt to the changes in scale between the modes associated with the three species and does a much better job of simultaneously smoothing the different features in the density.

Sain (2002) further experimented with such methods and demonstrated the potential of even this simple formulation of a multivariate sample-point estimator, in particular for picking out important structure and minimizing the number of false modes. However, Sain (2002) also showed that UCV was not as effective when a fully parameterized bandwidth matrix is used. Hazelton (2003) has explored the product kernel formulation using not a piecewise constant bandwidth structure as in the binning case but a linearly interpolated bandwidth function with some promising results.

3.5. Other estimators

Kernel estimators and orthogonal series density estimators were developed independently (Rosenblatt, 1956; Watson, 1969). It is well known that an orthogonal series estimator can be re-expressed as a kernel estimator. However, cross-validation algorithms are somewhat different (Wahba, 1981) and spline estimators are also available. More recently, wavelet bases have become available and fall into this category (Donoho et al., 1996). Wahba pioneered splines for density estimators; however, her representation places knots at each sample point. Kooperberg and Stone (1991) describe an alternative spline formulation on a log-scale. A new spline tool called P-splines has recently emerged that like the ASH model greatly reduces computation; see Eilers and Marx (1996) and Ruppert et al. (2003).

We remarked earlier that maximum likelihood had a role to play in the definition of histograms, but was limited in any role for defining smoothing parameters. This situation has changed in recent years with the development of local likelihood methods for density estimation as well as regression estimation. This promising family of estimators is surveyed by Loader (1999).

4. Mixture density estimation

An alternative to the kernel estimator is the so-called mixture model, where the underlying density is assumed to have the form

$$g(\mathbf{x}) = \sum_{i=1}^{k} p_i g_i(\mathbf{x}; \boldsymbol{\theta}_i). \tag{21}$$

The $\{p_i: i = 1, \ldots, k\}$ are referred to as mixing proportions or weights and are constrained so that $p_i > 0$ and $\sum_{i=1}^{k} p_i = 1$. The components of the mixture, $\{g_i: i = 1, \ldots, k\}$, are themselves densities and are parameterized by θ_i which may be vector valued. Often, the g_i are taken to be multivariate normal, in which case $\theta_i = \{\boldsymbol{\mu}_i, \boldsymbol{\Sigma}_i\}$.

Mixture models are often motivated by heterogeneity or the presence of distinct subpopulations in observed data. For example, one of the earliest applications of a mixture model (Pearson, 1894; see also McLachlan and Peel, 2000) used a two-component mixture to model the distribution of the ratio between measurements of forehead and body length on crabs. This simple mixture was effective at modeling the skewness in the distribution and it was hypothesized that the two-component structure was related to the possibility of this particular population of crabs evolving into two new subspecies.

The notion that each component of a mixture is representative of a particular subpopulation in the data has led to the extensive use of mixtures in the context of clustering and discriminant analysis. See, for example, the reviews by Fraley and Raftery (2002) and McLachlan and Peel (2000). It was also the motivation for the development of the multivariate outlier test of Wang et al. (1997) and Sain et al. (1999), who were interested in distinguishing nuclear tests from a background population consisting of different types of earthquakes, mining blasts, and other sources of seismic signals.

Often, a mixture model fit to data will have more components than can be identified with the distinct groups present in the data. This is due to the flexibility of mixture models to represent features in the density that are not well-modeled by a single component. Marron and Wand (1992), for example, give a wide variety of univariate densities (skewed, multimodel, e.g.) that are constructed from normal mixtures. It is precisely this flexibility that makes mixture models attractive for general density estimation and exploratory analysis.

When the number of components in a mixture is pre-specified based on some a priori knowledge about the nature of the subpopulations in the data, mixtures can be considered a type of parametric model. However, if this restriction on the number of components is removed, mixtures behave in nonparametric fashion. The number of components acts something like a smoothing parameter. Smaller numbers of components will behave more like parametric models and can lead to specification bias. Greater flexibility can be obtained by letting the number of components grow, although too many components can lead to overfitting and excessive variation. A parametric model is at one end of this spectrum, and a kernel estimator is at the other end. For example, a kernel estimator with a normal kernel can be simply considered a mixture model with weights taken to be $1/n$ and component means fixed at the data points.

4.1. Fitting mixture models

While mixture models have a long history, fitting mixture models was problematic until the advent of the expectation-maximization (EM) algorithm of Dempster et al. (1977). Framing mixture models as a missing data problem has made parameter estimation much easier, and maximum likelihood via the EM algorithm has dominated the literature on fitting mixture models. However, Scott (2001) has also had considerable success using the L₂E method, which performs well even if the assumed number of components, k, is too small.

The missing data framework assumes that each random vector $\mathbf{X}_i$ generating from the density (21) is accompanied by a categorical random variable Z_i where Z_i indicates the component from which $\mathbf{X}_i$ comes. In other words, Z_i is a single-trial multinomial with cell probabilities given by the mixing proportions $\{p_i\}$. Then, the density of $\mathbf{X}_i$ given Z_i is g_i in (21). It is precisely the realized values of the Z_i that are typically considered missing when fitting mixture models, although it is possible to also consider missing components in the $\mathbf{X}_i$ as well.

For a fixed number of components, the EM algorithm is iterative in nature and has two steps in each iteration. The algorithm starts with initial parameter estimates. Often, computing these initial parameter estimates involves some sort of clustering of the data, such as a simple hierarchical approach.

The first step at each iteration is the expectation step which involves prediction and effectively replaces the missing values with their conditional expectation given the data and the current parameter estimates. The next step, the maximization step, involves recomputing the estimates using both complete data and the predictions from the expectation step.

For normal mixtures missing only the realized component labels z_i, this involves computing in the expectation step

$$w_{ij} = \frac{\hat{p}_j^0 f_j(\mathbf{x}_i; \hat{\boldsymbol{\mu}}_j^0, \widehat{\boldsymbol{\Sigma}}_j^0)}{\sum_{j=1}^k \hat{p}_j^0 f_j(\mathbf{x}_i; \hat{\boldsymbol{\mu}}_j^0, \widehat{\boldsymbol{\Sigma}}_j^0)} \tag{22}$$

for $i = 1, \ldots, n$ and $j = 1, \ldots, k$ and where $\hat{p}_j^0$, $\hat{\boldsymbol{\mu}}_j^0$, and $\widehat{\boldsymbol{\Sigma}}_j^0$ are the current parameter estimates. The maximization step then updates the sufficient statistics:

$$T_{j1} = \sum_{i=1}^n w_{ij}, \qquad \mathbf{T}_{j2} = \sum_{i=1}^n w_{ij}\mathbf{x}_i, \qquad \mathbf{T}_{j3} = \sum_{i=1}^n w_{ij}\mathbf{x}_i\mathbf{x}_i'$$

for $j = 1, \ldots, k$ to yield the new parameter estimates:

$$\hat{p}_j^1 = T_{j1}/n, \qquad \hat{\boldsymbol{\mu}}_j^1 = \mathbf{T}_{j2}/T_{j1}, \qquad \widehat{\boldsymbol{\Sigma}}_j^1 = (\mathbf{T}_{j3} - \mathbf{T}_{j2}\mathbf{T}_{j2}'/T_{j1})/T_{j1}$$

for $j = 1, \ldots, k$. The process cycles between these two steps until some sort of convergence is obtained. The theory concerning the EM algorithm suggests that the likelihood is increased at each iteration. Hence, at convergence, a local maximum in the likelihood has been found.

A great deal of effort has been put forth to determine a data-based choice of the number of components in mixture models and many of these are summarized in McLachlan and Peel (2000). Traditional likelihood ratio tests have been examined but a breakdown in the regularity conditions have made implementation difficult. Bootstrapping and Bayesian approaches have also been studied. Other criterion such as Akaike's information criterion (AIC) have been put forth and studied in some detail. In many situations, however, it has been found that AIC tends to choose too many components. There is some criticism of the theoretical justification for AIC, since it violates the same regularity conditions as the likelihood ratio test. An alternative information criterion is the Bayesian information criterion (BIC) given by

$$\text{BIC} = -2\ell + d \log n$$

where ℓ is the maximized log-likelihood, d is the number of parameters in the model, and n is the sample size. While there are some regularity conditions for BIC that do not hold for mixture models, there is much empirical evidence that supports its use. For example, Roeder and Wasserman (1997) show that when a normal mixture model is used as a nonparametric density estimate, the density estimate that uses the BIC choice of the number of components is consistent.

Other sophisticated procedures for choosing the number of components in mixture models have also been explored. For example, Priebe and Marchette (1991, 1993) and Priebe (1994) discuss what the authors' refer to as "adaptive mixtures" that incorporate the ideas behind both kernel estimators and mixture models and that use a data-based method for adding new terms to a mixture. The adaptive mixture approach can at times overfit data and Solka et al. (1998) combine adaptive mixtures with a pruning step to yield more parsimonious models. These methods have also been shown, both theoretically and through simulation and example, to be effective at determining the underlying structure in the data.

4.2. An example

Figures 10 and 11 show an example of an application of a mixture model using bivariate data consisting of twenty-year averages of temperature and precipitation measured globally on a 5° grid (Covey et al., 2003; Wigley, 2003). An initial scatterplot of the measurements shows clearly the presence of multiple groupings in the data. It is hypoth-

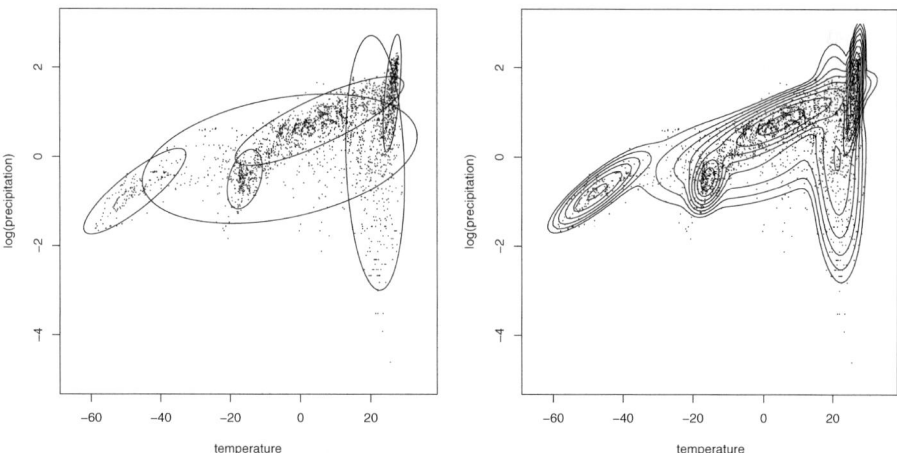

Fig. 10. Example using climate data. The best BIC model uses six components and these components are represented by ellipses in the left frame. The right frame shows a contour plot of the resulting density estimate.

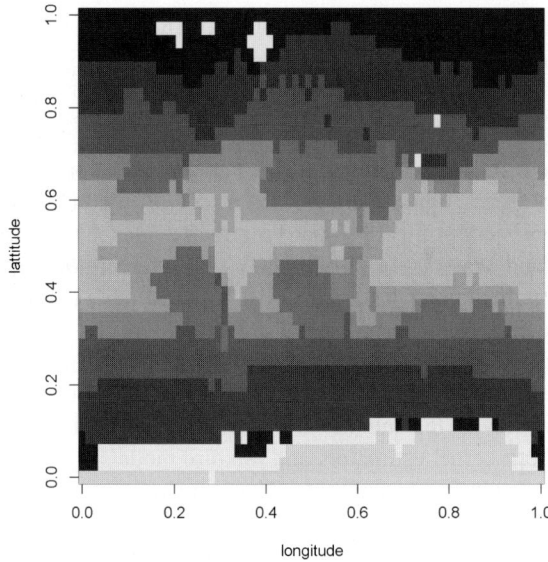

Fig. 11. An image plot displaying the results of the clustering based on the mixture estimate. The effects of land masses and latitude are clearly present in the clusters.

esized that this multimodality can be attributed to climatic effects as well as latitude and land masses across the globe.

A sequence of multivariate normal mixture models was fit to the data using various numbers of components. BIC suggested a six component model. The ellipses in the left frame of Figure 10 indicate location and orientation of the individual components while the right frame shows the contours of the resulting density overlaid on the data.

It seems clear from the contour plot that some components are present to model nonnormal behavior in the density. However, Figure 11 shows the result of classifying each observation as coming from one of the six components. This is done by examining the posterior probabilities as given by the w_{ij} in (22) at the end of the EM iterations. The groupings in the data do appear to follow latitude lines as well as the land masses across the globe.

5. Visualization of densities

The power of nonparametric curve estimation is in the representation of multivariate relationships. While univariate density estimates are certainly useful, the visualization of densities in two, three, and four dimensions offers greater potential in an exploratory context for feature discovery. Visualization techniques are described here.

We examine the zip code data described by Le Cun et al. (1990). Handwritten digits scanned from USPS mail were normalized into 16×16 grayscale images. Training and testing data (available at the U.C. Irvine data repository) were combined into one data

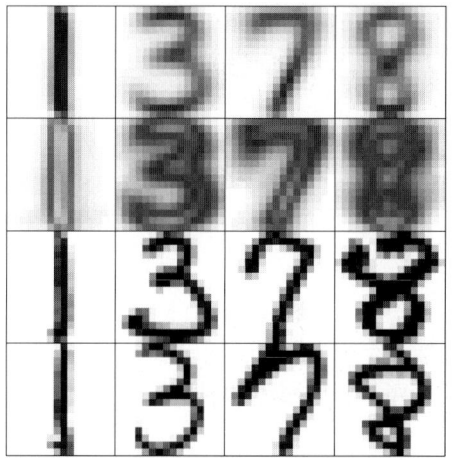 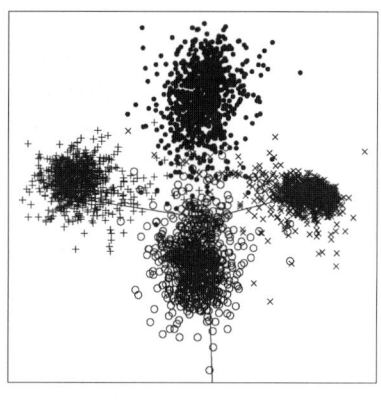

Fig. 12. (Left) Mean, standard deviation, and examples of zip code digits 1, 3, 7, and 8. (Right) LDA subspace of zip code digits 1 (×), 3 (•), 7 (+), and 8 (O).

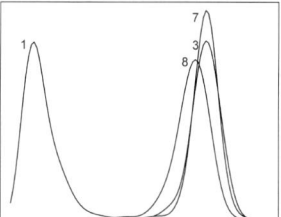 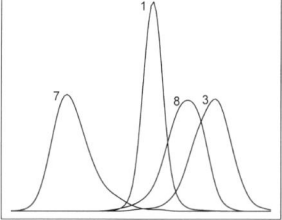 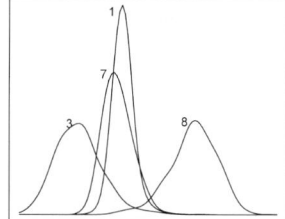

Fig. 13. ASH's for each of the 4 digits for the 1st, 2nd, and 3rd LDA variable (L–R).

set, and the digits 1, 3, 7, and 8 were extracted for analysis here (1269, 824, 792, and 708 cases, respectively). We selected these digits to have examples of straight lines (1 and 7) as well as curved digits (3 and 8). In Figure 12, some examples of the digits together with summary statistics are displayed. Typical error rates observed classifying these data are high, in the 2.5% range.

To analyze and visualize these data, we computed the Fisher linear discriminant analysis (LDA) subspace. We sphered the data using a pooled covariance estimate, and computed the LDA subspace as the three-dimensional span of the four group means. The right frame of Figure 12 displays a frame from XGobi (Swayne et al., 1991) and shows that the four groups are reasonably well-defined and separated in the LDA variable space.

If we examine averaged shifted histograms of each digit for each of the three LDA variables separately, we observe that the first LDA variable separates out digit 1 from the others; see the left frame Figure 13. In the middle frame, the second LDA variable separates digit 7 from digits 3 and 8. Finally, in the right frame, the third LDA variable almost completely separates digits 3 and 8 from each other (but not from the others).

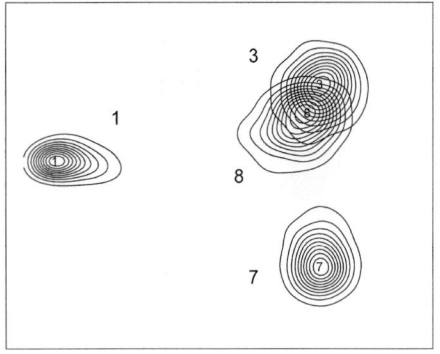 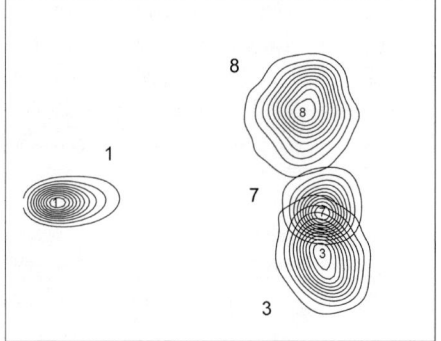

Fig. 14. Bivariate ASH's of the 4 digits using LDA variables (v_1, v_2) (left) and (v_1, v_3) (right).

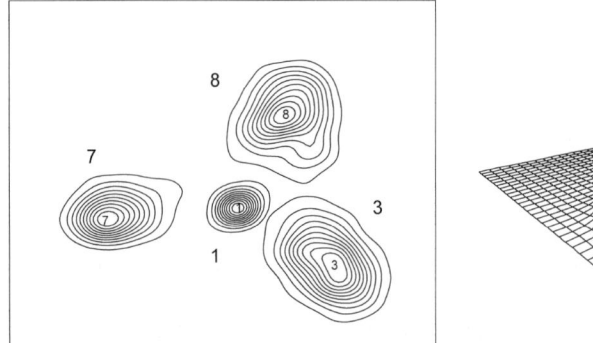 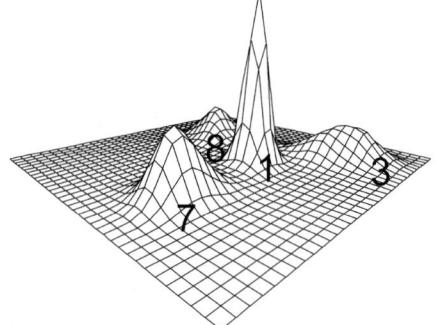

Fig. 15. ASH's for LDA variables (v_2, v_3).

We can obtain a less fragmented view of the feature space by looking at pairs of the LDA variables. In Figures 14 and 15, averaged shifted histograms for each digit were computed separately and are plotted. Contours for each ASH were drawn at 10 equally-spaced levels. The left frame in Figure 14 reinforces the notion that the first two LDA variables isolate digits 1 and 7. Digits 3 and 8 are separated by the first and third LDA variables in the right frame of Figure 14; recall that digit 7 can be isolated using the second LDA variables. Interestingly, in Figure 15, all four digits are reasonably separated by the second and third LDA variables alone. We also show a perspective plot of these ASH densities. (The perspective plot in Figure 14 does not display the full 100×100 mesh at this reduced size for clarity and to avoid overplotting the lines.)

Visualization of univariate and bivariate densities has become a fairly routine task in most modern statistical software packages. The figures in this chapter were generated using the Splus package on a Sun under the Solaris operating system. The ASH software is available for download at the ftp software link at author's homepage http://www.stat.rice.edu/~scottdw. The ASH software contains separate routines for the univariate and bivariate cases. Visualization of the *ash1* and *ash2* estimates was accomplished using the built-in Splus functions *contour* and *persp*.

A separate function, *ashn*, is also included in the ASH package. The *ashn* function not only computes the ASH for dimensions $3 \leqslant d \leqslant 6$, but it also provides the capability to visualize arbitrary three-dimensional contours of a level set of any four-dimensional surface. In particular, if $f_{\max}$ is the maximum value of an ASH estimate, $\hat{f}(x, y, z)$, and α takes values in the interval $(0, 1)$, then the α-th contour or level set is the surface

$$C_\alpha = \{(x, y, z): \hat{f}(x, y, z) = \alpha f_{\max}\}.$$

The mode of the density corresponds to the choice $\alpha = 1$. The *ashn* function can compute the fraction of data within any specified α-contour.

Some simple examples of C_α contours may be given for normal data. If the covariance matrix $\Sigma = I_d$, then contours are spheres centered at μ:

$$C_\alpha = \{(x, y, z): e^{-0.5((x-\mu_1)^2+(y-\mu_2)^2+(z-\mu_3)^2)} = \alpha\}$$

or $C_\alpha = \{(x, y, z): (x - \mu_1)^2 + (y - \mu_2)^2 + (z - \mu_3)^2 = -2\log \alpha\}$. For a general covariance matrix, the levels sets are the ellipses $C_\alpha = \{(x, y, z): (\mathbf{x}-\mu)'\Sigma^{-1}(\mathbf{x}-\mu) = -2\log \alpha\}$.

With a nonparametric density, the contours do not follow a simple parametric form and must be estimated from a matrix of values, usually on a regular three-dimensional mesh. This mesh is linearly interpolated, resulting in a large number of triangular mesh elements that are appropriately sorted and plotted in perspective. Since the triangular elements are contiguous, the resulting plot depicts a smooth contour surface. This algorithm is called marching cubes (Lorensen and Cline, 1987).

In Figure 16, a trivariate ASH is depicted for the data corresponding to digits 3, 7, and 8. (The digit 1 is well-separated and those data are omitted here.) The triweight kernel was selected with $m = 7$ shifts for each dimension. The contours shown correspond to the values $\alpha = 0.02, 0.1, 0.2, 0.35, 0.5, 0.7,$ and 0.9. The *ashn* function also permits an ASH to be computed for each of the digits separated and plotted in one frame. For these data, the result is very similar to the surfaces shown in Figure 16.

This figure can be improved further by using stereo to provide depth of field, or through animation and rotation. The *ashn* software has an option to output this static figure in the so-called *QUAD* format used by the *geomview* visualization package from the previous NSF Geometry Center in Minneapolis. This software is still available from http://www.geomview.org and runs on SGI, Sun, and Linux platforms (Geomview, 1998).

5.1. Higher dimensions

Scott (1992) describes extensions of the three-dimensional visualization idea to four dimensions or more. Here we consider just four-dimensional data, (x, y, z, t). The α-th contour is defined as above as

$$C_\alpha = \{(x, y, z, t): \hat{f}(x, y, z, t) = \alpha f_{\max}\}.$$

Since only a 3-dimensional field may be visualized, we propose to depict *slices* of the four-dimensional density. Choose a sequence of values of the fourth variable,

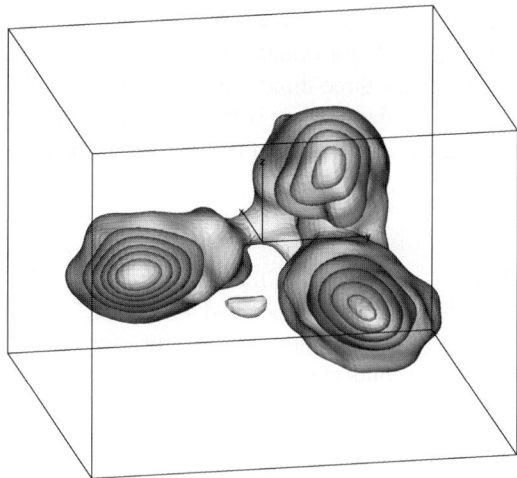

Fig. 16. Trivariate ASH of LDA variables (v_1, v_2, v_3) and digits 3, 7, and 8. The digit labels were not used in this plot. The digit 7 is in the left cluster; the digit 8 is in the top cluster; and the digit 3 in the lower right cluster.

$t_1 < t_2 < \cdots < t_m$, and visualize the sequence of slices

$$C_\alpha(k) = \left\{(x, y, z): \hat{f}(x, y, z, t = t_k) = \alpha f_{\max}\right\} \quad \text{for } k = 1, \ldots, m.$$

With practice, observing an animated view of this sequence of contours reveals the four-dimensional structure of the five-dimensional density surface. An important detail is that $f_{\max}$ is not recomputed for each slice, but remains the constant value of maximum of the entire estimate $\hat{f}(x, y, z, t)$. A possible alternative is viewing the conditional density, $\hat{f}(x, y, z \mid t = t_k)$; however, the renormalization destroys the perception of being in the low-density or tails of the distribution.

To make this idea more concrete, let us revisit the trivariate ASH depicted in Figure 16. This ASH was computed on a $75 \times 75 \times 75$ mesh. We propose as an alternative visualization of this ASH estimate $\hat{f}(x, y, z)$ to examine the sequence of slices

$$C_\alpha(k) = \left\{(x, y, z): \hat{f}(x, y, z = z_k)\right\} \quad \text{for } k = 1, \ldots, 75.$$

In Figure 17, we display a subset of this sequence of slices of the trivariate ASH estimate. For bins numbered less than 20, the digit 3 is solely represented. For bins between 22 and 38, the digit 7 is represented in the lower half of each frame. Finally, for bins between 42 and 62, the digit 8 is solely represented.

We postpone an actual example of this slicing technique for 4-dimensional data, since space is limited. Examples may be found in the color plates of Scott (1992). The extension to five-dimensional data is straightforward. The *ashn* package can visualize slices such as the contours

$$C_\alpha(k, \ell) = \left\{(x, y, z): \hat{f}(x, y, z, t = t_k, s = s_\ell) = \alpha \hat{f}_{\max}\right\}.$$

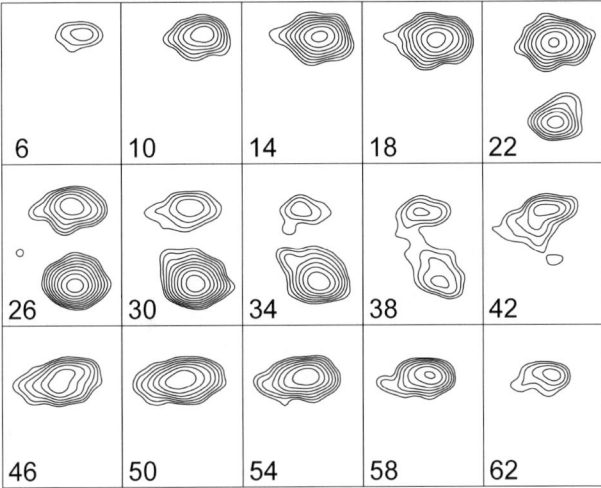

Fig. 17. A sequence of slices of the three-dimensional ASH of the digits 3, 7, and 8 depicted in Figure 16. The z-bin number is shown in each frame from the original 75 bins.

Scott (1986) presented such a visualization of a five-dimensional dataset using an array of ASH slices on the competition data exposition at the Joint Statistical Meetings in 1986.

5.2. Curse of dimensionality

As noted by many authors, kernel methods suffer from increased bias as the dimension increases. We believe the direct estimation of the full density by kernel methods is feasible in as many as six dimensions.

However, this does not mean that kernel methods are not useful in dimensions beyond six. Indeed, for purposes such as statistical discrimination, kernel methods are powerful tools in dozens of dimensions. The reasons are somewhat subtle. Scott (1992) argued that if the smoothing parameter is very small, then comparing two kernel estimates at the same point **x** is essentially determined by the closest point in the training sample. It is well known that the nearest-neighbor classification rule asymptotically achieves half of the optimal Bayesian misclassification rate. At the other extreme, if the smoothing parameter is very large, then comparing two kernel estimates at the same point **x** is essentially determined by which sample mean is closer for the two training samples. This is exactly what Fisher's LDA rule does in the LDA variable space. Thus, at the extremes, kernel density discriminate analysis mimics two well-known and successful algorithms. Thus there exist a number of choices for the smoothing parameter between the extremes that produce superior discriminate rules.

What is the explanation for the good performance for discrimination and the poor performance for density estimation? Friedman (1997) argued that the optimal smoothing parameter for kernel discrimination was much larger than for optimal density estimation. In retrospect, this result is not surprising. But it emphasizes how suboptimal density estimation can be useful for exploratory purposes and in special applications of nonparametric estimation.

6. Discussion

There are a number of useful references for the reader interested in pursuing these ideas and others not touched upon in this chapter. Early reviews of nonparametric estimators include Wegman (1972a, 1972b) and Tarter and Kronmal (1976). General overviews of kernel methods and other nonparametric estimators include Tapia and Thompson (1978), Silverman (1986), Härdle (1990), Scott (1992), Wand and Jones (1995), Fan and Gijbels (1996), Simonoff (1996), Bowman and Azzalini (1997), Eubank (1999), Schimek (2000), and Devroye and Lugosi (2001).

Scott (1992) and Wegman and Luo (2002) discuss a number of issues with the visualization of multivariate densities. Classic books of general interest in visualization include Wegman and De Priest (1986), Cleveland (1993), Wolff and Yaeger (1993), and Wainer (1997).

Applications of nonparametric density estimation are nearly as varied as the field of statistics itself. Research challenges that remain include handling massive datasets and flexible modeling of high-dimensional data. Mixture and semiparametric models hold much promise in this direction.

References

Abramson, I. (1982). On bandwidth variation in kernel estimates—a square root law. *Ann. Statist.* **10**, 1217–1223.

Banfield, J.D., Raftery, A.E. (1993). Model-based Gaussian and non-Gaussian clustering. *Biometrics* **49**, 803–821.

Bowman, A.W. (1984). An alternative method of cross-validation for the smoothing of density estimates. *Biometrika* **71**, 353–360.

Bowman, A.W., Azzalini, A. (1997). *Applied Smoothing Techniques for Data Analysis: the Kernel Approach with S-Plus Illustrations*. Oxford University Press, Oxford.

Breiman, L., Meisel, W., Purcell, E. (1977). Variable kernel estimates of multivariate densities. *Technometrics* **19**, 353–360.

Chaudhuri, P., Marron, J.S. (1999). SiZer for exploration of structures in curves. *J. Amer. Statist. Assoc.* **94**, 807–823.

Cleveland, W.S. (1993). *Visualizing Data*. Hobart Press, Summit, NJ.

Covey, C., AchutaRao, K.M., Cubasch, U., Jones, P.D., Lambert, S.J., Mann, M.E., Phillips, T.J., Taylor, K.E. (2003). An overview of results from the Coupled Model Intercomparison Project (CMIP). *Global Planet. Change* **37**, 103–133.

Dempster, A.P., Laird, N.M., Rubin, D.B. (1977). Maximum likelihood for incomplete data vi the EM algorithm (with discussion). *J. R. Statist. Soc. Ser. B* **39**, 1–38.

Devroye, L., Lugosi, T. (2000). Variable kernel estimates: On the impossibility of tuning the parameters. In: Gine, E., Mason, D. (Eds.), *High-Dimensional Probability*. Springer-Verlag, New York.

Devroye, L., Lugosi, T. (2001). *Combinatorial Methods in Density Estimation*. Springer-Verlag, Berlin.

Donoho, D.L., Johnstone, I.M., Kerkyacharian, G., Picard, D. (1996). Density estimation by wavelet thresholding. *Ann. Statist.* **24**, 508–539.

Duin, R.P.W. (1976). On the choice of smoothing parameters for Parzen estimators of probability density functions. *IEEE Trans. Comput.* **25**, 1175–1178.
Eilers, P.H.C., Marx, B.D. (1996). Flexible smoothing with B-splines and penalties. *Statist. Sci.* **11**, 89–102.
Eubank, R.L. (1999). *Nonparametric Regression and Spline Smoothing*. Dekker, New York.
Fan, J., Gijbels, I. (1996). *Local Polynomial Modelling and Its Applications*. Chapman and Hall, London.
Fisher, R.A. (1932). *Statistical Methods for Research Workers*, fourth ed. Oliver and Boyd, Edinburgh.
Fraley, C., Raftery, A.E. (2002). Model-based clustering, discriminant analysis, and density estimation. *J. Amer. Statist. Assoc.* **97**, 611–631.
Friedman, J.H. (1997). On bias, variance, 0/1-loss, and the curse-of-dimensionality. *Data Mining and Knowledge Discovery* **1**, 55–77.
Geomview (1998). http://www.geomview.org/docs/html, 1998.
Graunt, J. (1662). *Natural and Political Observations Made upon the Bills of Mortality*. Martyn, London.
Hall, P. (1983). On near neighbor estimates of a multivariate density. *J. Multivariate Anal.* **12**, 24–39.
Hall, P. (1992). On global properties of variable bandwidth density estimators. *Ann. Statist.* **20**, 762–778.
Hall, P., Marron, J.S. (1988). Variable window width kernel estimates. *Probab. Theory Rel. Fields* **80**, 37–49.
Hall, P., Hu, T.C., Marron, J.S. (1994). Improved variable window kernel estimates of probability densities. *Ann. Statist.* **23**, 1–10.
Härdle, W. (1990). *Smoothing Techniques with Implementations in S*. Springer-Verlag, Berlin.
Hart, J.D. (1984). Efficiency of a kernel density estimator under an autoregressive dependence model. *J. Amer. Statist. Assoc.* **79**, 110–117.
Hazelton, M.L. (1998). Bias annihilating bandwidths for local density estimates. *Statist. Probab. Lett.* **38**, 305–309.
Hazelton, M.L. (2003). Adaptive smoothing in bivariate kernel density estimation. Manuscript.
Hearne, L.B., Wegman, E.J. (1994). Fast multidimensional density estimation based on random-width bins. *Comput. Sci. Statist.* **26**, 150–155.
Jones, M.C. (1990). Variable kernel density estimates and variable kernel density estimates. *Austral. J. Statist.* **32**, 361–372.
Kanazawa, Y. (1992). An optimal variable cell histogram based on the sample spacings. *Ann. Statist.* **20**, 291–304.
Kogure, A. (1987). Asymptotically optimal cells for a histogram. *Ann. Statist.* **15**, 1023–1030.
Kooperberg, C., Stone, C.J. (1991). A study of logspline density estimation. *Comput. Statist. Data Anal.* **12**, 327–347.
Le Cun, Y., Boser, B., Denker, J., Henderson, D., Howard, R., Hubbard, W., Jackel, L. (1990). Handwritten digit recognition with a back-propagation network. In: Touretzky, D. (Ed.), *Advances in Neural Information Processing Systems, vol. 2*. Morgan Kaufmann, Denver, CO.
Loader, C. (1999). *Local Regression and Likelihood*. Springer-Verlag, New York.
Loftsgaarden, D.O., Quesenberry, C.P. (1965). A nonparametric estimate of a multivariate density. *Ann. Math. Statist.* **36**, 1049–1051.
Lorensen, W.E., Cline, H.E. (1987). Marching cubes: a high resolution 3D surface construction algorithm. *Comput. Graph.* **21**, 163–169.
Mack, Y., Rosenblatt, M. (1979). Multivariate k-nearest neighbor density estimates. *J. Multivariate Anal.* **9**, 1–15.
Marron, J.S., Wand, M.P. (1992). Exact mean integrated squared error. *Ann. Statist.* **20**, 536–712.
McKay, I.J. (1993). A note on the bias reduction in variable kernel density estimates. *Canad. J. Statist.* **21**, 367–375.
McLachlan, G., Peel, D. (2000). *Finite Mixture Models*. Wiley, New York.
Minnotte, M.C. (1997). Nonparametric testing of the existence of modes. *Ann. Statist.* **25**, 1646–1667.
Minnotte, M.C., Scott, D.W. (1993). The mode tree: A tool for visualization of nonparametric density features. *J. Comput. Graph. Statist.* **2**, 51–68.
Pearson, K. (1894). Contributions to the theory of mathematical evolution. *Philos. Trans. R. Soc. London* **185**, 72–110.
Pearson, K. (1902). On the systematic fitting of curves to observations and measurements. *Biometrika* **1**, 265–303.
Priebe, C.E. (1994). Adaptive mixtures. *J. Amer. Statist. Assoc.* **89**, 796–806.

Priebe, C.E., Marchette, D.J. (1991). Adaptive mixtures: Recursive nonparametric pattern recognition. *Pattern Recogn.* **24**, 1197–1209.
Priebe, C.E., Marchette, D.J. (1993). Adaptive mixture density estimation. *Pattern Recogn.* **26**, 771–785.
Roeder, K., Wasserman, L. (1997). Practical Bayesian density estimation using mixtures of normals. *J. Amer. Statist. Assoc.* **92**, 894–902.
Rosenblatt, M. (1956). Remarks on some nonparametric estimates of a density function. *Ann. Math. Statist.* **27**, 832–837.
Rudemo, M. (1982). Empirical choice of histograms and kernel density estimators. *Scand. J. Statist.* **9**, 65–78.
Ruppert, D., Carroll, R.J., Wand, M.P. (2003). *Semiparametric Regression*. Cambridge University Press.
Sain, S.R. (2001). Bias reduction and elimination with kernel estimators. *Commun. Statist.* **30**, 1869–1888.
Sain, S.R. (2002). Multivariate locally adaptive density estimation. *Comput. Statist. Data Anal.* **39**, 165–186.
Sain, S.R. (2003). A new characterization and estimation of the zero-bias bandwidth. *Austral. New Zealand J. Statist.* **45**, 29–42.
Sain, S.R., Scott, D.W. (1996). On locally adaptive density estimation. *J. Amer. Statist. Assoc.* **91**, 1525–1534.
Sain, S.R., Scott, D.W. (2002). Zero-bias bandwidths for locally adaptive kernel density estimation. *Scand. J. Statist.* **29**, 441–460.
Sain, S.R., Baggerly, K.A., Scott, D.W. (1994). Cross-validation of multivariate densities. *J. Amer. Statist. Assoc.* **89**, 807–817.
Sain, S.R., Gray, H.L., Woodward, W.A., Fisk, M.D. (1999). Outlier detection from a mixture distribution when training data are unlabeled. *Bull. Seismol. Soc. America* **89**, 294–304.
Schimek, M.G. (Ed.) (2000). *Smoothing and Regression*. Wiley, New York.
Scott, D.W. (1979). On optimal and data-based histograms. *Biometrika* **66**, 605–610.
Scott, D.W. (1985a). On optimal and data-based frequency polygons. *J. Amer. Statist. Assoc.* **80**, 348–354.
Scott, D.W. (1985b). Averaged shifted histograms: effective nonparametric density estimators in several dimensions. *Ann. Statist.* **13**, 1024–1040.
Scott, D.W. (1986). Data exposition poster. 1986 Joint Statistical Meetings.
Scott, D.W. (1992). *Multivariate Density Estimation: Theory, Practice, and Visualization*. Wiley, New York.
Scott, D.W. (2001). Parametric statistical modeling by minimum integrated square error. *Technometrics* **43**, 274–285.
Sheather, S.J., Jones, M.C. (1991). A reliable data-based bandwidth selection method for kernel density estimation. *J.R. Statist. Soc. Ser. B* **53**, 683–690.
Silverman, B.W. (1981). Using kernel density estimates to investigate multimodality. *J.R. Statist. Soc. Ser. B* **43**, 97–99.
Silverman, B.W. (1982). Algorithm AS176. Kernel density estimation using the fast Fourier transform. *Appl. Statist.* **31**, 93–99.
Silverman, B.W. (1986). *Density Estimation for Statistics and Data Analysis*. Chapman and Hall, London.
Simonoff, J.S. (1996). *Smoothing Methods in Statistics*. Springer-Verlag, Berlin.
Solka, J.L., Wegman, E.J., Priebe, C.E., Poston, W.L., Rogers, G.W. (1998). Mixture structure analysis using the Akaike information criterion and the bootstrap. *Statist. Comput.* **8**, 177–188.
Student (1908). On the probable error of a mean. *Biometrika* **6**, 1–25.
Sturges, H.A. (1926). The choice of a class interval. *J. Amer. Statist. Assoc.* **21**, 65–66.
Swayne, D., Cook, D., Buja, A. (1991). XGobi: Interactive dynamic graphics in the X window system with a link to S.. In: *ASA Proceedings of the Section on Statistical Graphics*. ASA, Alexandria, VA, pp. 1–8.
Tapia, R.A., Thompson, J.R. (1978). *Nonparametric Probability Density Estimation*. Johns Hopkins University Press, Baltimore.
Tarter, E.M., Kronmal, R.A. (1976). An introduction to the implementation and theory of nonparametric density estimation. *Amer. Statist.* **30**, 105–112.
Terrell, G.R., Scott, D.W. (1985). Oversmoothed nonparametric density estimates. *J. Amer. Statist. Assoc.* **80**, 209–214.
Terrell, G.R., Scott, D.W. (1992). Variable kernel density estimation. *Ann. Statist.* **20**, 1236–1265.
Wahba, G. (1981). Data-based optimal smoothing of orthogonal series density estimates. *Ann. Statist.* **9**, 146–156.
Wainer, H. (1997). *Visual Revelations*. Springer-Verlag, New York.
Wand, M.P., Jones, M.C. (1994). Multivariate plug-in bandwidth selection. *Comput. Statist.* **9**, 97–116.

Wand, M.P., Jones, M.C. (1995). *Kernel Smoothing*. Chapman and Hall, London.

Wang, S.J., Woodward, W.A., Gray, H.L., Wiechecki, S., Sain, S.R. (1997). A new test for outlier detection from a multivariate mixture distribution. *J. Comput. Graph. Statist.* **6**, 285–299.

Watson, G.S. (1969). Density estimation by orthogonal series. *Ann. Math. Statist.* **40**, 1496–1498.

Wegman, E.J. (1972a). Nonparametric probability density estimation I: A summary of available methods. *Technometrics* **14**, 513–546.

Wegman, E.J. (1972b). Nonparametric probability density estimation II: A comparison of density estimation methods. *J. Statist. Comput. Simulation* **1**, 225–245.

Wegman, E.J. (1975). Maximum likelihood estimation of a probability function. *Sankhya, Ser. A* **37**, 211–224.

Wegman, E.J., DePriest, D. (Eds.) (1986). *Statistical Image Processing and Graphics*. Dekker, New York.

Wegman, E.J., Luo, Q. (2002). On methods of computer graphics for visualizing densities. *J. Comput. Graph. Statist.* **11**, 137–162.

Wigley, T. (2003). MAGICC/SCENGEN 4.1: Technical manual. http://www.cgd.ucar.edu/cas/wigley/magicc/index.html, 2003.

Wolff, R., Yaeger, L. (1993). *Visualization of Natural Phenomena*. Springer-Verlag, New York.

Multivariate Outlier Detection and Robustness

Mia Hubert, Peter J. Rousseeuw and Stefan Van Aelst

1. Introduction

Many multivariate data sets contain outliers, i.e., data points that deviate from the usual assumptions and/or from the pattern suggested by the majority of the data. Outliers are more likely to occur in data sets with many observations and/or variables, and often they do not show up by simple visual inspection.

The usual multivariate analysis techniques (e.g., principal components, discriminant analysis, and multivariate regression) are based on arithmetic means, covariance and correlation matrices, and least squares fitting. All of these can be strongly affected by even a few outliers. When the data contain nasty outliers, typically two things happen:

- the multivariate estimates differ substantially from the 'right' answer, defined here as the estimates we would have obtained without the outliers;
- the resulting fitted model does not allow to detect the outliers by means of their residuals, Mahalanobis distances, or the widely used 'leave-one-out' diagnostics.

The first consequence is fairly well known (although the size of the effect is often underestimated). Unfortunately the second consequence is less well known, and when stated many people find it hard to believe or paradoxical. Common intuition says that outliers must 'stick out' from the classical fitted model, and indeed some of them may do so. But the most harmful types of outliers, especially if there are several of them, may affect the estimated model so much 'in their direction' that they are now well fitted by it.

Once this effect is understood, one sees that the following two problems are essentially equivalent:

- Robust estimation: find a 'robust' fit, which is similar to the fit we would have found without the outliers;
- Outlier detection: find all the outliers that matter.

Indeed, a solution to the first problem allows us to identify the outliers by their residuals, etc., from the robust fit. Conversely, a solution to the second problem allows us to remove or downweigh the outliers followed by a classical fit, which yields a robust result.

The approach we have followed in our research is to work on the first problem, and to use its results to answer the second. We prefer this to the opposite direction because, given the available tools, we found it easier (or rather, less difficult) to start by searching for *sufficiently many* 'good' data points than to start by finding *all* the 'bad' data points.

It turns out that most of the currently available highly robust multivariate estimators are difficult to compute, which makes them unsuitable for the analysis of large databases. Among the few exceptions is the minimum covariance determinant estimator (MCD) of Rousseeuw (1984, 1985). The MCD is a highly robust estimator of multivariate location and scatter that can be computed efficiently with the FAST-MCD algorithm of Rousseeuw and Van Driessen (1999).

Section 2 gives a brief description of the MCD estimator and discusses its computation and its main properties. Section 3 does the same for the least trimmed squares (LTS) estimator (Rousseeuw, 1984), which is an analog of MCD for multiple regression. Since estimating the covariance matrix is the cornerstone of many multivariate statistical methods, the MCD has also been used to develop robust and computationally efficient multivariate techniques. The chapter then goes on to describe MCD-based robust methods for multivariate regression (Section 4), classification (Section 5), principal component analysis (Section 6), principal component regression (Section 7), partial least squares regression (Section 8), and other settings (Section 9). Section 10 concludes with pointers to the available software for these techniques.

2. Multivariate location and scatter

2.1. The need for robustness

In the multivariate location and scatter setting, we assume that the data are stored in an $n \times p$ data matrix $X = (x_1, \ldots, x_n)'$ with $x_i = (x_{i1}, \ldots, x_{ip})'$ the ith observation. Hence n stands for the number of objects and p for the number of variables.

To illustrate the effect of outliers, we consider the following engineering problem. We thank Gertjan Otten, Herman Veraa and Frans Van Dommelen for providing the data and giving permission to publish the results. Philips Mecoma (The Netherlands) is producing diaphragm parts for TV sets. These are thin metal plates, molded by a press. When starting a new production line, nine characteristics were measured for $n = 677$ parts. The aim is to gain insight in the production process and to find out whether abnormalities have occurred and why.

A classical approach is to compute the Mahalanobis distance

$$\text{MD}(x_i) = \sqrt{(x_i - \hat{\mu}_0)' \widehat{\Sigma}_0^{-1} (x_i - \hat{\mu}_0)} \qquad (1)$$

of each measurement x_i. Here $\hat{\mu}_0$ is the arithmetic mean, and $\widehat{\Sigma}_0$ is the classical covariance matrix. The distance $\text{MD}(x_i)$ should tell us how far away x_i is from the center of the cloud, relative to the size of the cloud.

In Figure 1 we plotted the classical Mahalanobis distance versus the index i, which corresponds to the production sequence. The horizontal line is at the usual cutoff value $\sqrt{\chi^2_{9,0.975}} = 4.36$. Figure 1 suggests that most observations are consistent with the classical assumption that the data come from a multivariate normal distribution, except for a few isolated outliers. This should not surprise us, even in the first experimental run of

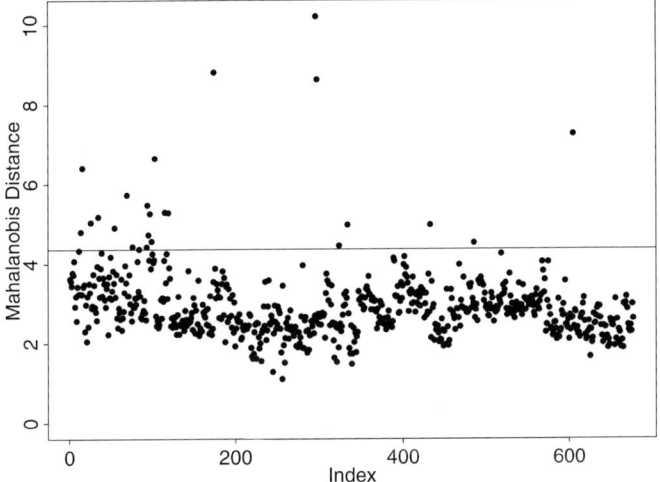

Fig. 1. Plot of Mahalanobis distances for the Philips data.

a new production line, because the Mahalanobis distances are known to suffer from the *masking effect*. That is, even if there were a group of outliers (here, deformed diaphragm parts) they would affect $\hat{\boldsymbol{\mu}}_0$ and $\widehat{\boldsymbol{\Sigma}}_0$ in such a way that they get small Mahalanobis distances $MD(x_i)$ and thus become invisible in Figure 1. To get a reliable analysis of these data, we need robust estimators $\hat{\boldsymbol{\mu}}$ and $\widehat{\boldsymbol{\Sigma}}$ that can resist possible outliers. For this purpose, we will use the MCD estimates described below.

2.2. Description of the MCD

The MCD method looks for the h observations (out of n) whose classical covariance matrix has the lowest possible determinant. The MCD estimate of location is then the average of these h points, whereas the MCD estimate of scatter is a multiple of their covariance matrix. The MCD location and scatter estimates are affine equivariant, which means that they behave properly under affine transformations of the data. That is, for a data set $X = (x_1, \ldots, x_n)'$ in $\mathbb{R}^p$ the MCD estimates $(\hat{\boldsymbol{\mu}}, \widehat{\boldsymbol{\Sigma}})$ satisfy

$$\hat{\boldsymbol{\mu}}(XA + \mathbf{1}_n v') = \hat{\boldsymbol{\mu}}(X)A + v, \qquad \widehat{\boldsymbol{\Sigma}}(XA + \mathbf{1}_n v') = A'\widehat{\boldsymbol{\Sigma}}(X)A$$

for all nonsingular $p \times p$ matrices A and $v \in \mathbb{R}^p$. The vector $\mathbf{1}_n = (1, 1, \ldots, 1)' \in \mathbb{R}^n$.

A useful measure of robustness in the data mining context is the finite-sample *breakdown value* (Donoho and Huber, 1983). The breakdown value $\varepsilon_n^*(\hat{\boldsymbol{\mu}}, X)$ of an estimator $\hat{\boldsymbol{\mu}}$ at the data set X is the smallest amount of contamination that can have an arbitrarily large effect on $\hat{\boldsymbol{\mu}}$. Consider all possible contaminated data sets $\widetilde{X}$ obtained by replacing *any* m of the original observations by *arbitrary* points. Then the breakdown value of a location estimator $\hat{\boldsymbol{\mu}}$ is the smallest fraction m/n of outliers that can take the estimate over all bounds:

$$\varepsilon_n^*(\hat{\boldsymbol{\mu}}, X) := \min_m \left\{ \frac{m}{n}; \ \sup_{\widetilde{X}} \|\hat{\boldsymbol{\mu}}(\widetilde{X}) - \hat{\boldsymbol{\mu}}(X)\| = \infty \right\}. \qquad (2)$$

For many estimators, $\varepsilon_n^*(\hat{\boldsymbol{\mu}}, X)$ varies only slightly with X and n, so that we can denote its limiting value (for $n \to \infty$) by $\varepsilon^*(\hat{\boldsymbol{\mu}})$. Similarly, the breakdown value of a covariance matrix estimator $\widehat{\boldsymbol{\Sigma}}$ is defined as the smallest fraction of outliers that can either take the largest eigenvalue $\lambda_1(\widehat{\boldsymbol{\Sigma}})$ to infinity or the smallest eigenvalue $\lambda_p(\widehat{\boldsymbol{\Sigma}})$ to zero. The MCD estimates $(\hat{\boldsymbol{\mu}}, \widehat{\boldsymbol{\Sigma}})$ of multivariate location and scatter have breakdown value $\varepsilon_n^*(\hat{\boldsymbol{\mu}}) = \varepsilon_n^*(\widehat{\boldsymbol{\Sigma}}) \approx (n-h)/n$. The MCD has its highest possible breakdown value when $h = [(n+p+1)/2]$ (see Lopuhaä and Rousseeuw, 1991).

2.3. The C-step

Rousseeuw and Van Driessen (1999) developed the FAST-MCD algorithm to efficiently compute the MCD. The key component is the C-step:

THEOREM 1. *Take $X = \{x_1, \ldots, x_n\}$ and let $H_1 \subset \{1, \ldots, n\}$ be an h-subset, that is $|H_1| = h$. Put $\hat{\boldsymbol{\mu}}_1 := \frac{1}{h} \sum_{i \in H_1} x_i$ and $\widehat{\boldsymbol{\Sigma}}_1 := \frac{1}{h} \sum_{i \in H_1} (x_i - \hat{\boldsymbol{\mu}}_1)(x_i - \hat{\boldsymbol{\mu}}_1)'$. If $\det(\widehat{\boldsymbol{\Sigma}}_1) \neq 0$, define the relative distances*

$$d_1(i) := \sqrt{(x_i - \hat{\boldsymbol{\mu}}_1)' \widehat{\boldsymbol{\Sigma}}_1^{-1} (x_i - \hat{\boldsymbol{\mu}}_1)} \quad \text{for } i = 1, \ldots, n.$$

Now take H_2 such that $\{d_1(i); i \in H_2\} := \{(d_1)_{1:n}, \ldots, (d_1)_{h:n}\}$ where $(d_1)_{1:n} \leq (d_1)_{2:n} \leq \cdots \leq (d_1)_{n:n}$ are the ordered distances, and compute $\hat{\boldsymbol{\mu}}_2$ and $\widehat{\boldsymbol{\Sigma}}_2$ based on H_2. Then

$$\det(\widehat{\boldsymbol{\Sigma}}_2) \leq \det(\widehat{\boldsymbol{\Sigma}}_1)$$

with equality if and only if $\hat{\boldsymbol{\mu}}_2 = \hat{\boldsymbol{\mu}}_1$ and $\widehat{\boldsymbol{\Sigma}}_2 = \widehat{\boldsymbol{\Sigma}}_1$.

If $\det(\widehat{\boldsymbol{\Sigma}}_1) > 0$, the C-step yields $\widehat{\boldsymbol{\Sigma}}_2$ with $\det(\widehat{\boldsymbol{\Sigma}}_2) \leq \det(\widehat{\boldsymbol{\Sigma}}_1)$. Note that the C stands for 'concentration' since $\widehat{\boldsymbol{\Sigma}}_2$ is more concentrated (has a lower determinant) than $\widehat{\boldsymbol{\Sigma}}_1$. The condition $\det(\widehat{\boldsymbol{\Sigma}}_1) \neq 0$ in the C-step theorem is no real restriction because if $\det(\widehat{\boldsymbol{\Sigma}}_1) = 0$ we already have the minimal objective value.

In the algorithm the C-step works as follows. Given the h-subset H_{old} or the pair $(\hat{\boldsymbol{\mu}}_{\text{old}}, \widehat{\boldsymbol{\Sigma}}_{\text{old}})$:

(1) compute the distances $d_{\text{old}}(i)$ for $i = 1, \ldots, n$;
(2) sort these distances, which yields a permutation π for which $d_{\text{old}}(\pi(1)) \leq d_{\text{old}}(\pi(2)) \leq \cdots \leq d_{\text{old}}(\pi(n))$;
(3) put $H_{\text{new}} := \{\pi(1), \pi(2), \ldots, \pi(h)\}$;
(4) compute $\hat{\boldsymbol{\mu}}_{\text{new}} := \text{ave}(H_{\text{new}})$ and $\widehat{\boldsymbol{\Sigma}}_{\text{new}} := \text{cov}(H_{\text{new}})$.

For a fixed number of dimensions p, the C-step takes only O(n) time (because H_{new} can be determined in O(n) operations without fully sorting all the $d_{\text{old}}(i)$ distances).

C-steps can be iterated until $\det(\widehat{\boldsymbol{\Sigma}}_{\text{new}}) = 0$ or $\det(\widehat{\boldsymbol{\Sigma}}_{\text{new}}) = \det(\widehat{\boldsymbol{\Sigma}}_{\text{old}})$. The sequence of determinants obtained in this way must converge in a finite number of steps because there are only finitely many h-subsets. However, there is no guarantee that the final value $\det(\widehat{\boldsymbol{\Sigma}}_{\text{new}})$ of the iteration process is the global minimum of the MCD objective function. Therefore an approximate MCD solution can be obtained by *taking many* initial choices of H_1, applying C-steps to each and keeping the solution with lowest determinant.

To construct an initial subset H_1, we draw a random $(p+1)$-subset J, and then compute $\hat{\boldsymbol{\mu}}_0 := \text{ave}(J)$ and $\widehat{\boldsymbol{\Sigma}}_0 := \text{cov}(J)$. (If $\det(\widehat{\boldsymbol{\Sigma}}_0) = 0$ then we extend J by adding another random observation, and we continue adding observations until $\det(\widehat{\boldsymbol{\Sigma}}_0) > 0$.) Then we compute the distances $d_0^2(i) := (\boldsymbol{x}_i - \hat{\boldsymbol{\mu}}_0)' \widehat{\boldsymbol{\Sigma}}_0^{-1} (X_i - \hat{\boldsymbol{\mu}}_0)$ for $i = 1, \ldots, n$. Sort them into $d_0(\pi(1)) \leqslant \cdots \leqslant d_0(\pi(n))$ and put $H_1 := \{\pi(1), \ldots, \pi(h)\}$. This method yields better initial subsets than by drawing random h-subsets directly, because the probability of drawing an outlier-free subset is much higher when drawing $(p+1)$-subsets than with h-subsets.

2.4. Computational improvements

Since each C-step involves the calculation of a covariance matrix, its determinant and the corresponding distances, using fewer C-steps would considerably improve the speed of the algorithm. To determine the necessary number of C-steps, we look at Figure 2 which traces the value of the determinant $\widehat{\boldsymbol{\Sigma}}_j$ obtained in successive C-steps for the first 100 random starts, for some data set with many outliers. The solid lines are the runs that led to a robust solution while the dashed lines correspond to nonrobust runs. We see that after two C-steps (i.e., at step (3)) many runs that will lead to the global minimum already have a considerably smaller determinant. Therefore, we can reduce the number of C-steps by applying only two C-steps on each initial subset and selecting the 10 different subsets with lowest determinants. Only for these 10 subsets we continue to take C-steps until convergence.

The resulting algorithm is very fast for small sample sizes n, but when n grows the computation time increases due to the n distances that need to be calculated in each

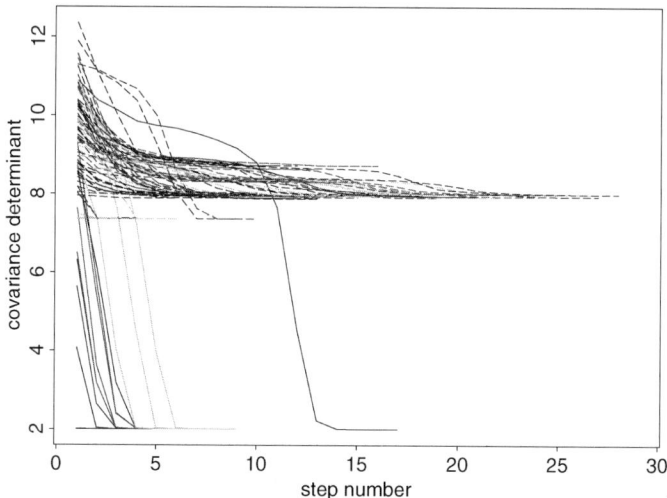

Fig. 2. Value of the determinant for successive C-steps. Each sequence stops when no further reduction is obtained.

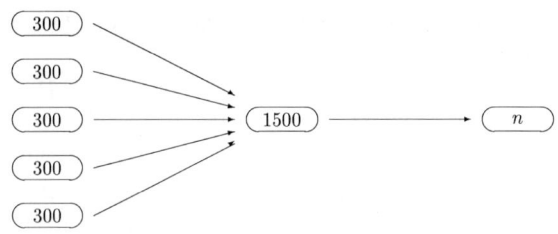

Fig. 3. Nested system of subsets generated by the FAST-MCD algorithm.

C-step. For large n, the FAST-MCD uses a special structure, which avoids doing all the calculations in the entire data. When $n > 1500$, the algorithm generates a nested system of subsets as shown in Figure 3, where the arrows mean 'is a subset of'. The algorithm draws 1500 observations without replacement out of the data set of size n. This subset is then partitioned into five non-overlapping subsets of size 300. The partitioning is very easy: the first 300 observations encountered by the algorithm in the random set of 1500 are put in the first subset, and so on. Because the subset of size 1500 is drawn at random, each subset of size 300 is roughly representative for the data set, and the merged set with 1500 cases is even more representative.

When $600 < n < 1500$, the data will be partitioned into at most 4 random subsets of size 300 or more observations, so that each observation belongs to a subset and such that the subsets have roughly the same size. For instance, 601 will be split as $300 + 301$ and 900 as $450 + 450$. For $n = 901$, the partitioning $300 + 300 + 301$ is used, and finally $1499 = 375 + 375 + 375 + 374$.

Whenever $n > 600$ (and whether $n < 1500$ or not), the FAST-MCD will take two C-steps from several starting subsamples within each subset, with a total of 500 starts for all subsets together. For every subset, the best 10 solutions are stored which gives a total of at most 50 available solutions. Pooling the subsets yields the merged set of at most 1500 observations. For each of the (at most 50) stored solutions ($\hat{\mu}_{sub}, \hat{\Sigma}_{sub}$), the algorithm now continues to take C-steps using all 1500 observations in the merged set. Again the best 10 solutions ($\hat{\mu}_{merged}, \hat{\Sigma}_{merged}$) are kept. These 10 solutions are extended to the full data set in the same way, and the best solution ($\hat{\mu}_{full}, \hat{\Sigma}_{full}$) is reported.

Since the final computations are carried out in the entire data set, they take more time when n increases. In the interest of speed, we can limit the number of initial solutions ($\hat{\mu}_{merged}, \hat{\Sigma}_{merged}$) and/or the number of C-steps in the full data set as n becomes large.

The numbers such as 5 subsets of 300 observations, 500 starts, 10 best solutions, and so on are defaults based on various empirical trials, but they can be changed by the user.

2.5. The FAST-MCD algorithm

To summarize all the aspects of the algorithm, we now give an overview in pseudocode of the complete FAST-MCD algorithm.

(1) The default subset size is $h = [(n+p+1)/2]$ which yields the maximal breakdown value $\varepsilon^*(MCD) \approx 50\%$, but any integer h between $[(n+p+1)/2]$ and n can be chosen. However, data usually contain less than 25% of contamination. Therefore,

putting $h = [0.75n]$ is a good compromise between breakdown value (ε^*(MCD) $\approx$ 25%) and statistical efficiency.
(2) If $h = n$ then the MCD location estimate $\hat{\mu}$ is the empirical mean of the full data set, and the MCD scatter estimate $\hat{\Sigma}$ is its empirical covariance matrix. Report these and stop.
(3) If $p = 1$ (univariate data), compute the MCD estimate $(\hat{\mu}, \hat{\Sigma})$ by the exact algorithm of Rousseeuw and Leroy (1987, pp. 171–172) in $O(n \log n)$ time; then stop.
(4) From here on, $h < n$ and $p \geqslant 2$. If n is small (say, $n \leqslant 600$) then
  - Repeat (say) 500 times:
    * construct an initial h-subset H_1 starting from a random $(p + 1)$-subset as outlined in Section 2.3;
    * carry out two C-steps as described by the pseudocode in Section 2.3.
  - Select the 10 results with lowest covariance determinant and carry out C-steps until convergence.
  - Report the solution $(\hat{\mu}, \hat{\Sigma})$ with lowest det($\hat{\Sigma}$).
(5) If n is larger (say, $n > 600$) then
  - Construct up to five disjoint random subsets of size n_{sub} (say, 5 subsets of size $n_{\text{sub}} = 300$).
  - Inside each subset, repeat $500/5 = 100$ times:
    * construct an initial subset H_1 of size $h_{\text{sub}} = [n_{\text{sub}}(h/n)]$;
    * carry out two C-steps, using n_{sub} and h_{sub};
    * keep the 10 best results $(\hat{\mu}_{\text{sub}}, \hat{\Sigma}_{\text{sub}})$.
  - Pool the subsets, yielding the merged set (say, of size $n_{\text{merged}} = 1500$).
  - In the merged set, repeat for each of the 50 solutions $(\hat{\mu}_{\text{sub}}, \hat{\Sigma}_{\text{sub}})$:
    * carry out two C-steps, using n_{merged} and $h_{\text{merged}} = [n_{\text{merged}}(h/n)]$;
    * keep the m_{full} (say $m_{\text{full}} = 10$) best results $(\hat{\mu}_{\text{merged}}, \hat{\Sigma}_{\text{merged}})$.
  - In the full data set, repeat for the m_{full} best results:
    * take several C-steps, using n and h;
    * keep the best final result $(\hat{\mu}_{\text{full}}, \hat{\Sigma}_{\text{full}})$.
  Here m_{full} and the number of C-steps (preferably, until convergence) depend on how large the data set is.

Note that the FAST-MCD algorithm is itself affine equivariant. FAST-MCD contains two more steps:

(6) In order to obtain unbiased and consistent estimates when the data come from a multivariate normal distribution, we put

$$\hat{\mu}_{\text{MCD}} = \hat{\mu}_{\text{full}} \quad \text{and} \quad \hat{\Sigma}_{\text{MCD}} = c_{h,n} \hat{\Sigma}_{\text{full}}$$

where the multiplication factor $c_{h,n}$ is the product of a consistency factor and a finite-sample correction factor (Pison et al., 2002).
(7) A one-step reweighted estimate is obtained by

$$\hat{\mu}_1 = \left(\sum_{i=1}^{n} w_i x_i \right) \bigg/ \left(\sum_{i=1}^{n} w_i \right),$$

$$\widehat{\Sigma}_1 = d_{h,n} \left(\sum_{i=1}^n w_i(x_i - \hat{\mu}_1)(x_i - \hat{\mu}_1)' \right) \Big/ \left(\sum_{i=1}^n w_i - 1 \right)$$

where $w_i = \begin{cases} 1 & \text{if } d_{(\hat{\mu}_{\text{MCD}}, \widehat{\Sigma}_{\text{MCD}})}(i) \leq \sqrt{\chi^2_{p,0.975}} \\ 0 & \text{otherwise.} \end{cases}$

Also here, $d_{h,n}$ consists of a consistency and finite-sample correction factor (Pison et al., 2002).

The one-step reweighted estimator computed in the last step has the same breakdown value as the initial MCD estimator but a much better statistical efficiency. For example, the asymptotic efficiency of the MCD scatter matrix with $h = 0.75n$ is 44% in 10 dimensions, while the one-step reweighted covariance matrix attains 83% of efficiency (Croux and Haesbroeck, 1999). In the special case where h or more observations of the data set lie on a hyperplane, the FAST-MCD algorithm yields the corresponding MCD location $\hat{\mu}$ and singular scatter matrix $\widehat{\Sigma}$. From this solution $(\hat{\mu}, \widehat{\Sigma})$ the program then computes the equation of the hyperplane that contains at least h of the data points. If the original data were in p dimensions but it turns out that most of the data lie on a hyperplane, we can apply FAST-MCD again to the projected data in this $(p-1)$-dimensional space.

2.6. Examples

Let us now re-analyze the Philips data. For each observation x_i, we now compute the robust distance (Rousseeuw and Leroy, 1987) given by

$$\text{RD}(x_i) = \sqrt{(x_i - \hat{\mu})' \widehat{\Sigma}^{-1} (x_i - \hat{\mu})} \qquad (3)$$

where $(\hat{\mu}, \widehat{\Sigma})$ are the MCD location and scatter estimates. Recall that the Mahalanobis distances in Figure 1 indicated no groups of outliers. On the other hand, the robust distances $\text{RD}(x_i)$ in Figure 4 show a strongly deviating group of outliers, ranging from index 491 to index 565. Something happened in the production process, which was not visible from the classical Mahalanobis distances due to the masking effect. Furthermore, Figure 4 also shows a remarkable change after the first 100 measurements. Both phenomena were investigated and interpreted by the engineers at Philips.

The second data set came from a group of Cal Tech astronomers working on the Digitized Palomar Sky Survey (see Odewahn et al., 1998). They made a survey of celestial objects (light sources) by recording nine characteristics (such as magnitude, area, image moments) in each of three bands: blue, red, and near-infrared. The database contains measurements for 27 variables on 137 256 celestial objects. Exploratory data analysis led to the selection of six variables (two from each band). Figure 5 shows the Mahalanobis distances (1) for a randomly drawn subset of 10 000 points. The cutoff is at $\sqrt{\chi^2_{6,0.975}} = 3.82$. Figure 5 reveals two groups of outliers with $\text{MD}(x_i) \approx 9$ and $\text{MD}(x_i) \approx 12$, and a few outliers even further away.

The astronomical meaning of the variables revealed that the outlying groups were all objects for which at least one measurement fell outside its physically possible range.

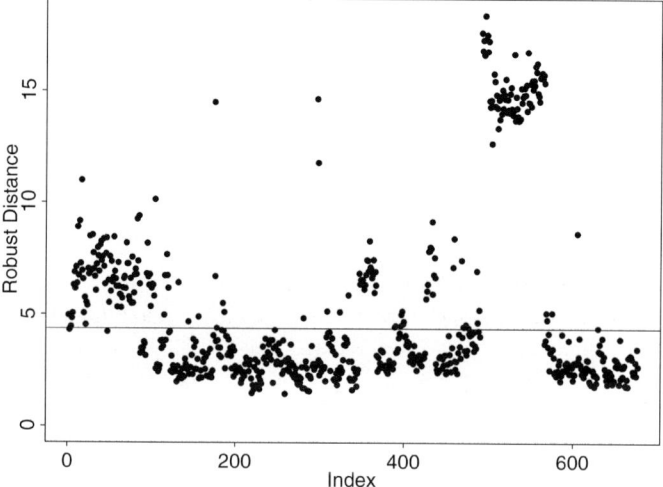

Fig. 4. Plot of robust distances for the Philips data.

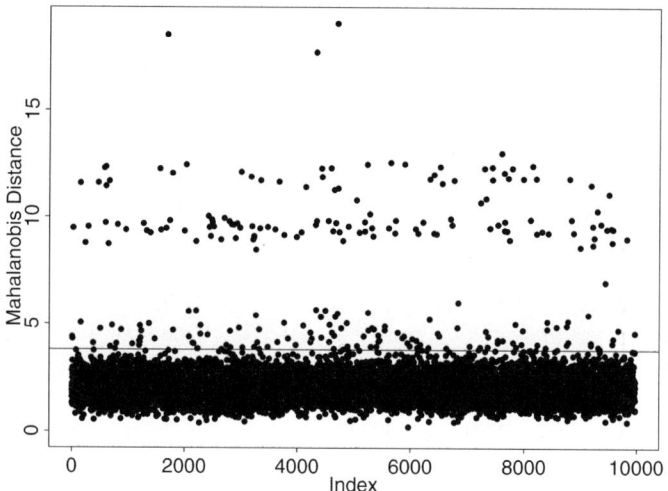

Fig. 5. Mahalanobis distances for the Digitized Palomar data.

Therefore, the data were cleaned by removing all objects with physically impossible measurements, yielding a reduced data set of size 132 402. The Mahalanobis distances of a random subset of the reduced data are shown in Figure 6(a).

This plot (and a Q–Q plot) suggests that the distances approximately come from the $\sqrt{\chi_6^2}$ distribution, as would be the case if the data came from a homogeneous population. The FAST-MCD algorithm now allows us to compute robust estimates for this database. Figure 6(b) plots the resulting robust distances. In contrast to the innocent-looking Mahalanobis distances, these robust distances reveal the presence of two groups. There is a

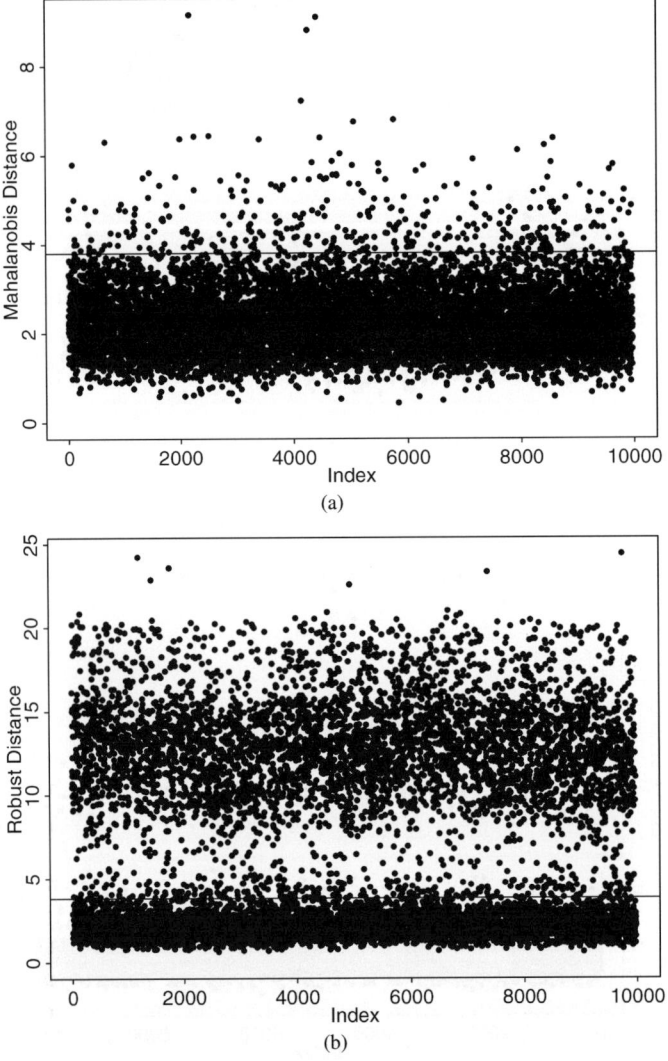

Fig. 6. Reduced Digitized Palomar data: (a) Mahalanobis distances; (b) Robust distances.

majority with $\mathrm{RD}(x_i) \leqslant \sqrt{\chi^2_{6,0.975}}$ and a group with $\mathrm{RD}(x_i)$ between 8 and 16. When reporting these results to the astronomers, it turned out that the lower group are mainly stars while the upper group are mainly galaxies.

3. Multiple regression

The multiple regression model assumes that also a response variable y is measured, which can be explained as an affine combination of the x-variables. More precisely, the

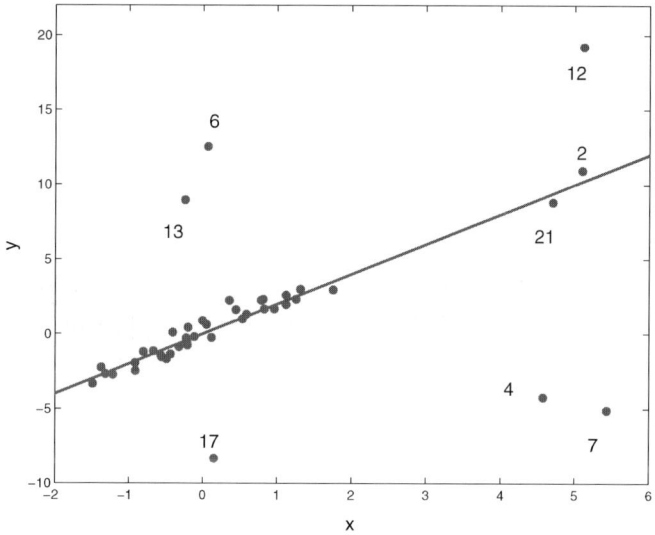

Fig. 7. Simple regression data with different types of outliers.

model says that for all observations (x_i, y_i) with $i = 1, \ldots, n$ it holds that

$$y_i = \theta_1 x_{i1} + \cdots + \theta_p x_{ip} + \theta_{p+1} + \varepsilon_i, \quad i = 1, \ldots, n, \qquad (4)$$

where the errors ε_i are assumed to be i.i.d. with zero mean and constant variance σ^2. The vector $\boldsymbol{\beta} = (\theta_1, \ldots, \theta_p)'$ is called the slope, and $\alpha = \theta_{p+1}$ the intercept. For regression without intercept, we require that $\theta_{p+1} = 0$. Denote $\boldsymbol{x}_i = (x_{i1}, \ldots, x_{ip})'$ and $\boldsymbol{\theta} = (\boldsymbol{\beta}', \alpha)' = (\theta_1, \ldots, \theta_p, \theta_{p+1})'$.

The classical least squares method to estimate $\boldsymbol{\theta}$ and σ is extremely sensitive to regression outliers, i.e., observations that do not obey the linear pattern formed by the majority of the data. In regression, we can distinguish between different types of points. This is illustrated in Figure 7 for simple regression. *Leverage points* are observations $(\boldsymbol{x}_i, y_i)$ whose $\boldsymbol{x}_i$ are outlying, i.e., $\boldsymbol{x}_i$ deviates from the majority in x-space. We call such an observation $(\boldsymbol{x}_i, y_i)$ a good leverage point if $(\boldsymbol{x}_i, y_i)$ follows the linear pattern of the majority, such as points 2 and 21. If, on the other hand, $(\boldsymbol{x}_i, y_i)$ does not follow this linear pattern, we call it a bad leverage point, like 4, 7 and 12. An observation whose $\boldsymbol{x}_i$ belongs to the majority in x-space but where $(\boldsymbol{x}_i, y_i)$ deviates from the linear pattern is called a vertical outlier, like the points 6, 13 and 17. A regression data set can thus have up to four types of points: regular observations, vertical outliers, good leverage points, and bad leverage points. Leverage points attract the least squares solution towards them, so bad leverage points are often masked in a classical regression analysis.

To illustrate this, we again consider the database of the Digitized Palomar Sky Survey. We now use the subset of 56 744 stars for which all the characteristics in the blue color (the F band) are available. On the vertical axis in Figure 8 the aperture magnitude (a brightness measure) of the star in the F band (MAperF) is plotted and the variable on the horizontal axis (called csfF) is based on spectroscopy. Figure 8 clearly shows

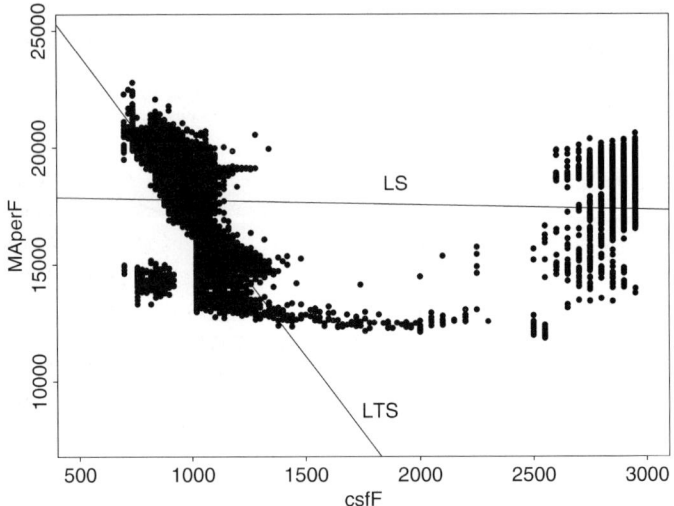

Fig. 8. Plot of aperture magnitude versus the quantity csf for the Digitized Palomar data.

that the classical least squares (LS) line is affected by the group of stars with outlying x-coordinate (on the right hand side of the plot).

In low dimensions, as in this example, visual inspection can be used to detect outliers and leverage points, but for large databases in higher dimensions this is not an option anymore. Therefore, we need robust and computationally efficient estimators that yield a reliable analysis of regression data. We consider the least trimmed squares estimator (LTS) of Rousseeuw (1984) for this purpose.

For the data set $Z = \{(x_i, y_i); \ i = 1, \ldots, n\}$ and for any $\theta \in \mathbb{R}^{p+1}$, denote the corresponding residuals by $r_i = r_i(\theta) = y_i - \beta' x_i - \alpha = y_i - \theta' u_i$ with $u_i = (x_i', 1)'$. Then the LTS estimator $\hat{\theta}$ is defined as

$$\hat{\theta} = \arg\min_{\theta} \sum_{i=1}^{h} (r^2)_{i:n} \tag{5}$$

where $(r^2)_{1:n} \leqslant (r^2)_{2:n} \leqslant \cdots \leqslant (r^2)_{n:n}$ are the ordered squared residuals (note that the residuals are first squared and then ordered). This is equivalent to finding the h-subset with smallest least squares objective function, which resembles the definition of the MCD method. The LTS estimate is then the least squares fit to these h points. The LTS estimates are regression, scale, and affine equivariant. That is, for any $X = (x_1, \ldots, x_n)'$ and $y = (y_1, \ldots, y_n)'$ it holds that

$$\hat{\theta}(X, y + Xv + 1_n c) = \hat{\theta}(X, y) + (v', c)',$$
$$\hat{\theta}(X, cy) = c\hat{\theta}(X, y),$$
$$\hat{\theta}(XA' + 1_n v', y) = \left(\hat{\beta}'(X, y) A^{-1}, \alpha(X, y) - \hat{\beta}'(X, y) A^{-1} v\right)'$$

for any $v \in \mathbb{R}^p$, any constant c and any nonsingular matrix $A \in \mathbb{R}^{p \times p}$. Again $1_n = (1, 1, \ldots, 1)' \in \mathbb{R}^n$.

The breakdown value of a regression estimator $\hat{\boldsymbol{\theta}}$ at a data set Z is the smallest fraction of outliers that can have an arbitrarily large effect on $\hat{\boldsymbol{\theta}}$. Formally, it is defined by (2) where X is replaced by (X, y). For $h = [(n + p + 1)/2]$, the LTS breakdown value equals $\varepsilon^*(\text{LTS}) \approx 50\%$, whereas for larger h we have that $\varepsilon_n^*(\text{LTS}) \approx (n - h)/n$. The usual choice $h \approx 0.75n$ yields $\varepsilon^*(\text{LTS}) = 25\%$.

When using LTS regression, the standard deviation of the errors can be estimated by

$$\hat{\sigma} = c_{h,n} \sqrt{\frac{1}{h} \sum_{i=1}^{h} (r^2)_{i:n}} \tag{6}$$

where r_i are the residuals from the LTS fit, and $c_{h,n}$ makes $\hat{\sigma}$ consistent and unbiased at Gaussian error distributions (Pison et al., 2002). Note that the LTS scale estimator $\hat{\sigma}$ is itself highly robust (Croux and Rousseeuw, 1992). Therefore, we can identify regression outliers by their standardized LTS residuals $r_i/\hat{\sigma}$.

To compute the LTS in an efficient way, Rousseeuw and Van Driessen (2000) developed a FAST-LTS algorithm similar to the FAST-MCD algorithm described in the previous section. The basic component of the algorithm is again the C-step, which now says that starting from an initial h-subset H_1 or an initial fit $\hat{\boldsymbol{\theta}}_1$ we can construct a new h-subset H_2 by taking the h observations with smallest absolute residuals $|r_i(\hat{\boldsymbol{\theta}}_1)|$. Applying LS to H_2 then yields a new fit $\hat{\boldsymbol{\theta}}_2$ which is guaranteed to have a lower objective function (5).

To construct the initial h-subsets, we now start from randomly drawn $(p + 1)$-subsets. For each $(p + 1)$-subset, the coefficients $\boldsymbol{\theta}_0$ of the hyperplane through the $(p + 1)$ points in the subset are calculated. If a $(p + 1)$-subset does not define a unique hyperplane, then it is extended by adding more observations until it does. The corresponding initial h-subset is then formed by the h points closest to the hyperplane (i.e., with smallest residuals). As was the case for the MCD, also here this approach yields much better initial fits than would be the case if random h-subsets were drawn directly.

For regression with an intercept, the accuracy of the algorithm can be improved further by intercept adjustment. This is a technique that decreases the value of the LTS objective function for any fit. For regression with an intercept, each initial $(p+1)$-subset, corresponding h-subset and each C-step yield an estimate $\hat{\boldsymbol{\theta}} = (\hat{\theta}_1, \ldots, \hat{\theta}_p, \hat{\theta}_{p+1})'$ where $\hat{\theta}_1, \ldots, \hat{\theta}_p$ are slopes and $\hat{\theta}_{p+1}$ is the intercept. The corresponding value of the LTS objective function is given by

$$\sum_{i=1}^{h} (r^2(\hat{\boldsymbol{\theta}}))_{i:n} = \sum_{i=1}^{h} ((y_i - x_{i1}\hat{\theta}_1 - \cdots - x_{ip}\hat{\theta}_p - \hat{\theta}_{p+1})^2)_{i:n}. \tag{7}$$

The adjusted intercept $\hat{\theta}'_{p+1}$ is now calculated as the exact univariate LTS location estimate of $\{t_i = y_i - x_{i1}\hat{\theta}_1 - \cdots - x_{ip}\hat{\theta}_p;\ i = 1, \ldots, n\}$. Hence,

$$\hat{\theta}'_{p+1} = \arg\min_{\mu} \sum_{i=1}^{h} ((t_i - \mu)^2)_{i:n}.$$

By construction, replacing $\hat{\theta}_{p+1}$ by $\hat{\theta}'_{p+1}$ yields the same or a lower value of the objective function (7).

For a given univariate data set, the LTS location estimator is the mean of the h-subset with the smallest variance (note that this coincides with the MCD location estimator). Let us sort the data. Clearly, the search can be limited to contiguous h-subsets $t_{j:n}, \ldots, t_{j+h-1:n}$. If we denote their mean by $\bar{t}^{(j)}$ and their sum of squares by

$$\mathrm{SQ}^{(j)} = \sum_{i=j}^{j+h-1} \left(t_{i:n} - \bar{t}^{(j)}\right)^2,$$

then the LTS location minimizes $\mathrm{SQ}^{(j)}$ over $j = 1, \ldots, n+1-h$. By updating $\bar{t}^{(j)}$ and $\mathrm{SQ}^{(j)}$ at each j, the whole univariate LTS location algorithm (Rousseeuw and Leroy, 1987, pp. 171–172) takes only $O(n)$ space and $O(n \log n)$ time.

The FAST-LTS algorithm can be summarized by the pseudocode given in Section 2.5 with the C-step described above, and includes the intercept adjustment for regression with an intercept. In step (2), if $h = n$ the least squares solution is now computed and in step (3), for regression with only an intercept the exact univariate LTS algorithm is used. The initial subsets in step (4) are generated as described above. In step (6) we put

$$\hat{\theta}_{\mathrm{LTS}} = \hat{\theta}_{\mathrm{full}} \quad \text{and} \quad \hat{\sigma}_{\mathrm{LTS}} = c_{h,n} \sqrt{\frac{1}{h} \sum_{i=1}^{h} \left(r(\hat{\theta}_{\mathrm{full}})^2\right)_{i:n}}.$$

Finally, in step (7) reweighted least squares estimates are given by

$$\hat{\theta}_1 = \left(\sum_{i=1}^{n} w_i u_i u_i'\right)^{-1} \left(\sum_{i=1}^{n} w_i u_i y_i\right), \qquad \hat{\sigma}_1 = d_{h,n} \sqrt{\frac{\sum_{i=1}^{n} w_i r_i(\hat{\theta}_1)^2}{\sum_{i=1}^{n} w_i - 1}}$$

where $u_i = (x_i', 1)'$ and $w_i = \begin{cases} 1 & \text{if } |r_i(\hat{\theta}_{\mathrm{LTS}})/\hat{\sigma}_{\mathrm{LTS}}| \leqslant 2.5, \\ 0 & \text{otherwise.} \end{cases}$

As before $d_{h,n}$ combines a consistency and a finite-sample correction factor (Pison et al., 2002). These reweighted estimates have the same breakdown value as the initial LTS estimates and a much better statistical efficiency. Moreover, from the reweighted least squares estimates we can obtain all the usual inferential output such as t-statistics, F-statistics, an R^2 statistic and the corresponding p-values. These p-values assume that the data with $w_i = 1$ come from the model (4) whereas the data with $w_i = 0$ do not.

In Figure 8 we see that the LTS line obtained by FAST-LTS yields a robust fit that is not attracted by the leverage points on the right hand side, and hence follows the pattern of the majority of the data. Of course, the LTS method is most useful when there are several x-variables.

To detect leverage points in higher dimensions, we must detect outlying x_i in x-space. For this purpose, we will use the robust distances RD_i based on the one-step reweighted MCD of the previous section. On the other hand, we can see whether a point (x_i, y_i) lies near the majority pattern by looking at its standardized LTS residual $r_i/\hat{\sigma}$. Rousseeuw and Van Zomeren (1990) proposed a diagnostic display which

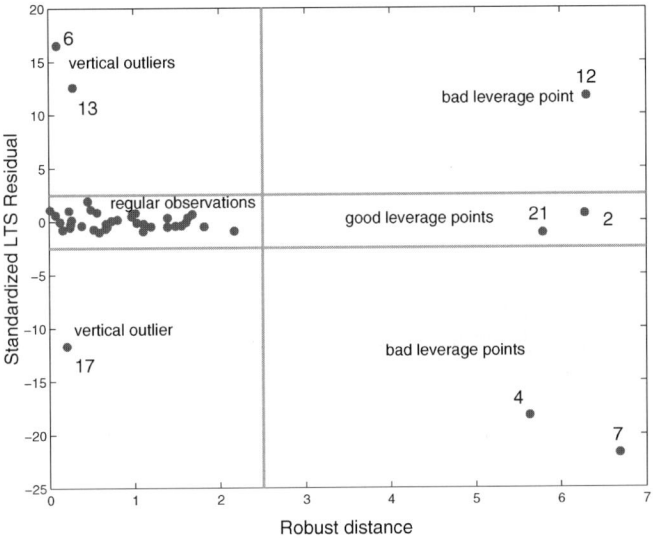

Fig. 9. Regression diagnostic plot for the artificial data of Figure 7.

plots robust residuals $r_i/\hat{\sigma}$ versus robust distances $RD(x_i)$, and indicates the corresponding cutoffs by horizontal and vertical lines. It automatically classifies the observations into the four types of data points that can occur in a regression data set. For the artificial data of Figure 7, the corresponding diagnostic plot is shown in Figure 9.

To illustrate this plot, let us return to the set of 56 744 stars with a complete F band in the Digitized Palomar Sky Survey database. The same response variable MaperF is now regressed against the other 8 characteristics of the color band F. These characteristics describe the size of a light source and the shape of the spatial brightness distribution in a source. Figure 10(a) plots the standardized LS residuals versus the classical Mahalanobis distances. Some isolated outliers in the y-direction as well as in x-space were not plotted to get a better view of the majority of the data. Observations for which the standardized absolute LS residual exceeds the cutoff 2.5 are considered to be regression outliers (because values generated by a Gaussian distribution are rarely larger than 2.5σ), whereas the other observations are thought to obey the linear model. Similarly, observations for which $MD(x_i)$ exceeds the cutoff $\sqrt{\chi^2_{8,0.975}}$ are considered to be leverage points. Figure 10(a) shows that most data points lie between the horizontal cutoffs at 2.5 and -2.5 which suggests that most data follow the same linear trend. On the other hand, the diagnostic plot based on LTS residuals and robust distances $RD(x_i)$ shown in Figure 10(b) tells a different story. This plot reveals a rather large group of observations with large robust residuals and large robust distances. Hence, these observations are bad leverage points. This group turned out to be giant stars, which are known to behave differently from other stars.

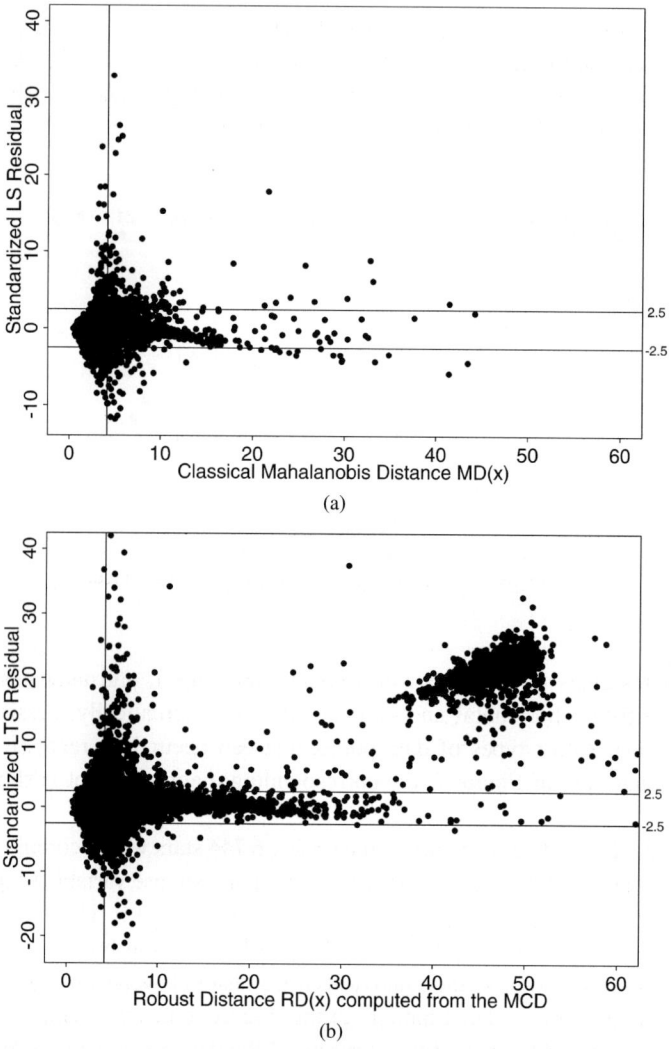

Fig. 10. Digitized Palomar Sky Survey data: regression of MAperF on 8 regressors. (a) Plot of LS residual versus Mahalanobis distance MD(x_i). (b) Diagnostic plot of LTS residual versus robust distance RD(x_i).

4. Multivariate regression

The regression model can be extended to the case where we have more than one response variable. For p-variate predictors $x_i = (x_{i1}, \ldots, x_{ip})'$ and q-variate responses $y_i = (y_{i1}, \ldots, y_{iq})'$, the multivariate regression model is given by

$$y_i = \mathcal{B}'x_i + \alpha + \varepsilon \tag{8}$$

where $\boldsymbol{B}$ is the $p \times q$ slope matrix, $\boldsymbol{\alpha}$ is the q-dimensional intercept vector, and the errors $\boldsymbol{\varepsilon} = (\varepsilon_1, \ldots, \varepsilon_q)'$ are i.i.d. with zero mean and with $\text{Cov}(\boldsymbol{\varepsilon}) = \boldsymbol{\Sigma}_\varepsilon$ a positive definite matrix of size q. Note that for $q = 1$ we obtain the multiple regression model of the previous section. On the other hand, putting $p = 1$ and $x_i = 1$ yields the multivariate location and scatter model. It is well known that the least squares solution can be written as

$$\widehat{\boldsymbol{B}} = \widehat{\boldsymbol{\Sigma}}_{xx}^{-1} \widehat{\boldsymbol{\Sigma}}_{xy}, \tag{9}$$

$$\widehat{\boldsymbol{\alpha}} = \widehat{\boldsymbol{\mu}}_y - \widehat{\boldsymbol{B}}' \widehat{\boldsymbol{\mu}}_x, \tag{10}$$

$$\widehat{\boldsymbol{\Sigma}}_\varepsilon = \widehat{\boldsymbol{\Sigma}}_{yy} - \widehat{\boldsymbol{B}}' \widehat{\boldsymbol{\Sigma}}_{xx} \widehat{\boldsymbol{B}} \tag{11}$$

where

$$\widehat{\boldsymbol{\mu}} = \begin{pmatrix} \widehat{\boldsymbol{\mu}}_x \\ \widehat{\boldsymbol{\mu}}_y \end{pmatrix} \quad \text{and} \quad \widehat{\boldsymbol{\Sigma}} = \begin{pmatrix} \widehat{\boldsymbol{\Sigma}}_{xx} & \widehat{\boldsymbol{\Sigma}}_{xy} \\ \widehat{\boldsymbol{\Sigma}}_{yx} & \widehat{\boldsymbol{\Sigma}}_{yy} \end{pmatrix}$$

are the empirical mean and covariance matrix of the joint (x, y) variables.

Vertical outliers and bad leverage points highly influence the least squares estimates in multivariate regression, and may make the results completely unreliable. Therefore, robust alternatives have been developed.

Rousseeuw et al. (2004) proposed to use the MCD estimates for the center $\boldsymbol{\mu}$ and scatter matrix $\boldsymbol{\Sigma}$ in expressions (9) to (11). The resulting estimates are called MCD regression estimates. It has been shown that the MCD regression estimates are regression, y-affine and x-affine equivariant. With $\boldsymbol{X} = (\boldsymbol{x}_1, \ldots, \boldsymbol{x}_n)'$, $\boldsymbol{Y} = (\boldsymbol{y}_1, \ldots, \boldsymbol{y}_n)'$ and $\widehat{\boldsymbol{\theta}} = (\widehat{\boldsymbol{B}}', \widehat{\boldsymbol{\alpha}})'$ this means that

$$\widehat{\boldsymbol{\theta}}(\boldsymbol{X}, \boldsymbol{Y} + \boldsymbol{X}\boldsymbol{D} + \boldsymbol{1}_n \boldsymbol{w}') = \widehat{\boldsymbol{\theta}}(\boldsymbol{X}, \boldsymbol{Y}) + (\boldsymbol{D}', \boldsymbol{w})', \tag{12}$$

$$\widehat{\boldsymbol{\theta}}(\boldsymbol{X}, \boldsymbol{Y}\boldsymbol{C} + \boldsymbol{1}_n \boldsymbol{w}') = \widehat{\boldsymbol{\theta}}(\boldsymbol{X}, \boldsymbol{Y})\boldsymbol{C} + (\boldsymbol{O}'_{pq}, \boldsymbol{w})', \tag{13}$$

$$\widehat{\boldsymbol{\theta}}(\boldsymbol{X}\boldsymbol{A}' + \boldsymbol{1}_n \boldsymbol{v}', \boldsymbol{Y}) = \left(\widehat{\boldsymbol{B}}'(\boldsymbol{X}, \boldsymbol{Y})\boldsymbol{A}^{-1}, \widehat{\boldsymbol{\alpha}}(\boldsymbol{X}, \boldsymbol{Y}) - \widehat{\boldsymbol{B}}'(\boldsymbol{X}, \boldsymbol{Y})\boldsymbol{A}^{-1}\boldsymbol{v}\right)' \tag{14}$$

where $\boldsymbol{D}$ is any $p \times q$ matrix, $\boldsymbol{A}$ is any nonsingular $p \times p$ matrix, $\boldsymbol{C}$ is any nonsingular $q \times q$ matrix, $\boldsymbol{v}$ is any p-dimensional vector, and $\boldsymbol{w}$ is any q-dimensional vector. Here $\boldsymbol{O}_{pq}$ is the $p \times q$ matrix consisting of zeroes.

MCD regression inherits the breakdown value of the MCD estimator, thus $\varepsilon_n^*(\widehat{\boldsymbol{\theta}}) \approx (n - h)/n$. To obtain a better efficiency, the reweighted MCD estimates are used in (9)–(11) and followed by the regression reweighting step described below. For any fit, $\widehat{\boldsymbol{\theta}}$ denote the corresponding q-dimensional residuals by $\boldsymbol{r}_i(\widehat{\boldsymbol{\theta}}) = \boldsymbol{y}_i - \widehat{\boldsymbol{B}}'\boldsymbol{x}_i - \widehat{\boldsymbol{\alpha}}$. Then the reweighted regression estimates are given by

$$\widehat{\boldsymbol{\theta}}_1 = \left(\sum_{i=1}^n w_i \boldsymbol{u}_i \boldsymbol{u}_i'\right)^{-1} \left(\sum_{i=1}^n w_i \boldsymbol{u}_i \boldsymbol{y}_i'\right), \tag{15}$$

$$\widehat{\boldsymbol{\Sigma}}_\varepsilon^1 = d_1 \frac{\sum_{i=1}^n w_i \boldsymbol{r}_i(\widehat{\boldsymbol{\theta}}_1) \boldsymbol{r}_i(\widehat{\boldsymbol{\theta}}_1)'}{\sum_{i=1}^n w_i} \tag{16}$$

where $\boldsymbol{u}_i = (\boldsymbol{x}'_i, 1)'$ and d_1 is a consistency factor. The weights w_i are given by

$$w_i = \begin{cases} 1 & \text{if } d^2(r_i(\hat{\boldsymbol{\theta}}_{\text{MCD}})) \leq \chi^2_{q,0.99} \\ 0 & \text{otherwise} \end{cases}$$

with $d^2(r_i(\hat{\boldsymbol{\theta}}_{\text{MCD}})) = r_i(\hat{\boldsymbol{\theta}}_{\text{MCD}})'(\hat{\boldsymbol{\Sigma}}_\varepsilon)^{-1} r_i(\hat{\boldsymbol{\theta}}_{\text{MCD}})$ the squared robust distances of the residuals corresponding to the initial MCD regression estimates $\hat{\boldsymbol{\theta}}_{\text{MCD}}$ and $\hat{\boldsymbol{\Sigma}}_\varepsilon$ based on the reweighted MCD. Note that these reweighted regression estimates (15), (16) have the same breakdown value as the initial MCD regression estimates.

To illustrate MCD regression, we analyze a data set obtained from Shell's polymer laboratory in Ottignies, Belgium by courtesy of Dr. Christian Ritter. To preserve confidentiality, all variables have been standardized. The data set consists of $n = 217$ observations with $p = 4$ predictor variables and $q = 3$ response variables. The predictor variables describe the chemical characteristics of a piece of foam, whereas the response variables measure its physical properties such as tensile strength. The physical properties of foam are determined by the chemical composition used in the production process. Therefore, multivariate regression is used to establish a relationship between the chemical inputs and the resulting physical properties of foam. After an initial exploratory study of the variables, a robust multivariate MCD regression was used. The breakdown value was set equal to 25%.

To detect leverage points and outliers, the diagnostic plot of Rousseeuw and Van Zomeren (1990) has been extended to multivariate regression. In multivariate regression, the robust distances of the residuals $r_i(\hat{\boldsymbol{\theta}}_1)$' are plotted versus the robust distances of the $\boldsymbol{x}_i$. Figure 11 shows the diagnostic plot of the Shell foam data. Observations 215 and 110 lie far from both the horizontal cutoff line at $\sqrt{\chi^2_{3,0.975}} = 3.06$ and the vertical cutoff line at $\sqrt{\chi^2_{4,0.975}} = 3.34$. These two observations can thus be classified as bad leverage points. Several observations lie substantially above the horizontal cutoff but not to the right of the vertical cutoff, which means that they are vertical outliers (their residuals are outlying but their x-values are not).

When this list of special points was presented to the scientists who had made the measurements, we learned that a fraction of the observations in Figure 11 were made with a different production technique and hence belong to a different population with other characteristics. These include the observations 210, 212 and 215. We therefore remove these observations from the data, and retain only observations from the intended population.

Running the method again yields the diagnostic plot in Figure 12. Observation 110 is still a bad leverage point, and also several of the vertical outliers remained. No chemical/physical mechanism was found to explain why these points are outliers, leaving open the possibility of some large measurement errors. But the detection of these outliers at least provides us with the option to choose whether or not to allow them to affect the final result.

Since MCD regression is mainly intended for regression data with random carriers, Agulló et al. (2001) developed an alternative robust multivariate regression method which can be seen as an extension of LTS to the multivariate setting. This multivariate least trimmed squares estimator (MLTS) can also be used in cases where the carriers are

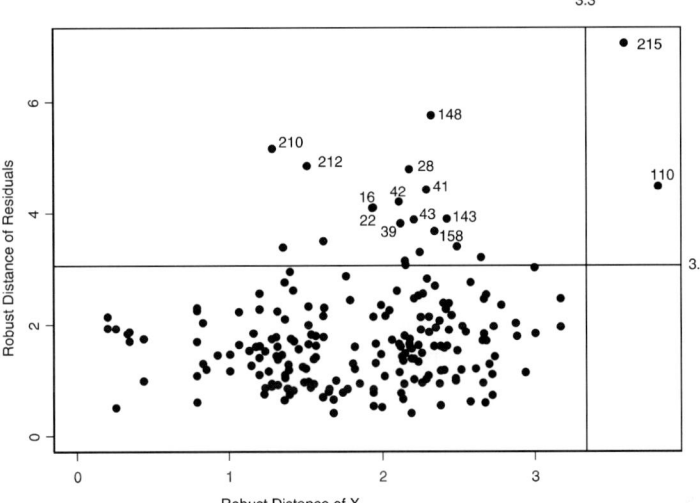

Fig. 11. Diagnostic plot of robust residuals versus robust distances of the carriers for the foam data.

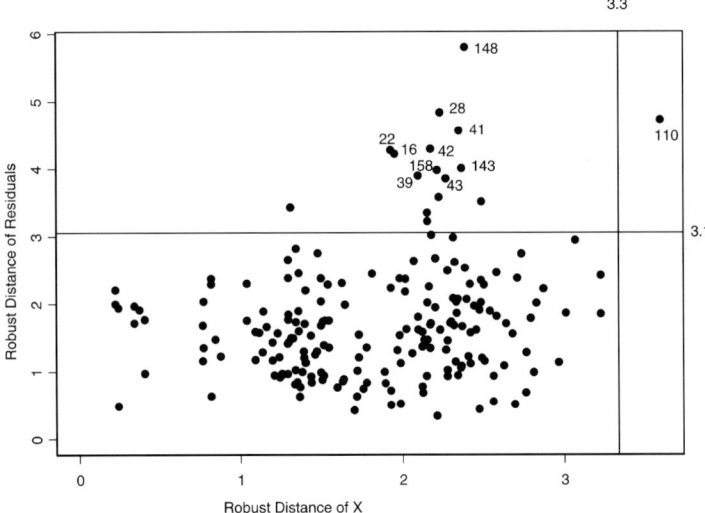

Fig. 12. Diagnostic plot of robust residuals versus robust distances for the corrected foam data.

fixed. The MLTS looks for a subset of size h such that the determinant of the covariance matrix of its residuals corresponding to its least squares fit is minimal. The MLTS satisfies $\varepsilon_n^*(\text{MLTS}) \approx (n-h)/n$ and the equivariance properties (12)–(14) expected from a multivariate regression estimator. The MLTS can be computed quickly with an algorithm based on C-steps as outlined in Section 2.5. To improve the efficiency while keeping the breakdown value, a one-step reweighted MLTS estimator can be computed using expressions (15), (16).

5. Classification

The goal of classification, also known as discriminant analysis or supervised learning, is to obtain rules that describe the separation between known groups of observations. Moreover, it allows to classify new observations into one of the groups. We denote the number of groups by l and assume that we can describe our experiment in each population π_j by a p-dimensional random variable X_j with distribution function (density) f_j. We write p_j for the membership probability, i.e., the probability for an observation to come from π_j. The maximum likelihood rule then classifies an observation $x \in \mathbb{R}^p$ into π_k if $\ln(p_k f_k(x))$ is the maximum of the set $\{\ln(p_j f_j(x)); \ j = 1, \ldots, l\}$. If we assume that the density f_j for each group is Gaussian with mean $\boldsymbol{\mu}_j$ and covariance matrix $\boldsymbol{\Sigma}_j$ then it can be seen that the maximum likelihood rule is equivalent to maximizing the discriminant scores $d_j^Q(x)$ with

$$d_j^Q(x) = -\frac{1}{2}\ln|\boldsymbol{\Sigma}_j| - \frac{1}{2}(x - \boldsymbol{\mu}_j)'\boldsymbol{\Sigma}_j^{-1}(x - \boldsymbol{\mu}_j) + \ln(p_j). \tag{17}$$

That is, x is allocated to π_k if $d_k^Q(x) > d_j^Q(x)$ for all $j = 1, \ldots, l$ (see, e.g., Johnson and Wichern, 1998).

In practice, $\boldsymbol{\mu}_j$, $\boldsymbol{\Sigma}_j$ and p_j have to be estimated. Classical Quadratic Discriminant Analysis (CQDA) uses the group's mean and empirical covariance matrix to estimate $\boldsymbol{\mu}_j$ and $\boldsymbol{\Sigma}_j$. The membership probabilities are usually estimated by the relative frequencies of the observations in each group, hence $\hat{p}_j = n_j/n$ with n_j the number of observations in group j.

A Robust Quadratic Discriminant Analysis (RQDA) is derived by using robust estimators of $\boldsymbol{\mu}_j$, $\boldsymbol{\Sigma}_j$ and p_j. In particular, we can apply the reweighed MCD estimator of location and scatter in each group. As a byproduct of this robust procedure, outliers (within each group) can be distinguished from the regular observations. Finally, the membership probabilities can be estimated robustly as the relative frequency of *regular* observations in each group. For an outline of this approach, see (Hubert and Van Driessen, 2004).

When all the covariance matrices are assumed to be equal, the quadratic scores (17) can be simplified to

$$d_j^L(x) = \boldsymbol{\mu}_j'\boldsymbol{\Sigma}^{-1}x - \frac{1}{2}\boldsymbol{\mu}_j'\boldsymbol{\Sigma}^{-1}\boldsymbol{\mu}_j + \ln(p_j) \tag{18}$$

where $\boldsymbol{\Sigma}$ is the common covariance matrix. The resulting scores (18) are linear in x, hence the maximum likelihood rule belongs to the class of *linear discriminant analysis*. It is well known that if we have only two populations ($l = 2$) with a common covariance structure and if both groups have equal membership probabilities, this rule coincides with Fisher's linear discriminant rule. Robust linear discriminant analysis based on the MCD estimator or S-estimators (Rousseeuw and Yohai, 1984; Davies, 1987) has been studied by Hawkins and McLachlan (1997), He and Fung (2000), Croux and Dehon (2001), and Hubert and Van Driessen (2004).

From Colin Greensill (Faculty of Engineering and Physical Systems, Central Queensland University, Rockhampton, Australia) we obtained a data set that contains the spectra of three different cultivars of the same fruit (cantaloupe – Cucumis melo L. Cantaloupensis). The cultivars (named D, M and HA) had sizes 490, 106 and 500, and all spectra were measured in 256 wavelengths. Our data set thus contains 1096 observations and 256 variables. First we applied a robust principal component analysis (as described in the next section) to reduce the dimension of the data space, and retained two components. For a more detailed description and analysis of these data, see (Hubert and Van Driessen, 2004).

We randomly divided the data into a training set and a validation set, containing 60% and 40% of the observations. Figure 13(a) shows the training data. In this figure, the cultivar D is marked with crosses, cultivar M with circles, and cultivar HA with diamonds. We see that cultivar HA has a cluster of outliers that are far away from the other observations. As it turns out, these outliers were caused by a change in the illumination system. We will use the model (18) with a common covariance matrix Σ. In the plot we have superimposed the robust tolerance ellipses for each group. Figure 13(b) shows the same data with the corresponding classical tolerance ellipses. Note how strongly the classical covariance estimator of the common Σ is influenced by the outlying subgroup of cultivar HA.

The effect on the resulting classical linear discriminant rules is dramatic for cultivar M. It appears that all the observations are badly classified because they would have to belong to a region that lies completely outside the boundary of this figure! The robust discriminant analysis does a better job. The tolerance ellipses are not affected by the outliers and the resulting discriminant lines split up the different groups more accurately. The misclassification rates are 17% for cultivar D, 95% for cultivar M, and 6% for cultivar HA. The misclassification rate of cultivar M remains very high. This is due to the intrinsic overlap between the three groups, and the fact that cultivar M has few data points compared to the others. (When we impose that all three groups are equally important by setting the membership probabilities equal to $1/3$, we obtain a better classification of cultivar M with 46% of errors.)

This example thus clearly shows that outliers can have a huge effect on the classical discriminant rules, whereas the robust version proposed here fares better.

6. Principal component analysis

6.1. Classical PCA

Principal component analysis is a popular statistical method which tries to explain the covariance structure of data by means of a small number of components. These components are linear combinations of the original variables, and often allow for an interpretation and a better understanding of the different sources of variation. Because PCA is concerned with data reduction, it is widely used for the analysis of high-dimensional

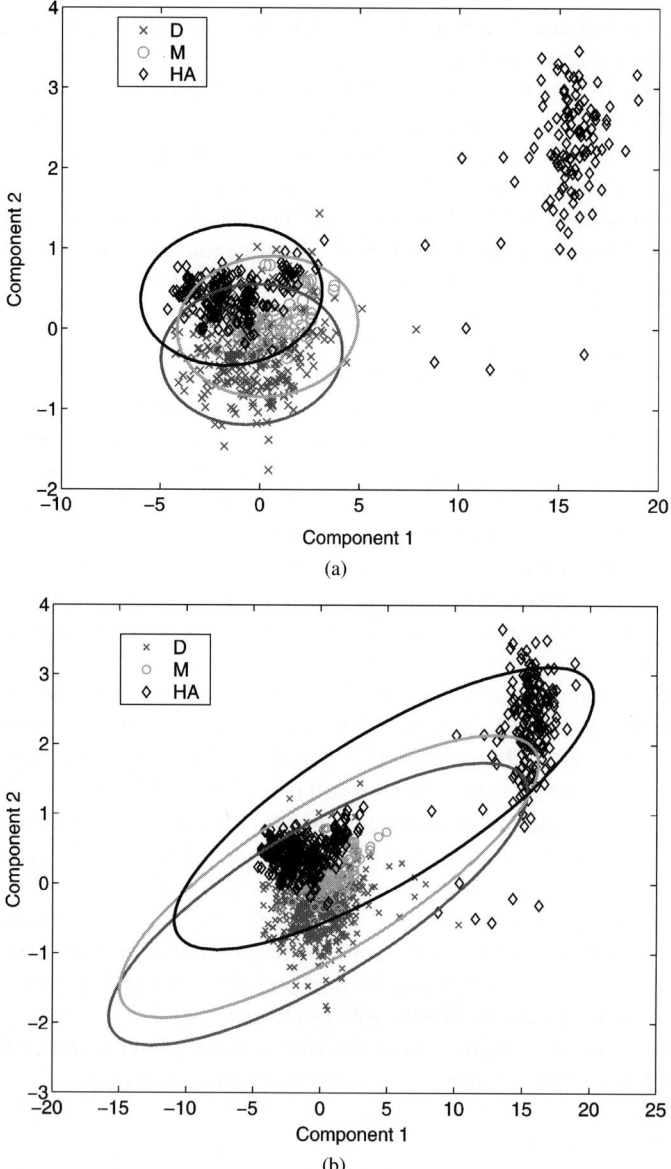

Fig. 13. (a) Robust tolerance ellipses for the fruit data with common covariance matrix; (b) classical tolerance ellipses.

data which are frequently encountered in chemometrics, computer vision, engineering, genetics, and other domains. PCA is then often the first step of the data analysis, followed by discriminant analysis, cluster analysis, or other multivariate techniques (see, e.g., Hubert and Engelen, 2005). It is thus important to find those principal components that contain most of the information.

In the classical approach, the first component corresponds to the direction in which the projected observations have the largest variance. The second component is then orthogonal to the first and again maximizes the variance of the data points projected on it. Continuing in this way produces all the principal components, which correspond to the eigenvectors of the empirical covariance matrix. Unfortunately, both the classical variance (which is being maximized) and the classical covariance matrix (which is being decomposed) are very sensitive to anomalous observations. Consequently, the first components are often attracted towards outlying points, and may not capture the variation of the regular observations. Therefore, data reduction based on classical PCA (CPCA) becomes unreliable if outliers are present in the data.

To illustrate this, let us consider a small artificial data set in $p = 4$ dimensions. The Hawkins–Bradu–Kass data set (listed in Rousseeuw and Leroy, 1987) consists of $n = 75$ observations in which 2 groups of outliers were created, labelled 1–10 and 11–14. The first two eigenvalues explain already 98% of the total variation, so we select $k = 2$. The CPCA scores plot is depicted in Figure 14(a). In this figure, we can clearly distinguish the two groups of outliers, but we see several other undesirable effects. We first observe that, although the scores have zero mean, the regular data points lie far from zero. This stems from the fact that the mean of the data points is a bad estimate of the true center of the data in the presence of outliers. It is clearly shifted towards the outlying group, and consequently the origin even falls outside the cloud of the regular data points. On the plot we have also superimposed the 97.5% tolerance ellipse. We see that the outliers 1–10 are within the tolerance ellipse, and thus do not stand out based on their Mahalanobis distance. The ellipse has stretched itself to accommodate these outliers.

6.2. Robust PCA

The goal of robust PCA methods is to obtain principal components that are not influenced much by outliers. A first group of methods is obtained by replacing the classical covariance matrix by a robust covariance estimator. Maronna (1976) and Campbell (1980) proposed to use affine equivariant M-estimators of scatter for this purpose, but these cannot resist many outliers. More recently, Croux and Haesbroeck (2000) used positive-breakdown estimators such as the MCD method and S-estimators. Let us reconsider the Hawkins–Bradu–Kass data in $p = 4$ dimensions. Robust PCA using the reweighted MCD estimator yields the score plot in Figure 14(b). We now see that the center is correctly estimated in the middle of the regular observations. The 97.5% tolerance ellipse nicely encloses these points and excludes all the 14 outliers.

Unfortunately the use of these affine equivariant covariance estimators is limited to small to moderate dimensions. To see why, let us, for example, consider the MCD estimator. If p denotes the number of variables in our data set, the MCD estimator can only be computed if $p < h$, otherwise the covariance matrix of any h-subset has zero determinant. Since $h < n$, the number of variables p may never be larger than n. A second problem is the computation of these robust estimators in high dimensions. Today's fastest algorithms (Woodruff and Rocke, 1994;

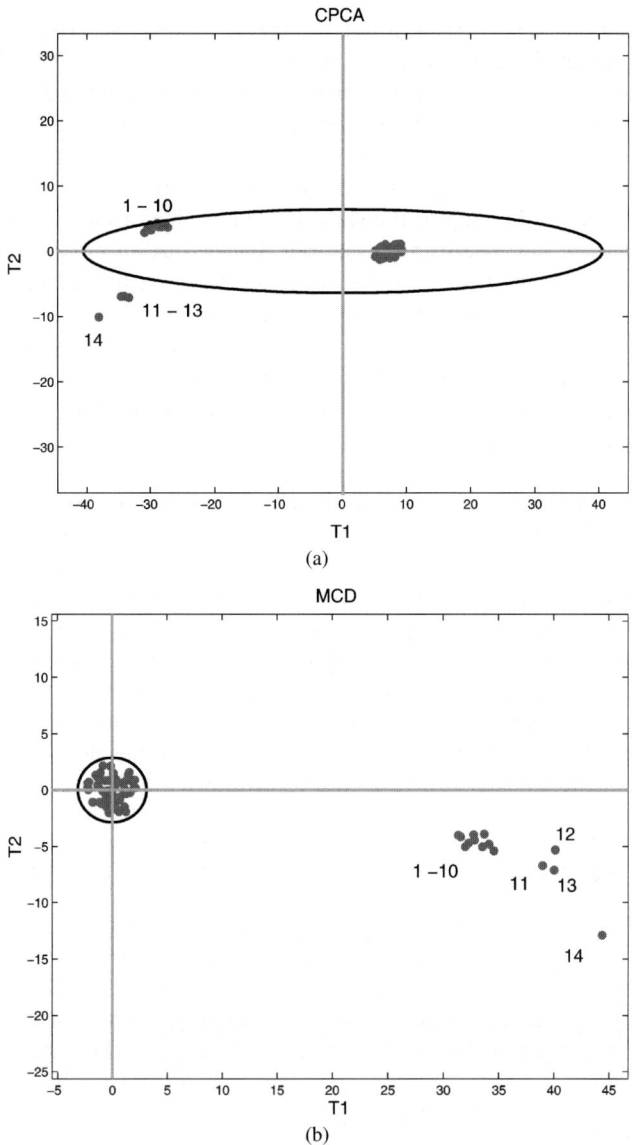

Fig. 14. Score plot and 97.5% tolerance ellipse of the Hawkins–Bradu–Kass data obtained with (a) CPCA; (b) MCD.

Rousseeuw and Van Driessen, 1999) can handle up to about 100 dimensions, whereas there are fields like chemometrics which need to analyze data with dimensions in the thousands.

A second approach to robust PCA uses *Projection Pursuit* (PP) techniques. These methods maximize a robust measure of spread to obtain consecutive directions on which the data points are projected. Hubert et al. (2002) presented a projection pursuit (PP) algorithm based on the ideas of Li and Chen (1985) and Croux and Ruiz-Gazen (1996). The algorithm is called RAPCA, which stands for *reflection algorithm for principal components analysis*. It has been successfully applied to detect outliers in large microarray data (Model et al., 2002).

If $p \geqslant n$, the method starts by reducing the data space to the affine subspace spanned by the n observations. This is done quickly and accurately by a singular value decomposition of $X_{n,p}$ (here, the subscripts indicate the dimensions of the matrix). Note that this singular value decomposition is just an affine transformation of the data. It is not used to retain only the first eigenvectors of the covariance matrix of $X_{n,p}$. This would imply that classical PCA is performed, which is of course not robust. Here, we merely represented the data set in its own dimensionality.

The main step of the algorithm is then to search for the direction in which the projected observations have the largest robust scale. To measure the univariate scale, the Q_n estimator of Rousseeuw and Croux (1993) is used. This scale estimator is essentially the first quartile of all distances between two data points. Croux and Ruiz-Gazen (2005) showed that using Q_n combines the highest breakdown value with a Gaussian efficiency of 67%.

After each projection a reflection is applied such that the dimension of the projected data points can be reduced by one. In this way, we do not need to do all the computations in $\mathbb{R}^p$. This is especially important for high dimensions p, since otherwise the algorithm can become numerically unstable (see Hubert et al., 2002, Section 2).

The pseudocode of RAPCA is as follows:

(1) Reduce the data space to the affine subspace spanned by the n observations.
(2) Center the data points by means of their L^1-median, which is robust and orthogonally equivariant.
(3) The first principal component ('eigenvector') is defined as the direction in which the projected observations have maximal robust scale. Search over all directions through the origin and a data point. The first 'eigenvalue' is defined as the squared robust scale of the projections on the first eigenvector.
(4) Reflect the data such that the first eigenvector is mapped onto the first basis vector.
(5) Project the data onto the orthogonal complement of the first eigenvector. This is simply done by omitting the first component of each (reflected) point. Note that this reduces the working dimension by one.
(6) Repeat steps (3) to (5) until all required eigenvectors and eigenvalues have been found.
(7) Transform each eigenvector back to $\mathbb{R}^p$ using the same reflections.

Note that the resulting algorithm is orthogonally equivariant, which is appropriate for a PCA method. Also note that it is not required to compute all eigenvectors, which would be very time-consuming for high p, but that the computations can be stopped as soon as the required number of components has been found. Asymptotic results about this approach are presented by Cui et al. (2003).

Another approach to robust PCA has been proposed by Hubert et al. (2005) and is called ROBPCA. This method combines ideas of both projection pursuit and robust covariance estimation. The PP part is used for the initial dimension reduction. Some ideas based on the MCD estimator are then applied to this lower-dimensional data space. Simulations by Hubert et al. (2005) have shown that this combined approach yields more accurate estimates than the raw PP algorithm. The complete description of the ROBPCA method is quite involved, so here we will only outline the main stages of the algorithm.

First, as in RAPCA, the x-data are preprocessed by reducing their data space to the affine subspace spanned by the n observations. As a result, the data are represented using at most rank($X_{n,p}$) variables (hence, at most $n-1$) without loss of information.

In the second step of the ROBPCA algorithm, a measure of outlyingness is computed for each data point. This is obtained by projecting the high-dimensional data points on many univariate directions. On every direction the univariate MCD estimator of location and scale is computed, and for every data point its standardized distance to that center is measured. Finally for each data point its largest distance over all the directions is considered. The h data points with smallest outlyingness are kept, and from the covariance matrix Σ_h of this h-subset we select the number k of principal components to retain.

The last stage of ROBPCA consist of projecting the data points onto the k-dimensional subspace spanned by the largest eigenvectors of Σ_h and of computing their center and shape by means of the reweighted MCD estimator. The eigenvectors of this scatter matrix then determine the robust principal components, and the MCD location estimate serves as a robust center.

Summarizing, the ROBPCA (and also the RAPCA) method applied to $X_{n,p}$ yields robust principal components which can be collected in a loading matrix $P_{p,k}$ with orthogonal columns, and a robust center $\hat{\mu}_x$. From here on the subscripts to a matrix serve to recall its size, e.g., $X_{n,p}$ is an $n \times p$ matrix and $P_{p,k}$ is $p \times k$. The robust scores are the $n \times 1$ column vectors

$$t_i = P'_{k,p}(x_i - \hat{\mu}_x).$$

The *orthogonal distance* measures the distance between an observation and its projection in the k-dimensional PCA subspace:

$$\text{OD}_i = \|x_i - \hat{\mu}_x - P_{p,k} t_i\|. \tag{19}$$

Let L denote the diagonal matrix which contains the eigenvalues l_j of the MCD scatter matrix, sorted from largest to smallest. The *score distance* of x_i with respect to $\hat{\mu}_x$, P and L is then defined as

$$\text{SD}_i = \sqrt{(x_i - \hat{\mu}_x)' P_{p,k} L^{-1}_{k,k} P'_{k,p}(x_i - \hat{\mu}_x)} = \sqrt{\sum_{j=1}^{k} \frac{t_{ij}^2}{l_j}}.$$

All the above mentioned methods are orthogonally and location equivariant. When we apply a translation v and/or an orthogonal transformation A (rotation, reflection) to the observations x_i, the robust center is also shifted and the loadings are rotated accordingly. To be precise, let $\hat{\mu}_x$ and P denote the robust center and loading matrix for the original x_i. Then the robust center and loadings for the transformed data $Ax_i + v$

are equal to $A\hat{\boldsymbol{\mu}}_x + \boldsymbol{v}$ and AP. Consequently the scores remain the same under these transformations:

$$t_i(Ax_i + v) = P'A'\big(Ax_i + v - (A\hat{\boldsymbol{\mu}}_x + v)\big) = P'(x_i - \hat{\boldsymbol{\mu}}_x) = t_i(x_i).$$

This property is not satisfied by the Orthogonalized Gnanadesikan–Kettenring estimator of Maronna and Zamar (2002).

We also mention the robust LTS-subspace estimator and its generalizations, introduced and discussed by Rousseeuw and Leroy (1987) and Maronna (2005). The idea behind these approaches consists in minimizing a robust scale of the orthogonal distances, similar to the LTS estimator and S-estimators in regression. For functional data, a fast PCA method is introduced by Locantore et al. (1999).

6.3. Diagnostic plot

The result of the PCA analysis can be represented by means of a diagnostic plot (Hubert et al., 2005). As in regression, this figure highlights the outliers and classifies them into several types. In general, we have defined an outlier as an observation which does not obey the pattern of the majority of the data. In the context of PCA, this means that an outlier either lies far from the subspace spanned by the k eigenvectors, and/or that the projected observation lies far from the bulk of the data within this subspace. This outlyingness can be expressed by means of the orthogonal and the score distances. These two distances define four types of observations, as illustrated in Figure 15(a). *Regular observations* have a small orthogonal and a small score distance. *Bad leverage points*, such as observations 2 and 3, have a large orthogonal distance and a large score distance. They typically have a large influence on classical PCA, as the eigenvectors will be tilted towards them. When points have a large score distance but a small orthogonal distance, we call them *good leverage points*. Observations 1 and 4 in Figure 11(a) can be classified into this category. Finally, *orthogonal outliers* have a large orthogonal distance, but a small score distance, as for example case 5. They cannot be distinguished from the regular observations once they are projected onto the PCA subspace, but they lie far from this subspace.

The diagnostic plot displays the OD_i versus the SD_i and hence classifies the observations according to Figure 15(b). On this plot, lines are drawn to distinguish the observations with a small and a large OD, and with a small and a large SD. For the latter distances, the cutoff value $c = \sqrt{\chi^2_{k,0.975}}$ is used. For the orthogonal distances, the approach of Box (1954) is followed. The squared orthogonal distances can be approximated by a scaled χ^2 distribution which in its turn can be approximated by a normal distribution using the Wilson–Hilferty transformation. The mean and variance of this normal distribution are then estimated by applying the univariate MCD to the $OD_i^{2/3}$ (see Hubert et al., 2005).

6.4. Example

We illustrate the diagnostic plot on a data set, which consists of spectra of 180 ancient glass pieces over $p = 750$ wavelengths (Lemberge et al., 2000). The measurements

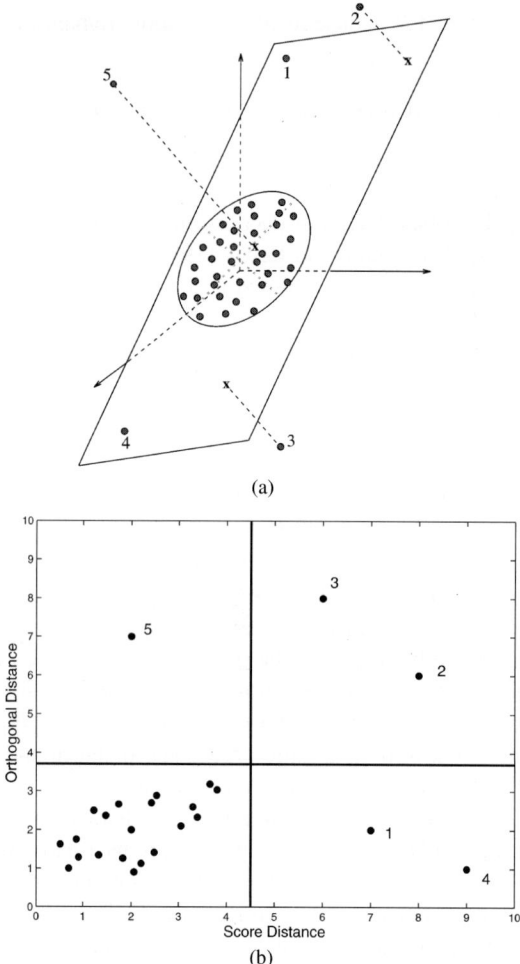

Fig. 15. (a) Different types of outliers when a three-dimensional data set is projected on a robust two-dimensional PCA-subspace, with (b) the corresponding diagnostic plot.

were performed using a Jeol JSM 6300 scanning electron microscope equipped with an energy-dispersive Si(Li) X-ray detection system. Three principal components were retained for CPCA and ROBPCA, yielding the diagnostic plots in Figure 16. From the classical diagnostic plot in Figure 16(a) we see that CPCA does not find big outliers. On the other hand the ROBPCA plot of Figure 16(b) clearly distinguishes two major groups in the data, a smaller group of bad leverage points, a few orthogonal outliers and the isolated case 180 in between the two major groups. A high-breakdown method such as ROBPCA treats the smaller group with cases 143–179 as one set of outliers. Later, it turned out that the window of the detector system had been cleaned before the last 38 spectra were measured. As a result of this less radiation (X-rays) is absorbed and more

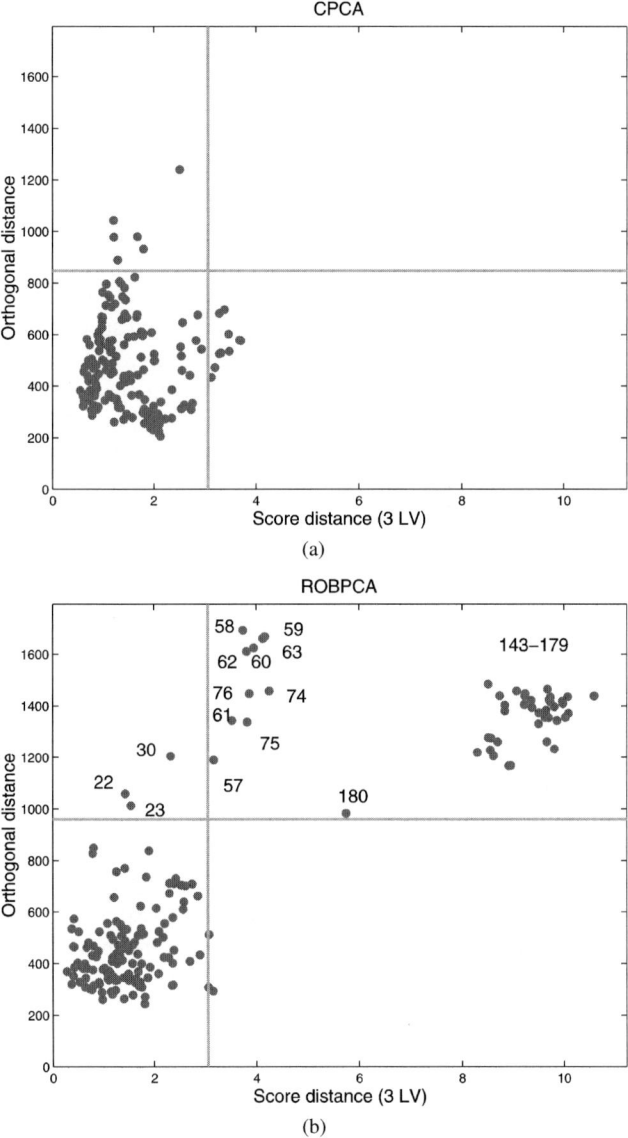

Fig. 16. Diagnostic plot of the glass data set based on three principal components computed with (a) CPCA; (b) ROBPCA.

can be detected, resulting in higher X-ray intensities. The other bad leverage points (57–63) and (74–76) are samples with a large concentration of calcic. The orthogonal outliers (22, 23 and 30) are borderline cases, although it turned out that they have larger measurements at the channels 215–245. This might indicate a larger concentration of phosphorus.

7. Principal component regression

7.1. Computation

Principal component regression is typically used for linear regression models (4) or (8) where the number of independent variables p is very large or where the regressors are highly correlated (this is known as multicollinearity). An important application of PCR is multivariate calibration in chemometrics, whose goal is to predict constituent concentrations of a material based on its spectrum. This spectrum can be obtained via several techniques such as Fluorescence spectrometry, Near InfraRed spectrometry (NIR), Nuclear Magnetic Resonance (NMR), Ultra-Violet spectrometry (UV), Energy dispersive X-Ray Fluorescence spectrometry (ED-XRF), etc. Since a spectrum typically ranges over a large number of wavelengths, it is a high-dimensional vector with hundreds of components. The number of concentrations on the other hand is usually limited to at most, say, five. In the univariate approach, only one concentration at a time is modeled and analyzed. The more general problem assumes that the number of response variables q is larger than one, which means that several concentrations are to be estimated together. This model has the advantage that the covariance structure between the concentrations is also taken into account, which is appropriate when the concentrations are known to be strongly intercorrelated with each other.

Classical PCR (CPCR) starts by replacing the large number of explanatory variables X_j by a small number of loading vectors, which correspond to the first (classical) principal components of $X_{n,p}$. Then the response variables Y_j are regressed on these components using least squares regression. It is thus a two-step procedure, which starts by computing scores t_i for every data point. Then the y_i are regressed on the t_i.

The robust PCR method proposed by Hubert and Verboven (2003) combines robust PCA for high-dimensional data with a robust multivariate regression technique such as MCD regression. For a univariate response variable, LTS regression can be used. The robust scores t_i obtained with ROBPCA thus serve as the independent variables in the regression model (4) or (8).

The proposed RPCR method is x-translation equivariant, x-orthogonally equivariant and y-affine equivariant. This means that the estimates satisfy Eqs. (13) and (14) for any orthogonal matrix A. These properties follow in a straightforward way from the orthogonal equivariance of the ROBPCA method. Robust PCR methods which are based on nonequivariant PCA estimators, such as those proposed by Pell (2000), therefore are also not x-equivariant. The y-affine equivariance is inherited from the LTS- and MCD-regression method and allows linear transformations of the y-variables.

7.2. Selecting the number of components

An important issue in PCR is the selection of the number of principal components, for which several methods have been proposed. A popular approach consists of minimizing the root mean squared error of cross-validation criterion $RMSECV_k$ which, for one response variable ($q = 1$), equals

$$\text{RMSECV}_k = \sqrt{\frac{1}{n} \sum_{i=1}^{n} (y_i - \hat{y}_{-i,k})^2} \qquad (20)$$

with $\hat{y}_{-i,k}$ the predicted value for observation i, where i was left out of the data set when performing the PCR method with k principal components. The goal of the RMSECV$_k$ statistic is twofold. It yields an estimate of the root mean squared prediction error $E(y - \hat{y})^2$ when k components are used in the model, whereas the curve of RMSECV$_k$ for $k = 1, \ldots, k_{\max}$ is a popular graphical tool to choose the optimal number of components.

This RMSECV$_k$ statistic is however not suited at contaminated data sets because it also includes the prediction error of the outliers in (20). Therefore a robust RMSECV measure is proposed by Hubert and Verboven (2003). Computing these R-RMSECV$_k$ values is rather time consuming, because for every choice of k they require the whole RPCR procedure to be performed n times. Therefore faster algorithms for cross-validation have recently been developed (Engelen and Hubert, 2005). They avoid the complete recomputation of resampling methods such as the MCD when one observation is removed from the data set.

7.3. Example

To illustrate RPCR, we analyze the Biscuit Dough data set of Osborne et al. (1984) preprocessed as by Hubert et al. (2002). This data set consists of 40 NIR spectra of biscuit dough with measurements every 2 nanometers, from 1200 nm up to 2400 nm. The responses are the percentages of 4 constituents in the biscuit dough: $y_1 = $ fat,

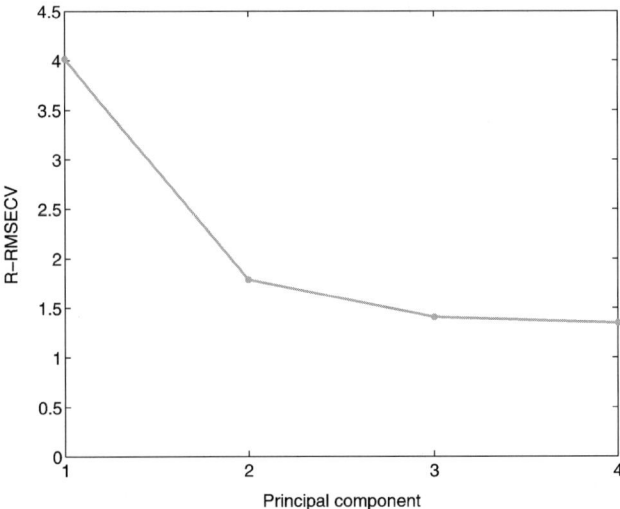

Fig. 17. Robust R-RMSECV$_k$ curve for the Biscuit Dough data set.

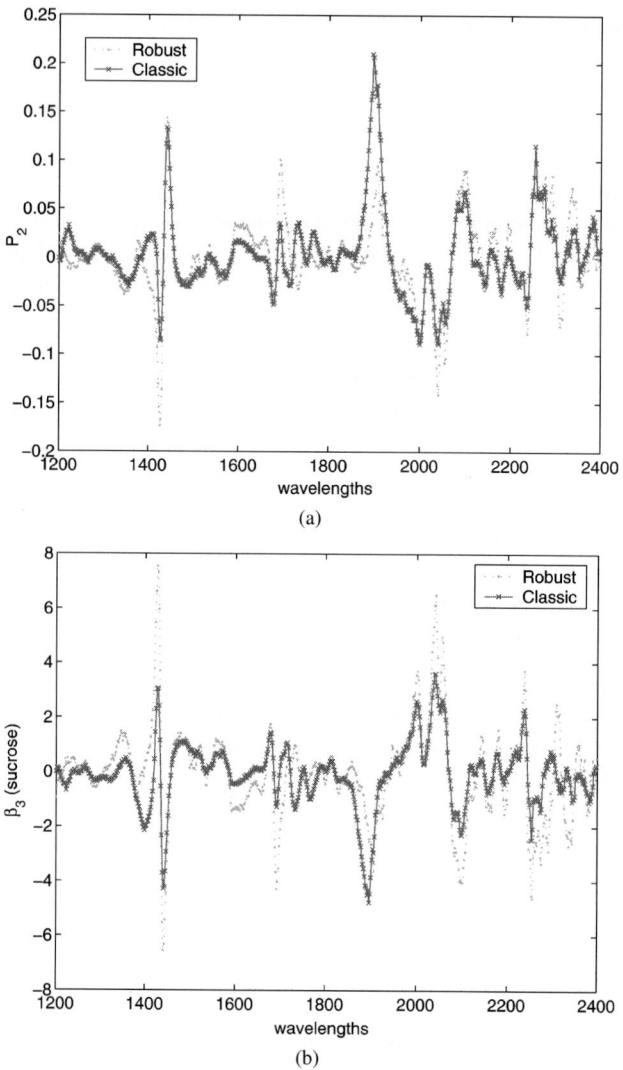

Fig. 18. Second loading vector and calibration vector of sucrose for the Biscuit dough data set, computed with (a) CPCR; (b) RPCR.

y_2 = flour, y_3 = sucrose and y_4 = water. Because there is a significant correlation among the responses, a multivariate regression is performed. The robust R-RMSECV$_k$ curve is plotted in Figure 17 and suggests to select $k = 2$ components.

Differences between CPCR and RPCR show up in the loading vectors and in the calibration vectors. Figure 18 shows the second loading vector and the second calibration vector for y_3 (sucrose). For instance, CPCR and RPCR give quite different results between wavelengths 1390 and 1440 (the so-called C-H bend).

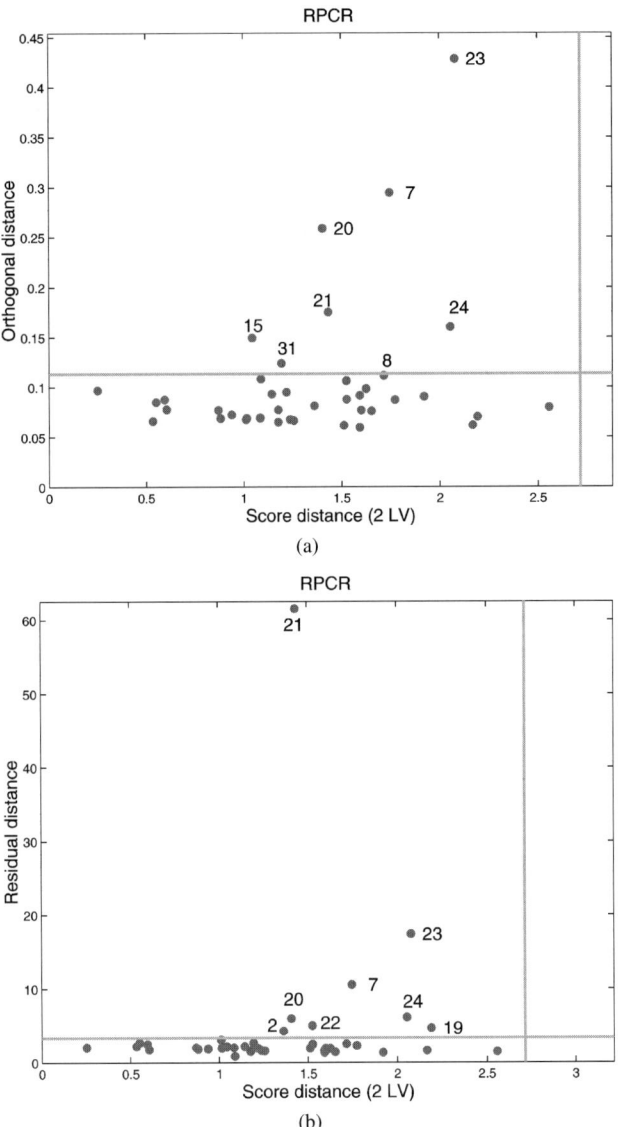

Fig. 19. (a) Score diagnostic plot when applying RPCR to the Biscuit Dough data set; (b) corresponding regression diagnostic plot.

Next, we can construct diagnostic plots as in Section 4 and in Section 6.3. ROBPCA yields the score diagnostic plot displayed in Figure 19(a). We see that there are no leverage points but there are some orthogonal outliers, the largest being 23, 7 and 20. The result of the regression step is shown in Figure 19(b). It exposes the robust distances of the residuals (or the standardized residuals if $q = 1$) versus the score distances.

RPCR shows that observation 21 has an extremely high residual distance. Other vertical outliers are 23, 7, 20, and 24, whereas there are a few borderline cases.

8. Partial Least Squares Regression

Partial Least Squares Regression (PLSR) is similar to PCR. Its goal is to estimate regression coefficients in a linear model with a large number of x-variables which are highly correlated. In the first step of PCR, the scores were obtained by extracting the main information present in the x-variables by performing a principal component analysis on them, without using any information about the y-variables. In contrast, the PLSR scores are computed by maximizing a covariance criterion between the x- and y-variables. Hence the first stage of this technique already uses the responses.

More precisely, let $\widetilde{X}_{n,p}$ and $\widetilde{Y}_{n,q}$ denote the mean-centered data matrices, with $\tilde{x}_i = x_i - \bar{x}$ and $\tilde{y}_i = y_i - \bar{y}$. The normalized PLS weight vectors r_a and q_a (with $\|r_a\| = \|q_a\| = 1$) are then defined as the vectors that maximize

$$\text{cov}(\widetilde{Y}q_a, \widetilde{X}r_a) = q_a' \frac{\widetilde{Y}'\widetilde{X}}{n-1} r_a = q_a' \widehat{\Sigma}_{yx} r_a \qquad (21)$$

for each $a = 1, \ldots, k$, where $\widehat{\Sigma}'_{yx} = \widehat{\Sigma}_{xy} = \widetilde{X}'\widetilde{Y}/(n-1)$ is the empirical cross-covariance matrix between the x- and the y-variables. The elements of the scores $\tilde{t}_i$ are then defined as linear combinations of the mean-centered data: $\tilde{t}_{ia} = \tilde{x}_i' r_a$, or equivalently $\widetilde{T}_{n,k} = \widetilde{X}_{n,p} R_{p,k}$ with $R_{p,k} = (r_1, \ldots, r_k)$.

The computation of the PLS weight vectors can be performed using the SIMPLS algorithm (de Jong, 1993). The solution of this maximization problem is found by taking r_1 and q_1 as the first left and right singular eigenvectors of $\widehat{\Sigma}_{xy}$. The other PLSR weight vectors r_a and q_a for $a = 2, \ldots, k$ are obtained by imposing an orthogonality constraint to the elements of the scores. If we require that $\sum_{i=1}^{n} t_{ia} t_{ib} = 0$ for $a \neq b$, a deflation of the cross-covariance matrix $\widehat{\Sigma}_{xy}$ provides the solutions for the other PLSR weight vectors. This deflation is carried out by first calculating the x-loading $p_a = \widehat{\Sigma}_x r_a/(r_a' \widehat{\Sigma}_x r_a)$ with $\widehat{\Sigma}_x$ the empirical variance-covariance matrix of the x-variables. Next an orthonormal base $\{v_1, \ldots, v_a\}$ of $\{p_1, \ldots, p_a\}$ is constructed and $\widehat{\Sigma}_{xy}$ is deflated as

$$\widehat{\Sigma}_{xy}^a = \widehat{\Sigma}_{xy}^{a-1} - v_a(v_a' \widehat{\Sigma}_{xy}^{a-1})$$

with $\widehat{\Sigma}_{xy}^1 = \widehat{\Sigma}_{xy}$. In general, the PLSR weight vectors r_a and q_a are obtained as the left and right singular vector of $\widehat{\Sigma}_{xy}^a$.

A robust method RSIMPLS has been developed by Hubert and Vanden Branden (2003). It starts by applying ROBPCA on the x- and y-variables in order to replace $\widehat{\Sigma}_{xy}$ and $\widehat{\Sigma}_x$ by robust estimates, and then proceeds analogously to the SIMPLS algorithm. Similar to RPCR, a robust regression method (ROBPCA regression) is performed in the second stage. Vanden Branden and Hubert (2004) proved that for low-dimensional data the RSIMPLS approach yields bounded influence functions for the weight vectors r_a

and $\boldsymbol{q}_a$ and for the regression estimates. Also the breakdown value is inherited from the MCD estimator.

Note that canonical correlation analysis tries to maximize the correlation between linear combinations of the x- and the y-variables, instead of the covariance in (21). Robust methods for canonical correlation are presented by Croux and Dehon (2002).

The robustness of RSIMPLS is illustrated on the octane data set (Esbensen et al., 1994), consisting of NIR absorbance spectra over $p = 226$ wavelengths ranging from 1102 nm to 1552 nm with measurements every two nm. For each of the $n = 39$ production gasoline samples, the octane number y was measured, so $q = 1$. It is known that the octane data set contains six outliers (25, 26, 36–39) to which alcohol was added. From the RMSECV values (Engelen et al., 2004) it follows that $k = 2$ components should be retained. The resulting diagnostic plots are shown in Figure 20.

The robust score diagnostic plot is displayed in Figure 20(a). We immediately spot the six samples with added alcohol. The SIMPLS diagnostic plot is shown in Figure 20(b). We see that the classical analysis only detects the outlying spectrum 26, which does not even stick out much above the border line. The robust regression diagnostic plot in Figure 20(c) shows that the outliers are good leverage points, whereas SIMPLS again only reveals spectrum 26.

9. Some other multivariate frameworks

Apart from the frameworks covered in the previous sections, there is also work in other multivariate settings. These methods cannot be described in detail here due to lack of space, but here are some pointers to the literature. Chen and Victoria-Feser (2002) address the problem of robust covariance matrix estimation with missing data. Also in the framework of multivariate location and scatter, an MCD-based alternative to the Hotelling test was provided by Willems et al. (2002). A technique based on robust distances was applied to the control of electrical power systems in (Mili et al., 1996). Highly robust regression techniques were extended to computer vision settings (e.g., Meer et al., 1991; Stewart, 1995). For logistic regression with a binary response and multiple regressors, Rousseeuw and Christmann (2003) proposed a robust fitting method using (among other things) robust distances on the x-variables. Boente and Orellana (2001) and Boente et al. (2002) introduced robust estimators for common principal components. Robust methods for factor analysis were proposed by Pison et al. (2003). Croux et al. (2003) fitted general multiplicative models such as FANOVA. Rocke and Woodruff (1999) proposed clustering methods based on robust scatter estimation. Of course, this short list is far from complete.

10. Availability

Stand-alone programs carrying out FAST-MCD and FAST-LTS can be downloaded from the website http://www.agoras.ua.ac.be, as well as Matlab versions. The FAST-MCD algorithm is already available in the package S-PLUS (as the built-in function

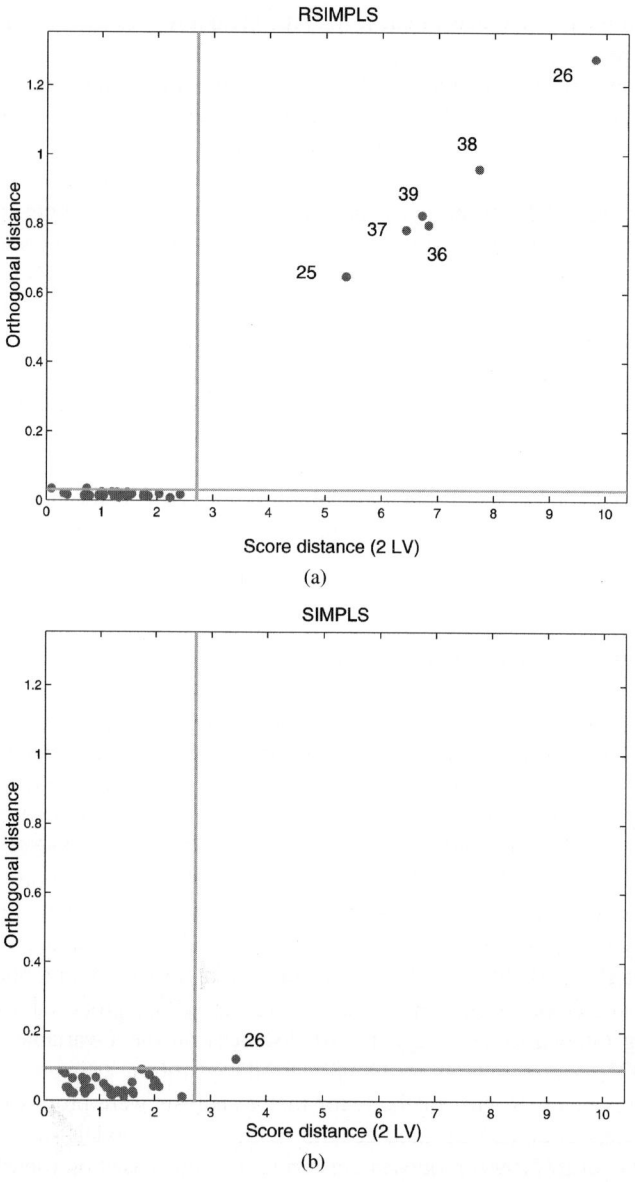

Fig. 20. (a) Score diagnostic plot of the octane data set obtained with RSIMPLS; (b) with SIMPLS; (c) regression diagnostic plot obtained with RSIMPLS; (d) with SIMPLS.

cov.mcd), in R (in *rrcov*), and in SAS/IML Version 7. It was also included in SAS Version 9 (in *PROC ROBUSTREG*). These packages all provide the one-step reweighted MCD estimates. The LTS is available in S-PLUS as the built-in function *ltsreg*, which uses a slower algorithm and has a low default breakdown value. The FAST-LTS algo-

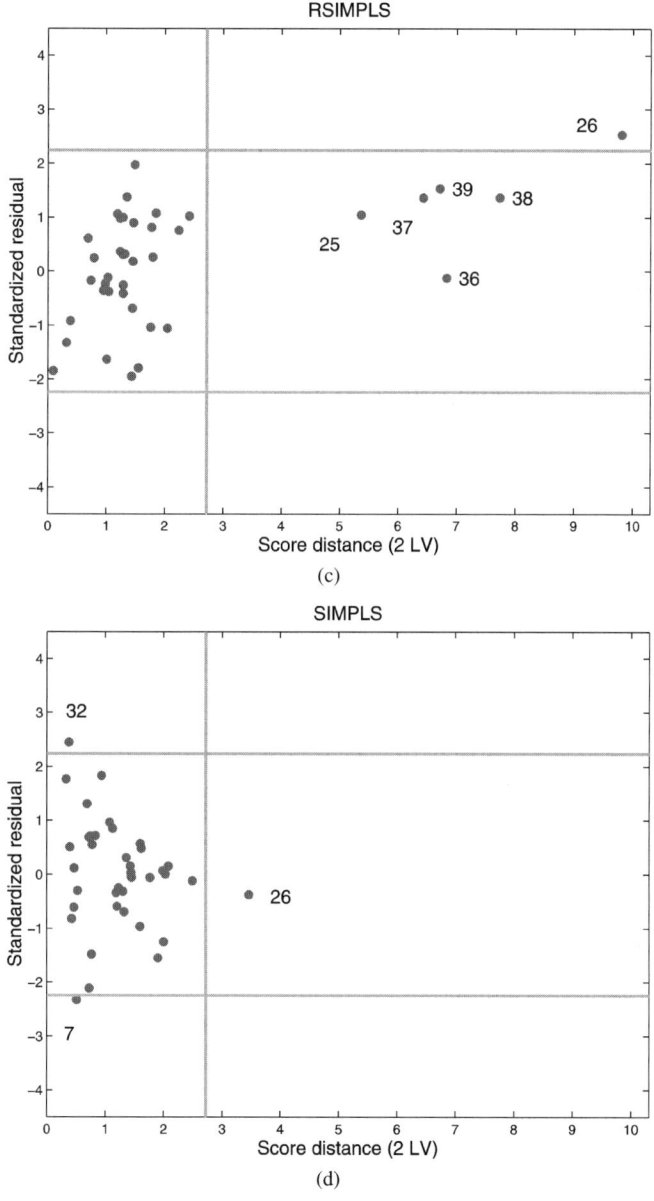

Fig. 20. Continued.

rithm is available in R (as part of *rrcov*) and in SAS/IML Version 7. In SAS Version 9 it is incorporated in *PROC ROBUSTREG*.

Matlab functions for most of the procedures mentioned in this chapter (MCD, LTS, MCD-regression, RQDA, RAPCA, ROBPCA, RPCR, and RSIMPLS) are part of LI-

BRA, a Matlab LIBrary for Robust Analysis (Verboven and Hubert, 2005) which can be downloaded from http://www.wis.kuleuven.ac.be/stat/robust.html.

Acknowledgements

We thank Sanne Engelen, Karlien Vanden Branden, and Sabine Verboven for help with preparing the figures of this chapter.

References

Agulló, J., Croux, C., Van Aelst, S. (2001). The multivariate least trimmed squares estimator. Submitted for publication.
Boente, G., Orellana, L. (2001). A robust approach to common principal components. In: Fernholz, L.T., Morgenthaler, S., Stahel, W. (Eds.), *Statistics in Genetics and in the Environmental Sciences*. Birkhäuser, Basel, pp. 117–146.
Boente, G., Pires, A.M., Rodrigues, I. (2002). Influence functions and outlier detection under the common principal components model: a robust approach. *Biometrika* **89**, 861–875.
Box, G.E.P. (1954). Some theorems on quadratic forms applied in the study of analysis of variance problems: effect of inequality of variance in one-way classification. *Ann. Math. Statist.* **25**, 33–51.
Campbell, N.A. (1980). Robust procedures in multivariate analysis I: Robust covariance estimation. *Appl. Statist.* **29**, 231–237.
Chen, T.-C., Victoria-Feser, M. (2002). High breakdown estimation of multivariate location and scale with missing observations. *British J. Math. Statist. Psych.* **55**, 317–335.
Croux, C., Dehon, C. (2001). Robust linear discriminant analysis using S-estimators. *Canad. J. Statist.* **29**, 473–492.
Croux, C., Dehon, C. (2002). Analyse canonique basée sur des estimateurs robustes de la matrice de covariance. *Revue Statist. Appl.* **2**, 5–26.
Croux, C., Haesbroeck, G. (1999). Influence function and efficiency of the minimum covariance determinant scatter matrix estimator. *J. Multivariate Anal.* **71**, 161–190.
Croux, C., Haesbroeck, G. (2000). Principal components analysis based on robust estimators of the covariance or correlation matrix: influence functions and efficiencies. *Biometrika* **87**, 603–618.
Croux, C., Rousseeuw, P.J. (1992). A class of high-breakdown scale estimators based on subranges. *Comm. Statist. Theory Methods* **21**, 1935–1951.
Croux, C., Ruiz-Gazen, A. (1996). A fast algorithm for robust principal components based on projection pursuit. In: *COMPSTAT 1996*. Physica, Heidelberg, pp. 211–217.
Croux, C., Ruiz-Gazen, A. (2005). High breakdown estimators for principal components: the projection–pursuit approach revisited. *J. Multivariate Anal.* In press.
Croux, C., Filzmoser, P., Pison, P., Rousseeuw, P.J. (2003). Fitting multiplicative models by robust alternating regressions. *Statist. Comput.* **13**, 23–36.
Cui, H., He, X., Ng, K.W. (2003). Asymptotic distributions of principal components based on robust dispersions. *Biometrika* **90**, 953–966.
Davies, L. (1987). Asymptotic behavior of S-estimators of multivariate location parameters and dispersion matrices. *Ann. Statist.* **15**, 1269–1292.
de Jong, S. (1993). SIMPLS: an alternative approach to partial least squares regression. *Chemometrics Intelligent Laboratory Systems* **18**, 251–263.
Donoho, D.L., Huber, P.J. (1983). The notion of breakdown point. In: Bickel, P., Doksum, K., Hodges, J.L. (Eds.), *A Festschrift for Erich Lehmann*. Wadsworth, Belmont.
Engelen, S., Hubert, M. (2005). Fast model selection for robust calibration methods. *Anal. Chimica Acta*. In press.

Engelen, S., Hubert, M., Vanden Branden, K., Verboven, S. (2004). Robust PCR and robust PLS: a comparative study. In: Hubert, M., Pison, G., Struyf, A., Van Aelst, S. (Eds.), *Theory and Applications of Recent Robust Methods*. In: *Statistics for Industry and Technology*. Birkhäuser, Basel, pp. 105–117.
Esbensen, K.H., Schönkopf, S., Midtgaard, T. (1994). *Multivariate Analysis in Practice*. Camo, Trondheim.
Hawkins, D.M., McLachlan, G.J. (1997). High-breakdown linear discriminant analysis. *J. Amer. Statist. Assoc.* **92**, 136–143.
He, X., Fung, W.K. (2000). High breakdown estimation for multiple populations with applications to discriminant analysis. *J. Multivariate Anal.* **72** (2), 151–162.
Hubert, M., Engelen, S. (2005). Robust PCA and classification in biosciences. *Bioinformatics* **20**, 1728–1736.
Hubert, M., Van Driessen, K. (2004). Fast and robust discriminant analysis. *Comput. Statist. Data Anal.* **45**, 301–320.
Hubert, M., Vanden Branden, K. (2003). Robust methods for partial least squares regression. *J. Chemometrics* **17**, 537–549.
Hubert, M., Verboven, S. (2003). A robust PCR method for high-dimensional regressors. *J. Chemometrics* **17**, 438–452.
Hubert, M., Rousseeuw, P.J., Verboven, S. (2002). A fast robust method for principal components with applications to chemometrics. *Chemometrics Intelligent Laboratory Systems* **60**, 101–111.
Hubert, M., Rousseeuw, P.J., Vanden Branden, K. (2005). ROBPCA: a new approach to robust principal components analysis. *Technometrics* **47**, 64–79.
Johnson, R.A., Wichern, D.W. (1998). *Applied Multivariate Statistical Analysis*. Prentice Hall, Englewood Cliffs, NJ.
Lemberge, P., De Raedt, I., Janssens, K.H., Wei, F., Van Espen, P.J. (2000). Quantitative Z-analysis of 16th–17th century archaeological glass vessels using PLS regression of EPXMA and μ-XRF data. *J. Chemometrics* **14**, 751–763.
Li, G., Chen, Z. (1985). Projection-pursuit approach to robust dispersion matrices and principal components: primary theory and Monte Carlo. *J. Amer. Statist. Assoc.* **80**, 759–766.
Locantore, N., Marron, J.S., Simpson, D.G., Tripoli, N., Zhang, J.T., Cohen, K.L. (1999). Robust principal component analysis for functional data. *Test* **8**, 1–73.
Lopuhaä, H.P., Rousseeuw, P.J. (1991). Breakdown points of affine equivariant estimators of multivariate location and covariance matrices. *Ann. Statist.* **19**, 229–248.
Maronna, R.A. (1976). Robust M-estimators of multivariate location and scatter. *Ann. Statist.* **4**, 51–67.
Maronna, R.A. (2005). Principal components and orthogonal regression based on robust scales. *Technometrics*. In press.
Maronna, R.A., Zamar, R.H. (2002). Robust multivariate estimates for high dimensional data sets. *Technometrics* **44**, 307–317.
Meer, P., Mintz, D., Rosenfeld, A., Kim, D.Y. (1991). Robust regression methods in computer vision: a review. *Internat. J. Comput. Vision* **6**, 59–70.
Mili, L., Cheniae, M.G., Vichare, N.S., Rousseeuw, P.J. (1996). Robust state estimation based on projection statistics. *IEEE Trans. Power Systems* **11**, 1118–1127.
Model, F., König, T., Piepenbrock, C., Adorjan, P. (2002). Statistical process control for large scale microarray experiments. *Bioinformatics* **1**, 1–9.
Odewahn, S.C., Djorgovski, S.G., Brunner, R.J., Gal, R. (1998). Data from the digitized Palomar sky survey. Technical report. California Institute of Technology.
Osborne, B.G., Fearn, T., Miller, A.R., Douglas, S. (1984). Application of near infrared reflectance spectroscopy to the compositional analysis of biscuits and biscuit dough. *J. Sci. Food Agriculture* **35**, 99–105.
Pell, R.J. (2000). Multiple outlier detection for multivariate calibration using robust statistical techniques. *Chemometrics Intelligent Laboratory Systems* **52**, 87–104.
Pison, G., Van Aelst, S., Willems, G. (2002). Small sample corrections for LTS and MCD. *Metrika* **55**, 111–123.
Pison, G., Rousseeuw, P.J., Filzmoser, P., Croux, C. (2003). Robust factor analysis. *J. Multivariate Anal.* **84**, 145–172.
Rocke, D.M., Woodruff, D.L. (1999). A synthesis of outlier detection and cluster identification. Submitted for publication.
Rousseeuw, P.J. (1984). Least median of squares regression. *J. Amer. Statist. Assoc.* **79**, 871–880.

Rousseeuw, P.J. (1985). Multivariate estimation with high breakdown point. In: Grossmann, W., Pflug, G., Vincze, I., Wertz, W. (Eds.), *Mathematical Statistics and Applications, vol. B*. Reidel, Dordrecht, pp. 283–297.

Rousseeuw, P.J., Christmann, A. (2003). Robustness against separation and outliers in logistic regression. *Comput. Statist. Data Anal.* **43**, 315–332.

Rousseeuw, P.J., Croux, C. (1993). Alternatives to the median absolute deviation. *J. Amer. Statist. Assoc.* **88**, 1273–1283.

Rousseeuw, P.J., Leroy, A.M. (1987). *Robust Regression and Outlier Detection*. Wiley–Interscience, New York.

Rousseeuw, P.J., Van Driessen, K. (1999). A fast algorithm for the minimum covariance determinant estimator. *Technometrics* **41**, 212–223.

Rousseeuw, P.J., Van Driessen, K. (2000). An algorithm for positive-breakdown regression based on concentration steps. In: Gaul, W., Opitz, O., Schader, M. (Eds.), *Data Analysis: Scientific Modeling and Practical Application*. Springer-Verlag, New York, pp. 335–346.

Rousseeuw, P.J., Van Zomeren, B.C. (1990). Unmasking multivariate outliers and leverage points. *J. Amer. Statist. Assoc.* **85**, 633–651.

Rousseeuw, P.J., Yohai, V.J. (1984). Robust regression by means of S-estimators. In: Franke, J., Härdle, W., Martin, R.D. (Eds.), *Robust and Nonlinear Time Series Analysis*. In: *Lecture Notes in Statistics*, vol. 26. Springer-Verlag, New York, pp. 256–272.

Rousseeuw, P.J., Van Aelst, S., Van Driessen, K., Agulló, J. (2004). Robust multivariate regression. *Technometrics* **46**, 293–305.

Stewart, C.V. (1995). Minpran: a new robust estimator for computer vision. *IEEE Trans. Pattern Anal. Machine Intelligence* **17**, 925–938.

Vanden Branden, K., Hubert, M. (2004). Robustness properties of a robust PLS regression method. *Anal. Chimica Acta* **515**, 229–241.

Verboven, S., Hubert, M. (2005). LIBRA: a Matlab Library for Robust Analysis. *Chemometrics Intelligent Laboratory Systems* **75**, 127–136.

Willems, G., Pison, G., Rousseeuw, P.J., Van Aelst, S. (2002). A robust Hotelling test. *Metrika* **55**, 125–138.

Woodruff, D.L., Rocke, D.M. (1994). Computable robust estimation of multivariate location and shape in high dimension using compound estimators. *J. Amer. Statist. Assoc.* **89**, 888–896.

Classification and Regression Trees, Bagging, and Boosting

Clifton D. Sutton

1. Introduction

Tree-structured classification and regression are alternative approaches to classification and regression that are not based on assumptions of normality and user-specified model statements, as are some older methods such as discriminant analysis and ordinary least squares (OLS) regression. Yet, unlike the case for some other nonparametric methods for classification and regression, such as kernel-based methods and nearest neighbors methods, the resulting tree-structured predictors can be relatively simple functions of the input variables which are easy to use.

Bagging and boosting are general techniques for improving prediction rules. Both are examples of what Breiman (1998) refers to as *perturb and combine* (P&C) methods, for which a classification or regression method is applied to various perturbations of the original data set, and the results are combined to obtain a single classifier or regression model. Bagging and boosting can be applied to tree-based methods to increase the accuracy of the resulting predictions, although it should be emphasized that they can be used with methods other than tree-based methods, such as neural networks.

This chapter will cover tree-based classification and regression, as well as bagging and boosting. This introductory section provides some general information, and briefly describes the origin and development of each of the methods. Subsequent sections describe how the methods work, and provide further details.

1.1. Classification and regression trees

Tree-structured classification and regression are nonparametric computationally intensive methods that have greatly increased in popularity during the past dozen years. They can be applied to data sets having both a large number of cases and a large number of variables, and they are extremely resistant to outliers (see Steinberg and Colla, 1995, p. 24).

Classification and regression trees can be good choices for analysts who want fairly accurate results quickly, but may not have the time and skill required to obtain them using traditional methods. If more conventional methods are called for, trees can still be helpful if there are a lot of variables, as they can be used to identify important variables and interactions. Classification and regression trees have become widely used among members of the data mining community, but they can also be used for relatively simple tasks, such as the imputation of missing values (see Harrell, 2001).

Regression trees originated in the 1960s with the development of AID (Automatic Interaction Detection) by Morgan and Sonquist (1963). Then in the 1970s, Morgan and Messenger (1973) created THAID (Theta AID) to produce classification trees. AID and THAID were developed at the Institute for Social Research at the University of Michigan.

In the 1980s, statisticians Breiman et al. (1984) developed CART (Classification And Regression Trees), which is a sophisticated program for fitting trees to data. Since the original version, CART has been improved and given new features, and it is now produced, sold, and documented by Salford Systems. Statisticians have also developed other tree-based methods; for example, the QUEST (Quick Unbiased Efficient Statistical Tree) method of Loh and Shih (1997).

Classification and regression trees can now be produced using many different software packages, some of which are relatively expensive and are marketed as being commercial data mining tools. Some software, such as S-Plus, use algorithms that are very similar to those underlying the CART program. In addition to creating trees using `tree` (see Clark and Pregibon, 1992), users of S-Plus can also use `rpart` (see Therneau and Atkinson, 1997), which can be used for Poisson regression and survival analysis in addition to being used for the creation of ordinary classification and regression trees. Martinez and Martinez (2002) provide Matlab code for creating trees, which are similar to those that can be created by CART, although their routines do not handle nominal predictors having three or more categories or splits involving more than one variable. The MATLAB Statistics Toolbox also has functions for creating trees. Other software, including some ensembles aimed at data miners, create trees using variants of CHAID (CHi-squared Automatic Interaction Detector), which is a descendant of THAID developed by Kass (1980). SPSS sells products that allow users to create trees using more than one major method.

The machine learning community has produced a large number of programs to create decision trees for classification and, to a lesser extent, regression. Very notable among these is Quinlan's extremely popular C4.5 (see Quinlan, 1993), which is a descendant of his earlier program, ID3 (see Quinlan, 1979). A newer product of Quinlan is his system See5/C5.0. Quinlan (1986) provides further information about the development of tree-structured classifiers by the machine learning community.

Because there are so many methods available for tree-structured classification and regression, this chapter will focus on one of them, CART, and only briefly indicate how some of the others differ from CART. For a fuller comparison of tree-structured classifiers, the reader is referred to Ripley (1996, Chapter 7). Gentle (2002) gives a shorter overview of classification and regression trees, and includes some more recent references. It can also be noted that the internet is a source for a large number of descriptions of various methods for tree-structured classification and regression.

1.2. Bagging and boosting

Bootstrap aggregation, or bagging, is a technique proposed by Breiman (1996a) that can be used with many classification methods and regression methods to reduce the variance associated with prediction, and thereby improve the prediction process. It is a relatively simple idea: many bootstrap samples are drawn from the available data, some prediction method is applied to each bootstrap sample, and then the results are combined, by averaging for regression and simple voting for classification, to obtain the overall prediction, with the variance being reduced due to the averaging.

Boosting, like bagging, is a committee-based approach that can be used to improve the accuracy of classification or regression methods. Unlike bagging, which uses a simple averaging of results to obtain an overall prediction, boosting uses a weighted average of results obtained from applying a prediction method to various samples. Also, with boosting, the samples used at each step are not all drawn in the same way from the same population, but rather the incorrectly predicted cases from a given step are given increased weight during the next step. Thus boosting is an iterative procedure, incorporating weights, as opposed to being based on a simple averaging of predictions, as is the case with bagging. In addition, boosting is often applied to *weak learners* (e.g., a simple classifier such as a two node decision tree), whereas this is not the case with bagging.

Schapire (1990) developed the predecessor to later boosting algorithms developed by him and others. His original method pertained to two-class classifiers, and combined the results of three classifiers, created from different learning samples, by simple majority voting. Freund (1995) extended Schapire's original method by combining the results of a larger number of weak learners. Then Freund and Schapire (1996) developed the *AdaBoost algorithm*, which quickly became very popular. Breiman (1998) generalized the overall strategy of boosting, and considered Freund and Schapire's algorithm as a special case of the class of *arcing* algorithms, with the term arcing being suggested by *adaptive resampling and combining*. But in the interest of brevity, and due to the popularity of Freund and Schapire's algorithm, this chapter will focus on AdaBoost and only briefly refer to related approaches. (See the bibliographic notes at the end of Chapter 10 of Hastie et al. (2001) for additional information about the development of boosting, and for a simple description of Schapire's original method.)

Some (see, for example, Hastie et al., 2001, p. 299) have indicated that boosting is one of the most powerful machine/statistical learning ideas to have been introduced during the 1990s, and it has been suggested (see, for example, Breiman (1998) or Breiman's statement in Olshen (2001, p. 194)) that the application of boosting to classification trees results in classifiers which generally are competitive with any other classifier. A particularly nice thing about boosting and bagging is that they can be used very successfully with simple "off-the-shelf" classifiers (as opposed to needing to carefully tune and tweak the classifiers). This fact serves to somewhat offset the criticism that with both bagging and boosting the improved performance comes at the cost of increased computation time. It can also be noted that Breiman (1998) indicates that applying boosting to CART to create a classifier can actually be much quicker than fitting a neural net classifier. However, one potential drawback to perturb and combine methods is that the final prediction rule can be appreciably more complex than what can be obtained using a method that does not combine predictors. Of course, it is often the case that simplicity has to be sacrificed to obtain increased accuracy.

2. Using CART to create a classification tree

In the general classification problem, it is known that each case in a sample belongs to one of a finite number of possible classes, and given a set of measurements for a case, it is desired to correctly predict to which class the case belongs. A classifier is a rule that assigns a predicted class membership based on a set of related measurements, $x_1, x_2, \ldots, x_{K-1}$, and x_K. Taking the measurement space $\mathcal{X}$ to be the set of all possible values of $(x_1, \ldots, x_K)$, and letting $\mathcal{C} = \{c_1, c_2, \ldots, c_J\}$ be the set of possible classes, a classifier is just a function with domain $\mathcal{X}$ and range $\mathcal{C}$, and it corresponds to a partition of $\mathcal{X}$ into disjoint sets, $B_1, B_2, \ldots, B_J$, such that the predicted class is j if $\mathbf{x} \in B_j$, where $\mathbf{x} = (x_1, \ldots, x_K)$.

It is normally desirable to use past experience as a basis for making new predictions, and so classifiers are usually constructed from a *learning sample* consisting of cases for which the correct class membership is known in addition to the associated values of $(x_1, \ldots, x_K)$. Thus statistical classification is similar to regression, only the response variable is nominal. Various methods for classification differ in how they use the data (the learning sample) to partition $\mathcal{X}$ into the sets $B_1, B_2, \ldots, B_J$.

2.1. Classification trees

Tree-structured classifiers are constructed by making repetitive splits of $\mathcal{X}$ and the subsequently created subsets of $\mathcal{X}$, so that a hierarchical structure is formed. For example, $\mathcal{X}$ could first be divided into $\{\mathbf{x} \mid x_3 \leqslant 53.5\}$ and $\{\mathbf{x} \mid x_3 > 53.5\}$. Then the first of these sets could be further divided into $A_1 = \{\mathbf{x} \mid x_3 \leqslant 53.5, x_1 \leqslant 29.5\}$ and $A_2 = \{\mathbf{x} \mid x_3 \leqslant 53.5, x_1 > 29.5\}$, and the other set could be split into $A_3 = \{\mathbf{x} \mid x_3 > 53.5, x_1 \leqslant 74.5\}$ and $A_4 = \{\mathbf{x} \mid x_3 > 53.5, x_1 > 74.5\}$. If there are just two classes ($J = 2$), it could be that cases having unknown class for which $\mathbf{x}$ belongs to A_1 or A_3 should be classified as c_1 (predicted to be of the class c_1), and cases for which $\mathbf{x}$ belongs to A_2 or A_4 should be classified as c_2. (Making use of the notation established above, we would have $B_1 = A_1 \cup A_3$ and $B_2 = A_2 \cup A_4$.) Figure 1 shows the partitioning of $\mathcal{X}$ and Figure 2 shows the corresponding representation as a tree.

While it can be hard to draw the partitioning of $\mathcal{X}$ when more than two predictor variables are used, one can easily create a tree representation of the classifier, which is easy to use no matter how many variables are used and how many sets make up the partition. Although it is generally harder to understand how the predictor variables relate to the various classes when the tree-structured classifier is rather complicated, the classifier can easily be used, without needing a computer, to classify a new observation based on the input values, whereas this is not usually the case for nearest neighbors classifiers and kernel-based classifiers.

It should be noted that when $\mathcal{X}$ is divided into two subsets, these subsets do not both have to be subsequently divided using the same variable. For example, one subset

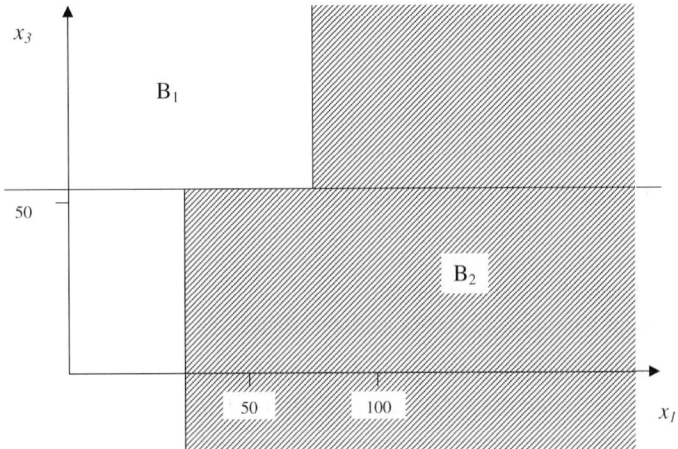

Fig. 1. A partition of $\mathcal{X}$ formed by orthogonal splits.

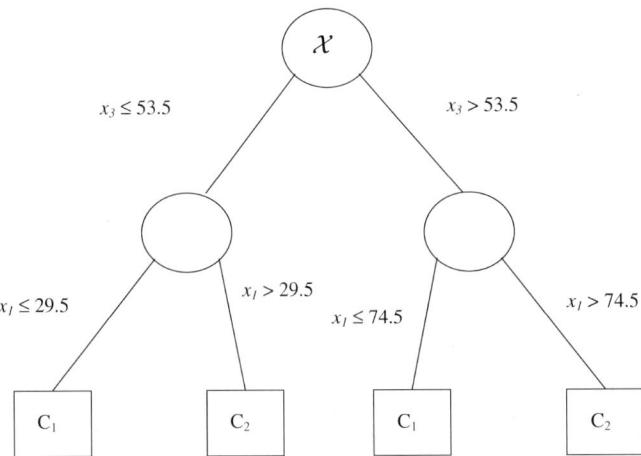

Fig. 2. Binary tree representation of the classifier corresponding to the partition shown in Figure 1.

could be split using x_1 and the other subset could be split using x_4. This allows us to model a nonhomogeneous response: the classification rules for different regions of $\mathcal{X}$ can use different variables, or be different functions of the same variables. Furthermore, a classification tree does not have to be symmetric in its pattern of nodes. That is, when $\mathcal{X}$ is split, it may be that one of the subsets produced is not split, while the other subset is split, with each of the new subsets produced being further divided.

2.2. Overview of how CART creates a tree

What needs to be determined is how to best split subsets of $\mathcal{X}$ (starting with $\mathcal{X}$ itself), to produce a tree-structured classifier. CART uses recursive binary partitioning to create a binary tree. (Some other methods for creating classification trees allow for more than two branches to descend from a node; that is, a subset of $\mathcal{X}$ can be split into more than two subsets in a single step. It should be noted that CART is not being restrictive in only allowing binary trees, since the partition of $\mathcal{X}$ created by any other type of tree structure can also be created using a binary tree.) So, with CART, the issues are:

(1) how to make each split (identifying which variable or variables should be used to create the split, and determining the precise rule for the split),
(2) how to determine when a node of the tree is a terminal node (corresponding to a subset of $\mathcal{X}$ which is not further split), and
(3) how to assign a predicted class to each terminal node.

The assignment of predicted classes to the terminal nodes is relatively simple, as is determining how to make the splits, whereas determining the right-sized tree is not so straightforward. In order to explain some of these details, it seems best to start with the easy parts first, and then proceed to the tougher issue; and in fact, one needs to first understand how the simpler parts are done since the right-sized tree is selected from a set of candidate tree-structured classifiers, and to obtain the set of candidates, the splitting and class assignment issues have to be handled.

2.3. Determining the predicted class for a terminal node

If the learning sample can be viewed as being a random sample from the same population or distribution from which future cases to be classified will come, and if all types of misclassifications are considered to be equally bad (for example, in the case of $J = 2$, classifying a c_1 case as a c_2 is given the same penalty as classifying a c_2 as a c_1), the class assignment rule for the terminal nodes is the simple *plurality rule*: the class assigned to a terminal node is the class having the largest number of members of the learning sample corresponding to the node. (If two or more classes are tied for having the largest number of cases in the learning sample with **x** belonging to the set corresponding to the node, the predicted class can be arbitrarily selected from among these classes.) If the learning sample can be viewed as being a random sample from the same population or distribution from which future cases to be classified will come, but all types of misclassifications are not considered to be equally bad (that is, classifying a c_1 case as a c_2 may be considered to be twice as bad, or ten times as bad, as classifying a c_2 as a c_1), the different misclassification penalties, which are referred to as *misclassification costs*, should be taken into account, and the assigned class for a terminal node should be the one which minimizes the total misclassification cost for all cases of the learning sample corresponding to the node. (Note that this rule reduces to the plurality rule if all types of misclassification have the same cost.) Likewise, if all types of misclassifications are to be penalized the same, but the learning sample has class membership proportions that are different from those that are expected for future cases to be classified with the classifier, weights should be used when assigning predicted classes to the nodes, so that the predictions will hopefully minimize the number of misclassifications for future cases. Finally, if all types of misclassifications are not to be penalized the same *and* the learning sample has class membership proportions that are different from those that are expected for cases to be classified with the classifier, two sets of weights are used to assign the predicted classes. One set of weights is used to account for different misclassification costs, and the other set of weights is used to adjust for the class proportions of future observations being different from the class proportions of the learning sample.

It should be noted that, in some cases, weights are also used in other aspects of creating a tree-structured classifier, and not just for the assignment of the predicted class. But in what follows, the simple case of the learning sample being viewed as a random sample from the same population or distribution from which future cases to be classified will come, and all types of misclassifications being considered equally bad, will be dealt with. In such a setting, weights are not needed. To learn what adjustments should be made for other cases, the interested reader can find some pertinent information in Breiman et al. (1984), Hastie et al. (2001), Ripley (1996).

2.4. Selection of splits to create a partition

In determining how to divide subsets of $\mathcal{X}$ to create two children nodes from a parent node, the general goal is to make the distributions of the class memberships of the cases of the learning sample corresponding to the two descendant nodes different, in such a way as to make, with respect to the response variable, the data corresponding to each of the children nodes purer than the data corresponding to the parent node. For example, in a four class setting, a good first split may nearly separate the c_1 and c_3 cases from the c_2 and c_4 cases. In such a case, uncertainty with regard to class membership of cases having **x** belonging to a specific subset of $\mathcal{X}$ is reduced (and further splits may serve to better divide the classes).

There are several different types of splits that can be considered at each step. For a predictor variable, x_k, which is numerical or ordinal (coded using successive integers), a subset of $\mathcal{X}$ can be divided with a plane orthogonal to the x_k axis, such that one of the newly created subsets has $x_k \leqslant s_k$, and the other has $x_k > s_k$. Letting

$$y_{k\,(1)} < y_{k\,(2)} < \cdots < y_{k\,(M)}$$

be the ordered distinct values of x_k observed in the portion of the learning sample belonging to the subset of $\mathcal{X}$ to be divided, the values

$$(y_{k\,(m)} + y_{k\,(m+1)})/2 \quad (m = 1, 2, \ldots, M-1)$$

can be considered for the split value s_k. So even if there are many different continuous or ordinal predictors, there is only a finite (albeit perhaps rather large) number of possible splits of this form to consider. For a predictor variable, x_k, which is nominal, having class labels belonging to the finite set D_k, a subset of $\mathcal{X}$ can be divided such that one of the newly created subsets has $x_k \in S_k$, and the other has $x_k \notin S_k$, where S_k is a nonempty proper subset of D_k. If D_k contains d members, there are $2^{d-1} - 1$ splits of this form to be considered.

Splits involving more than one variable can also be considered. Two or more continuous or ordinal variables can be involved in a *linear combination split*, with which a subset of $\mathcal{X}$ is divided with a hyperplane which is not perpendicular to one of the axes. For example, one of the created subsets can have points for which $2.7x_3 - 11.9x_7 \leqslant 54.8$, with the other created subset having points for which $2.7x_3 - 11.9x_7 > 54.8$. Similarly, two or more nominal variables can be involved in a *Boolean split*. For example, two nominal variables, gender and race, can be used to create a split with cases corresponding to white males belonging to one subset, and cases corresponding to males who are not white and females belonging to the other subset. (It should be noted that while Breiman et al. (1984) describe such Boolean splits, they do not appear to be included in the Salford Systems implementation.)

In many instances, one may not wish to allow linear combination and Boolean splits, since they can make the resulting tree-structured classifier more difficult to interpret, and can increase the computing time since the number of candidate splits is greatly increased. Also, if just single variable splits are used, the resulting tree is invariant with respect to monotone transformations of the variables, and so one does not have to consider whether say dose, or the log of dose, should be used as a predictor. But if linear combination splits are allowed, transforming the variables can make a difference in the resulting tree, and for the sake of simplicity, one might want to only consider single variable splits. However, it should be noted that sometimes a single linear combination split can be better than many single variable spits, and it may be preferable to have a classifier with fewer terminal nodes, having somewhat complicated boundaries for the sets comprising the partition, than to have one created using only single variable splits and having a large number of terminal nodes.

At each stage in the recursive partitioning, all of the allowable ways of splitting a subset of $\mathcal{X}$ are considered, and the one which leads to the greatest increase in node purity is chosen. This can be accomplished using an *impurity function*, which is a function of the proportions of the learning sample belonging to the possible classes of the response variable. These proportions will be denoted by $p_1, p_2, \ldots, p_{J-1}, p_J$.

The impurity function should be such that it is maximized whenever a subset of $\mathcal{X}$ corresponding to a node in the tree contains an equal number of each of the possible classes. (If there are the same number of c_1 cases as there are c_2 cases and c_3 cases and so on, then we are not able to sensibly associate that node with a particular class, and in fact, we are not able to sensibly favor any class over any other class, giving us that uncertainty is maximized.) The impurity function should assume its minimum value for a node that is completely pure, having all cases from the learning sample corresponding to the node belonging to the same class. Two such functions that can serve as the impurity function are the *Gini index of diversity*,

$$g(p_1, \ldots, p_J) = 2 \sum_{j=1}^{J-1} \sum_{j'=j+1}^{J} p_j p_{j'} = 1 - \sum_{j=1}^{J} p_j^2,$$

and the *entropy function*,

$$h(p_1, \ldots, p_J) = -\sum_{j=1}^{J} p_j \log p_j,$$

provided that $0 \log 0$ is taken to be $\lim_{p \downarrow 0} p \log p = 0$.

To assess the quality of a potential split, one can compute the value of the impurity function using the cases in the learning sample corresponding to the parent node (the node to be split), and subtract from this the weighted average of the impurity for the two children nodes, with the weights proportional to the number of cases of the learning sample corresponding to each of the two children nodes, to get the decrease in overall impurity that would result from the split. To select the way to split a subset of $\mathcal{X}$ in the tree growing process, all allowable ways of splitting can be considered, and the one which will result in the greatest decrease in node impurity (greatest increase in node purity) can be chosen.

2.5. Estimating the misclassification rate and selecting the right-sized tree

The trickiest part of creating a good tree-structured classifier is determining how complex the tree should be. If nodes continue to be created until no two distinct values of **x** for the cases in the learning sample belong to the same node, the tree may be overfitting the learning sample and not be a good classifier of future cases. On the other hand, if a tree has only a few terminal nodes, then it may be that it is not making enough use of information in the learning sample, and classification accuracy for future cases will suffer. Initially, in the tree-growing process, the predictive accuracy typically improves as more nodes are created and the partition gets finer. But it is usually the case that at some point the misclassification rate for future cases will start to get worse as the tree becomes more complex.

In order to compare the prediction accuracy of various tree-structured classifiers, there needs to be a way to estimate a given tree's misclassification rate for future observations, which is sometimes referred to as the *generalization error*. What does not work well is to use the *resubstitution estimate* of the misclassification rate (also known as the *training error*), which is obtained by using the tree to classify the members of the learning sample (that were used to create the tree), and observing the proportion that are misclassified. If no two members of the learning sample have the same value of **x**, then a tree having a resubstitution misclassification rate of zero can be obtained by continuing to make splits until each case in the learning sample is by itself in a terminal node (since the class associated with a terminal node will be that of the learning sample case corresponding to the node, and when the learning sample is then classified using the tree, each case in the learning sample will drop down to the terminal node that it created in the tree-growing process, and will have its class match the predicted class for the node). Thus the resubstitution estimate can be a very poor estimate of the tree's misclassification rate for future observations, since it can decrease as more nodes are created, even if the selection of splits is just responding to "noise" in the data, and not real structure. This phenomenon is similar to R^2 increasing as more terms are added to a multiple regression model, with the possibility of R^2 equaling one if enough terms are added, even though more complex regression models can be much worse predictors than simpler ones involving fewer variables and terms.

A better estimate of a tree's misclassification rate can be obtained using an independent *test sample*, which is a collection of cases coming from the same population or distribution as the learning sample. Like the learning sample, for the test sample the true class of each case is known in addition to the values for the predictor variables. The *test sample estimate* of the misclassification rate is just the proportion of the cases in the test sample that are misclassified when predicted classes are obtained using the tree created from the learning sample. If the cases to be classified in the future will also come from the same distribution that produced the test sample cases, the test sample estimation procedure is unbiased.

Since the test sample and the learning sample are both composed of cases for which the true class is known in addition to the values for the predictor variables, choosing to make the test sample larger will result in a smaller learning sample. Often it is thought that about one third of the available cases should be set aside to serve as a test sample, and the rest of the cases should be used as the learning sample. But sometimes a smaller fraction, such as one tenth, is used instead.

If one has enough suitable data, using an independent test sample is the best thing to do. Otherwise, obtaining a *cross-validation estimate* of the misclassification rate is preferable. For a V-fold cross-validation, one uses all of the available data for the learning sample, and divides these cases into V parts of approximately the same size. This is usually done randomly, but one may use stratification to help make the V cross-validation groups more similar to one another. V is typically taken to be 5 or 10. In a lot of cases, little is to be gained by using a larger value for V, and the larger V is, the greater is the amount of time required to create the classifier. In some situations, the quality of the estimate is reduced by making V too large.

To obtain a cross-validation estimate of the misclassification rate, each of the V groups is in turn set aside to serve temporarily as an independent test sample and a tree is grown, according to certain criteria, using the other $V - 1$ groups. In all, V trees are grown in this way, and for each tree the set aside portion of the data is used to obtain a test sample estimate of the tree's misclassification rate. Then the V test sample estimates are averaged to obtain the estimate of the misclassification rate for the tree grown from the entire learning sample using the same criteria. (If $V = 10$, the hope is that the average of the misclassification rates of the trees created using 90% of the data will not be too different from the misclassification rate of the tree created using all of the data.) Further details pertaining to how cross-validation is used to select the right-sized tree and estimate its misclassification rate are given below.

Whether one is going to use cross-validation or an independent test sample to estimate misclassification rates, it still needs to be specified how to grow the best tree, or how to create a set of candidate trees from which the best one can be selected based on their estimated misclassification rates. It does not work very well to use some sort of a *stop splitting rule* to determine that a node should be declared a terminal node and the corresponding subset of $\mathcal{X}$ not split any further, because it can be the case that the best split possible at a certain stage may decrease impurity by only a small amount, but if that split is made, each of the subsets of $\mathcal{X}$ corresponding to both of the descendant nodes can be split in ways to produce an appreciable decrease in impurity. Because of this phenomenon, what works better is to first grow a very large tree, splitting subsets in the current partition of $\mathcal{X}$ even if a split does not lead to an appreciable decrease in impurity. For example, splits can be made until 5 or fewer members of the learning sample correspond to each terminal node (or all of the cases corresponding to a node belong to the same class). Then a sequence of smaller trees can be created by *pruning* the initial large tree, where in the pruning process, splits that were made are removed and a tree having a fewer number of nodes is produced. The accuracies of the members of this sequence of subtrees—really a finite sequence of nested subtrees, since the first tree produced by pruning is a subtree of the original tree, and a second pruning step creates a subtree of the first subtree, and so on—are then compared using good estimates of their misclassification rates (either based on a test sample or obtained by cross-validation), and the best performing tree in the sequence is chosen as the classifier.

A specific way to create a useful sequence of different-sized trees is to use *minimum cost–complexity pruning*. In this process, a nested sequence of subtrees of the initial large tree is created by *weakest-link cutting*. With weakest-link cutting (pruning), all of the nodes that arise from a specific nonterminal node are pruned off (leaving that specific node as a terminal node), and the specific node selected is the one for which the corresponding pruned nodes provide the smallest *per node* decrease in the resubstitution misclassification rate. If two or more choices for a cut in the pruning process would produce the same per node decrease in the resubstitution misclassification rate, then pruning off the largest number of nodes is favored. In some cases (minimal pruning cases), just two children terminal nodes are pruned from a parent node, making it a terminal node. But in other cases, a larger group of descendant nodes are pruned all at once from an internal node of the tree. For example, at a given stage in the pruning process, if the increase in the estimated misclassification rate caused by pruning four nodes is no more than twice the increase in the estimated misclassification rate caused by pruning two nodes, pruning four nodes will be favored over pruning two nodes.

Letting $R(T)$ be the resubstitution estimate of the misclassification rate of a tree, T, and $|T|$ be the number of terminal nodes of the tree, for each $\alpha \geqslant 0$ the *cost–complexity measure*, $R_\alpha(T)$, for a tree, T, is given by

$$R_\alpha(T) = R(T) + \alpha|T|.$$

Here, $|T|$ is a measure of tree complexity, and $R(T)$ is related to misclassification cost (even though it is a biased estimate of the cost). α is the contribution to the measure for each terminal node. To minimize this measure, for small values of α, trees having a large number of nodes, and a low resubstitution estimate of misclassification rate, will be favored. For large enough values of α, a one node tree (with $\mathcal{X}$ not split at all, and all cases to be classified given the same predicted class), will minimize the measure.

Since the resubstitution estimate of misclassification rate is generally overoptimistic and becomes unrealistically low as more nodes are added to a tree, it is hoped that there is some value of α that properly penalizes the overfitting of a tree which is too complex, so that the tree which minimizes $R_\alpha(T)$, for the proper value of α, will be a tree of about the right complexity (to minimize the misclassification rate of future cases). Even though the proper value of α is unknown, utilization of the weakest-link cutting procedure described above guarantees that for each value for α (greater than or equal to 0), a subtree of the original tree that minimizes $R_\alpha(T)$ will be a member of the finite nested sequence of subtrees produced.

The sequence of subtrees produced by the pruning serves as the set of candidates for the classifier, and to obtain the classifier, all that remains to be done is to select the one which will hopefully have the smallest misclassification rate for future predictions. The selection is based on estimated misclassification rates, obtained using a test sample or by cross-validation.

If an independent test sample is available, it is used to estimate the error rates of the various trees in the nested sequence of subtrees, and the tree with the smallest estimated misclassification rate can be selected to be used as the tree-structured classifier. A popular alternative is to recognize that since all of the error rates are not accurately known, but only estimated, it could be that a simpler tree with only a slightly higher estimated error rate is really just as good or better than the tree having the smallest estimated error rate, and the least complex tree having an estimated error rate within one standard error of the estimated error rate of the tree having the smallest estimated error rate can be chosen, taking simplicity into account (and maybe not actually harming the prediction accuracy). This is often referred to as the *1 SE rule*.

If cross-validation is being used instead of a test sample, then things are a bit more complicated. First the entire collection of available cases are used to create a large tree, which is then pruned to create a sequence of nested subtrees. (This sequence of trees, from which the classifier will ultimately be selected, contains the subtree which minimizes $R_\alpha(T)$ for every nonnegative α.) The first of the V portions of data is set aside to serve as a test sample for a sequence of trees created from the other $V - 1$ portions of data. These $V - 1$ portions of data are collectively used to grow a large tree, and then the same pruning process is applied to create a sequence of subtrees. Each subtree in the sequence is the optimal subtree in the sequence, according to the $R_\alpha(T)$ criterion, for some range of values for α. Similar sequences of subtrees are created and the corresponding values of α for which each tree is optimal are determined, by setting aside, one at a time, each of the other $V - 1$ portions of the data, resulting in V sequences in all. Then, for various values of α, cross-validation estimates of the misclassification rates of the corresponding trees created using all of the data are determined as follows: for a given value of α the misclassification rate of the corresponding subtree in each of the V sequences is estimated using the associated set aside portion as a test sample, and the V estimates are averaged to arrive at a single error rate estimate corresponding to that value of α, and this estimate serves as the estimate of the true misclassification rate for the tree created using all of the data and pruned using this value of α. Finally, these cross-validation estimates of the error rates for the trees in the original sequence of subtrees are compared, and the subtree having the smallest estimated misclassification rate is selected to be the classifier. (Note that in this final stage of the tree selection process, the role of α is to create, for each value of α considered, matchings of the subtrees from the cross-validation sequences to the subtrees of the original sequence. The trees in each matched set can be viewed as having been created using the same prescription: a large tree is grown, weakest-link pruning is used to create a sequence of subtrees, and the subtree which minimizes $R_\alpha(T)$ is selected.)

2.6. Alternative approaches

As indicated previously, several other programs that create classification trees are apparently very similar to CART, but some other programs have key differences. Some can handle only two classes for the response variable, and some only handle categorical predictors. Programs can differ in how splits are selected, how missing values are dealt with, and how the right-sized tree is determined. If pruning is used, the pruning procedures can differ.

Some programs do not restrict splits to be binary splits. Some allow for *soft splits*, which in a sense divide cases that are very close to a split point, having such cases represented in both of the children nodes at a reduced weight. (See Ripley (1996) for additional explanation and some references.) Some programs use resubstitution estimates of misclassification rates to select splits, or use the result of a chi-squared test, instead of using an impurity measure. (The chi-squared approach selects the split that produces the most significant result when the null hypothesis of homogeneity is tested against the general alternative using a chi-squared test on the two-way table of counts which results from cross-tabulating the members of the learning sample in the subset to be split by the class of their response and by which of the children node created by the split they correspond to.) Some use the result of a chi-squared test to determine that a subset of $\mathcal{X}$ should not be further split, with the splitting ceasing if the most significant split is not significant enough. That is, a stopping rule is utilized to determine the complexity of the tree and control overfitting, as opposed to first growing a complex tree, and then using pruning to create a set of candidate trees, and estimated misclassification rates to select the classifier from the set of candidates.

Ciampi et al. (1987), Quinlan (1987, 1993), and Gelfandi et al. (1991) consider alternative methods of pruning. Crawford (1989) examines using bootstrapping to estimate the misclassification rates needed for the selection of the right-sized tree. Buntine (1992) describes a Bayesian approach to tree construction.

In all, there are a large number of methods for the creation of classification trees, and it is safe to state than none of them will work best in every situation. It may be prudent to favor methods, and particular implementations of methods, which have been thoroughly tested, and which allow the user to make some adjustments in order to tune the method to yield improved predictions.

3. Using CART to create a regression tree

CART creates a regression tree from data having a numerical response variable in much the same way as it creates a classification tree from data having a nominal response, but there are some differences. Instead of a predicted class being assigned to each terminal node, a numerical value is assigned to represent the predicted value for cases having values of **x** corresponding to the node. The sample mean or sample median of the response values of the members of the learning sample corresponding to the node may be used for this purpose. Also, in the tree-growing process, the split selected at each stage is the one that leads to the greatest reduction in the sum of the squared differences between the response values for the learning sample cases corresponding to a particular node and their sample mean, or the greatest reduction in the sum of the absolute differences between the response values for the learning sample cases corresponding to a particular node and their sample median. For example, using the squared differences criterion, one seeks the plane which divides a subset of $\mathcal{X}$ into the sets A and B for which

$$\sum_{i:\,\mathbf{x}_i \in A} (y_i - \bar{y}_A)^2 + \sum_{i:\,\mathbf{x}_i \in B} (y_i - \bar{y}_B)^2$$

is minimized, where $\bar{y}_A$ is the sample mean of the response values for cases in the learning sample corresponding to A, and $\bar{y}_B$ is the sample mean of the response values for cases in the learning sample corresponding to B. So split selection is based on making the observed response values collectively close to their corresponding predicted values. As with classification trees, for the sake of simplicity, one may wish to restrict consideration to splits involving only one variable.

The sum of squared or absolute differences is also used to prune the trees. This differs from the classification setting, where minimizing the impurity as measured by the Gini index might be used to grow the tree, and the resubstitution estimate of the misclassification rate is used to prune the tree. For regression trees, the error-complexity measure used is

$$R_\alpha(T) = R(T) + \alpha |T|,$$

where $R(T)$ is either the sum of squared differences or the sum of absolute differences of the response values in the learning sample and the predicted values corresponding to the fitted tree, and once again $|T|$ is the number of terminal nodes and α is the contribution to the measure for each terminal node. So, as is the case with classification, a biased resubstitution measure is used in the pruning process, and the $\alpha|T|$ term serves to penalize the overfitting of trees which partition $\mathcal{X}$ too finely. To select the tree to be used as the regression tree, an independent test sample or cross-validation is used to select the right-sized tree based on the mean squared prediction error or the mean absolute prediction error.

Some do not like the discontinuities in the prediction surface which results from having a single prediction value for each subset in the partition, and may prefer an alternative regression method such as MARS (Multivariate Adaptive Regression Splines (see Friedman, 1991)) or even OLS regression. However, an advantage that regression trees have over traditional linear regression models is the ability to handle a nonhomogeneous response. With traditional regression modeling, one seeks a linear function of the inputs and transformations of the inputs to serve as the response surface for the entire measurement space, $\mathcal{X}$. But with a regression tree, some variables can heavily influence the predicted response on some subsets of $\mathcal{X}$, and not be a factor at all on other subsets of $\mathcal{X}$. Also, regression trees can easily make adjustments for various interactions, whereas discovering the correct interaction terms can be a difficult process in traditional regression modeling. On the other hand, a disadvantage with regression trees is that additive structure is hard to detect and capture. (See Hastie et al. (2001, p. 274) for additional remarks concerning this issue.)

4. Other issues pertaining to CART

4.1. Interpretation

It is often stated that trees are easy to interpret, and with regard to seeing how the input variables in a smallish tree are related to the predictions, this is true. But the instability of trees, meaning that sometimes very small changes in the learning sample values can lead to significant changes in the variables used for the splits, can prevent one from reaching firm conclusions about issues such as overall variable importance by merely examining the tree which has been created.

As is the case with multiple regression, if two variables are highly correlated and one is put into the model at an early stage, there may be little necessity for using the other variable at all. But the omission of a variable in the final fitted prediction rule should not be taken as evidence that the variable is not strongly related to the response.

Similarly, correlations among the predictor variables can make it hard to identify important interactions (even though trees are wonderful for making adjustments for such interactions). For example, consider a regression tree initially split using the variable x_3. If x_1 and x_2 are two highly correlated predictor variables, x_1 may be used for splits in the left portion of a tree, with x_2 not appearing, and x_2 may be used for splits in the right portion of a tree, with x_1 not appearing. A cursory inspection of the tree may suggest the presence of interactions, while a much more careful analysis may allow one to detect that the tree is nearly equivalent to a tree involving only x_1 and x_3, and also nearly equivalent to a tree involving only x_2 and x_3, with both of these alternative trees suggesting an additive structure with no, or at most extremely mild, interactions.

4.2. Nonoptimality

It should be noted that the classification and regression trees produced by CART or any other method of tree-structured classification or regression are not guaranteed to be optimal. With CART, at each stage in the tree growing process, the split selected is the one which will immediately reduce the impurity (for classification) or variation (for regression) the most. That is, CART grows trees using a *greedy algorithm*. It could be that some other split would better set things up for further splitting to be effective. However, a tree-growing program that "looks ahead" would require much more time to create a tree.

CART also makes other sacrifices of optimality for gains in computational efficiency. For example, when working with a test sample, after the large initial tree is grown, more use could be made of the test sample in identifying the best subtree to serve as a classifier. But the minimal cost–complexity pruning procedure, which makes use of inferior resubstitution estimates of misclassification rates to determine a good sequence of subtrees to compare using test sample estimates, is a lot quicker than an exhaustive comparison of all possible subtrees using test sample estimates.

4.3. Missing values

Sometimes it is desired to classify a case when one or more of the predictor values is missing. CART handles such cases using *surrogate splits*. A surrogate split is based on a variable other than the one used for the primary split (which uses the variable that leads to the greatest decrease in node impurity). The surrogate split need not be the second best split based on the impurity criterion, but rather it is a split that mimics as closely as possible the primary split; that is, one which maximizes the frequency of cases in the learning sample being separated in the same way that they are by the primary split.

5. Bagging

Bagging (from bootstrap aggregation) is a technique proposed by Breiman (1996a, 1996b). It can be used to improve both the stability and predictive power of classification and regression trees, but its use is not restricted to improving tree-based predictions. It is a general technique that can be applied in a wide variety of settings to improve predictions.

5.1. Motivation for the method

In order to gain an understanding of why bagging works, and to determine in what situations one can expect appreciable improvement from bagging, it may be helpful to consider the problem of predicting the value of a numerical response variable, $Y_\mathbf{x}$, that will result from, or occur with, a given set of inputs, $\mathbf{x}$. Suppose that $\phi(\mathbf{x})$ is the prediction that results from using a particular method, such as CART, or OLS regression with a prescribed method for model selection (e.g., using Mallows' C_p to select a model from the class of all linear models that can be created having only first- and second-order terms constructed from the input variables). Letting μ_ϕ denote $\mathrm{E}(\phi(\mathbf{x}))$, where the expectation is with respect to the distribution underlying the learning sample (since, viewed as a random variable, $\phi(\mathbf{x})$ is a function of the learning sample, which can be viewed as a high-dimensional random variable) and not $\mathbf{x}$ (which is considered to be fixed), we have that

$$\begin{aligned}
&\mathrm{E}\big([Y_\mathbf{x} - \phi(\mathbf{x})]^2\big) \\
&= \mathrm{E}\big([(Y_\mathbf{x} - \mu_\phi) + (\mu_\phi - \phi(\mathbf{x}))]^2\big) \\
&= \mathrm{E}\big([Y_\mathbf{x} - \mu_\phi]^2\big) + 2\mathrm{E}(Y_\mathbf{x} - \mu_\phi)\mathrm{E}(\mu_\phi - \phi(\mathbf{x})) + \mathrm{E}\big([\mu_\phi - \phi(\mathbf{x})]^2\big) \\
&= \mathrm{E}\big([Y_\mathbf{x} - \mu_\phi]^2\big) + \mathrm{E}\big([\mu_\phi - \phi(\mathbf{x})]^2\big) \\
&= \mathrm{E}\big([Y_\mathbf{x} - \mu_\phi]^2\big) + \mathrm{Var}(\phi(\mathbf{x})) \\
&\geqslant \mathrm{E}\big([Y_\mathbf{x} - \mu_\phi]^2\big).
\end{aligned}$$

(Above, the independence of the future response, $Y_\mathbf{x}$, and the predictor based on the learning sample, $\phi(\mathbf{x})$, is used.) Since in nontrivial situations, the variance of the predictor $\phi(\mathbf{x})$ is positive (since typically not all random samples that could be the learning sample yield the sample value for the prediction), so that the inequality above is strict, this result gives us that if $\mu_\phi = \mathrm{E}(\phi(\mathbf{x}))$ could be used as a predictor, it would have a smaller mean squared prediction error than does $\phi(\mathbf{x})$.

Of course, in typical applications, μ_ϕ cannot serve as the predictor, since the information needed to obtain the value of $E(\phi(\mathbf{x}))$ is not known. To obtain what is sometimes referred to as the *true* bagging estimate of $E(\phi(\mathbf{x}))$, the expectation is based on the empirical distribution corresponding to the learning sample. In principle, it is possible to obtain this value, but in practice it is typically too difficult to sensibly obtain, and so the bagged prediction of $Y_\mathbf{x}$ is taken to be

$$\frac{1}{B} \sum_{b=1}^{B} \phi_b^*(\mathbf{x}),$$

where $\phi_b^*(\mathbf{x})$ is the prediction obtained when the base regression method (e.g., CART) is applied to the bth bootstrap sample drawn (with replacement) from the original learning sample. That is, to use bagging to obtain a prediction of $Y_\mathbf{x}$ in a regression setting, one chooses a regression method (which is referred to as the *base method*), and applies the method to B bootstrap samples drawn from the learning sample. The B predicted values obtained are then averaged to produce the final prediction.

In the classification setting, B bootstrap samples are drawn from the learning sample, and a specified classification method (e.g., CART) is applied to each bootstrap sample to obtain a predicted class for a given input, $\mathbf{x}$. The final prediction—the one that results from bagging the specified base method—is the class that occurs most frequently in the B predictions.

An alternative scheme is to bag class probability estimates for each class and then let the predicted class be the one with the largest average estimated probability. For example, with classification trees one has a predicted class corresponding to each terminal node, but there is also an estimate of the probability that a case having $\mathbf{x}$ corresponding to a specific terminal node belongs to a particular class. There is such an estimate for each class, and to predict the class for $\mathbf{x}$, these estimated probabilities from the B trees can be averaged, and the class corresponding to the largest average estimated probability chosen. This can yield a different result than what is obtained by simple voting. Neither of the two methods works better in all cases. Hastie et al. (2001) suggest that averaging the probabilities tends to be better for small B, but also includes an example having the voting method doing slightly better with a large value of B.

Some have recommended using 25 or 50 for B, and in a lot of cases going beyond 25 bootstrap samples will lead to little additional improvement. However, Figure 8.10 of Hastie et al. (2001) shows that an appreciable amount of additional improvement can occur if B is increased from 50 to 100 (and that the misclassification rate remained nearly constant for all choices of B greater than or equal to 100), and so taking B to be 100 may be beneficial in some cases. Although making B large means that creating the classifier will take longer, some of the increased time requirement is offset by the fact that with bagging one does not have to use cross-validation to select the right amount of complexity or regularization. When bagging, one can use the original learning sample as a test set. (A test set is supposed to come from the same population the learning sample comes from, and in bagging, the learning samples for the B predictors are randomly drawn from the original learning sample of available cases. A test set can easily be drawn as well, in the same way, although Breiman (1996a) suggests that one may use the original learning sample as a test set (since if a huge test set is randomly selected from the original learning sample of size N, each of the original cases should occur in the huge test set with an observed sample proportion of roughly N^{-1}, and so testing with a huge randomly drawn test set should be nearly equivalent to testing with the original learning sample).) Alternatively, one could use *out-of-bag* estimates, letting the cases not selected for a particular bootstrap sample serve as independent test sample cases to use in the creation of the classifier from the bootstrap sample.

5.2. When and how bagging works

Bagging works best when the base regression or classification procedure that is being bagged is not very stable. That is, when small changes in the learning sample can often result in appreciable differences in the predictions obtained using a specified method, bagging can result in an appreciable reduction in average prediction error. (See Breiman (1996b) for additional information about instability.)

That bagging will work well when applied to a regression method which is rather unstable for the situation at hand is suggested by the result shown in the preceding subsection. It can be seen that the difference of the mean squared prediction error for predicting $Y_{\mathbf{x}}$ with a specified method,

$$\mathrm{E}\bigl([Y_{\mathbf{x}} - \phi(\mathbf{x})]^2\bigr),$$

and the mean squared prediction error for predicting $Y_{\mathbf{x}}$ with the mean value of the predictor, $\mu_\phi = \mathrm{E}(\phi(\mathbf{x}))$,

$$\mathrm{E}\bigl([Y_{\mathbf{x}} - \mu_\phi]^2\bigr),$$

is equal to the variance of the predictor, $\mathrm{Var}(\phi(\mathbf{x}))$. The larger this variance is relative to the mean squared prediction error of the predictor, the greater the percentage reduction of the mean squared prediction error will be if the predictor is replaced by its mean. In situations for which the predictor has a relatively large variance, bagging can appreciably reduce the mean squared prediction error, provided that the learning sample is large enough for the bootstrap estimate of μ_ϕ to be sufficiently good.

When predicting $Y_{\mathbf{x}}$ using OLS regression in a case for which the proper form of the model is *known* and all of the variables are included in the learning sample, the error term distribution is approximately normal and the assumptions associated with least squares are closely met, and the sample size is not too small, there is little to be gained from bagging because the prediction method is pretty stable. But in cases for which the correct form of the model is complicated and not known, there are a lot of predictor variables, including some that are just noise not related to the response variable, and the sample size is not really large, bagging may help a lot, if a generally unstable regression method such as CART is used.

Like CART, other specific methods of tree-structured regression, along with MARS and neural networks, are generally unstable and should benefit, perhaps rather greatly, from bagging. In general, when the model is fit to the data adaptively, as is done in the construction of trees, or when using OLS regression with some sort of stepwise procedure for variable selection, bagging tends to be effective. Methods that are much more stable, like nearest neighbors regression and ridge regression, are not expected to greatly benefit from bagging, and in fact they may suffer a degradation in performance. Although the inequality in the preceding subsection indicates that μ_ϕ will not be a worse predictor than $\phi(\mathbf{x})$, since in practice the bagged estimate of μ_ϕ has to be used instead of the true value, it can be that this lack of correspondence results in bagging actually hurting the prediction process in a case for which the regression method is rather stable and the variance of $\phi(\mathbf{x})$ is small (and so there is not a lot to be gained even if μ_ϕ could be used). However, in regression settings for which bagging is harmful, the degradation of performance tends to be slight.

In order to better understand why, and in what situations, bagging works for classification, it may be helpful to understand the concepts of bias and variance for classifiers. A classifier ϕ, viewed as a function of the learning sample, is said to be *unbiased* at $\mathbf{x}$ if when applied to a randomly drawn learning sample from the parent distribution of the actual learning sample, the class which is predicted for $\mathbf{x}$ with the greatest probability is the one which actually occurs with the greatest probability with $\mathbf{x}$; that is, the class predicted by the Bayes classifier. (The Bayes classifier is an ideal classifier that always predicts the class which is most likely to occur with a given set of inputs, $\mathbf{x}$. The Bayes classifier will not always make the correct prediction, but it is the classifier with the smallest possible misclassification rate. For example, suppose that for a given $\mathbf{x}$, *class 1* occurs with probability 0.4, *class 2* occurs with probability 0.35, and *class 3* occurs with probability 0.25. The Bayes classifier makes a prediction of *class 1* for this $\mathbf{x}$. The probability of a misclassification for this $\mathbf{x}$ is 0.6, but any prediction other than *class 1* will result in a greater probability of misclassification. It should be noted that in practice it is rare to have the information necessary to determine the Bayes classifier, in which case the best one can hope for is to be able to create a classifier that will perform similarly to the Bayes classifier.) Letting $\mathcal{U}_\phi$ be the subset of the measurement space on which ϕ is unbiased, and $\mathcal{B}_\phi$ be the subset of the measurement space on which it is not unbiased, if we could apply ϕ to a large set of independently drawn replications from the parent distribution of the learning sample to create an ensemble of classifiers, and let them vote, then with high probability this aggregated classifier would produce the same predicted class that the Bayes classifier will (due to a law of large numbers effect), *provided that the $\mathbf{x}$ being classified belongs to $\mathcal{U}_\phi$*.

The *variance* of a classifier ϕ is defined to be the difference of the probability that $\mathbf{X}$ (a random set of inputs from the distribution corresponding to the learning sample) belongs to $\mathcal{U}_\phi$ and is classified correctly by the Bayes classifier, and the probability that $\mathbf{X}$ belongs to $\mathcal{U}_\phi$ and is classified correctly by ϕ. (To elaborate, here we consider a random vector, $\mathbf{X}$, and its associated class, and state that $\mathbf{X}$ is classified correctly by ϕ if $\phi(\mathbf{X})$ is the same as the class associated with $\mathbf{X}$, where $\phi(\mathbf{X})$ is a function of $\mathbf{X}$ *and* the learning sample, which is considered to be random. The situation is similar for $\mathbf{X}$ being classified correctly by the Bayes classifier, except that the Bayes classifier is not a function of the learning sample.) Since an aggregated classifier will behave very similarly to the Bayes classifier on the subset of the measurement space on which the base classifier is unbiased (provided that the ensemble of base classifiers from which the aggregated classifier is constructed is suitably large), the aggregated classifier can have a variance very close to zero, even if the base classifier does not. The key is that the voting greatly increases the probability that the best prediction for **x** (the prediction of the Bayes classifier, which is not necessarily the correct prediction for a particular case) *will be made* provided that **x** belongs to the part of the measurement space for which the base classifier will make the best prediction with a greater probability than it makes any other prediction. (Note that the probabilities for predicting the various classes with the base classifier are due to the random selection of the learning sample as opposed to being due to some sort of random selection given a fixed learning sample.) Even if the base classifier only slightly favors the best prediction, the voting process ensures that the best prediction is made with high probability. When bagging is applied using a particular base classifier and a given learning sample, the hope is that the learning sample is large enough so that bootstrap samples drawn from it are not too different from random replications from the parent distribution of the learning sample (which is assumed to correspond to the distribution of cases to be classified in the future), and that the base classifier is unbiased over a large portion of the measurement space (or more specifically, a subset of the measurement space having a large probability).

The *bias* of a classifier ϕ is defined to be the difference in the probability that **X** belongs to $\mathcal{B}_\phi$ and is classified correctly by the Bayes classifier, and the probability that **X** belongs to $\mathcal{B}_\phi$ and is classified correctly by ϕ. Thus the bias of ϕ pertains to the difference in the performance of ϕ and that of the Bayes classifier when **X** takes a value in the part of the measurement space on which ϕ is more likely to predict a class other than the class corresponding to the best possible prediction. (Note that the sum of the variance of ϕ and the bias of ϕ equals the difference in the misclassification rate of ϕ and the misclassification rate of the Bayes classifier. Breiman (1998, Section 2) gives more information about bias and variance in the classification setting, suggests that these terms are not ideal due to a lack of correspondence with bias and variance in more typical settings, and notes that others have given alternative definitions.) Just as the process of replication and voting makes a bagged classifier, with high probability, give the same predictions as the Bayes classifier on the subset of the measurement space on which the base classifier is unbiased, it makes a bagged classifier, with high probability, give predictions *different* from those given by the Bayes classifier on the subset of the measurement space on which the base classifier is biased. But if this subset of the measurement space is rather small, as measured by its probability, the amount by which the bias can increase due to aggregation is bounded from above by a small number. *So a key to success with bagging is to apply it to a classification method that has a small bias* with the situation at hand (which typically corresponds to a relatively large variance). It does not matter if the classifier has a large variance, perhaps due to instability, since the variance can be greatly reduced by bagging. (It should be noted that this simple explanation of why bagging a classifier having low bias works ignores effects due to the fact that taking bootstrap samples from the original learning sample is not the same as having independent samples from the parent distribution of the learning sample. But if the original learning sample is not too small, such effects can be relatively small, and so the preceding explanation hopefully serves well to get across the main idea. One can also see Breiman (1996a, Section 4.2), which contains an explanation of why bagging works in the classification setting.)

Generally, bagging a good classifier tends to improve it, while bagging a bad classifier can make it worse. (For a simple example of the latter phenomenon, see Hastie et al. 2001, Section 8.7.1.) But bagging a nearest neighbors classifier, which is relatively stable, and may be rather good in some settings, can lead to no appreciable change, for better or for worse, in performance. Bagging the often unstable classifier CART, which is typically decent, but not always great, can often make it close to being ideal (meaning that its misclassification rate is close to the Bayes rate, which is a lower bound which is rarely achieved when creating classifiers from a random sample). This is due to the fact that if a carefully created classification tree is not too small, it will typically have a relatively small bias, but perhaps a large variance, and bagging can greatly decrease the variance without increasing the bias very much. Results in Breiman (1996a) show that bagged trees outperformed nearest neighbors classifiers, which were not improved by bagging. Bagging can also improve the performances of neural networks, which are generally unstable, but like CART, tend to have relatively low bias.

Breiman (1996a) gives some indications of how well bagging can improve predictions, and of the variability in performance when different data sets are considered. With the classification examples examined, bagging reduced CART's misclassification rates by 6% to 77%. When the amount of improvement due to bagging is small, it could be that there is little room for improvement. That is, it could be that both the unbagged and the bagged classifier have misclassification rates close to the Bayes rate, and that bagging actually did a good job of achieving the limited amount of improvement which was possible. When compared with 22 other classification methods used in the Statlog Project (see Michie et al., 1994) on four publicly available data sets from the Statlog Project, Breiman (1996a) shows that bagged trees did the best overall (although it should be noted that boosted classifiers were not considered). With the regression examples considered by Breiman (1996a), bagging reduced CART's mean squared error by 21% to 46%.

6. Boosting

Boosting is a method of combining classifiers, which are iteratively created from weighted versions of the learning sample, with the weights adaptively adjusted at each step to give increased weight to the cases which were misclassified on the previous step. The final predictions are obtained by weighting the results of the iteratively produced predictors.

Boosting was originally developed for classification, and is typically applied to *weak learners*. For a two-class classifier, a weak learner is a classifier that may only be slightly better than random guessing. Since random guessing has an error rate of 0.5, a weak classifier just has to predict correctly, on average, slightly more than 50% of the time. An example of a weak classifier is a *stump*, which is a two node tree. (In some settings, even such a simple classifier can have a fairly small error rate (if the classes can be nearly separated with a single split). But in other settings, a stump can have an error rate of almost 0.5, and so stumps are generally referred to as weak learners.) However, boosting is not limited to being used with weak learners. It can be used with classifiers which are fairly accurate, such as carefully grown and pruned trees, serving as the *base learner*. Hastie et al. (2001) claim that using trees with between four and eight terminal nodes works well in most cases, and that performance is fairly insensitive to the choice, from this range, which is made. Friedman et al. (2000) give an example in which boosting stumps does appreciably worse than just using a single large tree, but boosting eight node trees produced an error rate that is less than 25 percent of the error rate of the single large tree, showing that the choice of what is used as the base classifier can make a large difference, and that the popular choice of boosting stumps can be far from optimal. However, they also give another example for which stumps are superior to larger trees. As a general strategy, one might choose stumps if it is suspected that effects are additive, and choose larger trees if one anticipates interactions for which adjustments should be made.

6.1. AdaBoost

AdaBoost is a boosting algorithm developed by Freund and Schapire (1996) to be used with classifiers. There are two versions of AdaBoost.

AdaBoost.M1 first calls for a classifier to be created using the learning sample (that is, the base learner is fit to the learning sample), with every case being given the same weight. If the learning sample consists of N cases, the initial weights are all $1/N$. The weights used for each subsequent step depend upon the weighted resubstitution error rate of the classifier created in the immediately preceding step, with the cases being misclassified on a given step being given greater weight on the next step.

Specifically, if $I_{m,n}$ is equal to 1 if the nth case is misclassified on the mth step, and equal to 0 otherwise, and $w_{m,n}$ is the weight of the nth case for the mth step, where the weights are positive and sum to 1, for the weighted resubstitution error rate for the mth step, e_m, we have

$$e_m = \sum_{n=1}^{N} w_{m,n} I_{m,n}.$$

The weights for the $(m + 1)$th step are obtained from the weights for the mth step by multiplying the weights for the correctly classified cases by $e_m/(1 - e_m)$, which should be less than 1, and then multiplying the entire set of values by the number greater than 1 which makes the N values sum to 1. This procedure downweights the cases which were correctly classified, which is equivalent to giving increased weight to the cases which were misclassified. For the second step, there will only be two values used for the weights, but as the iterations continue, the cases which are often correctly classified can have weights much smaller than the cases which are the hardest to classify correctly. If the classification method of the weak learner does not allow weighted cases, then from the second step onwards, a random sample of the original learning sample cases is drawn to serve as the learning sample for that step, with independent selections made using the weights as the selection probabilities.

The fact that the cases misclassified on a given step are given increased weight on the next step typically leads to a different fit of the base learner, which correctly classifies some of the misclassified cases from the previous step, contributing to a low correlation of predictions from one step and the next. Amit et al. (1999) provide some details which indicate that the AdaBoost algorithm attempts to produce low correlation between the predictions of the fitted base classifiers, and some believe that this phenomenon is an important factor in the success of AdaBoost (see Breiman's discussion in Friedman et al. (2000) and Breiman (2001, p. 20)).

Assuming that the weighted error rate for each step is no greater than 0.5, the boosting procedure creates M classifiers using the iteratively adjusted weights. Values such as 10 and 100 have been used for M. But in many cases it will be better to use a larger value, such as 200, since performance often continues to improve as M approaches 200 or 400, and it is rare that performance will suffer if M is made to be too large. The M classifiers are then combined to obtain a single classifier to use. If the error rate exceeds 0.5, which should not occur with only two classes, but could occur if there are more than 2 classes, the iterations are terminated, and only the classifiers having error rates less than 0.5 are combined. In either case, the classifiers are combined by using a weighted voting to assign the predicted class for any input $\mathbf{x}$, with the classifier created on the mth step being given weight $\log((1 - e_m)/e_m)$. This gives the greatest weights to the classifiers having the lowest weighted resubstitution error rates. Increasing the number of iterations for AdaBoost.M1 drives the training error rate of the composite classifier to zero. The error rate for an independent test sample need not approach zero, but in many cases it can get very close to the smallest possible error rate.

Breiman's discussion in Friedman et al. (2000) makes the point that after a large number of iterations, if one examines the weighted proportion of times that each of the learning sample cases has been misclassified, the values tend to be all about the same. So instead of stating that AdaBoost concentrates on the hard to classify cases, as some have done, it is more accurate to state that AdaBoost places (nearly) equal importance on correctly classifying each of the cases in the learning sample. Hastie et al. (2001) show that AdaBoost.M1 is equivalent to forward stagewise additive modeling using an exponential loss function.

AdaBoost.M2 is equivalent to AdaBoost.M1 if there are just two classes, but appears to be better than AdaBoost.M1 (see results in Freund and Schapire, 1996) if there are more than two classes, because for AdaBoost.M1 to work well the weak learners should have weighted resubstitution error rates less than 0.5, and this may be rather difficult to achieve if there are many classes. The weak learners used with AdaBoost.M2 assign to each case a set of *plausibility values* for the possible classes. They also make use of a loss function that can assign different penalties to different types of misclassifications. The loss function values used are updated for each iteration to place more emphasis on avoiding the same types of misclassifications that occurred most frequently in the previous step. After the iterations have been completed, for each $\mathbf{x}$ of interest the sequence of classifiers is used to produce a weighted average of the plausibilities for each class, and the class corresponding to the largest of these values is taken to be the prediction for $\mathbf{x}$. See Freund and Schapire (1996, 1997) for details pertaining to AdaBoost.M2.

6.2. Some related methods

Breiman (1998) presents an alternative method of creating a classifier based on combining classifiers that are constructed using adaptively resampled versions of the learning sample. He calls his method *arc-x4* and he refers to AdaBoost as *arc-fs* (in honor of Freund and Schapire), since he considers both arc-x4 and arc-fs to be arcing algorithms (and so Breiman considered boosting to be a special case of arcing). When the performances of these two methods were compared in various situations, Breiman (1998) found that there were no large differences, with one working a little better in some situations, and the other working a little better in other situations. He concludes that the *adaptive resampling is the key to success*, and not the particular form of either algorithm, since although both use adaptive resampling, with misclassified cases receiving larger weights for the next step, they are quite different in other ways (for example, they differ in how the weights are established at each step, and they differ in how the voting is done after the iterations have been completed, with arc-x4 using unweighted voting).

Schapire and Singer (1998) improve upon AdaBoost.M2, with their *AdaBoost.MH* algorithm, and Friedman et al. (2000) propose a similar method, called *Real AdaBoost*. Applied to the two class setting, Real AdaBoost fits an additive model for the logit transformation of the probability that the response associated with **x** is a particular one of the classes, and it does this in a stagewise manner using a loss function which is appropriate for fitting logistic regression models. Friedman et al. (2000) also propose *Gentle AdaBoost* and *LogitBoost* algorithms, which in some cases, but not consistently, give improved performance when compared to the other AdaBoost algorithms. Bühlmann and Yu, in their discussion of Friedman et al. (2000), suggest boosting bagged stumps or larger trees, and they give some results pertaining to this method, which they refer to as *bag-boosting*.

The boosting approach can be extended to the regression setting. Friedman (2001) proposes *TreeBoost* methods for function approximation, which create additive models, having regression trees as components, in a stagewise manner using a *gradient boosting* technique. Hastie et al. (2001) refer to this approach as *MART* (Multiple Additive Regression Trees), and they find that incorporating a shrinkage factor to encourage slow learning can improve accuracy.

6.3. When and how boosting works

When used with a rather complex base classification method that is unstable, such as carefully grown and pruned trees, boosting, and arcing in general, like bagging, in many cases can dramatically lower the error rate of the classification method. Stated simply, this is due *in part* to the fact that unstable methods have relatively large variances, and boosting decreases the variance without increasing the bias. Breiman (1998) shows that when applied to linear discriminant analysis (LDA), which is a fairly stable method of classification, neither bagging nor boosting has any appreciable effect. In some cases (see, for example, results reported by Freund and Schapire (1996) and Quinlan (1996), where C4.5 was compared to boosted C4.5), boosting *can* make a classifier appreciably worse, but this does not seem to happen in a large percentage of settings which have been investigated, and when it does happen, perhaps a small learning sample size is a contributing factor. Several studies have indicated that boosting can be outperformed when the classes have significant overlap. Nevertheless, it is often the case that some form of boosted classifier can do nearly as good or better than any other classifier.

As to how boosting typically achieves such a low misclassification rate, the opinions are many and varied. Friedman et al. (2000) point out that when used with a rather simple base classifier which is stable but perhaps highly biased, such as trees limited to a rather small number of terminal nodes, which is a common way of applying boosting, the success of boosting is due much more to bias reduction (see Schapire et al., 1998) than it is to variance reduction, and this makes boosting fundamentally different than bagging, despite the fact that both methods combine classifiers based on various perturbations of the learning sample. Schapire et al. (1998) offer the opinion that AdaBoost works because of its ability to produce generally high margins. But Breiman (1999) claims that their explanation is incomplete, and provides some explanation of his own. Some have attempted to explain boosting from a Bayesian point of view. Friedman et al. (2000) give support to their opinion that viewing boosting procedures as stagewise algorithms for fitting additive models greatly helps to explain their performance, and make a connection to maximum likelihood. In the discussion of Friedman et al. (2000), various authors offer additional explanation as to why boosting works. In particular, Breiman finds fault with the explanation provided by Friedman et al. (2000), since it does not explain the strong empirical evidence that boosting is *generally* resistant to overfitting. Friedman et al. (2000) provide one example of boosting leading to an overfit predictor. Also, Rätsch et al. (2001) conclude that overfitting can occur when there is a lot of noise, and they provide some pertinent references.) In his discussion, Breiman indicates that the key to the success of boosting is that it produces classifiers of reasonable strength which have low correlation. Breiman provided additional insight with his 2002 Wald Lecture on Machine Learning (see http://stat-www.berkeley.edu/users/breiman/wald2002-1.pdf).

While the preceding paragraph indicates that it is not fully understood why boosting works as well as it does, there is no question about the fact that it can work very well. In various comparisons with bagging, boosting generally, but not always, does better, and leading researchers in the field of machine learning tend to favor boosting. It should be noted that some of the comparisons of boosting to bagging are a bit misleading because they use stumps as the base classifier for both boosting and bagging, and while stumps, being weak learners, can work very well with boosting, they can be too stable and have too great a bias to work really well with bagging. Trees larger than stumps tend to work better with bagging, and so a fairer comparison would be to compare bagging fairly large trees, boosting both stumps and slightly larger trees, and using an assortment of good classifiers without applying boosting or bagging. Boosting and bagging classifiers other than trees can also be considered. From the results of some such studies, it can be seen that often there is not a lot of difference in the performances of bagged large trees and boosted large trees, although it appears that boosting did better more often than not.

Of course, the main interest should be in an overall winner, and not the comparison of boosting with bagging. However, various studies indicate that no one method works best in all situations. This suggests that it may be wise to be familiar with a large variety of methods, and to try to develop an understanding about the types of situations in which each one works well. Also, one should be aware that there are other types of methods for classification and regression, such as kernel-based methods and support vector machines, that have not been covered in this chapter. Indeed, one other class of methods, of the P&C variety, *random forests* with random feature selection, have been shown by Breiman (2001) to compare very favorably with AdaBoost. Furthermore, despite the fact that random forests use ensembles of classifiers created from independent identically distributed learning samples, as opposed to the adaptively produced learning samples used in boosting, Breiman (2001) claims that AdaBoost is similar to a random forest. So it seems premature to declare that boosting is the clear-cut best of the new methods pertaining to classification developed during the past 15 years, while at the same time it is safe to declare that boosting can be used to create classifiers that are very often among the best.

References

Amit, Y., Blanchard, G., Wilder, K. (1999). Multiple randomized classifiers: MRCL. Technical Report. Department of Statistics, University of Chicago.
Breiman, L. (1996a). Bagging predictors. *Machine Learning* **24**, 123–140.
Breiman, L. (1996b). Heuristics of instability and stabilization in model selection. *Ann. Statist.* **24**, 2350–2383.
Breiman, L. (1998). Arcing classifiers (with discussion). *Ann. Statist.* **26**, 801–849.
Breiman, L. (1999). Prediction games and arcing algorithms. *Neural Comput.* **11**, 1493–1517.
Breiman, L. (2001). Random forests. *Machine Learning* **45**, 5–32.
Breiman, L., Friedman, J.H., Olshen, R.A., Stone, C.J. (1984). *Classification and Regression Trees*. Wadsworth, Pacific Grove, CA.
Buntine, W.L. (1992). Learning classification trees. *Statist. Comput.* **2**, 63–73.
Ciampi, A., Chang, C.-H., Hogg, S., McKinney, S. (1987). Recursive partitioning: a versatile method for exploratory data analysis in biostatistics. In: MacNeil, I.B., Umphrey, G.J. (Eds.), *Biostatistics*. Reidel, Dordrecht.
Clark, L.A., Pregibon, D. (1992). Tree based models. In: Chambers, J.M., Hastie, T.J. (Eds.), *Statistical Models in S*. Wadsworth and Brooks/Cole, Pacific Grove, CA.
Crawford, S.L. (1989). Extensions to the CART algorithm. *Int. J. Man–Machine Stud.* **31**, 197–217.
Freund, Y. (1995). Boosting a weak learning algorithm by majority. *Inform. Comput.* **121**, 256–285.
Freund, Y., Schapire, R. (1996). Experiments with a new boosting algorithm. In: Saitta, L. (Ed.), *Machine Learning: Proceedings of the Thirteenth International Conference*. Morgan Kaufmann, San Francisco, CA.
Freund, Y., Schapire, R. (1997). A decision-theoretic generalization of on-line learning and an application to boosting. *J. Comput. System Sci.* **55**, 119–139.
Friedman, J. (1991). Multivariate adaptive regression splines (with discussion). *Ann. Statist.* **19**, 1–141.
Friedman, J. (2001). Greedy function approximation: a gradient boosting machine. *Ann. Statist.* **29**, 1189–1232.
Friedman, J., Hastie, T., Tibshirani, R. (2000). Additive logistic regression: a statistical view of boosting (with discussion). *Ann. Statist.* **28**, 337–407.
Gelfandi, S.B., Ravishankar, C.S., Delp, E.J. (1991). An iterative growing and pruning algorithm for classification tree design. *IEEE Trans. Pattern Anal. Machine Intelligence* **13**, 163–174.
Gentle, J.E. (2002). *Elements of Computational Statistics*. Springer-Verlag, New York.
Harrell, F.E. (2001). *Regression Modeling Strategies: with Applications to Linear Models, Logistic Regression, and Survival Analysis*. Springer-Verlag, New York.
Hastie, T., Tibshirani, R., Friedman, J. (2001). *The Elements of Statistical Learning: Data Mining, Inference, and Prediction*. Springer-Verlag, New York.
Kass, G.V. (1980). An exploratory technique for investigating large quantities of categorical data. *Appl. Statist.* **29**, 119–127.
Loh, W.-Y., Shih, Y.-S. (1997). Split selection methods for classification trees. *Statist. Sin.* **7**, 815–840.

Martinez, W.L., Martinez, A.R. (2002). *Computational Statistics Handbook with MATLAB*. Chapman and Hall/CRC, Boca Raton, FL.

Michie, D., Spiegelhalter, D.J., Taylor, C.C. (1994). *Machine Learning, Neural and Statistical Classification*. Horwood, London.

Morgan, J.N., Messenger, R.C. (1973). THAID: a sequential search program for the analysis of nominal scale dependent variables. Institute for Social Research, University of Michigan, Ann Arbor, MI.

Morgan, J.N., Sonquist, J.A. (1963). Problems in the analysis of survey data, and a proposal. *J. Amer. Statist. Assoc.* **58**, 415–434.

Olshen, R. (2001). A conversation with Leo Breiman. *Statist. Sci.* **16**, 184–198.

Quinlan, J.R. (1979). Discovering rules by induction from large classes of examples. In: Michie, D. (Ed.), *Expert Systems in the Microelectronic Age*. Edinburgh University Press, Edinburgh.

Quinlan, J.R. (1986). Induction of decision trees. *Machine Learning* **1**, 81–106.

Quinlan, J.R. (1987). Simplifying decision trees. *Int. J. Man–Machine Stud.* **27**, 221–234.

Quinlan, J.R. (1993). *C4.5: Programs for Machine Learning*. Morgan Kaufmann, San Mateo, CA.

Quinlan, J.R. (1996). Bagging, boosting, and C4.5. In: *Proceedings of AAAI '96 National Conference on Artificial Intelligence*. Morgan Kaufmann, San Francisco, CA.

Rätsch, G., Onoda, T., Müller, K.R. (2001). Soft margins for AdaBoost. *Machine Learning* **42**, 287–320.

Ripley, B.D. (1996). *Pattern Recognition and Neural Networks*. Cambridge Univ. Press, Cambridge.

Schapire, R. (1990). The strength of weak learnability. *Machine Learning* **5**, 197–227.

Schapire, R., Singer, Y. (1998). Improved boosting algorithms using confidence-rated predictions. In: *Proceedings of the Eleventh Annual Conference on Computational Learning Theory*. ACM Press.

Schapire, R., Freund, Y., Bartlett, P., Lee, W. (1998). Boosting the margin: a new explanation for the effectiveness of voting methods. *Ann. Statist.* **26**, 1651–1686.

Steinberg, D., Colla, P. (1995). *CART: Tree-Structured Nonparametric Data Analysis*. Salford Systems, San Diego, CA.

Therneau, T.M., Atkinson, E.J. (1997). An introduction to recursive partitioning using the rpart routine. Technical Report. Section of Statistics, Mayo Clinic.

Fast Algorithms for Classification Using Class Cover Catch Digraphs

David J. Marchette, Edward J. Wegman and Carey E. Priebe

Abstract

Non-parametric statistical pattern recognition algorithms, such as nearest neighbor algorithms, typically present a trade off between performance and complexity. Here "complexity" refers to computational complexity, which has implications for the size of data sets that can be processed. We present a methodology for the selection of prototypes for a graph-theoretical classifier, the class cover catch digraph. These prototypes, along with certain associated parameters, define the classifier. The methodology is illustrated on several simulated data sets and on two applications: the Sloan Digital Sky Survey and a text classification problem.

1. Introduction

Let $X = \{x_1, \ldots, x_n\}$ and $Y = \{y_1, \ldots, y_m\}$ be measurements taken from two distinct classes of objects. For example, these might be skull measurements from two species of ape, blood chemistry measurements from healthy individuals and patients with a specific disease, measurements taken from sonar returns from mines and from mine-like objects, etc. Objects (or individuals) of the class of the X observations are labeled class 0, and those of the same class as Y are labeled class 1. A classifier is a function $g(z; x_1, \ldots, x_n, y_1, \ldots, y_m)$ which, given the "training data" X and Y, and a new set of measurements Z from an individual of unknown class, returns a class label, 0 or 1.

One way to construct such a classifier is through nearest neighbors. The training observation closest to the new observation provides the class label. Similarly, one could vote amongst the k closest training observations (the k-nearest neighbor classifier). Other classifiers operate by estimating the probability densities of the different classes and using a likelihood ratio test to make the class determination for new observations. Finally, there are classifiers that try to estimate the decision region directly.

In this article we consider a novel classifier that comes from an estimate of the support of each class by a covering of a (relatively) small number of balls, centered at points in a subset of the training observations. This classifier, described in Section 2, takes a complete covering by all the observations and selects a subset to reduce the complexity of the cover. This has somewhat the flavor of a reduced nearest neighbor classifier, where one tries to reduce the number of training exemplars that need to be retained in order to obtain the desired level of performance of the classifier.

One problem with these types of classifiers, where a reduced set of exemplars is chosen from the training set, is that they are generally of order n^2 in complexity, where n is the size of the training sample. Thus, for very large data sets the computational resources needed to construct the classifier are prohibitive. (This assumes naive implementations; as we will see there are ways to reduce this complexity through clever algorithms; however, the complexity is still greater than linear in n.) What is needed are approximate algorithms that produce a good classifier in time closer to $O(n)$. In Section 5 we present such an algorithm for the class cover catch digraph (CCCD), and in Section 8 we illustrate its performance on several pedagogical examples. In Sections 9 and 10, we illustrate the algorithm's performance on two real-world problems.

2. Class cover catch digraphs

The class cover catch digraph (Priebe et al., 2001, 2003; Marchette and Priebe, 2003) is a random graph, which is constructed from observations in a classification problem. The vertices consist of the observations from one class, and the edges (arcs) are defined by a relationship to the other class. This construction results in a random graph of interest in its own right as well as a powerful classifier.

A directed graph is a pair $D = (V, A)$ where V is a set of vertices, and A is a set of ordered pairs of vertices, corresponding to the directed edges (sometimes called arcs) of the graph. If $vw \in A$ we say that w is adjacent to v. A random digraph is one whose arcs are defined by a random process. There are many models of random graphs and digraphs, the interested reader is directed to Bollobás (2001) for some of these. We will focus in this article on a class of *vertex random* graphs, for which the random component of the graph is inherited from the randomness of the vertices.

Define the open ball centered at a point $x \in \mathbb{R}^d$ by $B(x, r) = \{y \in \mathbb{R}^d : \rho(x, y) < r\}$, where ρ is a distance metric (usually Euclidean distance). Given observations $X = \{x_1, \ldots, x_n\}$, $Y = \{y_1, \ldots, y_m\}$, the class cover catch digraph (CCCD) for X on Y is defined as follows. Let the vertex set V be X and define the arc set A by $vw \in A \Leftrightarrow w \in B(v, \rho(v, Y))$. Here $\rho(v, Y)$ is the minimum distance from v to an element of Y, using the predefined distance ρ. The CCCD then provides a cover of the X observations by open balls centered at these observations, such that no ball covers a Y observation. These balls are maximal with respect to this constraint.

Given a CCCD, we can construct a classifier by observing whether a new observation is interior to one of the balls. More sophisticated (and accurate) classifiers are possible, for instance by covering both classes and using the radii of the balls in making the class determination. See (Priebe et al., 2003) and Section 3 for more information on these classifiers. We wish to reduce the computational complexity in this classifier by selecting a subset of the balls that contain the observations. To this end, we attempt to find a minimum dominating set for the digraph.

A dominating set for a digraph D is a set S such that every vertex of D is either an element of S or is adjacent to an element of S. A minimum dominating set is one with minimum cardinality.

In the CCCD, a dominating set corresponds to a set of balls which cover all the X observations. A minimum dominating set is then the smallest cardinality set of balls defined as above such that every observation in X is interior to at least one of the balls in the set. Figure 1 illustrates this with a pedagogical example. In this case, the dominating set is of size 2, a reduction by a factor of 5 in the number of balls needed to cover the X (dots) observations to the exclusion of the Y (triangles).

The dominating set provides us with a subset of the cover that has the same resubstitution error (zero) as the original cover. This can reduce over-fitting of the classifier as well as reducing its computational complexity. Note that the minimum dominating set is not necessarily unique. The choice of minimum dominating set may be important for certain applications, but in this article we are concerned only with finding one such set (or, more precisely, a dominating set which is approximately minimum).

While the selection of a minimum dominating set is known to be NP-hard, there are approximations that work quite well in practice. For example, in (Priebe et al., 2003; Marchette and Priebe, 2003), a greedy algorithm, presented below, is used to obtain approximate minimum dominating sets. Unfortunately, construction of the graph, as we have presented it here, is an $O(nm + n^2)$ operation (ignoring the dimension of the data), due to the fact that distances are required between all pairs of points. As will be seen in Section 6, this can be improved upon somewhat with clever preprocessing. However, it is still a daunting calculation for large values of n and m.

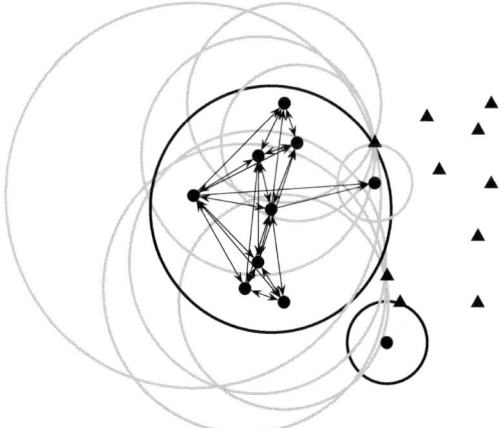

Fig. 1. An example of ball covering, CCCD, and a dominating set from the CCCD of the class indicated by dots. The balls associated with the dominating set are colored black, while those not selected are colored gray.

The greedy algorithm for finding an approximately minimum dominating set is:

NAIVE GREEDY ALGORITHM.

Set $\mathcal{J} = X, d = 0, l = 0$.
While $\mathcal{J} \neq \emptyset$
  $l = l + 1$.
  $x = \arg\max_{x' \in X} |B(x', \rho(x', Y)) \cap \mathcal{J}|$ (break ties arbitrarily).
  $W_l = \{x\}$.
  $r_l = \rho(x, Y)$.
  $\mathcal{J} = \mathcal{J} \setminus \{x' : \rho(x, x') < r_l\}$.
EndWhile
Set $d = l$.
Return ($\mathcal{W} = [W_1 \cdots W_d]', \mathcal{R} = [r_1 \cdots r_d]'$).

We refer to this algorithm as *naive* because it does not take advantage of any standard methods for computing nearest neighbors, as will be discussed below. The greedy algorithm is not guaranteed to find a minimum dominating set, but it does return a dominating set, and in many cases returns one that is quite close to minimum. See (Chvatal, 1979; Parekh, 1991) for more information on greedy heuristics. Note that the version of the algorithm presented here does not work on the graph directly, but on the inter point distance matrix; the arg max computes the degree using the cover, rather than using the degrees of the digraph. This means that the algorithm as stated requires $O((n^2 + nm)d)$ calculations to compute the inter point distance matrices on the d-dimensional data. It is clear how to modify this algorithm to operate on arbitrary directed graphs, which results in the standard algorithm for finding minimum dominating sets on graphs.

In (Wegman and Khumbah, 2001; Khumbah and Wegman, 2003), the issues of computations on large data sets are discussed, and it is shown that for data sizes typical of many real problems $O(n^2)$ algorithms are not feasible. Thus, for very large problems we need an algorithm which provides an approximation to the dominating set of the CCCD in time linear (or sub-linear) in the number of observations. Section 5 will discuss some algorithms designed to be faster than $O(n^2)$.

Note that the requirements that the balls cover all of the X observations and none of the Y observations can each be relaxed. This is discussed in some detail in (Marchette and Priebe, 2003; Priebe et al., 2003), and will be described briefly in Sections 3 and 4.

3. CCCD for classification

The CCCD can be used to define a classifier in a number of ways. Given X and Y, corresponding to classes 0 and 1 respectively, let $D_{X,Y}$ denote the CCCD on vertices X as defined by Y. Denote by S_X the set of balls corresponding to the elements of a dominating set of $D_{X,Y}$ (for example, that returned by the greedy algorithm), and S_Y

that corresponding to a dominating set of $D_{Y,X}$. Denote by $\mathring{D}_{X,Y} = \bigcup S_X$. Define

$$g(x; \mathring{D}_{X,Y}, \mathring{D}_{Y,X}) = \begin{cases} 0, & x \in \mathring{D}_{X,Y} \setminus \mathring{D}_{Y,X}, \\ 1, & x \in \mathring{D}_{Y,X} \setminus \mathring{D}_{X,Y}, \\ -1, & x \in \mathring{D}_{Y,X} \cap \mathring{D}_{X,Y}, \\ -2, & \text{else.} \end{cases} \quad (1)$$

The classifier g produces either a class label or a "no decision" as represented by the values -1 (indicating confusion) or -2 indicating that x is not in the region defined by the training data. In (Priebe et al., 2003) this is referred to as a *preclassifier*, due to the two types of "no decisions" it allows. This makes hard decisions based on the cover, and does not make a decision in the difficult overlap region. Thus, we define the classifier

$$g_{\text{CCCD}}(x) = I \left\{ \min_{z \in c(S_X)} \rho(x, z)/r(z) < \min_{z \in c(S_Y)} \rho(x, z)/r(z) \right\}, \quad (2)$$

where $c(S_X)$ is the set of centers for the balls in S_X. This makes the decision based on a scaled distance to the ball, scaled by the radius of the ball.

Figure 2 illustrates the CCCD classifier on the data from Figure 1. This simple example illustrates the potential for complex boundary regions, even in the case of simple representations. In this case the classifier is defined by four balls, two for each class. The training data are completely covered by their respective covers. Note that the decision boundary can be internal to the circles, and that points outside the circles are also given a classification.

It stands to reason that the rule for defining the balls may result in balls that are too small in regions of overlap. If we could relax the condition that the balls be pure, we

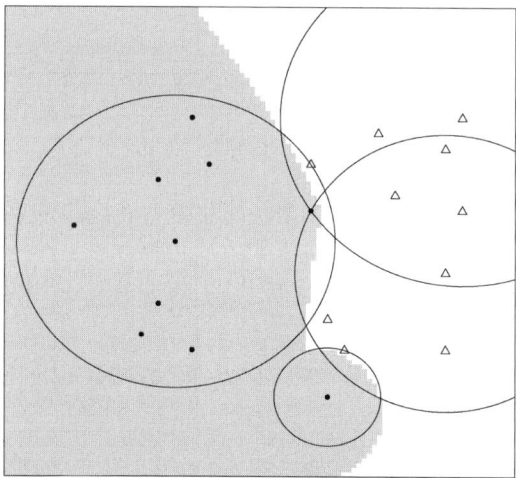

Fig. 2. A CCCD classifier on the data from Figure 1. The gray area represents the region classified as class 1 (dots), the white area represents the region classified as class 2 (triangles). The circles represent the covers defining the classifier.

could presumably produce a better classifier, especially in light of Eq. (2). The random walk is defined in order to solve this problem.

To determine the appropriate radius for an observation x, consider the following random walk: as the ball around x expands it will meet other observations, either from X or Y. If it meets an X observation it moves up by $1/n$. If it meets a Y it moves down by $1/m$. Figure 3 illustrates this idea.

Following the construction of the random walk, the radius $r(x)$ can be computed by choosing the value for which the walk was maximum (in the case of ties, choose the one with minimum radius). One may wish to first subtract a curve (such as the curve r^d) from the walk prior to selecting the radius, to penalize for large balls.

Figure 4 depicts four two-dimensional data sets, and Figure 5 illustrates the CCCD classifier run on these data. The complexity of the resulting classifiers are 10 and 15 balls for the circle and square without overlap (left column in the figures) and 48 and 49 for the overlapping cases (right column in the figures). In all cases, the points displayed in Figure 4 were used as training, and a grid of the square was used to evaluate the algorithms. The random walk version tends to fill in the holes, at the expense of the boundary, as can be seen in Figure 6. The resulting classifiers have 12 and 11 balls for the non-overlapping cases, and 25 and 16 for the overlapping cases.

It is important to note that since the CCCD operates only on inter point distances, it can be applied to arbitrary-dimensional data. Provided the information necessary for building an accurate model is contained in the inter point distance matrix (an as-

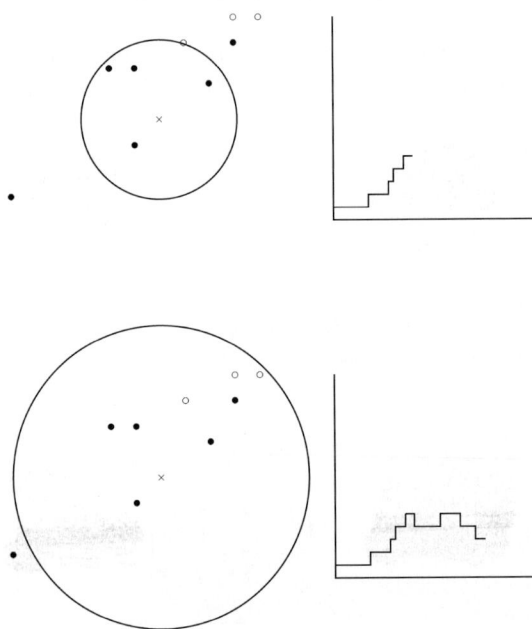

Fig. 3. Illustration of the random walk. The top shows the walk prior to encountering the first non-target point. The walk increases by a step at each target observation. In the bottom plot, the random walk has continued, decreasing a step at each non-target observation.

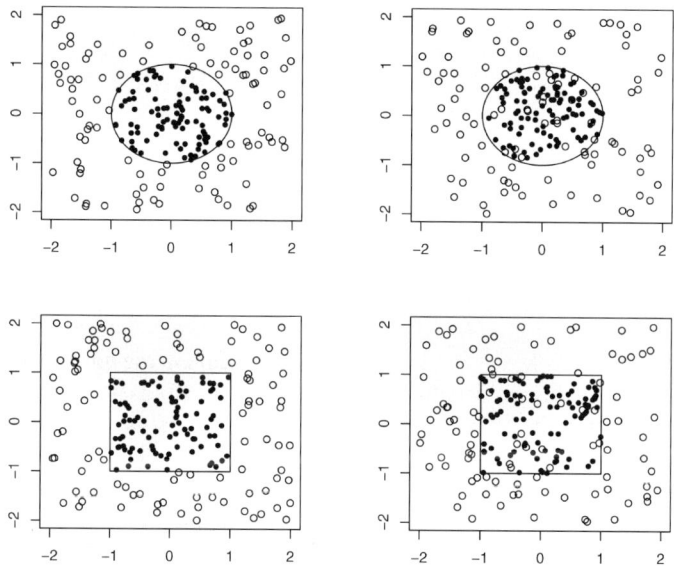

Fig. 4. Four instances of 100 observations drawn from each of two distributions. In the left column the supports of the classes do not overlap, and the data are uniform on their regions. In the right, one distribution is uniform over the entire region, while the other is uniform only within the circle (top) or square (bottom).

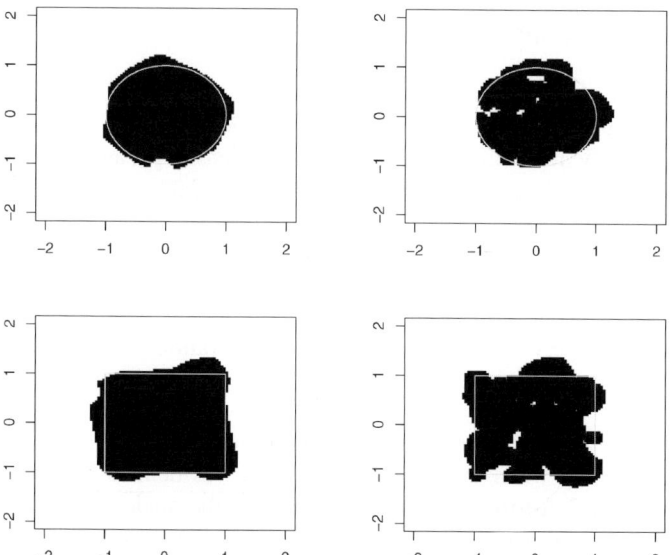

Fig. 5. The result of running the CCCD classifier on the data from Figure 4.

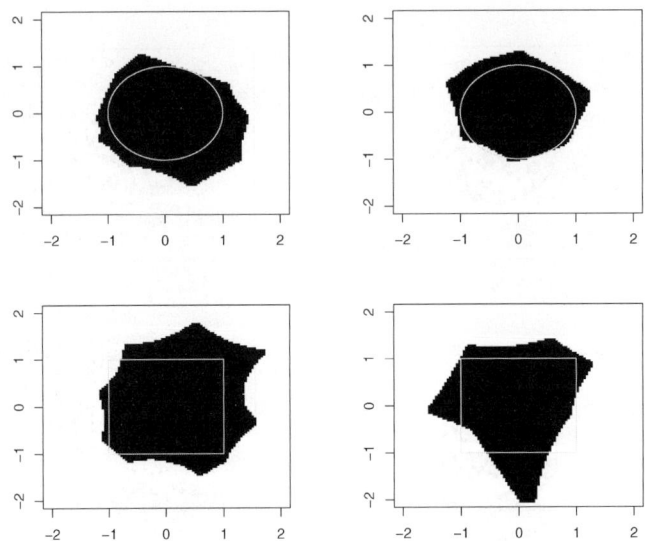

Fig. 6. The result of running the random walk version of the CCCD classifier on the data from Figure 4.

sumption that many classification algorithms, such as the nearest neighbor algorithms, rely upon), the CCCD can be used to produce good classifiers for high-dimensional data.

4. Cluster catch digraph

Catch digraphs can also be used to summarize data, and for clustering. In this case, we have observations $x_1, \ldots, x_n$, independent and identically distributed from some distribution F (we will assume there is an associated density f), and we wish to produce a summary picture of the data, to find groups within the data, or to estimate the support of the density f.

In the classification setting, we can use the distance to observations from other classes to define the balls, as described in the previous section. The problem of applying catch digraphs to this case is the one of defining the balls without reference to any "other class" observations. Instead, for each observation x_i we will define a radius r_i which will define the ball centered at x_i in terms of the "clusteredness" of the data around x_i.

The idea is to perform the random walk, as in Section 2, and compare it to a null distribution, stopping when the difference is "large" as defined by some test. If we denote the empirical distribution function defined by the random walk as $RW(r)$, and the null distribution as F_0, then the non-sequential Kolmogorov–Smirnov (K–S) test would suggest we should choose the radius to be

$$r^* = \arg\max_{r \geq 0} \{RW(r) - F_0(r)\}. \tag{3}$$

This selects the radius for which the data in the ball are farthest from the null hypothesis of complete spatial randomness. We require the deviation from "non-clustering" to be in the positive, or "more clustered" sense. Thus, we pick the radius so that the corresponding ball covers the most clustered group of observations possible, by the definition from the K–S test.

For the sequential version, define

$$l(r) = \min_{q \leq r}\{RW(q) - F_0(q)\},$$
$$u(r) = \max_{q \leq r}\{RW(q) - F_0(q)\}.$$

Let k_α be the value of the K–S statistic at which one would reject with level α. Let R be the first radius such that either $-l(R) \geq k_\alpha$ or $u(R) \geq k_\alpha$. Set R to be the first point at which rejection occurs due to $l(R)$ (if no such rejection occurs, then set R to be the maximum possible). Then set

$$r_x = \arg\max_{q < R}\{RW(q) - F_0(q)\}.$$

The K–S statistic associated with x would then be

$$k_x = RW(r_x) - F_0(r_x).$$

The idea is to pick the radius to be that which produces the largest positive difference from the null prior to any significant negative difference; we wish to find the highest density grouping prior to any region of less than expected density. Compare this with the random walk CCCD of Section 2.

The null distribution should be something related to the problem at hand: finding clusters. In this case "clusters" means regions of high density, and since we are using

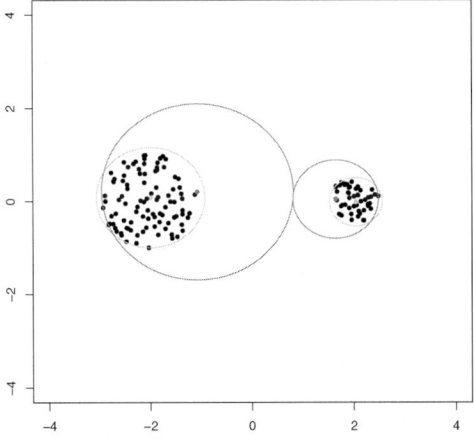

Fig. 7. The result of running the CCD on a simple data set. The small curves indicate the ball found for the observation nearest the center of the cluster, the large curves for those points within the cluster which are farthest from the center.

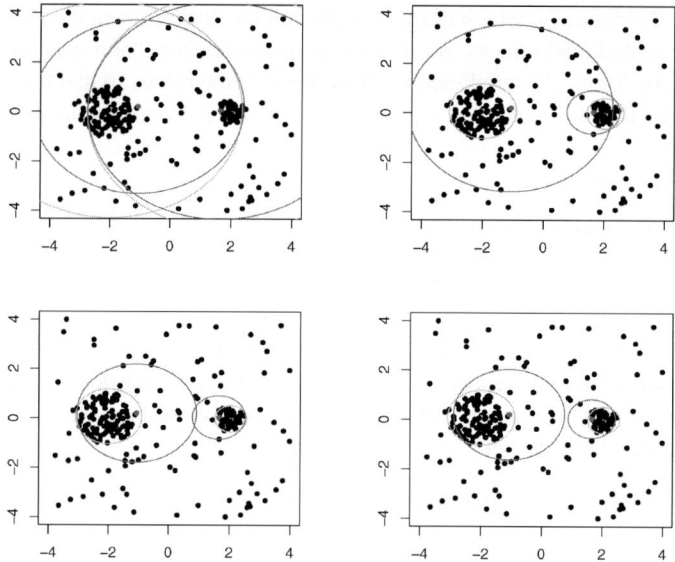

Fig. 8. The result of running the CCD on a simple data set with noise. The light curves indicate the ball found for the observation nearest the center of the cluster, the dark curves are for the point within the cluster which is farthest from the center. From left to right, top to bottom, the values for m were: 1, 1.5, 2, 2.5.

distances exclusively, appropriate null distributions are of the form mr^d where d is the dimension of the data. Changing d and m can be thought of as incorporating a penalty corresponding to the local "background density". Figure 7 shows the results for a simple problem, with four representative balls indicated. Figure 8 shows what happens when noise is added, as a function of the slope of the penalty, m. More information about both the CCCD and the CCD can be found in (Marchette, 2004).

5. Fast algorithms

Several methods are possible to speed up the normally $O((n^2 + nm)d)$ operation of constructing the CCCD and computing the (approximately minimum) dominating set when n is very large. The most obvious is to take a sample from the data of a manageable size. This has the drawback that interesting structure may be lost. In particular, regions of low density can be ignored, resulting in a sub-optimal class cover, and hence a sub-optimal classifier. Such an approach has no guarantee of selecting a dominating set of the CCCD digraph (on the full data set), and while this may seem to be ancillary to our desire to find a parsimonious cover, it is essential if we are to cover all the data, rather than the selected subset. Another approach to selecting the dominating set is detailed below.

The fast estimate proceeds as follows:

ALGORITHM 1.

Set P to a vector of n values of $1/n$,
   C to a vector of n zeros, *dists* to a vector of n values set to some very large number (larger than any likely interpoint distances), $D = \emptyset$.
While C contains a zero **Do**:
   Select x from X with probability given by P.
   Set $r = \min(d(x, Y))$.
   Mark with a 1 the positions of C corresponding to those X items covered by the ball $B(x, r)$.
      Note that this can be accomplished by computing only the distances between the new element and those X observations not already covered.
   Set *dists* to be the (pairwise) minimum of $d(x, X)$ and *dists*.
   Set $P = dists / \sum(dists)$.
   Set $D = D \cup \{x\}$.
Return D.

The idea of the fast algorithm is as follows. Given n d-dimensional observations from X and m observations from Y, the naive greedy algorithm requires $O((n^2 + nm)d + \gamma(n+m))$ operations, where γ is the size of the resultant dominating set. The dominant $(n^2 + nm)d$ comes from the interpoint distance calculation, while the secondary term comes from the greedy algorithm for the selection of the dominating set. We reduce this to $O(\gamma'(n+m)d)$ (where $\gamma' \geq \gamma$ is the size of the new dominating set) as follows. Select an observation at random, and compute its radius (distance to Y) and the X observations not covered, which takes $O(d(n - 1 + m))$ operations. The process repeats until all the X observations are covered.

If $\gamma \leq \gamma' \sim n$, then this algorithm has the same complexity as the greedy algorithm (including the cost of computing the digraph), and in fact will typically report a larger dominating set than the greedy algorithm. In practice, assuming the classes are reasonably well separated, we expect $\gamma \leq \gamma' \ll n$, in which case the algorithm may be significantly faster than the greedy algorithm.

This fast algorithm is guaranteed to produce a dominating set of the CCCD. The algorithm can be modified if desired to keep the random selection uniform instead of proportional to the distances to the X observations. Also, one can select from only those observations not already covered, thus ensuring that the new selection will increase the number of observations covered. None of these modifications significantly change the computational complexity of the algorithm, although they may have a significant impact for a given problem.

One can also check the γ' elements produced for redundancy to produce a minimal dominating set. This can be done either by eliminating balls that cover points covered by other balls within the set, or by eliminating balls that are contained within the other balls. This can be a combinatorial problem on its own, and so this cost must be traded off against the cost of retaining redundant elements.

The selection of the new observations with probability P is designed to ensure that the algorithm explores new territory, rather than areas with many points that may already be covered. One could instead ignore the probability vector P and always select the observations with equal probability, as noted above. Also, this could be "phased in", using a uniform probability early on to ensure that high density regions are well represented, then gradually changing the selection probability as desired.

Several modifications are possible to produce a further speedup, at the cost of relaxing the constraint that the result be a true dominating set of the CCCD. Subsampling has been mentioned. If one subsamples from X the result is a set which does not necessarily dominate. If one subsamples from Y the result is a set of balls which are not necessarily "pure" (that is, contain only X observations). Another approach would be to stop computation when D reaches a certain size, or after a prespecified amount of time. This will result in a set which does not necessarily dominate. The intuition is that, provided the number of points covered is a large percentage of the full data set, the observations missed are those which cover few observations and are therefore outliers. Removing these tends to have a positive, rather than negative, effect on the classifier (Priebe et al., 2003).

Other modifications can be made to the algorithm to attempt to make it more efficient in terms of γ', at the expense of computations. For example, one could insist, early on in the processes, that only balls covering at least a minimum number of observations will be retained, or balls of at least some minimum radius. This ensures that large balls are discovered earlier than they would otherwise be, and can eliminate small, unnecessary balls. Unfortunately, this comes at the expense of further iterations, and so results in a slower algorithm.

Note that one advantage the fast algorithm has over the naive greedy algorithm is the ability to stop the algorithm without complete loss. Since the greedy algorithm requires the calculation of the interpoint distance matrix, no dominating elements can be selected until this calculation is complete. The fast algorithm is a sequential algorithm. Thus, one can stop it at any time and obtain the dominating elements found to that point.

Another speedup that both algorithms can use to their advantage is the fact that the distance calculations need not be performed completely. For the calculation of the radius, we need to compute, for a given x, the minimum of $\rho(x, y)$ over all $y \in Y$. The d term sum for computing the distance between an x observation and a y observation can be terminated as soon as the sum exceeds the current minimum for the value of $\rho(x, Y)$. This same trick can be used when computing the interpoint distance matrix for X, since this matrix is needed only in order to determine which observations are covered by each ball. Thus, the observation can be marked "not covered" as soon as the sum exceeds the radius for the ball. This improvement effects only the factor d in the order calculations.

Rather than selecting the dominating elements at random, one could select a subset of the data, compute a dominating set, then repeat, combining the resultant dominating sets. A version of this is as follows.

ALGORITHM 2.

Set C to a vector of n zeros, and $D = \emptyset$.
Fix N at some manageable value less than n. Partition X into sets of size at most N.
While C contains a zero **Do**:
  Select the next N observations from X, call these X'.
  Compute the dominating set D' for (X', Y).
  Mark with a 1 the positions of C corresponding to those X items covered by a ball
    in D', and add those elements of D' which cover new observations to D.
Return D.

This algorithm stops when all the observations are covered. Like the previous algorithm, this is sequential, and thus can be stopped after a fixed amount of processing time, returning an approximate dominating set. This algorithm performs comparably to the random algorithm, and will not be discussed further in this chapter. More information can be found in Marchette (2004).

6. Further enhancements

The above complexity calculations assume that all interpoint distances need be calculated in order to find the nearest neighbor. There are nearest neighbor algorithms that do not require this, and these can easily be incorporated in any of the above algorithms. This does not alter the fact that fast algorithms are needed for sufficiently large data sets.

One of the most popular techniques utilizes kd-trees (see Bentley (1975), Friedman et al. (1977), Moore (1991) for more details). In its simplest form, a node in a kd-tree consists of the following fields:

- An observation. This defines the splitting threshold, and also corresponds to the value contained in a leaf node.
- A split variate. This is an integer corresponding to the variate on which to split.
- A left node. This is a pointer to a kd-tree.
- A right node. This is also a pointer to a kd-tree.

Other information can be stored at a node. For instance, one could store the defining corners of the bounding rectangle for the data below the node. One can also store summary statistics for the data below the node, the class or other attributes of the observation, etc.

Figure 9 depicts a small tree and the regions defined by the tree in the plane. Consider using this tree to find the nearest neighbor to the point $x = (6, 2)$ amongst these seven points. First we traverse the tree to find the region in which x lies: right, left, left. So we have done three comparisons of single values. We are now in the lower right section. We need to test this section, the one above, and the one to the left. This can be done via moving up and down the tree. The key point is that we never have to check the distance from x to $(7, 7)$. This trivial example illustrates the basic idea. Further efficiency, in

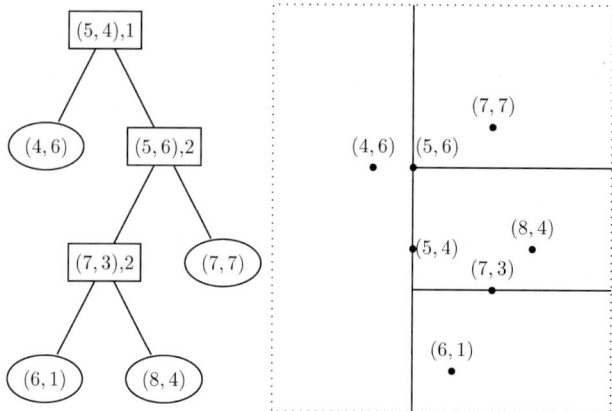

Fig. 9. A kd-tree for seven observations. Leaf nodes are denoted by ovals. Nodes are indicated by their observation and split, except for leaves which have no split. The regions defined by the tree are depicted on the right.

larger trees, can be obtained by geometric reasoning. In this simple example, by noting that the distance between x and (6, 1) is 1, and hence the circle is contained in the region of the leaf, we need not test any of the other regions. Thus, we have found the nearest neighbor with three floating point tests, a single distance calculation, and a containment test, in place of seven distance tests.

Searching the kd-tree for a nearest neighbor to a newly presented value x proceeds as follows. First, the tree is searched to find the leaf node into which x falls. The distance between x and this leaf is computed. The algorithm then proceeds back up the tree. At the parent node to the leaf, it is determined whether the other node could contain a nearer neighbor: if the current nearest distance is smaller than the split difference, then there cannot be a closer point down the other branch; otherwise the other branch is searched. The algorithm then proceeds up the tree, until the root node is reached, at which time the algorithm terminates.

By using kd-trees, the nearest neighbor calculations can be sped up significantly, in any of the above mentioned algorithms.

7. Streaming data

The above algorithms focused on finding a single cover. They also required the full data set to be available, although Algorithm 2 clearly could be used in the case where one of the classes was to be streaming: observations would be presented to the algorithm continuously. We now look at an algorithm for constructing CCCD-like classifiers on streaming data.

Assume that the observations will be obtained sequentially. Start with covers for the two classes (say, defined by a small data set, or defined a priori). As each observation is presented, determine whether it falls inside a current ball of its class or not. If not, it

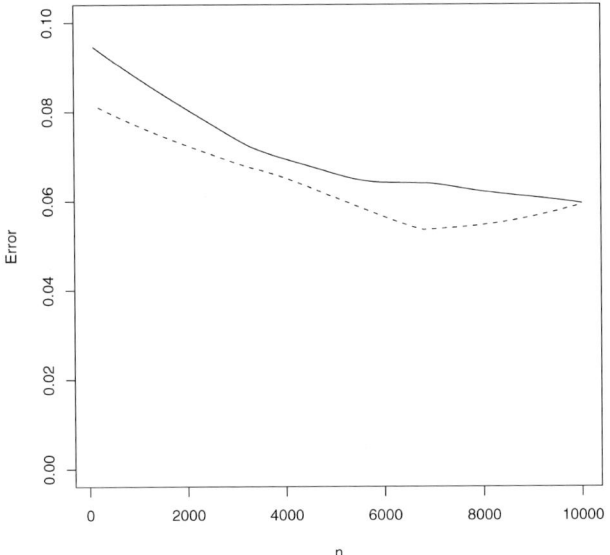

Fig. 10. Errors for the streaming version of the CCCD, Algorithm 3, applied to uniform data: $X \sim U[0, 1] \times U[-1, 1]^5$, $Y \sim U[-1, 0] \times U[-1, 1]^5$. The solid curve corresponds to the error for the CCCD, the dotted for a nearest neighbor classifier.

becomes the center of a new ball. The radius is set to the distance to the centers of the other class cover. The balls of the other cover are adjusted, if necessary, so that they do not cover the observation.

ALGORITHM 3.

Initialize covers S_X and S_Y.
Do
  Select the next observation z with class c_z.
  If there is no $B \in S_{c(z)}$ with $z \in B$, then $S_X = S_X \cup \{z\}$, with radius
    $r(z) = \rho(z, S_{!c(z)})$.
  Foreach $B \in S_{!c(z)}$ such that $\rho(z, B) < r(B)$, set $r(B) = \rho(z, B)$.

In this, $!c(z)$ is used to indicate the other class from that of z, and the distance between a point and a ball is the distance from the center of the ball to the point. The distance between a point and a set of balls is the minimum distance from the point to the centers of the balls. Once again, the covers can be reduced by eliminating balls wholly contained within other balls.

Let $X \sim U[0, 1] \times U[-1, 1]^5$ and $Y \sim U[-1, 0] \times U[-1, 1]^5$. Figure 10 shows the results of running Algorithm 3 on these data. For each observation, the current classifier is used to classify the observation, and a nearest neighbor classifier using all the data to date is also used. These allow the calculation of error curves which are subsequently smoothed and displayed in the figure. As can be seen, the streaming CCCD algorithm

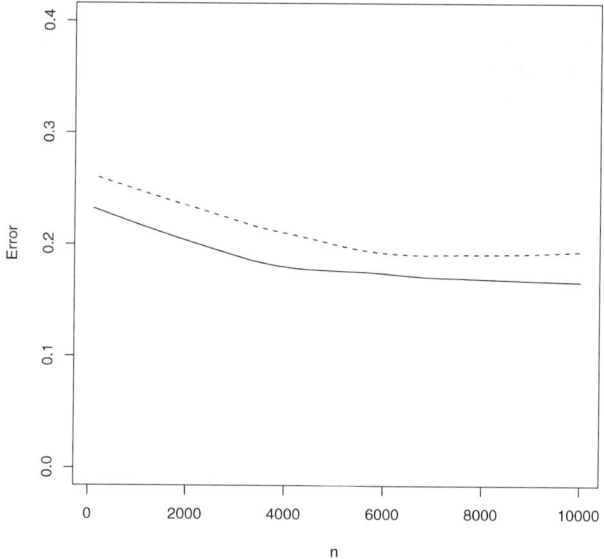

Fig. 11. Errors for the streaming version of the CCCD, Algorithm 3, applied to normal data. The solid curve corresponds to the error for the CCCD, the dotted for a nearest neighbor classifier.

performs quite well in this case. The final complexity after $n = 10\,000$ observations is $349 + 336 = 680$ balls in the cover.

Figure 11 depicts the same experiment with the data distributed as

$$X \sim N\left(\left(1, \frac{1}{2}, \frac{1}{3}, \ldots, \frac{1}{20}\right), I\right)$$

and

$$Y \sim N\left(\left(-1, -\frac{1}{2}, -\frac{1}{3}, \ldots, -\frac{1}{20}\right), I\right).$$

In the next sections we will consider applications of the fast algorithms discussed in Section 5 on simulated and real data.

8. Examples using the fast algorithms

A set of experiments were performed with simulated data in order to illustrate the fast algorithms of Section 5. Figure 12 shows the distributions used in the examples. The X observations were drawn from two squares, while the Y observations were drawn from the remainder of the square containing the X support. Squares were chosen (instead of circles) so that γ would grow as n and m increased.

Both the greedy algorithm and Algorithm 1 were coded in C, using the fast distance calculation algorithm mentioned above, where the calculation is stopped if it exceeds the current radius. The simulations were run on the same machine, a 750 MHz Athlon running RedHat Linux 7.1, with 512 Mb of RAM. The memory of the machine was

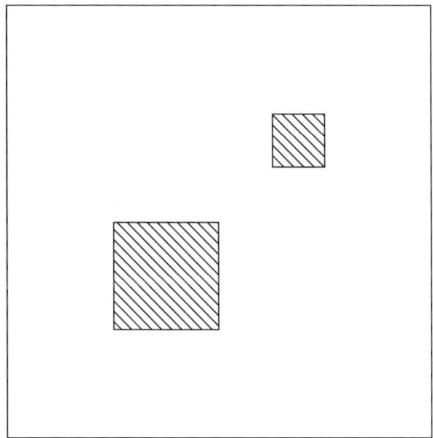

Fig. 12. Distribution of X (diagonal lines) and Y (clear) for the simulation examples.

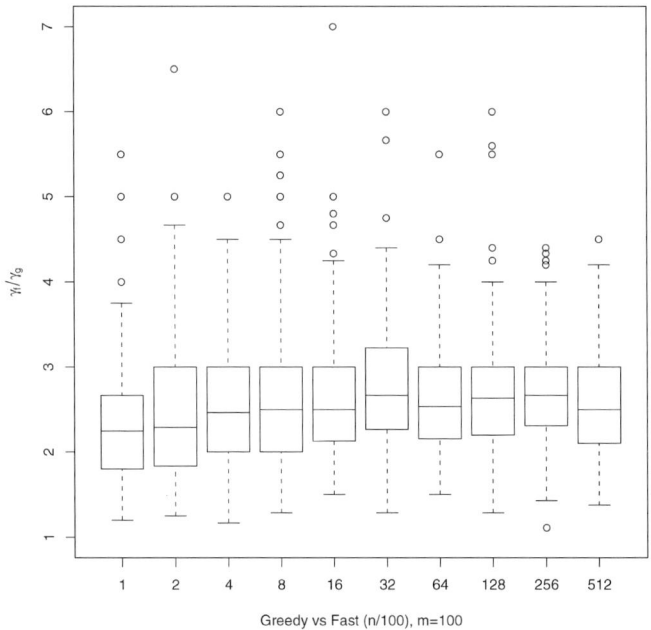

Fig. 13. Ratio of the estimated γs for $n = 100, 200, \ldots, 51\,200$ and $m = 100$. In each case 100 runs were made, and the greedy algorithm estimate (γ_g) is compared to that produced via the fast algorithm (γ_f).

sufficient that no swapping was necessary for either algorithm. We consider ratios of times, rather than absolute times, to reduce the dependence of the results on the actual hardware used. Only the actual time for the calculations of the dominating sets was computed, so that any time generating data or writing results to disk was ignored. The simulations were during a time in which no other programs, besides the operating system, were executing.

Figures 13 and 14 depict the results of a series of runs comparing the greedy and fast algorithms. In this experiment, the number of Y observations, m, is fixed at 100. Figure 13 shows a box plot of the ratio of the sizes of the dominating sets, showing that the fast algorithm produces larger dominating sets by a factor of slightly more than 2. In Figure 14 we see that the fast algorithm is significantly faster than the greedy algorithm in this case. This figure depicts the log of the ratios of the means of the time for the fast versus the greedy algorithm. As can be seen, the fast algorithm provides a considerable improvement in speed, with only about a factor of two increase in dominating set size over the greedy algorithm, in this example.

To see how the algorithm performs as a function of m, we consider two further experiments. Figures 15 and 16 depict the same experiment as above using $m = 1000$. Figures 17 and 18 show the results of the same basic experiment with $m = n$. The ratio of the γs is small, yet the speedup is significant.

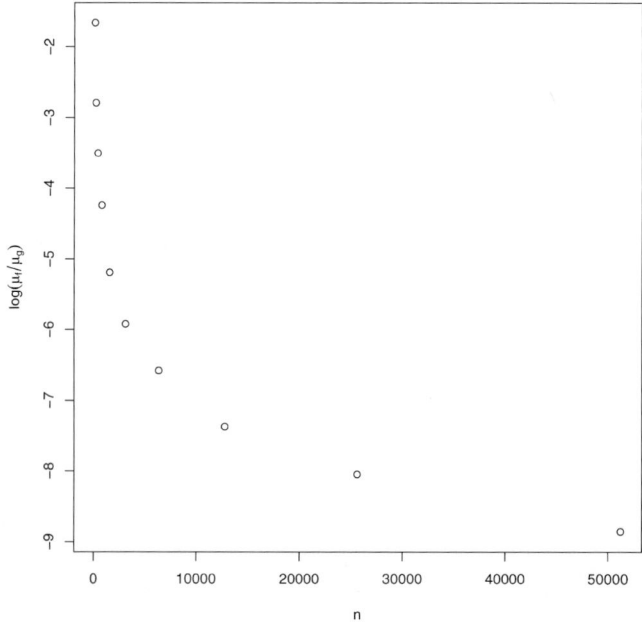

Fig. 14. Ratio of the run time of the greedy algorithm versus the fast algorithm for $n = 100, 200, \ldots, 51\,200$ and $m = 100$. In each case 100 Monte Carlo replications were performed.

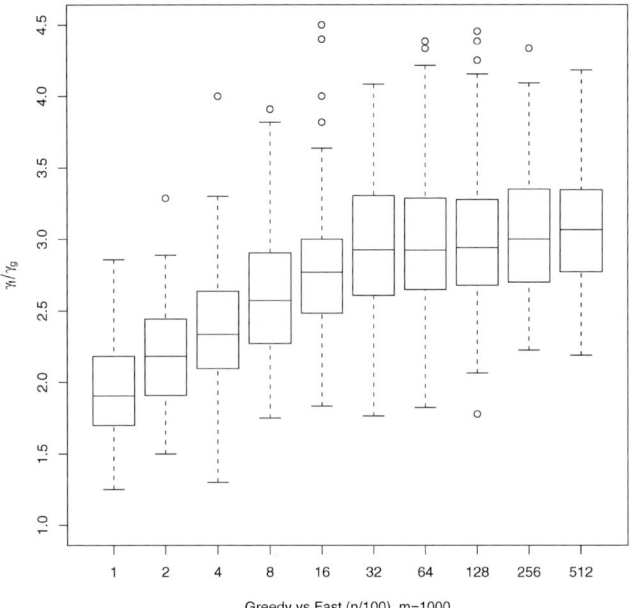

Fig. 15. Ratio of the estimated γs for $n = 100, 200, \ldots, 51\,200$ and $m = 1000$. In each case 100 Monte Carlo replications were performed, and the greedy algorithm estimate (γ_g) is compared to that produced via the fast algorithm (γ_f).

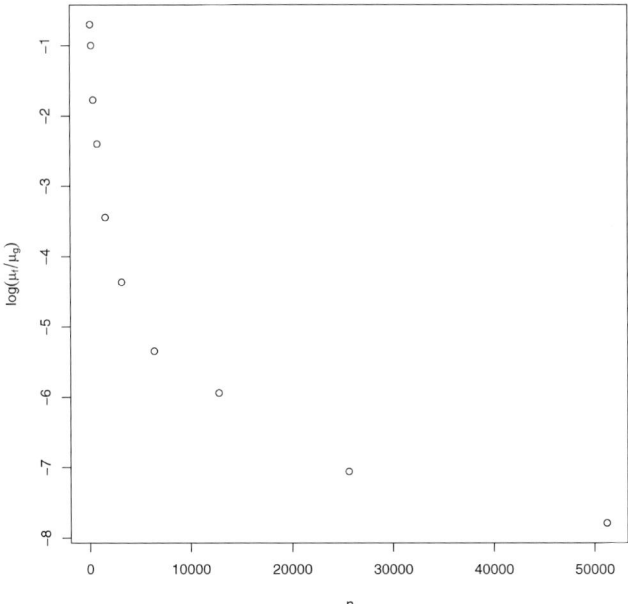

Fig. 16. Ratio of the run time of the greedy algorithm versus the fast algorithm for $n = 100, 200, \ldots, 51\,200$ and $m = 1000$. In each case 100 Monte Carlo replications were performed.

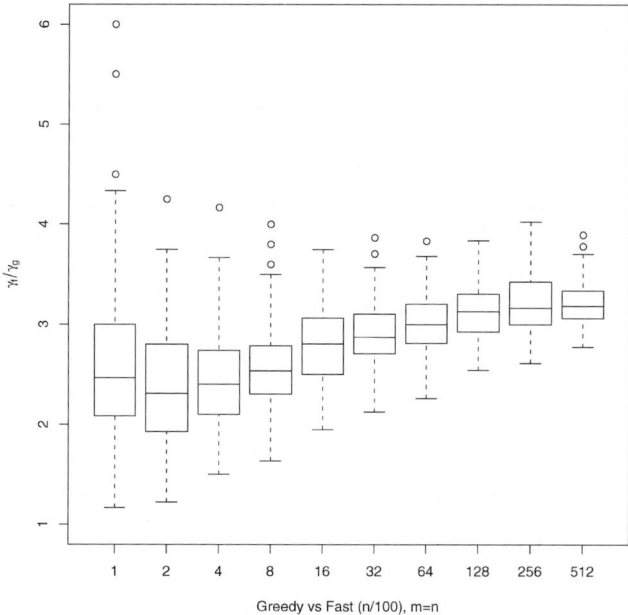

Fig. 17. Ratio of the estimated γs for $n = m = 100, 200, \ldots, 51\,200$. In each case 100 Monte Carlo replications were performed, and the greedy algorithm estimate (γ_g) is compared to that produced via the fast algorithm (γ_f).

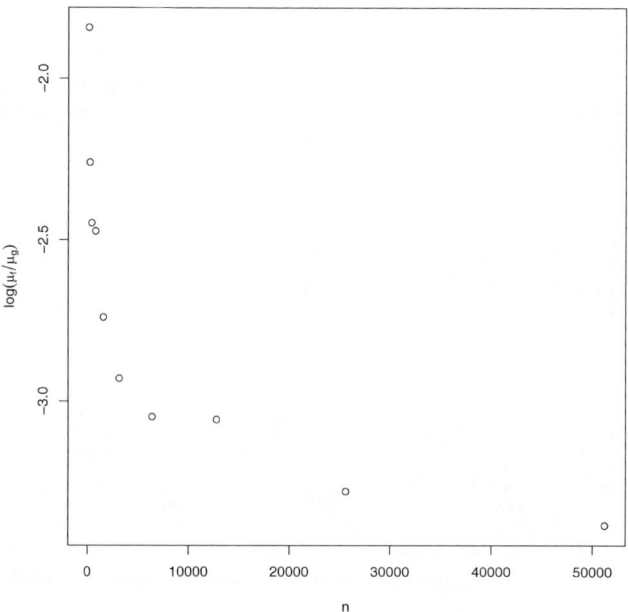

Fig. 18. Ratio of the run time of the greedy algorithm versus the fast algorithm for $n = m = 100, 200, \ldots, 51\,200$. In each case 100 Monte Carlo replications were performed.

9. Sloan Digital Sky Survey

A data set of astronomical objects measured by the Sloan Digital Sky Survey (www.sdss.org) was made available to us. The Sloan data consists of 2 091 444 X and 670 782 Y 5-dimensional observations. These five dimensions are the u, g, r, i, z magnitudes of the astronomical objects measured (see www.sdss.org for more information). For the first experiment we chose a random sample of 209 144 X and 335 391 Y observations. The algorithm was allowed to run until it reached a size of 100 for the dominating set. The results are depicted in Figures 19 and 20. 37 011 observations are covered, or 17.7%.

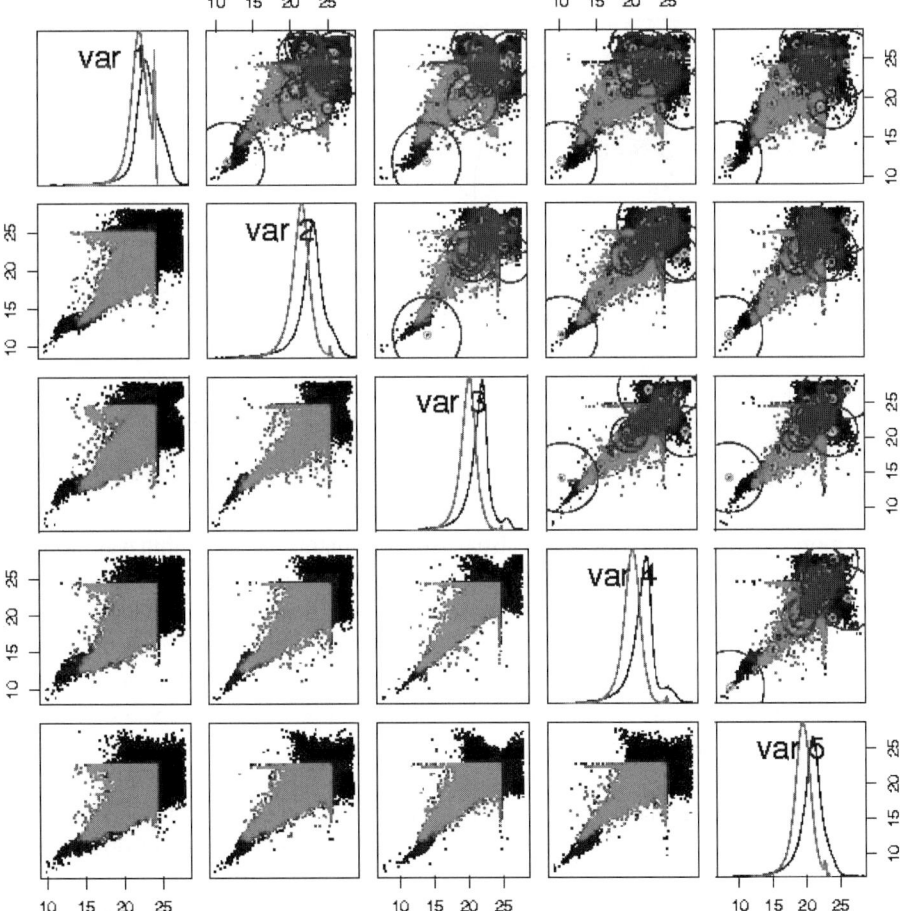

Fig. 19. Pairs plot of Sloan data. Circles are overlaid in blue in upper triangle, with centers indicated in green. Kernel densities of the two classes are plotted on diagonal. For a color reproduction of this figure see the color figures section, page 596.

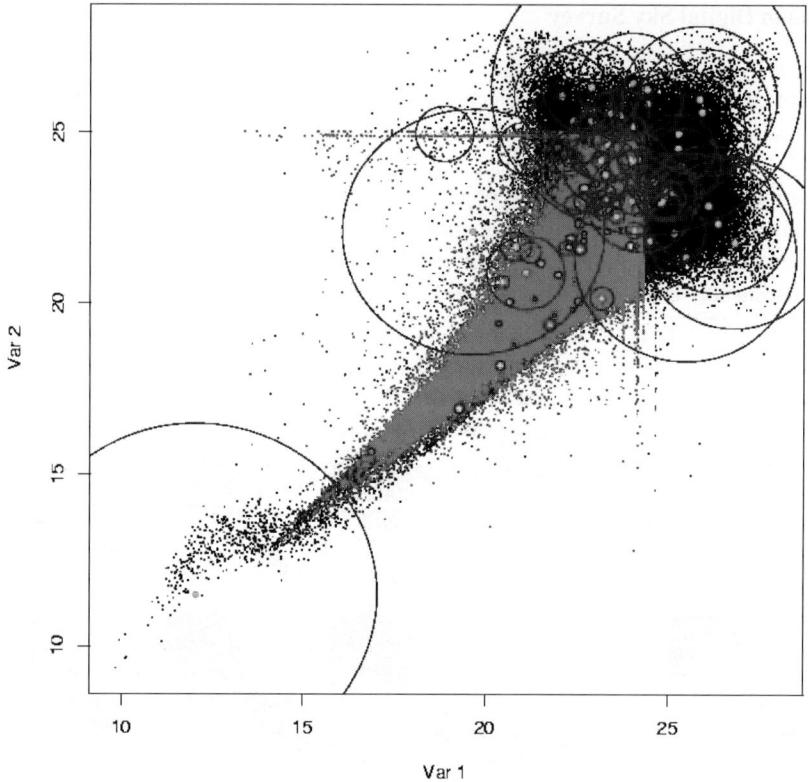

Fig. 20. Plot of Sloan data for variables 1 and 2. Circles are overlaid in blue, with their centers indicated in green. For a color reproduction of this figure see the color figures section, page 597.

A more extensive experiment was also performed. In this, the algorithm was modified to select uniformly from X at each stage of the algorithm, but in the early stages of the process, balls covering fewer than 100 observations were not accepted. This restriction was relaxed as the algorithm ran. All 2 091 444 X and 670 782 Y observations were used, and the algorithm was stopped when the D set had reached 1000 elements. These elements covered 1 183 440 observations, and so we refer to this as a sub-dominating set, emphasizing the fact that it does not dominate all the observations. The coverage of this sub-dominating set corresponds to slightly more than half of the X observations. Figure 21 shows a kernel estimator of the density of the radii for the 1000 elements in the set, and Figure 22 depicts a plot of the radius against the number of observations covered.

This is a particularly difficult problem for the CCCD, as a result of the nature of the data. The Y observations are a collection of "uncommon" objects, while the X observations are a data collection of a larger group of objections, containing the original Y observations. Thus the Y observations are very close to some of the X (since they are

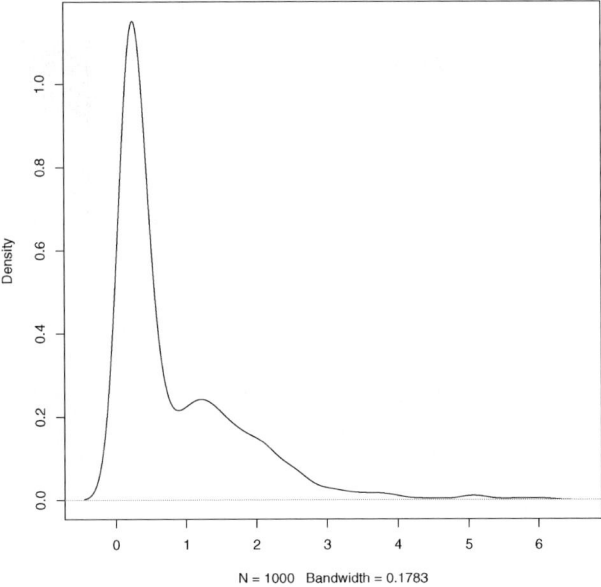

Fig. 21. Plot of the kernel density of the radii for the set of 1000 elements.

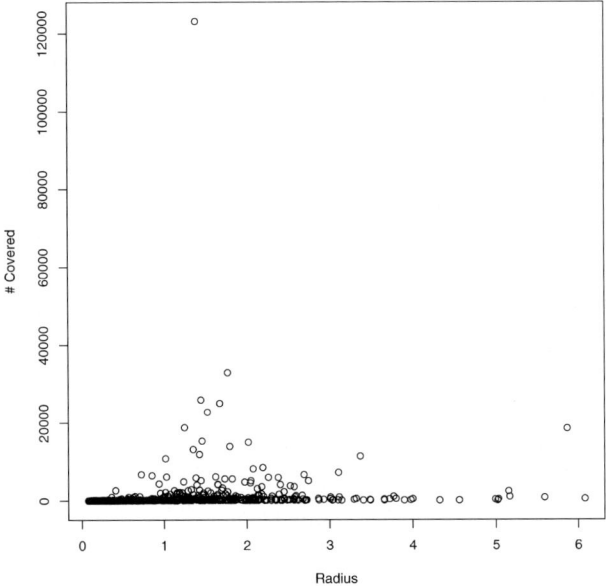

Fig. 22. Radius against number of observations covered for the 1000 element set.

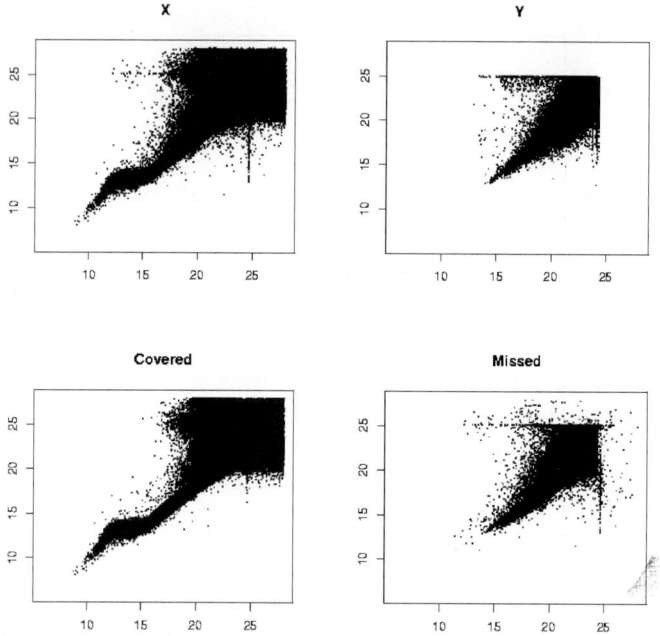

Fig. 23. Scatter plots of the first two variates for the Sloan data. Top: the original data; bottom: the data covered and missed by the 1000 point sub-dominating set.

a second sample of the same object). What is desired is a near-dominating set which covers most of the X observations.

Note that there is a group of over 120 000 observations contained in a ball of radius approximately 1.4, indicating that they are all a distance of less than 3 from the non-target class. There are 14 balls which contain more than 10 000 observations, 91 which contain more than 1000 observations. These 91 balls account for (cover) 593 613 observations, or approximately half of the observations. Note that these 91 balls cover four times as many observations as the 100 balls in the first experiment on these data. Clearly forcing the algorithm to choose balls that cover many observations has greatly improved the coverage of the set, at the expense of longer run time.

Figure 23 illustrates the coverage of the 1000 point sub-dominating set. For the first two variates, the full data set is plotted in the top two panels. The bottom two panels depict the observations covered by the sub-dominating set, and those missed. The observations covered do seem to be representative of the X observations, while those missed appear to closely match the Y observations, and hence are in regions of high overlap. This is what we expect from these data, since some of the X observations are resamples of the Y objects.

10. Text processing

Another area that generates huge, high-dimensional data is that of text processing. In this section we consider a problem of determining the topic of discussion of newsgroup postings.

The data consist of a subset of the data set of postings to 20 newsgroups available at the web site http://www.ai.mit.edu/people/jrennie/20_newsgroups/. We performed two experiments on subsets of these groups. In the first, the problem is to distinguish postings in religious discussion groups (alt.atheism, soc.religion.christian, talk.religion.misc) from those to political groups (talk.politics.guns, talk.politics.mideast, talk.politics.misc). In the second, we wish to distinguish religion from science (sci.crypt, sci.electronics, sci.med, sci.space).

There were 2424 posting to religious groups, 2624 to political groups, and 3949 to science groups. In both experiments the target set was the religious postings. A feature set was extracted based on word frequencies, and this resulted in a distance metric by which two documents could be compared. The fast algorithm was then used to obtain a dominating set. In the following, religions versus politics will be denoted RvP, and religion versus science RvS.

The dominating sets were fairly comparable in size. The dominating set for RvP was 612 objects, while that for RvS was 500. This is a reduction of complexity from 2424 to 500 or 612, corresponding to 21% or 25% of the observations needed. At the point in the algorithm at which the RvP sub-dominating set had 500 observations it covered $1956/2424 = 80.78\%$ of the documents. This indicates that the problem of discriminating between religion and science may be slightly easier than that of discriminating between religion and politics, but not by a great amount. The radii are depicted in Figure 24. The left histogram corresponds to RvP and the right to RvS. As can be seen, the radius distributions are comparable, indicating that the ball sizes are equally distributed

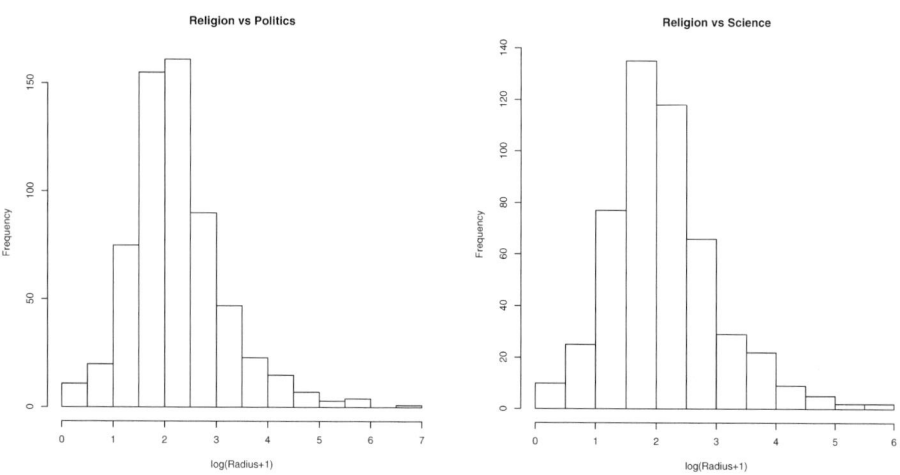

Fig. 24. Histograms of the log of the radii for the dominating elements for RvP (left) and RvS (right).

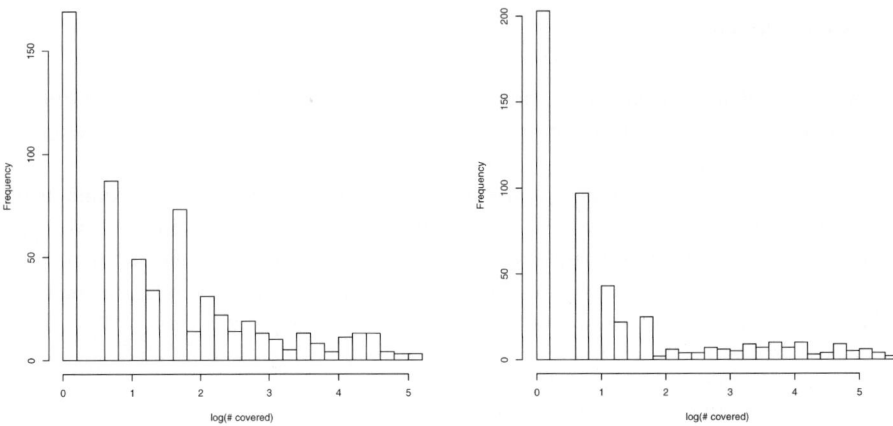

Fig. 25. Histograms of the log of the number of points covered by the dominating elements for RvP (left) and RvS (right).

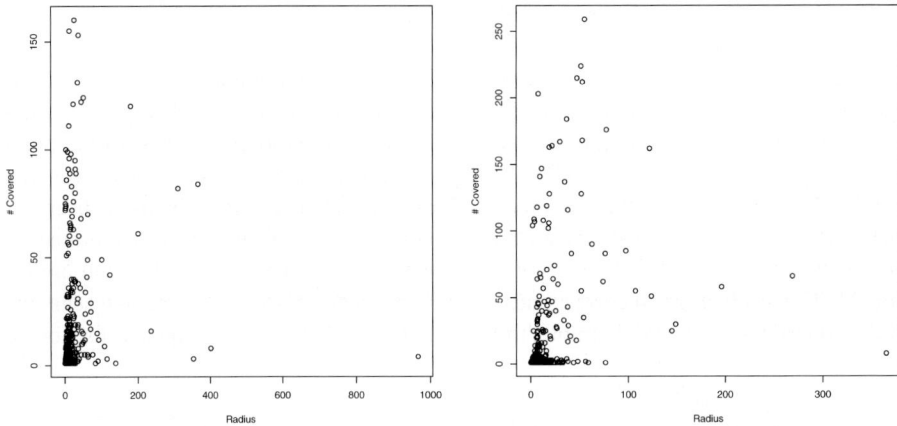

Fig. 26. Plots of radius against the number of points covered for the dominating elements for RvP (left) and RvS (right).

for the two dominating sets. A similar picture for the number of points covered is shown in Figure 25.

The plot of radius against number of observations covered for the two dominating sets is presented in Figure 26. Note that the balls covering the most observations are ones that have relatively small radii, indicating that a significant number of the observations tend to cluster near the discriminant boundary.

In Table 1 we consider the Y values that define the balls of the dominating sets (these are the Y observations closest to the center of the ball). Perhaps surprisingly, the political posting nearest to a religious posting is more likely to be in the guns newsgroup than the Mideast. The relatively few postings to the Middle East group may be a result of the fact that the religious newsgroups in the study represent primarily Christian, rather than

Table 1
The number of times an element of each newsgroup was the defining Y observation for a ball in the dominating set

RvP		RvS	
talk.politics.mideast	134	sci.crypt	106
talk.politics.misc	205	sci.electronics	109
talk.politics.guns	273	sci.space	123
		sci.med	162

Muslim, discussions. Less surprising is the fact that the nearest science posting is more likely to be in the space or medical newsgroup than the cryptography or electronics groups. The fact that there are any of the latter two in this set is in itself somewhat interesting.

This is a particularly difficult problem, since it uses only word frequencies to determine class. Note that the religion groups (particularly alt.atheism) often discuss scientific topics (such as the origin of life, the universe, and everything). Similarly, they also discuss many of the same topics discussed in political newsgroups.

11. Discussion

The problem of selecting a minimum dominating set for a digraph is hard. However, it is required for the CCCD classifier that a (nearly) minimum dominating set be found in order to reduce the complexity and potential for over-fitting of the classifier. Thus, for large data sets, we require a fast algorithm for the selection of a dominating set. In this article we have provided one such algorithm.

The fast algorithm works quite well in practice, as demonstrated by the simulations and examples. It tends to over estimate the size of the minimum dominating set, which is unavoidable. It has the attractive property that it can be stopped at any point to return a sub-dominating set, allowing one to specify a maximum run time for the algorithm while retaining a useful result should the time be exceeded.

Several modifications are possible, which may speed the algorithm in practice. The idea of selecting a subset of the data and using this along with the greedy algorithm to select the next ball to add to the set (Algorithm 2 above) is one that deserves further investigation. Modifying the subset selection method to select observations far from the region covered, as is done in the fast algorithm, may improve the algorithm substantially. A hybrid of the two methods that uses the second algorithm initially, then reverts to the first, might also improve performance.

Acknowledgements

The work of CEP was partially supported by Office of Naval Research Grant N00014-01-1-0011 and CEP and EJW were partially supported by Defense Advanced Research

Projects Agency Grant F49620-01-1-0395. David Marchette was partially supported by the Defense Advanced Research Projects Agency.

References

Bentley, J.L. (1975). Multidimensional binary search trees used for associative searches. *Comm. ACM* **18** (9), 509–517.
Bollobás, B. (2001). *Random Graphs*. Cambridge University Press, Cambridge.
Chvatal, V. (1979). A greedy heuristic for the set-covering problem. *Math. Oper. Res.* **4** (3), 233–235.
Friedman, J.H., Bentley, J.L., Finkel, R.A. (1977). An algorithm for finding best matches in logarithmic expected time. *ACM Trans. Math. Softw.* **3** (3), 209–226.
Khumbah, N.-A., Wegman, E.J. (2003). Data compression by geometric quantization. In: *Recent Advances and Trends in Nonparametric Statistics*. Elsevier, Amsterdam, pp. 35–48.
Marchette, D.J. (2004). *Random Graphs for Statistical Pattern Recognition*. Wiley, Hoboken, NJ.
Marchette, D.J., Priebe, C.E. (2003). Characterizing the scale dimension of a high dimensional classification problem. *Pattern Recognition* **36**, 45–60.
Moore, A.W. (1991). Efficient memory-based learning for robot control. Technical Report 209. University of Cambridge.
Parekh, A.K. (1991). Analysis of a greedy heuristic for finding small dominating sets in graphs. *Inform. Process. Lett.* **39**, 237–240.
Priebe, C., Marchette, D., Devinney, J., Socolinsky, D. (2003). Classification using class cover catch digraphs. *J. Classification*, 3–23.
Priebe, C.E., DeVinney, J.G., Marchette, D.J. (2001). On the distribution of the domination number for random class cover catch digraphs. *Statist. Probab. Lett.* **55** (3), 239–246.
Wegman, E.J., Khumbah, N.-A. (2001). Data compression by quantization. *Comput. Sci. Statist.* **33**. http://www.galaxy.gmu.edu/interface/I01/I2001Proceedings/EWegman/DataCompressionbyQuantization.pdf.

On Genetic Algorithms and their Applications

Yasmin H. Said

Abstract

Genetic computation is a rapidly growing field and can be applied to an array of subject areas, from optimization problems in medicine, data fitting and clustering in engineering, to acoustics in physics. Recently, it has been rapidly growing in aeronautics and space applications due to the fact that genetic algorithms provide optimal solutions to complex problems; the same in business, and much more. Further, genetic algorithms have provided trend spotting, scheduling, and path findings to name a few in a vast array of fields. Moreover, the objective of genetic algorithms is to optimize the best possible algorithm in any subject area. In this chapter, we describe the basic concepts and functionality of genetic computation.

Keywords: Evolutionary computation; Evolutionary algorithms; Evolutionary programming; Evolution strategies; Genetic algorithms

1. Introduction

Under the umbrella of evolutionary computation (EC) also referred to as evolutionary algorithms (EAs) are the areas of evolutionary programming (EPs) and evolution strategies (ESs). Each of these methods of evolutionary computation simulate the process of evolution through the mutation, selection, and/or reproduction processes and rely on perceived performance of individual structures assigned by the environment. Evolutionary algorithms support population structures, which progress to the rules of selection using genetic operators. Genetic operators determine which structures will move on to the next level and which will not. In essence, the individuals in the population obtain a degree of fitness from the environment. Reproduction concentrates on high fitness individuals.

As early as the 1950s and 1960s, researchers have been looking for ways to apply Darwinian theories of "Survival of the Fittest" and "Natural Selection" to machine learning. These researchers theorized that the natural evolutionary process, in its perfection, is a model that machine learning should logically follow. All of the language involved in evolutionary computation is derived from natural evolutionary processes. In fact, in studies detailing various applications of evolutionary computation, one is hard-pressed to remember that these processes are really only functioning for machines because the language is so vividly biological.

This very complex set of computational processes necessarily requires much more space than a synopsis such as this may allow. Therefore, this chapter will deal uniquely with processes directly involved with Genetic Algorithms (GAs) and its applications for scientists among different disciplines. This synopsis seeks to show the viability of GAs in an interdisciplinary sense. The applications are far reaching; computational scientists could work with political scientists and use GAs to hypothesize the evolutionary course cooperating nations might take in response to a threat; or they could work with biologists to predict patterns that cancer cells might follow in response to a new drug. In any case, GAs take into consideration all historical information involved in a problem, all hypotheses, which might be considered in solving it, as well as random information, which is added to keep the process as natural as possible.

2. History

In the middle of the twentieth century some computer scientists worked on evolutionary systems with the notion that this will yield to an optimization mechanism for an array of engineering queries (Mitchell, 1998). GAs were invented and developed by John Holland, his students and his colleagues at the University of Michigan. His team's original intentions were not to create algorithms, but instead to determine exactly how adaptation occurs in nature and then develop ways that natural adaptation might become a part of computer systems.

Holland's book *Adaptation in Natural and Artificial Systems* (1975) set forth the lexicon from which all further dialogue concerning GAs would be developed. In essence, his theoretical framework provided the point of reference for all work on genetic algorithms up until recently whereupon it has taken on a new direction, given new technology. GA, he stated, moves one population of bits (chromosomes and genes) to a new population using a type of "natural selection" along with genetic operators of crossover, mutation and inversion (all biological functions). These operators determine which chromosomes are the fittest and thus able to move on. Although some less fit chromosomes do move forward, on average the most fit chromosomes produce more offspring than their less fit counterparts. Biological recombination occurs between these chromosomes, and chromosomal inversion further completes the process of providing as many types as possible of recombination or "crossover".

This remarkable quality that genetic algorithms have of focusing their attention on the "fittest" parts of a solution set in a population is directly related to their ability to combine strings, which contain partial solutions. First, each string in the population is ranked to determine the execution of the tactic that it predetermines. Second, the higher-ranking strings mate in couplets. When the two strings assemble, a random point along the strings is selected and the portions adjacent to that point are swapped to produce two offspring: one which contains the encoding of the first string up to the crossover point, those of the second beyond it, and the other containing the opposite cross.

Biological chromosomes perform the function of crossover when zygotes and gametes meet, and so the process of crossover in genetic algorithms is designed to mimic its biological nominative. Successive offspring do not replace the parent strings; rather, they replace low-fitness ones, which are discarded information at each generation in order that the population size is maintained.

Although the mechanics of reproduction through recombination or crossover are relatively simple to understand, they are underestimated in terms of their power for computer applications. After all, it is chance that controls biological evolution and we have plenty of evidence that it is quite perfectly powerful, even in its simplicity. The mathematician, J. Hadamard, recognized that "there is an intervention of chance but a necessary work of unconsciousness ... [i]ndeed it is obvious that invention or discovery, be it in mathematics or anywhere else, takes place by combining ideas" (Hadamard, 1949).

Combinations of ideas, as has been discussed, occur additionally through some mutation of pieces (determined through chance) of information (genes) thus changing the values of some segments (alleles) in the bit strings (chromosomes). Mutation is often misunderstood in the computer/biological lexicon, partly because it is often seen as an event that is abnormal or destructive; visions of giant, radioactive tomatoes come to mind. However, mutation is needed because, even though reproduction and crossover are effective in searching out fit ideas and recombining randomly, unchecked, those systems might overpopulate population samples. Thus, mutations and mutation rates of data are also accounted for in the analysis of population data. Offspring of the mutations in genetic operations are carried on in equations. This recombination of differing parent hypotheses ensures against the evolutionary problem of overcrowding, where a very fit member of the population manages to succeed all other members of the population in creating progeny. This is important because in biological applications as in computational applications, when the diversity of the population is reduced, operational progress is slowed.

3. Genetic algorithms

Genetic algorithms are search and optimization tools that enable the fittest candidate among strings to survive and reproduce based on random information search and exchange imitating the natural biological selection. The production of new strings in the following generations depend on the initial "parent" strings, the offspring "child strings" are created using parts and portions of the parent strings; the fittest candidate, preserving the best biological features and thus improving the search process. The whole search process is not completely a random process; genetic algorithms utilize the chronological information about old strings in order to be able to produce new and enhanced ones (Goldberg, 1989).

Genetic algorithms are important because they maintain robustness through evolution. This is at the core of the search process and demonstrates the performance of the algorithm under various circumstances. Robustness in genetic algorithms ensures that simulated models are diverse. Designers can use this robustness to decrease high costs when artificial systems function efficiently, thus creating high level adaptive processes.

Computer system designers, engineering system designers, and software and hardware designers will see an advancement in the performance, usefulness and robustness of programs through use of the natural model. Significant properties of the natural system such as natural healing, control, management, leadership and reproduction are all benefits that the artificial model borrows from its natural parent model.

Studied theoretically and tested experimentally, genetic algorithms are justified not only because they mirror adaptation and survival but also because they prove themselves robust and efficient in complex systems. Through their effective and efficient role in solving search and optimization problems, genetic algorithms are becoming more popular among scientists and engineers and as such, the number of genetic algorithm applications is increasing dramatically. Genetic algorithms require simple computations and yet offer improved and powerful search methodology. Moreover, they do not impose restrictions on the domain of the search or optimization problem; restrictions found in calculus-based approaches such as continuity, smoothness of the function and the existence of multiple-derivatives.

Traditional search and optimization methods can be categorized into two main types: calculus-based and enumerative. Search and optimization methodologies are not necessarily compared at this point. Rather, it is more enlightening to examine the lack of robustness using these techniques in comparison to the effectiveness of genetic algorithms.

3.1. Calculus-based schemes

Perhaps the most popular optimization schemes are the calculus-based; they can be classified as direct search methods and indirect search methods. The latter optimizes an objective function subject to certain constraints (linear and/or nonlinear equations). The Lagrange methodology, which sets the gradient equal to zero, is a methodology commonly used. This approach is a generalization of the idea of searching for local extremal points taught in elementary calculus.

The procedure of finding local maxima or minima operates on a smooth and unconstrained function and starts by searching for those points that have derivatives equal to zero. Direct optimizing techniques depend on the value of the local gradient in order to direct searching to find local extremum. Although these techniques were modified, altered and enhanced to become more powerful, there are still some robustness concerns.

For one reason, this technique works best on the local scale; the search for a maximum or minimum is performed within the neighborhood of the present point. Optimizing the function around the local maximum point, i.e. searching for the point having a slope of zero, may overlook the absolute maximum point. Moreover, once the local maximum is attained, it may be harder to locate the absolute maximum unless one randomly initializes the procedure again or through a different approach. Ideal optimizing and search problems containing excellent and easily modeled objective functions with ideal constraints and continuous derivatives have attracted mathematicians in the past. However, real-world application problems are full of discontinuities, peaks, huge multimodal, and chaotic search domains making them difficult to model. See Figure 1.

Thus, conditions on continuity and the existence of derivatives necessary for optimization and search methodologies limit the use of those techniques to a certain domain

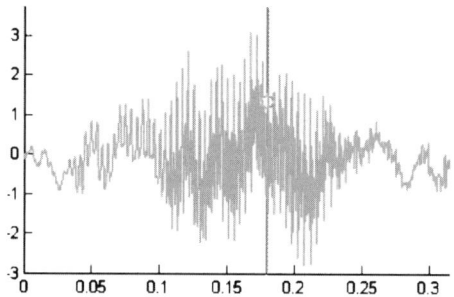

Fig. 1. A complex, multimodal search domain.

of problems. This brings us to the conclusion that calculus-based schemes lack the power to provide robust search through optimization techniques.

3.2. Enumerative-based optimization schemes

Enumerative techniques are less popular than calculus-based techniques. They depend on either searching a finite continuous highly focused region, or an infinite disconnected search region whereby the optimization algorithm evaluates every point in the region consecutively. Although the enumerative algorithm in and of itself is simple in nature and sounds like a reasonable process, it is inefficient and thus cannot compete for the robustness race. In reality, enumerative domains of search are considerably enormous or too absolute in focus to perform the search process successfully. Therefore, enumerative-based algorithms are not desirable.

3.3. Genetic algorithms – an example

Genetic algorithms are a new generation for optimization and search techniques. They have become very popular because of their efficiency, power, utility, and accessibility. There are several types of genetic algorithms (GAs) designed to deal with various simple and complicated problem domains that range from an array of disciplines, including engineering, information technology, medicine and business. Due to the broad scope of the GAs and their applications, they have also been adapted to decipher and solve simple and complex GAs, including and obtaining optimal solutions to fuzzy genetic algorithms and multi modal, and multiple optimal solutions to an array of problems (Deb, 1999).

Consider a cylindrical closed soup container (see Figure 2), which can be determined by the two variables, height h and radius r.

For simplicity, we may assume that the other parameters exist, but are nuisance factors including the thickness, shape and material characteristics. Without loss of generality, suppose the container has volume V of at least 0.1 liters. Our purpose is to minimize the material cost C of this container. So, now translate this information into an optimization model consisting of objective and constraint functions called the non-

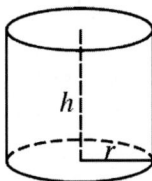

Fig. 2. Cylindrical container with radius r and height h.

linear programming problem. The problem is now:

$$\text{Minimize} \quad C(r, h) = \gamma(2\pi rh + 2\pi r^2) = 2\pi r\gamma(h + r),$$
$$\text{subject to} \quad V(r, h) = r^2 h\pi \geqslant 0.1 \quad \text{and}$$
$$\text{bounding variables:} \quad r_{\min} \leqslant r \leqslant r_{\max}, \quad h_{\min} \leqslant h \leqslant h_{\max},$$

where γ is the material cost per cm^2, r, h are the height and radius, respectively, in cm^2.

Optimize cost C, a function of h and r satisfying the constraint on V. Suppose we let the cost of the container be 5 units. To proceed with the genetic algorithm, we transform from decimal system to binary system, and hence we will work on binary strings. Suppose we have two 5-bit binary strings corresponding to the two variables h and r, which gives a total string length of 10. Also, assign the container a height of 7 cm and radius of 3 cm.

$$\underbrace{00011}_{r} \underbrace{00111}_{h}$$

The height and radius are given by $(r, h) = (3, 7)$ cm. The binary chromosomal "string" representation is given by 00011 00111. In the 5-bit representation, both the h and r have a lower limit of 0 and an upper limit of 31, which implies that there are $2^5 = 32$ distinct possible solutions. This gives the genetic algorithm the integer interval $[2^0 - 1, 2^5 - 1] = [0, 31]$ from which to select. On the other hand, the genetic algorithm is; in general, not limited to this set of integers and requires larger sets that contain both integer and non-integer quantities. This can be achieved simply by considering strings of longer lengths providing wider lower and upper limits. The following is a transformation to obtain the value of any parameter:

$$t_n = t_n^{\min} + \frac{t_n^{\max} - t_n^{\min}}{2^{k_n} - 1} \cdot \varphi(l_n)$$

where t_n is the nth parameter value, $t_n^{\min} = 0$ is the minimum value the nth parameter can hold, $t_n^{\max} = 31$ is the maximum value the nth parameter can hold, $k_n = 5$ is the length of the nth string, $\varphi(l_n)$ is the decoding value of the string l_n. This transformation has the subsequent properties:

(i) Variables can hold positive values or negative values.
(ii) A well-defined string length can be used to obtain a variable with specified degree of accuracy.
(iii) Various string lengths can be used to obtain different accuracies for unlike variables.

The coding stage of the variables allows us to obtain a pseudo-chromosomal description of the model in a binary string. In this problem that is being modeled, one may biologically represent the container with radius $r = 3$ cm and height $h = 7$ cm using the 10-bit chromosome "string". In the case of the soup can, the container is the phenotypic representation consisting of 10 genes, which build the simulated chromosome. The study the leftmost bit "gene" of the chromosome of radius r allows one to observe how the shape "phenotype" of the container is determined by the 10 genes. If the value of this bit, perhaps the most important bit "gene", is zero then the radius can hold values that range between 0 cm and 15 cm. If the value is one, then the radius can hold values that range between 16 cm and 31 cm. As a result, this gene "bit" controls the narrowness of the container. Thus, the zero value means the container is narrow and a one value means the container is wide. Any permutation of the other genes may result in the different characteristics of the container, some of which are desirable properties and others might not be that useful. Hence, one has transformed the produced children into a string representation. Furthermore, one proceeds in the same process applying genetic operations to produce improved and desirable offspring, which requires us to define and assign fitness values of goodness; in the form of a string, accompanied with each child, offspring.

Once the genetic algorithm finds a solution to the optimization problem, it is essential to rerun the algorithm using this solution as the new values of the variables. It is recommended to evaluate the objective function and the constraint using the produced string "solution". If the optimization problem has no constraints, then one assigns the fitness of the string a function value that corresponds to the solution of the objective function. In general, this value equals to the value of the objective function. As an illustration, consider the above example where the container is characterized by the 10-bit string, which has a fitness value of $f(l) = 2 \cdot \pi \cdot 3 \cdot 0.0254 \cdot (7 + 3) = 5$, where $\gamma = 0.0254$. Again, the goal is to minimize cost of the objective function. Therefore, when the fitness value is small, a better solution is guaranteed.

3.4. Operational functionality of genetic algorithms

Genetic algorithms function in a totally different way compared to the traditional search and optimization algorithms. The following steps encapsulate the whole process of genetic algorithms. First, the genetic algorithm starts off with an arbitrary initial solution set in order to perform the search process, as an alternative to the one solution set in the classical search. In the container problem, the initial solution set is a random set of binary strings. The algorithm proceeds by randomly generating a population of strings to be evaluated in the nonlinear programming problem. After that, the algorithm checks the validity whether a given termination condition is fulfilled or not. If that condition is satisfied the algorithm stops; however, if not, it generates a new adapted population of strings with the aid of genetic operators finishing the first iteration and declaring the first generation of strings.

There are three major operators: the *reproduction operator*, the *crossover operator*, and the *mutation operator*.

3.4.1. The reproduction operator
The reproduction operator functions by considering promising offspring "strings" to be potential candidates for reproduction and discarding the weak ones without changing the range of the population. Therefore, the reproduction operator operates in the following manner:

(i) Among the diverse strings in the population, the operator detects only potential candidates that qualify for reproduction.
(ii) The operator then creates manifold versions of these candidates.
(iii) It replaces the superior candidates for the potentially less or inferior candidates.

Reproduction preserves the size of the population. There are some frequently used techniques in identifying good candidates for reproduction, which include *rank grouping*, *proportionate grouping*, and *tournament grouping*.

The purpose of ranking is to prevent rapid convergence. According to Baker (1987), rank selection involves individuals in a particular population to be positioned according to their expected value and fitness, but not absolute fitness. The main reason we do not have to scale the fitness is due to the fact that the absolute fitness are often times masked. Hence, there are pros and cons to this issue. One advantage to this removal of the absolute fitness is when using it, one may obtain convergence problems, while a disadvantage maybe vital to identify that an individual is far better than its closest competitor. In addition, positioning eludes in providing the farthest and greatest distribution of the progeny to a smaller cluster of extremely qualified individuals. Therefore, this diminishes the selection burden upon an excessive fitness variance. Moreover, it furthers the selection burden upon a minimal fitness variance. Further, it does not matter if the absolute fitness is elevated or dilapidated, the ratio of the expected value will remain to be the equivalent. The ranking technique (i.e. the fitness proportionate selection) necessitates two elapses for each origination. One of the elapses is when a computation of the expected value of the fitness is preformed for each individual. In addition, it lapses to perform the mean fitness. This technique necessitates the arranging the full population by order. This is an extremely time-consuming technique.

The technique of tournament selection is very much like ranking. The advantage of using this technique is that it is more effective in parallel implementation. Tournament selection does not consume much time, it is much more efficient than the ranking method with regards to the time computation and it is more acquiescent. A selection of two individuals is made from the indiscriminate population. An unbiased number, n is selected from population, such that n is anywhere between zero and one, and m is a parameter. The better of the two selections is chosen to be the parent and it is defined by $n < m$. If this is not the case, then the reduced amount of fit of the individual is chosen. Subsequently, the two individuals are taken back to the initial population, so they maybe reselected (Deb and Goldberg, 1989).

In tournament grouping, two strings compete for a spot in the new generation and off course, the winner string superior candidate takes the spot. The process is repeated again with another two strings to fill another spot under the condition that each string may not participate in more than two tournaments. In this manner, a superior string will have the opportunity to enter the competition twice and thus guarantee two spots.

In the same fashion, an inferior string will also have the chance to compete and lose twice. Therefore, every string will either have 0, 1, or 2 representations in the new generation. It has been shown that tournament grouping for replication (reproduction) operator conjoins at a highly rapid pace and needs less processes than the other operator.

EXAMPLE. Consider a sample population of seven strings "containers" with corresponding fitness values "cost" of {23, 26, 29, 31, 38, 41, 42}. Suppose that 23, 29, 31, 38 and 42 enter the tournament. The rule here is that every string has two chances to compete, also if say 29 and 31 compete; 29 occupies the spot in the new generation because the objective is to minimize the cost. Here is a random tournament. Without loss of generality, let $\otimes$ represent the reproduction operator:

$$29 \otimes 31 \to 29, \quad 41 \otimes 42 \to 41, \quad 38 \otimes 23 \to 23, \quad 29 \otimes 38 \to 29,$$
$$23 \otimes 26 \to 23, \quad 26 \otimes 42 \to 26, \quad 31 \otimes 38 \to 31.$$

Hence, the winners of this tournament that generates the breeding pool consists of the following set of strings {23, 23, 26, 29, 29, 31, 41}.

3.4.2. The crossover operator

The second operator to be discussed is the crossover operator. This operator functions on the breeding pool. While the reproduction operator lacks the ability to produce new nonexistent strings, the crossover operator is used when dealing with this issue. The operator selects two random parent strings from the breeding pool to perform random swaps between cross sections of the strings. There are several types of crossovers that include single crossover also known as one-point crossover, two-point crossover, uniform crossover among others.

EXAMPLE. Consider the following two containers with their corresponding genotype strings:

Fitness	(r, h)	Binary representation
23	(8, 10)	0100:0 01010
29	(7, 13)	0011:1 01101

If one applies the crossover operator to the above parent strings one obtains the following two child strings. Notice the creation of the totally new containers with new fitness values:

Binary representation	(r, h)	Fitness
0100:1 01101	(9, 13)	32
0011:0 01010	(6, 10)	15

The above one-point crossover operator produced new children "containers", one of them has a better fitness value, "cost" of 15 units compared to both parent strings. It is worth mentioning that it is not always necessary to obtain strings with better fitness.

However, the process of having better offspring is not random and more probable since the parent strings are involved in the crossover process, which are the same winner parents of the tournament, in fact, they are expected to produce more enhanced breeds. In the container examples, this means that the containers produced have a higher chance of being near the optimal solution.

3.4.3. The mutation operator

The crossover operator and the mutation operator are both used by the genetic algorithm for the purposes of identifying and obtaining the optimal solutions. However, instead of exchanging cross-sections of a given two strings, the mutation operator selects one binary string and randomly interchanges a set of ones and zeros. The main objective of the mutation operator is to produce a variety of different strings.

EXAMPLE. The container with fitness value of 41 has the following binary representation:

Fitness	(r, h)	Binary representation
41	(7, 30)	001:1:1 11110

The below mutation has produced the child strings "container" with fitness value of 28.

Binary representation	(r, h)	Fitness
001:0:1 11110	(5, 30)	28

Note that 41 is the output obtained from the reproduction process, which was already a good solution, but now 28 is even better. Nevertheless, it is now the better case. Consider the following example.

EXAMPLE. The container with fitness value of 15 has the following binary representation:

Fitness	(r, h)	Binary representation
15	(6, 10)	0011:0: 010:1 :0

The following mutation has produced the child strings "container" with fitness value of 17:

Binary representation	(r, h)	Fitness
0011:1: 010:0:0	(7, 8)	17

Mutation is not an accidental operation, which would have a major impact on the creation of new strings because mutated strings are going to be within the same range as their parents from which they come.

Reproduction, crossover, and mutation operators are the main commonly used operators in genetic algorithms among others. In the reproduction process, only superior strings survive by making several copies of their genetic characteristics. There are some reproduction techniques such as the tournament selection. In the crossover process, two cross sections of parent strings with good fitness values are exchanged to form new chromosomes. Conversely, in the mutation process, only parts of a given string are altered in the hope of obtaining genetically a better child. These operators limit the chances of inferior strings to survive in the successive generations. Moreover, if superior strings existed or are created during the process, they will likely make it to the subsequent generations.

3.5. Encryption and other considerations

Genetic algorithms operate differently than classical search and optimization algorithms. Genetic algorithms depend on encrypted version of the parameters, rather than performing direct calculations on these parameters. Binary genetic algorithms function on discrete functional space despite the fact that the original model might be defined over a continuous space. Because the calculations involve discrete solutions, genetic algorithms use a discrete function to bypass this issue, which gives the algorithms the ability to deal with multiple types of modeling problems. Moreover, genetic algorithm operators take advantage of similarities in strings arrangements for more powerful optimized search. It is not clear which encryption method gives better optimization results, however, generally an encryption that preserves the process of building blocks is recommended. Therefore, the challenge that faces the genetic algorithm is to determine the proper encryption method.

Another significant disparity between genetic algorithms and classical search and optimization algorithms is the population issue. Unlike the classical techniques, which in most cases depend on one solution, genetic algorithms function on a space of solutions. This characteristic is expected because genetic algorithms perform several processes simultaneously on strings to modify or create one string. Nonetheless, there are a few classical algorithms that are population oriented, but these algorithms do not seize the diversity of information available.

Genetic algorithms assign probability values to strings depending on their fitness in order to direct the search process. In most classical search algorithms, the transition process between phases is determined by fixed conditions limiting the range of problems that can be solved, which raises the concern that if for some reason an error occurs during the search process it would be difficult to repair it. Genetic algorithms start the search process with a random sample of the population of which each member is assigned a probability value to direct the search process. In the early stages of the process, no major changes in strings formation are adopted, therefore, any miscalculation will not have direct impact on the search process. Once the search converges to a certain solution narrowing down the search space, an optimal solution is more probable. Genetic algorithms are adaptive to various types of changes; this unique characteristic allows the algorithm correct any unexpected fault during the creation of strings. It also makes the algorithm applicable to a wide range of optimization and search problems.

Genetic algorithms can be executed with parallel computations and can function on parallel computers independently and efficiently. For example, two processors can be used simultaneously to perform the tournament selection on two random strings. Parallel computations can also be applied to carry out strings that crossover and strings that mutate where more than one string is required for the process. The main advantage of parallel computations is they cut down the total computing time significantly since the calculations are done independently.

The primitive operators discussed earlier which strings are involved in duplicating, substrings swapping, and bit alteration, which are the foundation of genetic algorithms. They provide the basic search tools. Although the three operators reproduction, crossover, and mutation operators may appear trivial, their overall functionality and contribution in search process is highly complex. In fact, genetic algorithms are extremely nonlinear, multidimensional yet stochastic algorithms. A number of studies have shown that to examine the time and solution convergence of the nonlinear system; they considered Markov chain analysis to investigate appropriate probability values that determine the changeover in search phases. In a typical optimization problem the size of the population and the number of processes phases grows enormously, which requires more calculations and fast processors to enhance the performance and convergence of the genetic algorithm.

3.6. Schemata

To learn more about the working principle of genetic algorithms, consider the optimization model (see Figure 3):

$$f(\theta) = \cos\left(\theta - \frac{\pi}{2}\right), \quad 0 \leqslant \theta \leqslant \pi.$$

The objective here is to maximize $f(\theta)$ with the bounding variable θ being between 0 and π.

Recall the 5-bit binary representation of the radius and height of the container. The variable θ is going to have the same representation in the interval $[0, \pi]$ with the strings

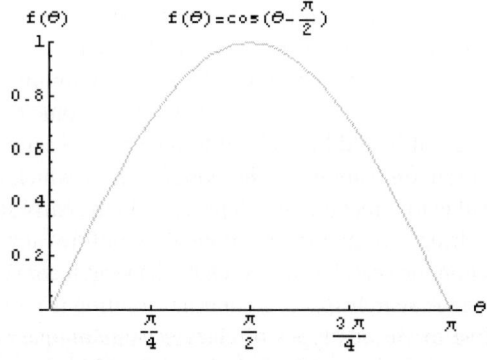

Fig. 3. Optimization model $f(\theta) = \cos(\theta - \pi/2), 0 \leqslant \theta \leqslant \pi$.

00000; meaning $\theta = 0$, and 11111; meaning $\theta = \pi$, while the rest of the 30 strings are mapped in the interval $(0, \pi)$. Also, suppose for simplicity that the population has size of 4, proportionate selection, one-point crossover probability $P_x = 1$, and bit wise mutational probability $P_m = 0.01$.

To proceed with the algorithm, initialize a random space of population, then perform strings evaluation and finally optimize the function using the three operators as follows:

Initial generation							New generation				
Parent string	φ	θ	$f(\theta)$	$\bar{f}(\theta)$	$\dfrac{f}{\bar{f}}$	Breeding pool	Crossover child string	φ	θ	$f(\theta)$	$\bar{f}(\theta)$
10001	17	1.723	0.988	0.575	1.7	10|0|01	10|1|01	21	2.128	0.849	0.684
00101	5	0.507	0.485	0.575	0.8	00|1|01	00|0|01	1	0.101	0.101	0.684
11110	30	3.040	0.101	0.575	0.2	100|0|1	100|1|1	19	1.925	0.938	0.684
10111	23	2.331	0.725	0.575	1.3	101|1|1	101|0|1	21	2.128	0.849	0.684

Initial generation							New generation			
Parent string	φ	θ	$f(\theta)$	$\dfrac{f}{\bar{f}}$	Breeding pool	Crossover child string	φ	θ	$f(\theta)$
10001	17	1.723	0.988	1.72	10|0|01	10|1|01	21	2.128	0.849
00101	5	0.507	0.485	0.84	00|1|01	00|0|01	1	0.101	0.101
11110	30	3.040	0.101	0.18	100|0|1	100|1|1	19	1.925	0.938
10111	23	2.331	0.725	1.26	101|1|1	101|0|1	21	2.128	0.849
		$\bar{f}(\theta) = 0.575$							$\bar{f}(\theta) = 0.684$	

There are several things that are achieved by the end of the first iteration. First, the average performance of the model has been improved entirely when the string was duplicating the operator and substring-swapping operator were used. Also, the bit similarities in the chromosomes between the two generations can be observed as well. The three genetic algorithm operators discussed earlier play an enormous role in the process of increasing the bits similarities and thus introducing what is known in the genetic algorithm terms schema or schemata (a collection of schema.) A schema is a procedure of binary coding to unfold similarities of subsets of strings at a particular position or positions. Genetic algorithms are not defined over the space of solutions rather than on the space of schemata.

DEFINITION. Let v be the string $(v_1, v_2, \ldots, v_j)$, then the length of v denoted by $|v|$ is j. Given a binary alphabet $\Sigma = \{0, 1, *\}$. A schema over Σ is the string $(v_1, v_2, \ldots, v_i, \ldots, v_j)$, where $v_i \in \Sigma$ and $1 \leqslant i \leqslant j$.

Let $S_1 = (10***)$ and $S_2 = (11***)$ be schemata representing $2^3 = 8$ distinct chromosomes each starting with 1, 0 in the first and second positions of S_1 while 1, 1 in the first and second positions of S_2. S_1 contains the two chromosomes from the initial generation 10001, 10111 and the two strings 10101, 10011 from the new generation. Although one chromosome is included in S_2 from the initial generation, S_2 does not contain any chromosome from the new generation.

Because schemata contain strings with similar bit position(s), they may be considered as a region representation in the domain search. Recall $f(\theta) = \cos(\theta - \pi/2)$, $0 \leq \theta \leq \pi$. In this illustration, S_1 represents chromosomes carrying θ values in the interval $[1.621, 2.331]$ that correspond to $f(\theta)$ in the range $[0.725, 0.998]$; yet, S_2 represents chromosomes carrying θ values in the interval $[2.533, 3.141]$ that correspond to $f(\theta)$ in the range $[0, 0.572]$. The main goal of the model is to maximize $f(\theta)$, therefore an optimizer working on this problem would be more interested in considering chromosomes from S_1 than S_2, in fact duplicates of strings from S_1 will help this purpose. This is exactly what one is looking for, which is when manipulated strings from the domain of search in the hope of obtaining optimal solution without the concept of schema competitions. Simple calculations show that the near optimum string in this example is actually the chromosome 10000, this string is contained in S_1.

S_1 and S_2 both have two fixed positions, closely arranged, and have three variable positions. Short, low-ordered and above-average schemata grow exponentially in subsequent generations turning into *building blocks*. The genetic algorithm operators use these building blocks to construct larger building blocks that converge to the optimum solution. Building blocks are behind the success of genetic algorithm, which relates to the Building Block Hypothesis.

While setting up the container's model, certain conditions and restrictions that might be imposed into the problem were avoided. This issue was handled by the simple penalty technique, which focused more on penalizing strings from the infeasible region. A more generalized and efficient method is to be introduced next.

4. Generalized penalty methods

A typical optimization model is stated in the form of nonlinear programming problem, which will be presented in the following manner.

$$\text{Maximize and minimize} \quad f(\theta),$$
$$\text{subject to} \quad k_n(\theta) \geq 0, \quad n = 1, 2, 3, \ldots, N,$$
$$l_m(\theta) = 0, \quad m = 1, 2, 3, \ldots, M,$$
$$\theta_j^{\min} \leq \theta_j \leq \theta_j^{\max}, \quad j = 1, 2, 3, \ldots, J,$$

where $f(\theta)$ is the objective function, N, M the numbers of inequality and equality constraints, respectively, J the number of variable or parameters. In the container's model, $J = 2, N = 1, M = 0$.

The latter representation of the nonlinear model can be translated into the following unconstrained optimization model using the method of simple penalty function to penalize strings from the infeasible region:

$$U(\theta, I, E) = f(\theta) + \sum_{n=1}^{N} I_n \cdot (k_n(\theta))^2 + \sum_{m=1}^{M} E_m \cdot (l_m(\theta))^2$$

where I_n is the inequality penalty parameter, E_m is the equality penalty parameter.

Good results are achieved by choosing the values of these parameters properly; they should be well picked so that the value of the objective function and their values are within a reasonable neighborhood. If however, this method assigns penalties to constraints that are fairly huge compared to each other, then the algorithm will produce solutions that contravene with some of the constraints, which may lead to premature convergence to undesirable solutions.

There have been an array of experiments to determine a reasonable working set of penalty parameters that can be used in the optimization models and it has been discovered that there are some promising values. A good way to reduce the number of parameters used in the problem is to normalize all constraints so that one penalty parameter represents the constraints. Let take the constraint $V(r, h) = r^2 h \pi \geqslant 0.1$ from the container example. After normalization, the constraint becomes

$$k_1(\theta) = \frac{r^2 h \pi}{0.1} - 1 \geqslant 0.$$

As mentioned earlier, genetic algorithms operate on a space of solutions rather than a single solution. The implementation of the penalty method in genetic algorithms would be by exploiting a one-to-one penalty comparison while the tournament selection is processing. There are standards that need to be imposed. Feasible solutions will have advantage on infeasible ones in terms of fitness. Also, feasible solutions guarantee better fitness. When comparing two feasible solutions, it will depend only on the values of their objective functions. When comparing two infeasible solutions, it will depend on the number violations each constraint has made. Genetic algorithms were enhanced to solve the more complicated search and optimization models, which include the real coded genetic algorithms.

Unlike the binary genetic algorithm, real-coded genetic algorithms were introduced to deal with models, where the real-valued parameters cannot be utilized directly. Real coded genetic algorithm takes care of problems obtaining random decision variable accuracy. The operators associated with the binary genetic algorithms are also valid for the real coded genetic algorithm; however, the difference encompasses the use of advanced crossover and mutation operators.

The real-coded crossover operator functions on one variable at a time. The operator picks two parent strings randomly from a variable say the nth variable with corresponding string values of x_n^1 and x_n^2 to produce the two child strings y_n^1 and y_n^1 by applying the following probability density function (pdf):

$$\phi(\theta) = \begin{cases} \frac{1}{2}(\tau + 1)\theta^\tau, & 0 < \theta \leqslant 1, \\ \frac{1}{2}(\tau + 1)\theta^{-(\tau+2)}, & \theta > 1, \end{cases}$$

where θ is the magnitude of the ratio of the parent's difference to the children's difference and is given by

$$\theta = \left| \frac{y_n^2 - y_n^1}{x_n^2 - x_n^1} \right|.$$

The operator proceeds with the process in the following manner:

(i) Selects a random number α in the interval $[0, 1]$.
(ii) Evaluates the integral $\int_0^{\bar{\theta}} \phi(\theta)\, d\theta = \alpha$ to obtain $\bar{\theta}$.
(iii) Creates the children strings using the formulas:

(a) $y_n^1 = \dfrac{(x_n^1 + x_n^2) - \bar{\theta}|x_n^1 - x_n^2|}{2}$,

(b) $y_n^2 = \dfrac{(x_n^1 + x_n^2) + \bar{\theta}|x_n^1 - x_n^2|}{2}$.

The purpose of the probability density function is to create child solutions similar in characteristics to their parents having relatively higher probability compared to other solutions. The variable τ represents how much biologically similar child strings to their parent strings. The higher the value of τ the more alike parents and children are, yet the smaller the value of τ the less similar parents and children are. Generally, when $\tau = 2$, it ends up working perfectly.

The real-coded crossover operator can adapt to any circumstances, this advantage is a result of θ being dimensionless. During the early phases of the optimizing process, strings are randomly scattered and therefore parent chromosomes are not likely to be similar at this early stage, the function of the crossover operator is to gather these scattered strings using a uniform probability function and thus converge to an optimal solution. Chromosomes are closer to each other now than before; the operator will then limit the creation of any divergent solution and promote the creation of convergent solution allowing the genetic algorithm to produce solutions having random accuracy. Further, probability distributions function properly on bounding variables.

A great number of the natural models have multiple optimum solutions, which are all valid solutions. The concept of having more than one solution directs the optimizer to consider different but equivalent solutions, which increases the probability of finding an optimal solution in the search domain. In some classical methods such as the point-by-point technique, one suggestion to obtain all optimum solutions is to execute the optimization algorithm several times using different initial values. The searcher needs to have bare idea about where the equilibrium points occur, otherwise running the algorithm so many times might lead to the same one as being the optimal solution. There is a higher chance to obtain multiple optimum coexisting points simultaneously when executing genetic algorithms because they operate on a population of solutions rather than one solution as mentioned earlier.

The simple genetic algorithm needs to be adapted to handle coexisting optimum solutions. One way to treat this issue is to introduce sharing function among coexisting solutions. When two species coexist in a certain environment they share the available resources, same idea applies here. The genetic algorithm investigates how much sharing of resources is needed by two strings during the optimization process to assign sharing values accordingly. Linear sharing function can be used if $D_{nm} = \|l_m - l_n\|$ is the distance; difference norm in the mathematical terms, between strings l_n and l_m:

$$\Psi(D_{nm}) = \begin{cases} \dfrac{D_{\max} - D_{nm}}{D_{\max}}, & D_{nm} < D_{\max}, \\ 0, & D_{nm} \geqslant D_{\max}, \end{cases}$$

where $D_{\max}$ is a constant known in advance representing the maximum shared distance between any given two strings. The multimodal algorithm proceeds in the following manner. First, the algorithm selects at random a sample population of strings to evaluate the sample function for those strings and this is done to each string belongs to the population. The sum of the values of the sharing function is taken to find out the position count, which is $w_n = \sum_m \Psi(D_{nm})$. The algorithm then computes the fitness value of the nth chromosome as follows: $\bar{f}_n = f_n/w_n$. The reproduction operator uses this value as the shared fitness in place of the value of the objective function f_n. This process makes the concept of having multiple coexisting local and global optima acceptable. A sample population from any given generation with few optimal chromosomes will have smaller position count in comparison with other optimal chromosomes, while the value of the share fitness corresponds to these chromosomes will be high. Chromosomes having high fitness values imply that the reproduction operator of the genetic algorithm will produce as many duplicates of these chromosomes (optimal solutions) as the situation permits promoting superior candidates. The sharing technique preserves optimum strings from decreasing and decaying thus providing a prominent solution to the multimodal optimization problem.

4.1. Multi-objective optimization

Multi-objective optimization methods operate on several objective functions simultaneously. The purpose of this technique is to represent the multiple objective functions as just single objective function; some optimizers use the average of the multi-objective functions to be the function, which will turn into the normal one objective optimization model. The solution in this case will be a set of optimum solutions found by optimizing a number of single objective functions called the Pareto-optimal optimization.

DEFINITION. A state α, which is the set of object parameters is called a Pareto optimal, when there exists no other state β such that the dominating state α with respect to a set of objective functions. A state α dominates a state β, if and only if α is much superior than β, in at least one objective function, and may not be inferior with respect to all other objective functions.

The genetic algorithm to investigate Pareto-optimal solutions uses non-dominated sorting of the population. The method basically depends on a single fitness value of a given solution from the search space, where the fitness values corresponding to every objective function are compared in order to be assigned a new fitness value based on the result of the comparison. The concept of string non-dominated to assign fitness values to solutions is used. Solutions that are better optimizers of one or more objective function are called non-dominated solutions. The non-dominated solutions go through the following process upon obtaining an optimal solution. The algorithm sorts the population and then selects a collection of non-dominated solutions in order to assign them high artificial fitness.

To obtain Pareto-optimal solutions, the algorithm applies the sharing method discussed earlier to the non-dominated solution in order to produce strings with new shared-fitness values. A different non-dominated collection is established while the first set is saved in a temporary place. Fitness values are assigned to the strings in this new set that are significantly smaller in value than strings having the least shared fitness contained in the first set.

The algorithm repeats (reiterates) the sharing process on the current non-dominated collection of strings. This goes until all strings are arranged in a descending order. After the sorting process is completed, the algorithm executes the reproduction phase using the assigned fitness values. In the last step, the algorithm runs both the crossover operator and the mutation operator for purification purposes.

4.2. Fuzzy logic controller

Fuzzy logic controller is a methodology used to solve optimal control models, it consists two primary components, which control the strategy and a set of action plans. The optimal fuzzy controller does the following activities:

(i) Defines the control and action parameters.
(ii) Calculates the best possible association functions corresponding to those parameters.
(iii) For the control and action parameters, it defines the best rules accordingly.

Figure 4 presents a classical fuzzy logic problem. In the example, (t_1, t_2, t_3) represent the control/action parameters for the membership functions with the options (*below average, moderate, above average*). In general, the optimizer sets the values of (t_1, t_2, t_3). Also the values of the membership function are normalized to be in the interval [0, 1]. Genetic algorithms can be applied to obtain the optimal rule base for a given control strategy problem.

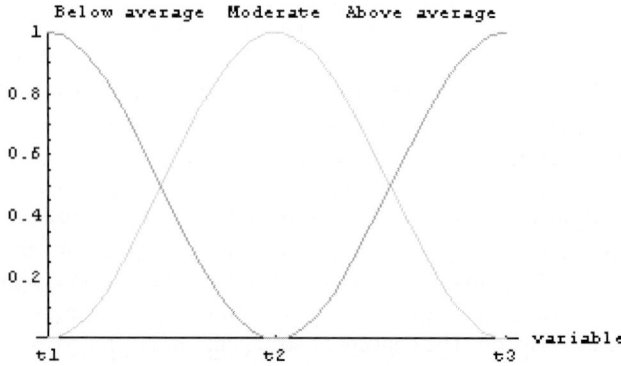

Fig. 4. Classical fuzzy logic problem.

Consider a model consists of two control parameters (t_1 = *height*, t_2 = *weight*) with corresponding options (*below average, moderate, above average*), respectively, and one action parameter (t_3 = *food consumption*). The number of combinations of control parameters is $3 \times 3 = 9$ different possibilities. If we independently consider the control parameter's individual effect, then in this case the number of combinations of control parameters is $(4 \times 4) - 1 = 15$ different possibilities. Need to mention here that the fourth option is the (*no action*) choice. To find the optimal rule base for this problem, every control parameter's combination needs to attain one option of the action parameter. Suppose we are given the following string in ternary coding of length 16:

2 0 2 1 0 2 3 0 3 3 2 3 0 1 2 3

In this representation, every bit in the string corresponds to a grouping of action parameter (t_3 = *food consumption*). Zero implies no action, one implies below average, two implies moderate, and three implies above average. Some action variables may have no combination represented in the rule base, in this case (*no action*) is added to the already existing options. The rule base for the latter string has 12 action variable rules containing nonzero actions and four action variable rules containing no action. In the following table the two control parameters (t_1 = *height*, t_2 = *weight*) are in bold while the action parameter (t_3 = *food consumption*) is set to be regular style.

Weight\height	Below average	Moderate	Above average	No preference
Below average	Moderate	No action	Above average	No action
Moderate	No action	Moderate	Above average	Below average
Above average	Moderate	Above average	Moderate	Moderate
No preference	Below average	No action	Above average	Above average

If the above rule base is not optimized, then genetic algorithms can be used to obtain such an optimal rule base. The genetic algorithm uses the string to determine the rules that belong to the rule base, while the optimizer defines the membership functions to initiate the optimizing process. Then the genetic algorithm proceeds by evaluating the objective function and applying the operators discussed earlier starting with the reproduction operator followed by the crossover and the mutation operators.

It is possible to represent this combination of action parameter using binary strings as a substitute to the ternary string. In such a case, a 30-bit string is needed to represent the rule base with two bits representing each option. The process of investigating an optimized rule base and optimized number of rules is considered to be unique because the foundation of genetic algorithms requires parameters to be discrete and solutions to be in string representation.

Genetic algorithms can utilize the membership function and the rule base simultaneously during the optimization process by concatenating their corresponding codes (strings); a mechanism similar in some sense to parallel computation techniques. Part of the code (string) will be used for the control parameter; while the other remaining bits will be saved for the rules, which will then be exploited by the genetic algorithm to determine the fitness values.

5. Mathematical underpinnings

Genetic algorithm can be enhanced for better convergence and performance using some advanced techniques. A mathematical analysis of the genetic algorithm with the aid of schemata will be discussed.

5.1. Mathematical analysis

Schemata are the collection of schema procedures used to unfold similarities of subsets of strings at a particular position(s). Genetic algorithms are not defined over the space of solutions, but rather on the space of schemata (Biesbroek, 2000).

DEFINITION. Let v be the string $(v_1, v_2, \ldots, v_j)$, then the length of v denoted by $|v|$ is j. Given a binary alphabet $\Sigma = \{0, 1, *\}$. A schema over Σ is the string $(v_1, v_2, \ldots, v_i, \ldots, v_j)$, where $v_i \in \Sigma$ and $1 \leqslant i \leqslant j$.

Here the star symbol $*$ means no preference, which implies accept either zero or one. A schema that has as many of these stars becomes indefinite (i.e. it depicts more than one string). Suppose this schema contains n stars, subsequently note that the above alphabet contains three symbols, one can then determine that the number of strings in this schema will be 2^n.

EXAMPLE. Consider the schema (101*10*). This schema describes the set

$$\{(1010100), (1011100), (1010101), (1011101)\}.$$

A schema provides an optimized method to describe similar strings that have finite length defined over a finite alphabet Σ. By knowing the fitness of a certain string, one can predict the formation of other strings with analogous structure.

Let k be the length of a string v, and m be the cardinality of an alphabet Σ. Then the total number of distinct schemata is $(m + 1)^k$, which gives the schemata space. Note that m^k is the cardinality of the original space of solutions. The $*$ characters, which can be part of the schema, does not participate in the process of the actual strings genetic algorithms. In fact, they function as a tool to describe the string structure and similarity in a more optimized matter. Therefore, the space of schemata is larger than the search space.

5.2. Schema Theorem

The Schema Theorem is the fundamental theorem in genetic algorithms. John Holland was the first to initiate the Schema Theorem (Holland, 1975). He first introduced the concept of "implicit parallelism", which explains the process of genetic algorithm. Consider a binary string with 2^k different degrees of freedom; subsequently this string includes 2^k possible schemata. Hence, one is able to obtain promising information about the schema and its fitness. Genetic algorithms can be thought of as a competition among schemata with 2^k distinct competitions.

The Schema Theorem explains the increase of a schema from a particular generation to the subsequent one. The Schema Theorem is frequently explained such that concise, low-order, short schemas whose mean fitness stays beyond the average will be given exponential increasing number(s) of samples beyond the instance of time. The quantity of the assessed occurrence of the schemas that are not disrupted and which stay above the mean in fitness is enhanced by a factor of $f(S)/f_{\text{pop}}$ for every generation, where $f(S)$ is the fitness mean value of schema S, and f_{pop} is the fitness mean value of the population.

The Schema Theorem deals with only the damaging effects of recombination and mutation and is a lower bound. On the other hand, recombination is the chief source of the genetic algorithm's strength. It has the ability to put together situations with decent schemas, which to develop various occurrences of equally decent or more enhanced higher-order schemas. The procedure by which genetic algorithms operate is known as the Building Block Hypothesis. The low-ordered and above-average schemata grow exponentially in subsequent generations are referred to as *building blocks*.

DEFINITION. Given a schema S defined over an alphabet Σ. The *order* of S denoted by $O(S)$ is defined to be the number of fixed positions.

EXAMPLE. The *order* of the schema (101∗10∗) is 5.

DEFINITION. Given a schema S defined over an alphabet Σ. The *length* of S denoted by $L(S)$ is defined to be the distance between the outermost fixed positions.

EXAMPLE. The *length* of the schema (101∗10∗) is $6 - 1 = 5$.

SCHEMA THEOREM.

$$e(S, n+1) = e(S, n)\frac{f(S)}{f_{\text{pop}}}\left(1 - P_{\text{x}} \cdot \frac{L(S)}{1-1} - O(S) \cdot P_{\text{m}}\right)$$

such that $e(S, n+1)$ is the expected number of instances of schema S in generation $n+1$, $f(S)$ is the fitness mean value of schema S, f_{pop} is the fitness mean value of the population, P_{x} is the recombination probability, P_{m} is the mutation probability, $L(S)$ is the length of schema S, $O(S)$ is the order of schema S.

In conclusion, implicit parallelism is the fundamental foundation for genetic algorithms. The ability to determine all the possible solutions concurrently is perhaps the most significant quality genetic algorithms can provide. Strings, in general, contain an array of building blocks, which are evaluated immediately using implicit parallelism, however, genetic algorithm measures these blocks simultaneously when determining string's fitness. The process of transferring the building blocks to the following generation pioneers when the algorithm starts searching for patterns and similarities inside the string. Unlike traditional algorithms, this procedure enhances the performance of genetic algorithms significantly.

5.3. Hybridization

Hybridization is the integration of genetic algorithm in association with other optimization algorithms, taking only the constructive aspects of those algorithms. Optimization can be achieved in several ways.

(i) If the existing algorithm is fast, the hybrid genetic algorithm can add the solution it generates to the initial population. This guarantees that the offspring algorithm will be at least be of the same efficiency compared to the parent algorithm. Overall, crossing the results of the existing algorithm over with each other or with other results leads to improvements.

(ii) If the existing algorithm carries out consecutive transformations on encodings, the hybrid genetic algorithm can incorporate those transformations into the operator set of the algorithm.

(iii) If the existing algorithm interprets encoding in a wise fashion, the crossbreeding process of the genetic algorithm is able to exploit its functionality into the decoding stage from the assessed method.

Generally, algorithms that frequently have co-adapted encoding strategies are frequently tailored to a problem domain. This standard might possibly lead to the use of features of the existing deciphering techniques.

The above techniques are valid with the assumption that the only existing algorithm is the one operating; however, if there are several algorithms operating, the above techniques should be altered accordingly.

5.4. Genetic algorithm fitness

5.4.1. Scaled fitness

In linear scaling, the raw fitness f and the scaled fitness f_p are linearly proportional as given by the following equation:

$$f_p = \alpha \cdot f + \beta, \quad \text{where}$$

$$\alpha = \frac{C_{\text{mult}} - 1}{\Delta} \cdot f_{\text{pop}}, \quad \beta = \frac{f_{\text{max}} - f_{p\_\text{max}}}{\Delta} \cdot f_{\text{pop}},$$

$$f_{p\_\text{max}} = C_{\text{mult}} \cdot f_{\text{pop}}, \quad \Delta = f_{\text{max}} - f_{\text{pop}}.$$

Hence, both of the scaled fitness mean and the raw fitness mean are equal. The number of predicted copies wanted for the best population member C_{mult} controls the number of offspring with optimal raw fitness set to the population member. In general, a population of 50 to 100 members proved to have promising C_{mult} results of 1.2 to 2. Furthermore, the population with low fitness proved to have negative values.

5.4.2. Windowing technique

The next technique to be addressed is windowing, which is applied to improve the low fitness of vulnerable members of the population by simply assigning fitness to the desired population of chromosomes. Windowing inhibits a certain group of chromosomes from dominating the whole population and also cuts off the elimination of the weaker members. The process of assigning fitness values to chromosomes is as follows:

(i) Locate the member in the population that has minimum fitness call it $f_{\min}$.
(ii) Assign each member of the population a fitness value that exceeds $f_{\min}$.

5.4.3. Linear normalization technique
The third technique is linear normalization. Linear normalization is similar to windowing, but differs in the sense that it is biased when it comes to determining and assigning values. The process for linear normalization is:

(i) Arrange the chromosomes in a descending order with respect to their fitness.
(ii) Obtain a fixed fitness value from the user.
(iii) Assign the chromosomes a fitness value that decreases linearly based on the given fixed value.

5.4.4. Fitness technique
When applying these techniques, one must use care not to run into a constraint violation by forgetting to set the maximum fitness values. Linear normalization deals with this problem during the sorting process, however, windowing and linear scaling lack a concrete method to solve this issue. There are three suggested solutions to constraint violation.

High penalty. This method works in such a way that penalizes violators excessively. An advantage of this approach is that it controls violators, strings may produce, but on a limited basis. The disadvantage, however, encompasses the population that has high fraction of violators, which may lead to undesirable solution and creating what is called population stagnation. Population stagnation occurs when a small percentage of the legal stings successively dominate the generations, while violators that represent the high percentage are left out.

Moderate penalty. Unlike high penalty, moderate penalty assigns modest penalty to violators in which enhances the likelihood of violators to produce, which minimizes the population stagnation. However, the concern rises when the ratio of violators is greater than the ratio of certified strings, because one does not expect violators to produce certified strings.

Elimination. This method looks for restricted violators among strings prior to assigning the fitness through the eradication process. There are some disadvantages for this approach. First, the elimination may exclude valid solutions in the sense that violating strings may produce non-violating strings. Moreover, implementation of the elimination technique through this means would be too costly.

6. Techniques for attaining optimization

6.1. Elitism
Elitism is a technique used to avoid loosing members of the population and maximizing the best values by replicating several copies of these chromosomes into the following

generation. When a new generation is produced there is a probability that one loses some of the best characteristics from the original population. Due to the fact that when performing the algorithm, one has a slight chance of losing the optimal solution, hence, one would have to use elitism, because elitism would minimize such occurrences (i.e. avoid deleting the optimal solution). If the optimal strings are eliminated from the reproduced population, it may slow down convergence and lead to instability of the genetic algorithm, hence, saving these strings from decomposition provides the chance for a fast and stable dominance. As a result, elitism helps to improve the performance of the genetic algorithm.

6.2. Linear probability

Linear probability is another technique for obtaining and maximizing the optimal members for higher survival rate and assigning high probability values to the best ones, while, assigning inferior members low probability values. This can be done with the aid of a linear probability curve, which scales the fitness value. Usually, superior members receive a probability value of 0.95, which implies that they have a high likelihood of survival, and inferior members receive a probability value of 0.05, which implies that they have a small likelihood of survival. By doing this, one guarantees not all healthy members would be allowed to reproduce indefinitely and not all weak members would be prevented from reproduction and thus be extinct. Technically, linear probability preserves all types of inherited characteristics from disappearing. Note that a chromosome with a probability value of one would definitely make it to the next generation similar to elitism.

Occasionally undesirable consequence of genetic algorithm happens when parent string are replaced by their progeny in its entirety. This occurrence yields to negative consequences. The reason for the negative consequence is that some superior members may not be able to reproduce, hence, will lose their genes. In addition, applying the mutation or crossover method with this technique will yield to some of the characteristics found in the original chromosomes, which maybe altered and thus lost forever.

6.3. Steady-state technique

This undesirable consequence can be avoided using steady-state reproduction. There are two forms of this technique. The steady state reproduction technique is achieved by creating a copy of the solution during the production process, a copy of which is inserted into the parents' position in the original population. The two weak strings in the population are replaced with the new solution through repeated replications until the number of iterations (solutions) in the population equals the number of new solutions added to the population from the last generation. Moreover, the steady state technique without replication throws out all offspring that have been simulated in the existing population instead of inserting these copies in the population.

6.4. Advanced crossover techniques

The demand for more advanced crossover techniques resulted from the fact that one-point crossover has some disadvantages when applied to genetic algorithms. For example, one-point crossover fails to pass a combination of encoded characteristics of the chromosomes to the new generation. Therefore, schemata that have outsized lengths maybe dislocated. Moreover, some particular strings maybe allowed to exist only once. There are four different types of advanced crossovers, which include:

(i) two-point crossover,
(ii) uniform crossover,
(iii) partially mixed crossover,
(iv) uniform order-based crossover.

6.4.1. Two-point crossover

The most popular among the different types of crossover techniques is perhaps the two-point crossover. Unlike the one-point crossover, two-point crossover performs two cuts simultaneously and at random, instead of one cut, then exchanges the information between the two cuts.

EXAMPLE (The two-point crossover operator).

α $\beta\gamma|\delta\varepsilon|\zeta\eta\theta$	α $\beta\gamma|\mu\nu|\zeta\eta\theta$
ι $\kappa\lambda|\mu\nu|\xi o\pi$	ι $\kappa\lambda|\delta\varepsilon|\xi o\pi$
Before crossover	After crossover

6.4.2. Uniform crossover

Uniform crossover is the second crossover technique; it performs and swaps more than two cuts simultaneously and at random as follows. It selects two parents and two produced children. Then randomly pick the bit position on the parent and the bit position on the first child to hold the desired characteristics. Repeat this process for the second child. Now, create a mask to indicate the bit on the parent that is going to be donated to the corresponding child; the same mask is going to be used for both children.

EXAMPLE (The uniform crossover operator).

α $\beta\gamma\delta\varepsilon\zeta\eta\theta$	α $\kappa\lambda\mu\nu\xi\eta\pi$
ι $\kappa\lambda\mu\nu\xi o\pi$	ι $\beta\gamma\delta\varepsilon\zeta o\theta$
mask: 1 0 0 1 1 0 1 0	
Before crossover	After crossover

6.4.3. Partially mixed crossover

Another crossover technique is the partially mixed crossover, which operates on two strings; the procedure is as follows: from the two given strings; string 1 and string 2, select a uniform cross section at random, and then that cross section becomes the matching section. This matching section will be used for position-wise swaps. In the example below, we select the cross sections "$\zeta\eta\alpha$" from string one and "$\gamma\kappa\iota$" from string two. Then swap the letters "ζ" and "γ", "η" and "κ", "α" and "ι".

EXAMPLE (The partially mixed crossover operator).

ι	$νξο|ζηα|γμκ$	α	$νξο|γκι|ζμη$
ν	$ηαμ|γκι|οξζ$	ν	$κιμ|ζηα|οξγ$
Before crossover		After crossover	

6.4.4. Uniform order-based crossover

The last technique is the uniform order-based crossover, often used for the Traveling Salesman Problem, which is when an individual, specifically, a salesman is set to pass through a city and the objective of the salesman is to minimize the total distance traveled in that particular city.

Genetic algorithms have shown much potential concerning a particular problem known as the 'traveling salesman problem' in which a salesperson has the task of visiting a number of clients, located in different cities, and returns home. The distance traveled by the salesman must be minimized. With this problem, the strings of the genetic algorithm represent the order of the cities. Precautions must be taken when a certain item (a city to be visited) appears only once in a string. Like uniform order-based crossover, *uniform order-based mutation* is a way to deal with this problem. This operator selects a section of the chromosome and mixes up the elements within the section.

Before mutation	After mutation
$ι \quad δε|ζηαγ|βκ$	$ι \quad δε|γηζα|βκ$

This technique generates a mask string of a length that equals the length of the parent string to maintain the order of bits when crossing over. Then it fills part of the progeny's bit positions with the same bits from parent one wherever the mask holds "1". In the next step, it creates a list of the bits of parent one wherever the mask holds "0" to permute them so that the order of these bits is maintained as they exist in parent two. To finish, we fill in the empty positions within the progeny the permuted bits with the same order created in the last step.

$$\begin{array}{c} αβγδ\boxed{εζηθ} \\ ικλμ\boxed{νξοπ} \end{array} \Rightarrow \begin{array}{c} αβγδ\boxed{νξοπ} \\ ικλμ\boxed{εζηθ} \end{array}$$
Before crossover After crossover
$$\begin{array}{c} αβγ\boxed{δε}ζηθ \\ ικλ\boxed{μν}ξοπ \end{array} \Rightarrow \begin{array}{c} αβγ\boxed{μν}ζηθ \\ ικλ\boxed{δε}ξοπ \end{array}$$

6.5. Mutation

6.5.1. Uniform order-based mutation

Mutation works in a way similar to the hypothetical traveling salesman, where an object journeys between terminals with a minimum travel distance under the condition that no one terminal is visited twice. Applying this concept to the genetic algorithm, we have strings representing terminals and the problem becomes finding the optimum solution provided that no one bit is being used twice. During the process, the uniform order-based mutation's operator chooses a certain string cross section and randomly mixes the bits up.

EXAMPLE (The uniform order-based mutation operator).

$v \quad \eta\|\alpha\mu\gamma\kappa\iota\|o\xi\zeta$	$v \quad \eta\|\kappa\iota\mu\alpha\gamma\|o\xi\zeta$
Before mutation	After mutation

6.5.2. Advanced mutation

As previously noted, crossover techniques are decent methods by which to arrange and promote fit members of the population onto successive levels of reproduction. However, over-population of these fit members gathers and converges to decrease the chances that crossover will operate again. Therefore, the call for a different technique to handle the new case is necessary. Mutation thrown into the mix is the technique used to hunt for better members among the existing elite members. This operation will increase the fitness's binary mutation while the algorithm is operating. Mutation focuses on similarities between the parents of a given child; the more similarities there are the higher the chance for a mutation. This process cuts down the probability of having premature convergence.

6.6. Genetic algorithm parameters

Many researchers, including DeJong (1975), have worked on the parameters of the operators of mutation and one-point crossover and found out that the following values provide desirable results. DeJong set the population size to be 50, the crossover probability P_x to be 0.6, the mutation probability P_m to be 0.001, elitism to be 2 and no windowing.

6.6.1. Multi-parameters

To obtain parameters with optimal results, one may opt for an integer from the interval $[0, 15]$, which is equivalent to $[0, 2^k - 1]$ where k is the length of the string. In this case, a parameter of 4 bits is used for encoding. However, a new approach can be used to encode optimization problems with real multi-parameters. This technique is called the concatenated, multi-parameter, mapped, fixed-point coding (Goldberg, 1989). Instead of working on $[0, 2^k - 1]$, the method maps $[0, 2^k]$ to the interval $[I_{\min}, I_{\max}]$ using a linear transformation. This allows us to monitor the decoded parameter including its precision and range; the following formula can be used to compute the precision R:

$$R = \frac{I_{\max} - I_{\min}}{2^k - 1}.$$

The procedure to build the multi-parameter code is straightforward. Consider a single-parameter code, to proceed with the process take one of these code blocks which has a length, $I_{\min}$ and $I_{\max}$ then start concatenating them until reaching the desired point.

6.6.2. Concatenated, multi-parameter, mapped, fixed-point coding

Consider a single-parameter code I_n of length $k_n = 3$. This parameter has corresponding $I_{\min} = 000$ and $I_{\max} = 111$. Then a linear transformation of a multi-parameter code consisting of 9 parameters would be

$$001011 \cdots 100111$$
$$I_1 I_2 \cdots I_8 I_9.$$

6.7. Exploitable techniques

Although, inversion, addition and deletion are exploitable techniques their impact is still uncertain. Inversion is basically taking a cross section of a given string then rearranging the characters of this section in a reverse order. The concept of inversion guided researchers to both the uniform ordered-base and the partially mixed crossovers, which discussed earlier.

6.7.1. Inversion operator

ν $\eta\|\alpha\mu\gamma\kappa\iota\|o\xi\zeta$	ν $\eta\|\iota\kappa\gamma\mu\alpha\|o\xi\zeta$
Before inversion	After inversion

6.7.2. Addition operator

ν $\eta\alpha\mu\gamma\kappa\iota o\xi\zeta$	ν $\eta\alpha\mu\gamma$ M $\kappa\iota o\alpha\xi\zeta$
Before addition	After addition

6.7.3. Deletion operator

ν $\eta\alpha\mu\gamma\kappa\iota o\xi\zeta$	ν $\alpha\mu\gamma\kappa\iota o\xi$
Before deletion	After deletion

One performs the function of deletion by randomly selecting and deleting one or more characters of the targeted cross section of a given string, however, addition can be done by randomly making copy of a certain permitted character and then inserting this character randomly into a targeted position. It is noteworthy to mention that the addition and deletion operators have direct impact on the length of the string, which calls for adapting our techniques including the fitness.

7. Closing remarks

Genetic algorithms are robust, useful, and are most powerful apparatuses in detecting problems in an array of fields. In addition, genetic algorithms unravel and resolve an assortment of complex problems. Moreover, they are capable of providing motivation for their design and foresee broad propensity of the innate systems. Further, the reason for these ideal representations is to provide thoughts on the exact problem at hand and to examine their plausibility. Hence, employing them as a computer encode and identify how the propensities are affected from the transformations with regards to the model (Mitchell, 1998). Without genetic algorithms, one was not able to solve real world issues. The genetic algorithm methods may permit researchers to carry out research, which was not perceivable during this evolutionary technological era. Therefore, one is able to replicate this phenomenon, which would have been virtually impossible to obtain or analyze through traditional methods or through the analysis of certain equations.

Because GAs form a subset field of evolutionary computation, optimization algorithms are inspired by biological and evolutionary systems and provide an approach to learning that is based on simulated evolution. Given a basic understanding of biological evolution and Darwin's ideas of the survival of the fittest or most successful members of a biological population, one is able to comprehend the computational applications of such a theory wherein a computer analyzes populations of data and "learns" which data strings to follow and repeat based on the most "fit" or successful data in a population (Baldwin, 1896).

Through successive hypotheses, "fitness" can be ascertained through training examples in games such as chess. Mitchell (1997), in *Machine Learning*, states that the computer's task in a game of chess would be to learn the strategy for the game. Fitness could be defined, he states, "as the number of games won ... when playing against other individuals in the current population." This simple example shows how groups of hypotheses known as populations are continuously updated so that the fittest data to date is evaluated to form the next most likely population.

Genetic algorithms have proven time and again that they are excellent means to solve complex real world problems.

Acknowledgements

The author is in great debt to Dr. Edward J. Wegman for providing impeccable support, encouragement, guidance, and for introducing me to the genetic algorithm field. The author is also grateful to Dr. Clifton D. Sutton, for his willingness to provide critical and insightful comments.

References

Baker, J.E. (1987). Reducing bias and inefficiency in the selection algorithm. In: Grefenstette, J.J. (Ed.), *Genetic Algorithms and Their Applications*, Proceedings of the Second International Conference on Genetic Algorithms, Erlbaum.

Baldwin, J.M. (1896). A new factor in evolution. *Amer. Natur.* **30**, 441–451, 536-553. http://www.santafe.edu/sfi/publications/Bookinfo/baldwin.html.

Biesbroek, R. (2000). Genetic algorithm tutorial. http://www.estec.esa.nl/outreach/gatutor/history_of_ga.htm.

Deb, K. (1999). An introduction to genetic algorithms. Preprint. http://www.iitk.ac.in/kangal/papers/sadhana.ps.gz.

Deb, K., Goldberg, D.E. (1989). An investigation of niche and species formation in genetic function optimization. In: Schaffer, J.D. (Ed.), *Proceedings of the Third International Conference on Genetic Algorithms*. Morgan Kaufmann, San Mateo, CA.
DeJong, K.A. (1975). An analysis of the behavior of a class of genetic adaptive systems. Doctoral dissertation. University of Michigan. *Diss. Abstr. Internat.* **36** (10), 5140B.
Goldberg, D.E. (1989). *Genetic Algorithms in Search, Optimization and Machine Learning*. Addison–Wesley, Boston.
Hadamard, J. (1949). *The Psychology of Invention in the Mathematical Field*. Princeton University Press, Princeton, NJ.
Holland, J.H. (1975). *Adaptation in Natural and Artificial Systems*. University of Michigan Press, Ann Arbor.
Mitchell, M. (1998). *An Introduction to Genetic Algorithms*. MIT Press, Cambridge, MA.
Mitchell, T.M. (1997). *Machine Learning*. WCB/McGraw–Hill, Boston.

Further reading

Ackley, D., Littman, M. (1994). A case for Lamarckian evolution. In: Langton, C. (Ed.), *Artificial Life III*. Addison–Wesley, Reading, MA.
Back, T. (1996). *Evolutionary Algorithms in Theory and Practice*. Oxford University Press, Oxford, UK.
Barberio-Corsetti, P. (1996). Technical assistance for genetic optimization of Trellis codes. Final report for the Communications System Section (XRF) of ESTEC. Eurospacetech Frequency 94.02, Rev. 1.
Belew, R. (1990). Evolution, learning, and culture: Computational metaphors for adaptive algorithms. *Complex Systems* **4**, 11–49.
Belew, R.K., Mitchell, M. (Eds.) (1996). *Adaptive Individuals in Evolving Populations: Models and Algorithms*. Addison–Wesley, Reading, MA.
Beyer, H.-G. (2002). Glossary: Evolutionary Algorithms – Terms and Definitions. http://ls11-www.cs.uni-dortmund.de/people/beyer/EA-glossary/node102.html.
Bledsoe, W. (1961). The use of biological concepts in the analytical study of systems. In: *Proceedings of the ORSA–TIMS National Meeting*, San Francisco.
Booker, L.B., Goldberg, D.E., Holland, J.H. (1989). Classifier systems and genetic algorithms. *Artificial Intelligence* **40**, 235–282.
Box, G. (1957). Evolutionary operation: A method for increasing industrial productivity. *J. Roy. Statist. Soc.* **6** (2), 81–101.
Cohoon, J.P., Hegde, S.U., Martin, W.N., Richards, D. (1987). Punctuated equilibria: A parallel genetic algorithm. In: *Genetic Algorithms and Their Applications*, Proceedings of the Second International Conference on Genetic Algorithms, Erlbaum, pp. 148–154.
Davidor, Y. (1991). *Genetic Algorithms and Robotics: A Heuristic Strategy for Optimization*. World Scientific, Singapore.
Davis, L. (1991). *Handbook of Genetic Algorithms*. Van Nostrand Reinhold, New York.
Dawn, T. (1995). Nature shows the way to discover better answers. *Sci. Comput. World* **6**, 23–27.
Deb, K., Goldberg, D.E. (1993). Analyzing deception in trap function. In:, L.D. Whitley (Ed.), *Foundations of Genetic Algorithms 2*. Morgan Kaufmann, San Mateo, CA.
DeJong, K.A., Spears, W.M., Gordon, D.F. (1993). Using genetic algorithms for concept learning. *Machine Learning* **13**, 161–188.
Dorigo, M., Maniezzo, V. (1993). Parallel genetic algorithms: Introduction and overview of current research. In: *Parallel Genetic Algorithms: Theory and Applications*, pp. 5–42.
Folgel, L.J., Owens, A.J., Walsh, M.J. (1966). *Artificial Intelligence through Simulated Evolution*. Wiley, New York.
Folgel, L.J., Atmar, W. (Eds.) (1993). *Proceedings of the Second Annual Conference on Evolutionary Programming*. Evolutionary Programming Society.
Forrest, S. (1993). Genetic algorithms: Principles of natural selection applied to computation. *Science* **261**, 872–878.

French, R., Messinger, A. (1994). Genes, phenes, and the Baldwin effect: Learning and evolution in a simulated population. In: Brooks, R., Maes, P. (Eds.), *Artificial Life IV*. MIT Press, Cambridge, MA.

Goldberg, D. (1994). Genetic and evolutionary algorithms come of age. *Comm. ACM* **37** (3), 113–119.

Goldberg, D.E., Deb, K., Korb, B. (1991). Don't worry, be messy. In: Belew, R., Brooker, L. (Eds.), *Proceedings of the Fourth International Conference on Genetic Algorithms*. Morgan Kaufmann, San Mateo, CA, pp. 24–30.

Grant, K. (1995). An introduction to genetic algorithms. *C/C++ Users J.* (March), 45–58.

Green, D.P., Smith, S.F. (1993). Competition based induction of decision models from examples. *Machine Learning* **13**, 229–257.

Grefenstette, J.J. (1991). Lamarchian learning in multi-agent environments. In: Belew, R., Brooker, L. (Eds.), *Proceedings of the Fourth International Conference on Genetic Algorithms*. Morgan Kaufmann, San Mateo, CA.

Grefenstette, J.J. (1998). Credit assignment in rule discovery systems based on genetic algorithms. *Machine Learning* **3**, 225–245.

Hart, W., Belew, R. (1995). Optimization with genetic algorithm hybrids that use local search. In: Below, R., Mitchell, M. (Eds.), *Adaptive Individuals in Evolving Populations: Models and Algorithms*. Addison–Wesley, Reading, MA.

Harvey, I. (1993). The puzzle of the persistent question mark: A case study of genetic drift. In: Forrest, S. (Ed.), *Proceedings of the Fifth International Conference of Genetic Algorithms*. Morgan Kaufmann, San Mateo, CA.

Heitkoetter, J., Beasley, D. (1994). The Hitch-Hiker's Guide to Evolutionary Computation: A List of Frequently Asked Questions (FAQ). USENET: comp.ai.genetic. Available via anonymous FTP from ftp://rtfm.mit.edu/pub/usenet/news.answers/ai-faq/genetic/.

Hinton, G.E., Nowlan, S.J. (1987). How learning can guide evolution. *Complex Systems* **1**, 495–502.

Holland, J.H. (1962). Outline for a logical theory of adaptive systems. *J. ACM* **3**, 297–314.

Holland, J.H. (1986). Escaping brittleness: The possibilities of general-purposed learning algorithms applied to a parallel rule-based systems. In: Michalski, R., Carbonell, J., Mitchell, T. (Eds.), *Machine Learning: An Artificial Intelligence Approach, vol. 2*. Morgan Kaufmann, San Mateo, CA.

Holland, J.H. (1989). Searching nonlinear functions for high values. *Appl. Math. Comput.* **32**, 255–274.

Holland, J.H. (1992). *Adaptation in Natural and Artificial Systems*, 2nd ed. MIT Press, Cambridge, MA.

Janikow, C.Z. (1993). A knowledge-intensive GA for supervised learning. *Machine Learning* **13**, 189–228.

Koza, J. (1992). *Genetic Programming: On the Programming of Computers by Means of Natural Selection*. MIT Press, Cambridge, MA.

Koza, J.R. (1994). *Genetic Programming II: Automatic Discovery of Reusable Programs*. MIT Press, Cambridge, MA.

Koza, J.R., Bennett III, F.H., Andre, D., Keane, M.A. (1996). Four problems for which a computer program evolved by genetic programming is competitive with human performance. In: *Proceedings of the 1996 IEEE International Conference on Evolutionary Computation*. IEEE Press, pp. 1–10.

Michalewicz, Z. (1999). *Genetic Algorithms + Data Structures = Evolution Programs*, 3rd ed. Springer-Verlag, Berlin.

O'Reilly, U.-M., Oppacher, R. (1994). Program search with a hierarchical variable length representation: Genetic programming, simulated annealing, and hill climbing. In: Davidor, Y. et al. (Ed.), *Parallel Problem Solving from Nature*, PPSN III, Lecture Notes in Comput. Sci., vol. 866. Springer-Verlag, New York.

Parker, B.S. (1992). Demonstration of Using Genetic Algorithm Learning. Manual of DOUGAL. Information Systems Teaching Laboratory.

Rechenberg, I. (1965). Cybernetic solution path of an experimental problem. Ministry of Aviation, Royal Aircraft Establishment, UK.

Rechenberg, I. (1973). *Evolutionsstrategie: Optimierung Technischer Systeme nach Prinzipien der biologischen Evolution*. Frommann–Holzboog, Stuttgart.

Schwefel, H.P. (1975). Evolutionsstrategie und numerische Optimierung. PhD dissertation. Technical University of Berlin.

Schwefel, H.P. (1977). *Numerische Optimierung von Computer-Modellen Mittels der Evolutionsstrategie*. Birkhäuser, Basel.

Schwefel, H.P. (1995). *Evolution and Optimum Seeking*. Wiley, New York.

Smith, S. (1980). A learning system based on genetic adaptive algorithms. PhD dissertation. Computer Science Department, University of Pittsburgh.

Spiessens, P., Maderick, B. (1991). A massively parallel genetic algorithm: Implementation and first analysis. In: Belew, R., Brooker, L. (Eds.), *Proceedings of the Fourth International Conference on Genetic Algorithms*. Morgan Kaufmann, San Mateo, CA, pp. 279–286.

Stender, J. (Ed.) (1993). *Parallel Genetic Algorithms*. IOS Press, Amsterdam.

Tanese, R. (1989). Distributed genetic algorithms. In: Schaffer, J.D. (Ed.), *Proceedings of the Third International Conference on Genetic Algorithms*. Morgan Kaufmann, San Mateo, CA, pp. 434–439.

Teller, A., Veloso, M. (1994). PADO: A new learning architecture for object recognition. In: Ikeuchi, K., Veloso, M. (Eds.), *Symbolic Visual Learning*. Oxford Univ. Press, Oxford, UK, pp. 81–116.

Turnkey, P.D. (1995). Cost-sensitive classification: Empirical evaluation of a hybrid genetic decision tree induction algorithm. *J. Artificial Intelligence Res.* **2**, 369–409. http://www.cs.washington.edu/research/jair/home.html.

Turnkey, P.D., Whitley, D., Anderson, R. (Eds.) (1997). Special Issue: The Baldwin Effect. *Evolut. Comput.* **4** (3). http://www-mitpress.mit.edu/jrnlscatalog/evolution-abstracts/evol.html.

Whitley, L.D., Vose, M.D. (Eds.) (1995). *Foundations of Genetic Algorithms 3*. Morgan Kaufmann, San Mateo, CA.

Wright, S. (1977). *Evolution and the Genetics of Populations; vol. 4: Variability within and among Natural Populations*. University of Chicago Press, Chicago, IL.

Computational Methods for High-Dimensional Rotations in Data Visualization

Andreas Buja, Dianne Cook, Daniel Asimov and Catherine Hurley

Abstract

There exist many methods for visualizing complex relations among variables of a multivariate dataset. For pairs of quantitative variables, the method of choice is the scatterplot. For triples of quantitative variables, the method of choice is 3D data rotations. Such rotations let us perceive structure among three variables as shape of point scatters in virtual 3D space.

Although not obvious, three-dimensional data rotations can be extended to higher dimensions. The mathematical construction of high-dimensional data rotations, however, is not an intuitive generalization. Whereas three-dimensional data rotations are thought of as *rotations of an object* in space, a proper framework for their high-dimensional extension is better based on *rotations of a low-dimensional projection* in high-dimensional space. The term "data rotations" is therefore a misnomer, and something along the lines of "high-to-low dimensional data projections" would be technically more accurate.

To be useful, virtual rotations need to be under interactive user control, and they need to be animated. We therefore require projections not as static pictures but as movies under user control. Movies, however, are mathematically speaking one-parameter families of pictures. This article is therefore about *one-parameter families of low-dimensional projections in high-dimensional data spaces*.

We describe several algorithms for dynamic projections, all based on the idea of smoothly interpolating a discrete sequence of projections. The algorithms lend themselves to the implementation of interactive visual exploration tools of high-dimensional data, such as so-called grand tours, guided tours and manual tours.

1. Introduction

Motion graphics for data analysis have long been almost synonymous with 3D data rotations. The intuitive appeal of 3D rotations is due to the power of human 3D perception and the natural controls they afford. To perform 3D rotations, one selects triples of data variables, spins the resulting 3D pointclouds, and presents a 2D projection thereof to the viewer of a computer screen. Human interfaces for controlling 3D data rotations follow natural mental models: Thinking of the pointcloud as sitting inside a globe, one enables the user to rotate the globe around its north–south axis, for example, or to pull an axis into oblique viewing angles, or to push the globe into continuous motion with a sweeping motion of the hand (the mouse, that is).

The mental model behind these actions proved natural to such an extent that an often asked question became vexing and almost unanswerable: How would one generalize 3D rotations to higher dimensions? If 3D space presents the viewer with one hidden backdimension, the viewer of $p > 3$ dimensions would face $p - 2 > 1$ backdimensions and would not know how to use them to move the p-dimensional pointcloud!

Related is the fact that in 3D we can describe a rotation in terms of an axis around which the rotation takes place, while in higher than 3D the notion of a rotation axis is generally not useful: If the rotation takes place in a 2D plane in p-space, the "axis" is a $(p - 2)$-dimensional subspace of fixed points; but if the rotation is more general, the "axis" of fixed points can be of dimensions $p - 4, p - 6, \ldots$, and such "axes" do not determine a unique 1-parameter family of rotations. A similar point was made by Tukey (1987, Section 8).

The apparent complexity raised by these issues was answered in a radical way by Asimov's notion of a grand tour (Asimov, 1985): Just like 3D data rotations expose viewers to dynamic 2D projections of 3D space, a grand tour exposes viewers to dynamic 2D projections of higher dimensional space, but unlike 3D data rotations, a grand tour presents the viewer with an automatic movie of projections with no user control. A grand tour is by definition a movie of low-dimensional projections constructed in such a way that it comes arbitrarily close to any low-dimensional projection; in other words, a grand tour is a space-filling curve in the manifold of low-dimensional projections of high-dimensional data spaces. The grand tour as a fully automatic animation is conceptually simple, but its simplicity leaves users with a mixed experience: that of the power of directly viewing high dimensions, and that of the absence of control and involvement.

Since Asimov's original paper, much of our work on tours has gone into reclaiming the interactive powers of 3D rotations and extending them in new directions. We did this in several ways: by allowing users to restrict tours to promising subspaces, by offering a battery of view-optimizations, and by re-introducing manual control to the motion of projections. The resulting methods are what we call "guided tours" and "manual tours".

At the base of these interactively controlled tours is a computational substrate for the construction of paths of projections, and it is this substrate that is the topic of the present article. The simple idea is to base all computations on the continuous interpolation of discrete sequences of projections (Buja and Asimov, 1986). An intuitive depiction of this interpolation substrate is shown in Figure 1, and a realistic depiction in Figure 2. Interpolating paths of projections are analogous to connecting line segments that interpolate points in Euclidean spaces. Interpolation provides the bridge between continuous animation and discrete choice of sequences of projections:

- Continuous animation gives viewers a sense of coherence and temporal comparison between pictures seen now and earlier. Animation can be subjected to controls such as start and stop, move forward or back up, accelerate or slow down.

Computational methods for high-dimensional rotations in data visualization 393

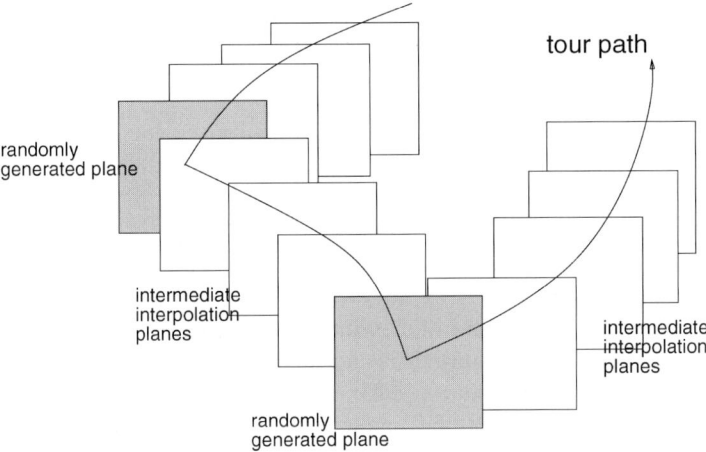

Fig. 1. Schematic depiction of a path of projection planes that interpolates a sequence of randomly generated planes. This scheme is an implementation of a grand tour as used in XGobi (Swayne et al., 1998).

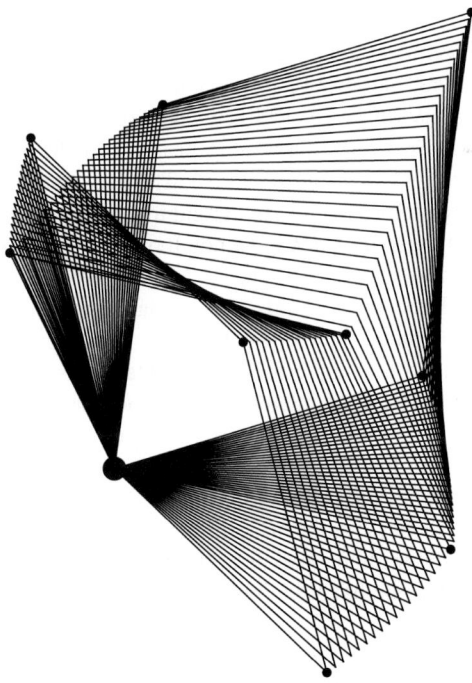

Fig. 2. XGobi looking at XGobi: This figure shows a random projection of a path of projection planes that interpolates three randomly generated projections. The planes are represented as squares spanned by their projection frames. The origin is marked by a fat dot, and the three random planes by small dots. The figure was created in XGobi's tour module with data generated from a sequence of XGobi projection frames.

- Discrete sequences offer freedom of use for countless purposes: they can be randomly selected, systematically constrained, informatively optimized, or manually directed. In other words, particular choices of discrete sequences amount to implementations of grand tours, guided tours, and manual tours.

For a full appreciation of the second point, which reflects the power of the proposed interpolation substrate, we list a range of high-level applications it can support. Most of these applications are available in the XGobi software (Swayne et al., 1998) and its successor, the GGobi software. Sequences of projections can be constructed as follows:

- *Random choice*: If the goal is to "look at the data from all sides", that is, to get an overview of the shape of a p-dimensional pointcloud, it may be useful to pick a random sequence of projections. This choice yields indeed a particular implementation of the *grand tour* (Asimov, 1985), defined as an infinite path of projections that becomes dense in the set of all projections. Denseness is trivially achieved by interpolating a discrete set of projections that is dense in the set of projections, the so-called Grassmann manifold.
- *Precomputed choice*: At times one would like to design paths of data projections for specific purposes. An example is the *little tour* (McDonald, 1982) which follows a path that connects all projections onto pairs of variables. The little tour is the animated analog of a scatterplot matrix which also shows plots of all pairs of variables, but without temporal connection. Another example is what may be called the *packed tour*: One places a fixed number of projections as far as possible from each other – that is, one finds a packing of fixed size of the Grassmann manifold – and places a shortest (Hamiltonian) path through these projections. Such Grassmann packings have been computed by Conway et al. (1996). The packed tour is essentially an improvement over the random-choice based grand tour under the stipulation that one is only willing to watch the tour for a fixed amount of time (Asimov, 1985, Remark R1, p. 130).
- *Data-driven choice*: These are methods we summarily call *guided tours* because the paths are partly or completely guided by the data. A particular guided tour method uses *projection pursuit*, a technique for finding data projections that are most structured according to a criterion of interest, such as a clustering criterion or a spread criterion. Our implementation of interactive projection pursuit follows hill-climbing paths on the Grassmann manifold by interpolating gradient steps of various user-selected projection pursuit criteria (Cook et al., 1995). Alternation with the grand tour provides random restarts for the hill-climbing paths. Another guided tour method uses *multivariate analysis* rather than projection pursuit: One restricts paths of projections for example to principal component and canonical variate subspaces (Hurley and Buja, 1990).
- *Manual choice*: Contrary to common belief, it is possible to manually direct projections from higher than three dimensions. Intuitively, this can be done by "pulling variables in and out of a projection". Generally, any direction in p-space can serve as a "lever" for pulling or pushing a projection. The notion is thus to move a projection plane with regard to immediately accessible levers such as variable directions in data space. Unlike customary 3D controls, this notion applies in arbitrarily many

dimensions, resulting in what we may call a *manual tour*. See (Cook and Buja, 1997) for more details and for an implementation in the XGobi software. A related technique has been proposed by Duffin and Barrett (1994).

While our tendency has been away from random grand tours towards greater human control in guided and manual tours, the emphasis of Wegman (1991) has been to remove the limitation to 1D and 2D projections with a powerful proposal of arbitrary k-dimensional tours, where k is the projection dimension. Such projections can be rendered by parallel coordinate plots and scatterplot matrices, but even the full-dimensional case $k = p$ is meaningful, when there is no dimension reduction, only dynamic p-dimensional rotation of the basis in data space.

Another area of extending the idea of tours by Wegman and co-authors (Wegman et al., 1998; Symanzik et al., 2002) is in the so-called image tour, in which high-dimensional spectral images are subjected to dynamic projections and reduced to 1- or 2-dimensional images that can be rendered as time-varying gray-scale or false-color images.

A word on the relation of the present work to the original grand tour paper by Asimov (1985) is in order: That paper coined the term "grand tour" and devised the most frequently implemented grand tour algorithm, the so-called "torus method". This algorithm parametrizes projections, constructs a space-filling path in parameter space, and maps this path to the space of projections. The resulting space-filling path of projections has some desirable properties, such as smoothness, reversibility and ease of speed control. There are two reasons, however, why some of us now prefer interpolation methods:

- The paths generated by the torus method can be highly non-uniformly distributed on the manifold of projections. As a result, the tour may linger among projection planes that are near some of the coordinate axes but far from others. Such non-uniformities are unpredictable and depend on subtle design choices in the parametrization. Although the paths are uniformly distributed in parameter space, their images in the manifold of projections are not, and the nature of the non-uniformities is hard to analyze. By contrast, tours based on interpolation of uniformly sampled projections are uniformly distributed by construction.

  Wegman and Solka (2002) make an interesting argument that the problem of non-uniformity is less aggravating for so-called full-dimensional tours (Wegman, 1991), in which data space is not projected but subjected to dynamic basis rotations and the result shown in parallel coordinate plots or scatterplot matrices. This argument sounds convincing, and the above criticism of the torus method therefore applies chiefly to tours that reduce the viewing dimension drastically. Still, non-uniformity is a defect even for full-dimensional tours (although more aesthetic than substantive), and there is no reason why one should put up with a defect that can be remedied.
- In interactive visualization systems, for which the grand tour is meant, users have a need to frequently change the set of active variables that are being viewed. When such a change occurs, a primitive implementation of a tour simply aborts the current tour and restarts with another subset of the variables. As a consequence there is discontinuity. In our experience with DataViewer (Buja et al. 1986, 1988) and XGobi (Swayne et al., 1998) we found this very undesirable. In order to fully grant the benefits of

continuity – in particular visual object and shape persistence – a tour should remain continuous at all times, even when the subspace is being changed. Therefore, when a user changes variables in XGobi, the tour interpolates to the new viewing space in a continuous fashion. The advantages of interpolation methods were thus impressed on us by needs at the level of user perception.

The above argument applies of course only to tours that keep the projection dimension fixed, as is the case for X/GGobi's 1D and 2D tours. Wegman (1991) k-dimensional tours, however, permit arbitrary projection dimensions, and when this dimension is changed, continuous transition is more difficult to achieve because projection dimensions are added or removed. However, when the dimension of data space is changed but the projection dimension is preserved, continuous transitions are possible.

This article is laid out as follows: Section 2 gives an algorithmic framework for computer implementations, independent of the particular interpolating algorithms. Section 3 describes an interpolation scheme for projections that is in some sense optimal. Section 4 gives several alternative methods that have other benefits. All methods are based on familiar tools from numerical linear algebra: real and complex eigendecompositions, Givens rotations, and Householder transformations.

Free software that implements dynamic projections can be obtained from the following sites:

- *GGobi* by Swayne, Temple-Lang, Cook, and Buja for Linux and MS Windows™:

    http://www.ggobi.org/.

    See (Swayne et al. 2002, 2003).

- *XGobi* by Swayne et al. (1998) for Unix® and Linux operating systems:

    http://www.research.att.com/areas/stat/xgobi/.

    See (Swayne et al., 1998). The tour module of XGobi implements most of the numerical methods described in this paper. A version of the software that runs under MS Windows™ using a commercial X™ emulator has been kindly provided by Brian Ripley:

    http://www.stats.ox.ac.uk/pub/SWin/.

- *CrystalVision* for MS Windows™ by Wegman, Luo and Fu:

    ftp://www.galaxy.gmu.edu/pub/software/.

    See (Wegman, 2003). This software implements among other things k-dimensional (including full-dimensional) tours rendered with parallel coordinates and scatterplot matrices.

- *ExploreN* for SGI Unix® by Luo, Wegman, Carr and Shen:

    ftp://www.galaxy.gmu.edu/pub/software/.

    See (Carr et al., 1996).

- *Lisp-Stat* by Tierney contains a grand tour implementation:

    http://lib.stat.cmu.edu/xlispstat/.

    See (Tierney, 1990, Chapter 10).

Applications. This article is about mathematical and computational aspects of dynamic projections. As such it will leave those readers unsatisfied who would like to see dynamic projection methods in use for actual data analysis. We can satisfy this desire partly by providing pointers to other literature: A few of our own papers give illustrations (Buja et al., 1996; Cook et al., 1995; Furnas and Buja, 1994; Hurley and Buja, 1990). A wealth of applications with interesting variations, including image tours, is by Wegman and co-authors (Wegman, 1991; Wegman and Carr, 1993; Wegman and Shen, 1993; Wegman et al., 1998; Symanzik et al., 2002; Wegman and Solka, 2002; Wegman, 2003).

3D data rotations are used extensively in the context of regression by Cook and Weisberg (1994).

It should be kept in mind that the printed paper has never been a satisfactory medium for conveying intuitions about motion graphics. Nothing replaces live or taped demonstrations or, even better, hands-on experience.

Terminology. In what follows we use the terms *"plane"* or *"projection plane"* to denote subspaces of any dimension, not just two.

2. Tools for constructing plane and frame interpolations: orthonormal frames and planar rotations

We outline a computational framework that can serve as the base of any data visualization with dynamic projections. We need notation for the linear algebra that underlies projections:

- Let p be the high dimension in which the data live, and let $x_i \in \mathbb{R}^p$ denote the column vector representing the ith data vector ($i = 1, \ldots, N$). The practically useful data dimensions are p from 3 (traditional) up to about 10.
- Let d be the dimension onto which the data is being projected. The typical projection dimension is $d = 2$, as when a standard scatterplot is used to render the projected data. However, the extreme of $d = p$ exists also and has been put to use by Wegman (1991) in his proposal of a full-dimensional grand tour in which the rotated p-space is rendered in a parallel coordinate plot or scatterplot matrix (this is a special case of his general k-dimensional tour; note we use d where he uses k).
- An "orthonormal frame" or simply a "frame" F is a $p \times d$ matrix with pairwise orthogonal columns of unit length:

    $$F^T F = I_d,$$

    where I_d is the identity matrix in d dimensions, and F^T is the transpose of F. The orthonormal columns of F are denoted by f_i ($i = 1, \ldots, d$).

- The projection of x_i onto F is the d-vector $y_i = F^T x_i$. This is the appropriate notion of projection for computer graphics where the components of y_i are used as coordinates of a rendering as scatterplot or parallel coordinate plot on a computer screen.

    (By comparison, d-dimensional projections in the mathematical sense are linear idempotent symmetric rank-d maps; they are obtained from frames F as $P = FF^T$. In this sense the projection of x_i is $P x_i = F y_i$, which is a p-vector in the column space of F. Two frames produce the same mathematical projection iff their column spaces are the same.)
- Paths of projections are given by continuous *one-parameter families* $F(t)$ where $t \in [a, z]$, some real interval representing essentially time. We denote the starting and the target frame by $F_a = F(a)$ and $F_z = F(z)$, respectively.
- The animation of the projected data is given by a path $y_i(t) = F(t)^T x_i$ for each data point x_i. At time t, the viewer of the animation sees a rendition of the d-dimensional points $y_i(t)$, such as a scatterplot if $d = 2$ as in XGobi's 2D tour, or a parallel coordinate plot if $d = p$ as in Wegman's full-dimensional tour, a special case of his general k-dimensional tour (we use d where he uses k).
- We need notation for the dimensions of various subspaces that are relevant for the construction of paths of projections. We write span(...) for the vector space spanned by the columns contained in the arguments. Letting

$$d_S = \dim\bigl(\mathrm{span}(F_a, F_z)\bigr) \quad \text{and} \quad d_I = \dim\bigl(\mathrm{span}(F_a) \cap \mathrm{span}(F_z)\bigr),$$

it holds

$$d_S = 2d - d_I.$$

Some special cases:
- We have $d_S = 2d$ whenever $d_I = 0$, that is, the starting and target plane intersect only at the origin, which is the generic case for a 2D tour in $p \geqslant 4$ dimensions.
- Tours in $p = 3$ dimensions are just 3D data rotations, in which case generally $d_S = 3$ and $d_I = 1$.
- In the other extreme, when the two planes are identical and the plane is rotated within itself, we have $d_S = d_I = d$, which is the generic case for a full-dimensional tour with $p = d_S = d_I$.

2.1. Minimal subspace restriction

The point of the following considerations is to reduce the problem of path construction to the smallest possible subspace and allow for particular bases in which interpolation can be carried out simply.

A path of frames $F(t)$ that interpolates two frames F_a and F_z should be parsimonious in the sense that it should not traverse parts of data space that are unrelated to F_a and F_z. For example, if both these frames exist in the space of variables $x_1, \ldots, x_5$, then the path $F(t)$ should not make use of variable x_6. In general terms, $F(t)$ should live in the joint span span(F_a, F_z). The restriction of the frame path $F(t)$ to this subspace requires some minor infrastructure which is set up in a step that we call "preprojection".

Preprojection is carried out as follows: Form an arbitrary orthonormal basis of span(F_a, F_z), for example, by applying Gram–Schmidt to F_z with regard to F_a. Denote the resulting basis frame of size $p \times d_S$ by

$$B = (\boldsymbol{b}_1, \boldsymbol{b}_2, \ldots, \boldsymbol{b}_{d_S}).$$

Note that when span(F_a) = span(F_z) = span(F_a, F_z) and hence $d_S = d_I = d$, the plane is rotated within itself, yet preprojection is *not* vacuous: Most interpolation algorithms require particular (for example canonical) bases in order to simplify interpolation.

We can now express the original frames in this basis:

$$F_a = BW_a \quad \text{and} \quad F_z = BW_z,$$

where $W_a = B^{\mathsf{T}} F_a$ and $W_z = B^{\mathsf{T}} F_z$ are orthonormal frames of size $d_S \times d$. The problem is now reduced to the construction of paths of frames $W(t)$ that interpolate the preprojected frames W_a and W_z. The corresponding path in data space is

$$F(t) = BW(t)$$

and the viewing coordinates of a data vector $\boldsymbol{x}_i$ are

$$\boldsymbol{y}_i(t) = F(t)^{\mathsf{T}} \boldsymbol{x}_i = W^{\mathsf{T}}(t) B^{\mathsf{T}} \boldsymbol{x}_i.$$

If one anticipates projections from very high dimensions, it might be useful to preproject the data to $\boldsymbol{\xi}_i = B^{\mathsf{T}} \boldsymbol{x}_i$ and lessen the computational expense by computing only $\boldsymbol{y}_i(t) = W^{\mathsf{T}}(t) \boldsymbol{\xi}_i$ during animation. When the data dimension p is below 10, say, the reduction in computational cost may not be worth the additional complexity.

2.2. Planar rotations

The basic building blocks for constructing paths of frames are planar rotations, that is, rotations that have an invariant 2D plane with action corresponding to

$$\begin{pmatrix} c_\tau & -s_\tau \\ s_\tau & c_\tau \end{pmatrix}$$

and an orthogonal complement of fixed points. ($c_\tau = \cos(\tau)$ and $s_\tau = \sin(\tau)$.)

If the action is in the plane of variables i and j, we denote the rotation by $R_{ij}(\tau)$, which is then also called a *Givens rotation*. Note that the order of i and j matters: $R_{ij}(\tau) = R_{ji}(-\tau)$. For efficiency, $R_{ij}(\tau)$ is never stored explicitly; matrix multiplications involving $R_{ij}(\tau)$ are directly computed from i, j, and τ. See (Golub and Van Loan, 1983, Section 3.4) for computational details (note that their $J(i, k, \theta)$ is our $R_{ik}(-\theta)$).

The methods we introduce in the following sections are based on the composition of a number of Givens rotations in a suitable coordinate system. The basic step is always the construction of a composition that maps the starting frame onto the target frame. Writing $R_\mu(\tau_\mu)$ for $R_{i_\mu j_\mu}(\tau_\mu)$, this is

$$W_z = R_m(\tau_m) \ldots R_2(\tau_2) R_1(\tau_1) W_a.$$

in the preprojection. We arrive at an interpolating path of frames by simply inserting a time parameter t into the formula:

$$W(t) = R_m(\tau_1 t) \ldots R_2(\tau_2 t) R_1(\tau_1 t) W_a,$$

where $0 \leqslant t \leqslant 1$. Obviously, $W(0) = W_a$ and $W(1) = W_z$. In compact notation we write

$$R(\tau) = R_m(\tau_m) \ldots R_2(\tau_2) R_1(\tau_1), \quad \tau = (\tau_1, \ldots, \tau_m).$$

Interpolating paths based on rotations are never unique. For one thing, if $W_z = R(\tau) W_a$, then any other $\tilde{\tau}$ with $\tilde{\tau}_j = \tau_j + k_j \cdot 2\pi$ (where k_j are integers) also satisfies $W_z = R(\tilde{\tau}) W_a$. Among all these candidates, one usually selects the τ that is closest to the vector of zero angles.

Now the raw parameter $t \in [0, 1]$ is of course not a good choice for creating an animation: if one were to move in equi-distant steps $t_i = \Delta \cdot i$ from 0 to 1, one would move at differing speeds, depending on how far the starting and target frames are apart. The speed would be slow for nearby frames and fast for distant ones. In addition, there are some subtle issues of non-constancy of speed when the sequence of Givens rotations is complex.

Because speed is an issue that is general and independent of the choice of path of frames, we deal with it before we describe specific choices of paths.

2.3. Calculation and control of speed

Assuming that a speed measure for moving frames has been chosen, we can step along paths of frames in such a way that the motion has constant speed with regard to the chosen measure, thereby providing constancy of motion. What measure of speed should be chosen? We cannot discuss here the full answer, but an outline is as follows.

Speed is distance traveled per unit of time. For paths of frames $F(t)$ the speed at time t is therefore some function of the derivative $F'(t)$. The question is what the natural speed measures are. A natural requirement is certainly invariance under orthogonal coordinate transformations: a rotated path should have the same speed characteristics as the unrotated path. Surprisingly this requirement alone is powerful enough to constrain the speed measures to the following family:

$$g_F(F') = \alpha_p \cdot \|F'\|_{\text{Frob}}^2 + (\alpha_w - \alpha_p) \cdot \|F^\mathsf{T} F'\|_{\text{Frob}}^2, \qquad (1)$$

where $\alpha_p > 0$ and $\alpha_w \geqslant 0$, and $\|A\|_{\text{Frob}}^2 = \sum_{i,j} A_{i,j}^2 = \text{trace}(A^\mathsf{T} A) = \text{trace}(AA^\mathsf{T})$ is the Frobenius norm of matrices. In terms of differential geometry, these are all possible rotation-invariant Riemannian metrics on the so-called Stiefel manifold of frames. See (Buja et al., 2004, Theorem 2). We cannot go into the details, but here is some intuition: FF^T being the orthogonal projection onto span(F) and $I - FF^\mathsf{T}$ onto its orthogonal complement, the quantity $\|(FF^\mathsf{T})F'\|_{\text{Frob}}^2 = \|F^\mathsf{T} F'\|_{\text{Frob}}^2$ measures how fast the path rotates *within* the projection plane, whereas $\|(I - FF^\mathsf{T})F'\|_{\text{Frob}}^2 = \|F'\|_{\text{Frob}}^2 - \|F^\mathsf{T} F'\|_{\text{Frob}}^2$ measures how fast the path rotates *out of* the current projection plane. In this light, Eq. (1) takes the more interpretable form

$$g_F(F') = \alpha_p \cdot \|(I - FF^\mathsf{T}) F'\|_{\text{Frob}}^2 + \alpha_w \cdot \|(FF^\mathsf{T}) F'\|_{\text{Frob}}^2. \qquad (2)$$

According to this formula any squared speed measure can be obtained as a positive linear combination of squared out-of-plane and within-plane speeds.

Even with the reduction to the above family of speed measures, we are still left with many choices. Measures that differ only by a constant positive factor are equivalent, however, hence the remaining choice is that of the ratio α_w/α_p. We (Buja et al., 2004) single out two choices for which arguments can be given: $\alpha_p = \alpha_w = 1$, which is the simplest in algebraic terms, and $\alpha_p = 2$, $\alpha_w = 1$, which is mathematically most meaningful. The latter gives a greater weight to out-of-plane speed than the former. Either choice works reasonably well in practice.

In order to control speed in computer implementations, the following considerations are elementary: Computer animations are generated by updating the display at a fast but *constant* rate (at least 5–10 per second). This implies that animation speed is not usually controlled by varying the update rate but by varying the step size along the motion path: Wider steps produce faster motion.

For dynamic projections, this implies that a path of frames is discretized, and speed is controlled by proper choice of the step size of the discretization. Incrementally, this means that at time t, one steps from $F(t)$ to $F(t + \Delta)$, amounting to a distance $\int_t^{t+\Delta} g_F(F'(\tau))^{1/2} \, d\tau$. Approximate constancy of speed can be provided as follows. Using the first order approximation

$$\text{StepSize} = \int_t^{t+\Delta} g_{F(\tau)}(F'(\tau))^{1/2} \, d\tau \approx g_{F(t)}(F'(t))^{1/2} \cdot \Delta,$$

we can choose the increment Δ as

$$\Delta = \text{StepSize}/g_{F(t)}(F'(t))^{1/2},$$

where StepSize is a user-chosen speed parameter that expresses something akin to "degrees of motion between updates". Typically, the user does not need to know actual values of StepSize because this quantity is controlled interactively through graphical gauges. (As a recommendation to implementors, gauges such as sliders should not represent StepSize linearly: At slow speeds it is important to be offered very precise control, while for high speeds it is more important to be offered a large range than high precision. Letting StepSize be proportional to the square of the value read from a slider works well.)

We note that all speed calculations can be carried out in the preprojection: Because $B^T B = I_{d_S}$ we have $g_F(F'(t)) = g_W(W'(t))$ for the invariant speed measures of Eq. (1).

Abbreviating the concatenated planar rotations of Section 2.2 and their derivatives by $R = R(\tau t)$ and $R' = d/dt\, R(\tau t)$, respectively, we have $W' = R'W_a$, and the speed measures of Eq. (1) in the form given above become

$$g_W(W') = \alpha_p \cdot \|R'W_a\|_{\text{Frob}}^2 + (\alpha_w - \alpha_p) \cdot \|W_a^T R^T R' W_a\|_{\text{Frob}}^2.$$

When R is a complex concatenation of many planar rotations, it can be impossible or at least messy to obtain R' analytically. It may then be advantageous to approximate

R' numerically by actually computing the matrix $R(\tau t)$ for two close values of t, and calculating a finite difference quotient,

$$R' \approx \big(R\big(\tau(t+\delta)\big) - R(\tau t)\big)/\delta$$

which can be substituted for R'. In some cases, such as the optimal paths of Sections 3 and 4.1, there exist explicit formulas for speed.

2.4. Outline of an algorithm for interpolation

In this section we give the schematic outline of an interpolation algorithm. Given starting and target frames F_a and F_z, the outline assumes that we know how to construct a preprojection basis B and a sequence of planar rotations $R(\tau) = R_m(\tau_m) \ldots R_1(\tau_1)$ that map $W_a = B^T F_a$ to $W_z = B^T F_z$. The construction of B and $R(\tau)$ is dependent on the interpolation method, and in the following sections we give several such constructions.

ALGORITHM.

1. Given a starting frame F_a, create a new target frame F_z.
2. Initialize interpolation:

    - Construct a preprojection basis B of span(F_a, F_z), where B has d_S columns.
    - Check: If starting and target plane are the same: $d = d_S$, and
      if starting and target frame have opposite orientations: $\det(F_a^T F_z) = -1$,
      then interpolation within the span is impossible.
      Possible remedy: flip one frame vector, $\boldsymbol{f}_{z,d} \leftarrow -\boldsymbol{f}_{z,d}$.
    - Preproject the frames: $W_a = B^T F_a$, $W_z = B^T F_z$.
    - Construct planar rotations in coordinate planes:

    $$R(\tau) = R_m(\tau_m) \ldots R_1(\tau_1), \quad \tau = (\tau_1, \ldots, \tau_m),$$

    such that $W_z = R(\tau) W_a$.
    - Initialize: $t \leftarrow 0$.

3. Execute interpolation; iterate the following:
    $t \leftarrow \min(1, t)$
    $W(t) \leftarrow R(\tau t) W_a$
    $F(t) \leftarrow B W(t)$ (render frame)
    $\boldsymbol{y}_i(t) \leftarrow F(t)^T \boldsymbol{x}_i$, $i = 1, \ldots, N$ (render data)
    If $t = 1$: break iterations.
    Else: $\Delta \leftarrow \text{StepSize}/g_W(W')^{1/2}$, $t \leftarrow t + \Delta$, do next iteration.
4. Set $F_a \leftarrow F_z$ and go to beginning.

In the following sections, we will only specify the construction of B and $R(\tau)$ and refer the reader to the above algorithm.

When the algorithm is integrated in a larger interactive visualization system such as XGobi or GGobi, a number of additional measures have to be taken at each iteration because interactions by the user may have reset some of the parameters:

- Read the speed parameter StepSize.
- Check whether a new subspace of data space has been selected. Most often a new subspace is specified in terms of a new subset of data variables.
- Check whether another method of target generation has been selected.

If one of the last two checks is positive, the current path has to be interrupted and a new path has to be initialized with the proper changes.

The two lines marked "*render*" imply execution of the rendering mechanisms for generating a new graphical scene from the data projection, such as drawing a new scatterplot or parallel coordinate plot, and for giving the viewer feedback about the position of the current projection, such as drawing a p-pod or generalized tripod that represents the projection of the canonical variable basis onto the current plane.

In what follows we describe the details for algorithms that interpolate frames F_a and F_z (Section 4), but prior we describe an algorithm that interpolates planes (Section 3) in the sense that only span(F_z) is specified and the particular frame F_z within the target plane is determined by the algorithm. The paths produced by the algorithm have the desirable property that their within-plane rotation is zero: $F^T F' = 0$.

3. Interpolating paths of planes

We describe paths of frames that optimally interpolate planes in the sense that they are locally shortest (geodesic) with regard to the metrics discussed in Section 2.3. The interpolation scheme based on these paths is appropriate for rendering methods that are visually invariant under changes of orientation (Buja et al., 2004). The construction of the paths is based on the following:

FACT. *Given two planes of dimension d, there exist orthonormal d-frames G_a and G_z that span the planes and for which $G_a^T G_z$ is diagonal.*

Optimal paths of planes essentially rotate the columns of G_a into the corresponding columns of G_z. Denoting the columns of G_a and G_z by $g_{a,j}$ and $g_{z,j}$, respectively, we see that the planes spanned by $g_{a,j}$ and $g_{z,j}$ are mutually orthogonal for $j = 1, 2, \ldots, d$, and they are either 1-dimensional (if $g_{a,j} = g_{z,j}$) or 2-dimensional (otherwise). The motion is carried out by a moving frame $G(t) = (g_1(t), \ldots, g_d(t))$ as follows: If $g_{a,j} = g_{z,j}$, then $g_j(t) = g_{a,j}$ is at rest; otherwise $g_j(t)$ rotates from $g_{a,j}$ to $g_{z,j}$ at constant speed proportional to the angle between $g_{a,j}$ and $g_{z,j}$.

The columns of the frames G_a and G_z are called "principal directions" of the pair of planes. Without loss of generality, we can assume that the diagonal elements of $G_a^T G_z$ are (1) non-negative and (2) sorted in descending order: $1 \geq \lambda_j \geq \lambda_{j+1} \geq 0$. (For non-negativity, multiply a column of G_a with -1 if necessary.) The diagonal elements λ_j of $G_a^T G_z$, called "principal cosines", are the stationary values of the cosines of angles between directions in the two planes. In particular, the largest principal cosine describes the smallest angle that can be formed with directions in the two planes. The angles $\alpha_j = \cos^{-1} \lambda_j$ are called principal angles ($0 \leq \alpha_1 \leq \alpha_2 \leq \cdots \leq \alpha_d \leq \pi/2$). Note

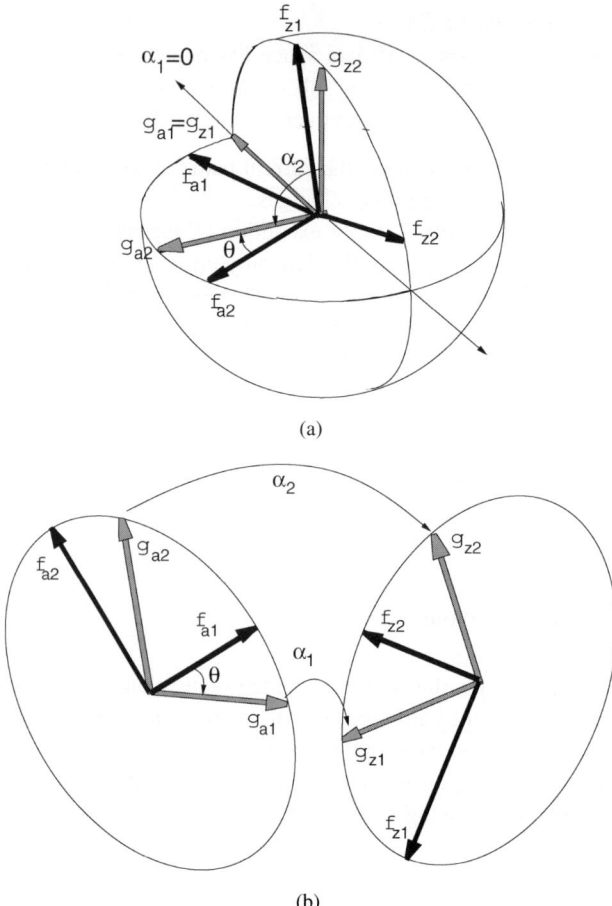

Fig. 3. Relative positions of given orthonormal bases, $(f_{a,1}, f_{a,2})$, $(f_{z,1}, f_{z,2})$, and pairs of principal directions, $(g_{a,1}, g_{a,2})$, $(g_{z,1}, g_{z,2})$, of the two 2-planes in (a) 3 dimensions, (b) 4 dimensions (the origin is pulled apart to disentangle the picture).

that for two planes with a nontrivial intersection of dimension $d_I > 0$, there will be d_I vanishing principal angles: $\alpha_1 = \cdots = \alpha_{d_I} = 0$. In particular, two 2-planes in 3-space always have at least an intersection of dimension $d_I = 1$, and hence $\lambda_1 = 1$ and $\alpha_1 = 0$. The geometry of principal directions is depicted in Figure 3.

(The principal angles are the essential invariants for the relative position of two planes to each other. This means technically that two pairs of planes with equal principal angles can be mapped onto each other with a rotation (Halmos, 1970).)

Principal directions are easily computed with a singular value decomposition (SVD). Given two arbitrary frames F_a and F_z spanning the respective planes, let

$$F_a^T F_z = V_a \Lambda V_z^T$$

be the SVD of $F_a^T F_z$, where V_a and V_z are $d \times d$ orthogonal matrices and Λ is diagonal with singular values λ_j in the diagonal. The frames

$$G_a = F_a V_a \quad \text{and} \quad G_z = F_z V_z$$

satisfy $G_a^T G_z = \Lambda$. Hence the singular values are just the principal cosines, and the orthogonal transformations V_a and V_z provide the necessary rotations of the initial frames (Bjorck and Golub, 1973; Golub and Van Loan, 1983, Section 12.4).

The following construction selects an interpolating path with zero within-plane spin that is not only locally but globally shortest. The path is unique iff there are no principal angles of 90 degrees.

PATH CONSTRUCTION.

1. Given a starting frame F_a and a preliminary frame F_z spanning the target plane, compute the SVD

$$F_a^T F_z = V_a \Lambda V_z^T, \quad \Lambda = \text{diag}(\lambda_1 \geqslant \cdots \geqslant \lambda_d),$$

and the frames of principal directions:

$$G_a = F_a V_a, \quad G_z = F_z V_z.$$

2. Form an orthogonal coordinate transformation U by, roughly speaking, orthonormalizing (G_a, G_z) with Gram–Schmidt. Due to $G_a^T G_z = \Lambda$, it is sufficient to orthonormalize $g_{z,j}$ with regard to $g_{a,j}$, yielding $g_{*,j}$. This can be done only for $\lambda_j < 1$ because for $\lambda_j = 1$ we have $g_{a,j} = g_{z,j}$, spanning the d_I-dimensional intersection $\text{span}(F_a) \cap \text{span}(F_z)$. (Numerically, use the criterion $\lambda_j > 1 - \epsilon$ for some small ϵ to decide inclusion in the intersection.)

Form the preprojection basis

$$B = (g_{a,d}, g_{*,d}, g_{a,d-1}, g_{*,d-1}, \ldots,$$
$$g_{a,d_I+1}, g_{*,d_I+1}, g_{a,d_I}, g_{a,d_I-1}, \ldots, g_{a,1}).$$

The last d_I vectors $g_{a,d_I}, g_{a,d_I-1}, \ldots, g_{a,1}$ together span the intersection of the starting and target plane. The first $d - d_I$ pairs $g_{a,j}, g_{*,j}$ span each a 2D plane in which we perform a rotation:

3. The sequence of planar rotations $R(\tau)$ is composed of the following $d - d_I$ planar rotations:

$$R_{12}(\tau_1) R_{34}(\tau_2) \ldots,$$
$$\text{where } \tau_j = \cos^{-1} \lambda_{d+1-j} \text{ for } j = 1, 2, \ldots, d - d_I.$$

The resulting path moves G_a to G_z, and hence $F_a = G_a V_a^T$ to $G_z V_a^T$. The original frame $F_z = G_z V_z^T$ in the target plane is thus replaced with the target frame $G_z V_a^T = F_z V_z V_a^T$. The rotation $V_z V_a^T$ maps F_z to the actual target frame in $\text{span}(F_z)$.

For these paths, it is possible to give an explicit formula for speed measures, of which there exists essentially only one: In the preprojection basis, the moving frame is of the form

$$W(t) = R(\tau t)W_a = \begin{pmatrix} c_{\tau_1 t} & 0 & \cdots \\ s_{\tau_1 t} & 0 & \cdots \\ 0 & c_{\tau_2 t} & \cdots \\ 0 & s_{\tau_2 t} & \cdots \\ \cdots & \cdots & \cdots \end{pmatrix},$$

hence $g_W(W') = \|W'\|^2 = \tau_1^2 + \tau_2^2 + \cdots$. In the formula for speed measures, Eq. (2), the second term for within-plane rotation vanishes due to $F^T F' = 0$. The speed measure is therefore the same for all choices of α_w. The speed measure $g_W(W')$ is constant along the path and therefore needs to be computed only once at the beginning of the path.

In the most important case of $d = 2$-dimensional projections, the SVD problem is of size 2×2, which can be solved explicitly: The eigenvalues of the 2×2 symmetric matrix $(F_a^T F_z)(F_a^T F_z)^T$ are the squares of the singular values of $F_a^T F_z$ and can be found by solving a quadratic equation. We give the results without a proof:

LEMMA 1. *The principal directions in a 2D starting plane are given by*

$$\boldsymbol{g}_{a,1} = \cos\theta \cdot \boldsymbol{f}_{a,1} + \sin\theta \cdot \boldsymbol{f}_{a,2}, \qquad \boldsymbol{g}_{a,2} = -\sin\theta \cdot \boldsymbol{f}_{a,1} + \cos\theta \cdot \boldsymbol{f}_{a,2},$$

where

$$\tan(2\theta) = 2\frac{(\boldsymbol{f}_{a,1}^T \boldsymbol{f}_{z,1}) \cdot (\boldsymbol{f}_{a,2}^T \boldsymbol{f}_{z,1}) + (\boldsymbol{f}_{a,1}^T \boldsymbol{f}_{z,2}) \cdot (\boldsymbol{f}_{a,2}^T \boldsymbol{f}_{z,2})}{(\boldsymbol{f}_{a,1}^T \boldsymbol{f}_{z,1})^2 + (\boldsymbol{f}_{a,1}^T \boldsymbol{f}_{z,2})^2 - (\boldsymbol{f}_{a,2}^T \boldsymbol{f}_{z,1})^2 - (\boldsymbol{f}_{a,2}^T \boldsymbol{f}_{z,2})^2}. \qquad (3)$$

The principal cosine λ_j is obtained by projecting $\boldsymbol{g}_{a,j}$ onto the target plane: $\lambda_j^2 = (\boldsymbol{g}_{a,j}^T \boldsymbol{f}_{z,1})^2 + (\boldsymbol{g}_{a,j}^T \boldsymbol{f}_{z,2})^2$.

Caution is needed when using Eq. (3): It has four solutions in the interval $0 \leqslant \theta < 2\pi$ spaced by $\pi/2$. Two solutions spaced by π yield the same principal direction up to a sign change. Hence the four solutions correspond essentially to two principal directions.

The denominator of the right hand side of Eq. (3) is zero when the principal angles of the two planes are identical, in which case all unit vectors in the two planes are principal.

The principal 2-frame in the target plane should always be computed by projecting the principal 2-frame of the starting plane. If both principal angles are $\pi/2$, any 2-frame in the target plane can be used for interpolation. If only one of the principal angles is $\pi/2$, one obtains $\boldsymbol{g}_{z,2}$ as an orthogonal complement of $\boldsymbol{g}_{z,1}$ in the target plane.

4. Interpolating paths of frames

Frame interpolation – as opposed to plane interpolation – is necessary when the orientation of the projection matters, as in full-dimensional tours. In addition, frame interpolation can always be used for plane interpolation when the human cost of implementing paths with zero within-plane spin is too high. Some frame interpolation schemes are indeed quite simple to implement. They have noticeable within-plane spin – a fact which XGobi users can confirm by playing with interpolation options in the tour module.

The methods described here are based on (1) decompositions of orthogonal matrices, (2) Givens decompositions, and (3) Householder decompositions. The second and third of these methods do not have any optimality properties, but the first method is optimal for full-dimensional tours in the sense that it yields geodesic paths in $SO(p)$.

4.1. Orthogonal matrix paths and optimal paths for full-dimensional tours

The idea of this interpolation technique is to augment the starting frame and the target frame to square orthogonal matrices and solve the interpolation problem in the orthogonal group ($SO(d_S)$, to be precise). The implementation is quite straightforward. The suboptimality of the paths is obvious from the arbitrariness of the way the frames are augmented to square orthogonal matrices. This is why full-dimensional paths are optimal: They do not require any augmentation at all. Strictly speaking, the requirement for optimality is not "full dimensionality" in the sense $d = p$, but simply that the starting plane and the target plane are the same, that is, $d_S = d$, in which case a simple rotation of the resting space takes place. For lower-dimensional frames whose space is not at rest ($d < d_S$), this arbitrariness can be remedied at least for the metric defined by $\alpha_w = 1$ and $\alpha_p = 2$: The trick is to optimize the augmentation (Asimov and Buja, 1994). The method is based on the following:

FACT. For any orthogonal transformation A with $\det(A) = +1$, there exists an orthogonal matrix V such that

$$A = V R(\tau) V^T \quad \text{where } R(\tau) = R_{12}(\tau_1) R_{34}(\tau_2) \ldots.$$

That is, in a suitable coordinate system, every orthogonal transformation with $\det = +1$ is the composition of planar rotations in mutually orthogonal 2D planes:

$$A(v_1, v_2) = (v_1, v_2) \begin{pmatrix} c_\phi & -s_\phi \\ s_\phi & c_\phi \end{pmatrix},$$

where v_1 and v_2 form an orthonormal basis of an invariant plane. See, for example, (Halmos, 1958, Section 81).

Invariant planes can be found with a complex eigendecomposition of A (as implemented, for example, in subroutine "dgeev" in LAPACK, available from *netlib.lucent.com/netlib/lapack*). If $v = v_r + i v_i$ is a complex eigenvector of A with eigenvalue $e^{i\phi}$, then the complex conjugate $\bar{v} = v_r - i v_i$ is an eigenvector with eigenvalue $e^{-i\phi}$, hence

$$A(v_r, v_i) = (v_r, v_i) \begin{pmatrix} c_\phi & s_\phi \\ -s_\phi & c_\phi \end{pmatrix},$$

which implies that $-\phi$ is the rotation angle in the invariant plane spanned by the frame (v_r, v_i). (The vectors v_r and v_i need to be normalized as they are not usually of unit length when returned by a routine such as "dgeev".)

PATH CONSTRUCTION.

1. Form a preliminary basis frame $\tilde{B}$ for preprojection by orthonormalizing the combined frame (F_a, F_z) with Gram–Schmidt. The preprojection of F_a is

$$\tilde{W}_a = \tilde{B}^T F_a = E_d = ((1, 0, 0, \ldots)^T, (0, 1, 0, \ldots)^T, \ldots).$$

The target frame $\tilde{W}_z = \tilde{B}^T F_z$ is of a general form.

2. Expand $\tilde{W}_z$ to a full orthogonal matrix A with $\det(A) = +1$, for example, by appending random vectors to $\tilde{W}_z$ and orthonormalizing them with Gram–Schmidt. Flip the last vector to its negative if necessary to ensure $\det(A) = +1$. Note that

$$\tilde{W}_z = A \tilde{W}_a,$$

because $\tilde{W}_a$ extracts the first d columns from A, which is just $\tilde{W}_z$.

3. Find the canonical decomposition $A = V R(\tau) V^T$ according to the above.
4. Change the preprojection basis: $B = \tilde{B} V$, such that $W_a = B^T F_a = V^T \tilde{W}_a$ and $W_z = B^T F_z = V^T \tilde{W}_z$. From the canonical decomposition follows

$$W_z = R(\tau) W_a.$$

The above is the only method described in this paper that has not been implemented and tested in XGobi.

4.2. Givens paths

This interpolation method adapts the standard matrix decomposition techniques based on Givens rotations. While we use Givens rotations for interpolation, Asimov's original grand tour algorithm (Asimov, 1985) uses them to (over-)parametrize the manifold of frames. Wegman and Solka (2002) call methods based on Givens rotations "winding algorithms". At the base of all uses of Givens rotations is the following:

FACT. In any vector u one can zero out the i'th coordinate with a Givens rotation in the (i, j)-plane for any $j \neq i$. This rotation affects only coordinates i and j and leaves all coordinates $k \neq i, j$ unchanged.

For example, to zero out coordinate x_2 in the (x_1, x_2)-plane, use a rotation $\begin{pmatrix} c_\tau & -s_\tau \\ s_\tau & c_\tau \end{pmatrix}$ with $c_\tau = x_1/(x_1^2 + x_2^2)^{1/2}$ and $s_\tau = -x_2/(x_1^2 + x_2^2)^{1/2}$. That is, τ is the angle from (x_1, x_2) to $(1, 0)$. See Figure 4 for a depiction. For computational details, see (Golub and Van Loan, 1983, Section 3.4).

Sequences of Givens rotations can be used to map any orthonormal d-frame F in p-space to the standard d-frame $E_d = ((1, 0, 0, \ldots)^T, (0, 1, 0, \ldots)^T, \ldots)$ as follows:

- Apply a sequence of $p - 1$ Givens rotations to zero out coordinates $2, \ldots, p$ of the first vector f_1. Examples of suitable sequences are rotations in the variables $(1, 2), (1, 3), (1, 4), \ldots, (1, p)$, or $(p - 1, p), \ldots, (3, 4), (2, 3), (1, 2)$, where the second variable is the one whose coordinate is being zeroed out. Care should be taken

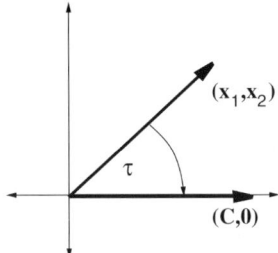

Fig. 4. Rotation to make subvector of matrix coincide with one of the coordinate axes.

that the last rotation chooses among the suitable angles τ and $\tau + \pi$ the one that results in the first coordinate $= +1$. The resulting frame will have $e_1 = (1, 0, 0, \ldots)^T$ as its first vector.
- Apply to the resulting frame a sequence of $p - 2$ Givens rotations to zero out the coordinates $3, \ldots, p$. Do not use coordinate 1 in the process, to ensure that the first vector e_1 remains unchanged. A suitable sequence of planes could be $(2, 3), (2, 4), \ldots, (2, p)$, or else $(p - 1, p), \ldots, (3, 4), (2, 3)$. Note that the zeros of the first column remain unaffected because $(0, 0)$ is a fixed point under all rotations.
- And so on, till a sequence of $(p - 1) + (p - 2) + \cdots + (p - d) = pd - \binom{d}{2}$ Givens rotations is built up whose composition maps F to E_d.

PATH CONSTRUCTION.

1. Construct a preprojection basis B by orthonormalizing F_z with regard to F_a with Gram–Schmidt:
$$B = (F_a, F_*).$$

2. For the preprojected frames
$$W_a = B^T F_a = E_d \quad \text{and} \quad W_z = B^T F_z$$
construct a sequence of Givens rotations that map W_z to W_a:
$$W_a = R_m(\tau_m) \ldots R_1(\tau_1) W_z.$$
Then the inverse mapping is obtained by reversing the sequence of rotations with the negative of the angles:
$$R(\boldsymbol{\tau}) = R_1(-\tau_1) \ldots R_m(-\tau_m), \qquad W_z = R(\boldsymbol{\tau}) W_a.$$

We made use of Givens rotations to interpolate projection frames. By comparison, the original grand tour implementation proposed in (Asimov, 1985), called "torus method", makes use of Givens decompositions in a somewhat different way: Asimov parametrizes the Stiefel manifold $V_{2,p}$ of 2-frames with angles $\boldsymbol{\tau} = (\tau_1, \tau_2, \ldots)$ provided by Givens rotations, and he devises infinite and uniformly distributed paths on the space (the "torus") of angles $\boldsymbol{\tau}$. The mapping of these paths to $V_{2,p}$ results in dense paths of frames, which therefore satisfy the definition of a grand tour. The non-uniformity problem mentioned at the end of the introduction stems from this mapping: The path of frames is not uniformly distributed, although its pre-image in the space of angles is. The non-uniformity may cause the tour to spend more time in some parts of the Stiefel manifold than in others. It would be of interest to better understand the mapping from the torus of angles to the manifold of frames. In particular, it would be interesting to know which of the many ways of constructing Givens decompositions lead to mappings with the best uniformity properties.

4.3. Householder paths

This interpolation method is based on reflections on hyperplanes, also called "Householder transformations". See, for example, (Golub and Van Loan, 1983, Section 3.3). A reflection at the hyperplane with normal unit vector r is given by

$$H = I - 2 \cdot rr^T.$$

The usefulness of reflections stems from the following:

FACTS.

1. Any two distinct vectors of equal length, $\|w\| = \|w_*\|$, $w \neq w_*$, can be mapped onto each other by a uniquely determined reflection H with

$$r = (w - w_*)/\|w - w_*\|.$$

2. Any vector orthogonal to r is fixed under the reflection.
3. The composition of two reflections H_1 and H_2 with vectors r_1 and r_2, respectively, is a planar rotation in the plane spanned by r_1 and r_2, with an angle double the angle between r_1 and r_2.

See Figure 5 for a depiction. We illustrate the technique for $d = 2$. We use a first reflection H_1 to map $f_{a,1}$ to $f_{z,1}$. Subsequently, we use a second reflection H_2 to map $H_1 f_{a,2}$ to $f_{z,2}$. Under H_2, the vector $H_1 f_{a,1}$ is left fixed because both $H_1 f_{a,2}$ and $f_{z,2}$, and hence their difference, are orthogonal to $H_1 f_{a,1} = f_{z,1}$. Thus

$$F_z = H_2 H_1 F_a,$$

where $H_2 H_1$ is a planar rotation according to the third fact above. The computational details are somewhat more involved than in numerical analysis applications because we must make sure that we end up with exactly two reflections that amount to a planar rotation:

PATH CONSTRUCTION (*for $d = 2$*).

1. Find a first reflection vector r_1 such that $H_1 f_{a,1} = f_{z,1}$: If $f_{a,1} \neq f_{z,1}$, use $r_1 = (f_{a,1} - f_{z,1})/\|f_{a,1} - f_{z,1}\|$, and $r_1 \perp f_{a,1}$ otherwise.
2. Map $f_{a,2}$ with this reflection: $f_{a,2+} = H_1 f_{a,2} = f_{a,2} - 2 \cdot (f_{a,2}^T r_1) r_1$.

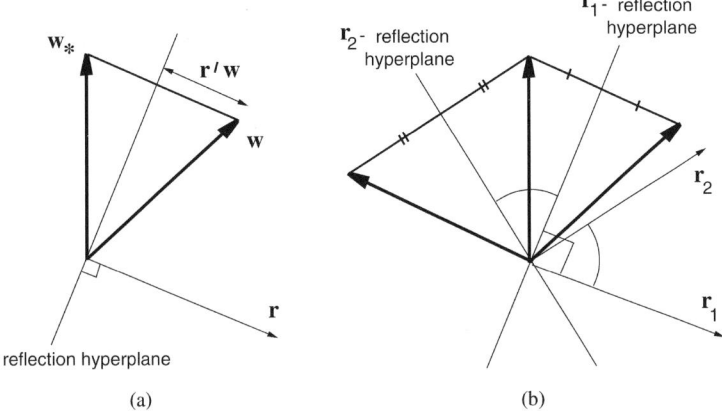

Fig. 5. (a) Reflection of a vector; (b) composition of two reflections to yield a planar rotation.

3. Find a second reflection vector r_2 such that $H_2 f_{a,2+} = f_{z,2}$: If $f_{a,2+} \neq f_{z,2}$, use $r_2 = (f_{a,2+} - f_{z,2})/\|f_{a,2+} - f_{z,2}\|$, and $r_2 \perp f_{a,2+}, \perp f_{z,2}$ otherwise.
4. Form a preprojection basis B: Orthonormalize r_2 with regard to r_1 to yield r_*; expand (r_1, r_*) to an orthonormal basis of $\mathrm{span}(F_a, F_z)$.
5. Rotation: $R_{12}(\tau)$ with $\tau = 2\cos^{-1}(r_1^T r_2)$.

The generalization of Householder paths to d-frames for $d > 2$ is quite obvious for d even: The process generates d reflections that can be bundled up into $d/2$ planar rotations. For d odd, some precautions are in order: If $\mathrm{span}(F_a) \neq \mathrm{span}(F_z)$, one has to introduce one additional dummy reflection that leaves F_z fixed, using an r_{d+1} in $\mathrm{span}(F_a, F_z)$ orthogonal to F_z; if $\mathrm{span}(F_a) = \mathrm{span}(F_z)$, the last reflection was not necessary because $H_{d-1} \ldots H_1 f_{a,d} = \pm f_{z,d}$ automatically, hence $(d-1)/2$ rotations result. If the spans are identical, the frames have to have the same orientation in order to allow continuous interpolation; hence it may be necessary to change the sign of $f_{z,d}$.

A peculiar aspect of the Householder method is that it uses fewer planar rotations than any of the other methods; as we have seen it transports 2-frames onto each other with a single planar rotation. Yet it does not produce paths that are optimal in any sense that we know of. It would be of interest to better understand the geometry of the Householder method and possibly produce criteria for which it is optimal. For example, we do not even know whether Householder paths are geodesic for one of the invariant metrics mentioned in Section 2.3.

5. Conclusions

The goal of this paper was to give an algorithmic framework for dynamic projections based on the interpolation of pairs of projections. The notion of steering from target projection to target projection makes for a flexible environment in which grand tours, interactive projection pursuit and manual projection control can be nicely embedded. Finally, we proposed several numerical techniques for implementing interpolating paths of projections.

References

Asimov, D. (1985). The grand tour: a tool for viewing multidimensional data. *SIAM J. Sci. Statist. Comput.* **6** (1), 128–143.
Asimov, D., Buja, A. (1994). The grand tour via geodesic interpolation of 2-frames. In: *Visual Data Exploration and Analysis, Symposium on Electronic Imaging Science and Technology*. IS&T/SPIE (Soc. Imaging Sci. Technology/Internat. Soc. Optical Engrg.).
Bjorck, A., Golub, G.H. (1973). Numerical methods for computing angles between linear subspaces. *Math. Comp.* **27** (123), 579–594.
Buja, A., Asimov, D. (1986). Grand tour methods: an outline. In: *Proceedings of the 17th Symposium on the Interface*. In: *Computing Science and Statistics*. Elsevier, Amsterdam, pp. 63–67.
Buja, A., Hurley, C., McDonald, J.A. (1986). A data viewer for multivariate data. In: *Proceedings of the 18th Symposium on the Interface*. In: *Computing Science and Statistics*. Elsevier, Amsterdam.
Buja, A., Asimov, D., Hurley, C., McDonald, J.A. (1988). Elements of a viewing pipeline for data analysis. In: Cleveland, W.S., McGill, M.E. (Eds.), *Dynamic Graphics for Statistics*. Wadsworth, Belmont, CA, pp. 277–308.
Buja, A., Cook, D., Swayne, D.F. (1996). Interactive high-dimensional data visualization. *J. Comput. Graph. Statist.* **5**, 78–99.
Buja, A., Cook, D., Asimov, D., Hurley, C. (2004). Theory of dynamic projections in high-dimensional data visualization. Submitted for publication, can be downloaded from www-stat.wharton.upenn.edu/~buja/.
Carr, D.B., Wegman, E.J., Luo, Q. (1996). ExploreN: design considerations past and present. Technical Report 129. Center for Computational Statistics, George Mason University, Fairfax, VA 22030.
Conway, J.H., Hardin, R.H., Sloane, N.J.A. (1996). Packing lines, planes, etc.: packings in Grassmannian spaces. *J. Experiment. Math.* **5**, 139–159.
Cook, D., Buja, A. (1997). Manual controls for high-dimensional data projections. *J. Comput. Graph. Statist.* **6**, 464–480.
Cook, D., Weisberg, S. (1994). *An Introduction to Regression Graphics*. Wiley, New York, NY.
Cook, D., Buja, A., Cabrera, J., Hurley, H. (1995). Grand tour and projection pursuit. *J. Comput. Graph. Statist.* **2** (3), 225–250.
Duffin, K.L., Barrett, W.A. (1994). Spiders: a new user interface for rotation and visualization of N-dimensional point sets. In: *Proceedings Visualization '94*. IEEE Computer Society Press, Los Alamitos, CA, pp. 205–211.
Furnas, G.W., Buja, A. (1994). Prosection views: dimensional inference through sections and projections. *J. Comput. Graph. Statist.* **3**, 323–385.
Golub, G.H., Van Loan, C.F. (1983). *Matrix Computations*, second ed. The Johns Hopkins University Press, Baltimore, MD.
Halmos, P.R. (1958). *Finite-Dimensional Vector Spaces*. Springer, New York, NY.
Halmos, P.R. (1970). Finite-dimensional Hilbert spaces. *Amer. Math. Monthly* **77** (5), 457–464.
Hurley, C., Buja, A. (1990). Analyzing high-dimensional data with motion graphics. *SIAM J. Sci. Statist. Comput.* **11** (6), 1193–1211.
McDonald, J.A. (1982). Orion I: interactive graphics for data analysis. In: Cleveland, W.S., McGill, M.E. (Eds.), *Dynamic Graphics for Statistics*. Wadsworth, Belmont, CA.
Swayne, D.F., Cook, D., Buja, A. (1998). XGobi: interactive dynamic data visualization in the X window system. *J. Comput. Graph. Statist.* **7** (1), 113–130.
Swayne, D.F., Temple-Lang, D., Buja, A., Cook, D. (2002). GGobi: evolving from XGobi into an extensible framework for interactive data visualization. *J. Comput. Statist. Data Anal.*
Swayne, D.F., Buja, A., Temple-Lang, D. (2003). Exploratory visual analysis of Graphs in GGobi. In: *Proceedings of the Third Annual Workshop on Distributed Statistical Computing*. DSC 2003, Vienna. AASC.

Symanzik, J., Wegman, E., Braverman, A., Luo, Q. (2002). New applications of the image grand tour. In: *Computing Science and Statistics*, vol. 34, pp. 500–512.

Tierney, L. (1990). *Lisp-Stat*. Wiley, New York, NY.

Tukey, J.W. (1987). Comment on 'Dynamic graphics for data analysis' by Becker et al. *Statist. Sci.* **2**, 355–395. Also in: Cleveland, W.S., McGill, M.E. (Eds.), *Dynamic Graphics for Statistics*. Wadsworth, Belmont, CA, 1988.

Wegman, E.J. (1991). The grand tour in k-dimensions. In: *Proceedings of the 22nd Symposium on the Interface*. In: *Computing Science and Statistics*. Elsevier, Amsterdam, pp. 127–136.

Wegman, E.J. (2003). Visual data mining. *Statist. Medicine* **22**, 1383–1397, plus 10 color plates.

Wegman, E.J., Carr, D.B. (1993). Statistical graphics and visualization. In: Rao, C.R. (Ed.), *Computational Statistics*. In: *Handbook of Statistics*, vol. 9. Elsevier, Amsterdam, pp. 857–958.

Wegman, E.J., Shen, J. (1993). Three-dimensional Andrews plots and the grand tour. In: *Computing Science and Statistics*, vol. 25, pp. 284–288.

Wegman, E.J., Solka, J.L. (2002). On some mathematics for visualizing high dimensional data. *Sanhkya Ser. A* **64** (2), 429–452.

Wegman, E.J., Poston, W.L., Solka, J.L. (1998). Image grand tour. In: *Automatic Target Recognition VIII*. In: *Proc. SPIE*, vol. 3371. SPIE, pp. 286–294.

Some Recent Graphics Templates and Software for Showing Statistical Summaries

Daniel B. Carr

1. Introduction

The rendering, interaction with and evaluation of statistical summaries are key elements of visual analytics. This paper describes the three graphical templates for showing statistical summaries. The templates in this paper are linked micromap (LM) plots, conditioned choropleth (CC) maps, and self-similar coordinate (SC) plots. The first two templates feature geospatially indexed statistics. The third template might more aptly be called an approach. The statistics are indexed by sequences of letters that in the examples represent amino acid sequences. The approach generates coordinates as the basis for rendering. The software in this case, GLISTEN, illustrates a generalization of geographic information system software. The paper also describes selected interactive and dynamic facets of the software implementing the templates.

In the face of huge data sets, the graphics used in this paper can seem very modest in the sense that the basic layouts were initially conceived as one page layouts for showing a small number of cases and variables. Coupling the layouts with interactive and dynamic methods, that include scrolling, variable selection, drill down and alternative views, reduces some limitations, but still many practical limitations remain. These include the limitations of visual real estate and human cognition. The pixel real estate available for pixel displays (see Keim, 1996) provides a bound on number of variables values that can be shown in one screen flash. Pragmatic considerations related to providing readable labels and human cognition lead to much stronger limitations. However, statistical summaries can make the study of very large and even massive data feasible. A single case in a graph can be a surrogate or summary for a huge number of cases. Each "value" of a single variable could be a box plot summary of a huge number of observations. Modest graphics can stretch along way when applied to statistical summaries, especially in an interactive setting.

The human–computer interface mantra is "overview first and then detail on demand." The task of creating good overviews is difficult. When dealing with large or massive data sets, heavy statistical summarization is often necessary to make the structure perceptually and cognitively accessible. Decisions must be made about what is most important to show, what is relegated to the background and what is totally omitted. Informed decisions about constructing an overview imply having used *algorithms* to work with the *data*. An informed decision implies having the strong background in terms of the choices available for summarizing *statistics*. It can also be crucial to have a strong background about the *phenomena* being described and the policy ramifications. Since such a breadth of knowledge is rarely present at one individual, the development of quality summaries may require a substantial collaboration and suppose the creating of analytic teams.

Since the target audience consists of people, making informed decisions about summaries and the underlying structure for a deeper analysis implies using knowledge about the graphic design based on *human perception and cognition*. Effective overviews should be tailored to the task (communication, discovery, hypothesis generation) and to the audience's expertise and cultural biases.

Because of the intellectual investment and the scope of knowledge required for developing visual analytic frameworks from the overview down, overview design has received a little attention. It is easier to develop yet another clustering algorithm. At least historically, being associated with mere graphics seems to reduce prestige even though the scope of knowledge required for producing quality overviews lies near the top of the knowledge pyramid. Pressing national security needs may change the appreciation level for quality quantitative graphics where crucial details, like uncertainty, do not get forgotten. The realization is dawning that visual analytics are required to cope with the flood of information that is sometimes of life and death importance. Quantitative visual analysts are important and this growing awareness will lead to better tools and working environments.

Like many papers, this chapter avoids the host of difficult summarization issues associated with large or massive data sets. What the chapter provides are some practical graphics templates suitable for the study of modest-sized summaries. Summaries can evolve to address increasingly complex situations. Looking at some summaries can motivate the production of others.

This chapter focuses on the layouts and encodings used in the graphics and briefly discusses selected interactive and dynamic features of the corresponding software. The linked micromap (LM) plots follow very sound design guidelines. Section 2 briefly indicates the selected design guidelines. Section 3 presents the basic LM plot template and indicates numerous variations. LM plots can serve for either communication or data discovery.

Conditioned choropleth (CC) maps encode statistics using colors. By itself, the color encoding has a poor perceptual accuracy of extraction (Cleveland and McGill, 1984). However, as Section 4 suggests, dynamic sliders help to shore up this weakness and make CC maps an interesting environment for both hypothesis generation and education.

The self-similar coordinate (SC) plots are newer and have been subject to little use and evaluation. Nonetheless, they address a class of problems, statistics indexed by repetitive letter sequences, which warrant attention. Section 5 indicates the construction of coordinates sequences for rendering statistical summaries and briefly discusses an organized set of widgets that provides control of 3D rendering and filtering.

Section 6 ends by turning attention back to the larger framework of visual analytics for potentially huge sets.

2. Background for quantitative graphics design

This section draws attention to a few basic points related to the design of quantitative graphics. This provides motivation for the LM plot designs appearing in Section 3 and suggests some initial weaknesses in the subsequent graphics.

Several books serve as important resources for those seriously interested in quantitative graphics design. In terms of human perception and cognition, Ware's (2004) book on perception for design is required reading. To have this knowledge implemented in statistical graphics software along with the rendering capabilities developed in computer graphics would be a blessing. Kosslyn (1994) provides good links into the perception and cognition literature and many easy to understand (but small data set) examples. Cleveland (1985, 1993a, 1993b) provides results on perceptual accuracy of extraction of univariate encodings and connections to statistics graphics applications. Tufte (1983, 1990, 1997) provides a wonderful set of examples. More encyclopedic breadth is available in Bertin's (1983) classic book and in Harris (1996). Wilkinson's (1999) book shows breath of knowledge about quantitative graphics and translates this into a graphic algebra that provides a coherent frame for software implementation. These books are, of course, just a starting place. There is much published literature and occasional gems appear in published application examples.

Of course papers and web sites provide additional resources. For example Brewer (1997) uses experiments to assess the strong opinions on spectral color schemes and provides an excellent web site (http://www.colorbrewer.org) for selecting color schemes that include schemes suitable for major categories of color blindness.

The simple starting point here comes from Cleveland and McGill (1984) and Cleveland (1985). In terms of high perceptual accuracy of extraction, position along a scale provides the best encoding for a continuous univariate variable. The decoding is like reading values against a ruler. The left panel in Figure 1 shows an example.

The middle panel correctly suggests that increased distance from the scale degrades decoding accuracy. Distance also degrades comparison accuracy. The horizontal

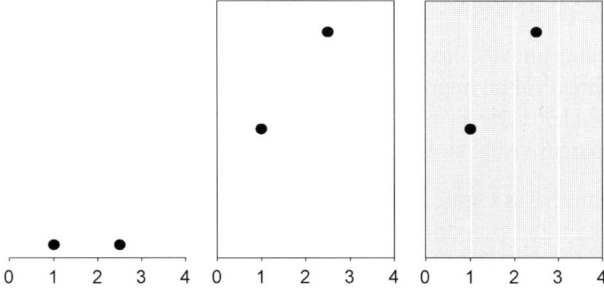

Fig. 1. Positioning along a scale and grid lines.

distance between points is harder to assess in the middle panel. The right panel use grid lines to extend the axis tick marks as basis for comparison. Weber's law covers the visual phenomena that (background) grid lines help to making more accurate comparisons. Cleveland's experiments indicated that length and angle encodings work pretty well. Most other encodings such as area, volume, color, and density are poor by comparison.

The relationships of values to a distribution and of distributions to other distributions provide frameworks for interpreting quantitative relationships. Statistical graphics tasks include *describing and comparing distributions*.

The heart of quantitative graphics is comparison. This includes comparison of statistics against reference values, modeled values, or statistics from other distributions. Wilkinson (1999) defines a schema as a diagram that includes both general and particular features in order to represent a distribution. An example schema includes a line segment with an overplotted dot showing a confidence interval and a mean, a line segment or a bar showing the range of the distribution (the minimum and the maximum or a local low or high value). The box plot is an increasingly well-known schema or distributional caricature. The version of the box plot here shows the median, the quartiles, the adjacent values and outliers (Cleveland, 1993a). Adjacent values are used in the CC map sliders in Section 4. Note that all these schema use position along a scale to encode schema components.

Statistical graphics also address tasks related to stochastic model building and model criticism. A common practice is to parameterize the mean of a distribution so that it represents a functional relationship whose parameters can be estimated. The functional relationship does not have to fit the data perfectly since it is only the mean of the distribution. The random variation in such stochastic models accommodates the residuals, the departures from the functional relationship. Many statistical graphics involve *presenting functional relationships in the context of uncertainty*. This is little discussed here, but dynamic scatterplot smooth suggest functional relationships are a part of the CC maps described in Section 4.

2.1. General guidance

Graphical challenges arise when distributions are multivariate, contexts are multivariate, and functional relations become increasingly complex. A host of guidelines have been developed to help in addressing complex situations. The guidelines can be grouped into four sometimes conflicting categories. These are:

- enable accurate comparisons;
- provide a context for an appropriate interpretation;
- involve the reader; and
- strive for apparent visual simplicity.

It would seem that accurate comparison would be important. However some popular graphics, like tilted pseudo-perspective 3D pie charts, are not the best in terms of accurate comparisons. Providing context can be a big challenge. Questions to address include who, what, when, where, why, how, and how well. In bioinformatics, the volume of context providing metadata can easily exceed the volume of data. Providing an extensive context can conflict with simple appearance. Involving the reader includes producing appealing graphics and providing interactive re-expression tools. Striving for apparent visual simplicity is really problematic when the relationships are complex. Still there is plenty what can be done. John Tukey's notions of simplifying the reference relationship are often helpful (see Cleveland, 1988). Graphics design guidance in terms of layouts, visual layering, gentle use of color (see http://www.colorbrewer.org for useful color schemes), and minimizing memory burdens can all come into play.

2.2. Challenging convention

A big challenge is to successfully promote a break with the tradition. Conventional graphics guidance says go with convention. Cleveland and McGill (1984) described the great merits of dot plots in terms of high perceptual accuracy of extraction and simplicity. Yet to this day, dot plots are hard to find in US federal publications! The graphics in Section 3 were substantially motivated by the desire to use and promote dot plots with their easily read horizontal labels. Software convenience and tradition are powerful opponents.

A serious graphics designer will encounter situations in which the preferable design goes against convention. The designer can usually make arguments for defying convention-based on personal experiments. If the arguments are correct, there may very well be supportive perception and cognition literature. It can come as a relief to read Ware (2004) as he calls attention to design conflicts in his discussion of sensory and arbitrary cultural representations. Sensory aspects of visualization derive their expressive power from being well designed to stimulate the visual sensory system. In contrast, arbitrary, cultural conventions for visualizations derive their power from how well they are learned.

As Ware indicates, these two aspects are conceptual extremes. In practice, encodings run the gamut but never quite reach pure extremes. Cultural encodings require learning, may slow processing, and lead to cross-cultural communication inefficiencies. In some cases, it is worth the cultural battle to make the arbitrary cultural conventions more compatible with sensory representations.

Of course, quantitative designs need to adapt to both the task and to the physical and educational constraints of the targeted audience. Federal graphics accessibility regulations now require that physical constraints be addressed. Ideally quantitative software could contain educational components to address the educational constraints.

2.3. Templates and GUIs

The templates to follow provide a framework for panel layout, visual linking, layering, and organizing the visual elements in the graphics. In dynamic graphics, templates can include the placement and appearance of graphical user interfaces (GUIs) as well as the options provided. Templates here apply to many different data sets but maintain a regularity that leads to familiarity with repeated use. A template may afford some flexibility in terms of relocatable GUIs and panels, but must have a clearly definable structure.

GUIs can control many things such as color choice, sorting, filtering, brushing, rotation, animation, and alternative layouts. Here the naming of a template derives from the symbolization, layout and layering of visual elements used in the template and not from GUI components. While templates can be dynamic, this is not typically reflected in the names.

With new templates in hand, there is a need to educate people about their merits and use. John Tukey promoted the box plot schema. The median, quartiles, adjacent values and outlines provide a much richer and more robust description of a distribution than a mean with a confidence interval. It took considerable education before staff in the US Food and Drug Administration became comfortable with this schema. The story is probably no different at other agencies and institutions. Today, simplified box plots appear in US elementary grade school curriculum. Education and change are possible.

3. The template for linked micromap (LM) plots

Linked micromap plots have now been described in a series of papers including (Carr and Pierson, 1996; Carr et al., 1998, 2000a, 2000b, 2002a; Carr, 2001, 2003; Wang et al., 2002). The study units in LM plots have names, map locations, and summary statistics. Figure 2 shows an LM plot example with the study units being states, the locations are polygon regions on a map and the statistical summaries are box plots of each state's constituent county mortality rates.

Encodings that use position along a scale to represent geospatially indexed statistics motivated the development of LM plots. Cartographers tended to use the best encoding, position along a scale, to show region boundaries. The statistics of interest would typically receive an inferior encoding even though they were often of greater interest than the precise location of political boundaries. LM plots sacrifice the region boundary resolution to increase the perceptual accuracy of extraction for the statistics. In Figure 2 the map is a caricature. It was adapted from a state visibility map (Monmonier, 1993) that increased the visibility of low area states.

The template for LM plots has four key features. First, LM plots include at least three parallel sequences of panels with constituent sub-panels that are linked by the position. As illustrated in Figure 2, one column of panels consists of micromaps, one column consists of linking names, and one column of statistical graphics, in this case box plots. The template can include more sequences of panels as long as it has these three types of panels. The second feature includes sorting the study units, such as the states in Figure 2. The third feature partitions the study units into panels to focus attention on a few units at a time. This produces the sub-panels in Figure 2. The sub-panels are not outlined in the first two columns but the perceptual grouping is evident. The fourth feature links the visual elements of each study unit across the corresponding sub-panels. In Figure 2 color links the state's visual elements across all three panels and position also links the state name to the box plot.

The perceptual grouping of sorted study units in LM plots serves several purposes. First, it creates small perceptual groups that facilitate focus on a manageable number of visual elements. Kosslyn (1994) cites literature indicating that four items as a good

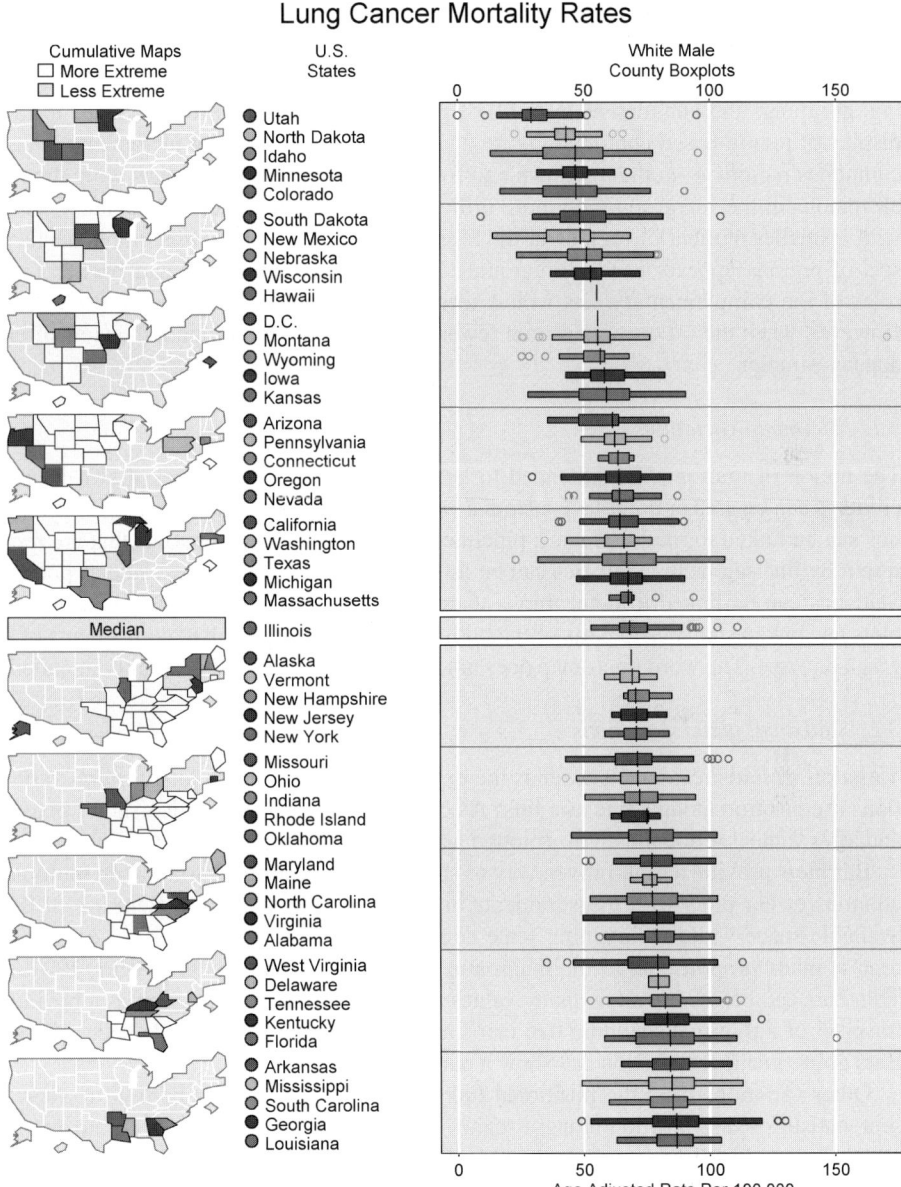

Fig. 2. An LM plot with a box plot schema. For a color reproduction of this figure see the color figures section, page 598.

number for direct comparison. Counting by five motivates the grouping in Figure 2. Second, with five study units in a group, five distinct hues suffice to clearly indicate which graphical elements belong to each study unit. Third, visual grouping makes long lists less intimidating. The edges of the groups are more salient and provide visual entry points into the list. Finding an interesting study unit signals the beginning of increased reader involvement, so increasing visual entry points is usually good. Finally, the sorting, grouping, and cumulative color fill of previously highlighted regions leads to the display of region-based contours. Figure 2 shows an experiment that incrementally accumulates regions from the top down and from the bottom up. Micromaps with almost all regions in the foreground are more difficult to study than the complementary view with a smaller number of regions in the foreground. When one of two complementary sets of regions appears in the foreground, reversing the foreground and background roles of the complementary sets would seem to results in logically equivalent views. However, when the set of regions with fewer edges is in the foreground, the view often appears simpler.

3.1. Micromap variations

The notion of micromaps is intended to be quite general. The features shown in micromaps can be arcs or point layers from a GIS rather the polygonal regions. The statistics can be linked to roads, streams, pipelines or communication networks that are represent by line segments. Statistics can be linked to cities that are represented as points. Statistics can be linked to other things than geospatial features. One variation of LM plots has linked statistics to parts of the body. The statistics can link to components of a graph or a tree. The word micromap does not necessarily convey the generality intended.

3.2. Statistical panel variations

Variation of statistical panels extends the expressive range of LM plots. Several variations use position along a line as a link. A common statistical application shows before and after values. A dot and an overplotted arrow glyph show this nicely.

Bar plots provide a time-proven way to show values. A simple variation on multiple columns of bar plots with measurements in the same units makes efficient use of the available space while maintaining scale comparability. The panel width for each column is made proportional to the maximum bar length of each column (Carr, 1994a). Thin bars can be used to show many values and multivariate sorting, such as a breadth traversal of a minimal spanning tree can simplify appearance (Carr and Olsen, 1996). Bars going in either direction can show a percentage change.

Other variations drop the positional link, so the *y*-axis can also be used to represent statistics using position along a scale. This leads to time-series plots (Carr et al., 1998), binned bivariate scatterplots involving hundreds of thousands of points (Carr et al., 2000a) and the display of rank order scatterplot smoothes and residuals (Carr, 2001).

The time-series variation typically has limited vertical resolution and suffers a bit from overplotting. With no missing data and five or fewer items in a group (four is better), the hidden points and lines can usually be inferred. In some cases, plotting back to front with decreasing point size and line thickness helps. In an interactive setting, any panel can potentially be expanded to provide increased resolution.

The study of Omernik ecoregions (Carr et al., 2000a) led to a variant of LM plots that stretches the previous LM plot definition by leaving the column of names. Similarly, in

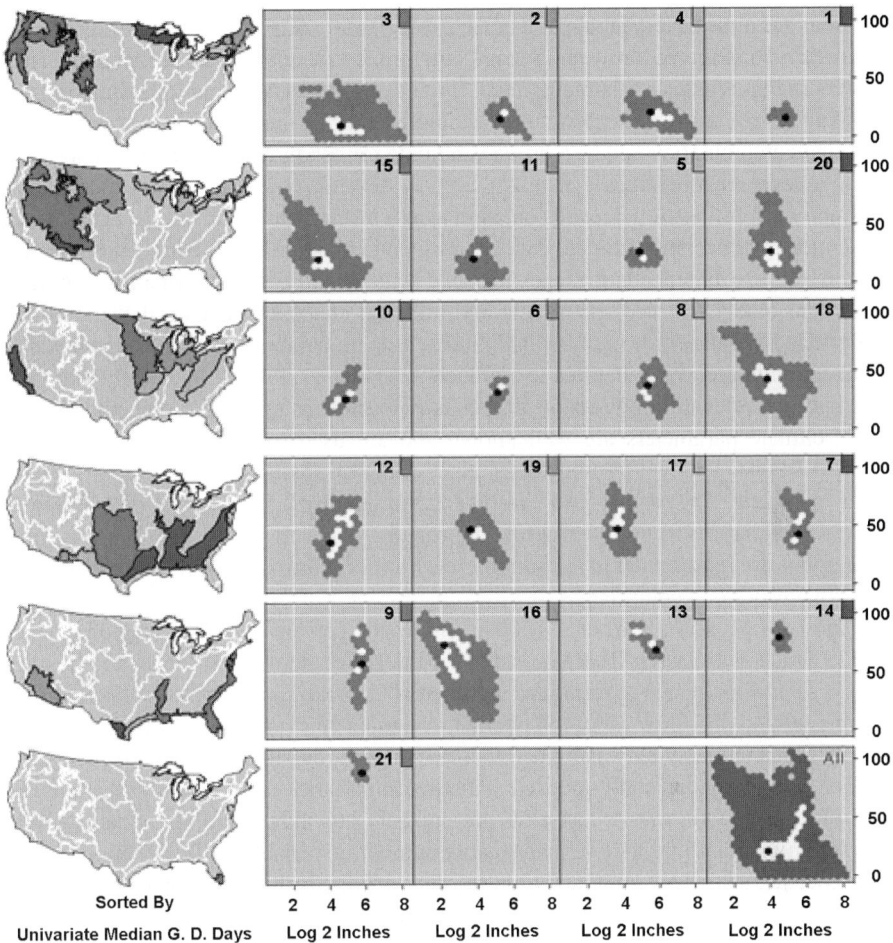

Fig. 3. LM bivariate boxplots. 1961–1990 precipitation (x) versus growing degree days/100 (y). For a color reproduction of this figure see the color figures section, page 599.

Figure 3 example, the names (numbers) of the twenty-one level-2 ecoregions do not appear in a separate column. In Figure 3 the ecoregions have been sorted by the median growing degrees days computed for grid cells covering the ecoregion. As described below, the grid has about 1/2 million cells for the continental US. The grid cell variables, growing degree days and precipitation, were used in the bivariate binning. The dark gray hexagons show hexagon cells with low counts. The yellow light when in gray-level hexagon cells have high counts that taken together constitute roughly 1/2 of all counts. The yellow footprint is analogous to showing the interquartile range in a univariate box plot. The black cell is the last cell left from a gray level erosion of the yellow hexagon cells. This cell can be considered as one kind of bivariate median. See Small (1990) for other kinds of bivariate medians. The binned scatterplot then provides a bivariate box plot that has omitted the details of adjacent values and outliers. Note the bifurcation for Ecoregion 16 in the fifth row and second column. A first guess at an explanation is that this ecoregion crosses a mountain range. If the two climatic variables, precipitation and growing degree days, are important for ecoregion delineation at this level of spatial resolution ecoregion 16 would be a prime candidate for subdivision when producing more detailed level 3 ecoregions.

Ecoregion delineation depends on climate as a macro forcing function. Figure 3 uses the nationally consistent climate data set developed using PRISM (Parameter-elevation Regression and Independent Slopes Model). Daly et al. (1994) describe PRISM as an analytical model that uses point data and a digital elevation model (DEM) to generate gridded estimates for monthly and annual climate parameters. PRISM models data from individual climate stations to produce a regular grid based on locally weighted precipitation /elevation regression functions. Orographics effects on precipitation and other climate parameters are well known and are used to define the local regions. PRISM has been shown to outperform other common spatial interpolation procedures such as kriging. The regular grid used is a 2.5 minute by 2.5 minute latitude/longitude grid, a nominal 5 km by 5 km grid. The regular grid helps to ensure that the entire ecoregions are represented in the summaries. Further information can be obtained from http://www.ocs.orst.edu/prism/.

Another LM plot in Carr et al. (2000a) shows the percent of each ecoregion devoted to 159 landcover classes. The data came for the Loveland et al. (1995) who classified roughly 8 million pixels covering the continental US based on multiple AVHRR temporal images and digital elevation. Showing Loveland's 159 variables on a page was not a problem, but the labeling was limited to a few major groupings of land classes.

The sizes of geospatially indexed statistical summaries that are readily handled by LM plots are quite limited. LM plots are not going to show a million variables without summarization into a more manageable number of variables. What is impressive is the power of statistical summarization to reduce the information so that LM plots become applicable. Many schemata such as box plots and bivariate bin cells are $O(n)$ and scale to large data sets. Showing an overview of mortality patterns based on billions of person–years of experience is not a problem.

3.3. Name panel variations

The development of single page of LM plots showing the counties for almost all states provided the challenge of showing 120 counties on a single page (Carr, 2001). The readability of names led to use of a double column format. Another solution is scrollable interactive graphics.

3.4. Interactive extensions

An LM plot applet is now a part of the National Cancer Institute's State Cancer Profiles web site (Carr et al., 2002b, 2003). The link to the LM plots is http://www.statecancerprofiles.cancer.gov/micromaps. Figure 4 shows an example.

The web site provides a "quick stop" for cancer-related statistics for planners, policymakers, and epidemiologists. It was developed by the National Cancer Institute (NCI)

Some recent graphics templates and software for showing statistical summaries 425

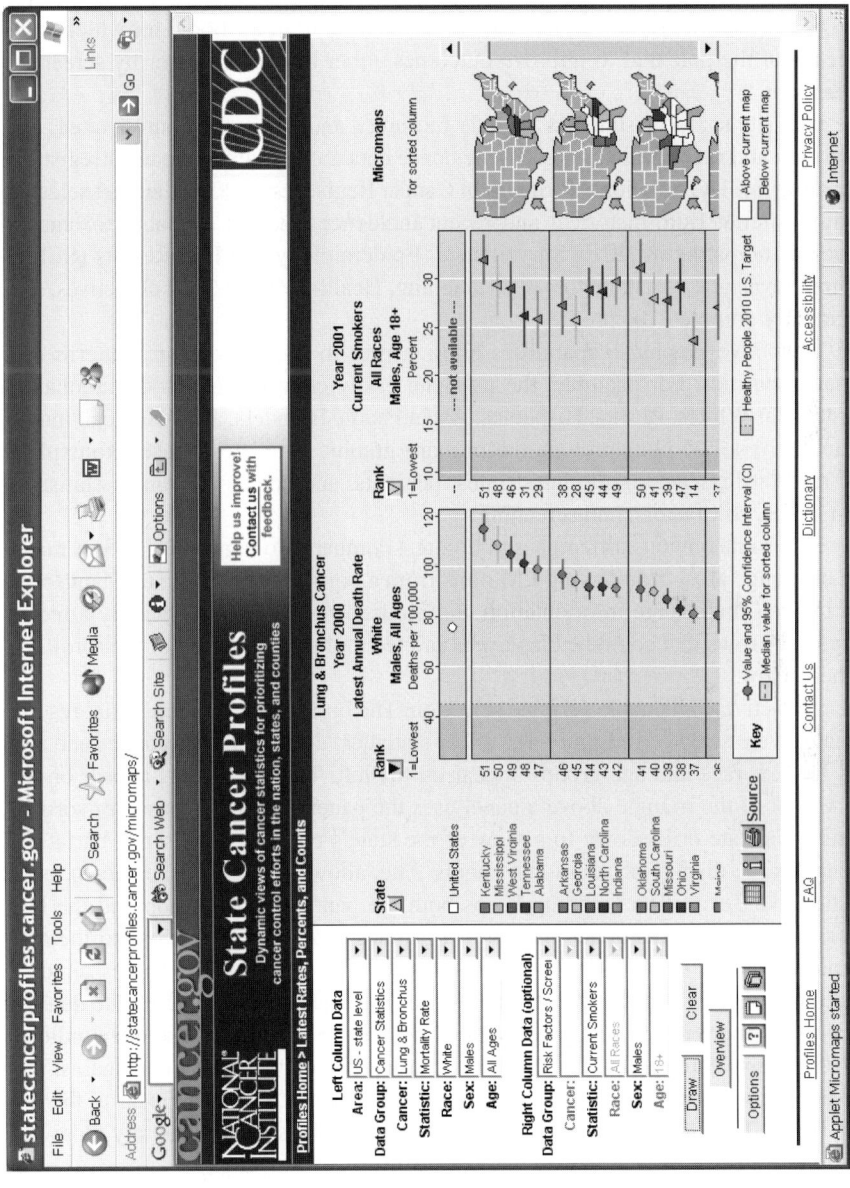

Fig. 4. Web-based LM plots with many interactive and dynamic features. For a color reproduction of this figure see the color figures section, page 600.

in collaboration with the Centers for Disease Control and Prevention (CDC) and is an integral component of NCI's Cancer Control PLANET, a web portal that links to resources for comprehensive cancer control. The web site provides national, state, and county views of cancer statistics collected and analyzed in support of annual federal reports. The web site focus is on eight major cancer sites (part of the body) for which there is evidence of the potential to improve outcomes either by prevention or by screening and treatment.

The web site's cancer statistics include mortality and incidence counts, rates, and trends by sex and race/ethnicity. Recent incidence data are available for cancer registries participating in CDC's National Program of Cancer Registries (NPCR) that met selected eligibility criteria. Both historical and recent incidence data are available for cancer registries participating in NCI's Surveillance, Epidemiology and End Results (SEER) program. Prevalence of risk factors and screening, Healthy People 2010 objectives, and demographics complete the profiles.

The interactive graphics capabilities allow a user to quickly explore patterns and potential disparities. For example, the user can easily compare graphs of national or state trends for Whites, Blacks, Hispanics, Asian Pacific Islanders, and American Indian Alaskan Natives. LM plots provide the primary graphic template for users to explore spatial relationships among the latest rates, percents, and counts for cancer statistics, demographics, risk factors, and screening.

There are some obvious differences in Figure 4 compared to Figures 2 and 3. A major design challenge arose from taking LM plots from a portrait print image to a landscape screen image. The change in orientation and reduced resolution significantly affected the number of rows that could be displayed, and led to the introduction of the scroll bar described below.

The interactive GUI enable variable selection. The option to select geospatial resolution (states were selected in Figure 4), cancer statistics, demographic statistics and risk factors for sex, race and age groups appear on the left. The micromaps appear on the right. A click of the triangle above a panel uses the panel values as a basis for sorting. Clicking on a state drills down to a view of the state's counties. The scroll bar at the right enables the display of additional panels. This allows counties to be shown, even for a state like Texas. The scrolling keeps both the panel axis at the top and the legend below in view, so maintains the context. Values can be brought close to the axis for a more accurate visual assessment and mouseovers can show the values directly. The estimates appear along with their confidence intervals to indicate estimate quality. When viewing the web site, the green regions with a reference line at the edge indicate the 2010 Health People goal. The state rates can immediately be compared against this goal. Many people are interested in the rank of their state. The ranks appear to the left of each panel. The Java applet has many capabilities such as printer friendly plots and help features.

The Java applet has been through usability assessment. It is in the process of being moved to other federal agencies including the National Center for Health Statistics and the Bureau of Labor Statistics. A lot can and will be done with the LM plot template.

4. Dynamically conditioned choropleth maps

Choropleth maps have provided a popular way to represent geospatially indexed statistics even thought the word "choropleth" is not well known. One sources says the word derives from two words meaning value and place. In a choropleth map, the value of each region, such as a state, is encoded by the color (or shading) used to fill the region's polygon(s).

Researchers have identified several problems with choropleth maps (Dent, 1990) and their use has evolved accordingly. The use of a continuous color scheme to encode continuous values leads to problems in color matching so classed choropleth maps became the standard. With classed maps, continuous region values are first transformed into an ordered categorical variable with a few levels or classes. A few easily distinguished colors then represent the classed values. Over the years, many papers were written on how the intervals should be determined. The Brewer and Pickle (2002) study of several map reading tasks indicates that quintile classes (20% in each class) are generally preferable for static maps. Additional well-noted problems have also been addressed. A persistent problem was the influence of the nuisance variable, region area, on human perception and interpretation. Showing statistics that were heavily confounded with region areas was not very helpful, so common practice focused primarily on the display of rates. For example, mortality rates and disease incidents rates could be low or high irrespective of a region area. The award winning atlas of Pickle et al. (1996) addressed several important issues including the choice of colors suitable for two common types of color blindness. One of the atlas' many positive features, the listing of known risk factors, became the point of departure for developing dynamically conditioned choropleth maps.

In the context of exploratory analysis and hypothesis generation, John Tukey (1976) said that the unadjusted plot should not be made. In simple words, it is difficult to mentally remove known patterns, envision the residuals, note the new patterns, and think about possible explanations. Common mortality and diseases incidence mapping has evolved to control for the common risk factors of sex and race by producing separate maps for the specific combinations. Producing age-specific maps or age-adjusted maps controls for the differences in age distribution across regions. However, published maps such as in mortality atlas rarely control for other known risk factors even though they are listed. Maps provide a wonderful context for generating hypothesis about geospatial patterns since many people organize their knowledge geospatially. People know where it is warm in the winter, where the traffic is heavy, where crops are grown, where pesticides are used, and so on. Thus it is highly desirable to produce maps that control variation due to known sources. Statistical modeling can control for unwanted variation by at least partially removing it. The approach has some problems since models are inevitably imperfect, and the data on risk factors may be poor and are sometimes confounded with other variables. A common problem is that many interested people with useful knowledge do not have the background to engage formal statistical modeling.

Carr et al. (2000a) proposed conditioned choropleth maps as an accessible way to control the variation of risk factors. This approach partitions regions into homogeneous groups based on risk factors. Geospatial patterns that remain within homogeneous groups call for explanation. The patterns across groups are also of interest especially when the partitioning variables where not anticipated to be risk factors or factors that would help to control variation. The dynamic software called CCmaps (Carr et al., 2000b, 2002a, 2003, 2005) also extends conditioning/partitioning approach from Cleveland et al. (1992) to choropleth maps. The shareware is available via http://www.galaxy.gmu.edu/~dcarr/ccmaps. Figure 5 provides an example.

Figure 5 is an enlarged single-panel view that shows three dynamic partitioning sliders at the top, bottom and right. A variables menu allows different variables to be attached to each slider. The primary variable of interest, called the dependent variable, attaches to the top slider that converts the values into three classes, low, middle and high and assigns three colors.

In Figure 5 the variable is the lung cancer mortality rate for White males ages 65–72. The analyst can dynamically change the two slider thresholds between the three classes and immediately focus on very high and very low value regions that would remain mixed with others in a static plot. Mouseovers provide direct access to region values. In Figure 5 the upper slider boundary was moved to put regions with highest rates and containing 10% of the population in the high class. Similarly, the lower slider boundary was set to put regions with the lowest rates and containing about 10% of the population into the low class. The other two sliders have the boundaries set to include all regions in the middle class and appear hidden.

Figure 6 shows all results of using all three sliders in the conditioned view. The sliders setting put regions with roughly 20% of the population in the extreme classes. Many regions with low mortality rates (blue) appear in the left column of panels. The left column highlights regions with low precipitation, with the threshold set by the bottom slider. Many regions with high rates appear in the top right panel. The highlighted regions in the map indicate a fairly strong association of the partitioning variables (precipitation and percent of households below the poverty level) with lung cancer mortality. Population-weighted averages in the top right of each panel back the impression up with numbers. The average rate in the lower left panel is 317 deaths per 100 000 and the average rate in the upper right panel is 539 deaths per 100 000. The difference of 222 deaths per 100 000 is huge.

Why would precipitation be associated with lung cancer mortality? Looking at the high rates in the Southeastern US and remembering the strong historical presence of the tobacco industry suggests the hypothesis that precipitation in confounded with cigarette smoking. Looking more deeply requires cigarette smoking data. The relevant exposure data would provide a description in terms of packs per day and cover the relevant previous 20 years. Sadly, such data is not available at this and possibly any geospatial level of resolution. The roughly 800 regions shown are health service areas. These are counties or aggregates of counties based on where people get their hospital care.

While CCmaps is still evolving, it has features to help find and assess patterns, and to keep track of the analysis process. The following indicates several of the interesting features.

When there are outliers as defined by the box plot approach, the slider is modified to allocate a little space for tails that show outlier intervals. The values increase more quickly as the mouse moves across the tail. This avoids the loss of slider resolution in the body of the data.

Some recent graphics templates and software for showing statistical summaries 429

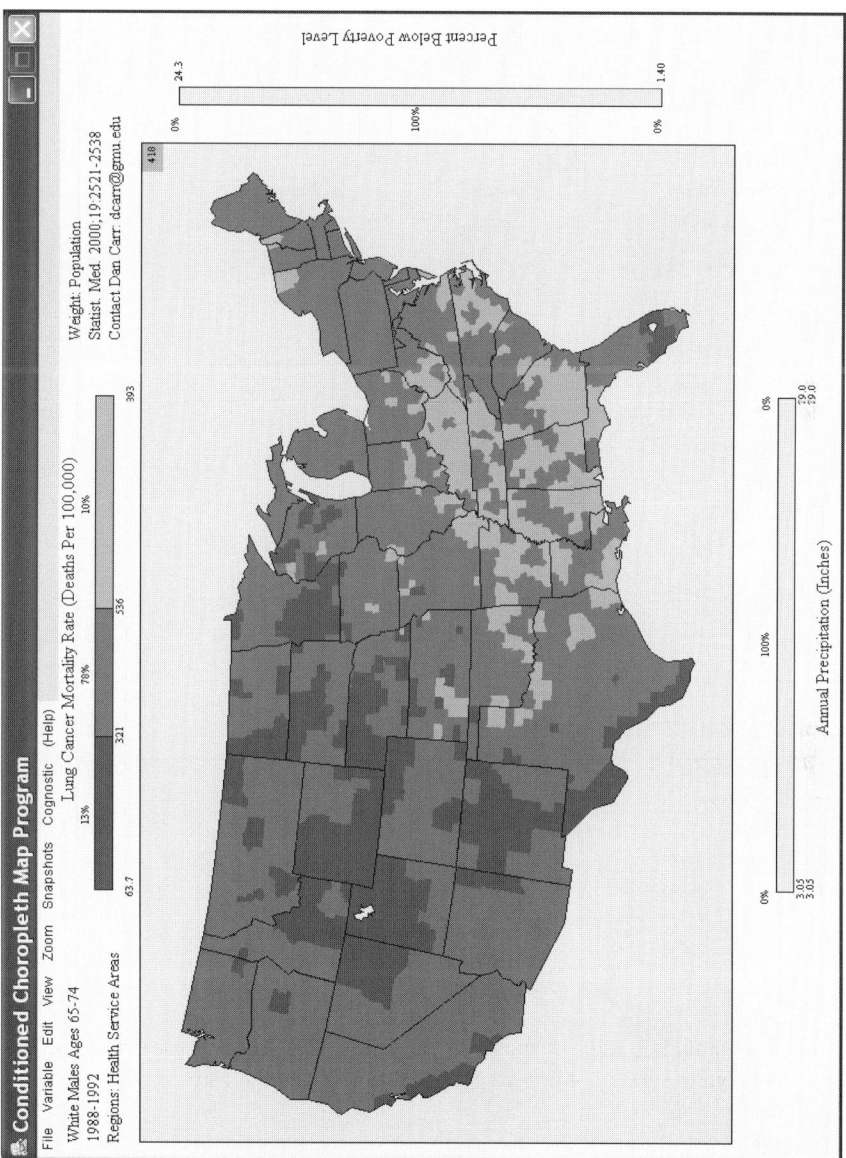

Fig. 5. CCmaps: an enlarged view of one conditioned panel with the conditioning constraints removed For a color reproduction of this figure see the color figures section, page 601.

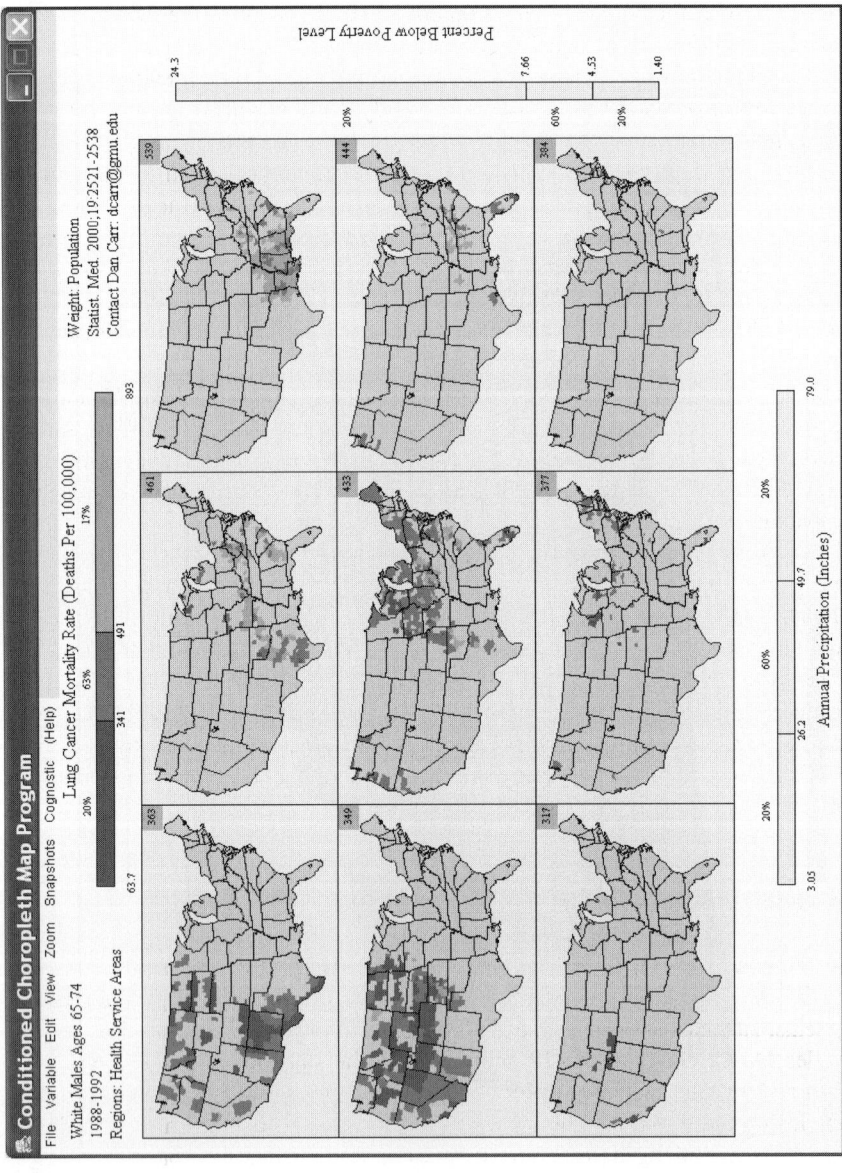

Fig. 6. CCmaps: a dynamic two-way partitioning of regions. For a color reproduction of this figure see the color figures section, page 602.

CCmaps has a search algorithm that looks for interesting slider setting. The analyst can step through the ordered suggested settings. Similarly, an algorithm to be added will search prioritizes suggestive pairs of partitioning variables.

Another non-standard mapping feature shows weighted means for the panels to help refine the rapid visual assessment based on color. Weights address the fact that not all summarizing statistics are of equal importance. Figures 5 and 6 weight the morality rates by the population size. The weighted average has a direct interpretation as deaths divided by person–years at risk. There is also a dynamic weighted QQplot view that compares the distributions in the panels and not just their weighted means.

An alternative scatterplot smoother view suggests functional relationships outside of the spatial context. This view helps to assess the impact of degrading two of the variables into ordered categorical variables in the map view.

A view of two-way tables of means and effects provide a quantitative approach that helps in assessing whether or not group differences are of practical importance. The table also includes a permutation test that helps to assess the non-randomness of apparent associations.

There is a snapshot capability that supports annotation and keeps track of the program states. The program is live when the snapshots are replayed. All the sliders, mouseovers, panel enlargement, zooming and other features are immediately available for restored views. This is the beginning of a more elaborate visual analysis management capability.

CCmaps works with other kinds of region boundaries files and data. The boundary files may correspond to elementary school districts or to regular grids for environmental assessment. The shareware comes with education, health and environment examples. CCmaps data handling accommodates analyst-provided comma-delimited files. The methodology can extend to the study of linear features such as multivariate networks.

Hypothesis generation is a risky activity. Tools help to find interesting patterns and can help to weed out apparent patterns that are not worthy of further study. Simply representing statistics on maps helps to promote hypothesis generation and to discount patterns likely associated with confounded variables.

5. Self-similar coordinates plots

Working with maps provides a good context for thinking about relationships. However, many data sets come without geospatial coordinates. In the study of gene regulation and proteins, statistics are often indexed by sequences of letters standing for the four nucleotides (A, C, G, T) or the 20 amino acids. The statistics in publications often appear in a short table with one column listing the letter sequences and one or more columns listing the statistics. The table may be sorted by one column of statistics or by a multivariate ranking based on several columns. This does not provide much of an overview. A sequence of length nine (a 9-mer) for amino acids involves over 1/2 trillion possibilities. There are a few graphical alternatives as sequence logo displays (Schneider and Stephens, 1990). Such sequence logo display is roughly equivalent to showing a sequence of one-dimensional margin letter-indexed stacked bars. The limited

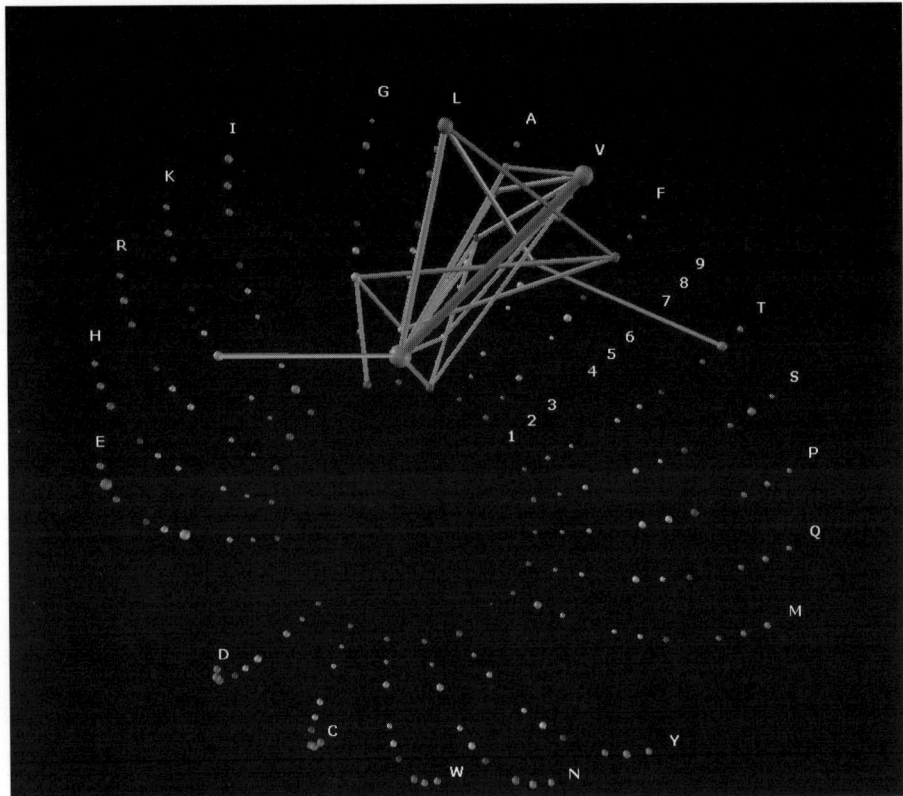

Fig. 7. GLISTEN: capless hemisphere coordinates with perceptual grouping alterations. Path shows immune system molecule docking statistics. The large count (thick path) is for Leucine in position 2 and Valine in position 9. For a color reproduction of this figure see the color figures section, page 603.

goal here is to demonstrate two different sets of self-similar coordinates and a multi-layer display that supports the rendering of statistics from higher-dimensional tables.

The self-similar coordinate plots here can be thought of as a generalization of parallel coordinates plots. Figure 7 uses capless hemisphere coordinates as the provided coordinates for plotting. The statistics are associated with amino acid 9-mers. Latitude represents the position in the sequence. Longitude represents the amino acid. The size of spherical glyph at a coordinate indicates the marginal counts (or other statistic) for the given position and amino acid. The plot of points by themselves is conceptually similar to sequences logos displays. However, this forms just one glyph (or map) layer. The thickness and color of a single line segment (or 3D appearing tube) between two points with different positions can encode the 2D margin counts for a pair of positions. The counts derive from $\binom{9}{2} = 36$ different 20×20 tables. Such statistics are rarely examined. A single-segment tube glyph is a part of the second layer that can be shown simultaneously in the display. Figure 7 is heavily filtered to reveal some of the dominant patterns. Consider the tube between L2 and V9. Note that the hemisphere curvature

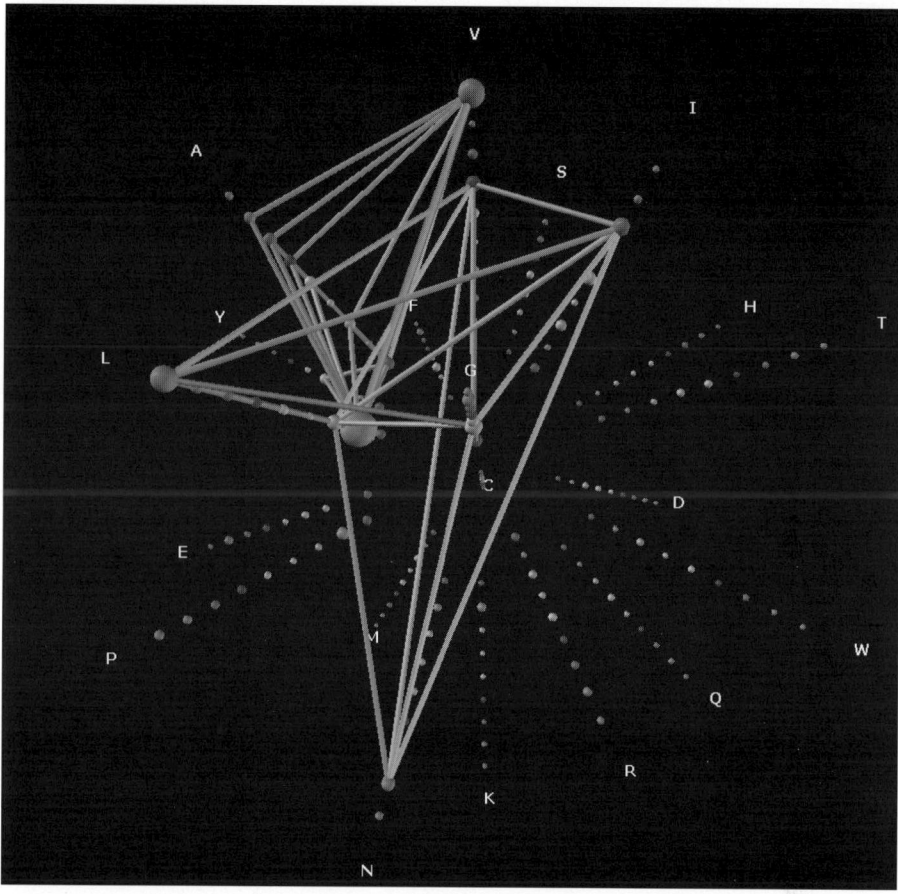

Fig. 8. GLISTEN: icosahedron face-center axes. High two-way table counts for different positions along amino acid 9-mers. For a color reproduction of this figure see the color figures section, page 604.

enables drawing the segment between these non-consecutive positions without going through the axes of intermediate positions. This does not work for parallel coordinates in the plane. Extending the approach to three-segment tubes allows one to encode statistics from $\binom{9}{4} = 126$ tables with 20^4 cells. Again, heavy filtering is required to see the dominant patterns. A two three-segment tube example in Lee et al. (2002) shows a plot – a bit of a surprise – from the 4D tables. Several high count cells have Glycine (G) in position 4. Figure 7 does not clearly reveal the more prominent role of G4 in the high-count 4D cells. This certainly would not be expected from the one-dimensional counts.

Software called GLISTEN did the rendering for Figure 7. GLISTEN was patterned after geographic information systems both in supporting the merging of more than one information source and allowing the choice of difference coordinates. GLISTEN does not provide the full analogue of a library of different map projections but allows substitution of new coordinate files. The statistics files remain unchanged. Figure 8 provides examples where the amino acid axes follow a vector from the origin to the 20-face center of an icosahedron. The first position along each axis starts a ways from the origin. The different geometry provides more sorting opportunities. For example, it is often desirable for hydrophobic amino acid axes to be neighbors. This tends to minimize high-count tube length. The different view can reveal different facets of the structure.

The use of connecting lines or 3D tubes is a bit controversial since they use so many pixels and since multi-segment paths (tubes) through the same point are ambiguous. Still, connectedness is a strong perceptual organizational principle, so a path can reveal structure. Some analysts are adverse to 3D visualization despite the visual appeal of a very simple lighting models and rendered glyphs. A weakness drawback of the glyphs chosen is that the encoding are not high in terms of perception accuracy of extraction. This increases the dependence of widgets for making fine distinctions. The problems of rendering in more than three dimensions are not easily addressed. Definitely, the 3D coordinates provide arrangement opportunities that can be advantageous, and the use of paths or tubes provides an encoding that extends to higher-dimensional tables.

Figures 7 and 8 concern the docking of foreign peptides on human immune system HLA A2 molecules. Docking indicates the cell that is contaminated and should be destroyed. The data comes from Brusic et al. (1998). Carr and Sung (2004) provide a further discussion of both data and coordinates. Examples including the use of the self-similar fractal coordinates for encoding nucleotides statistics.

The rendering software is called GLISTEN. This stands for geometric letter-indexed statistical table encoding. The GLISTEN GUI has evolved to provide many widgets that control the encoding of statistics as glyphs, filtering, view manipulation and even widget organization. The GUI warrants a separate paper.

6. Closing remarks

The graphics in this chapter illustrate different templates and the discussion addresses dynamics graphics. Some methods are being accepted by US federal agencies and some are still experimental. However useful the graphics are, focusing on specific graphics may miss the bigger picture.

- Lower-level statistical summarization can go a long way towards making large data sets available at a level where visual analysis can be conducted in a broader context using slower algorithms.
- Consideration of human perception and cognitive factors is essential to extending and enhancing analyst capabilities. The opinion here is that analyst training, the analyst–computer interface and analyst communications continue to be a weak links in the analysis process. Simply off loading petty short-term memory consuming tasks by providing visual analysis management environments could be a big help.
- Quantitative pattern finding, pattern evaluation, and decision making require a full complement of searching, analysis and reasoning tools.

- Much is already known that could be employed. A historical constraint to developing better tools and analysis environments has been small analysis budgets relative to the size of data gathering and data base construction budgets.
- Pressing world problems are beginning to raise the status of visual analysts and visual analytics teams.

Huge progress been made in the last decade, in terms of computer (cpu) and graphics (gpu) processing capabilities, increased computer memory and fast in memory algorithms, improved graphics rendering and projection capabilities, better understanding of human perception, cognition and communication, and better human–computer interfaces and graphics layouts. Massive data sets and streaming data sets will continue to outstrip analysis capabilities. This makes it imperative to focus on developing serious visual analytic environments and to help analysts to extend their reach as far as humanly possible.

Acknowledgements

NSF cooperative grant #9983461 supported this research. The many collaborators listed in the coauthored papers provided important contributions. Those primarily involved in software developments deserve explicit mention. Bill Killam played a major role in the NCI web site usability assessment and David Eyerman played a major role in the Java implementation of LM plots. Duncan MacPherson implemented the dynamic CCmaps prototype. Yuguang Zhang made the major contributions in the current version of CCmaps. Yaru Li and Chunling Zhang are responsible for the recent additions of dynamic QQplots and scatterplots, and the table of means, respectively. Duoduo Liao implemented the prototype versions of GLISTEN. The current version of GLISTEN is the work of Yanling Liu.

References

Bertin, J. (1983). *Semiology of Graphics*. Translated by W.J. Berg. University of Wisconsin Press, Madison.

Brewer, C.A. (1997). Spectral schemes: controversial color use on maps. *Cartogr. Geogr. Inform. Systems* **24** (4), 203–220.

Brewer, C.A., Pickle, L. (2002). Evaluation of methods for classifying epidemiological data on choropleth maps in series. *Ann. Assoc. Amer. Geogr.* **92** (4), 662–681.

Brusic, V., Rudy, G., Harrison, L.C. (1998). MHCPEP, a database of MHC-binding peptides: update 1997. *Nucl. Acids Res.* **26** (1), 368–371.

Carr, D.B., (1994a). Converting plots to tables. Technical Report No. 101. Center for Computational Statistics, George Mason University, Fairfax, VA.

Carr, D.B. (2001). Designing linked micromap plots for states with many counties. *Statist. Med.* **20**, 1331–1339.

Carr, D.B., Olsen, A.R. (1996). Simplifying visual appearance by sorting: an example using 159 AVHRR classes. *Statist. Comput. Graph. Newsletter* **7** (1), 10–16.

Carr, D.B., Sung, M.-H., (2004). Graphs for representing statistics indexed by nucleotide or amino acid sequences. In: *Proceedings of CompStat2004*. In press.

Carr, D.B., Pierson, S. (1996). Emphasizing statistical summaries and showing spatial context with micromaps. *Statist. Comput. Graph. Newsletter* **7** (3), 16–23.
Carr, D.B., Olsen, A.R., Courbois, J.P., Pierson, S.M., Carr, D.A. (1998). Linked micromap plots: named and described. *Statist. Comput. Graph. Newsletter* **9** (1), 24–32.
Carr, D.B., Olsen, A.R., Pierson, S.M., Courbois, J.P. (2000a). Using linked micromap plots to characterize Omernik ecoregions. *Data Min. Knowl. Discov.* **4**, 43–67.
Carr, D.B., Wallin, J., Carr, D.A. (2000b). Two new templates for epidemiology applications. Linked micromap plots and conditioned choropleth maps. *Statist. Med.* **19**, 2521–2538.
Carr, D.B., Chen, J., Bell, B.S., Pickle, L., Zhang, Y. (2002a). Interactive linked micromap plots and dynamically conditioned choropleth maps. In: *Proceedings of the Second National Conference on Digital Government*. Digital Government Research Center, pp. 61–67.
Carr, D.B., Zhang, Y., Li, Y. (2002b). Dynamically conditioned choropleth maps: shareware for hypothesis generation and education. *Statist. Comput. Graph. Newsletter* **13** (2), 2–7.
Carr, D.B., Bell, S., Pickle, L., Zhang, Y., Li, Y. (2003). The state cancer profiles web site and extensions of linked micromap plots and CCmaps. In: *Proceedings of the Third National Conference on Digital Government Research*. Digital Government Research Center, pp. 269–273.
Carr, D.B., White, D., MacEachren, A.M. (2005). Conditioned choropleth maps and hypothesis generation. *Ann. Assoc. Amer. Geogr.* **95** (1), 32–53.
Cleveland, W.S. (1985). *The Elements of Graphing Data*. Hobart Press, Summit, NJ.
Cleveland, W.W. (Ed.) (1988). *The Collected Works of John W. Tukey*. Wadsworth & Brooks/Cole, Pacific Grove, CA.
Cleveland, W.S. (1993a). *Visualizing Data*. Hobart Press, Summit, NJ.
Cleveland, W.S. (1993b). Display methods of statistical graphics. *J. Comput. Graph. Statist.* **2** (4), 323–343.
Cleveland, W.S., McGill, R. (1984). Graphical perception: theory, experimentation, and application to the development of graphical methods. *J. Amer. Statist. Assoc.* **79**, 531–554.
Cleveland, W.S., Grosse, E., Shyu, W.M. (1992). Local regression models. In: Chambers, J.M., Hastie, T.J. (Eds.), *Statistical Models in S*. Wadsworth and Brooks/Cole, Pacific Grove, CA.
Daly, C., Neilson, R.P., Phillips, D.L. (1994). A statistical–topographic model for mapping climatological precipitation over mountainous terrain. *J. Appl. Meteor.* **33**, 140–158.
Dent, B.D. (1990). *Cartography—Thematic Map Design*. Brown, Dubuque, IA.
Harris, R.L. (1996). *Information Graphics. A Comprehensive Illustrated Reference*. Management Graphics, Atlanta, GA.
Keim, D.A. (1996). Pixel-oriented visualization techniques for exploring very large databases. *J. Comput. Graph. Statist.*, 58–77.
Kosslyn, S.M. (1994). *Elements of Graph Design*. Freeman, New York, NY.
Lee, J.P., Carr, D., Grinstein, G., Kinney, J., Saffer, J. (2002). The next frontier for bio- and cheminformatics visualization. Rhine, T.-M. (Ed.), *IEEE Computer Graphics and Applications* (Sept/Oct), 6–11.
Loveland, T.R., Merchant, J.W., Reed, B.C., Brown, J.F., Ohlen, D.O., Olson, P., Hutchinson, J. (1995). Seasonal land cover regions of the United States. *Ann. Assoc. Amer. Geogr.* **85**, 339–355.
Monmonier, M. (1993). *Mapping It Out*. University of Chicago Press, Chicago, IL.
Pickle, L.W., Mungiole, M., Jones, G.K., White, A.A. (1996). *Atlas of United States Mortality*. National Center for Health Statistics, Hyattsville, MD.
Schneider, T.D., Michael Stephens, R. (1990). Sequence logos: a new way to display consensus sequences. *Nucl. Acids Res.* **18**, 6097–6100.
Small, C.G. (1990). A survey of multidimensional medians. *Int. Statist. Rev.* **58**, 263–277.
Tufte, E.R. (1983). *The Visual Display of Quantitative Information*. Graphics Press, Cheshire, CT.
Tufte, E.R. (1990). *Envisioning Information*. Graphics Press, Cheshire, CT.
Tufte, E.R. (1997). *Visual Explanations*. Graphics Press, Cheshire, CT.
Tukey, J.W. (1976). Statistical mapping: what should not be plotted. In: *Proceedings of the 1976 Workshop on Automated Cartography and Epidemiology*. March 18–19, US Department of Health, Education, and Welfare, Arlington, VA, pp. 18–26.
Wang, X., Chen, J.X., Carr, D.B., Bell, B.S., Pickle, L.W. (2002). Geographics statistics visualization: web-based linked micromap plots. In: Chen, J. (Ed.), *Computing Science and Engineering*, pp. 92–94.
Ware, C. (2004). *Information Visualization, Perception for Design*. Morgan Kaufmann, New York.
Wilkinson, L. (1999). *The Grammar of Graphics*. Springer-Verlag, New York.

Interactive Statistical Graphics: the Paradigm of Linked Views

Adalbert Wilhelm

1. Graphics, statistics and the computer

The modern history of statistical graphs as we know them goes back to the middle of the 18th century and is closely related with state statistics or Staatenkunde. Mostly being used to visualize results and to present conclusions, there have always been remarkable examples of exploratory approaches achieved by statistical plots. The excellent discussions by Tufte (1983, 1990, 1997) give many examples of this fact. Influenced by the work of the late John W. Tukey, statistical graphics have experienced a strong resurgence as a cornerstone in exploratory data analysis, see (Nagel et al., 1996; Velleman and Hoaglin, 1981). The switch from using graphics as a result presenting device to an analytic tool was accompanied by the invention of new complex plots and techniques, for example *mosaic plots, parallel coordinate plots* and *projection pursuit*. The electronic revolution that has taken place in the last quarter of a century fostered this paradigmatic change. Dynamic plots like *3D rotating plots* and the *Grand Tour*, could not have been imagined without the rapid developments in computer science.

It was only in the last twenty years that human interaction with graphical displays could be put into practice. As a consequence, plots changed their character from formerly being a final product to now being a temporary tool that can be modified and adapted according to the situation by simple mouse clicks or keyboard commands. Information overload that would prevent perception can be hidden at the first stage and made available on demand by responding to interactive user queries. Various selection modes that can even be combined to a sequence help in choosing a specific set of data points to assign interest to them. Unusual observations can be easily identified and isolated for special treatment.

Linking is the basic concept giving graphical methods the capability of multivariate analysis. Linking builds up a relation between different observations and between different displays. It propagates the changes and selections made in one plot to all other connected plots that deal with the same database. We encounter the standard case in multivariate statistics where we deal with individuals of a population on which different variables have been observed. In this case, all observations made on one and the same individual are linked together in a natural way.

1.1. Literature review

The resurgence of statistical graphics since the early seventies was made possible by the event of powerful computer technology, but it was greatly enhanced by the rise of methods of exploratory data analysis that John W. Tukey pioneered. In (Tukey, 1972) as well as in the monograph *Exploratory Data Analysis* (Tukey, 1977), he introduced a rich palette of data displays that form a critical part of the methodology for exploring data. Fisherkeller et al. (1975) started the first software project for interactive statistical graphics, PRIM-9, at a time when user interaction and graphical representations were still only possible on expensive and specialized computer hardware.

The hallmarks of the development of statistical computer graphics in the 1980s have been laid down in several conference proceedings and collections. Of particular importance are (Wegman and DePriest, 1986; Cleveland and McGill, 1988; Buja and Tukey, 1991).

The work shown in all these publications on statistical graphics can be roughly classified into the following four types: "ad hoc" procedures that have been mainly inspired by the analysis of a particular data set and its specific needs (e.g., Nagel, 1994; Wilhelm and Sander, 1998); mathematically oriented procedures like parallel coordinate plots (Inselberg, 1985; Wegman, 1990), Andrew curves (Andrews, 1972; Wegman and Shen, 1993), and biplots (Gower and Hand, 1996); work that mainly deals with problems of perception and how to support the human visual system (e.g., Rogowitz and Treinish, 1996; Tremmel, 1995); and software and implementational issues with a strong relationship to programming and computer science (e.g., Hurley and Oldford, 1991; Tierney, 1990; Unwin et al., 1996).

Some monographs on statistical graphics (e.g., Tufte, 1983, 1990, 1997; Cleveland, 1993; Geßler, 1993; Nagel et al., 1996) have been published during the years, most of them giving an enlightening description of the variety of data displays and their peculiarities. Although they provide guidance for good taste, good practice, and good graphic design they do not attempt to provide an underlying general theory.

The recent books on statistical graphics neglect the interactive statistical graphics capabilities and reflect mainly the state-of-the-art at the end of the eighties. Dynamic and interactive graphics from this period, however, are better described in the previous noted article collections edited by Wegman and DePriest (1986), Cleveland and McGill (1988), and Buja and Tukey (1991).

A good introduction to statistical graphics, covering a broad range from mathematical transformations, projections and hidden surfaces, over rendering, transparency and visual perception to multidimensional visualization techniques is (Wegman and Carr, 1993). A very rich source of the latest research results are the proceedings of the annual meeting of the Interface Foundation of North America. A special issue of *Computational Statistics* (Swayne and Klinke, 1999) has been devoted to interactive graphical data analysis. The papers in that issue cover a broad range of current research, from proposing general requirements for interactive graphics software over interactive features of older statistical environments to descriptions of special interactive systems. In addition, several interesting applications of interactive graphical systems to data analysis problems are included.

The rise of new data displays soon raised the demand for a theoretic foundation of statistical graphics, see (Cox, 1978). More than ten years earlier, Jacques Bertin had already written his famous semiological theory of quantitative graphics (Bertin, 1967). Written in French it needed the English translation (Bertin, 1983) to disseminate his comprehensive discussion of quality and efficiency of graphic transcriptions. By then, however, map and graphics creation had been computerized and Bertin's semiology fell behind. Nevertheless, Bertin's work is still one of the most cited books on creating and utilizing diagrams, networks, and maps. His ideas still apply for static graphics and some aspects can also be adopted to the world of virtual reality.

Some isolated attempts can be found to give a rigorous description on what kind of peculiarities in the data can be successfully revealed by certain graphics types. Furnas and Buja (1994) develop how scatterplot brushing and low-dimensional views can be used to resolve the dimensionality of high-dimensional loci. Sasieni and Royston (1996) write in the traditional style from the point of view of static displays and describe the use of isolated one- and two-dimensional dotplots for finding clusters and outliers. Wilkinson (1999a) considers dotplots from a density-estimation point of view and formulates an algorithm that produces plots that are close to hand-drawn dotplots. Sawitzki (1990) discusses tools and concepts that are used in data analysis and tries to judge them by questioning validity and perception of graphical displays. For presenting univariate distributions, some specific answers are given by Sawitzki (1994). Hofmann (1997) treats the reliability of graphics.

Wilkinson (1997) started a new attempt to formalize statistical graphics. This article is inspired by the work of Bertin (1983) and establishes rules for the mathematical construction and the aesthetic layout of graphs. These ideas are presented in depth in the recently published book *A Grammar of Graphics* (Wilkinson, 1999b). There is an accompanying graphics library developed by Dan Rope (1998) and written in Java. The main target of this approach are static plots, although it should be possible to adopt some of Wilkinson's definitions for interactive graphics.

Some basic requirements for interactive statistical graphics have been formulated by Unwin (1995, 1999). The thesis by Theus (1996) gives an overview on current state-of-the-art interactive graphics without formalizing concepts. A broad range of interactive techniques and graphical displays is presented, but the discussion is often rather cursory and does not establish a unifying theory. This work includes many new ideas for interactive graphics and good examples of exploratory analyses done with interactive graphics.

Recently, visualization techniques have met interest in the data base and data mining community, where they have been extensively used to represent results of dynamic data base queries (Shneiderman, 1994; Keim, 1995; Rogowitz et al., 1996; Rogowitz and Treinish, 1996). The parameters of the query are visualized by sliders each representing the range of one query parameter. The user can change the sliders interactively and the query results are shown in a linked graphic. Different methods to represent the distances between the query and the data items have been proposed in the literature: pixel-oriented techniques (Keim, 1997), different intensities of the highlighting color (Tweedie and Spence, 1998), or the standard linked views approach using a $\{0, 1\}$-distance (Derthick et al., 1997).

Special visualization techniques for particular situations and data have been proposed by Shneiderman (1992), Eick (1994), Wills (1999b) and others. Linking views has become extensively used in the exploration of spatial data, see (Anselin, 1996; Wills, 1992) for a comprehensive discussion. Here a map of the geographic locations has been introduced in addition to the statistical displays. Linking is possible in both directions, either by selecting a subset of locations in the map or by selecting a subset of data in a non-spatial view. A close relationship exists between interactive statistical graphics and Geographic Information Systems (GIS). Exploratory spatial data analysis uses methodologies from both the interactive statistical visualization and the GIS community. There is a vast literature available covering topics from cartography (e.g., Kraak and Ormeling, 1996; Monmonier, 1996), over software implementational issues (Anselin and Bao, 1997; Symanzik et al., 1996) to particular topics in spatial analysis (Hofmann and Unwin, 1998). Moreover, some special issues of statistical journals have been devoted to this subject. An issue of *The Statistician* covered recent developments in exploratory spatial data analysis with local statistics (Unwin and Unwin, 1998). A special issue of *Computational Statistics* focused on the aspect of computer-aided analysis of spatial data (Ostermann, 1996).

Linked scatterplots have been widely used in the development of static displays, see (Tufte, 1983; Diaconis and Friedman, 1983). Linking graphics using a computer environment was first introduced by McDonald (1982) to connect observations from two scatterplots. The most popular form of linking is scatterplot brushing as promoted by Becker et al. (1987). Furnas and Buja (1994) use brushing as a conditioning technique to obtain sections of high-dimensional loci with thickened versions of affine subspaces. Although there are some software packages available, e.g., LISPSTAT (Tierney, 1990), DATA DESK (Velleman, 1995), or MANET (Unwin et al., 1996) that offer linking for all kind of plots, the power of the linking paradigm is not recognized by the leading statistical software vendors.

Linking has been commonly used in the form of one of the following two concepts: linking as an attempt to escape 'the two-dimensional poverty of endless flatlands of paper and computer screen' (Tufte, 1990) and as a tool to show interesting or anomalous areas of one view of the data in the context of other views of the data. Therefore, interactive linking techniques have widely been seen as subsetting or conditioning techniques.

Pointing out different characteristics of linking, various authors have classified the linking procedures into diverse groups. Young et al. (1992) coin the term 'empirical linking' and differentiate this linking type into linking by observation and linking by variables. Linking by observation is the common linking scheme of connecting data points that correspond to the same individuals. Linking by variables is used in displays that contain more than one plot. Every plot has its style – scatter, bar, histogram, etc. – and this style remains the same while the variables displayed alter according to the user interaction. Opposed to the empirical linking is 'algebraic linking' that is linking between data values and derived functions, see (Young et al., 1992) and also (Nagel et al., 1996). By Velleman (1995) linking is classified into three different states: cold objects, warm objects and hot objects. Cold objects are those that need to be completely rebuild to react to the linking request. Warm objects are yet capable to reflect the linking but the user has to invoke the update. Hot objects immediately react to the linking request and they are automatically updated.

Subsetting is the most established form of interactive graphical analyses and is variously described as "multiple views" (McDonald, 1988), "brushing" (Becker and Cleveland, 1987) and "painting" (Stuetzle, 1987).

1.2. Software review

In recent years, a lot of software packages for visual data analysis and statistics have been made available. A fair comparison of all the packages is impossible because they differ in their aims as well as in the size of their supporting development teams. A first attempt to judge interactive graphic facilities offered by statistical software has been made by Wilhelm et al. (1996).

The leading commercial statistical software packages like SAS, SPSS, BMDP, MINITAB, SYSTAT, STATISTICA are very limited in their interactive graphical analysis capabilities. This is mainly to their historical development originating from mainframe computer packages based on a command language and just recently established with a graphical user interface. The available plot types range from bar charts, histograms, dot- and boxplots to scatterplot matrices. More specific plots for categorical data like spine plots or mosaic plots are usually not provided. Also user interaction with the graphics is very limited such that an interactive graphical analysis in the sense specified in later sections is not possible with these tools.

The above mentioned programs are designed to serve the needs of a wide range of users, incorporating a plethora of statistical analysis techniques. To meet special needs, they include extensible programming languages, but typically the software itself is so complex that user-defined enlargement of the code cannot be done easily.

One statistical software package that offers some of the greatest flexibility with regard to the implementation of user defined functions and the customization of one's environment is S-PLUS. S-PLUS is more a high-level programming language than a software package. Many add-on modules are commercially available and user-written macros can be downloaded from the world wide web. S-PLUS has been designed for the easy implementation of statistical functions. Statistical and user defined operations are highly supported and a wide range of graphics capabilities is present. More specific plots can be included by user definition, but the graphics device does not encourage the extensive use of interactive statistical graphics. The system offers many features to create plots that fulfill the specific needs of an analyst, but these features have to be specified in advance of drawing. Interactive changes of plot parameters are not supported. Instead of updating the displays to user changes, plots are completely redrawn making it impossible to see effects of user interaction instantaneously and to compare the new picture with the previous one.

Since a couple of years, a non-commercial alternative to S-PLUS exists under the name R (see http://www.cran.org). For R, a consortium of many researchers worldwide contribute code under the auspices of a core development team and make it available to a wide user community. The functionalities are similar to the one of S-PLUS, however more innovations are currently to be expected from the R-team. For interactive graphics, there is an ongoing project to develop a new library called *i-plots* of interactive statistical plots to be used within R.

In the group of commercially available software packages, only some products are available that offer a broad palette of interactive graphical tools: DATA DESK, SAS INSIGHT and JMP. DATA DESK was one of the earliest interactive statistics packages. It is a Macintosh-based product authored by Paul Velleman from Cornell University. The current release is 6.0 and it is the first that is also available for Windows operating systems. DATA DESK stores the data in relations and within one relation all data points having the same index are linked. In addition, there is a link between input and output windows in such a way that changes made in the input window are immediately taken into account in output windows. By this method, it is possible to see the effects of excluding data points from certain analyses or the effects of changing regression parameters. But linking is also possible between separate relations by defining a link function. For some plots, it is possible to link other information than the index, for example the scale. But this feature is not generally available in all plot types.

SAS INSIGHT and JMP are SAS products that are highly visualization oriented. JMP is a stand alone product for PC and Macintosh platforms. It keeps the data in a matrix arrangement and it offers more analytic and modeling tools than DATA DESK but less linked interactive graphics.

A new product called ADVIZOR has been put on the market by *Visual Insights*, a spin-off company of *Lucent Bell Labs*. This is an interactive data visualization-based decision support application. The central visualization component is a 'perspective', a combination of linked plots that are placed together in a frame.

XPLORE is a statistical computing environment with an high-level programming language that enables the creation of user-defined menus. It offers a wide range of smoothing methods but only limited brushing and linking tools. For example, only transient brushing is possible due to the data structure chosen. The current version of XPLORE is designed for use within the world wide web.

XGOBI (Swayne et al., 1998) is a dynamic graphics program restricted to an array of scatterplot tools. It focuses on dynamic analyses based on projection pursuit and the grand tour. XGOBI provides many choices for the projection pursuit index as well as linked scatterplot brushing and is especially intended for the analysis of continuous variables in two to ten dimensions.

The intention of XGOBI was to offer state-of-the-art methods for display and manipulation of scatterplots and linking them to the "S" system. Recent work with XGOBI has concentrated on interlacing it with ARCVIEW, the most popular geographic information system, see (Cook et al., 1996) and with the computing environment XPLORE. XGOBI has been extended to a program for interactive data visualization with multidimensional scaling, called XGVIS.

XLISP-STAT is a freeware alternative to S-PLUS. It is an object-oriented programming environment for statistical computing and dynamic graphics. As well as the "S" system it was motivated by the thought that research on innovative computational methods for statistics can only be done in an extensible system. XLISP-STAT is written by Luke Tierney, School of Statistics, University of Minnesota, and is available for all platforms. XLISP-STAT provides a set of high-level tools to develop new dynamic graphics techniques. Besides the more widespread possibility of empirical linking, that is all data points having the same index are linked, functions can be evaluated and plotted for highlighted points. The latter is known as algebraic linking. However, there is no rigorous differentiation between nominal and continuous variables and the structure of the linking implementation is based on the index vector, thus, resulting in uninformative highlighting in histograms and bar charts.

XLISP-STAT is widely used and there are some other programs based on it which are designed for particular problems. The most important are VISTA and R-CODE. VISTA – the Visual Statistics System – is an open and extensible visual environment for statistical visualization and data analysis. It is primarily designed as teachware and provides visual guidance for novices in data analysis and authoring tools for experts to create guidances for novices. But it can also be used as an interface to the complete XLISP-STAT system that is available to program and extend Vista's capabilities. R-CODE is designed for the concise treatment of regression and residual diagnostics (Cook and Weisberg, 1994).

QUAIL (Oldford et al., 1998) is another LISP-based system that offers interactive graphics. The linking structure has been extended to handle not only the standard is-equal linking but also is-part-of linking (Hurley, 1993). QUAIL has a brushing tool that is appropriate for multilevel and multiway data (Hurley, 1991).

EXPLORN is a statistical graphics and visualization package designed for the exploration of high-dimensional data. According to the authors, Qiang Luo, Edward J. Wegman, Daniel B. Carr and Ji Shen, it has a practical limit of 100 or so dimensions. The software combines stereo ray glyph plots, previously implemented in EXPLOR4 (Carr and Nicholson, 1988), with parallel coordinates, see (Wegman, 1990) and a d-dimensional grand tour (Wegman, 1991). Three forms of multidimensional displays are available: scatterplot matrices, stereo ray glyph plots and parallel coordinate plots. Subsets of the data can be brushed with color and the color saturation may be varied. The alpha-channel feature of the Silicon Graphics workstations is used to add saturation levels. Thus, for massive data sets the color saturation will show the amount of overplotting and it will result in a color-coded density plot, see (Wegman and Luo, 2002).

XMDVTOOL (Ward, 1994) is another public-domain software package for multivariate data visualization. It integrates scatterplot matrices, parallel coordinate plots, star glyphs, and dimensional stacking. The concept of n-dimensional brushes with ramped boundaries and the glyph brush tool are implemented within XMDVTOOL (Ward, 1998).

MANET is software for interactive statistical graphics running on a Macintosh computer. It has been originated from the attempt to consistently represent missing values graphically, but has evolved over time to a more general platform for innovative methods of interactive statistical graphics. Visualization of categorical data has become a second direction of research in this project. MANET requires the data to be given in a standard data matrix format and all displays dealing with the same data matrix are fully linked. So far, linking of two or more data sets is not possible. User interaction in MANET is not only restricted to selection and linked highlighting but sorting and grouping data is also supported. MANET also includes some special features for the visual exploration of spatial data. Polygon maps are available as well as 'tracing', a form of generalized brushing (Craig et al., 1989).

Based on MANET's philosophy is MONDRIAN a 100% pure JAVA application (Theus, 1999). Beside the standard statistical plots MONDRIAN offers some special plots for categorical and geographical data, like scrollable bar charts, mosaic plots, and polygon plots. MONDRIAN also offers graphical aid for log-linear models. In MONDRIAN, a version of multiple brushes is implemented. Whereas in most interactive graphical statistics software only one selection is allowed in each display, here an arbitrary number of selections is possible within one display. Those selections can be combined using one of the standard logical operations XOR, OR, AND, NOT.

Visual analysis by interactive methods is also well known in geography. Modern Geographic Information Systems (GIS) include interactive features to query and investigate maps. There exist also systems that are particularly designed for the exploration of spatially referenced data, for example CDV (Dykes, 1998) and REGARD (Unwin, 1994). In both systems, linking between maps and statistical plots is possible. REGARD's special feature is the possibility of working with several layers of data. In the latest version, only data within one layer can be linked. An earlier version of REGARD experimented with linking of various layers. It also included tools to visualize network flows and extended linking structures for nodes and edges. CDV is a more cartographic based software in which various neighborhood definitions and statistical quantities like Moran's I and Geary's c can be calculated. Beside the basic linking of regions in the map to points in scatterplots or lines in parallel coordinate plots a 1-to-m linking is included for showing local Moran's I coefficients for all neighbors of some region.

Several software packages for interactive statistical graphics exist. Each has its own special features and is based on slightly different paradigms. A user who wants to draw full benefit from all interactive tools currently available has either to build his/her own system or has to work with a lot of different environments.

2. The interactive paradigm

The increasing power of computers created a demand for more powerful, more intuitive and easy-to-use software tools. Interactivity has evolved to be one of the most desirable characteristics of an up-to-date statistical software package. Almost all developers claim that their packages are highly interactive. But there is confusion and disagreement about the definitions and meaning of interactivity, even within the statistical graphics community. Swayne and Klinke (1999) reported the results of a questionnaire that had been launched within the community about the use of interactive statistical graphics and they have expressed surprise about the different understanding of this term. They suggest using the terms "direct and indirect manipulation of graphs" for describing the work in that field. In the following section, I discuss several basic principles of interactivity and propose a classification of interactive methods.

The first step to interactive graphical computing was laid in the PRIM-9 project back in 1973, but till now only scatterplot brushing and 3D-rotation have become a fixed part of standard statistical software. These two concepts, however, cover only a very small aspect of interactive graphics, a variety of features has been proposed in the literature and made available as prototypes. Work with such prototypes has shown that interactive graphics are indispensable for acquisition of qualitative insights into the data sets, for studying model residuals and for revealing quantitative results.

The *Dictionary of Computing* (1991) defines '*interactive*' as "*a word used to describe a system or a mode of working in which there is a response to operator instructions as they are input. The instructions may be presented via an input device such as a keyboard or light pen, and the effect is observable sufficiently rapidly that the operator can work almost continuously.*" In the *Computer Dictionary* (1994), interactive graphics is defined as "*a form of computer use in which the user can change and control graphic displays, often with the help of a pointing device such as a mouse or a joystick. Interactive graphics is used in a range of computer products from games to computer-aided design (CAD) systems.*" Thus, two main characteristics of interactive graphics systems are the speed in which the system reacts to user instructions and the direct user control over the graphic displays. These two characteristics have been the ingredients for the definition of dynamic graphical methods given by Cleveland and McGill (1988): "direct manipulation of graphical elements on a computer screen and virtually instantaneous change of elements." Speed is a necessary feature of an interactive system but it is in no way sufficient. Almost all software tools that are currently available react almost instantaneously to actions caused by the user and interactivity in this sense has become a standard requirement for any modern software. To base a decision on whether a software system is interactive or not only on technical speed measurements ignores the fact that human users adjust the amount of time that they are willing to wait for a response to the difficulty of the desired action. While asking for simple graphical changes a user will typically want the update within a small portion of a second. For complex tasks, he/she will accept a longer response time. It is important that the reaction comes fast enough so that users do not have to interrupt their train of thought.

Huber (1988) corrected the term dynamic graphic to high-interaction graphic. Highly interactive statistical graphics are not only the result of a technical development in computer science they are also the product of research and experience of statisticians and data analysts. They allow the user to grab the data, to ask questions as they arise and to search through a body of data to find interesting relationships and information.

The second characteristic of interactive graphics involves the choice of a user-interface. Although the choice of user-interface is mainly determined by the hardware used – and choosing the hardware is often more a philosophical question than a matter of quality and power – there is the general trend to unify user interfaces. More and more graphical user interfaces (GUI) are replacing command-line interfaces. Programming and batch interfaces are no longer asked for because they hamper interaction. High-interaction graphics are in majority based on GUIs but using a command-line interface does not exclude interaction per se.

To specify the general demand for speed and direct user control for interactive statistical displays, we require that highly interactive statistical graphics software must be able to immediately respond to the following change and control commands created by the user:

- *Scaling*: Perception of graphical displays strongly depends on the scale. Since there are no unique choices, statistical software should provide the user with tools to flexibly change plot scales.

- *Interrogation*: Graphics should not be overloaded. On demand additional information must be available directly from the graphic.
- *Selection*: Selecting subgroups and focusing on specific data points help to reveal structure in the data set. A wide variety of tools to select groups of points from graphical representations is needed to perform sophisticated analyses.
- *Projection views*: Paper and screen are unfortunately restricted to two dimensions, and the human vision system is trained only for the three-dimensional world. Dimension reduction techniques are applied to produce low-dimensional views. A rapid, dynamic and smooth change of projection views is then needed to show as much of the multivariate structure as possible.
- *Linking*: Full interactivity is only achieved when selection is not restricted to a single display but propagated to other plots. This means that all displays are connected and that each view of the data shows each case consistently. Linking is the key concept of interactive statistical graphics, it builds up a relation between measurements of various variables, between different graphical representations as well as between raw data and models. These links can also perform different functions – the standard one is highlighting, others are color encoding or hiding.

Interactivity not only means that the user can interact with the data, but also that the results from the changes made by the user can be seen instantaneously. A rapid and responsive interaction facilitates active exploration in a manner that is inconceivable with static displays. Users can start to pose "What if" queries spontaneously as they work through a task. Therefore, interactive displays not only offer the possibility of comparing resulting static views of different aspects of the data, they even encourage to draw conclusions from the way things are changing. Unfortunately, this aspect cannot be shown in a written paper, this can only be seen live and on-line with a computer.

3. Data displays

Statistical graphics can be grouped into two classes: presentation displays and data analysis displays. The former address the issue of showing results and of underlining conclusions and explanations and they often incorporate more artistic features. The latter are usually as plain as possible. In this thesis, I will focus on data analysis displays. I will use the term 'data display' in a very general way, not only for classical graphical displays like histograms, scatterplots, or bar charts but also for text lists of observations, for contingency tables, or any output of modeling tools. What I call a data display is named a 'data view' by Wills (1999a). His definition of a data view as "a representation the user can look at and study to help understand relationships and determine features of interest in the data they are studying" (Wills, 1999a, p. 20).

3.1. General definition

Based on the graph algebra (Wilkinson, 1997) and using object oriented terminology, I define data displays in the following way.

DEFINITION 1. A data display is a class object comprising the following objects:

(1) A *frame*: Displays are typically drawn on paper or in a computer environment shown in a window on the screen. Thus the frame will often be a two-dimensional Euclidean region. There are some attempts to extend displays beyond the 2-dimensional Euclidean space. Three-dimensional spaces are used in Virtual Reality Labs, and some programs map data to pseudo-three-dimensional cubes. In general, we think of a frame as a set of pixels (or voxels). Every frame owns a size which limits the amount of graphical observations that can be displayed in it. We denote the frame by $\mathcal{F}$. The frame is not restricted by the size of the window in which it is contained. The frame might even be larger than any possible window on the screen such that only part of the frame is visible. Using windows with scroll bars the visible part of the frame can be modified. On the other hand more than one frame can be put into one window.

(2) A *type*: A type is an object made up from two other objects: a set of data representing graphical elements $\mathcal{G} = \{G_1, \ldots, G_g\}$ and a set of scale representing axes $s_\mathcal{G}$ that are both embedded in the frame. The data representing elements are geometric forms or their deformations, such as bars, curves, point sets, glyphs, text representations or numbers. Various types of displays we can choose from are bar charts, histograms, scatterplots, dotplots, contingency tables, or variable displays. The axes determine how many graphical objects will be put in the frame. Axes can be explicitly included in the type or only be present intrinsically. Every graphical element owns additional attributes like size, shape, and color. According to Bertin (1983) there are eight attributes or 'visual variables' as he called them: two planar dimensions, size, value, texture, color, orientation, and shape (Bertin, 1983, p. 42).

(3) A *model*: A model is a symbol for a variable term and its scale $s_\mathcal{X}$. The variable term is a transformation of a set of distinguishable observations $\mathcal{X} = \{x_1, \ldots, x_k\}$. The model describes exactly what we would like to visualize if our graphical facilities were unlimited. It is the goal of the visualization step to give a graphical representation of the model.

(4) A *sample population*: A sample population Ω is a set of cases, individuals, or experimental units on which the variables have been observed. Often they are only intrinsic in the definition of a data display as a set of indices $\Omega = \{1, \ldots, n\}$. Sometimes additional variables with names or other identifications are present. When dealing with categorical data levels of a categorical variable or cross-classifications of two or more might be regarded as identifying cases.

The size of the frame, the axes and the scale are usually considered as associated with the corresponding sets of pixels, graphical elements and observations. Thus, they are often not explicitly stated when describing a data analysis display. It is therefore convenient to define a data analysis display $\mathcal{D}$ as a quadruple consisting of a frame $\mathcal{F}$, a type and its associated set of graphical elements $\mathcal{G}$, a model $\mathcal{X}$, and a sample population Ω, i.e., $\mathcal{D} = (\mathcal{F}, \mathcal{G}, \mathcal{X}, \Omega)$. The pair $(\mathcal{X}, \Omega)$ is the data part and $(\mathcal{F}, \mathcal{G})$ the plotting part.

Associated with each graph level is an identifier that determines which elementary objects are distinguishable on this level. On the frame level the pixels are the elementary objects and the size of the frame defines the grid of pixels that can be identified. The graphical elements define an identifier on the type level. The set of observations described in the model yields an identifier for the model level and the sample population is an identifier itself by definition.

Drawing a graph within a computer system means defining a frame and representing the observations by some graphical elements that have to be placed within that frame. The minimum set-up for a data display consists of a set of observations with a corresponding set of graphical elements that are put together in a frame. The frame determines the available range for plotting and thus sets limits for the number of graphical elements that can be shown. It also has a strong impact on the resolution of the graph and hence the visibility of any structure.

The plot only becomes meaningful when the graphical elements are related to the observations in such a way that a human analyst can perceive the encoded information. The core visualization step is now to build up a structure connecting the various elements of a data analysis display. The way this is performed by statistical software depends on the available space in the frame, the plot type, the axes scaling, the observations, their scale, and the underlying data structure. The conventional solution provided in software packages is to use a fixed dependence structure between data structure and plot type. The data is required to be in a particular format such that the desired graphical representation can be created. The optimal solution is a bijective mapping between the sample population and the set of graphical elements via the model. However, in many situations it is impossible to find such a mapping due to the restricted resolution of the frame and the dimension reduction of the model.

Working with a model level in the display and not with the variables themselves liberates us from the specification of a particular data storage format. Extending Good's "first approximation, that a body of data can be defined as a collection of vectors or k-tuples that all relate to a single topic" (Good, 1983, p. 285), it is assumed in this thesis that a data set is accessible as a list of variables. Every variable is a collection of distinct observations and their corresponding counts that indicate how often a particular value has been observed. The observations might be real measurements as well as ordinal or nominal categories.

DEFINITION 2. A variable A is a symbol for a set of $\{A_1, \ldots, A_a\}$ that are all distinct, their corresponding counts count(A_i), $i = 1, \ldots, A$, and the scale s_A on which the values are displayed. Associated with a variable A is its identifier Ω_A.

It should be kept in mind that this definition does not require that the actual data storage is in that form. It is for notational reasons that we need a basic definition of variables to start with. All known data formats for variables can be converted to this form. Most variables will be univariate and higher-dimensional variables are created as models of the univariate variables, but it might well be that pairs of measurements are stored as a two-dimensional variable. Moreover, geometric areas like polygons, etc. might be stored as one variable. There is no need that all variables are of the same length because they all have their own identifiers. Multilevel data is included as well as multiway data. The advantage of introducing the model level is that variables can be transformed and combined on the fly without requiring that derived variables have to be physically created for drawing a plot. Typically, data sets are augmented while an analysis proceeds. Transformations of variables, model coefficients, model residuals, simulated values and statistical figures might be added as new variables to the data set.

The relational dependence between plotting part and data part is the basic structure that is used for linking and is thus controlling construction of graphics and the interactive procedures. Before I turn to the interrelation of the components of a data display, I will discuss the parts of classical data displays as isolated objects in the next sections. To illustrate some of the definitions, some simple examples will be given. Those examples will be based on two data sets. The first data set contains passenger data of the Titanic disaster, as presented by Dawson (1995). This data, as Andreas Buja (1999, p. 32), pointed out "seem to become to categorical data analysis what Fisher's Iris data are to discriminant analysis." For the second data set, a couple of variables on land usage, population and population movement have been recorded for the 96 counties (Landkreise) and cities (kreisfreie Städte) in Bavaria. The counties and cities in Bavaria are administratively organized in seven regions: Lower Franconia, Upper Franconia, Central Franconia, Upper Palatinate, Swabia, Lower Bavaria, and Upper Bavaria. One data set contains the original observations taken at the counties level, a second data set has been created for the regions by aggregating the original values.

3.2. Sample population

The sample population is usually hidden in the data base architecture in use. Statisticians usually think that their multivariate data is arranged in form of a data matrix and thus assume that the rows of the matrix are used for identification and that every data value having the same index belongs to the same case. But routinely other data storage situations arise like multilevel or multiway data. Therefore the sample population is introduced as an essential part of the display.

To illustrate the sample population, a common visualization task – drawing a bar chart – is described for some statistical packages and their favorite data format. In DATA DESK, for example, the input for all graphics is required to be in a case by case format as it is common in multivariate analysis. But quite often the data is given in a different format, for example as a contingency table. Table 1 shows information on the people on board the Titanic for its maiden voyage. In this table, we have 32 cells of which eight are empty since there were no children in the crew and all children traveling in first or second class survived.

To make this data plottable for DATA DESK, it must be converted to case by case variables. The result is shown in Table 2. However, this data format contains a lot of redundancy and as long as there is no further information to distinguish the individual passengers it is quite unusual to list them individually. The knowledge in the present data set is mainly information on the various groups that traveled on the Titanic but not on the individual passengers. Actually, more information on the passengers of the Titanic is available, but researchers have never universally agreed either on the passenger listings or on the numbers of lives lost. Also for reasons of data protection the contingency data format might be more appropriate for personal data. To cope with various

Table 1
Contingency table of persons on board the Titanic cross-classified according to class, age, sex and surviving

Class	Gender	Survived			
		No		Yes	
		Age		Age	
		Child	Adult	Child	Adult
First	Male	0	118	5	57
	Female	0	4	1	140
Second	Male	0	154	11	14
	Female	0	13	13	80
Third	Male	35	387	13	75
	Female	17	89	14	76
Crew	Male	0	670	0	192
	Female	0	3	0	20

Table 2
A part of the Titanic data matrix

Class	Age	Sex	Survived
First	Adult	Male	Yes
First	Adult	Male	Yes
First	Adult	Male	Yes
First	Adult	Male	Yes
First	Adult	Male	Yes
First	Adult	Male	Yes
First	Adult	Male	Yes
First	Adult	Male	Yes
First	Adult	Male	Yes

Each row corresponds to a passenger on the Titanic's maiden voyage. The variables are 'Class' (first, second third, or crew), 'Age' (adult or child), 'Sex' (male or female), and 'Survived' (yes or no).

data formats, MANET offers two different types of bar charts, the standard one for the case by case format, see Figure 1, and the weighted bar chart for a modified contingency table format, see Figure 2. The modified contingency table format for MANET contains a list for all non-empty cells in the contingency table. Every row in the data matrix lists the corresponding categories for the four variables 'Class', 'Age', 'Sex', and 'Survived'. A fifth column is appended that contains the counts that fall into this cell of the contingency table.

In JMP, case by case variables can be represented as a bar chart by drawing a histogram for them. Use of the bar chart command on the other hand requires a data vector that just contains one data row for each category of a variable and the corresponding counts. So neither the case by case format nor the contingency table format can be used as basis for a bar chart in JMP. Here it is necessary to pre-group the data prior to plotting to obtain the required data structure, see Figure 3. To plot all four variables of the Titanic data set, JMP requires either one data set for each variable or a combined data set that contains all ten marginal counts. In the case by case format, we have 2201 individuals

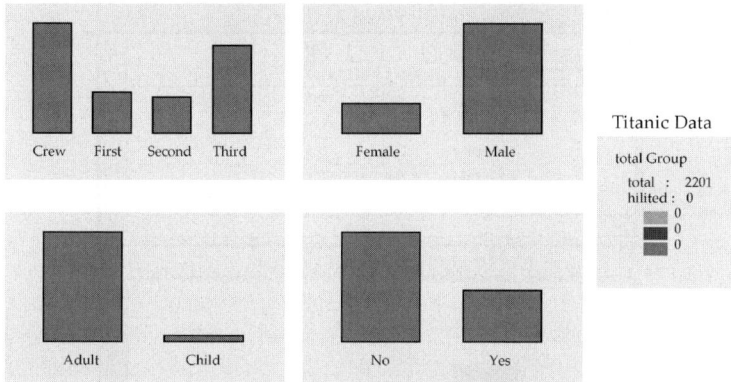

Fig. 1. Bar charts showing the marginal distributions of all variables in the Titanic data set. As can be seen in the group window on the right there are 2201 individuals in the data set. The data are stored in a case by case format which can be seen from the fact that there is no weighting variable indicated in the plot (cf. Figure 2). For a color reproduction of this figure see the color figures section, page 605.

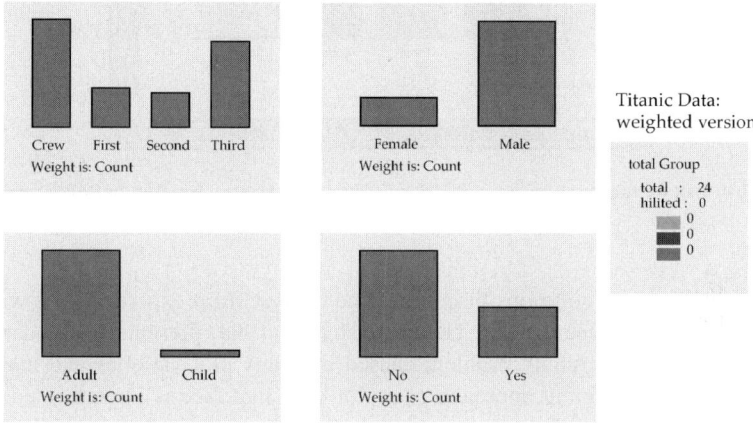

Fig. 2. Weighted bar charts showing the marginal distributions according to class, age, sex and survival of those on board the Titanic. As can be seen in the group window on the right only 24 items are indicated here that correspond to the non-empty cells in the contingency table. For a color reproduction of this figure see the color figures section, page 605.

that can be individually selected and used in the linking process. In the modified contingency table format, we have 24 items. So the kind of data storage has an enormous impact on how many cases or individuals can be separated in the selection and linking process.

The close relationship between data format and possible graphical representation burdens the analyst with either converting the data to the prescribed format, keeping various data formats of the same data set, building conversion functions for every combination of plot and data set type, or providing variations of the same plot type for every

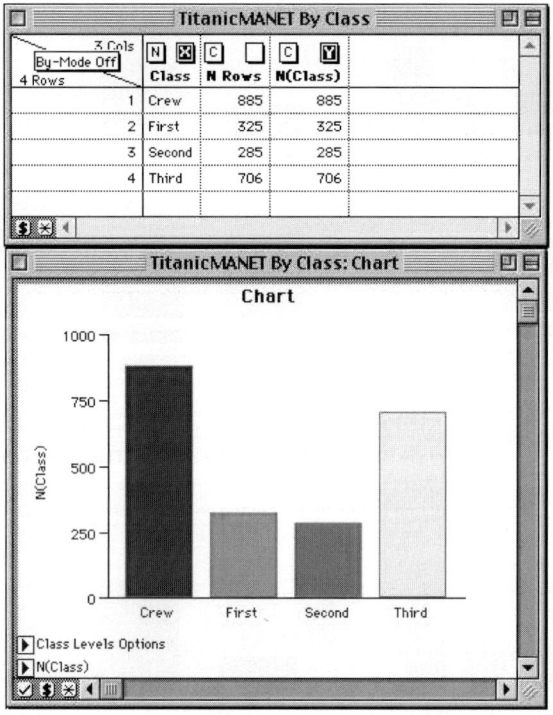

Fig. 3. A bar chart for the class variable in the Titanic data set. To create a bar chart in JMP, a data matrix is needed in which there are not any identical rows.

data format. All these solutions discourage the analyst from experimenting with new plot types or data representations. Therefore Hurley (1993) presents a plot–data interface to avoid the multiplicity problem caused by many plot varieties and many data representations. A different approach of a plot–data interface is implemented in MICROSOFT EXCEL where the user is guided through the various choices while creating a plot. Unfortunately, the user choices offered in the EXCEL system are not yet flexible enough to encourage experimentation.

In interactive statistical graphics, the described relationship between data format and plot types is not only inconvenient for the user but it also limits so far the possibilities of linking, since the linking scheme of software is completely based on the intrinsic case to data relation.

3.3. Model operations

Models are symbols for variable terms and define the set of observations or an appropriate transformation of them that will be represented in the display. The model includes two objects: the set of observations, respectively their transformed values, and an object that contains information on the scales on which the observations have been taken. The model is the central part of our data display and describes exactly the amount of information that shall be visualized. Since statistical visualization aims at portraying distributions and not at showing the variable's values only, as much distributional information as possible is included in the model. In the following, the aim is to allocate the necessary objects in the model formulae to describe the complete information. Future research might lead to enhanced model formulae and hence to enhanced type descriptions since additional objects in the model will require additional graphical representations.

Let A and B be two variables with values $(A_1, \ldots, A_a)$ and $(B_1, \ldots, B_b)$ respectively. According to our variable definition there is one variable entry for every distinct value and a corresponding weight vector that indicates how often each value has been observed.

3.3.1. Identity model

The simplest model is the identity model that just takes a variable and its scale as a model:

$$\mathcal{X} = A = \big(A_i, \ldots, A_a; \text{count}(A)\big),$$

$$s_\mathcal{X} = \Big(\min(A), \max(A), \text{prec}(A), \pi_a, \max_{i=1,\ldots,a} \big(\text{count}(A_i)\big)\Big).$$

The description of the model $\mathcal{X}$ is obvious. How can the elements of the scale be explained? That a scale contains the minimum and the maximum of the observations is rather obvious. For continuous data, $\text{prec}(A)$ defines the available precision of the data and is necessary to indicate which values are distinguishable in the model. For categorical data, the precision is not defined. π_a is an ordering of the a values of A. For quantitative data, this will usually be the natural ordering induced by the values, however, there is at least the choice to order the values increasingly or decreasingly. For qualitative variables, this can be any ordering, for example the lexicographic one. Only integers are possible as counts for a value of the variable A. Thus, the minimum number of counts must be 1 and only the maximum number of counts is stored in the scale to indicate the range.

Let us take a look at the Bavaria data set aggregated for the regions. For the variable 'population density', the identity model has the following form:

$$\mathcal{X} = \big(A_1 = 219.53, \ A_2 = 105.79, \ A_3 = 105.29, \ A_4 = 149.90,$$
$$A_5 = 223.79, \ A_6 = 149.24, \ A_7 = 165.67; \ 1, 1, 1, 1, 1, 1, 1\big),$$
$$s_\mathcal{X} = \big(105.29, 223.37, 0.01, (A_3, A_2, A_6, A_4, A_7, A_1, A_5), 1\big).$$

3.3.2. Variable transformations

Not all features of data are apparent in the way the data are given. It is often convenient and effective to re-express the data to stretch or alter the shape of their distribution. Mosteller and Tukey (1977, p. 89) point out the importance of data transformations for exploratory data analysis stating that "Numbers are primarily recorded or reported in a form that reflects habit or convenience rather than suitability for analysis. As a result, we often need to re-express data before analyzing it."

The most common variable transformations are the powers and the logarithm but any other algebraic expression might also be used. The components of a variable transformation model correspond to those of the identity model. The main impact of a transformation can typically be seen in the change of precision.

The square transformation of the variable 'population density' yields the model description

$$\mathcal{X} = (A_1 = 48193.4209, \ A_2 = 11191.5241, \ A_3 = 11085.9841,$$
$$A_4 = 22470.01, \ A_5 = 50081.9641, \ A_6 = 22272.5776,$$
$$A_7 = 27446.5489; \ 1, 1, 1, 1, 1, 1, 1),$$
$$s_\mathcal{X} = (11085.9841, 50081.9641, 0.0001, (A_3, A_2, A_6, A_4, A_7, A_1, A_5), 1).$$

3.3.3. Pair operator

The classical operation to combine two variables is to pair them. For this operation, a relation between the variables is needed. Since the relation is usually not defined directly on the values but in most cases on the underlying sample population, this model is defined on arbitrary identifiers for these variables. Let Ω_A and Ω_B be identifiers for the variables A and B and let $r : \Omega_A \to \Omega_B$ be a relation. The counts for the resulting pairs (A_i, B_j) are obtained as number of elements in the image of an identifier element i under r. The pair operator is then defined via

$$\mathcal{X} = (A, B)_r = (\{(A_i, B_j) : i \in \Omega_a, \ j \in r(i)\}; (\text{count}(A_i, B_j))),$$
$$s_\mathcal{X} = \Big(\min(A), \max(A), \min(B), \max(B), \text{prec}(A), \text{prec}(B), \pi_{(A,B)},$$
$$\max_{i=1,\dots,A, \ j \in r(i)} (\text{count}((A_i, B_j)))\Big).$$

The components of the pair model are quite clear from the process of pairing two variables. Since r is only a relation and not a function, there might be some elements in Ω_a that are mapped to more than one point in Ω_b. In these cases, the definition ensures that all pairs are listed by repeating the first entry.

Let us look at some examples:

(1) The standard case in multivariate statistics is to have a number of attributes that have been all observed for the same set of individuals. This can be formalized by $r = \text{id}$, $\Omega_A = \Omega_B$, $(A, B) = \{(A_i, B_i) : i \in \Omega_A\}$. In the aggregated Bavaria data set, pairing the variables 'total area' and 'population' yields seven pairs of values. The pair model of two variables A and B is used as input for scatterplots for example.

(2) If we have a variable with missing values the rules of relational data bases and data matrices require the use of a special symbol for the missings. Data storage would be more efficient if we could just list those entries that are not missing. If only one variable, let us say variable B, has missings values a complete-cases-only analysis is based on

$$\Omega_A \subset \Omega_B : r|_{\Omega_A} = \text{id}; \quad (A, B)_r = \{(A_i, B_i) : i \in \Omega_A\}.$$

An analysis based on all cases would then result in

$$\Omega_A \subset \Omega_B \colon r|_{\Omega_A} = \mathrm{id};$$
$$(A, B)_r = \{(A_i, B_i) \colon i \in \Omega_A\} \sqcup \{B_j \colon j \in \Omega_B \setminus \Omega_A\};$$

⊔ denotes the disjoint union of two sets.
(3) Pairing a variable, e.g., 'population change' from the original Bavaria data set with a variable, e.g., 'population density' from the aggregated Bavaria data set, can be done by using the administrative inclusion relation, i.e., $r(i) = j$ if the county i lies in the region j. With this relation the variable 'population change' that has 92 entries is paired with the variable 'population density' that has seven entries. The resulting paired model consists of 95 entries since there are two counties having the same population change and lying in the same region.

The order in which the variables are written in the pairing term define which relations between variables are used.

The pairing operator can be used for more than two variables and is used to define the models for 3D rotating plots or parallel coordinate plots.

3.3.4. Split operator

For categorical data, one often needs a cross-classification of two or more variables. The split operator provides this by combining every category of variable A with every category of variable B.

$$A \mid B = \left(\{(A_i, B_j) \colon j = 1, \ldots, b, \ i = 1, \ldots, a\}; \mathrm{count}(A_i, B_j)\right),$$
$$s_{\mathcal{X}} = \left((\pi_A, \pi_B), \max_{j=1,\ldots,b,\ i=1,\ldots,a} \mathrm{count}(A_i, B_j)\right).$$

As an example, let us look at the cross-classification of the variables 'Class' and 'Age' of the Titanic data:

$$\text{Class} \mid \text{Age} = \big(\{(\text{crew, child}), (\text{crew, adult}), \ldots, (\text{third, child}),$$
$$(\text{third, adult})\}; (0, 885, \ldots, 79, 627)\big),$$
$$s_{\mathcal{X}} = \big((\text{crew, first, second, third}), (\text{child, adult}), 885\big).$$

Note here that knowledge about the variables A and B is not sufficient to create the split operator. Information on the counts of the split classes is necessary and it is assumed that this information is available via the identifiers.

The split model can be visualized by mosaic plots.

3.3.5. Weight operator

Any vector with real non-negative values can be taken as a weight vector to replace the counts of each observation. Let A be a variable and let w be a vector of the same length a that has non-negative real values. Then the model of A weighted by w is

$$A \sqcup w = (A_1, \ldots, A_a; w_1, \ldots, w_a),$$
$$s_{\mathcal{X}} = \left(\min(A), \max(A), \pi_a, \max_{i=1,\ldots,a}(w_i)\right).$$

Weighting is useful for multivariate regional data, especially, when data is expressed in percentages. Let A be a variable that represents percentages for each region and let B be a related variable of absolute numbers. An informative weight vector w can then be obtained by taking $w_i = \sum_{j \in r(i)} B_j$. Weighted models can be best visualized by displays that aggregate the data, like histograms or bar charts.

3.3.6. Categorize operator
Any vector $C = (C_0, \ldots, C_c)$ of real values which segment the range of a variable can be used as categorization.

$$A \boxplus C = \Bigg([C_0, C_1], (C_1, C_2], \ldots, (C_{c-1}, C_c];$$
$$\operatorname{count}(AC) := \Bigg(\sum_{i:\, C_0 \leqslant A_i \leqslant C_1} \operatorname{count}(A_i), \ldots,$$
$$\sum_{i:\, C_{c-1} < A_i \leqslant C_c} \operatorname{count}(A_i) \Bigg) \Bigg),$$
$$s_{\mathcal{X}} = \big(C, \pi_c, \max(\operatorname{count}(AC))\big).$$

3.3.7. Projection operator
In multivariate exploratory data analysis, the task is to identify functional relationships between the data. A central point in achieving this goal is a sufficient dimension reduction in both the number of variables and the number of cases. Projection is a widely used task to reduce high-dimensional data to a low-dimensional space. Given a model $\tilde{\mathcal{X}}$ in p dimensions it can be projected onto a d-dimensional space $\mathcal{B}$ for $1 \leqslant d \leqslant p$ with basis $b_1, \ldots, b_d$. The projection x_i for the ith data point is then given by the coefficients $x_{i1}, \ldots, x_{id}$ of the new basis $b_1, \ldots, b_d$.

$$\mathcal{X} = (x_1, \ldots, x_n; w_1, \ldots, w_n),$$
$$s_{\mathcal{X}} = \Big(\min(b_1), \max(b_1), \ldots, \min(b_d), \max(b_d), \pi_{\mathcal{X}}, \max_{i=1,\ldots,n}(w_i)\Big).$$

3.3.8. Linear models
Linear models form the core of classical statistics. Simple linear regression with a univariate response depending on a univariate predictor can be displayed by a regression line. A common tool to visualize the dependency of a univariate response from a couple of predictors are *partial response plots*. They show the response variable against each of the predictors individually.

$$\mathcal{X} = (y \sim x),$$
$$s_{\mathcal{X}} = \big(\min(x), \max(x), \min(y), \max(y), \alpha, \beta\big).$$

α and β are the parameters in the linear regression equation $y = \alpha + \beta x$. Linear models also generate several new variables like residuals, leverages, or Cook's distances. These individual new variables are included in the notation $y\tilde{x}$.

3.3.9. Smoothing models

Instead of using parametric models like regression in which the formal structure of the regression curve is pre-specified up to some unknown parameters, smoothing methods can be used. Smoothing works with the relatively weak assumption that the regression curve is a smooth function. The true pattern is obtained from the data. As with regression models smoothers can be visualized by a single graphical element: the smooth curve. Smoothers can be seen as interpolating methods and are thus defined via their segments between two adjacent support points.

$$\mathcal{X} = (x_{s_1}, y_{s_1}, \ldots, x_{s_s}, y_{s_s}),$$
$$s\mathcal{X} = (x_{s_1}, x_{s_s}, y_{s_1}, y_{s_s}).$$

Here x_{s_i} denotes the ith support point and y_{s_i} the smoothed value at this point.

Also other statistical models that are applied in data analysis can be visually represented by a single graphical element. Such models should also be kept in mind for this formalization, although they are not explicitly defined here.

3.3.10. Models with missing values

Real data sets often confront the analyst with incomplete data. Missing values cause difficulties with standard confirmatory techniques as well as with contemporary graphical techniques. To avoid these problems, standard statistical software either restricts the analysis to the complete cases only or assigns a fictitious number (like -9999) to the missing values. Both paths get into troubles with multivariate data, since with increasing number of variables the chance that the majority of cases are incomplete goes up as well as the chance that the fictitious number comes close to regular values for at least one variable. In addition, when missings are not at random they can tell us a lot about the structure of the data set. The variable format that underlies our data model does not know about missing values because we do not use a data matrix format and there is no need for variables to have the same length. To incorporate missing values in the model, the set of observations $\mathcal{X}$ is joint with a list of counts indicating how many values are missing in any variable that is included in the model.

For example, the pair model of two variables A and B will show the missings for A and the missings for B. Those cases that are missing in both can then be obtained by taking the intersection of the corresponding identifiers.

I have just presented the basic models that arise in statistical graphics. These models can be used iteratively and they can be combined.

For example, (A, B, C) yields a triple of observations and $(A, B) \mid C$ a separate pair model of A and B for every value of C.

3.4. Types of graphics

In the following section, the standard plot types used in exploratory data analysis are defined as data displays and their corresponding formal descriptions are given. In the description, all parts of a data display are handled independently but to yield a useful plot in practice it is necessary to observe the interrelations between the various plot parts. The definitions are made to comply with the model definitions of the previous section. They therefore differ sometimes from the standard implementations in statistical software. The forthcoming descriptions aim at defining graphics that display all of the corresponding model information.

3.4.1. Style: point

The simplest model is the identity model and the corresponding display for non-categorical data is the dotplot. Other point plots are the scatterplot and 3D rotating plots. All point plots consist of a point cloud and a set of axes. The axes can be either visible or invisible and they might be added or removed interactively. This is a matter of taste and of implementational consistency. In the canonical case, every point in the plot corresponds to a single observation. The position of a point is fully determined by the observed values it represents and thus changing the position of a point affects the underlying data. For educational purposes, such interactive changes are desirable but they should not be allowed for analytical tools. Other parameters like color, size and shape are attributes of the points and can be explicitly defined as subobjects of them. In many programs, changing these attributes interactively is the only kind of user interaction that is supported.

3.4.1.1. Dotplot.
The formal description of a dotplot is given by

$$\mathcal{G} \mapsto \{\text{point}_i, \text{intens}_i, \text{color}_i, \text{size}_i, \text{shape}_i: i = 1, \ldots, g\},$$
$$s_\mathcal{G} = \big(\text{lower}, \text{upper}, \text{prec}, \text{max}(\text{intens})\big).$$

Here max(intens) is the count of observations for which a point shows brightest.

The max(intens) parameter is used for tonal highlighting. The use of tonal highlighting offers a possibility of representing overplotted points. It means that the brightness of a point is proportional to the frequency of its occurrence. Standard implementations of the dotplot set the parameter max(intens) equal to infinity and hence plot each point with the same brightness. Thus, such plots do not include any information on the amount of overplotting or tied values.

To illustrate the definition, let us look at some examples. The simplest case is a dotplot for a small data set. In Figure 4 a dotplot for the total area of the seven regions is drawn. Here we obtain:

$$\mathcal{G} = \{\text{point}_1, \text{point}_2, \text{point}_3, \ldots, \text{point}_6\},$$
$$s_\mathcal{G} = (723129, 1589227, 4).$$

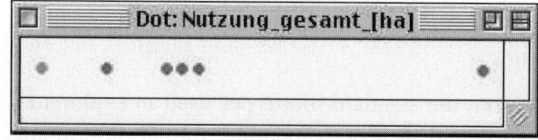

Fig. 4. A dotplot showing the total area of the seven regions in Bavaria. In theory, each observation is represented by a single dot but we can only see six dots because two observations fall too close to be visually separated.

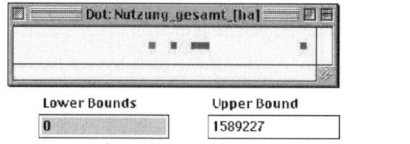 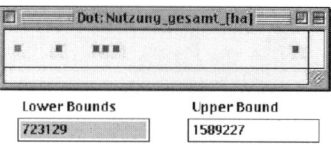

Fig. 5. By default, the axis in a dotplot is scaled to the range of the data that is to be represented. Other choices for the upper and lower bound of the dotplot axis might lead to completely different pictures.

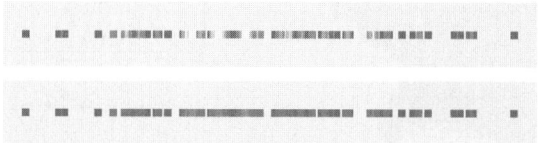

Fig. 6. Tonal highlighting is used to show overplotting of points. Changing the parameter max(intens) changes the intensity of overplotted points.

The axis of the display s_G defines the subset of the model domain that will be shown in the plot as long as the size of the frame is large enough to provide the necessary resolution. By default, most programs will adjust the axis s_G to the range of the model by setting lower $= \min(A)$, upper $= \max(A)$ (see Figure 5).

In Figure 6 tonal highlighting is used to show the amount of overplotting. The dotplot at the top uses the default value 4 for max(intens), for the dotplot at the bottom this value has been set to 30. This value is far too high for the Bavaria data set in which only 96 counties are present, so that no tonal highlighting is apparent.

3.4.1.2. Scatterplot. The scatterplot is the most well-known graphical display for two-dimensional data. It is excellent for displaying relationships between two continuous variables.

$$\mathcal{G} = \{\text{point}_i, \text{intens}_i, \text{color}_i, \text{size}_i, \text{shape}_i : i = 1, \ldots, g\},$$

$$s_\mathcal{G} = \left(\text{lower}_{\text{abs}}, \text{upper}_{\text{abs}}, \text{lower}_{\text{ord}}, \text{upper}_{\text{ord}}, \text{prec}_{\text{abs}}, \text{prec}_{\text{ord}}, \max(\text{intens})\right).$$

As in the dotplot tonal highlighting is used to visualize dense regions in the scatterplot. Tonal highlighting is similar to brushing with hue and saturation. The latter, however, is based on a hardware component of the SGI, the α-channel. Tonal highlighting adds the brightness of overlapping points: the brighter white a region in the scatterplot appears, the more dense lie the points in that region.

3.4.1.3. Trace plot. Tracing is a special kind of generalized brushing proposed by Craig et al. (1989). It computes statistics like the mean, the range, or the standard deviation, of specified variables for all points covered by the brush rectangle. As the brush moves over the window the computations are updated and the results are displayed in a new view in a time series plot format (see Figure 7). Tracing is particularly useful for classifying variables and detecting spatial dependence. Although it shows lines in the display the inner structure of the trace plot is a point plot. This is due to the fact that the line segments are just connecting the points to make the trend more visible. But interactive

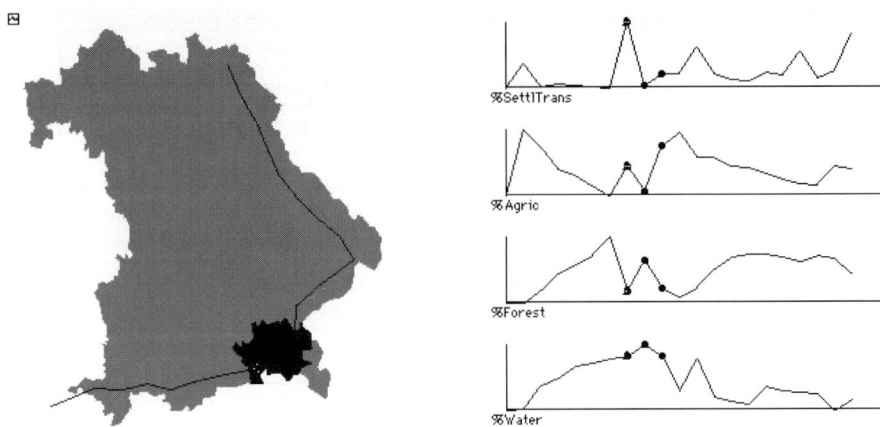

Fig. 7. On the left hand is a polygon map of the counties in Bavaria; on the right hand is the trace plot for four variables on type of land usage.

operations cannot be performed with the lines, they are restricted to the points only.

$$\mathcal{G} = \{\text{point}_i, \text{intens}_i, \text{color}_i, \text{size}_i, \text{shape}_i \colon i = 1, \ldots, g\},$$

$$s_\mathcal{G} = (\text{lower}_{\text{abs}}, \text{upper}_{\text{abs}}, \text{lower}_{\text{ord}}, \text{upper}_{\text{ord}}, \text{prec}_{\text{abs}}, \text{prec}_{\text{ord}}).$$

3.4.2. Style: area
Point plots aim at displaying every observation individually. Area plots aggregate the data and show counts or frequencies of groups of observations.

3.4.2.1. Bar charts and pie charts. The pie chart and the bar chart represent counts by area and both charts reduce this two-dimensional information to a visual comparison of scalars by simply keeping one dimension fixed for all categories. Thus, in the bar chart counts in each category are represented by a rectangle (bar, tile) with common width, and in the pie chart the area of a circle is divided into various segments (or slices or wedges) such that the areas of the slices only vary with the angles at the circle's center. It is much easier for most people to compare heights of bars than to compare different angles of segments. Especially, small differences in the angles of the pie widgets are hard to discern. Hence the bar chart is preferable whenever the sizes of the groups are of greater interest than the division of the whole into subgroups. On the other hand, by dividing one circle into segments we do not get only information on the counts of each individual category but also information on the union of all categories and all univariate proportions. In Figure 8 bar and pie charts for the four variables of the Titanic data are given. In the bar chart, all bars aligned on the baseline and it is therefore easy to sort the categories. In the pie chart, this is much harder as can be seen in the one for the variable 'Class' where it is hard to tell whether there are more people in first or in second class. On the other hand a glance on the pie chart of 'Gender' shows that three quarters of the people on board of the Titanic are males.

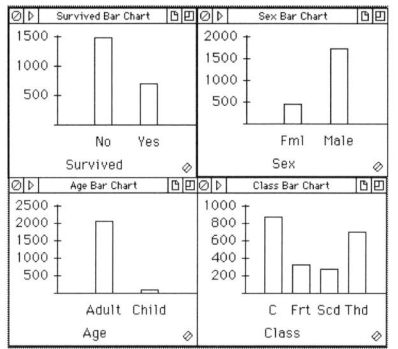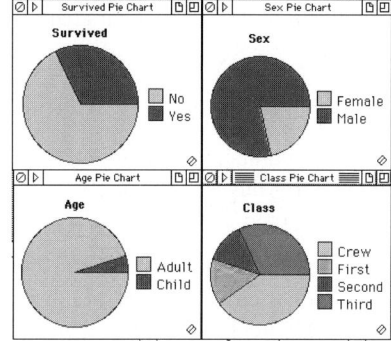

Fig. 8. Bar charts and pie charts for the four variables in the Titanic data set. For a color reproduction of this figure see the color figures section, page 606.

Formalization of a bar chart:

$$\mathcal{G} = \{\text{bar}_i, \text{intens}_i, \text{color}_i, \text{width}_i, \text{height}_i : i = 1, \ldots, g\},$$
$$s_\mathcal{G} = \bigl(\max(\text{width}), g, \pi_g, \max(\text{height})\bigr).$$

Here g is the number of bars and π_g reflects the ordering of the bars and max(height) is the height of the highest bar that can be displayed in the frame.

Formalization of a pie chart:

$$\mathcal{G} = \{\text{wedge}_i, \text{intens}_i, \text{color}_i : i = 1, \ldots, g\},$$
$$s_\mathcal{G} = (\text{radius}, g, \pi_g).$$

Again g is the number of wedges and π_g reflects the ordering of the wedges.

An alternative to pie charts are stacked bar charts (see Figure 9). Their main idea is to draw one bar for all cells and split this bar horizontally in slices according to the counts in the categories. Shading or coloring is used to distinguish the resulting blocks. As well as in pie charts it is difficult to compare proportions of two groups that only differ a little. Also the type of shading or coloring influences the human perception and might lead to wrong conclusions. The splitting idea of stacked bar chart is recursively extended in mosaic plots to higher dimensions.

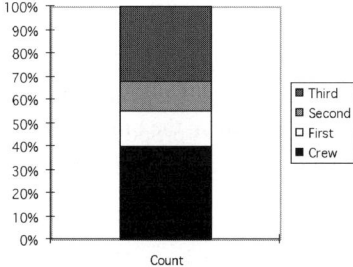

Fig. 9. Stacked bar charts representing a two-way classification between 'Gender' and 'Class'.

Formalization of a stacked bar chart:

$$\mathcal{G} = \{\text{slice}_{ij}, \text{intens}_{ij}, \text{color}_{ij}, \text{height}_{ij}, \text{width}_i : i = 1, \ldots, g, \ j = 1, \ldots, k\},$$
$$s_\mathcal{G} = \big(\max(\text{width}), g, \pi_g, k, \pi_k, \max(\text{height})\big).$$

Here g is the number of categories for the first variable and k represents the number of categories for the second variable. The corresponding ordering of the categories are reflected by π_g and π_k.

3.4.2.2. Spine plot. The spine plot (Hummel, 1996) is a modification of the bar chart. In a standard bar chart, the area of a bar represents the frequency of the corresponding category, and typically, bars are drawn with a common width such that the frequency can easily be read from the height of the bar. The disadvantage of this procedure appears with the use of linked subsetting. If selected cases are highlighted in bar charts it is usually hard to compare relative proportions with each other. In a spine plot, each bar is chosen with a constant height, so that now the width varies according to the number of data points in a category. Comparison of highlighted proportions is now straightforward.

$$\mathcal{G} = \{\text{spine}_i, \text{intens}_i, \text{color}_i, \text{width}_i : i = 1, \ldots, g\},$$
$$s_\mathcal{G} = \big(\text{height}, g, \pi_g, \max(\text{width})\big).$$

3.4.2.3. Histogram. Histograms are the basic method to display the distribution of numeric values. They are sometimes confused with bar charts. Both show frequencies as bars but the histogram is based on an arbitrary division of a continuous interval while the bar chart uses predefined groups that usually do not cover a continuous range. As a consequence the histogram bars are sequentially arranged without gaps. In histograms, we view the density of the distribution which is portrayed in the overall appearance of the plot.

$$\mathcal{G} = \{\text{bin}_i, \text{intens}_i, \text{color}_i, \text{width}_i, \text{height}_i : i = 1, \ldots, g\},$$
$$s_\mathcal{G} = \big(\text{lower}, \text{upper}, \max(\text{height})\big).$$

3.4.2.4. Mosaic plot. Mosaic plots as introduced by Hartigan and Kleiner (1981) are a graphical analogue to multivariate contingency tables. They show the frequencies in a contingency table as a collection of rectangles whose areas represent the cell frequencies. The construction of mosaic plots is hierarchically (see Figure 10) and resembles the way multiway tables are often printed. For tables, rows and columns will be recursively split to include more variables. In the same way, we split up horizontal and vertical axes of the mosaic plot recursively to obtain tiles that represent the cells in the contingency table. The area of each tile is chosen to be proportional to the observed cell frequency. Thus, mosaic plots are a multidimensional extension of divided bar charts. The width of each column of tiles in a mosaic plot is proportional to the marginal frequency. The formal description of the mosaic plot is an extension of the stacked bar chart.

$$\mathcal{G} = \{\text{tile}_{ij\ldots}, \text{intens}_{ij\ldots}, \text{color}_{ij\ldots}, \text{height}_{ij\ldots}, \text{width}_{ij\ldots} :$$
$$i = 1, \ldots, g, \ j = 1, \ldots, k, \ \ldots\},$$
$$s_\mathcal{G} = \big(\max(\text{width}), g, \pi_g, k, \pi_k, \ldots, \max(\text{height})\big).$$

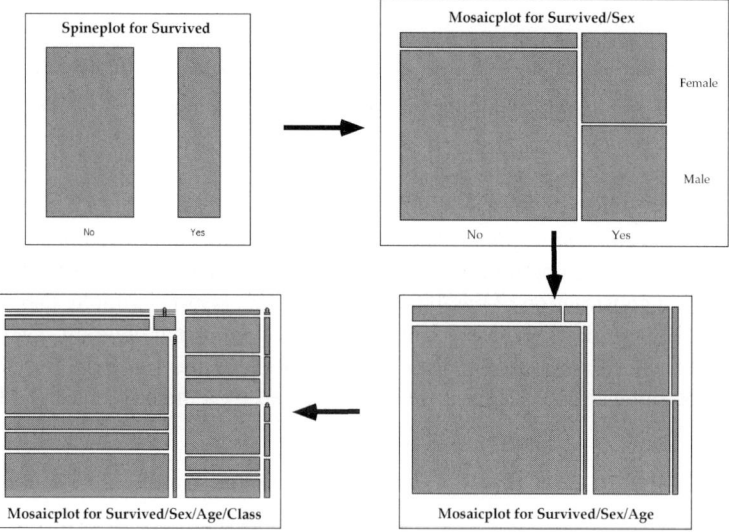

Fig. 10. Construction of a mosaic display.

3.4.2.5. Polygon map. The key tool to enhance interactive graphics software to the analysis of spatial data is to add a map tool. Data with point locations can simply be handled by drawing a scatterplot for longitude and latitude, but many spatial data sets are based on regions. To deal with such data, a polygon plot or map is necessary.

$$\mathcal{G} = \{\text{polygon}_i, \text{intens}_i, \text{color}_i, \text{size}_i, \text{shape}_i : i = 1, \ldots, g\},$$
$$s_{\mathcal{G}} = \big(\text{lower}_{\text{abs}}, \text{upper}_{\text{abs}}, \text{lower}_{\text{ord}}, \text{upper}_{\text{ord}}, \max(\text{intens})\big).$$

3.4.3. Style: curves, lines

In many cases, points shown in a graphical display are connected with lines to focus on the shape of the visualized pattern. Instead of histograms, frequency polygons or smooth density curves are used to portray distributional properties. Regression lines or curves are drawn to visualize relationships between variables.

All these situations have in common that there is a single graphical element – the curve – displayed in a two-dimensional layout.

$$\mathcal{G} = \{\text{curve, color, width}\},$$
$$s_{\mathcal{G}} = (\text{lower}_{\text{abs}}, \text{upper}_{\text{abs}}, \text{lower}_{\text{ord}}, \text{upper}_{\text{ord}}, \text{prec}_{\text{abs}}, \text{prec}_{\text{ord}}).$$

3.4.4. Style: hybrid

A lot of data analysis displays contain more than one kind of graphical elements. When a display contains several kinds of graphical elements then every set of graphical elements can define its own identifier.

3.4.4.1. Boxplot. The boxplot is a paper and pencil method introduced by Tukey. It is a view that aggregates the data to the five letter summary and displays them as boxes and whiskers. All observations that fall off the bulk of the data are represented as individual points.

$$\mathcal{G} = \{\text{point}_{\text{out}\,l1}, \ldots, \text{point}_{\text{out}\,ll}, \text{whisker}_{\text{low}}, \text{box}_{\text{low}}, \text{median}, \text{box}_{\text{upper}},$$
$$\text{whisker}_{\text{upper}}, \text{point}_{\text{out}\,h1}, \ldots, \text{point}_{\text{out}\,hh}, \text{intens}_i, \text{color}_i, \text{size}_i, \text{shape}_i :$$
$$i = 1, \ldots, l+h\},$$
$$s_{\mathcal{G}} = \bigl(\text{lower}, \text{upper}, \text{prec}, \max(\text{intens})\bigr).$$

3.4.4.2. Parallel coordinate plots. The parallel coordinate plot is a geometric device for displaying multivariate continuous measurements, in particular, for dimensions greater than three. The construction of parallel coordinate plots is rather simple. Draw the axes for all variables parallel to each other in order to obtain a planar diagram. Then plot the measurements for each variable on the corresponding axis and connect those points that belong to the same case. So, each case is represented by a broken line. In addition, all the measurements are present as points on the axes.

$$\mathcal{G} = \{\text{point}_{ai}, \ldots, \text{point}_{hi}, i\,\text{line}_i, \text{intens}_i, \text{color}_i, \text{width}_i : i = 1, \ldots, g\},$$
$$s_{\mathcal{G}} = (\text{lower}_{\text{axis}_a}, \text{upper}_{\text{axis}_a}, \text{lower}_{\text{axis}_b}, \text{upper}_{\text{axis}_b}, \ldots, \text{lower}_{\text{axis}_h}, \text{upper}_{\text{axis}_h},$$
$$\text{prec}_a, \ldots, \text{prec}_h).$$

A third identifier can be included in parallel coordinate plots. Due to the point–line duality an intersection point of lines in the parallel coordinate plot corresponds to straight lines in a Cartesian coordinate system.

3.4.4.3. 3D rotating plot. Plot rotation is a widely used tool to study relationships among three or more continuous variables. Rotation enhances the familiar scatterplot to give the illusion of a third dimension. Rotating plots are available in many statistical and graphics packages. Rotating plots are often combined with projection techniques like the grand tour and projection pursuit. To get a hint on which projection is currently displayed, rotating plots also contain a projection of the canonical axes.

$$\mathcal{G} = \{\text{point}_i, \text{intens}_i, \text{color}_i, \text{size}_i, \text{shape}_i : i = 1, \ldots, g\} \cup \{\text{axis}_1, \ldots, \text{axis}_\ell\},$$
$$s_{\mathcal{G}} = \bigl(\text{lower}_{\text{proj}_1}, \text{upper}_{\text{proj}_1}, \text{lower}_{\text{proj}_2}, \text{upper}_{\text{proj}_2}, \text{prec}_{\text{proj}_1}, \text{prec}_{\text{proj}_2},$$
$$\max(\text{intens})\bigr).$$

Here ℓ denotes the dimensionality of the data and proj_i the two projection axes.

3.4.4.4. Biplot (PCA). The biplot (Gabriel, 1971) is another method that represents both the individuals and the variables in one plot. It is similar to a specific projection in a 3D rotating plot: the projection onto the first principle components. But in addition to the rotating plot, the displayed axes in the biplot show additional information on the associations of the variables.

$$\mathcal{G} = \{\text{point}_i, \text{intens}_i, \text{color}_i, \text{size}_i, \text{shape}_i : i = 1, \ldots, g\}$$
$$\cup \{\text{axis}_j, \text{color}_j, \text{length}_j : j = 1, \ldots, \ell\},$$
$$s_{\mathcal{G}} = \bigl(\min_{pc1}, \max_{pc1}, \min_{pc2}, \max_{pc2}, \text{prec}_{pc1}, \text{prec}_{pc2}, \max(\text{intens}),$$
$$\max(\text{length})\bigr).$$

A book-length treatment of biplots is given by Gower and Hand (1996). A version of interactive biplots is described by Hofmann (1998). There exists also a biplot version for displaying categorical data. There the projection is based on a multivariate correspondence analysis.

Many other hybrid graphs can be created by blending a scatterplot with other plot types, in particular, by blending with smoothers and regression lines. Here the individual descriptions of the plots are brought together by simple taking the union of the individual sets.

3.4.5. Panel plots
Panel plots are space saving arrangements of the same plot type for different variables. Panel plots enhance comparison between the distributions of variables and groups. In some software packages, interactive techniques like brushing and linked highlighting are limited to panel plots.

3.4.5.1. Scatterplot matrix. A scatterplot matrix for k variables presents all $\binom{k}{2}$ scatterplots of variable i versus variable j, $i \neq j$. Some implementations reduce the full scatterplot matrix to a triangular matrix by ignoring all symmetric plots. Every scatterplot in the matrix behaves as a standard scatterplot and can be independently modified by the user. Linking is in some implementations (cf. S-PLUS) limited to scatterplot matrices and is not available for other displays.

3.4.5.2. Conditional plots. Coplots are sophisticated arrangements to show conditional relations among variables. A coplot is an array of plots, each of these plots drawn for different subsets of a conditioning variable. The core of the plot is a one-, two-, or even three-dimensional plot of the axis variables, sometimes called the panel function. Any kind of plot can be used for these panel functions. Each panel in the array shows the partial associations among the variables presented in the panel. The collection of all these panels in a matrix arrangement shows how these vary as the given variables change. An important point in the construction of these plots is the scaling of each panel function. It must be guaranteed that all plots have the same scale to enable comparison of the plots. Coplots are a good arrangement for showing specific results of an analysis of multivariate categorical data. But one has to know beforehand on which associations the impact will be laid on. So, coplots are a good presentation tool but with only limited use for analysis purposes. They are not very flexible and hard to extend. (See Figure 11.)

3.4.6. Style: lists
A list of strings or numerical values can also be seen as a data display. Typically, such lists represent a variable.

3.4.6.1. Text list or variable list.

$$\mathcal{G} = \{\text{string}_i, \text{intens}_i, \text{color}_i, \text{size}_i : i = 1, \ldots, g\},$$

$$s_\mathcal{G} = \pi_g.$$

Fig. 11. A coplot for the three variables 'Class', 'Age', and 'Gender' of the Titanic data.

3.4.7. Style: tables

Contingency tables are a useful representation of cross-classifications as provided by the split operator.

$$\mathcal{G} = \{\text{cell}_{ij}, \text{intens}_{ij}, \text{color}_{ij}, \text{size}_{ij}: i = 1, \ldots, g, \ j = 1, \ldots, k\},$$
$$s_{\mathcal{G}} = (g, k, \pi_g, \pi_k, \text{height}, \text{width}).$$

Here g and k are the number of categories for the row and column variables. Height and width gives the total layout area of the table.

3.4.8. Extensions for missing values

When missing values occur in the model description we augment the set $\mathcal{G}$ with graphical elements that indicate the numbers of missings. Since the model contains the counts of missing values for every single variable as well as for the model itself, we need $p + 1$ additional graphical elements when p is the number of variables in the model. We thus

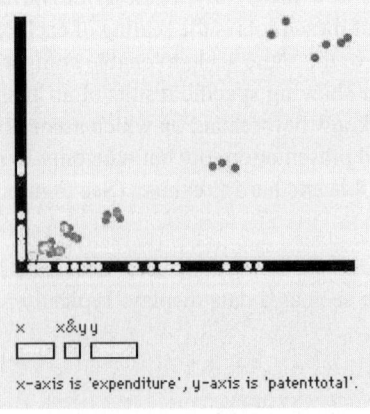
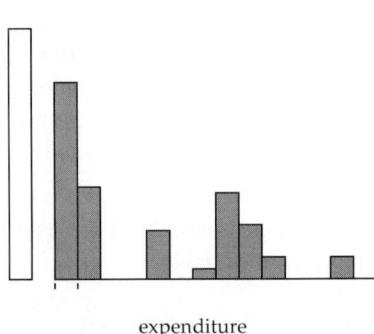

Fig. 12. Visualization of incomplete data in MANET. In the histogram, a separate bar is drawn representing the number of incomplete data. The white circles on the axes of the scatterplot represent all those pair of observations for which only one component has been measured.

obtain
$$\mathcal{G} = \mathcal{G}_{\text{nomissing}} \cup \{\text{bar}_{\text{missing}_1}\} \cup \cdots \cup \{\text{bar}_{\text{missing}_{p+1}}\}.$$

Figure 12 shows a scatterplot from MANET in which all complete observations are plotted in a standard way and the incomplete data points are shown in two ways: first, in three boxes underneath the plot indicating which variables have not been observed for the individual points, and secondly, those data points having one variable observed and the other missing are plotted as white dots on the corresponding axes.

4. Direct object manipulation

The big disadvantage of computer generated graphics is that the user loses control over the construction. Internally defined default values are used for the construction of the display and the user is not involved in the process of choosing these values. While creating a graph in statistical software the user makes only two choices:

(1) Which model shall be represented?
(2) Which style of plot is to be used (points, areas, lines, etc.)?

All other necessary input that is needed for constructing a display and transforming the model into a visible object is taken from the implemented default values – values that are chosen by the program designer and not by the user. For example, there is an upper limit for the size of the frame and this value will also limit the number of graphical elements that can be successfully placed within that frame. Also a lower limit exists, since computer graphics work on a discrete pixel grid and therefore graphical elements cannot be smaller than one pixel in size.

Not all users are aware of these restrictions and not every software tells the user that the created graph might show something different than the model specified by the user. An interactive system gives the control back to the user and provides means for interactively changing the default values. Since the possible gain of varying a display is not known in advance, it is important that the user has not to put too much effort into the variation. Authors of hand-drawn graphs know exactly where the displays deviate from their models. With computer-generated graphics this knowledge is lost and the four levels of a display – frame, type, model, and sample population – have their individual behavior.

User interaction is a two-fold procedure. It comprises an interaction request issued by the user and the response of the software. This response can either be the requested one, or a warning, or the requested response enhanced by a warning. An intelligent system should give just the right amount of additional hints and warnings. Hofmann (1997) discusses the issue of graphical reliability and shows some important warnings. Theus (1996) listed warnings, besides highlighting, linking, and querying, as one of the basic elements of interactive systems. I agree that interaction is a two-way procedure and that warnings are part of the interaction process, but I would like to stress the point that a warning is one kind of response of the system to user interactions and not an interactive user operation itself.

From an abstract viewpoint user interaction can be defined as manipulation of graph objects. What can the user interactively change? The user can

- select objects,
- query objects,
- group objects,
- change attributes of objects,
- modify objects,
- replace objects by other objects,
- change relations between objects.

The amount of corresponding changes in the visualization varies with the object at which the user interaction is directed. The range of user interaction effects goes from changing color of points to replacing the current model or plot type. However, there are some invariants: first of all the data set is an invariant. All interactive operations require that we are still working with the same data set. This is to be understood in an extended sense of data frames in S-PLUS. All user interactions are limited by the current data frame. However, within that frame we can interactively change the data by defining subsets, transforming variables, deriving new variables, or fitting statistical models. Given the data there are semi-invariants at every graph level. For example, if a histogram shall be displayed for a variable A the finest possible type is a histogram that contains one bar for every distinct value of the variable A. To this histogram corresponds the model

$$A \mid \tilde{A}, \quad \text{with } \tilde{A}_i = \begin{cases} A_{i\uparrow} + \text{prec}, & \text{for } i = 1, \ldots, a, \\ A_{1\uparrow} & \text{for } i = 0. \end{cases}$$

(Here $A_{i\uparrow}$ denotes the ith increasingly ordered value of A.) On the other hand there are a coarsest model and a coarsest type where all observations fall into one bin. The definition of the coarsest model (covering just the range of the variable A) is given by

$$A \mid \tilde{A}, \quad \text{with } \tilde{A}_0 = \min(A) \text{ and } \tilde{A}_1 = \max(A) + \text{prec}.$$

For every model and type of a data display, there exist corresponding finest choices which are denoted by $\mathcal{G}^*$ and $\mathcal{X}^*$, the coarsest choices are denoted by $\mathcal{G}^0$ and $\mathcal{X}^0$. The most common and also the most powerful user interactions just vary model or type of the display between the two extremes. In a similar way, defining subsets of individuals corresponds to partitioning the sample population Ω.

Which objects of a data display can receive user interaction requests? In principle, all objects could be interactively accessible but not all objects are directly visible. How should the user directly manipulate objects that cannot be seen? It seems reasonable to restrict direct manipulation to objects that have a graphical representation. Therefore, the following axiom is postulated.

AXIOM 1. *Only objects that are graphically represented can be directly manipulated.*

How can the users issue their interaction requests? Current implementations use mainly three ways of user interaction. The first one is to evoke an action by drawing, dragging or clicking with the mouse while some action keys may be pressed to alter the operation. The second one is to use keystrokes to manipulate the active display. The third method is to open dialog boxes in which the user can insert the desired manipulations. Dialog boxes then constitute visual representations in the sense of Axiom 1. While the first two methods are faster, the last one offers a much finer control. Thus, these methods are not alternatives but complement each other. It depends on the kind of user interaction and also the purpose of the user which way is more appropriate for a given situation. An excellent way to guide the user through an interactive analysis is either to use hyperview menus as is done in DATA DESK or to use cues integrated in the graph interface. By cues I mean pop-up triggers that are made visible by changing the mouse pointer's shape when moved over an area where a new feature is accessible. This approach is successfully implemented in MANET.

Whichever implementation of interactive operations is used, the user interaction starts with selecting those graphical objects that are to be interactively modified.

5. Selection

Selection is not only a common manipulation in visual data analysis systems but selecting objects is ubiquitous in human–computer interactions, for example, in a desktop GUI where icons have to be selected, in drawing programs, or in geographic information systems.

Within an interactive computer environment, it is quite natural to allow the user to select a certain subgroup of interest directly in the plot. To reflect the selection, the appearance of the graphical elements is changed – usually highlighted. According to the basic assumption that only objects with a graphical representation can be directly accessed by user interactions selection mainly takes place in the frame or in the type level.

Selection can be classified with respect to the graph level at which the selection is triggered. This can be

- selection of frame regions,
- selection of graphical elements,
- selection of axes.

Selection triggered in the frame does not aim to select pixels but aims to select a subset of the type. In our framework, the graph type level consists of two different objects leading to two selection concepts: selecting graphical elements and selecting axes.

DEFINITION 3. A selection that transforms the region of interest indicated by the user in the frame to a set of selected graphical elements is called a graphical selection.

A selection that transform this region of interest into conditions on the axes of the display is called an axis based selection.

When transforming the region selected in the frame into a selection at the type level the kind of selection tool used can be taken to distinguish between graphical and axes based selection.

5.1. Selection tools

A selection tool is a mechanism to define an object or an area of interest in the display. It can also be seen as a tool to define restrictions that the points of interest have to fulfill.

DEFINITION 4. A selection tool is a mapping S from the set $\mathcal{P}(\mathcal{F})$ of all possible subsets of the frame to the interval [0, 1]. It thus defines a partition $\mathcal{F}_S$ of the frame $\mathcal{F}$.

All current selection implementations do only offer binary selections, and then S will only take values 0 or 1. The definition above includes m-nary selections. With m-nary selections a continuous degree of selection can be represented. Such tools are often necessary in geographical analysis systems. For example, one might want to select objects according to their distance from the center of selection, and thus the closer an object is to that center the higher the intensity should be.

Another use of m-nary selections is to mimic multiple selections within one display. On first glance multiple selections within one plot seem not to be allowed since two blocks of a partition are disjoint by definition and it is thus impossible that a graphical object is both selected and unselected. In such a selection system, multiple selections are only possible when the selections are combined using logical operations. To display a multiple selection using a m-nary selection scheme, we simply add the selection flags for every observation and display the relative frequency of selection for each observation. In the following, we will restrict ourselves mainly to binary selections since those are most common.

User interaction takes place on the screen and since common screens are only two-dimensional, the selection region in the frame can only have dimension less than or equal to 2. The frame is of minor interest for data analysis, so the selected region of interest is automatically translated into useful information on the graph type level. To ensure a smooth working environment and to avoid wrong conclusions, this transition has to be as intuitive as possible. The dimensionality of a selection tool that results on the graph type level can be completely different from the one in the frame level. It can vary from one dimension up to as many dimensions as there are in the data set. Typically, the way in which selected regions are translated into data subsets is defined by the selection tool used and the style of the active plot. The most common selection tool is a rectangular brush. Pointer, lasso and slicers are others. Wills (1996) distinguishes data dependent and data independent selection tools. The latter are the more widespread used ones and with them the region of interest is taken as drawn and not modified by the data. The data dependent tools adapt the region of interest to certain restrictions that are evaluated for the currently selected points. An example is a selection that contains always the same proportion of the data, thus, the size of the selection region has to be adjusted to cover the pre-specified amount of data points.

Recent research in virtual reality might enable a broader class of users to draw benefit from 3D immersive virtual reality environments like the C2 (Nelson et al., 1999). In addition to the already mentioned selection tools, we have access to a cube or some cylinder in a three-dimensional environment.

Data dependent selection tools are common in geographic information systems where neighboring regions of the currently selected one are chosen. A typical selection operation will then be based on a neighborhood structure and select all neighbors up to the chosen order. In a map with n regions, the amount of selected points can then vary from 0 to $n - 1$.

5.1.1. Zero-dimensional selection tools
As the name suggests the pointer just picks out a single point on the screen. This tool is available in many software systems and its purpose is to focus on a certain observation point, or, if the action is performed consecutively on various points, to select a noncontiguous but clearly described subset of the data. Embedded in the pointer's purpose is already the way a data subset is selected. For pointers, the frame level does not really matter, since the user will aim at directly selecting elements of the graphic, like points, bins, bars or polygons. The selection of graphical elements with a certain amount of extension is straightforward, more problems occur with selecting individual points, since it requires a certain training with the mouse to let it work smoothly. Moreover, points that lie close together can hardly be separated, and individual selection of tied observations is impossible. To make the work flow smooth, selection with the pointer should be robust in the sense that a point is selected whenever the pointer is activated within a certain distance of that point. This threshold may be chosen individually for each display. It may be great for sparse scatterplots but it should be rather small for dense plots.

5.1.2. One-dimensional selection tools
The slicer is drawn as a line parallel to one of the axes to cut out the interval between the origin and this line without posing any restrictions on the other dimensions. Arbitrary intervals that are not necessarily starting at the origin can be defined by using two slicing lines at the same time. Drawing such a slicing line can be interpreted as a one-dimensional selection. A generalization of that would be to allow the drawing of arbitrary straight lines or curves on the screen. For arbitrary curve and line segments, there is little useful interpretation, but straight lines that cut the data space into two pieces constitute a helpful tool. In scatterplots, this would offer easy selection of phenomena that are not parallel to the coordinate axes but show along the diagonal for example.

5.1.3. Two-dimensional selection tools
The drag box is the most common selection tool and it is available in any graphical user interface. A specific instance of the drag box is the brush, a drag box with fixed, predefined size, that can be moved dynamically over a plot. A drag box is mostly a rectangular shaped two-dimensional region, but as pointed out by many authors, a lot of other brush shapes are helpful. The most general two-dimensional selection tool is the lasso. With the lasso one draws an arbitrary curve on the frame. Start and end point of this curve will be automatically connected and the enclosed region will be selected. Questions arise when the curve intersects itself or when a straight line is drawn. The lasso is best suited when the region of interest can be bounded by a smooth curve whose shape is simple enough so that a user can draw it precisely with a standard mouse pointing device.

Other two-dimensional shapes are often convenient: circles, ellipses, triangles. Especially, when dealing with distances such tools can be used to select all those graphical elements that are a certain distance away from a reference point. Depending on the

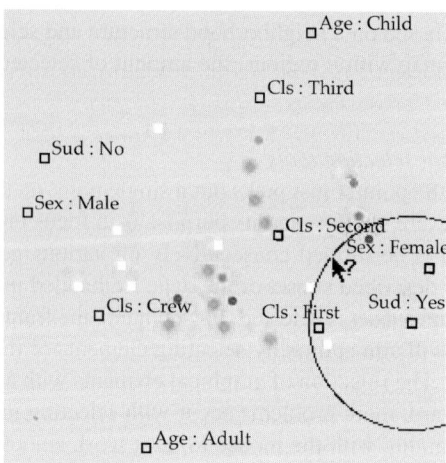

Fig. 13. Distances reflect the amount of association between data (small dots) and level points (squares). The closer a data point lies, the more is suggested that a connection exists.

metric used various shapes have to be provided. A prototype of this selection tool is implemented in MANET in biplots visualizing multiple correspondence analysis. There, option-clicking a category level point generates a distance neighborhood that can be shrunk and expanded by dragging the mouse. So far the operational behavior of this prototype is on the query level and is used to indicate which projected points fall within a certain distance from the selected category level point, see Figure 13. Extending the action to also select those points that fall in this neighborhood is straightforward but not yet implemented.

5.2. Selection memory

Besides *selection tools*, Wills (1996) used the notions *selection memory* and *selection operation* to propose a taxonomy of selection systems. A memoryless system only knows the current selection and neither stores the way that lead to the current state nor does it keep records of other selections. A system with memory keeps track of a series of selections and combines the individual selections to a final selection. Selection sequences (Theus and Hofmann, 1998) as implemented in MANET are an example here. Selection sequences make a system very forgiving to user errors and enable an easy update of complex selection processes. Thus, easy comparison between subgroups of the individuals is enhanced. The implementation in MANET stores the *selection area*, the *logical selection mode*, and the *ordering of selections*. Any selection in the sequence can then be changed interactively and the final selection is determined by considering all steps in the sequence.

5.3. Selection operation

For a system with memory, the calculus of selections is important. Memoryless systems just replace the current selection by the next one. So only regions that can be specified by a single selection tool can be selected. To avoid such restrictions, current systems provide set operators to combine a series of selections. The following four selection operations can be considered as standard for an interactive selection system: replace, intersect, add, toggle. Different approaches can be pursued according to the graph level on which the selections are combined.

5.4. Graphical selection

As stated above the dimensionality of selection tools can be used to assign a particular selection concept to a selection tool. I propose that a graphical selection is generated by either drawing a two-dimensional region on the frame or by simply clicking in one pixel of the frame. In the latter case, the zero-dimensional selection tool leads to selecting the graphical element that contains the selected pixel or the one that falls closest to it. The standard way to translate two-dimensional regions into subsets of graphical elements is to use an inclusion relation between the displayed quantities and the specified region: either all graphical elements are selected that have a non-empty intersection with the region specified in the frame or only those elements that are fully included in the selection region. The latter is a quite natural solution for plots where each observation is drawn separately, as is the case in scatterplots or parallel coordinate plots. For displays with area representations, it seems more natural to select all bins that have a non-empty intersection with the specified region of interest.

Let R be the region of interest specified in the frame, i.e., $R \subseteq \mathcal{F}$ and $S(R) = 1$. Using a graphical selection only unions of elements in $\mathcal{G}$ can be selected. The graphical elements $\mathcal{G}$ for that display define a topology for the frame. The two variants mentioned above correspond then to either choosing the interior $int(R)$ or the closure $cl(R)$ of the specified region with respect to this topology. Under both variants, however, it can happen that the combinations of selections do not conform with the internal linking structure, see Figure 14. Let R_1 and R_2 be two selection regions in the frame, and G_1 and G_2 their corresponding sets of graphical elements. Then working with the closure

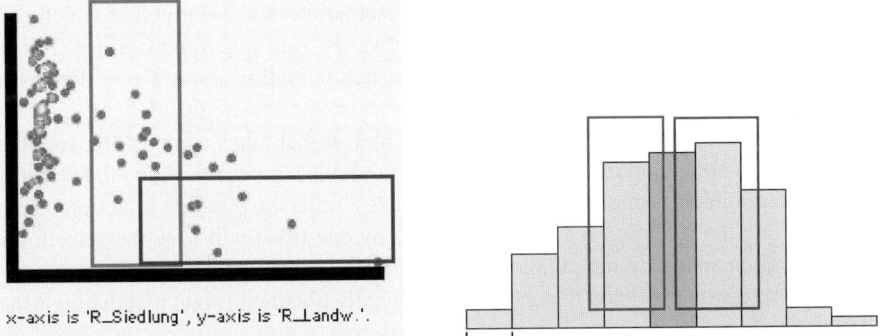

Fig. 14. Selection regions that overlap in the frame might have an empty intersection at the type level (scatterplot on the left). On the other hand regions that do not overlap in the frame for aggregating displays might result in a non-empty intersection at the type level.

and combining the selections on the frame level yields the selection cl($R_1 \cup R_2$) whereas combining the selections at the type level results in $G_1 \cup G_2$ which equals cl(R_1) $\cup$ cl(R_2). It is obvious that cl($R_1 \cup R_2$) is generally a subset of cl(R_1) $\cup$ cl(R_2) and that there are settings in which the two sets do not coincide.

When working with the interior relative to the topology induced by the graphical elements the reverse inclusion holds. There, combining the selections on the frame level yields int($R_1 \cup R_2$), and combining on the type level gives int(R_1) $\cup$ int(R_2). Here, the latter is a subset of the former.

After the selection in the frame has been transformed to a corresponding selection at the type level, this set of selected points can be propagated to the model and sample population level via the internal linking structure. These propagations are usually straightforward. However, the graph level that has been taken to create a selection might be important when selections performed in various displays are combined.

5.5. Axes based selection

One-dimensional selection tools should operate on the axes, building restrictions on one variable, independent of the graphical elements used in the particular plot. Although, current implementations of slicers do not always work in that way, this seems to be a natural choice for that type of selection tool.

Axes based selection is a direct graphical representation for data queries. In some packages, it is the standard selection mode for certain graphics. In EDV (Wills, 1997), for example, selection in histograms or density estimation plots is entirely based on the observation values and is neither affected by the model nor by the graphical elements in use. This selection type operates on axes, observations, and variables and not on the model representing objects. Depending on the precision of the scale the user-specified frame region will be transformed into conditions on the variables that are represented by the chosen axes. Then the boundaries of the selection region are transformed to restrictions on the observation space and all those observations that fulfill the resulting conditions are selected.

In XMDVTOOL (Ward, 1998), a glyph brush tool is available for creating selections. The glyph brush tool contains an enlarged representation of a star glyph in which each dimension is mapped as a line that radiates outwards from a common center. The selected subset is displayed as a filled polygonal structure on this glyph. Every dimension axis behaves like a slider and can be dragged to resize the selected subset.

Axes based selection works with its own model. It just takes some of the variables shown in the display and determines the finest model $\mathcal{X}^*$ for them. Since the selection is based on graphical tools, the precision of the axes limits the possible selection boundaries. Varying the selection region in the frame by one pixel will vary the selection in the observation space by the smallest unit as defined by the precision of the axes s_G. The finest selection precision that can be theoretically obtained is one that distinguishes all distinct observations.

Axes based selection is very useful for graphics having only a very limited amount of graphical elements. Selection in boxplots, for example, can only be defined in a meaningful way through this method. Graphical selection in boxplots would yield only a couple of graphical elements to choose from: five graphical elements for the center of the data plus some individual points for the outlying observations. Also for density curves the graphical selection does not make much sense because there is only one graphical element to be selected.

5.6. Data queries

Although there are only two different types of graphical objects in a data display – graphical elements and axes – an additional method for selection is available in most software programs. By using parameter displays in which the user can type exact values for the selection boundaries or even long-winded data base queries, the user can access directly the model level and detour the graph type level. This is the standard form of data query and is well known in data base visualization. The advantage is the possibility of making precise statements on which observations are currently selected. This preciseness is paid for by a huge loss of speed and intuition.

Parameter displays give a visual or textual representation of plot and model parameters. They are linked to their corresponding data display and follow the same internal linking structures as the corresponding data display (at least for the objects that are referenced from both). Parameter displays must be accessed through their corresponding data display. This can be done by using some menu option while the data display is active or by special cues that are placed within the data frame. Cues appear to be more elegant and also faster to use because a single click is sufficient to create the parameter display and the cursor has not to be moved out of the current data display.

6. Interaction at the frame level

The frame itself does not have too many subobjects so there are only a few candidates at which user interaction can be targeted. For a single frame, changing the entire frame and changing the size and color of the frame are the only user interactions that I find useful. Although these operations are of minor interest from the knowledge discovery aspect, they are sometimes very helpful and make interactive analyses rather elegant.

6.1. Changing frame

The most well-known form of changing frames is to switch from a horizontal layout to a vertical layout. Especially for bar charts this switch seems essential. In the vertical layout, the categories' names can be printed in full length – at least in most cases. On the other hand the horizontal layout is more common and people are used to compare the frequencies of categories via their heights and not via their widths.

Not only the bar chart profits from changing the frame. To ease comparisons or to arrange a greater number of plots within one window, changing the direction of the frame are also helpful tools for dotplots and boxplots. The context of the data set to be analyzed might also propose a particular arrangement of the frames.

Trellis displays and conditional plots are also an area in which frame changing might help to extract information from the display.

For parallel coordinate plots, there is no natural layout of the frame. Wilhelm et al. (1999, p. 126) recommend the horizontal layout in which the parallel axes are drawn as horizontal lines because of the wider aspect ratio. Changing the frame to the users' preference is appreciated here.

6.2. Resizing frame

As well as changing the direction of the frame it should be possible to increase or decrease the size of the frame. Resizing the frame can either be implemented to only affect the frame level but it seems to be more reasonable to link the frame size to the plot type in such a way that whenever the size of the frame is altered the plot size is changed as well.

6.3. Changing frame color

The background color of the frame does not bear any information but is usually chosen to let the data points stand out. Depending on the number and kind of colors that are used to represent the graphical elements different background colors might be more appropriate to make the patterns more visible. When the output occurs on a different medium – like printing on paper – it might be worthwhile to change the frame color once again.

This short list covers all currently available frame interactions. We turn to the user interaction on the plot type level in the next section.

7. Interaction at the type level

Plotting actions are operations that are sent to the plotting part of a data display. Operations that are received by the frame and transmitted to the type level have already a high impact on the interpretation and deduction of statistical graphics.

7.1. Operations on the graphical elements

The graph type definitions $\mathcal{G}$ as presented in Section 3.4 consist of graphical elements and their attributes. So two groups of operations can be distinguished: changes of the type by using other graphical elements and changes of the type by modifying the attributes of the graphical elements.

7.1.1. Changing graphical elements

Some models can be visualized by different types and in these cases it is convenient to have a simple option to switch between different graphical representations. The more similar the two types of displays are the less distraction is created when switching the type. When the two types differ more it might be necessary to update also the frame, an operation that might be too interruptive.

Dotplots and boxplots are closely related graphs and in DATA DESK it is possible to change a boxplot into a dotplot by removing the box or vice versa by adding the box.

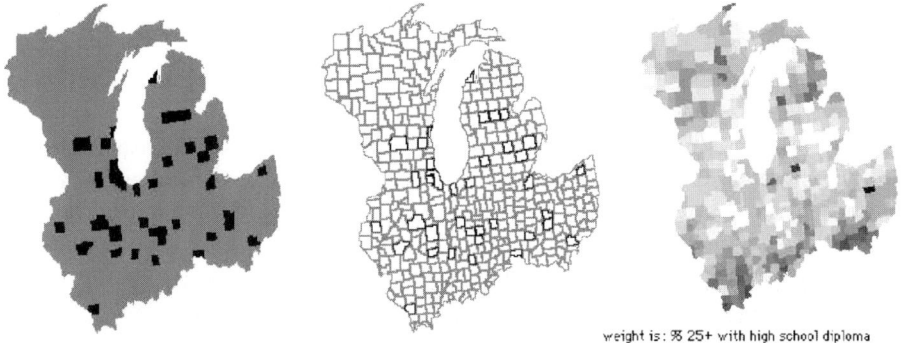

weight is: % 25+ with high school diploma

Fig. 15. The default polygon map is a filled one (left). Changing plot parameters draws only the boundaries (center). Dragging and dropping a variable over the plot changes it to a choropleth map (right).

Close connections also exist between histograms and density curves. An intermediate step is provided by frequency polygons. Replacing the rectangular bins by connecting the midpoints of the top line of each bin yields a piecewise linear curve that represents the same model as the original histogram but uses a different graph type. Similar operations can be imagined for changing a scatterplot into parallel dot or boxplots, but this is a switch that needs to be accompanied by an update of the frame.

The standard plot for displaying categorical variables is the bar chart that draws one bar for each category. The area of the bar represents the number of cases that fall in this category. By convention, the width of all bars is the same and the height of each bar is proportional to the count in that category. Alternatively, we can keep the height of the bars fixed and vary the width according to the number of cases in that category and we obtain what is called a spine plot (Hummel, 1996).

How can we make a polygon map interactive? Changing projection views could be interpreted as using different projection techniques to map the three-dimensional locations into planar coordinates. A less sophisticated approach is to allow different representations of the same map. This is done in MANET where the user can switch between a filled polygon map, a hollow polygon map and a choropleth map, see Figure 15. Switching from one representation to another is easy and can be performed by simple mouse clicks or drag and drop operations. The filled representation is more effective in showing the global spatial distribution but it is highly affected by the size of the regions.

7.1.2. Changing attributes of graphical elements
Looking again at the definitions of types in Section 3.4 we can see that for most types the graphical elements own the attributes intensity, color, size, and shape. Changing these attributes interactively is an important step towards multivariate analyses via linked highlighting, see Section 18. When the attributes are changed only the type level is affected while the underlying model remains unchanged. Change of one attribute can force other attributes to update. For example, increasing or decreasing the size of points in a dotplot changes the intensity of the points because the intensity represents the

amount of overplotting. At the same time the model remains the same since the counts of the observations do not change.

7.1.3. Adding or removing graphical elements

Statistical analyses yield new quantities which can often be represented visually. Fitting linear regressions or smoothing point clouds are examples in which scatterplots can be used to visualize the raw data and a regression line or a density curve respectively can be included to present an image of the model. In such cases, an additional identifier will be defined for the new graphical elements.

7.2. Axes operations

Plot axes in data analysis displays in an interactive environment are often of secondary importance and are, as a result, often invisible. There is no need to immediately show the particular values, since our interest lies more in the overall pattern. Although the axes might be invisible and only be obtainable on demand their information is present at any time and can be interactively modified.

Most displays have two axes: a vertical one and a horizontal one. Plots like dotplots or boxplots have only one axis, either a horizontal or a vertical one depending on the chosen frame layout. According to the definitions in Section 3 axes consist typically of the following objects: start and end points for each axis, increments that indicate the difference in values that is represented by one pixel. Also an order of the graphical elements is included in the axes information, either by stating that the numeric values have been ordered increasingly or decreasingly, or by listing the different strings for non-numeric data in a certain order. If additional information is encoded by attributes of the graphical elements, the axes can contain the necessary information to decode the attributes. For example, when tonal highlighting is used to visualize overlapping points, the number of overlaying points that correspond to the brightest color might be given. Another example is the color scheme that is used in chorochromatic and/or choropleth maps.

7.2.1. Zooming

By default, axes are scaled to cover the complete range of observed values from the minimum to the maximum of the corresponding variable values. To focus on a certain region, it might be necessary to change the scaling of the axes and by doing so to vary the type level. The most common reason for changing scales will be the desire to focus on a certain region of interest, typically a region with a high density of observations. The default choice for the axes parameters produces often heavy overplotting and rather uninformative displays for such distributions. Changing the end points of the axes will use the current window size for the specified region and will thus resize the graphical elements that lie within the specified region to fit the available window size. Zooming thus selects a subset $\mathcal{A}$ of $\mathcal{G}$ and updates the display to just showing the subset $\mathcal{A}$.

Also a valid comparison of different variables or groups is only possible when the plots under consideration show a comparable scaling. Depending on the context there might be some standard scale to compare with. This scale should then be used with less regard to the values actually observed. But for the purpose of comparing some plots it is recommended to link the axes information (see Section 15.2). Special commands are requested for returning to the original scale ('home scale').

7.2.2. Changing brightness
Graphical elements overlap when the values they represent fall to close together on the scale used in the current display. Brightness of the graphical elements can be used to visualize the extent of overplotting. The more graphical elements overlap the brighter they are drawn. One implementation of this is tonal highlighting as provided for dot- and scatterplots in MANET. In EXPLORN, brushing with hue and saturation is available for parallel coordinates using the α-channel of the Silicon Graphics workstation. Although the two implementations differ in their technical realization they have in common that the level of saturation (or brightness) can be controlled by the user. For sparse plots, brightness should change already with little overplotting, for dense plots a stronger overplotting is needed to cause a change in brightness, see Figure 16.

7.2.3. Changing color schemes
Perception of spatial distributions shown in choropleth maps depends heavily on the scale used. In MANET, two sliders are available to transform the data towards a more appropriate scaling. A preview of the color distribution makes it easy to find a transformation that results in a more informative choropleth map, see Figure 17.

The choropleth map is not restricted to continuous variables. Discrete variables are handled in the same way resulting in what is often called a chorochromatic or k-color map.

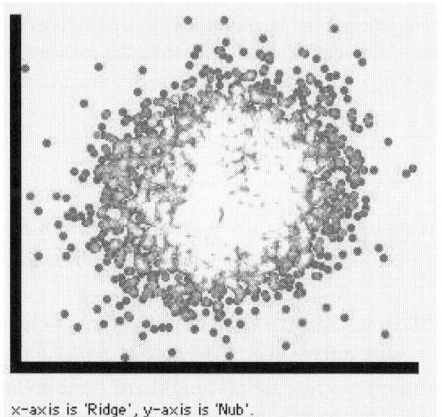 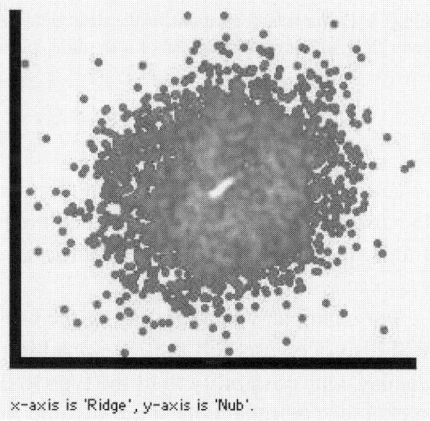

Fig. 16. Tonal highlighting is used to show the amount of overplotting. Increasing the brightness parameter with the overall density of the plot points towards areas with relatively high overplotting. The left plot shows the default setting. The plot on the right with increased brightness parameter clearly shows a pencil shape region in the center with a very high density.

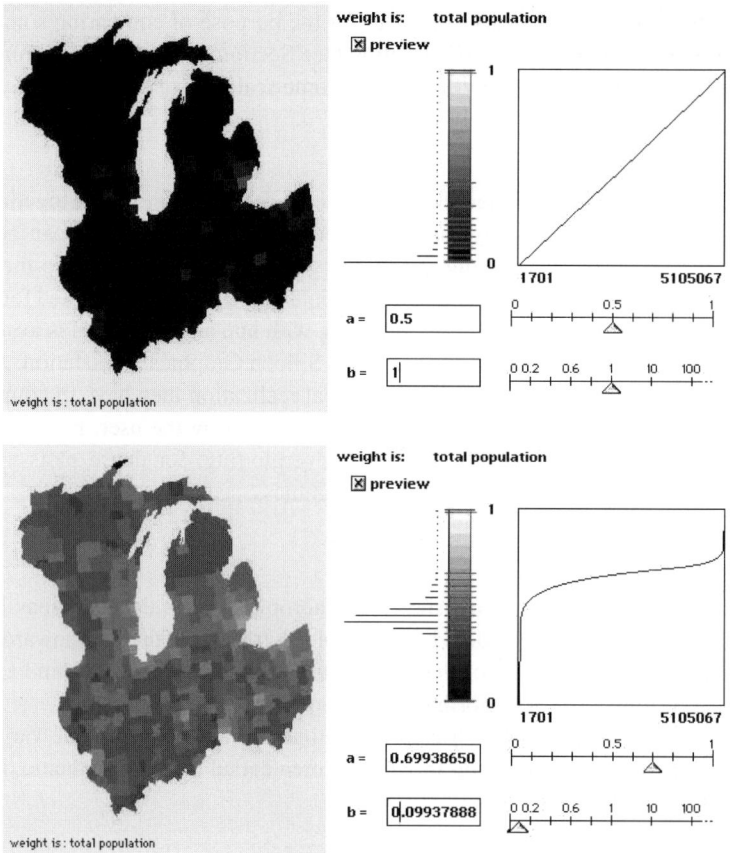

Fig. 17. Power transformations of the data yield new color schemes for choropleth maps. A preview of the color distribution is used instead of a simultaneous update of the color scheme. So unnecessary calculations are avoided.

7.2.4. Reformatting type

A common problem after updating the model or changing the plot type is that some of the graphical elements either exceed the frame size or cover only a small portion of the frame. In those cases, it is helpful to have an option for updating the type such that it fits exactly into the frame.

For example, when decreasing the bin width in a histogram the counts of each class will also decrease and thus the resulting bins will only cover the bottom part of the display. One possibility to fit the type to the frame is to rescale all heights of the bins in such a way that the highest bin fits into the frame. Since the axes of the histogram bear the information which count is the maximum possible in that frame, this operation is fitting the axes to the frame. Histograms are density estimators so focus lies on the area represented by the diagram. An alternative approach of reformatting would therefore be to rescale the bin heights in such a way that the area of all bins together is kept constant.

Let h denote the height of the highest possible bin in that frame and max(count) the maximum number of counts that can be represented in the highest bin, n the number of observations and w_0 the width of one bin. By default, the height h corresponds to the maximum number of counts max(count). Thus, a_0, the area per case in the original histogram, can be calculated as

$$a_0 = \frac{h}{\max(\text{count})_0} \cdot w_0.$$

Varying the bin width to the new value w_1 changes the area per case to

$$a_1 = \frac{h}{\max(\text{count})_0} \cdot w_1.$$

To keep the area constant, $\max(\text{count})_0$ must be replaced by

$$\max(\text{count})_1 = \frac{w_1}{w_0} \max(\text{count})_0.$$

An example with these choices is shown in Figure 18.

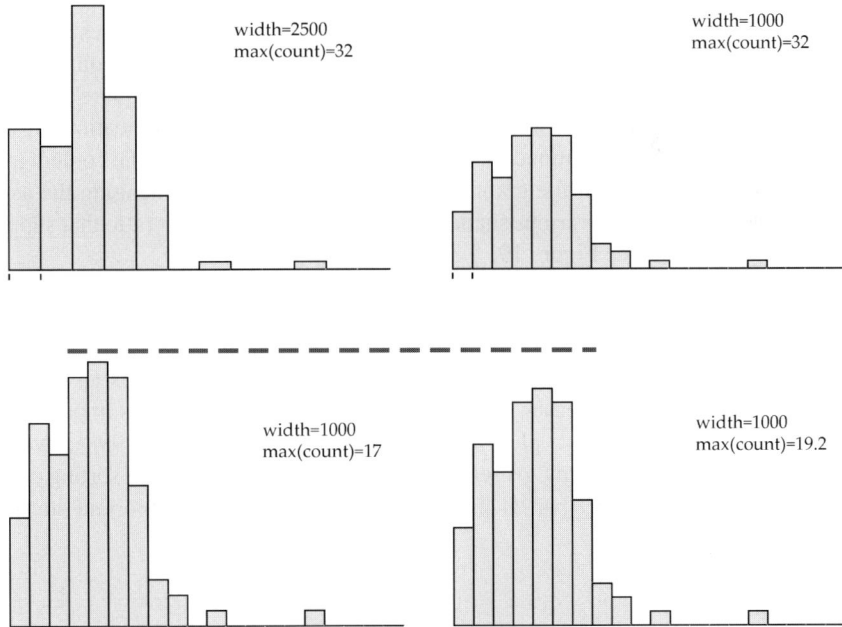

Fig. 18. Top row: A histogram (on the left) has been interactively modified. The resulting plot leaves a lot of space empty, see display on the right. Bottom row: The plot on the left is obtained by triggering the reformat cue in MANET. The plot on the right has been created by specifying the parameters. This histogram covers the same area as the original one in the top row on the left. The dashed line is included to ease comparison of the plots.

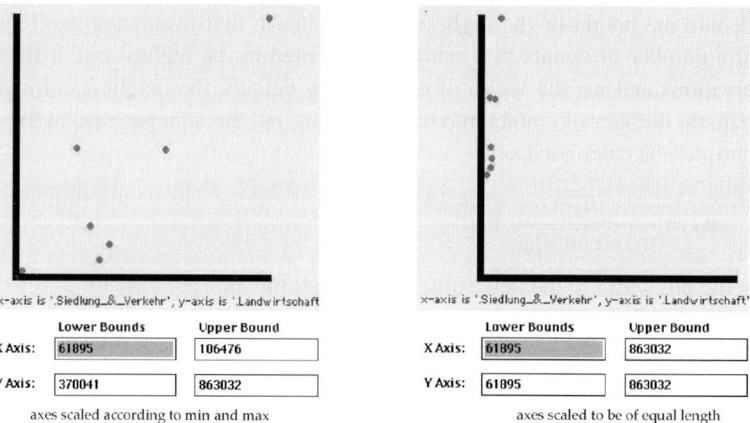

Fig. 19. In the scatterplot, we have two axes and both can have their individual scaling. Upper and lower bounds can be changed to any value but most often a switch to equal scaling or to a particular aspect ratio is most useful.

7.2.5. Changing aspect ratio

The aspect ratio of a display is the ratio of the precision of the vertical axis to the precision of the horizontal axis. Often this ratio is chosen to make the ranges of the vertical and horizontal axes fit in the default frame size. Wilkinson (1999b) points out that "many conventional prescriptions for aspect ratios in a plot ("make it square", "make it a Greek golden rectangle", etc.) have no empirical or theoretical justification. Instead, aspect ratios should be determined by the perceptual considerations of the content in the frame, namely, the shape of the graphics." An interactive tool for changing the aspect ratios offers exploration of various shapes and helps finding an aspect ratio that supports proper interpretation. See Figure 19.

7.3. Sorting data representing objects

When dealing with more than two variables the order of the variables is an important question because it is much easier to compare adjacent variables. This is in particular a problem for parallel coordinate plots or similar panel plots. Carr et al. (1998) have used sorting and perceptual grouping to create linked micromaps. A variety of sorting criteria can be used. The following partial list names some of the more important statistical numbers:

(1) median,
(2) range,
(3) high hinge,
(4) low hinge,
(5) upper hinge,
(6) whiskers' lower limit,
(7) whiskers' upper limit.

8. Interactions at the model level

8.1. Changing the model

8.1.1. Model parameters
Many models used for graphical representations are based on parameters that are needed for creating the display. These parameters are chosen by default and the user gains more control when there is a feature for interactively changing these values. Various graphical representations exist for model parameters. An easy to handle, rapid, and fully graphical method is provided by sliders. Sliders are graphical representations of scalar values and offer dynamic control of model parameters.

Interactively modifying the bin-width and the anchor point in histograms allows rapid exploration of a large set of alternative versions of this density estimator. This alleviates the shortcomings that single static histograms have. In some implementations, many models are overparametrized. In such cases, care has to be taken of how changes of one parameter conflict with other parameter settings and how those conflicts are resolved. In MANET, the definition of histograms with equal bin widths is based on four parameters: the offset, the bin width, the number of bins, and the upper limit. Actually, setting any three out of these four parameters uniquely defines a histogram. The anchor point and the number of bins are not updated when the other parameters are changed. Changing the upper limit forces the bin width to update, changing the bin width causes a corresponding change of the upper limit.

8.1.2. Inclusion/exclusion of variables
Displays that are generalizable to more dimensions should allow for adding or removing variables from the model description. Multivariable rotating plots, for example, offer such interactive changes. Also mosaic plots are generalizable to theoretically any dimension and an interactive option to include variables in the model or to take variables away should be provided. In MANET, it is possible to include and exclude variables from the mosaic plot model by simple key strokes.

In displays that cannot be extended to more dimensions, adding variables to the model will typically replace a variable that has already been in the model. Such drag and drop changes are implemented in DATA DESK.

8.1.3. Reordering variables
Scatterplots and rotating plots offer the possibility of interchanging the dimensions that are shown in the plot, i.e., instead looking at (a, b) we switch to (b, a).

Points in which lines in a parallel coordinate plot intersect each other correspond to straight lines in a conventional scatterplot and indicate that for these individuals the adjacent variables are negatively correlated. A positive correlation can be detected from lines that are parallel to each other. The human eye is much better in detecting intersection points than parallel lines. It is helpful to switch the direction of some axes to replace positive correlation by negative correlation.

In mosaic plots, various models can be investigated by running through all possible permutations of the variables. For parallel coordinate plots, similar operations are required.

8.1.4. Grouping categories

Sometimes during an analysis we might realize that the originally given categorization for discrete data is not optimum and that it might be better to join two or more categories. This feature is especially helpful when dealing with variables that have a large number of categories. Then often some categories include only a few cases and one would like to focus on the more prominent categories. But also for variables with a few categories there might be good reasons to combine categories under the aspect of their meaning.

8.1.5. Weighting

Spatially referenced area data is typically based on politically defined regions. Bar charts or histograms that reflect the number of regions falling in a particular class are often misleading and bear low information. Instead, some demographic figures, like total population or total area, might be more appropriate to use as weight for each bin area than the bare number of regions. In MANET, weighting is possible for bar charts, histograms and mosaic plots and it is performed by simply multiplying each case of the displayed variable with the corresponding value of the weight variable.

Weighted versions are most suitable for displays that aggregate the data, namely histograms, bar charts and mosaic plots. The plotted areas do then not reflect the number of counts in each class, as would be done by the standard plots of this type, but the areas reflect the amount of another variable measured at the objects that fall in a class, see Figure 20. In many surveys, such weighted plots help to adjust results and to avoid false impressions that are mainly caused by a specific structure in the underlying sample space. A drag and drop option to change standard plots into weighted ones or to change the weighting of plots would stimulate the experimentation with weightings. Choropleth maps, chorochromatic maps, and cartograms are special forms of visualizing a weight variable in a map display.

8.1.6. Adding model information

Adding graphical elements to an existent display has been discussed above. In most cases, we do not want to add only graphical elements but we want to add model information. Having created a scatterplot we might like to calculate the least squares regression and then add the regression equation to the scatterplot model. Then we can visualize

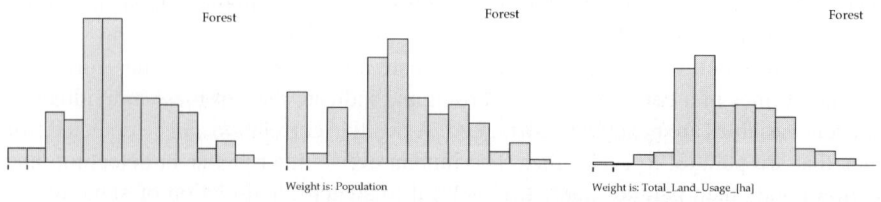

Fig. 20. On the left hand we show a standard histogram for the percentage of land that is covered by woods. Here the area of the bins represents number of counties in Bavaria. The other two histograms are weighted versions. In the middle, the area of the bins reflects the total population in those counties, on the right, the area represents the total area in the counties.

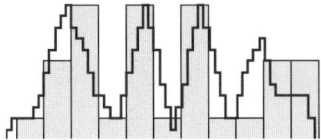

Fig. 21. An average shifted histogram has been calculated and laid over a histogram.

the regression by a regression line. Instead of regression lines, data-driven smoothing methods can be used in the same way.

To compensate for the dependency of histograms on the choice of anchor points, it is recommended to compute average shifted histograms and to lay them over the histogram, see Figure 21.

8.2. Changing scales

8.2.1. Re-ordering scales
Sorting is a very elementary method for exploratory data analysis. It is often routinely performed at the beginning of an analysis to increase numerical stability of calculations as well as to increase operational speed. For numerical values, the question of sorting comes to just the choice of sorting in ascending or descending order. Other orderings that contradict the natural order of numbers are seldom informative and should only be used with care. For categorical data, there is nothing like a natural ordering. A common default is to use the alphabetical order of the categories but this depends strongly on the language used and does not reflect the intrinsic order given by the meaning of the categories. When the context can be used to order categories it is very helpful to allow for manual rearrangements of the graphical objects.

For example, Figure 22 on the left shows a bar chart in which the categories are sorted alphabetically. By grabbing a bar in the bar chart and releasing it at its new place, we can change the left-to-right order of the categories to give it a meaningful ordering, see Figure 22 on the right. Reordering bars interactively is more helpful than renaming the categories which would be the alternative offered by standard software packages. Renaming might be done once but interactive reordering stimulates by its ease the exploration of quite different orders that might be of interest.

There are situations, especially when datasets become larger, where manual ordering is not enough. To compare classes, reordering the categories according to their counts is often desired. Here automated sorting algorithms, where the user can (interactively

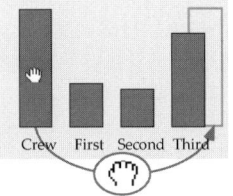 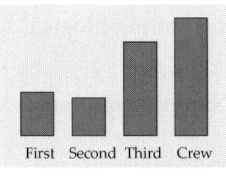

Fig. 22. Re-ordering categories in a bar chart "manually". This is more helpful than renaming the categories.

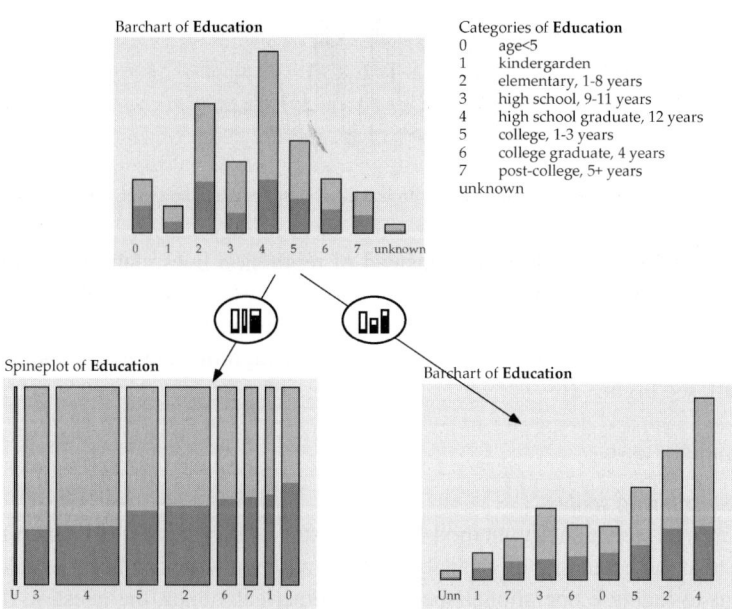

Fig. 23. Automatic sorting of bars in a spine plot (left-hand side) and a bar chart (right-hand side).

and graphically) construct the sorting criterion, have proved to be very useful in the analytical process. Figure 23 shows two different kinds of ordering mechanisms. In the bar chart, the sorting reflects the increasing order of highlighted counts. In the spine plot, the automatic sorting produces an order with increasing highlighted proportions. The different results of the sorting procedure comply with the different usage of bar charts and spine plots.

8.3. Logical zooming

Zooming selects a subset $\mathcal{A}$ of $\mathcal{G}$ and shows this in magnification. Logical zooming as proposed by Unwin (1999) in contrast shows this subset in more detail and therefore creates a new model $\mathcal{X}_\mathcal{A}$. Logical zooming might even lead to an extended sample population $\Omega_\mathcal{A}$. Logical zooming would be valuable for all statistical displays that aggregate the data, such as bar charts, histograms, mosaic plots, boxplots, and contingency tables. One could then investigate how additional variables or different model parameters would affect the selected section. Hot selection as provided in DATA DESK can be seen as an elementary form of logical zooming.

9. Interaction at sample population level

9.1. Selecting individuals

Linking can show structure only when combined with selection. If all observations in a plot are drawn in the same way, linking can not give any additional information. With linking, we draw our conclusions from comparing the distribution of the highlighted points with the distribution of either all data points or the ones that are not highlighted. In a dynamic environment, we can also compare the current selection with our mental image of the previous selection. Within an interactive computer environment, it is quite natural to allow the user to select a certain subgroup of interest directly in the plot. To reflect the selection, the appearance of the points is changed – usually highlighted – and the same operation will be performed in all connected plots. Other operations might be performed with the selected points like masking, deleting, or taking them as input for statistical models. But whatever these actions do with the selected cases the principle remains the same.

9.2. Grouping

Selecting individual cases often aims at finding homogeneous subgroups in the data. Having identified such a subgroup we would like to define the corresponding individuals as members of the group. Conceptual there are no problems in handling arbitrary groupings, including overlapping and not exhaustive groups. But technically we run into difficulties. How do we store the grouping for all individuals? How do we know which group is selected when an individual is chosen that belongs to two or more groups? Do we choose the alphabetically first, second, or last group, or all groups together. A restrictive method of working with groups is to forbid overlapping groups. Each individual can then just be a member of one group. Such restrictive approaches are used in DATA DESK or XGOBI. The advantage is that we can easily add a group variable to indicate which individual belongs to which group. If we allow individuals to be a member of more than one group we will have to deal with a rapid increase of possible subgroups. If we have specified k different groups and if we allow the groups to overlap we will have to generate a couple of group variables. Combining these group variables yields a total of $2^k - 1$ possible groups. To visualize a user-defined group, we will change the attributes of the corresponding graphical elements. For k different groups, we will therefore need an attribute that can take on $2^k - 1$ different values. The issue of visualizing groups is discussed in more detail in Section 19.

10. Indirect object manipulation

The discussion of user interactions so far relies on the ability to visually represent each data display object. Otherwise an object can not receive an interaction request due to Axiom 1 in Section 4. This restriction makes user interaction rather sensitive to implementational conventions and ideas. The implementation must ensure that there are either visual representations for all objects which can be manipulated by the user or that user interactions can be directed also to invisible objects. The main interest of statistical analysis is laid on deducing statements about the sample population. One could possibly create a list of all individuals to interact with the sample population level directly, however, in most cases one will work with a visual display that shows observations reported on those individuals. Thus, we need a method to propagate interaction requests from a visible graph level to an invisible one. For such actions, we distinguish receiver and addressee of a user request. The receiver is the object whose visual representation is accessed by the user interaction, the addressee is the object to which the user interaction is targeted. An interaction request issued at a visible object is passed through the levels of a data display and automatically propagated to the correct addressee of the message.

AXIOM 2. Interactive user request can be passed to objects without graphical representation using the internal linking structure.

On the other hand many of the user interactions that aim at the model or sample population level force the type and the frame level to update correspondingly. When we change a model we like to see the display react immediately. In summary, an internal linking structure is needed that controls the transmission of interaction messages between the graph levels within one display.

DEFINITION 5. Internal linking is the entire process of propagating a message from one object of a data display to another object in the same data display.

11. Internal linking structures

In the previous sections, the components of a data display have been viewed in isolation. But to constitute an interpretable image the graphical components have to be related to each other. The levels of a display are connected by relations between the set of identifiers and manipulating one object will cause other objects to update properly. It would be possible to investigate all possible relations between the graph levels but it seems appropriate to limit the relations to those that obey the hierarchical structure of the graph levels.

The basic structuring of the elements of a data display used here is chosen in such a way that the resulting hierarchy is equivalent to the order of visibility of the graph levels. The most visible part of a data display is the frame followed by the type of the display with its graphical elements and the scales. This choice has already been incorporated into the definition of data displays.

Whereas the frame must be physically present to show any information, not all information contained in the type must be visible at first sight. Scales for example might only be drawn on demand. The model is only visible via the plot type but the better the data display representation is the more information of the model is visible in the type. The sample population is usually completely hidden, but in fact, the aim of any data analysis is to draw conclusions for the sample population.

Within the above specified hierarchy we need to define relations between all neighboring graph levels. The sample population is directly connected only to the model level. The model splits into the set of observations and the scales. The latter are in close relation to the axes. The observations are connected with the set of graphical elements. Graphical elements and axes interact and both have to be embedded in the frame. Thus, the basic internal linking structure can be summarized in the diagram that is shown in Figure 24.

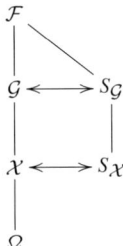

Fig. 24. Creation of a data display and user interactions are controlled by the internal linking structure. In this diagram, all possible links are included within the hierarchy derived from visibility.

Any object of a data display can be the target of a selection process but Axiom 1 concedes direct access only to objects that have a graphical representation. The internal linking structure is used to pass the user interaction message from the first recipient of a user interaction to the final addressee. This basic linking structure diagram is now to be filled with specific relations.

DEFINITION 6. *An internal linking structure is the complete set of relations that are present between the objects of a data display* $\mathcal{D} = (\Omega, \mathcal{X}, \mathcal{G}, \mathcal{F})$.

We begin the discussion with the classical linking structure that underlies the construction of conventional static plots. This structure is dominated by two graph levels: the frame and the model. Given a pre-specified frame it is the goal to transform the given model into an appropriate visualization without loosing information.

Thus, on the one hand the frame dominates the type of a data display and in particular restricts the amount of plottable graphical elements. This restriction that the frame can be actually too small to display the chosen graph type is included in the type information by the set $\mathcal{G}^*$ which describes the finest possible set of graphical elements of the current style that fits into the maximum frame. For a univariate model of a quantitative variable and a display of the histogram style, for example we expect the set $\mathcal{G}^*$ to contain exactly one single bar for every distinct observation.

On the other hand the observations in the model set are mapped to some graphical elements. This is done by introducing a function

$$G : \mathcal{X} \to \mathcal{G}, \quad x_i \mapsto G_j.$$

The observations are internally related to the scale which is mapped to the axes via

$$g : s_\mathcal{X} \to s_\mathcal{G}.$$

The connection between sample population and model is determined by the model type that has been chosen. The mapping

$$X : \Omega \to \mathcal{X}, \quad \omega \mapsto x_i,$$

provides this relation. The relations between observations and scale as well as the one between graphical elements and axes are intrinsic to the model and type definitions

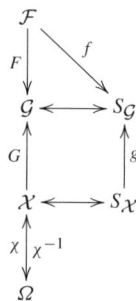

Fig. 25. A completely specified internal linking structure as it is present in the standard creation of displays. In this diagram, the structure is governed by the frame and the model.

and will not be further investigated here. The situation of the classic internal linking structure can be summarized in a modification of the above diagram (see Figure 25).

A class of data displays is defined by setting up a linking structure for a sample population, a model, a type, and a frame.

DEFINITION 7. An internal linking structure that is defined on a display $\mathcal{D} = (\Omega, \mathcal{X}, \mathcal{G}, \mathcal{F})$, its finest possible type $\mathcal{G}^*$, and its finest possible model $\mathcal{X}^*$ constitute a subclass of data displays called the canonical class of data displays for $\mathcal{D}$.

Using some default values the computer will generate an instance of this class. Other instances can be created by interactive modification of the current instance. All these various instances will bear the same internal linking structure. The default values are important and in most cases a plethora of intuitive correct and sensible default values exist. Usually, every statistical software offers only one of these possible values. Working in an interactive computer environment liberates us from the need to specify optimal parameter values – whenever such optimal values do exist. But instead it offers the possibility to easily modify the display and to check out a whole lot of useful views.

Conflicts for visualization might arise when the image of $\mathcal{X}$ under G is not included in the set $\mathcal{G}^*$. In the following, these possible conflicts are neglected assuming that the size of the frame can be chosen large enough to display all graphical elements with a sensible resolution within the frame. However then, the display can not be seen entirely in one image, and the user will not gain the complete overview of the display but instead see only a part of the display. Interactive operations help to modify the visible frame part and to overcome these limitations. Restrictions that can not be surmounted by user interaction will occur when the frame size demanded by the mapping G exceeds the upper limit of possible frame sizes.

By this assumption, we can neglect the influence of the frame level and the most important part of the internal linking structure results to be the relation between the sample population, the model, and the graph type.

11.1. 1-to-1 linking

The simplest case of an internal linking structure is a one-to-one correspondence between the identifiers of the sample population, the model, and the type. Historically there has been no need to investigate internal linking structures or the relationships among sample population, model, and graph type because brushing and linking have been restricted to scatterplot matrices and have been used mainly for small data sets without ties. This ensured that all sample units could be individually represented by points in the scatterplot and thus the sample population Ω could be bijectively mapped to the model $\mathcal{X}$ and the set of graphical elements $\mathcal{G}$. Under these circumstances the three graph levels are just different realizations of the same underlying phenomenon and we can easily switch from one level to the other. Yet already medium sample sizes or highly dispersed data will create tied values or overplotting each of which destroys the bijective property of the canonical mappings. Modifications of the sets Ω, $\mathcal{X}$, and $\mathcal{G}$ have been used to ensure bijectivity between the sample population and the plot type. For example, jittering can be used in point type diagrams to obtain one plot point for every individual in the sample population Ω. In area-based diagrams, the area could be segmented to assign a specific part of the area to a single individual. Three different approaches to achieve bijectivity of sample population and plot type can be pursued based on either the sample population, the model, or the graph type.

For the time being, we neglect axes and scales and focus on the data representing elements and discuss the various extended sets. The three possible bijective internal linking structures can be summarized in the following diagram:

$$\begin{array}{ccc}
\mathcal{G}^\Omega = \{G^{\omega_1}, \ldots, G^{\omega_n}\} & \mathcal{G}^{\mathcal{X}} = \{G^{\mathcal{X}_1}, \ldots, G^{\mathcal{X}_k}\} & \mathcal{G} = \{G_1, \ldots, G_g\} \\
\uparrow & \uparrow & \downarrow \\
\mathcal{X}^\Omega = \{x^{\omega_1}, \ldots, x^{\omega_n}\} & \mathcal{X} = \{x_1, \ldots, x_k\} & \mathcal{X}^{\mathcal{G}} = \{x^{\mathcal{G}_1}, \ldots, x^{\mathcal{G}_g}\} \\
\uparrow & \downarrow & \downarrow \\
\Omega = \{\omega_1, \ldots, \omega_n\} & \Omega^{\mathcal{X}} = \{\omega^{\mathcal{X}_1}, \ldots, \omega^{\mathcal{X}_k}\} & \Omega^{\mathcal{G}} = \{\omega^{\mathcal{G}_1}, \ldots, \omega^{\mathcal{G}_g}\} \\
\text{sample pop. based} & \text{model based} & \text{graph type based}
\end{array}$$

On the diagonal are the canonical sets for each graph level, the off-diagonals are quantities induced by the other levels to bijectively map the canonical set in each column to the other two graph levels. The advantage of a bijective internal linking structure is obvious. With a bijective structure it does not matter on which graph level the user interaction has been triggered. Interaction messages can be propagated to all related objects without loss of information. However, every one of the three possible bijective choices has its shortcomings because only the information available at the defining level is incorporated in the bijective structure. The other two levels will show a discrepancy between the canonical specification and the induced ones.

In standard situations, the sample population defines the finest sets leading to the ordering

$$\Omega^{\mathcal{G}} \subseteq \Omega^{\mathcal{X}} \subseteq \Omega,$$
$$\mathcal{X}^{\mathcal{G}} \subseteq \mathcal{X} \subseteq \mathcal{X}^* \subseteq \mathcal{X}^\Omega,$$

$$\mathcal{G} \subseteq \mathcal{G}^* \subseteq \mathcal{G}^{\mathcal{X}} \subseteq \mathcal{G}^{\Omega}.$$

The finest model $\mathcal{X}^*$ is included in $\mathcal{X}^{\Omega}$ because there might be tied values which can be distinguished in $\mathcal{X}^{\Omega}$ but not in $\mathcal{X}^*$. The set $\mathcal{G}^{\mathcal{X}}$ includes a graphical element for each distinct observation and is thus larger than $\mathcal{G}^*$ which is additionally restricted by the frame size and the drawing conventions.

With spatial data it might happen that the graphical elements induce more elements than the sample population because some regions are split into various polygons. When displaying a map of the European countries for example, France is split into the mainland and the isle of Corsica. But whenever the data relates to France only as a whole, the sample population will only include one element named France whereas there are two graphical elements yielding the inclusion relation $\Omega \subseteq \Omega^{\mathcal{G}}$.

The classical scatterplot brushing assumes that at every level the three sets coincide, i.e.,

$$\Omega^{\mathcal{G}} = \Omega^{\mathcal{X}} = \Omega,$$
$$\mathcal{X}^{\mathcal{G}} = \mathcal{X} = \mathcal{X}^{\Omega},$$
$$\mathcal{G} = \mathcal{G}^{\mathcal{X}} = \mathcal{G}^{\Omega}.$$

But already when dealing with aggregated displays, like histograms, boxplots, or bar charts, the equalities will no longer hold.

The choice for one of the internal linking approaches is important for the message passing of interactive user requests. Especially, it has an impact on how the selection region on the frame is transformed into a selection of the sample population. The advantage of sample linking lies in the fact that every single case can be selected individually. It is then also possible to inherit attributes of the graphical elements to the individuals and definition of brushing or linked highlighting is straightforward. XLISPSTAT uses a sample population based linking structure. In Figure 26, some examples of linked views created with XLISPSTAT are shown. In the top row, a rectangle has been selected in the left plot. Via linked highlighting various slices of the bins corresponding to the selected cases are highlighted in the right plot. In the bottom row, some cases have been selected individually by shift clicking in the bins of the left plot. The corresponding highlighting is also present in the right plot.

The disadvantages are the need of sorting the graphical elements according to selected and unselected elements. Otherwise linked highlighting results in uninterpretable views as can be seen in Figure 26. Even worse is the fact that selections will be unclear. Having one graphical element for each case in an aggregated display like a histogram or bar chart would allow to select individual cases within one bar of these diagrams. But the user would only know that a case falling in the corresponding class was selected but not exactly which one. Such a selection is not only counter-intuitive, it is also dangerous since it opens a very wide field of misinterpretation. Also technical problems arise since not all graphical elements can be usefully divided into individual parts. For example, it is impossible to intuitively subdivide density curves into segments for each observation.

Model linking allows individual selection of any observed value and thus displays the maximum information that is locally present in the current graphic. But as well as in the

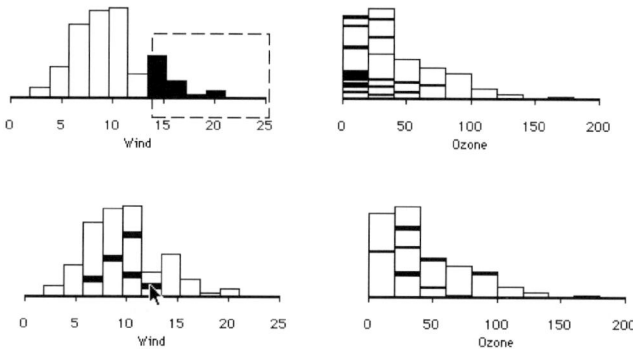

Fig. 26. Some examples of linked views for a sample population based internal linking structure. In both rows, the left plot is the active plot in which selections have been specified: in the top row a rectangle has been drawn, in the bottom row individual slices have been selected by shift clicking.

sample based linking selections are unclear in aggregated displays. Also this method is hampered by the difficulties associated with the subdivision of some graphical elements.

Graphical linking reflects exactly the visible information in the display. It has the great advantage that selections are directly visible and that the user will always know what has been selected. However, the possibilities of selection are reduced, because only those subsets of the individuals are selectable that correspond to a union of the induced sets $\omega_i^{\mathcal{G}}$. In many cases, this is an unnecessary reduction and summarizing of the information that could be otherwise drawn from the display.

As a result we have to leave the haven of bijective internal linking structures and we must consider more general linking relations. So we discard the induced sets $\Omega^{\mathcal{X}}$, $\Omega^{\mathcal{G}}$, $\mathcal{X}^{\Omega}$, $\mathcal{X}^{\mathcal{G}}$, $\mathcal{G}^{\Omega}$, $\mathcal{G}^{\mathcal{X}}$ and work with the canonical sets Ω, $\mathcal{X}$, and $\mathcal{G}$ of a display and the corresponding partitions of these sets induced by the inverse mappings.

Let $\mathcal{X}(\mathcal{G}) = (G^{-1}(G_1), \ldots, G^{-1}(G_g))$ be the partition of the set of observations $\mathcal{X}$ induced by G^{-1} and $\mathcal{G}$. One should keep in mind that G is not necessarily an invertible function. By G^{-1} we denote the set valued mapping that returns the set of points in the domain of G that are mapped to an element in its image.

This partition induces a related partition $\Omega(\mathcal{G})$ on the sample population Ω via the inverse mapping X^{-1}, i.e., $\Omega(\mathcal{G}) = X^{-1}(\mathcal{X}(\mathcal{G}))$. At the same time the model itself describes a partition $\Omega(\mathcal{X}) = X^{-1}(\mathcal{X})$ of the sample population by grouping all cases that show the same value in the model.

Thus, any interactive operation that is triggered on one of the canonical sets Ω, $\mathcal{X}$, or $\mathcal{G}$ leads to a corresponding update of the induced partitions at the other levels.

11.2. 1-to-n linking

Displays that aggregate the data yield to a 1-to-n linking scheme. This means that one graphical element represents n individuals at the sample population level. Propagating user interactions from the type level down to the sample population is straightforward because the message is passed from the one graphical element to all individuals represented by it. As long as we are dealing with just one display there are no problems with 1-to-n linking.

In general, 1-to-n linking is defined by a relation r between two identifiers I_1 and I_2 with

$$r : I_1 \to I_2 \quad \text{with} \ |r(i)| \geq 1 \ \forall i \in I_1, \text{ and } |r(i)| > 1 \text{ for at least one } i.$$

11.3. m-to-1 linking

This form of linking goes the other way round. For the first identifier, m objects are needed to make up for one object of the second identifier. The formal definition is give by a relation

$$r : I_1 \to I_2 \quad \text{with} \ |r^{-1}(j)| \geq 1 \ \forall j \in I_2, \text{ and } |r^{-1}(j)| > 1 \text{ for at least one } j.$$

Here $r^{-1}(j)$ is defined by $r^{-1} = \{i \in I_1 : r(i) = j\}$.

m-to-1 linking between type level and sample population requires a decision on how the system reacts to selection of one out of the m objects of the type identifier. Do we select the corresponding case or do we select it only partially, for example to the amount of $1/m$. With m-nary selection systems such problems of m-to-1 linking can be handled, at least theoretically. Whether it can also be correctly visualized depends on the number m and also on the display style. m-to-1 linking will typically occur in cases in which the set $\Omega^{\mathcal{G}}$ is larger than the sample population Ω.

12. Querying

The traditional aim of graphics was to show in one display as much information as possible. Interactive graphics make available as much information as possible but do not display it all at once. Information is hidden to make the plots easily readable and understandable. Only information that is fundamentally necessary for interpreting the data view is included by default. Any additional information needed by the user can be directly interrogated from the graphics. Interactive queries triggered at graphical elements give information on the data. They do not only take information from the type level but they combine and gather information of all levels. Actually, querying is inversely related to visibility in the sense that querying should provide more information on those graph levels that are less visible.

Identifying points in scatterplots is the basic and most widespread querying feature. In the beginning of interactive graphics, it was quite impressive to be able to find out the ID number of a point in the scatterplot. You could then look up further informations like names, values, etc. that belonged to this ID in a spreadsheet. In general, we want to be able to query all graphical elements of a plot, for example points, bars, lines, scales. Querying is usually listed as one of the interactive operations but it can also be seen as a special linking scheme between the active display and an information window that might either be permanently shown or just temporarily drawn as long as the corresponding key is hold. While moving from one graphical object to another the displayed information in this info window is automatically updated. Such a feature is not only present in interactive statistical graphics but also in some text editors. For example, Claris Homepage, a WYSIWYG editor for creating HTML-pages, opens a link editor to store the URL which the selected text refers to. Moving to another text passage updates the link editor automatically.

12.0.1. Querying a single graphical element
Interrogation is a context sensitive operation. The result of the interrogation depends not only on the plot type, but also on the part of the plot where the interrogation has been triggered and whether any other key has been pushed in addition or not. Which information do we expect to gain by querying? The general strategy for graphical elements is to show the most important information that is connected to it. A standard choice includes the exact values represented by the plot symbol, the number of cases that are aggregated in that graphical element, and how many of these cases are currently selected. Pushing additional keys might alter the amount of information provided. In MANET, shift-option-clicking changes the default interrogation information to the corresponding information taken from all variables that are currently selected in the variable window. For example, option-clicking a point in a scatterplot shows the values of the variables displayed in the scatterplot; shift-option-clicking shows the values of all selected variables.

Querying graphical elements that aggregate the data should only give the most important summaries and should yield the same information at any part of the graphical element. For example, clicking any part of the box in a boxplot should give the five-letter summary for that distribution: median, upper and lower hinges, and upper and lower adjacent values.

Current implementations do not offer querying for statistical model representations. Querying a regression line should result in a concise description of the regression model.

12.0.2. Querying two or more graphical elements
There are possibly different expectations on what we would like to know when we interrogate two or more graphical elements at the same time. One possibility is that we want to have the standard information of the individual graphical elements aggregated. Another goal might be to compare between subgroups of the sample population that are represented by the graphical elements. Such a query system is asked for when visualizing statistical models. Imagine we have a scatterplot showing two subset regressions via their regression lines. Querying the two lines should not show the complete information of the individual regressions but should offer tests or other criteria for comparing the models. Another example is querying two bars in a bar chart or spine plot. When there is no highlighting present in the plots then the standard numbers are to be shown. But when highlighting indicates a selected subgroup then it is more likely that the analyst wants to know whether amount of highlighting is independent of the categories in the bar chart. So, it makes more sense to display the results of a test for independence.

12.0.3. Interrogating axes
Information on the axes and the scales can be derived by querying the corresponding graphical representations – the axes. There is not much doubt how to query most of the

quantities represented by axes. Some axes information, like the ordering of categories or the brightness parameter, might not have a direct representation that can be interrogated. Here cues might provide a solution.

There is lots of other information that is worth to be derived from a plot by interrogation. Querying the frame for example could possibly result in a general description of the current plot – which model is to be represented, which data belong to it. Also user-defined comments made when creating the plot should be stored and made available by querying. Querying is an open field for new implementations and experiments.

13. External linking structure

User interactions can be local in the sense that they are limited to just one plot or they can be global in the sense that they are immediately propagated to all other linked plots. We have discussed the user interactions in the local form in the previous chapter. This chapter describes the extensions that are necessary for the global form. To pass user interactions to other plots, these plots have to be connected – or as it is called *linked*. The external linking structure controls the communications between various plots and is the basis for multivariate analysis using linked views. From an abstract viewpoint linking of two graphs D_1 and D_2 means exchanging information between them.

DEFINITION 8. External linking is the entire process of propagating a message from one object in the active data display – that is the one in which the interaction is triggered – to corresponding objects in the linked data displays.

Based on the definition of a data display in Section 3 and assuming that relations can only exist between corresponding graph levels of two displays, four types of external linking can be distinguished: *linking frames*, *linking types*, *linking models*, and *linking sample populations*. In addition, it is possible to distinguish between data representing objects and scale representing objects within the type and model level.

DEFINITION 9. An external linking structure is the relational structure that is present between objects at the same graph level of two or more linked displays.

The diagram in Figure 27 illustrates the linking schemes for two displays $D_1 = (\Omega_1, \mathcal{X}_1, \mathcal{G}_1, \mathcal{F}_1)$ and $D_2 = (\Omega_2, \mathcal{X}_2, \mathcal{G}_2, \mathcal{F}_2)$. Linking schemes that include the passing of messages between distinct graph levels may be theoretically possible but have not been tried in practice.

AXIOM 3. Interactive operations triggered in the active plot can be passed to objects at the same display level using the external linking structure.

The most widely used external linking structure is sample population linking. Besides the graph level based classification of linking schemes sometimes linking is categorized from a technical viewpoint into *cold*, *warm*, or *hot linking* (cf. Velleman, 1995).

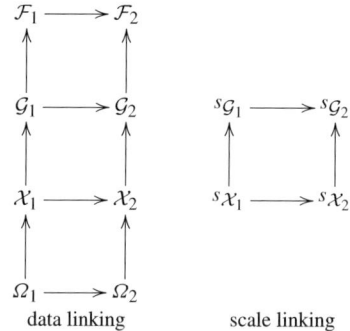

Fig. 27. External linking scheme for two displays.

Cold linking means that the user has to evoke a command that sends the message to the linked graphics. This is the standard case for non-interactive systems in which the user creates a new display to visualize the changes that have been performed in the active plot. Warm linking forms an intermediate step: the linked graphics receive the message from the active plot but they do not automatically update. Good software solutions indicate that changes have been made but are not yet incorporated into all linked displays. An example is the exclamation mark in DATA DESK's hyperview. Hot linking leads to an immediate update of all linked displays. This ensures consistency of all data displays currently shown. The disadvantage of hot linking is that the user can not focus on all updates at the same time. Cold linking hampers the workflow of an analysis and requires too much action of the analyst until a variation of the display can actually be seen. The preference between warm and hot linking is not as clear-cut. It depends on the situation and the kind of analysis step that is performed. I recommend that the user should get the option to change the general preference of warm and hot linking for parts of a session. Here, I focus on the classification according to the graph level and at each level linking can be realized in any of the three technical categories. As default I will assume hot linking since this mode raises the most visualization problems.

The above diagram shows one-directional linking only in which a message is sent from the active display to all linked displays. When linking two or more sample populations the question arises whether the displays should be treated on an equal basis or whether the linking relation should reflect a system of preferences. If the displays are treated with the same preference then we do not only propagate a message from the active display $\mathcal{D}_1$ to display $\mathcal{D}_2$, but also vice versa. In a bi-directional linking structure, the message is not only sent from $\mathcal{D}_1$ to $\mathcal{D}_2$, but it is also automatically returned. The linked plot $\mathcal{D}_2$ interprets the message in its context and the returning message can be either the same as the original one or it can be a modified version of it. If it is a modification it will then ask the active plot $\mathcal{D}_1$ to update accordingly. Conflicts might arise at this point and it is therefore reasonable to take one-directional linking as the standard scheme. One can choose bi-directional linking when the situation asks for it and the technical realization is possible without conflicts. Tracing is an example in which bi-directional linking has been successfully implemented.

Even more complex situations have to be handled when multidimensional linking structures are investigated, i.e., modified messages are not only returned to the active plot but also exchanged between all linked plots or when the series of sending and returning messages is continued until the resulting selection is stable. These situations require more research. In particular, it is necessary to gain more experience with linking different data matrices. For the time being, attention is restricted to one-directional linking only.

Most available software restricts linking on the aspect of multivariate analysis with the aim of showing each case consistently. However, some first attempts for a general linking implementation do exist. Linking at the frame, the type, and the model level usually results in a 1-to-1 linking and it is straightforward to implement.

14. Linking frames

The size of the frame could be adjusted by linking the frames directly. This is not only relevant for a screen-space saving layout of displays but also to establish correct comparisons of graphical displays. The use of different frame sizes might distract the analyst and lead to wrong conclusions. Also other attributes of frames like background color or other frame options like printing black-on-white, or white-on-black could be controlled simultaneously for all displays.

14.1. Arranging frames

Working with a couple of graphics at the same time can lead to a messy desktop. To keep control over the variety of aspects given for the data, it is convenient to have a command for arranging graphics. It should be possible to arrange frames in a common window (for example as is offered for layouts in DATA DESK) or on the desktop (as is done by the arrange command in MANET).

Putting a horizontal dot- or boxplot below the x-axis of a scatterplot and a vertical one beside the y-axis enhances the scatterplot that shows the two-dimensional distribution with the corresponding marginal distributions.

14.2. Linking frame size

Working with modern interactive systems usually results in a lot of windows scattered over the screen and overlapping each other. Arranging the windows to make efficient use of the available screen space is therefore possible in some systems. The command 'tile' is commonly used for this task and this yields to a common size for all windows and a screen-covering layout starting from one of the window corners. The result of linking frames is the same frame size for all linked views. There are two contrasts to the tile method: linking frame size does not change the location of frames and the currently active frame in which the command has been issued does not change its size. Formally linking frames is setting $s_{\mathcal{F}_i} = s_{\mathcal{F}_1}, \forall i \neq 1$. Linking frame sizes is rarely done without any other user interaction. Identical frame sizes are needed to make the results of other interactive operations comparable.

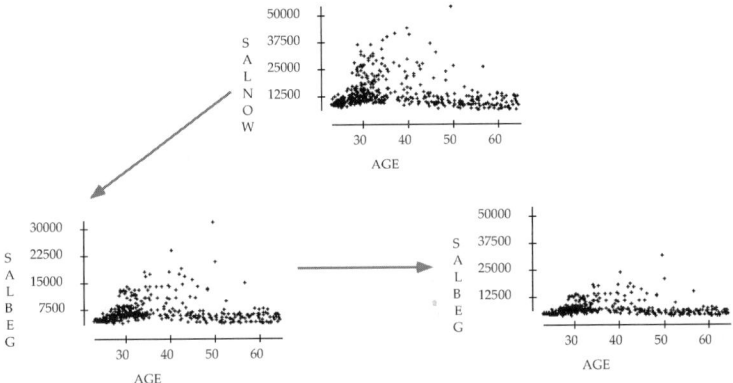

Fig. 28. Dragging the same scaling icon of the scatterplot at the top inside the left scatterplot in the bottom row and dropping it there adjusts the scale of this scatterplot to the scale of plot at the top. The resulting new plot (bottom row at the right) is now fully comparable.

15. Linking types

When information on the type level is exchanged it is helpful to match also the frame sizes so that a proper comparison between the displays is ensured.

15.1. Linking graphical elements

Attributes of graphical elements are used to visualize selection and grouping of cases. Therefore they cannot display other information. Thus, one should not take into account the direct linking of graphical elements without the connection to the underlying sample population.

15.2. Linking axes

Linking axes information comes typically to the fact that both displays use the same lower and upper limits for the axes. DATA DESK for example allows the scale of scatterplots to be adopted from other scatterplots by simply dragging and dropping an icon, see Figure 28. This feature should be made available for all plot types. A question that arises is whether the two linked plots should match in all axes limits or whether linking is also possible for one axis only. In histograms and bar charts, for example, it is helpful to adjust only the vertical axis such that the heights of the bins are comparable. Note that linking axes does not affect the scales of the model. So, linking the bin-widths for histograms is not a linking of axes but a linking of scales.

16. Linking models

For the model building step, a number of parameters are needed that are set by default. Many interaction requests aim at modifying these parameters to result in a model that displays the data better. Such adaption of the default values can easily be shared with other displays by linking the models together. Two models $\mathcal{X}_1$ and $\mathcal{X}_2$ can be linked by simply using the same objects in both models. Such a linking can be performed in a variety of ways depending on the number of objects that are involved in the linking process. We do not require that model linking forces all objects to be currently linked. The minimum requirement for linking models is that some objects at the model levels exchange messages. The closer the two model descriptions are the more objects can be linked. The set of objects that are contained in a model varies with the model. At least observational objects and scale objects can be distinguished, see Section 3. Technically, linking models is also a from of 1-to-1 linking. One model object in $\mathcal{X}_1$ is related to exactly one object in $\mathcal{X}_2$.

16.1. Linking observations

Models are based on variables and a possible link would be to use the same variables in both displays. Young et al. (1993) have used this empirical linking via variables in windows that contain various views of the same data set. In their system, one plot – typically a scatterplot matrix – controls which variables are displayed in all the other plots contained in one layout. Clicking a cell in the scatterplot matrix shows marginal views for these variables, for example. The aim of such a system is to create multiple views that act as a single visualization of the data space.

An extension of this approach is easily imagined. Why should the plots be arranged in a pre-specified layout? It would be more appropriate for the user interactive paradigm to let the user decide which plots shall be linked via variables and which plot shall play the role of the control panel.

Young et al. (1993) have also introduced algebraic linking in spreadplots. A spreadplot is the graphical equivalent of a spreadsheet in which plots play the role of cells. The linkage is not directly based on observations or variables but on equations. Empirical and algebraic linking can both exist simultaneously in one layout. Young et al. (1993) use algebraically linked plots in the context of tour plots. In an example, three plots are created: one rotating plot and two scatterplots. The projections used in the scatterplots are based on the residual information obtained from the projection given in the rotating plot. Spinning the rotating plot to a new projection and sending an update request to the algebraically linked scatterplots changes these plots to the new residual information and projections.

In MANET, algebraic linking is intrinsically present in biplots. Creating a biplot automatically opens two residual displays showing scatterplots for the residuals against the first and second principal components. However, the biplot in MANET cannot be rotated and thus the algebraic link cannot be used for updating.

Algebraic linking has also been used as a name for another form of linking by equations, see (Tierney, 1990; Nagel et al., 1996). In this form, a functional equation is evaluated for the selected subset of observations and the graph of this function is included in the current plot. This form is again based on observations and does thus not pass the criterion for algebraic linking as set up by Young et al. (1993). Instead, it is a special form of empirical linking where the usual highlighting performed on the selected points is replaced by a function equation. Since linked highlighting could also be seen as overlaying the same plot for the selected observations, there is only a small difference between the standard empirical linking and this form of "algebraic linking". Especially the implementation of standard linked highlighting in boxplots comes pretty close to that. Algebraic linking is also present in DATA DESK templates.

Another form of linking variables not yet implemented in software is to use the content of a variable. When working with multiple data matrices it might occur that variables are present in the individual data matrices that are related by their meaning. For example, one data matrix might contain a variable time that is measured in years. In another data matrix, a variable time measured in weeks may be included. If the two variables cover an identical time span it would be appropriate to link the two data matrices using these two variables. Similar other derived variables can be used for linking.

Another attempt is the use of weighted plots in MANET. Weighted versions are available for histograms, bar charts and mosaic plots. The plotted areas do not reflect the number of counts in each class, as would be done by the standard plots of this type, but the areas reflect the amount of another variable measured at the objects that fall in a class. An example of weighted plots has already been presented in Figure 20. In many surveys, such weighted plots help to adjust results and to avoid false impressions that are mainly caused by a specific structure in the underlying sample space.

16.2. Linking scales

A very widespread use of dynamic graphics is to work with sliders. Sliders are one-dimensional graphical representations of model parameters. The user can dynamically change the slider and thus the value of the model parameter. Then this parameter is automatically propagated to the plot that displays the model. One application for such sliders is interactively controlling Box-Cox-Transformations or checking various lags for time-series analyses.

MANET offers linking scales for a variety of plots. Linking scales in histograms adapts upper and lower limits, the bin width, the number of bins, and also the vertical axis. The user has to choice of resizing the frame as well. In Figure 29, three histogram scales are linked. The plot in the center is the active plot that propagates its scale to the others. The plot at the left adjusts also the frame size while the plot on the right keeps the original frame size and changes only the scale parameters.

The order of categories is also a scale information that can be linked. This is especially useful when working with mosaic plots. Bar charts can then be used to change the order of categories and this change is not a locally restricted action, but all other plots showing the same discrete random variable will immediately update their category ordering correspondingly, see Figure 30. In particular, the arrangement of mosaic plots can be matched to the structure of some of the variables. This procedure not only supports detection of patterns in mosaic plots it also makes patterns more interpretable. Also the grouping of categories can be linked to other displays. In Figure 30 (middle and bottom row), all classes for passengers have been grouped together. The scale information on how many and which categories are available for the variable 'Class' is linked to the scale of the mosaic plot and the same grouping is used.

Having seen these first attempts, it is quite obvious to consider linking of interactively performed scale changes to all relevant plots. It is also worthwhile to think about

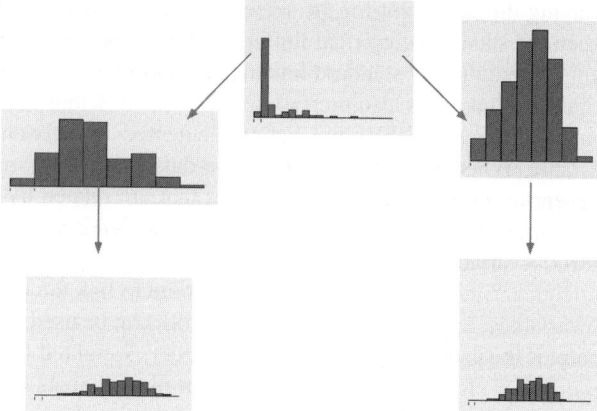

Fig. 29. Three histograms with linked scales. The plot in the center is the one that determines the scaling for all plots. The plot on the right is only linked for the scales but not for the frame size. The bottom row shows the updated plots.

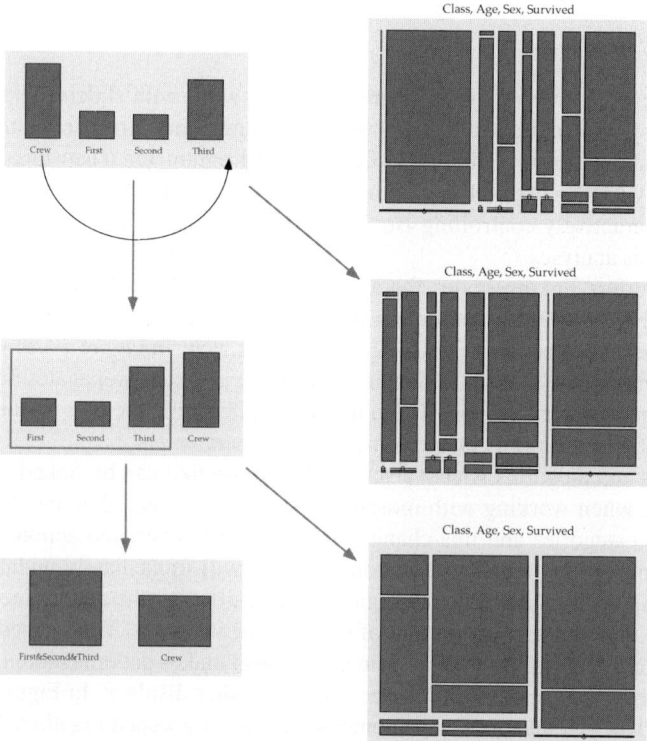

Fig. 30. A bar chart of the variable 'Class' of the Titanic data and a mosaic plot of all variables in this data set. Changing the order of categories in the bar chart affects also the mosaic plot (displays at the top and in the middle). Grouping some categories to one single group is also propagated to the mosaic plot.

mechanisms that only link part of the scales and not necessarily all scale information. For example, one could think of propagating changes of starting point and bin width of a histogram to all other histograms. Because different variables are measured on completely different scales, bin width and anchor point should not be linked as absolute quantities but in a relative sense providing the command for a proportional change. Let h_1 be the bin width in the active plot and let the user change this value to $h_1 - a$ then a possible choice is to change the bin width of a linked histogram from h_2 to $\frac{h_1-a}{h_1} h_2$. Modern computer systems will have enough power to do the necessary update of graphics instantaneously.

17. Linking sample populations

Linking sample populations is the origin of the linking paradigm. In multivariate statistics, it seems quite natural to visualize the connection between observations that have been taken at the same individual or case. In general, sample population based linking for two data displays $\mathcal{D}_1$ and $\mathcal{D}_2$ is defined by a relation $r: \Omega_1 \to \Omega_2$. As with internal linking we encounter 1-to-1 linking, 1-to-n linking and m-to-1 linking.

17.1. Identity linking

The easiest case of sample population based linking is empirical linking via observations in which both graphics rely on the same sample population Ω and the identity mapping is used as relation between the two sets, $\text{id}: \Omega \to \Omega$. This is a special case of 1-to-1 linking. The typical identity linking situation is as follows: a number of variables have been measured on a set of individuals Ω and a two-dimensional graphical representation of one or more of these measurements is given as graphic on a computer screen. The user draws a region R on the screen and according to the present plot type corresponding parts $\mathcal{G}_R \subset \mathcal{G}_1$ of the graphical display $\mathcal{D}_1$ are highlighted. These parts correspond to a subset $A = X^{-1}(G^{-1}(\mathcal{G}_R)) \subset \Omega$ of individuals. This selected subset of individuals is then kept constant for all plots and the corresponding subset model $Y(A)$ has to be displayed in the linked plot. Visualization is straightforward if the resulting image $Y(A)$ fits the internal linking structure in the display $\mathcal{D}_2$ in the sense that the model $Y(A)$ can be mapped to the graphical elements $\mathcal{G}_2$ without loss of information. This is the case for multivariate data where every plot shows each case individually – like scatterplots for untied and non-overlapping points for example.

Identity linking is not necessarily restricted to identical sample populations. Whenever two variables have the same length they can be bind together in a single data matrix and then all software programs will treat the variables as if they have been observed at the same individuals. However, one has to be careful when interpreting such artificially linked variables.

Identity linking renders superfluous the distinction between one-, bi- or multidirectional linking. Identity linking will not modify the message and thus the returning message is the same as the one sent. Hence no conflicts arise.

Identity linking is the natural choice for displays that are based on the same sample population. So the standard cases in statistics in which the data set consists of an $n \times k$

data matrix use this type of linking. Complex data sets however consist of multiple data matrices and therefore other linking relations are required.

17.2. Hierarchical linking

Databases to be analyzed in practice are a combination of data matrices stemming from different sources and relating to different sample population. But quite often the sample populations are related to each other in a certain way. A common situation is that one sample population is hierarchically on a higher level than the other. Such situations often arise in geographical investigations where one sample population is a finer geographical grid than the other. Divisions of countries in administrative regions are a typical example as shown in Figure 31.

This is a special case of 1-to-n linking. In a hierarchical linking scheme, selection of one object at a higher level automatically selects all related objects at the lower hierarchy levels. For example, selecting Germany in Figure 31 selects all federal states of Germany as well as all their districts and counties.

Hierarchical linking relations can also be used from the finer level to the coarser levels yielding a m-to-1 linking scheme, but then visualization problems might occur, see Section 18.

17.3. Distance and neighborhood linking

For spatial data, distance or neighborhood information are a natural choice to build up a linking relation. Distance will in most cases be Euclidean distance. A variety of neighborhood definitions are used in spatial data analysis each one leading to a somewhat different linking relation. Such linking relations can be based on distance and neighborhood within one sample population but they can also be established from one sample population to another whenever neighborhoods or distances between elements of the two sample populations can be defined. These are all special cases of 1-to-n linking. Different linking relations can be established by either calculating the distances individually or by taking the distance to the centroid of the selected subset. Inverse distance

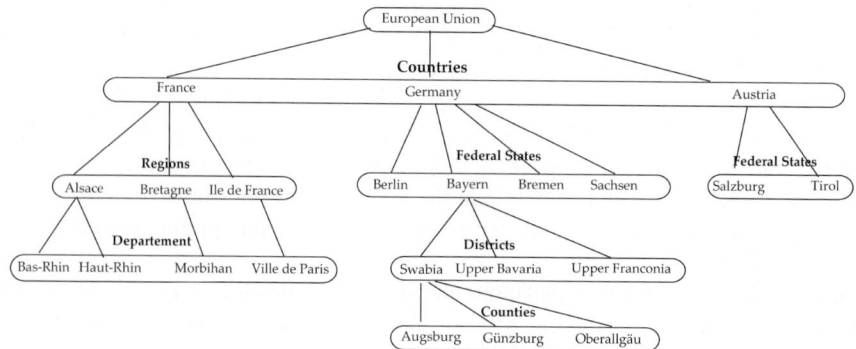

Fig. 31. An example of a hierarchical linking tree showing parts of the administrative division of the European Union.

weighting – in the sense that the closer a point is to the reference location, the higher is its weight – leads to an m-nary selection.

A linking scheme can also consist of all neighbors of one location – first order neighbors or neighbors up to a certain order. Incorporating the different orders of neighborhood results in an m-nary selection.

18. Visualization of linked highlighting

Brushing is a widely used technique for exploring multivariate relationships of quantitative data. It was first implemented for scatterplot matrices but can be generally used for all displays. Unfortunately, there are only a couple of commercial software products that offer brushing for displays other than the scatterplot matrix. To make clear that we use the brushing technique not only for the scatterplot matrix but also for other displays, we use the term "linked highlighting" instead of brushing.

18.1. Attributive highlighting

Only graphical elements in a plot can be directly selected via graphical selection. This selection is typically visualized in the active plot by coloring the selected graphical elements with the system's highlight color which means that the color attribute of the graphical elements has to be changed. Visualization of linked highlighting means that the highlight attribute must be shared by a couple of graphical elements in different displays. Linking of attributes of graphical elements can be passed along one of the following paths

(1) $\mathcal{G}_1 \to \mathcal{X}_1 \to \Omega_1 \to \Omega_2 \to \mathcal{X}_2 \to \mathcal{G}_2$,
(2) $\mathcal{G}_1 \to \mathcal{X}_1 \to \mathcal{X}_2 \to \mathcal{G}_2$,
(3) $\mathcal{G}_1 \to \mathcal{G}_2$.

Visualization is technically straightforward when the composed mapping from $\mathcal{G}_1 \to \mathcal{G}_2$ is a function. Then linking can be attributively visualized by setting the same attributes for all linked graphical elements. If path (1) is used attributive linking requires 1-to-1 internal linking structures in both displays and a 1-to-1 external sample population linking structure. This is rather unlikely to occur because we have discarded bijective internal linking structures. Even when the external linking is 1-to-1 the internal linking structures will not be. Attributive linking via path (2) requires an ideal visualization of the two models and a 1-to-1 external model linking. Path (3) requires a 1-to-1 graphical linking.

18.2. Overlaying

Axis selection is not based on graphical elements and can therefore not use their attributes. In boxplots, axis selection is recommended as standard selection method. DATA DESK uses attributive highlighting and limits the highlighting in boxplots to the outliers only. To visualize axis selection, overlaying can be used instead of attributive highlighting. This means that an additional plot is created for the selected subgroup and is

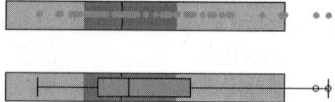

Fig. 32. Two different ways of visualizing selected points in a boxplot. The selection in the boxplot at the top is visualized by drawing a dotplot for the selected points. In the boxplot at the bottom, the same type of graphic – a boxplot – is drawn for the selected points.

placed on top of the current plot. While it is possible to change the graph type for the selected subgroup one should follow the principle of representing the selected subgroup by a graphic of the same type, i.e., highlighting within a boxplot is drawn as a boxplot, see Figure 32. Unfortunately, the overlaid plot covers part of the original one and for some plots this makes comparisons harder. MANET therefore draws the boxplot without highlighting in a non-standard way to make the standard boxplot used for the selected subgroup more visible.

Overlaying works in the active plot as well as in all linked plots no matter what kind of linking scheme is used.

18.3. Proportional highlighting

Whenever the complete linking scheme is not a 1-to-1 linking highlighting in the linked displays cannot be performed by attributive highlighting. Overlaying can be used but for some situations the technique of proportional highlighting is more appropriate. Proportional highlighting can visualize m-to-1 linking schemes. Assume that only some, let us say k, out of m objects are selected. If all m objects would have been selected then linking would cause the one corresponding object to be selected as well. How can we represent the fact that only a fraction k/m of this object has been selected? Here, proportional highlighting can be used to indicate the amount of selection. Proportional highlighting is in its results similar to overlaying and in many cases it can actually be implemented in the form of overlaying. Proportional highlighting is especially convenient for graphics that display area. Proportional highlighting is performed in the following way: every graphical element represents a certain number of cases. When a selection is performed the corresponding proportion of selected cases is calculated for every graphical element and the resulting proportion of the total area is colored. This method is best realized for displays which have their graphical elements aligned on a baseline, like bar charts or histograms. For histograms, bar charts, and mosaic plots the results of overlaying and proportional highlighting are identical and without deeper knowledge of the software in use, one cannot decide which version of linking is implemented. Proportional highlighting is in close connection to conditional probabilities which are further discussed in Section 23.

Instead of coloring proportions of the area, one can use different intensities to represent the selected fractions. This is recommended for displays with graphical elements that have a non-rectangular layout. Figure 33 shows a map of Bavaria, the left map shows all 96 counties, the right map the 7 regions. In the left map, some counties are selected and this selection is linked to the right map in which the corresponding regions

Interactive statistical graphics: the paradigm of linked views 507

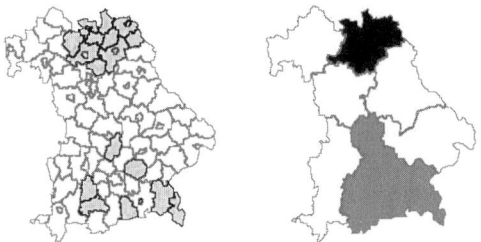

Fig. 33. Visualization of *m*-to-1 linking in maps. Intensity of the highlighting color in the map on the right reflects proportion of selected counties in each region.

are highlighted. Intensity of the highlighting color reflects how many counties are selected in each region.

18.4. Juxtaposition

Another alternative not yet available in any of the standard software tools is juxtaposition. This means placing the plot for the selected subgroup not directly on the original plot but close at the side to it (see Figure 34). This avoids masking important features of the original plot and still allows easy comparison between the two representations. However, for some plots like bar charts the overlaid plot has to be placed inside the original plot to get a reference for a proper judgment of conditional probabilities. Juxtaposition is well known for static plots but has not gained acceptance for interactive graphics. Current computer power makes the interactive creation of juxtaposed plots possible but the problem of creating too much distraction still remains. However, for comparing subgroups juxtaposed plots show the information much clearer.

The principle of juxtaposition can be straightforwardly extended from graphical displays to statistical model displays. Having calculated a statistical model for the whole sample we might be interested in particular subsets. Interactively specifying a subset of the sample population should then run the model for the selected subgroup only. The

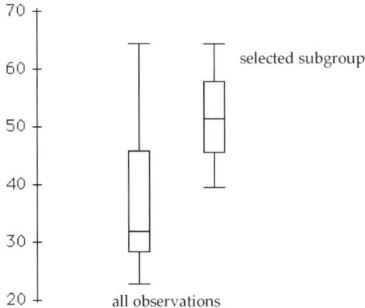

Fig. 34. Instead of overlaying the plot for the selected subgroup, it is placed next to the original one such that no overlapping occurs.

new results could then be juxtaposed with the original ones to allow for easy model comparison.

19. Visualization of grouping

In real data sets, the traditional assumption of statistics that the data are homogeneous is often invalid. Exploration of data often reveal clusters of data that are interesting and should be further investigated. In many cases, it comes to comparisons between two clusters or groups. To distinguish two or more groups of data in an analysis, the corresponding graphical elements are typically colored to reflect the affiliation to a group. There are different approaches to working with colors and groups. One of them is, as in XGOBI or DATA DESK, that each group is represented by one color and each data point may belong to only one group. In MANET, the concept is less restrictive in that one data point may belong to different groups. That implies, that if we have k different groups, we may want to deal with a further 2^{k-1} groups, which represent all possible intersections between the initial k groups. For a start, we therefore restrict the maximal number of different groups, dealing only with two groups and their intersection. Several problems result from this: overplotting in dot charts, e.g., which cannot be easily resolved (Wegman and Luo (1997) add colors by means of the α-channel). For bin-based representations, there is no unique or optimal solution (as well as no solution which could readily be generalized to more than two different groups), as the following discussion shows (see Figure 35):

- Stacked color bar charts have the advantage that the sum of colored observations is easy to compare vs. highlighted numbers or the totals, yet, comparisons between groups are not easy.
- Overlayed bar charts on the other hand, make comparisons between groups easy, though, comparisons involving sums of colored observations again become inaccessible.

20. Linking interrogation

Interrogation of plots is in many systems restricted to simply show the case number or its name if appropriate. In MANET, all plot classes can be interrogated to either

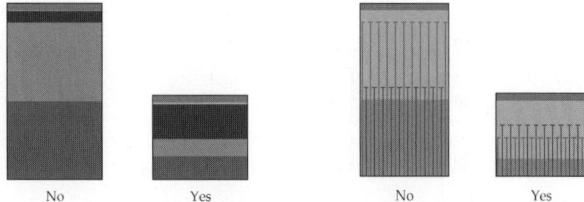

Fig. 35. Two different ways of coloring subgroups (and their intersection) in a bar chart. On the left-hand side the colors are *stacked*, on the right-hand side, colored bar charts are *overlayed*. For a color reproduction of this figure see the color figures section, page 606.

show boundaries, class names or class numbers or to show the relevant content of all currently selected variables. Linking interrogation could mean that the values of all currently displayed variables are displayed or equivalently that interrogating one plot would lead to showing the corresponding information in all other plots. However, since the human visual system is not trained to keep track of many plots at a time a bundled form of this information might be preferable. One special form of linked interrogations is implemented in MANET. When the shift-key is pressed in addition querying shows not only information on the selected individual for the currently displayed variables but also for all variables that are currently highlighted in the variable window. So, this selection information is linked from the variable window to all other displays.

21. Bi-directional linking in the trace plot

The trace plot has been introduced in Section 2 as one form of generalized brushing. The linking scheme that is used there is a combination of 1-to-n linking and m-to-1 linking. When a region in the map is selected then all points in the trace plot are highlighted that depend on the value observed at the selected region. When a point in the trace plot is selected then all regions that contribute to this value are highlighted. This highlighting action is then returned to the trace plot and all points are highlighted that depend on the currently highlighted regions.

One can argue whether bi-directional linking is useful in this case since when selecting in the trace plot one knows which point has been selected and it is therefore misleading to change the user specification on the system. The same argument can also be used the other way round. Since the user knows which point has been selected, the system can give the additional information that the currently selected regions contribute also to all the other highlighted points. I tend more towards the second argument and it should be noted that in trace plots bi-directional linking can be realized without conflicts because the trace plot represents the calculated quantities by points that are displayed along the horizontal axes such that no overplotting can occur.

22. Multivariate graphical data analysis using linked low-dimensional views

Historically, graphical methods have been seen as a part of exploratory data analysis (EDA). Although this relationship still holds true, there is an increasing number of areas where the use of graphics is more closely related to confirmatory analysis.

Graphical data analysis can not replace confirmatory analyses, but in many cases it can help to make confirmatory investigations easier to understand and more reliable. The use of graphics as a means of detecting outliers and unusual observations seems overemphasized. Not all data sets have outliers and these are by far not the only types of patterns that graphics can help to detect. In this chapter, we focus on multivariate dependencies and we assume familiarity with the use of graphics for univariate exploratory analysis. In multivariate exploratory data analysis, the task is to identify relationships between the variables. Statistically that means we seek information on the conditional distribution $F(Y \mid X)$. A central point in achieving this goal is a sufficient dimension reduction in the number of variables. A couple of analytical, like Principal Components Analysis, and graphical methods, like high-dimensional projection, exist to reduce dimensions. Once structure has been detected in the data the user wants to understand it. While complex techniques might be necessary to detect the structure it is often more convenient to use simple and low-dimensional views to interpret the structure. Extracting an easily understandable statement from some high-dimensional data projected on two-space is typically hard to achieve. It is, therefore, useful to give a description in terms of the original variables. Linking univariate views of individual quantities enhances this interpretation step.

The underlying theme of linked views is to show each case consistently. When an analyst has encountered some observations that show peculiarities in one view, it is quite natural to ask how these cases perform in all the other plots. In many cases, the expectation based on association assumptions will be met and a homogeneous subset of the data can be extracted. Sometimes the results will contradict the expectations, hence forcing a search for explanations for this unexpected behavior.

Linking views is not only used to see one-dimensional structure found in one plot in the light of other variables. By systematically subsetting the sample points, we attempt to detect structure in two or more dimensions.

The advantage of this approach compared to just using one single graphic for visualization of the dependence structure is that a single 2D or 3D plot can only show the whole structure if the structure is 2- or 3-dimensional, respectively. A graphic capable of doing this is called an "ideal summary plot" by Cook and Weisberg (1994).

Stuetzle (1991) distinguished two ways of linked brushing for finding associations depending on the different visual judgments that are required by the two techniques: static and dynamic brushing.

Stuetzle (1991) described brushing in an abstract way for the linking of scatterplots. I extend his description to plots that do not show each observation individually. Let us assume we have a sample of i.i.d. observations of a pair of one-dimensional random variables (X, Y) recorded for individuals in the set Ω. Suppose that the distribution of each variable is portrayed in a histogram and that the histogram of X is the active plot. The standard selection techniques available in interactive graphics offer only the possibility of selecting subsets of the population Ω whose observed values for X fall in a union of classes determined by the bins. Let $\mathcal{A} = (A_i)_{i \in I}$, I a finite index set, denote the partition of the set Ω that is induced by the choice of the bin size for the histogram of X. Assume further that we have currently selected a set of bins in the active plot; that means we have selected a subset $\mathcal{X}_A$ of the image $\mathcal{X}$ of the random variable X, or equivalently, a subset $A = \bigcup_{j=1}^{a} A_j$, $\{1, \ldots, a\} \in I$ of the underlying sample population Ω. We then superimpose a histogram for the selected subset A in all connected plots, that is we draw a histogram for $\mathcal{Y}_A = \{y(\omega): \omega \in A\}$ on top of the histogram for Y. If X and Y are independent, the conditional distribution $P^{Y \mid X \in A}$ is identical to the unconditional distribution of Y. So for any measurable sets $A \subset \Omega$ and $B \subset \mathcal{Y}$ we have the following independence properties:

$$P(Y \in B \mid X \in \mathcal{X}_A) = P(Y \in B).$$

In practice, however, we are not able to visualize all measurable sets; the only sets we can see are based on the partitions of Ω induced by the various plot scales. Let $\mathcal{B} = (B_j)_{j \in J}$ denote the partition induced by the histogram of Y. What we are actually judging are the distributions

$$P_Y(B_j \mid A_i) = P_Y(B_j), \quad \forall i \in I, j \in J.$$

In static painting, we fix the index i of the conditioning event, and our eyes run over the linked plots, mentally passing through the index set J, searching for non-uniformities in the relation between the original histogram and the overlaid part.

In dynamic painting, while we move the brush over the active plot, we pass through the index set I and try to detect changes in the distribution of the highlighted objects. So, in this case, we look simultaneously at sets of the partitions $\mathcal{A}$ and $\mathcal{B}$ and focus on the differences between two subsequent graphs, comparing the current image with our mental copy of the previous image. Actually, in many cases the judgment is more qualitative than quantitative and will be mainly based on the magnitude of changes. As Stuetzle (1991) already pointed out, there is no need to draw the plot for the entire population in dynamic painting, since we just focus on the various subset selections.

What kind of structure is easily detected by such an approach? In the dynamic case, uniformity of the various images is the main characteristic. In particular, the human eye is able to distinguish between a change of location and a change of shape.

23. Conditional probabilities

Although categorical variables occur frequently in real world data sets their proper handling is currently only supported by a few statistical software packages. In many cases, categories have to be recoded into integer values in such a way that standard methods for continuous data can be applied to them. This is true for graphical as well as for numerical methods, although the bar chart – the basic plot for univariate categorical data – was already used by Playfair two hundred years ago. But instead of generalizing univariate plots to two and more dimensions, recent enhancements of such graphical tools have focused on the artistic side, for example by introducing misleading three-dimensional aspects in two-dimensional graphs. These attempts neither improve the ease of interpretation nor do they provide an extension to higher dimensions.

An important goal in the design of visualization methods for categorical data is to provide a flexible tool that can serve various needs that arise during the analysis. What needs are there? First of all we want to gain a simple overview of the data and we wish to see how frequently certain categories occur. When analyzing two or more categorical variables we are not only interested in the frequencies of each cell, but we also want to know how the variables relate to each other. We want to check whether proportions are the same for all categories or whether there are dependencies between the variables. When the analysis proceeds certain models will be examined to find some acceptable model that explains the observed data sufficiently well. For categorical data, loglinear models are by now a well-established method. Such models can be evaluated and improved by looking at the residuals. Large residuals indicate which higher interaction terms should be added to the model. It is therefore necessary that the three major aspects of categorical data, namely counts, proportions and interactions, can be successfully visualized. The natural visual representation of counts (or frequencies) is as areas (Friendly, 1995). Proportions or fractions can then be visualized as subareas, and it is required that a graph to be powerful must offer an easy and correct comparison of two or more subareas. Interactions, or more generally associations, between variables are the most difficult to visualize.

A couple of univariate plots exist for categorical data that are designed either for counts or for proportions. Switching from one aspect to the other is usually not straightforward. As a first step in analyzing categorical data we want to get some information on the univariate marginal distributions. So we want to know how many categories there are, what relative size each category has and whether there are any particularly dominant or particularly sparse categories. For such univariate questions, bar charts and pie charts are often used.

Visualization of multivariate contingency tables can be done in three different ways. First, we can create new specific plots that are capable of portraying multidimensional categorical data. Second, we can create special arrangements of univariate plots that incorporate the multidimensional structure. Third, we can bring univariate plots together via linked highlighting.

For some particular types of multivariate contingency tables, graphical methods have been developed, e.g., the fourfold display for $2 \times 2 \times k$ tables (Fienberg, 1975; Friendly, 1994). Such plots are not readily available in standard software, and they are not yet well known. Also matrix and array layouts have been proposed to extend uni- or bivariate displays to higher dimensions, e.g., trellis displays. By conditioning, we partition the entire sample into subsamples and show a series of similar displays which we want to compare. In principle, this strategy can be applied to any univariate display for categorical data. The quality of such matrices will then depend on how easy it is to make good comparisons between the cells of the matrix. A matrix of pie charts for example would only be a space saving arrangement without any multidimensional information. A comparison between two pie charts would be hard because the angles are varied, and, thus, corresponding categories might be drawn at different positions within the circle. A shortcoming which is not shared by other displays. When analyzing two or more categorical variables the marginal distributions will be of less interest and the major focus will be on the relationship between the variables. A rather common question will be whether the proportions of some variable are the same in all cells. Proportions are a rough estimate for conditional probabilities and thus conditional plots are often used.

Selecting a subset of the data and highlighting it is one of the basic methods used with interactive statistical graphics. How does linked highlighting and conditioning look in plots for categorical data. Let us start with bar charts. The standard view of a bar chart shows the counts of cases that fall into each class. Usually, the bars in a bar chart have all the same width thus displaying the counts not only by the area but also by the height of the bars. A bar chart hence shows the unnormalized probability density function for a variable, see Figure 36.

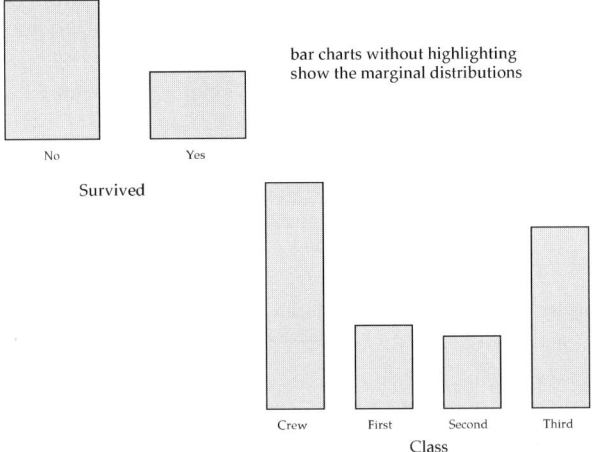

Fig. 36. Unnormalized probability density functions.

If we now start to condition on a subgroup (for example let us select the survivors in the Titanic disaster), then the highlighted bar chart shows four different distributions at the same time (see Figure 37). First of all the total heights of the bars are still displaying the number of cases that fall into that particular class, i.e., the marginal distribution. The heights of the highlighted areas reflect the counts for the cross-classification in the variable 'Class', i.e., they show $|\{\omega: \text{Class}(\omega) = \cdot, \text{Surv}(\omega) = \text{yes}\}|$. Dividing these counts by the total number of survivors yields an estimate for the conditional probabilities $P(\text{Class} \mid \text{Survival} = \text{yes})$. Since the denominator is constant for all classes, it can be ignored whence the counts and the highlighted areas respectively can be taken as representation of the conditional distribution $P(\text{Class} = \cdot \mid \text{Survival} = \text{yes})$. At the same time we can interpret all areas (highlighted parts and those that are not highlighted) as graphical representations of the joint distribution $P(\text{Class}, \text{Survival})$. This can be seen

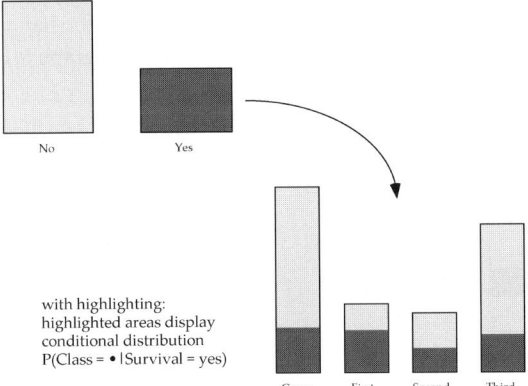

Fig. 37. Unnormalized conditional probability density functions for class given survival status.

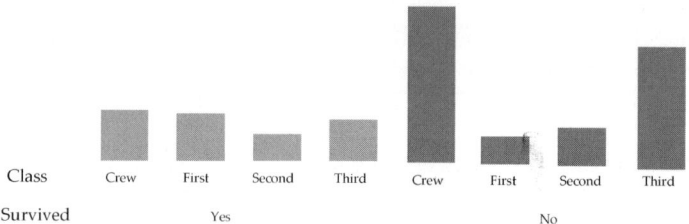

joint distribution: visualized by highlighted and not highlighted segments

Fig. 38. Rearranging the areas of a bar chart with highlighting yields a graphical representation of the joint distribution.

more clearly when we rearrange the areas to obtain a standard bar chart for the 8 classes of the cross-classification, see Figure 38. A third view in highlighted bar charts arises when one restricts oneself to an individual bar and compares the highlighted segment of the bar with the one that is not highlighted. This proportion can now be seen as an estimate for the conditional probability $P(\text{Survived} \mid \text{Class})$ for every category of the variable 'Class'. Usually, one does not want to simply look at an individual proportion but to check whether the highlighted proportion is the same in all cells. Let us look again at the Titanic data set in Figure 39. By selecting the survivors in the bar chart, we get via linked highlighting the corresponding cross-classification in the variable 'Class'. Unfortunately, the underlying frequencies in the various classes are rather different such that a comparison of the survival rates is hard to achieve. So it is not clear from the linked bar charts how the survival rate in the crew behave compared to the survival rates in second and third class. But when we modify our bar chart and use the same height for each bar and vary the width according to the number of counts, we can simply compare the proportions by just looking at the heights of the highlighted areas. Such plots are called spine plots (Hummel, 1996). In terms of conditional probabilities, we compare in Figure 39 $P(\text{Survived} = \text{yes} \mid \text{Class} = \text{crew})$ to $P(\text{Survived} = \text{yes} \mid \text{Class} = \text{first})$ and $P(\text{Survived} = \text{yes} \mid \text{Class} = \text{second})$ and $P(\text{Survived} = \text{yes} \mid \text{Class} = \text{third})$. To get this graphically, we have to select the category 'yes' in the bar chart for 'Survived' and switch the 'Class' bar chart to a spine plot. From the graph we conclude that the

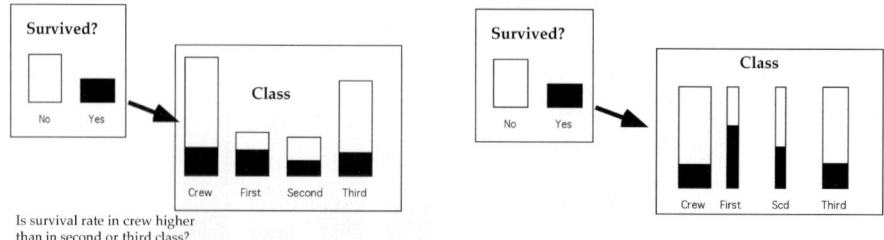

Fig. 39. In the linked bar charts on the left, it is hard to compare survival rates. Switching the bar chart for variable 'Class' to a spine plot eases the comparison substantially.

survival rate was highest for the first class, substantially lower for the second class, and much the same for third class and crew.

What happens when we change the model interactively and add more variables? Let us add the variable 'Gender'. Still we are interested in the survival rate, but now we have to compare the rates for each class cross-classified by gender. It is not possible to handle this with linked bar charts or spine plots. The sometimes proposed way of building the intersection of two selections, one in a bar chart for 'Class' and one in a bar chart for 'Gender' does not yield the correct result. Proceeding as proposed we would end up comparing for example the conditional probabilities P(Gender = female and Class = crew | Survived = yes) and P(Gender = female and Class = crew | Survived = no). The desired answer can only be obtained by linking spine plots and mosaic plots.

The construction of a mosaic plot results from a straightforward application of conditional probabilities. We first portray a single categorical variable in a spine plot. We then subdivide each bar in the spine plot in proportion to the conditional probabilities of the second variable given the first. Hence the area of each rectangle is proportional to the observed cell frequency which we can also obtain via the multiplication rule for conditional probabilities,

$$P(A = i, B = j, C = k, D = \ell, \ldots)$$
$$= P(A = i) P(B = j \mid A = i) P(C = k \mid B = j, A = i)$$
$$\times P(D = \ell \mid C = k, B = j, A = i) \cdots$$

We can now add a condition by linking a spine plot and performing a selection in it. Thus, we obtain in the mosaic plot the conditional probability of the selected category given the cross-classification as defined in the mosaic plot. In Figure 40, we have the full Titanic data, in the mosaic plot a cross-classification of the explanatory variables 'Class', 'Age' and 'Gender', and a spine plot for 'Survived'. In the spine plot, the survivors have been selected and in the mosaic plot the conditional probabilities P(Survived = yes | Class = i, Age = j, Gender = k) can be derived from the highlighted subareas. In the shown display, impact lies on the comparison between survival rates of the two genders for given age and class. This is due to the fact that the variable

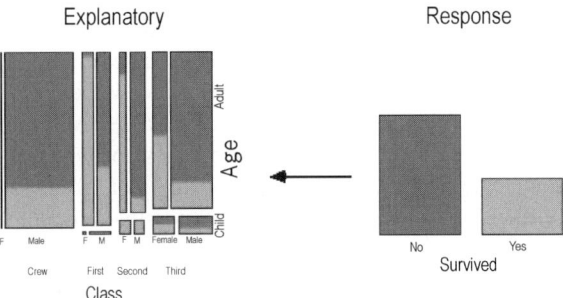

Fig. 40. Comparison of survival rates for females and males given the other variables.

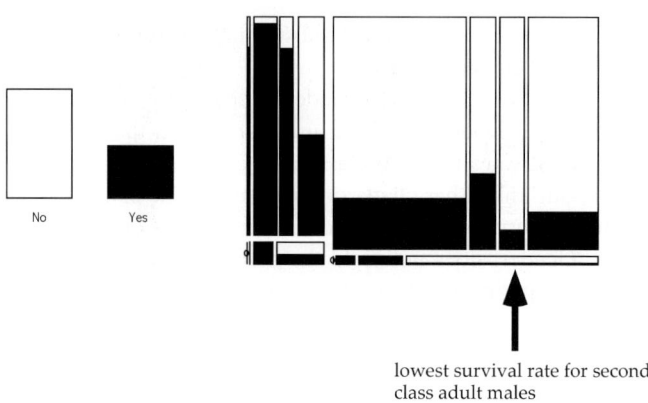

Fig. 41. Comparison of survivals rates for each class given age and gender.

'Gender' was the last to enter the construction of the mosaic plot. We can see that in all class and age groups survival rates for females are substantially higher than those for males. But we also see that females in third class had a much lower chance of survival than those in first or second class. This can be seen from the highlighted areas, but it can be derived more precisely from looking at the non-highlighted areas. These areas are pending from the ceiling and thus have a common base line which enables a proper comparison by judging height (or in this case depth) of the bars. Remarkably, in Figure 40 is the high survival rate for female crew members which is mainly caused by the small number of female crew members. The conclusion from this display is that within all age/class groups the rule "women first" was obeyed.

But what is the influence of the economic status? To obtain a suitable plot, we just have to switch the order of the variables in the mosaic plot. In Figure 41, the variable 'Class' has been the last to enter. Now comparisons of survival rates for each class given age and gender are readily available. Surprisingly, the lowest survival rate is observed for adult males in second class. It is much lower than the rate for crew or third class men.

Comparisons of highlighted subareas should only be made within groups that are aligned. In the above example, it would be not advisable to compare the survival rate of second class females with second class males, since the tiles are not based on the same level. It is always the last variable that enters the mosaic plot which can be used for proper comparisons. So, graphically the variables in the definition of the conditioning set play different role and they cannot be interchanged without changing the visual information. As a general rule we can state that the variable displayed in the spine plot shows up in the formula in front of the conditioning sign, whereas the variables on which we condition are in reverse order in the mosaic plot. To get all possible cross-classifications, we therefore have to run through all possible orders of variables for the mosaic plot.

To support the proper comparison of subareas, it is important to draw the highlighting in the direction of the next possible split of the mosaic plot axes. So far all our mosaic plots had the highlighted areas drawn from bottom to top, as in the spine plot. This

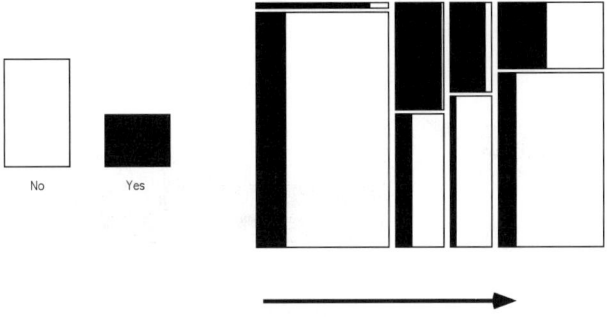

Highlighted areas are drawn from left to right
to ease comparison within each column.

Fig. 42. Comparison of survival rates for males and females within each class (age aggregated). Note that the highlighting is now from left to right.

came with the odd number of variables in the mosaic plot. Whenever there is an even number of variables in the mosaic plot then the highlighting will be drawn from left to right. In Figure 42, survival rates for females can be compared with those for males separated for each class and aggregated over age. In the top row are the females, the bottom row shows the males. Highlighting is from left to right and makes the intended comparisons straightforward. Changing the frame by a 90 degrees counter-clockwise rotation yields a mosaic plot in which the highlighting is again in the more common bottom–up direction.

Simpson's paradox is a well-known phenomenon when analyzing contingency tables. How can we find such a phenomenon with linked bar charts and mosaic plots. For the last time, we look at the Titanic data (see Figure 43) and draw a one-dimensional mosaic plot for variable 'Class'. Using the interactive rebinning method the number of classes is cut down to two: crew and passengers. Using linked highlighting we see that the survival rate is higher for passengers than for crew members. We now add one more variable – 'Gender'. Reordering the variables such that 'Class' is the last to enter the mosaic plot we can compare the survival rates for crew and passengers given the gender. For both males and females, the crew has a higher survival rate than the passengers.

Also for continuous data linked highlighting can be used to display conditional distributions. For categorical data, conditional probabilities are in the main focus, they do not play such an important role in the graphical analysis of continuous data. Histograms are the only graphical displays that yield to estimation of conditional probabilities for continuous data. In histograms, the bin heights are used as reference and are put into relation with the heights of the highlighted sections. To compare proportions, it is more convenient to transform the histogram bins to equal heights and let the bin widths vary with the counts. This leads to a continuous analogue of the spine plot which is implemented in MANET. Figure 44 shows a histogram with highlighting and its spine version in which all bins have the same height and the frequencies are represented by

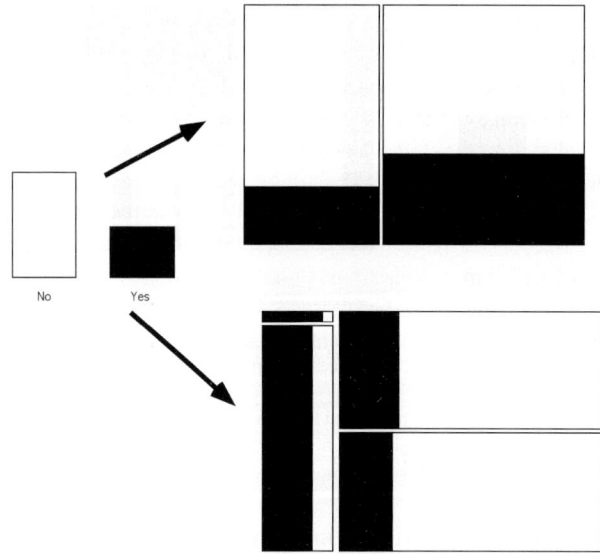

Fig. 43. Simpson's paradox can also be found in the Titanic data.

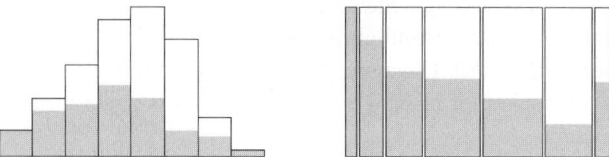

Fig. 44. Histograms show conditional probabilities. But changing the bars in the histogram to spines that have the same height and represent the counts by their widths makes the exploration of conditional probabilities for continuous data much easier.

the widths of the bins. As can be seen conditional probabilities can be easier deduced from the spine version than from the original histogram.

An important point to remember here is that whenever we want to deduce conditional distributions from a display we must have frequencies represented in the plot and the highlighting must be part of the original graphical elements. So only overlaying and proportional highlighting work for this purpose. Another consequence is that we can use juxtaposition whenever our goal does not include perception of conditional distributions. This is especially true for most explorations of continuous data.

24. Detecting outliers

Graphical exploratory techniques may be used for a range of goals. Outlier detection is often named first. This originated from the very beginning of statistical graphics and exploratory data analysis when Tukey and his acolytes stressed this point to motivate and promote their work. Unfortunately, this point has been overemphasized and prevented more widespread usage of graphics. The human eye will primarily classify as outliers, or more general as unusual observations, all data points that are separated from the main point cloud. Gaps in the distribution or low density areas that naturally arise at the boundaries of the domain of the distribution may falsely indicate outliers. The human vision system encounters problems when judging how far away points are from the center of the distribution and whether this is unusual for this kind of data, especially in the multivariate case. So, many points that seem to be suspicious at first glance will not be classified as outliers by more solid investigation. But nevertheless, gaps, large nearest neighbor distances and other apparent patterns in the data will first attract the attention of the analyst. Highlighting these cases and looking at how they perform in other plots allows an instantaneous check of the dimensionality of the pattern. Assuming correlation between the data, we expect that these unusual points also show a striking pattern in most of the other plots. Even if these points might not prove to be outliers, focusing on them gives us a hint on the correlation structure of the data and helps us in detecting high and low correlated variables.

25. Clustering and classification

How can linked views be used to detect clusters? Working with linked views is a highly interactive approach and depends heavily on the experience of the user. It includes a rapid back and forth between various views of the data. Selecting subsets is the core interactive procedure for linked views. The results we will obtain depend on whether we are able to find those subsets that define a good classification. An integral part of that is to extract those variables that bear most of the information that can be used for classification. Obviously, one-dimensional clusters will show up in the dotplots and by selecting one of these clusters we can check in all the other linked plots whether this cluster extends to more dimensions. This is an easy first step to explain some structure and especially to find influential variables. But this approach will not suffice to detect higher-dimensional clusters that are not present in marginal dimensions. For bivariate structure, we have to create subsets of one variable systematically and then to look in the other plots whether there shows up some clustering in the highlighted points. Higher-dimensional structure can be found if we combine selections made in different plots using logical operations. Mainly we will work with the intersection mode to define complex selections.

In low-dimensional views, we can easily recognize low-dimensional clusters, but the use of selection sequences (Theus et al., 1998) enables us to specify more elaborate queries to the database and, hence, we can detect also clusters in higher dimensions. Linking bar charts and histograms is an effective way to analyze data sets with both discrete and continuous variables.

Single univariate views can only show clusters that can be discriminated by a line perpendicular to the plot axis. The standard example for this limitation are clusters that split along the diagonal in two dimensions. Those can easily be depicted in a scatterplot but they can not be seen in any of the standard projections. Linking also enables us to

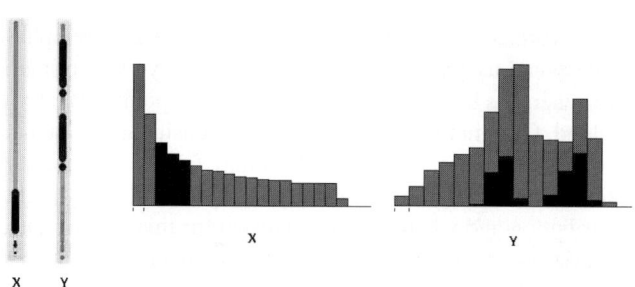

Fig. 45. Linking shows two-dimensional clusters that can not be seen in the projections.

see such clusters in the marginal views. While brushing over one of the variables the highlighted points in the other variable will split and show the clusters. Dotplots as well as histograms can show this effect (see Figure 45). But histograms smooth the data and do not show small gaps between the clusters as clearly as dotplots do. A separation between clusters can only be shown in a histogram if the distance between the clusters in this variable is greater than the bin width. It also depends on which variable we choose for brushing. For two-dimensional data, clusters can be seen best if we choose that variable for brushing that has the smaller one-dimensional distance between the clusters. Since then, the greater distance can be used to show the clusters. Also the smaller the bin width in the histogram the more clusters we can see. There is certainly a limit beyond which too many cluster artifacts are created. Dotplots, in contrast, are only slightly affected by change of the scale, but the highlighting unfortunately covers the original plot.

How do we proceed in searching for three-dimensional clusters that can not be seen in any two-dimensional projection? We then have to combine selections and to condition on two variables searching for a subset that produces a sufficient gap within the highlighted points of the third variable.

26. Geometric structure

Visualization of point clouds in high-dimensions aims particularly at identifying the existence of lower-dimensional structure. Carr and Nicholson (1988) translated dimensionality into terms of constraints on the display variables. They investigated three petals generated by

$$x = \cos(u) * \cos(3u) + \text{noise},$$
$$y = \sin(u) * \sin(3u) + \text{noise}.$$

In a scatterplot, the structure can be seen immediately (Figure 46). Do we have a chance to reconstruct the two-dimensional picture by linking one-dimensional marginal plots? To produce the petals, we have used 1000 equidistant points in the interval $[0, \pi]$. In one-dimensional scatterplots, much of the structure would remain obscure due to the heavy overplotting of points. We therefore work with linked marginal histograms. While

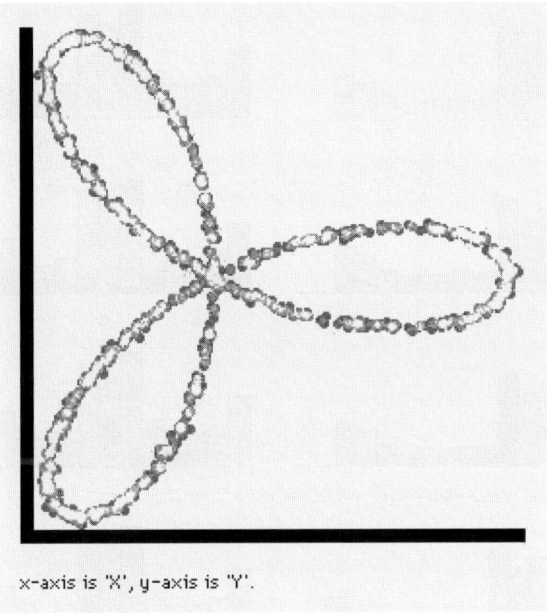

Fig. 46. An ideal summary plot for petal structure.

we brush over the x-values in the histogram the constraints in the y-variable become apparent (Figure 47). The holes in the petals become visible as well as the intersection point of the three petals in the center. It is easy to recognize four branches for low x-values in contrast to two branches for high x-values. With some effort we are able to reconstruct the entire structure of the petals from one-dimensional marginal views. In higher dimensions, the procedure works in the same way, with the only difference that there is no perfect two-dimensional summary plot to compare with – as the scatterplot is one for two-variate data.

By focusing on high density regions, we might be able to detect homogeneous clusters and local structure. To demonstrate this feature, we use a synthetic data set about the geometric features of pollen grains. The data set was made up by David Coleman of RCA Labs and was distributed in 1986 as ASA Data Exposition dataset. (The data is available from STATLIB at http://www.stat.cmu.edu/datasets/.) The dataset consists of 3648 observations in five dimensions. In some scatterplots (see Figure 48), a pencil shape high density region can be found in the center of the point cloud. Selecting one of these regions shows that these are two-dimensional projections of the same region. Zooming in – by interactively changing the boundaries of the plot – the hidden structure becomes apparent (see Figure 49). Six clusters of points forming the letters of the word EUREKA had been inscribed in a five-dimensional sphere of random normal data.

In both examples, projection alone could not reveal the structure. Linking adds the necessary component to mentally reconstruct the structure.

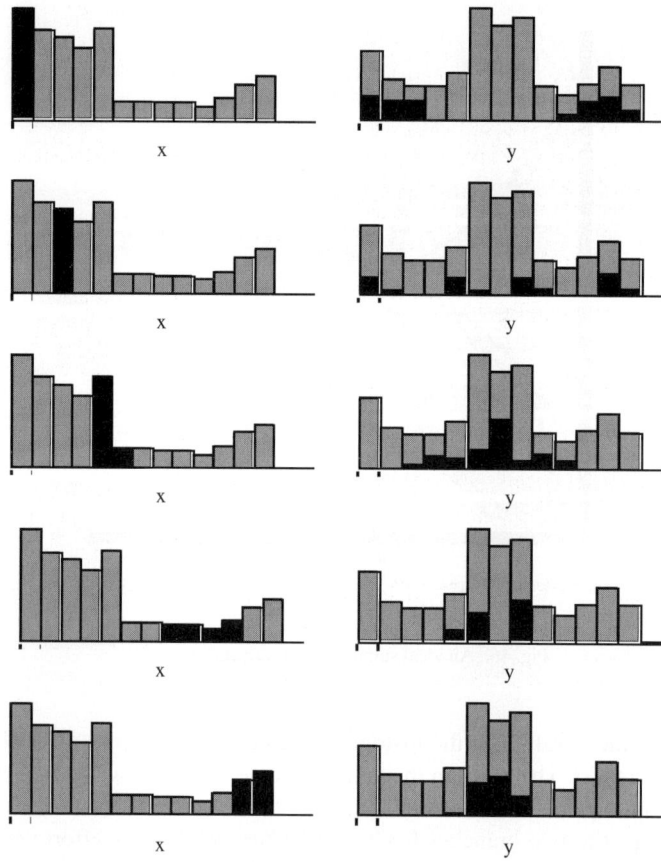

Fig. 47. Linked marginal histograms to depict two-dimensional structure.

27. Relationships

Many statistical problems can be formulated as regression models. Historically, most work on regression dealt with numerical procedures. Graphics have been used as diagnostic plots in the last two decades. Anscombe (1973) showed the relevance of looking at the data with four artificial data sets that have the same numerical regression parameters, but have a completely different view in a scatterplot. Thus, graphics are widely used for checking the model assumptions and the type of regression function to be used. But graphics can do much more than only diagnose departures from the model assumptions. Graphics are particularly helpful in determining which variables should be included in the modeling step. For some data sets, the resulting patterns are so striking that a formal analysis is no longer needed.

The classic attempt to portray functional dependencies between variables is to use a scatterplot. For two variables, this works excellently, and for medium sized data sets

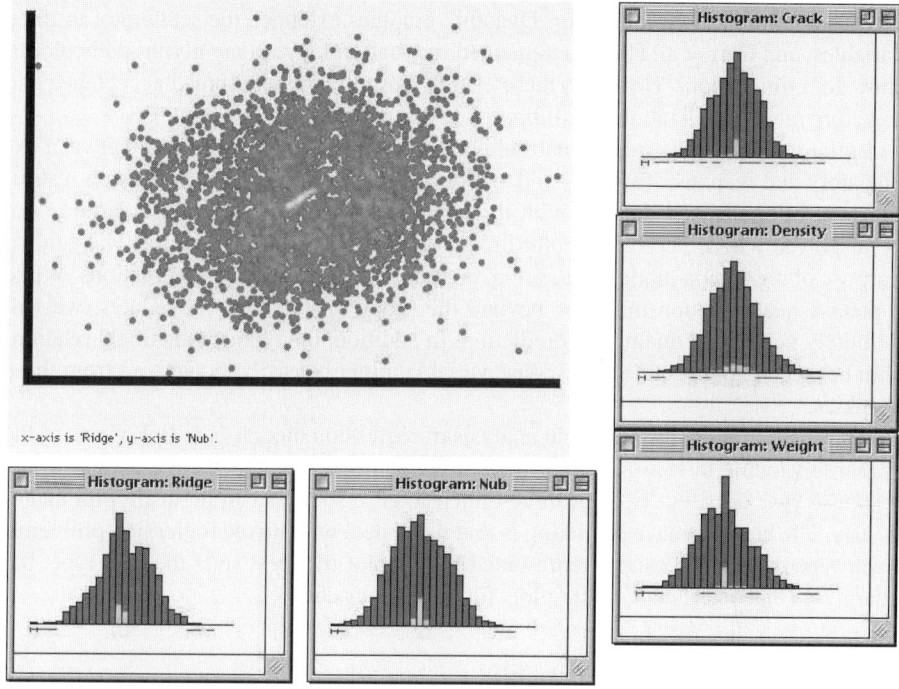

Fig. 48. By tonal highlighting hidden structure appears as a high density region in the scatterplot. This region lies in the center of all marginal distributions.

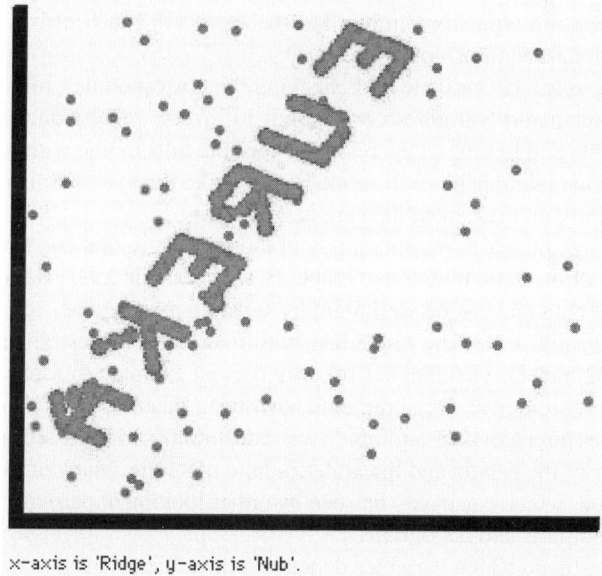

Fig. 49. Zooming easily reveals the hidden structure.

there is not much more to ask for. Dynamic graphics extended the scatterplot to three variables, and Carr et al. (1987) augmented the scatterplot with ray glyphs and color to show four dimensions. However, these efforts did not gain widespread acceptance and therefore most visualization is still limited to two or three dimensions.

Linking seems to be a quite natural choice to portray the relationship between explanatory and response variables and also to overcome some of the dimension restrictions. A 4-dimensional data set with three explanatory and one response variable can be displayed with a 3D rotating plot linked to a one-dimensional dotplot. The general purpose of regression models is to set up prediction rules for future observations. While a precise quantification might be beyond the scope of graphical procedures, we will definitely get a good qualitative prediction. In addition, the type of functional relationship between explanatory and response variable might be easily recognized from these graphics.

The basic strategy for graphical analysis of regression models is to link views of the response variable to views of the explanatory variables. We demonstrate the univariate response case only, but two- or three-dimensional responses can be dealt with analogously. The big advantage of linking is that it can deal with mixed regression problems. Each type of variable can be represented by the plot that best suits the data type: bar charts for categorical data, scatterplots for continuous data.

27.1. Models with continuous response

Continuous response models are the main theme in most introductory statistics courses. However, they do not occur as often in practice. To visualize a one-dimensional continuous response variable, we can choose either a dotplot, a histogram or a boxplot. Although the boxplot typically summarizes the data very much, it is a useful tool in assessing functional relationships.

Brushing the response variable and checking the corresponding highlighting in the plots for the explanatory variables corresponds to inverse regression; brushing an explanatory variable and linking to the response variable falls in line with partial response plots. A monotone relationship will immediately strike the eye, but also other patterns that depend mainly on the ranked data are easy to assess.

As the data size increases it gets more and more unlikely that one single regression function fits all of the data. Interactive graphical subsetting is a very helpful tool to find partitions of the data that can be well fitted by separate regressions.

The following illustrates the procedure with data on 202 Australian athletes (100 females and 102 males). The data has been used as an example by Cook and Weisberg (1994). Five hemotological measurements have been taken at each person as well as six physical measurements. In addition, two categorical variables are included in the data set: gender of the person and the athlete's field of sports. Many of the variables are highly correlated, which can easily be seen by either looking at pairwise scatterplots or brushing over dotplots and histograms.

First we investigate which variables depend on the athlete's gender. Figure 50 shows in the top row a bar chart of sex with the females selected and boxplots of all continuous variables. The boxplots show actually two distributions at a time: the distributions for all

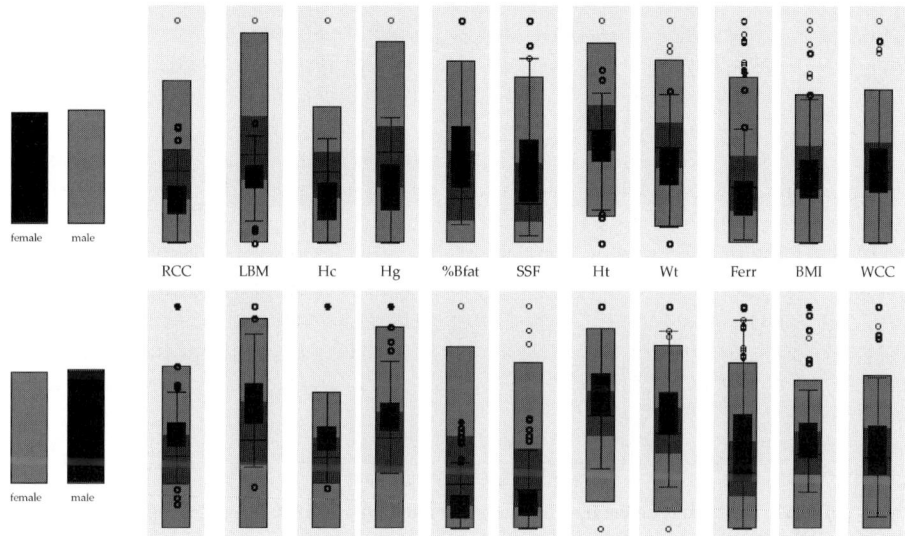

Fig. 50. Searching for correlation with gender.

athletes as gray boxplots in the background and the distributions for females depicted as black boxplots overlaid on the others. First and second order dependencies can be detected by comparing the medians and the interquartile ranges of the two overlaid boxplots. Judging how much the interquartile range of the female group overlaps with the interquartile range of all athletes we can roughly classify the 11 variables into two groups: the first group consists of 'RCC', 'LBM', 'Hc', 'Hg', 'SSF', '%Bfat', 'Ht', 'Wt' and is highly correlated with sex; the second group consisting of 'Ferr', 'BMI', 'WCC' indicates no correlation with sex. By judging the overlap of interquartile ranges, we made use of the static form of painting. In addition, we can use dynamic painting by toggling between females and males in the bar chart. The dynamic of changes in the highlighted boxplots is then the basis for judgment. Comparing top and bottom rows in Figure 50 is the static form of that. The variables correlated with sex follow mixture distributions. By looking at the boxplots for male and female separately, two different forms of mixtures can be distinguished: mixtures of rather separated distributions for males and females, e.g., 'RCC' and 'LBM'; mixtures of two distribution of which one resembles the overall distribution and the other a subgroup, e.g., '%Bfat' and 'SSF'.

27.2. Models with discrete response

Discrete data arise commonly in many surveys in the social and behavioral sciences. For the modeling of such data, loglinear and logistic models have been designed. How can interactive graphics be used to make the structure more visible and to support the numerical analysis?

To illustrate the capabilities of MANET, a partial analysis of a real data set is presented in the following. The data stems from (Küchenhoff and Ulm, 1997) and consists

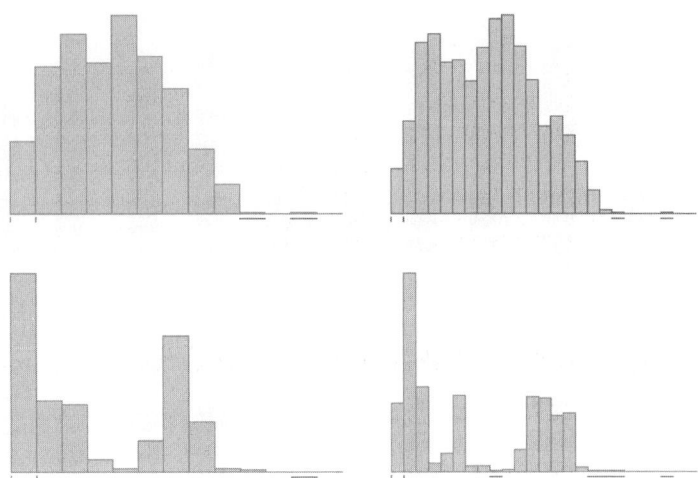

Fig. 51. Histograms for duration (top) and dust (bottom) with 13 (left) and 25 (right) bins.

of measurements taken on 1246 workers at a single plant. The variables are 'bronchitis', 'smoker', 'dust' and 'duration'. 'Bronchitis' is the response variable, this is a binary variable and just tells whether a person suffers from bronchitis or not. 'Smoker' is also a binary variable indicating whether a person smokes regularly or not. The variable 'dust' gives the logarithmic transformed amount of dust to which the worker is exposed at his or her working place. The fourth variable 'duration' shows how long this person has been exposed to pollution by dust at work.

The main goal for the analysis was to determine a maximum safety value for the amount of dust exposure at the workplace.

As a first step the dependence structure of the three explanatory variables will be examined.

One-dimensional structure of the continuous variables can be summarized by histograms. The two variables react quite differently to changes of the anchor point and the bin width. While the shape of the histogram for duration remains almost the same for all choices, the histogram for dust depends heavily on these parameters, see Figure 51. Sturges' rule would ask for 12 bins, as well as the Freedman–Diaconis rule. The oversmoothed histogram (Scott, 1992, Section 3.3.1) has 14 bins. Thus, it seems reasonable to use at least 15 bins. Using even more bins result in a quite stable histogram shape for 'dust' indicating a tri-modal distribution. Boxplots of the two continuous variables indicate just one possible outlier in the duration variable and do not show any other interesting structure.

To visualize the dependency between the two continuous variables, a scatterplot is the first choice. The gestalt in Figure 52 indicates that there is no structural relationship between the two variables. Amount of dust and duration of exposure seem to be rather uncorrelated. There is no hint of a dust exposure adjusted workplace assignment procedure in the form of exchanging workers in high dust places more often than others. Quite

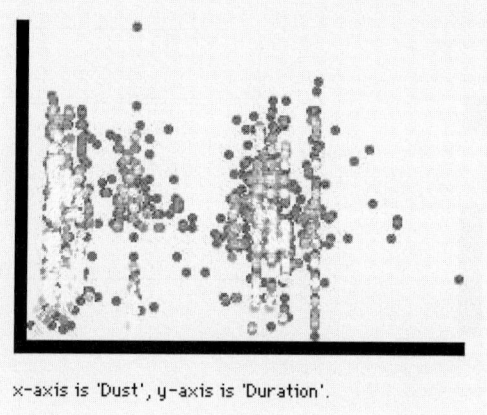

Fig. 52. Low correlation between dust and duration.

striking is the partitioning into three clusters according to the dust variable, that again supports the choice of smaller bin widths in the histogram. In some areas, the scatterplot shows a kind of grid structure. This is due to the precision of the dust measurement and the integer units of the duration variable. The original dust variable (before being log transformed) has been observed with an accuracy of two decimal digits.

As one would expect, there seems to be no relation between duration, dust exposure and the smoking behavior of the workers. Selecting the smoker category shows quite similar distributions for the selected and the entire sample population for variables 'dust' and 'duration'. Also, conditioning on 'dust' or 'duration' always shows similar proportions for smokers and non-smokers.

To investigate the relationship between response variable and explanatory variables, we first look at the partial response plots. For the discrete explanatory variable 'smoker', we can use either one mosaic or two linked spine plots. It is clear from Figure 53 that there is an interaction between the two variables stating that smokers are more likely to suffer from bronchitis than non-smokers. The corresponding χ^2-test has a value of 14.69 and indicates significance of the interaction at the 0.01% level.

Brushing the histogram of 'dust' shows no increased bronchitis risk for the lower two clusters. The high dust cluster, however, strongly indicates an increasing risk for developing bronchitis, see Figure 54. That the first two clusters of 'dust' do not show any influence on bronchitis but the third cluster does gives us a strong hint that there is a threshold above which the dust exposure has a negative impact on the health status of the workers. At the same time we can see in that figure that there are no interactions between the variable 'dust' and the other covariables.

Brushing the histogram of duration shows increased bronchitis risk for employees with a long working history (Figure 55). In contrast to the dependence between dust and bronchitis, there is a continuously increasing risk for developing bronchitis depending on the duration of exposure.

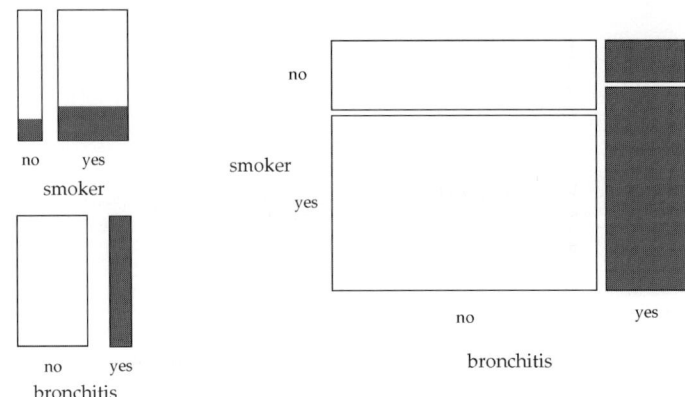

Fig. 53. Mosaic plot and linked spine plots for smoker and bronchitis.

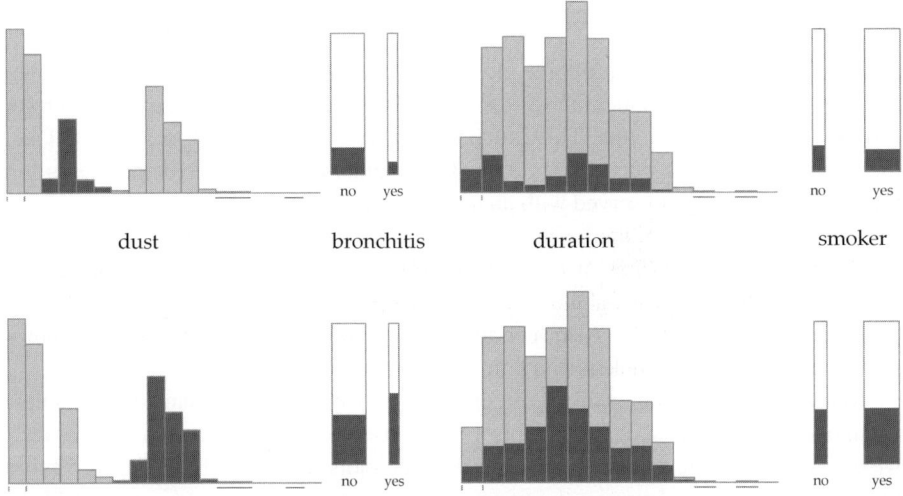

Fig. 54. Influence of amount of dust exposure on bronchitis.

Two last questions are to be answered in this study. Although there was no apparent interaction between the covariates in the previous plots, does the amount of dust really increase the bronchitis risk? And second, what is a reasonably maximum safety value for the dust exposure at the working place? The first question is answered in the affirmative in Figure 56, where in the top row the interactive selection sequence techniques of MANET are used to select smoking workers with high duration but low dust exposed working places. The resulting bar chart for 'bronchitis' shows almost no difference in the proportions. However, when we change the third criterion in the sequence to high dust exposition (see bottom row of plots) we get a clear distinction between workers who suffer from bronchitis and those who do not.

Interactive statistical graphics: the paradigm of linked views 529

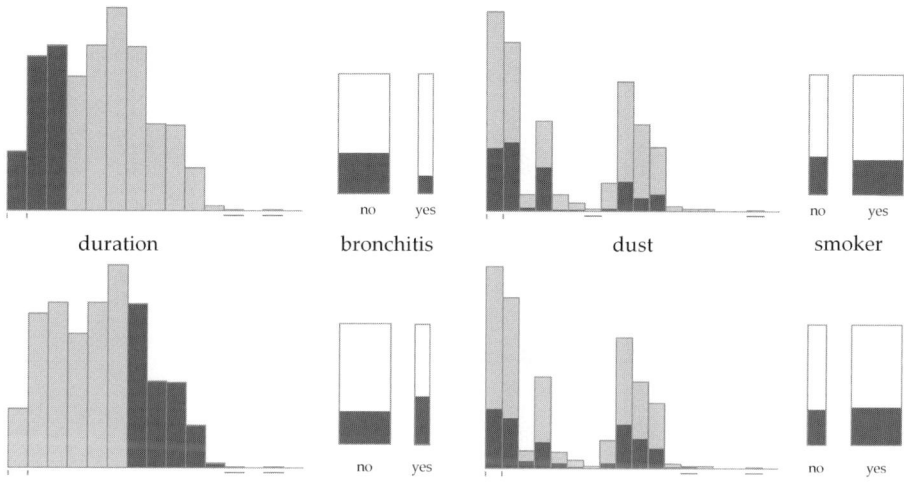

Fig. 55. High duration leads to an increased bronchitis risk. Proportion of smokers are independent of duration of dust exposure.

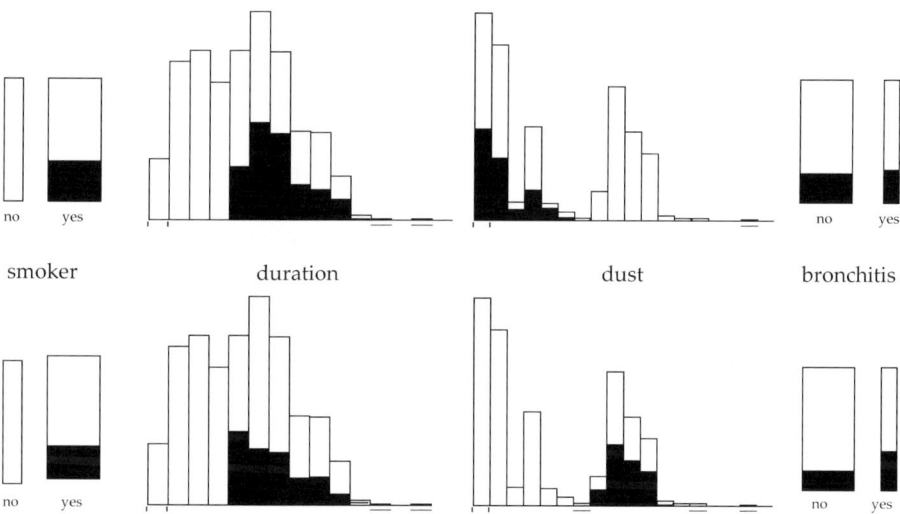

Fig. 56. Smokers that have been for a long time on their working place with high dust exposure are more affected with bronchitis than those with low dust exposure.

To answer the second question which was also the main purpose of the analysis, we might go back to brushing a histogram for 'dust' and looking when the proportions in the spine plot for 'bronchitis' start to differ. Taking the lower limit of this brush is a good candidate for maximum safety value. Here the interactive query tools become essential. By querying the graphical elements in Figure 57, we derive the value 1.32 as a reasonable candidate for the maximum dust exposure allowed at a working place.

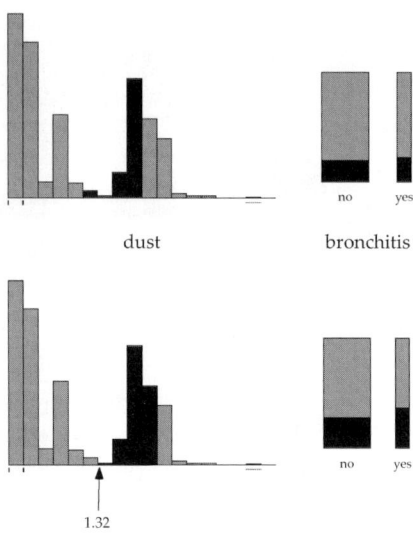

Fig. 57. Graphical determination of a threshold value for dust exposure.

27.3. Independence models

The mosaic plot is not only used for the raw data it proves also useful when dealing with loglinear models. If we assume that we have data from an independence model then all tiles in the mosaic plot would align, because the sides of the tiles would be determined by the marginal counts. Thus, deviation from an aligned pattern shows deviation from the independence model. Let us look at a two-dimensional example. In Figure 58, a mosaic plot for 'Gender' and 'Survived' is drawn. If the two variables are independent then the tiles in each row will align. For the displayed variable, this is not the case, indicating that there is a relationship between gender and the survival status.

With more variables more complex models and a variety of independence structures may arise. Each model-type – mutual independence, conditional independence, partial independence – shows a different and particular shape in the mosaic plot. A detailed

Fig. 58. The mosaic plot for the variables 'Survived' and 'Gender' shows also clear association between the variables, since tiles in a row do not align.

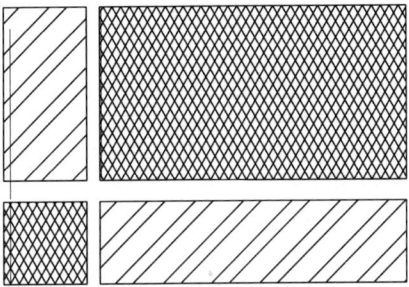

Fig. 59. The direction of the deviation from the independence model is portrayed by different grids.

description of this is given by Theus and Lauer (1997). Colors or shading can be used to add residual information for such models either to the mosaic plot of the observed counts or to the mosaic plot of the expected counts. We thus get a clear hint which interaction terms should be added to the model. When the mosaic plot is drawn for the observed counts coloring cells with count zero masks the actual observed value. In Figure 59, a mosaic plot for the independence model between variables 'Gender' and 'Survived' is shown. There the observed counts for cells with the denser grid exceed the expected count whereas the inequality goes in the opposite direction in the sparser grids.

An important problem in exploratory modeling is to find the most suitable model. Suitable in the sense that it is as parsimonious as possible while explaining as much as possible of the deviation from mutual independence. For this purpose, stepwise procedures are often used that are based on χ^2 and G^2 statistics to judge adequacy of the model. A visual analogue to that can be created with mosaic plots in the form of a graphical backward selection as well as in the form of a graphical forward selection (Theus and Lauer, 1997).

Mosaic plots have been proven to be very helpful to analyze categorical data. They are the only graphical tool that allows analysis of the three main aspects of categorical data simultaneously: counts, proportions, and interactions. A highly interactive implementation is required to draw full benefit from mosaic displays. Due to the recursive construction labeling of mosaic plots is only sensible for up to four variables. With good interactive querying tools we can escape these problems. But mosaic plots alone are not sufficient for a successful exploration of categorical data. It is necessary that we have the mosaic plot integrated in a highly interactive system that allows linked highlighting with other plot types, especially bar charts and spine plots. But not only the sample population based linking scheme is important for this display. Linking on the model level and linking scales are needed because we have to be aware that the hierarchy in drawing a mosaic plot determines the kind of easy comparison within the plot. It is not just a matter of the alphabetical order of variables in the data set or in which order they have been typed in. The user has to be able to specify the order of variables he or she wants to see. Therefore, interactive features like rebinning or reordering variables must be included. Linking models for superposing residual information makes the mosaic plot even more powerful.

28. Conclusion

User interactions are an essential ingredient of modern statistical software. Exploratory data analysis needs such highly interactive tools and benefits from them. Graphical displays that can be directly modified in a simple and intuitive manner encourage the user to experiment with the data and to try out different approaches. Visualization is not only an important part of exploratory data analysis in the classical sense it also offers much help for the process of discovering knowledge in databases. Having found something of interest we want to explain it. Linked highlighting allows us to use simple and well-known views that can be easily interpreted.

Modern data sets are complex and go beyond the traditional statistical point of view of representing data sets as data matrices. Data warehousing and data mining are two key words for modern data analysis. The data model for interactive statistical graphics that I have presented in the previous chapters gives a general specification of an interactive system that can handle some of the complexities of modern data sets. It covers the currently existing implementations and points towards new developments. The distinction of the display levels – frame, type, model, and sample population – clarifies the plot–data relationship and presents an appropriate classification of user interactions. New interactive techniques can be derived from generic methods for the various objects and new and enhanced displays can be generated by the model and graph type operators.

The four levels of a display define different levels of interaction and also different possibilities of linking. Linking sample populations by identity linking is fairly known by now. Other linking relations, like spatial inclusion or distance, can be defined. Problems of such m-to-1 or 1-to-n linking schemes arise with linked highlighting. Overlaying, proportional highlighting, and juxtaposition can often solve the visualization problems. Juxtaposition is a standard technique for static plots but has not been used in interactive statistical graphics packages. For continuous data, only interpretation of histograms benefits from overlaying. In all other plot types, selected subgroups and complete samples can be efficiently and effectively compared by juxtaposition as long as the same scales and axes are used. At the very least, juxtaposition should be available as an interactive option, in particular, to prepare an interesting view for printing or other output in a static format. Not all possible techniques have yet been implemented and tested. Further research is necessary.

Throughout the previous chapters some principles of interactive statistical graphics have been pointed out. Some of them have been explicitly formulated, some others are present implicitly. It seems appropriate to summarize them at this point and to postulate their content in form of axioms.

AXIOM 0. Construction of data displays aims at finding good graphical representations of the model.

AXIOM 1. Only objects that are graphically represented can be directly manipulated.

AXIOM 2. Interactive operations are passed to other objects within a plot using the internal linking structure.

AXIOM 3. Interactive operations triggered in the active plot are passed to corresponding objects in linked plots using the external linking structure.

AXIOM 4. Selections can be combined by logical operations. The combinations are resolved on the lowest available graph level. Usually this is the sample population level.

AXIOM 5. Linked highlighting should avoid excessive visual effects but it should also provide the relevant changes in a perceptible way. Overlaying works well as a default but juxtaposition is better for comparing subgroups and can be used whenever the original graphical elements are not needed as reference for the highlighted ones.

To fulfill the demands in Axiom 5, the selected subset should be represented by the same type of plot as has been originally used.

Future work

In which direction should future research aim? What we need is an integrated package that offers all of these interactive tools. The model that has been laid out here can be used to build such a consistent system for interactive statistical visualization. Currently a prototype for linking various data matrices is under construction.

A common point of criticism of interactive graphical methods is that it is hard to store the analysis path and the results. The structure given by the model presented here could be used as a basis for a journaling system, but many details would have to be worked out. The model also leads to a structure that can be used for teaching and communicating future research developments.

Data exploration based on selecting subgroups and using linked highlighting depends on the knowledge and the experience of the user. Slightly increasing or decreasing the area of a selection rectangle might sometimes cause a significant change in the resulting graphics. To illustrate these problems, it is useful to have an automatic update that shows neighboring selections. Also, to guide novices through the possibilities of interactive data analysis it is helpful to automatically run through a set of possible and interesting selections. These two methods fall under the heading of animation.

Selections can be animated by working through a list of prespecified subsets. Which subsets are more interesting depends on the problem at hand. To make comparisons between sequential selections more sound, one should try to keep the number of selected points stable. The more the number of selected points varies, the more artifacts from visual perception will show up.

Graphical methods complement the statistician's toolbox, they do not replace traditional analysis methods like hypothesis testing and estimation. The relation between interactive graphics and statistical tests or models deserves more study. Interactive statistical graphics is still a young research area. Many promising techniques and methods are available but they are only loosely integrated. In this chapter, we tried to lay the foundation for a common framework and consistent description of interactive statistical graphics. Hopefully, further extensions will be made on the theoretical side as well as on the practical side to provide us with intuitive and easy to use methods that encourage sound analyses of data.

References

Andrews, D. (1972). Plots of high-dimensional data. *Biometrics* **28**, 125–136.
Anscombe, F. (1973). Graphics in statistical analysis. *Amer. Statist.* **27**, 17–21.
Anselin, L. (1996). Interactive techniques and exploratory spatial data analysis, Technical Report. Regional Research Institute, West Virginia University.
Anselin, L., Bao, S. (1997). Exploratory spatial data analysis linking SpaceStat and ArcView. In: Fischer, M., Getis, A. (Eds.), *Recent Developments in Spatial Analysis – Spatial Statistics, Behavioural Modelling and Neurocomputing*. Springer-Verlag, Berlin, pp. 35–59.
Becker, R.A., Cleveland, W.S. (1987). Brushing scatterplots. *Technometrics* **29**, 127–142.
Becker, R.A., Cleveland, W.S., Wilks, A.R. (1987). Dynamic graphics for data analysis. *Statist. Sci.* **2**, 355–395. With discussion.
Bertin, J. (1967). *Sémiologie Graphique*. Mouthon–Gauthiers–Villars, Paris.
Bertin, J. (1983). *Semiology of Graphics*. University of Wisconsin Press, Madison, WI. Translation of (Bertin, 1967).
Buja, A. (1999). A word from the editor of JCGS. *Statist. Comput. Graphics* **10** (1), 32–33.
Buja, A., Tukey, P.A. (Eds.) (1991). *Computing and Graphics in Statistics. IMA Vols. Math. Appl.*, vol. 36. Springer-Verlag.
Carr, D.B., Nicholson, W. (1988). Explor4: a program for exploring four-dimensional data using stereo-ray glyphs, dimensional constraints, rotation and masking. In: *Dynamic Graphics for Statistics*. Wadsworth, pp. 309–329.
Carr, D.B., Littlefield, R., Nicholson, W., Littlefield, J. (1987). Scatterplot matrix techniques for large N. *J. Amer. Statist. Assoc.* **82**, 424–436.
Carr, D.B., Olsen, A.R., Curbois, J., Pierson, S.M., Carr, D. (1998). Linked micromap plots: named and described. *Statist. Comput. Graph. Newslett.* **9** (1), 24–32.
Cleveland, W.S. (1993). *Visualizing Data*. Hobart Press, Summit, NJ.
Cleveland, W.S., McGill, M.E. (Eds.) (1988). *Dynamic Graphics for Statistics*. Wadsworth and Brooks/Cole, Pacific Grove, CA.
Computer Dictionary (1994), 2nd ed. Microsoft Press, Redmond, WA.
Cook, R.D., Weisberg, S. (1994). *An Introduction to Regression Graphics*. Wiley, New York.
Cook, D., Majure, J., Symanzik, J., Cressie, N. (1996). Dynamic graphics in a GIS: exploring and analyzing multivariate spatial data using linked software. *Comput. Statist.* **11**, 467–480.
Cox, D.R. (1978). Some remarks on the role in statistics of graphical methods. *Appl. Statist.* **27**, 4–9.
Craig, P., Haslett, J., Unwin, A.R., Wills, G. (1989). Moving statistics – an extension of brushing for spatial data. In: *Proceedings of the 21st Symposium on the Interface*. In: *Computing Science and Statistics*. Interface Foundation of North America, pp. 170–174.
Dawson, R.J.M. (1995). The "unusual episode" data revisited. *J. Statist. Education* **3** (3).
Derthick, M., Kolojejchick, J., Roth, S.F. (1997). An interactive visualization environment for data exploration. Technical Report. Carnegie Mellon University, Pittsburgh, PA.
Diaconis, P.N., Friedman, J.H. (1983). M and N plots. In: Riszi, M., Rustagi, J., Siegmund, D. (Eds.), *Recent Advances in Statistics: Papers in Honor of Herman Chernoff on His Sixtieth Birthday*. Academic Press, New York, pp. 425–447.
Dictionary of Computing (1991), 3rd ed. Oxford University Press, New York.
Dykes, J. (1998). Cartographic visualization: exploratory spatial data analysis with local indicators of spatial association using Tcl/Tk and cdv. *Statist.* **47**, 485–497.
Eick, S.G. (1994). Data visualization sliders. In: *Proc. ACM UIST*. ACM Press, pp. 119–120.
Fienberg, S.E. (1975). Perspective Canada as a social report. *Soc. Indicators Res.* **2**, 153–174.
Fisherkeller, M.A., Friedman, J.H., Tukey, J.W. (1975). Prim-9: an interactive multidimensional data display and analysis system. In: *Data: Its Use, Organization and Management*. ACM Press, New York, pp. 140–145.

Friendly, M. (1994). Mosaic displays for multi-way contingency tables. *J. Amer. Statist. Assoc.* **89**, 190–200.
Friendly, M. (1995). Conceptual and visual models for categorical data. *Amer. Statist.* **49**, 153–160.
Furnas, G.W., Buja, A. (1994). Prosection views: dimensional inference through sections and projections. *J. Comput. Graph. Statist.* **3**, 323–385. With discussion.
Gabriel, K. (1971). The biplot graphical display of matrices with application to principal component analysis. *Biometrika* **58**, 453–467.
Geßler, J.R. (1993). *Statistische Graphik*. Birkhäuser, Basel.
Good, I. (1983). The philosophy of exploratory data analysis. *Philos. Sci.* **50**, 283–295.
Gower, J., Hand, D.J. (1996). *Biplots*. Chapman and Hall.
Hartigan, J.A., Kleiner, B. (1981). Mosaics for contingency tables. In: Eddy, W. (Ed.), *Proceedings of the 13th Symposium on the Interface*. In: *Computing Science and Statistics*. Springer-Verlag, New York, pp. 268–273.
Hofmann, H. (1997). Graphical stability of data analysing software. In: Klar, R., Opitz, O. (Eds.), *Classification, Data Analysis and Knowledge Organization*. Springer-Verlag, Berlin, pp. 36–43.
Hofmann, H. (1998). Interactive biplots. In: *NTTS 98 – New Techniques and Technologies in Statistics*. EuroStat, Sorrento, pp. 127–136.
Hofmann, H., Unwin, A.R. (1998). New interactive graphic tools for exploratory analysis of spatial data. In: Carver, S. (Ed.), *Innovations in GIS 5 – Selected Papers from the Fifth National Conference on GIS Research UK (GISRUK)*. Taylor and Francis, London.
Huber, P.J. (1988). Comment in: Cleveland, W.S., McGill, M.E. (Eds.), *Dynamic Graphics for Statistics*. Wadsworth and Brooks/Cole, Pacific Grove, CA, pp. 55–57.
Hummel, J. (1996). Linked bar charts: analysing categorical data graphically. *Comput. Statist.* **11**, 36–44.
Hurley, C. (1991). Applications of constraints in statistical graphics. In: *ASA Proceedings on Statistical Graphics*. Amer. Statist. Assoc.
Hurley, C. (1993). The plot–data interface in statistical graphics. *J. Comput. Graph. Statist.* **2**, 365–379.
Hurley, C.B., Oldford, R. (1991). A software model for statistical graphics. In: Buja, A., Tukey, P.A. (Eds.), *Computing and Graphics in Statistics*. In: *IMA Vols. Math. Appl.*, vol. 36. Springer-Verlag, pp. 77–94.
Inselberg, A. (1985). The plane with parallel coordinates. *Visual Comput.* **1**, 69–91.
Keim, D.A. (1995). Enhancing the visual clustering of query-dependent data visualization techniques using screen-filling curves. In: *Proc. Int. Workshop on Database Issues in Data Visualization*. Springer-Verlag.
Keim, D.A. (1997). Visual techniques for exploring databases. In: *Tutorial Notes: Third International Conference on Knowledge Discovery and Data Mining*. AAAI Press, pp. 1–121.
Kraak, M.-J., Ormeling, F. (1996). *Cartography: Visualization of Spatial Data*. Longman, Harlow.
Küchenhoff, H., Ulm, K. (1997). Comparison of statistical methods for assessing threshold limiting values in occupational epidemiology. *Comput. Statist.* **12**, 249–264.
McDonald, J.A. (1982). Interactive graphics for data analysis. PhD thesis. Stanford University.
McDonald, J.A. (1988). Orion I: interactive graphics for data analysis. In: *Dynamic Graphics for Statistics*. Wadsworth and Brooks/Cole.
Monmonier, M.S. (1996). *How to Lie with Maps*. University of Chicago Press.
Mosteller, F., Tukey, J.W. (1977). *Data Analysis and Regression*. Addison–Wesley, Reading, MA.
Nagel, M. (1994). Interactive analysis of spatial data. *Comput. Statist.*, 295–314.
Nagel, M., Benner, A., Ostermann, R., Henschke, K. (1996). *Graphische Datenanalyse*. Fischer, Stuttgart.
Nelson, L., Cook, D., Cruz-Neira, C. (1999). XGobi vs the C2: results of an experiment comparing data visualization in a 3-D immersive virtual reality environment with a 2-D workstation display. *Comput. Statist.* **14**, 39–51.
Oldford, R.W., Hurley, C.B., Bennett, G., Desvignes, G., Anglin, D., Lewis, M., Bennet, N., Poirier, P., Chipman, H., Whimster, C., White, B. (1998). Quantitative analysis in Lisp. Technical Report. University of Waterloo. http://setosa.uwaterloo.ca/~ftp/Quail/Quail.html.
Ostermann, R. (1996). Computer-aided analysis of spatial data: Editorial. *Comput. Statist.* **11**, 383–385.
Rogowitz, B.E., Treinish, L.A. (1996). How not to lie with visualization. *Comput. Phys.* **10**, 268–273.
Rogowitz, B.E., Rabenhorst, D.A., Gerth, J.A., Kalin, E.B. (1996). Visual cues for data mining. Technical Report. IBM Research Division, Yorktown Heights, NY.
Rope, D.J. (1998). A Java-based graphics production class library. In: *Computing Science and Statistics*, vol. 29, p. 145.

Sasieni, P.D., Royston, P. (1996). Dotplots. *Appl. Statist.* **45**, 219–234.
Sawitzki, G. (1990). Tools and concepts in data analysis. In: Faulbaum, F., Haux, R., Jöckel, K.-H. (Eds.), *SoftStat'89 – Fortschritte der Statistik Software 2*, pp. 237–248.
Sawitzki, G. (1994). Diagnostic plots for one-dimensional data. In: Dirschedl, P., Ostermann, R. (Eds.), *Computational Statistics*. Physica, Heidelberg, pp. 237–258.
Scott, D.W. (1992). *Multivariate Density Estimation*. Wiley, New York.
Shneiderman, B. (1992). Tree visualization with Treemaps: a 2D space-filling approach. *ACM Trans. Graphics* **11**, 92–99.
Shneiderman, B. (1994). Dynamic queries for visual information seeking. *IEEE Software* **11** (6), 70–77.
Stuetzle, W. (1987). Plot windows. *J. Amer. Statist. Assoc.* **82**, 466–475.
Stuetzle, W. (1991). Odds plots: a graphical aid for finding associations between views of a data set. In: Buja, A., Tukey, P.A. (Eds.), *Computing and Graphics in Statistics*. In: *IMA Vols. Math. Appl.*, vol. 36. Springer-Verlag, pp. 207–217.
Swayne, D.F., Klinke, S. (1999). Introduction to the special issue on interactive graphical data analysis: what is interaction? *Comput. Statist.* **14**, 1–6.
Swayne, D.F., Cook, D., Buja, A. (1998). XGobi: interactive dynamic data visualization in the X window system. *J. Comput. Graph. Statist.* **7**, 113–130.
Symanzik, J., Majure, J.J., Cook, D. (1996). Dynamic graphics in a GIS: a bidirectional link between ArcView 20 and XGobi. In: *Computing Science and Statistics*, vol. 27, pp. 299–303.
Theus, M. (1996). Theorie und Anwendung interaktiver statistischer Graphik. PhD thesis. Universität Augsburg.
Theus, M. (1999). MONDRIAN – Interactive statistical graphics in Java. Manuscript.
Theus, M., Hofmann, H. (1998). Selection sequences in Manet. *Comput. Statist.* **13**, 77–87.
Theus, M., Lauer, S.R. (1997). Visualizing loglinear models. Manuscript.
Theus, M., Hofmann, H., Wilhelm, A.F. (1998). Selection sequences – interactive analysis of massive date sets. In: *Proceedings of the 29th Symposium on the Interface*. In: *Computing Science and Statistics*. Interface Foundation of North America, pp. 439–444.
Tierney, L. (1990). *LispStat – An Object-oriented Environment*. Wiley, New York.
Tremmel, L. (1995). The visual separability of plotting symbols in scatterplots. *J. Comput. Graph. Statist.* **4**, 101–112.
Tufte, E.R. (1983). *The Visual Display of Quantitative Information*. Graphics Press, Cheshire, CT.
Tufte, E.R. (1990). *Envisioning Information*. Graphics Press, Cheshire, CT.
Tufte, E.R. (1997). *Visual Explanations: Images and Quantities, Evidence and Narrative*. Graphics Press, Cheshire, CT.
Tukey, J.W. (1972). Some graphic and semigraphic displays. In: Bancroft, T. (Ed.), *Statistical Papers in Honor of George W. Snedecor*. Iowa State University, pp. 293–316.
Tukey, J.W. (1977). *Explorative Data Analysis*. Addison–Wesley, Reading, MA.
Tweedie, L., Spence, R. (1998). The prosection matrix: a tool to support the interactive exploration of statistical models and data. *Comput. Statist.* **13**.
Unwin, A.R. (1994). Regarding geographic data. In: Dirschedl, P., Ostermann, R. (Eds.), *Computational Statistics*. Physica, Heidelberg, pp. 315–326.
Unwin, A.R. (1995). Interaktive statistische Grafik – eine übersicht? In: *Appl. Statistics – Recent Developments*. Vandenhoeck and Rupprecht, Göttingen, pp. 177–183.
Unwin, A.R. (1999). Requirements for interactive graphics software for exploratory data analysis. *Comput. Statist.* **14**, 7–22.
Unwin, A., Unwin, D. (1998). Exploratory spatial data analysis with local statistics. *Statist.* **47**, 415–421.
Unwin, A.R., Hawkins, G., Hofmann, H., Siegl, B. (1996). Manet – missings are now equally treated. *J. Comput. Graph. Statist.* **5**, 113–122.
Velleman, P.F. (1995). *DataDesk 5.0 User's Guide*. Ithaca, NY.
Velleman, P.F., Hoaglin, D.C. (1981). *Applications, Basics, and Computing of Exploratory Data Analysis*. Duxbury Press, Boston, MA.
Ward, M.O. (1994). Xmdvtool: integrating multiple methods for visualizing multivariate data. In: *Visualization'94*. Washington, DC. IEEE, pp. 326–336.

Ward, M.O. (1998). Creating and manipulating n-dimensional brushes. In: *Proceedings of the ASA Section on Statistical Graphics*. Amer. Statist. Assoc.

Wegman, E.J. (1990). Hyperdimensional data analysis using parallel coordinates. *J. Amer. Statist. Assoc.* **85**, 664–675.

Wegman, E.J. (1991). The grand tour in k-dimensions. In: *Proceedings of the 22nd Symposium on the Interface*. In: *Computing Science and Statistics*. Interface Foundation of North America, pp. 127–136.

Wegman, E.J., Carr, D.B. (1993). Statistical graphics and visualization. In: Rao, C. (Ed.), *Handbook of Statistics*, vol. 9. Elsevier, pp. 857–958.

Wegman, E.J., DePriest, D.J. (Eds.) (1986). *Statistical Image Processing and Graphics*. Marcel Dekker.

Wegman, E.J., Luo, Q. (1997). High dimensional clustering using parallel coordinates and the Grand Tour. In: Klar, R., Opitz, O. (Eds.), *Classification and Knowledge Organization*. Springer-Verlag, pp. 93–101.

Wegman, E.J., Luo, Q. (2002). Visualizing densities. *J. Comput. Graph. Statist.* **11** (1), 137–162.

Wegman, E.J., Shen, J. (1993). Three-dimensional Andrews plots and the grand tour. In: *Computing Science and Statistics*.

Wilhelm, A., Sander, M. (1998). Interactive statistical analysis of dialect features. *Statist.* **47**, 445–455.

Wilhelm, A., Unwin, A.R., Theus, M. (1996). Software for interactive statistical graphics – a review. In: Faulbaum, F., Bandilla, W. (Eds.), *SoftStat'95 – Advances in Statistical Software 5*. Lucius and Lucius, Stuttgart, pp. 3–12.

Wilhelm, A.F., Wegman, E.J., Symanzik, J. (1999). Visual clustering and classification: the Oronsay particle size data set revisited. *Comput. Statist.* **14** (1), 109–146.

Wilkinson, L. (1997). A graph algebra. In: *Computing Science and Statistics*, vol. 28, pp. 341–351.

Wilkinson, L. (1999a). Dot plots. *Amer. Statist.* **53**, 276–281.

Wilkinson, L. (1999b). *A Grammar for Graphics*. Springer-Verlag.

Wills, G. (1992). Spatial data: exploration and modelling via distance-based and interactive graphics methods. PhD thesis. Trinity College, Dublin.

Wills, G. (1996). 524 288 ways to say this is interesting. In: *Proceedings of Visualization'96*. IEEE.

Wills, G. (1997). How to say 'this is interesting'. In: *Proceedings of Section of Stat. Graphics*. IEEE, pp. 25–31.

Wills, G. (1999a). Linked data views. *Statist. Comput. Graphics* **10** (1), 20–24.

Wills, G. (1999b). Nicheworks – interactive visualization of very large graphs. *J. Comput. Graph. Statist.* **8** (3).

Young, F.W., Faldowski, R.A., McFarlane, M.M. (1992). Vista: a visual statistics research and development testbed. In: *Graphics and Visualization, Proceedings of the 24th Symposium on the Interface*. In: *Computing Science and Statistics*. Interface Foundation of North America, pp. 224–233.

Young, F.W., Faldowski, R.A., McFarlane, M.M. (1993). Multivariate statistical visualization. In: Rao, C. (Ed.), *Handbook of Statistics*, vol. 9. Elsevier, pp. 959–998.

Data Visualization and Virtual Reality

Jim X. Chen

1. Introduction

We present a survey of basic principles and advanced technologies in data visualization, with a focus on Virtual Reality (VR) hardware and software. In order to understand visualization and VR, we should understand computer graphics first. Most data visualization today relies on computer graphics to present data in innovative ways, so that obscured abstract or complex information become obvious and easy to comprehend through images and/or animations. VR, also called VE (Virtual Environment) in some applications, is an extension to computer graphics to include other human senses, in addition to eyes, that perceive data, phenomena, or simulated virtual world from different perspectives. VR extends the 3D graphics world to include stereoscopic, acoustic, haptic, tactile, and other kinds of feedbacks to create a sense of immersion.

In the rest of the chapter, we first concisely introduce principles of computer graphics and graphics software tools (Chen, 2002, Foley et al., 1996), then provide some data visualization technologies and a few examples (Chen and Wang, 2001, Wang et al., 2002, Wegman and Carr, 1993, Wegman and Luo, 1997), after that discuss VR and software (Burdea and Coiffet, 2003, Stanney, 2002, Vince, 1995, Vince, 1998, Wegman, 2000, Wegman and Symanzik, 2002), and finally list some applications that exploit an integration of visualization and VR.

2. Computer graphics

A graphics *display device* is a drawing board with an array of fine points called pixels. At the bottom of a graphics system is a mechanism named *magic pen* here, which can be moved to a specific pixel position, and tune the pixel into a specific color (an RGB vector value). This magic pen can be controlled directly by hand through an input device (mouse, keyboard, etc.) as a simple paintbrush. In this case, we can draw whatever we imagine, but it has nothing to do with computer graphics and data visualization. Computer graphics, or simply graphics, is more about controlling this magic pen automatically through programming to generate images or animation: given input data (geometric objects, computational results, etc.), the application program will move the magic pen to turn the abstract data into objects in a 2D or 3D image. Here, we will explain what we need to provide to a graphics system and how a graphics system works in general.

2.1. Shape

In computer graphics, unlike drawing by hand, we need to provide accurate description of the shape of the object to be drawn: a list of vertices or a geometric description (such as a specific curve equation) which can be used to find all the points of the object. Therefore, instead of just having a picture, we have a *model* that describes the object as initial data saved in the memory. A *graphics library* or *package* provides a set of graphics commands or output subroutines. Graphics commands can specify primitive geometric models that will be digitized and displayed. Here "primitive" means that only certain simple shapes (such as points, lines, polygons) can be accepted by a graphics library. A very complex shape needs an application program to assemble small pieces of simple shapes. As you know, we have the magic pen that can draw a pixel. If we can draw a pixel (point), we can draw a line, a polygon, a curve, a block, a building, an airplane, etc. A general application program could be included into a graphics library as a command to draw a complex shape.

A graphics system digitizes a specific model into a set of discrete color points saved in a piece of memory called *frame buffer*. This digitization process is called *scan-conversion*. The color points in the frame buffer will be sent to the corresponding pixels in the display device by a piece of hardware called *video controller*. Therefore, whatever in the frame buffer corresponds to an image on the screen. The display device indicates both the frame buffer and the screen. The application program accepts user input, manipulates the model (create, store, retrieve, and modify the descriptions), and produces the picture through the graphics system. The display device is also a window for us to manipulate the image as well as the model behind the image through the application program. Most of the programmer's task concerns creating and editing the model, and handling user interaction in a programming language with a graphics library.

2.2. Transformation

After we have the model (description of the object), we may move it around or even transform the shape of the model into a completely different object. Therefore, we need to specify the rotation axis and angle, translation vector, scaling vector, or other manipulations to the model. The ordinary *geometric transformation* is a process of mathematical manipulations of each vertices of a model through matrix multiplication, and the graphics system then displays the final transformed model. The transformation can be predefined, such as moving along a planed trajectory, or interactive depending on user input. The transformation can be permanent, which means that the coordinates of the vertices are changed and we have a new model replacing the original one, or just temporarily. In many cases a model is transformed in order to display it at a different position/orientation, and the graphics system discard the transformed model after scan-conversion. Sometimes, all the vertices of a model go through the same transformation and the shape of the model is preserved; sometimes, different vertices go through different transformations, and the shape is dynamic.

In a movie theater, motion is achieved by taking a sequence of pictures and projecting them at 24 frames per second on the screen. In a graphics system, as we assumed, the magic pen moves at lightning speed. Therefore, we can first draw an object and keep it on display for a short period of time, then erase it and redraw the object after a small step of transformation (rotation, translation, etc.), and repeat this process. The resulting effect is that the object is animated on display. Of course, in order to achieve animation for complex objects, we need not only a fast pen, but also other high performance algorithms and hardware that carry out many graphics functions (including transformation and scan-conversion) efficiently. In order to achieve smooth transformation, two frame buffers are used in the hardware. While one frame buffer is used for displaying, the other is used for scan-conversion, and then they are swapped so the one that was displaying is used for scan-conversion, and the one that was used for scan-conversion is now for displaying. This hardware mechanism allows parallelism and smooth animation.

2.3. Viewing

The display device is a 2D window, and our model may be in 3D. When we specify a model and its transformations in a coordinate system, we need to consider the relationship between this coordinate system and the coordinates of the display device (or its corresponding frame buffer.) Therefore, we need a *viewing* transformation, the mapping of a part of the model's coordinate scene to the display device coordinates. We need to specify a viewing volume, which determines a projection method (*parallel* or *perspective*) – how 3D model is projected into 2D. The viewing volume for parallel projection is like a box. The result of parallel projection is a less realistic view, but can be used for exact measurements. The viewing volume for perspective projection is like a truncated pyramid, and the result looks more realistic in many cases, but does not preserve sizes on display – objects further away are smaller.

Through the graphics system, the model is clipped against the 3D viewing volume and then projected into a 2D plane corresponding to the display window. This transformation process is analogous to taking a photograph with a camera. The object in the outside world has its own coordinate system. Our film in the camera has its own coordinate system. By pointing and adjusting the zoom, we have specified a viewing volume and projection method. If we aim at objects very far away, we can consider the projection is parallel, otherwise perspective.

2.4. Color and lighting

We can specify one color for a model, or different colors for different vertices. The colors of the pixels between the vertices are interpolated by the graphics system.

On the other hand, we can specify color indirectly by setting light sources of different colors at different positions in the environment and material properties, so the model will be lit accordingly. When we specify light sources, we need also specify the material properties (*ambient*, *diffuse*, and *specular* parameters) of the model, so the model can have overall brightness, dull reflections, and shiny spots. The graphics system achieves *lighting* (namely, *illumination* or *shading*) by calculating the color, or more precisely, the RGB vector of each individual pixel in the model. It uses empirical methods to integrate the material properties, the normal of the pixel, the position of the viewer, and the colors and positions of the light sources. The normal of a pixel tells the system which direction the pixel is facing. It can be calculated/interpolated by the normals of the surrounding vertices of the pixel in the model. It can be either provided or calculated from its corresponding surface equation.

A more expensive method, which is generally not provided in a low-level graphics library, is to calculate lightness by *ray-tracing*. A ray is sent from the view point through a point on the projection plane to the scene to calculate the recursive intersections with objects, bounces with reflections, penetrations with refractions, and finally the color of the point by accumulating all fractions of intensity values from bottom up. The drawback is that the method is time consuming.

2.5. Texture mapping

When calculating the color of a pixel, we can also use or integrate a color retrieved from an image. To do this, we need to provide a piece of image (array of RGB values) called texture. Therefore, *texture mapping* is to tile a surface of a model with existing texture. At each rendered pixel, selected texture pixels (*texels*) are used either to substitute or scale one or more of the surface's material properties, such as its diffuse color components. Textures are generated from real images scanned into memory or by any other means.

2.6. Transparency

When we have transparent objects, or even want to visualize the contents of an opaque object, we need to specify the *transmission coefficients* for overlapping objects. A final pixel's R, G, or B value depends on the blending of the corresponding R, G, or B values of the pixels in the overlapping objects through the transmission coefficients: $I_\lambda = \alpha_1 I_{\lambda 1} + \alpha_2 I_{\lambda 2}$, where λ is R, G, or B; $I_{\lambda 1}$ the color component of the pixel in the first object, α_1 the transmission coefficient of the pixel in the first object, and similarly for the second object.

2.7. Graphics libraries

A graphics system includes a graphics library and the corresponding hardware. A graphics library includes most of the basic components we have described in this section. Some of the functions are more or less implemented in hardware, otherwise they would be very slow. Some of the functions requires us to provide input, as a list of vertices for drawing a polygon, and others we do not, as removing hidden surfaces. In a graphics library, like controlling the pen by hand, we may draw pixels directly into frame buffer for display without going through transformation and viewing. In general, we provide vertices for primitives (points, lines, or polygons), and they all go through similar operations: transformation, viewing, lighting, scan-conversion, etc. OpenGL is currently the most popular low-level graphics library. Direct3D and PHIGS are two other well-known graphics libraries.

3. Graphics software tools

A low-level graphics library or package is a software interface to graphics hardware, namely the application programmer's interface (API). All graphics tools or applications are built on top of a low-level graphics library, such as visualization, rendering, and VR tools (http://cs.gmu.edu/~jchen/graphics/book/index.html), which we discuss in more detail in this section (Chen, 2002). A taxonomy of graphics tools can be found at http://cs.gmu.edu/~jchen/graphics/book/index_tools_catagories.html.

3.1. Visualization

Visualization employs graphics to make pictures that give us insight into certain abstract data and symbols. The pictures may directly portray the description of the data, or completely present the content of the data in an innovative form. Many visualization tools have multiple and overlapping visualization functions. Some tools include capabilities of interactive modeling, animation, simulation, and graphical user interface construction as well. In the following, we briefly introduce several visualization tools.

AVS/Express, IRIS Explorer, Data Explorer, MATLAB, PV-WAVE, Khoros, and Vtk are multiple purpose visualization commercial products that satisfy most of the visualization needs. *AVS/Express* has applications in many scientific areas, including engineering analysis, CFD, medical imaging, and GIS (Geographic Information Systems). It is built on top of OpenGL and runs on multiple platforms. *IRIS Explorer* includes visual programming environment for 3D data visualization, animation and manipulation. IRIS Explorer modules can be plugged together, which enable users to interactively analyze collections of data and visualize the results. IRIS Explorer is build on top of *OpenInventor*, an interactive 3D object scene management, manipulation, and animation tool. OpenInventor has been used as the basis for the emerging Virtual Reality Modeling Language (VRML). The rendering engine for IRIS Explorer and OpenInventor are OpenGL. IBM's *Data Explorer (DX)* is a general-purpose software package for data visualization and analysis. OpenDX is the open source software version of the DX Product. DX is build on top of OpenGL and runs on multiple platforms. *MATLAB* was originally developed to provide easy access to matrix software. Today, it is a powerful simulation and visualization tool used in a variety of application areas including signal and image processing, control system design, financial engineering, and medical research. *PV-WAVE* integrates charting, volume visualization, image processing, advanced numerical analysis, and many other functions. *Khoros* is a software integration, simulation, and visual programming environment that includes image processing and visualization. *Vtk* is a graphics tool that supports a variety of visualization and modeling functions on multiple platforms. In Vtk, applications can be written directly in C++ or in Tcl (an interpretive language).

Volumizer, 3DVIEWNIX, ANALYZE, and VolVis are 3D imaging and volume rendering tools. *Volume rendering* is a method of extracting meaningful information from a set of volumetric data. For example, a sequence of 2D image slices of human body can be reconstructed into a 3D volume model and visualized for diagnostic purposes or for planning of treatment or surgery. NIH's Visible Human Project (http://www.nlm.nih.gov/research/visible/visible_human.html) creates anatomically detailed 3D representations of the human body. The project includes effort of several universities and results in many imaging tools.

StarCD, FAST, pV3, FIELDVIEW, EnSight, and Visual3 are CFD (Computational Fluid Dynamics) visualization tools. Fluid flow is a rich area for visualization applications. Many CFD tools integrate interactive visualization with scientific computation of turbulence or plasmas for the design of new wings or jet nozzles, the prediction of atmosphere and ocean activities, and the understanding of material behaviors.

NCAR, Vis5D, FERRET, Gnuplot, and SciAn are software tools for visual presentation and examination of data sets from physical and natural sciences, often requiring the integration of terabyte or gigabyte distributed scientific databases with visualization. The integration of multi-disciplinary data and information (e.g. atmospheric, oceanographic, and geographic) into visualization systems will help and support cross-disciplinary explorations and communications.

3.2. Modeling and rendering

Modeling is a process of constructing a virtual 3D graphics object (namely, computer model, or simply model) from a real object or an imaginary entity. Creating graphics models require a significant amount of time and effort. Modeling tools make creating and constructing complex 3D models easy and simple. A graphics model includes geometrical descriptions (particles, vertices, polygons, etc.) as well as associated graphics attributes (colors, shadings, transparencies, materials, etc.), which can be saved in a file using certain standard (3D model) formats. Modeling tools help create virtual objects and environments for CAD (computer-aided design), visualization, education, training, and entertainment. *MultigenPro* is a powerful modeling tool for 3D objects and terrain generation/editing. *AutoCAD* and *MicroStation* are popular for 2D/3D mechanical designing and drawing. *Rhino3D* is for free-form curve surface objects.

Rendering is a process of creating images from graphics models (Watt and Watt, 1992). 3D graphics models are saved in computer memory or hard-disk files. The term *rasterization* and *scan-conversion* are used to refer to low-level image generation or drawing. All modeling tools provide certain drawing capabilities to visualize the models generated. However, in addition to simply drawing (scan-converting) geometric objects, rendering tools often include lighting, shading, texture mapping, color blending, ray-tracing, radiosity, and other advanced graphics capabilities. For example, *RenderMan* Toolkit includes photo-realistic modeling and rendering of particle system, hair, and many other objects with advanced graphics functions such as ray-tracing, volume display, motion blur, depth-of-field, and so forth. Some successful rendering tools were free (originally developed by excellent researchers at their earlier career or school years), such as POVRay, LightScape, Rayshade, Radiance, and BMRT. *POVRay* is a popular ray-tracing package across multiple platforms that provides a set of geometric primitives and many surface and texture effects. *LightScape* employs radiosity and ray-tracing to produce realistic digital images and scenes. *Rayshade* is an extensible system for creating ray-traced images that includes a rich set of primitives, CSG (constructive solid geometry) functions, and texture tools. *Radiance* is a rendering package for the analysis and visualization of lighting in design. It is employed by architects and engineers to predict illumination, visual quality and appearance of design spaces, and by researchers to evaluate new lighting technologies. *BMRT* (Blue Moon Rendering Tools) is a RenderMan-compliant ray-tracing and radiosity rendering package. The package contains visual tools to help users create RenderMan Input Bytestream (RIB) input files.

Many powerful graphics tools include modeling, rendering, animation, and other functions into one package, such as Alias|Wavefront's Studio series and Maya, SoftImage, 3DStudioMax, LightWave, and TrueSpace. It takes serious course training to use these tools. Alias|Wavefront's *Studio* series provide extensive tools for industrial design, automotive styling, and technical surfacing. Its *Maya* is a powerful and productive 3D software for character animation that has been used to create visual effects in some of the hottest recent film releases, including A Bug's Life and Titanic. *SoftImage3D* provides advanced modeling and animation features such as NURBS, skin, and particle system that are excellent for special effects and have been employed in many computer games and films, including stunning animations in Deep Impact and Airforce One. *3DStudioMax* is a popular 3D modeling, animation, and rendering package on Window 95/NT platform for game development. Its open plugin architecture makes it an idea platform for third party developers. *LightWave* is another powerful tool that has been successfully used in many TV feature movies, games, and TV commercials. *TrueSpace* is another popular and powerful 3D modeling, animation, and rendering package on Win95/NT platforms.

3.3. Animation and simulation

In a movie theater, animation is achieved by taking a sequence of pictures, and then projecting them at 24 frames per second on the screen. Videotape is shown at 30 frames per second. Computer animation is achieved by refreshing the screen display with a sequence of images at 60–100 frames per second. Of course, we can use fewer than 60 images to refresh 60 frames of screen in a second, so that some images will be used to refresh the display more than one time. The fewer the number of different images displayed in a second, the jerkier the animation will be. Keyframe animation is achieved by pre-calculate keyframe images and in-between images, which may take significant amount of time, and then display (play-back) the sequence of generated images in real time. Keyframe animation is often used in visual effects in films and TV commercials, which no interactions or unpredictable changes are necessary. Interactive animation, on the other hand, is achieved by calculating, generating, and displaying the images at the same time on the fly. When we talk about real-time animation, we mean the virtual animation happens in the same time frames as in real world behavior. However, for graphics researchers, real-time animation often simply implies the animation is smooth or interactive. Real-time animation is often used in virtual environment for education, training, and 3D games. Many modeling and rendering tools are also animation tools, which are often associated with simulation.

Simulation, on the other hand, often means a process of generating certain natural phenomena through scientific computation. The results of the simulations may be large datasets of atomic activities (positions, velocities, pressures and other parameters of atoms) or fluid behaviors (volume of vectors and pressures). Computer simulation allows scientists to generate the atomic behavior of certain nanostructured materials for understanding material structure and durability and to find new compounds with superior quality (Nakano et al., 1998). Simulation integrated with visualization can help pilots to learn to fly and aid automobile designers to test the integrity of the passenger compartment driving crashes. For many computational scientists, simulation may not be related to any visualization at all. However, for many graphics researchers, simulation often means simply animation. Today, graphical simulation, or simply simulation, is an animation of certain process or behavior that are generated often through scientific computation and modeling. Here we emphasize an integration of simulation and animation – the simulated results are used to generate graphics model and control animation behaviors. It is far easier, cheaper, and safer to experiment with a model through simulation than with a real entity. In fact, in many situations, such as training space-shuttle pilots and studying molecular dynamics, modeling and simulation are the only feasible method to achieve the goals. *Real-time simulation* is an overloaded term. To computational scientists, it often means the simulation time is the actual time in which the physical process (under simulation) should occur. In automatic control, it means the output response time is fast enough for automatic feedback and control. In graphics, it often means that the simulation is animated at an interactive rate of our perception. The emphasis in graphics is more on responsiveness and smooth animation rather than strictly accurate timing of the physical process. In many simulations for training applications, the emphasis is on generating realistic behavior for interactive training environment rather than strictly scientific or physical computation.

IRIS Performer is a toolkit for real-time graphics simulation applications. It simplifies development of complex applications such as visual simulation, flight simulation, simulation-based design, virtual reality, interactive entertainment, broadcast video, CAD, and architectural walk-through. *Vega* is MultiGen-Paradigm's software environment for real-time visual and audio simulation, virtual reality, and general visualization applications. It provides the basis for building, editing, and running sophisticated applications quickly and easily. *20-sim* is a modeling and simulation program for electrical, mechanical, and hydraulic systems or any combination of these systems. *VisSim/Comm* is a Windows-based modeling and simulation program for end-to-end communication systems at the signal or physical level. It provides solutions for analog, digital, and mixed-mode communication system designs. *SIMUL8* is a visual discrete event simulation tool. It provides performance measures and insights into how machines and people will perform in different combinations. *Mathematica* is an integrated environment that provides technical computing, simulation, and communication. Its numeric and symbolic computation abilities, graphical simulation, and intuitive programming language are combined with a full-featured document processing system. As we discussed earlier *MATLAB*, *Khoros*, and many other tools contains modeling and simulation functions.

3.4. File format converters

Currently, 3D model file formats and scene description methods are numerous and different. Different modeling tools create models with a variety of attributes and parameters, and use their own descriptions and formats. It happens that we are unable to use certain models because we cannot extract certain information from the files. There are 3D model and scene file format converting tools available (Chen and Yang, 2000), such as *PolyTran*, *Crossroads*, *3DWin*, *InterChange*, *Amapi3D*, *PolyForm*, *VIEW3D*, and *Materialize3D*. Some attributes and parameters unique to certain formats will be lost or omitted for simplicity in the conversions.

3.5. Graphics user interfaces

Graphical user interface tools, such as *Motif*, *Java AWT*, *Microsoft Visual C++ MFC/Studio*, *XForms*, and *GLUI*, are another category of graphics tools. Computer graphics covers a wide range of applications and software tools. The list of software tools is growing exponentially when we try to cover more. We list all the software tools covered in this chapter at http://cs.gmu.edu/graphics/.

4. Data visualization

Computer graphics provides the basic functions for generating complex images from abstract data. Visualization employs graphics to give us insight into abstract data. Apparently, a large variety of techniques and research directions exist in data visualization. Choosing a specific technique is mostly determined by the type and the dimension of the data to be visualized. The generally discussed research areas include volume visualization for volumetric data, vector fields and flow visualization for computational fluid dynamics, large data-set visualization for physical and natural sciences, information visualization for text libraries, and statistical visualization for statistical data analysis and understanding. Many visualization techniques deal with multidimensional multivariate data sets.

4.1. Data type

A visualization technique is applicable to data of certain types (discrete, continual, point, scalar, or vector) and dimensions (1D, 2D, 3D, and multiple: ND). *Scatter Data* represent data as discrete points on a line (1D), plane (2D), or in space (3D). We may use different colors, shapes, sizes, and other attributes to represent the points in high dimensions beyond 3D, or use a function or a representation to transform the high-dimensional data into 2D/3D. *Scalar Data* have scalar values in addition to dimension values. The scalar value is actually a special additional dimension that we pay more attention. 2D diagrams like histograms, bar charts, or pie charts are 1D scalar data visualization methods. Both histograms and bar charts have one coordinate as the dimension scale and another as the value scale. Histograms usually have scalar values in confined ranges, while bar charts do not carry this information. Pie charts use a slice area in a pie to represent a percentage. 2D contours (iso-lines in a map) of constant values, 2D images (pixels of x–y points and color values), and 3D surfaces (pixels of x–y points and height values) are 2D scalar data visualization methods. Volume and iso-surface rendering methods are for 3D scalar data. A *voxel* (volume pixel) is a 3D scalar data with

(x, y, z) coordinates and an intensity or color value. *Vector Data* include directions in addition to scalar and dimension values. We use line segments, arrows, streamlines, and animations to present the directions.

4.2. Volumetric data – volume rendering

Volume rendering or visualization is a method of extracting meaningful information from a set of 2D scalar data. A sequence of 2D image slices of human body can be reconstructed into a 3D volume model and visualized for diagnostic purposes or for planning of treatment or surgery. For example, a set of volumetric data such as a deck of Magnetic Resonance Imaging (MRI) slices or Computed Tomography (CT) can be blended into a 2D X-ray image by firing rays through the volume and blending the voxels along the rays. This is a rather costly operation and the blending methods vary. The concept of volume rendering is also to extract the contours from given data slices. An iso-surface is a 3D constant intensity surface represented by triangle strips or higher order surface patches within a volume. For example, the voxels on the surface of bones in a deck of MRI slices appear to be the same intensity value.

4.3. Vector data – fluid visualization

From the study of turbulence or plasmas to the design of new wings or jet nozzles, flow visualization motivates much of the research effort in scientific visualization. Flow data are mostly 3D vectors or tensors of high dimensions. The main challenge of flow visualization is to find ways of visualizing multivariate data sets. Colors, arrows, particles, line convolutions, textures, surfaces, and volumes are used to represent different aspects of fluid flows (velocities, pressures, streamlines, streaklines, vortices, etc.)

4.4. Large datasets – computation and measurement

The visual presentation and examination of large data sets from physical and natural sciences often require the integration of terabyte or gigabyte distributed scientific databases with visualization. Genetic algorithms, radar range images, materials simulations, atmospheric and oceanographic measurements are among the areas that generate large multidimensional/multivariate data sets. The variety of data comes with different data geometries, sampling rates, and error characteristics. The display and interpretation of the data sets employ statistical analyses and other techniques in conjunction with visualization. The integration of multi-disciplinary data and information into visualization systems will help cross-disciplinary explorations and communications.

4.5. Abstract data – information visualization

The field of information visualization includes visualizing retrieved information from large document collections (e.g., digital libraries), the World Wide Web, and text databases. Information is completely abstract. We need to map the data into a physical space that will represent relationships contained in the information faithfully and efficiently. This could enable the observers to use their innate abilities to understand through spatial relationships the correlations in the library. Finding a good spatial representation of the information at hand is one of the most challenging tasks in information visualization.

Many forms and choices exist for the visualization of 2D or 3D data sets, which are relatively easy to conceive and understand. For data sets that are more than 3D, visualization is still the most promising as well as challenging research area.

4.6. Interactive visualization

Interactive visualization allows us to visualize the results or presentations interactively in different perspectives (e.g., angles, magnitude, layers, levels of detail, etc.), and thus help us to understand the results on the fly better. GUI is usually needed for this interaction. Interactive visualization systems are most effective when the results of models or simulations have multiple or dynamic forms, layers, or levels of detail. Thus interactive visualization helps users interact with visual presentations of scientific data and understand the different aspects of the results. Interactive visualization makes the scientist an equal partner with the computer to manipulate and maneuver the 3D visual presentations of the abstract results.

4.7. Computational steering

Many software tools, packages, and interactive systems have been developed for scientific visualization. However, most of these systems are limited to post-processing of datasets. Any changes and modifications made to the model, input parameters, initial conditions, boundary conditions, or numerical computation considerations are accomplished off-line, and each change made to the model or input parameters causes most of the other steps and computation processes to repeat. In many cases the computations are done after hours of numerical calculations, only to find the results do not help solving the problems or the initial value of a parameter has been wrong. We would prefer to have a system in which all these computational components were linked so that, parameters, initial and boundary conditions, and all aspects of the modeling and simulation process could be steered, controlled, manipulated, or modified graphically through interactive visualization.

The integration of computation, visualization, and control into one tool is highly desirable, because it allows users to interactively "steer" the computation. At the beginning of the simulation before there is any result generated, a few important feedbacks will significantly help choose correct parameters and initial values. One can visualize some intermediate results and key factors to steer the simulation in the right direction. With computational steering, users are able to modify parameters in their systems as the computation progress, and avoid something being wrong or uninteresting after long hours of tedious computation. Therefore, computational steering is an important tool to adjust uncertain parameters, to steer the simulation in the right direction, and to fine tune the results. Implementation of a computational steering framework requires a successful integration of many aspects of visual supercomputing, including numerical computation, system analysis, and interactive visualization, all of which need to be effectively coordinated within an efficient computing environment.

4.8. Parallel coordinates

Parallel coordinates method represents a d-dimensional data as values on d coordinates parallel to the x-axis equally spaced along the y-axis (Figure 1, or the other way around rotating 90 degrees). Each d-dimensional data corresponds to the line segments between the parallel coordinates connecting the corresponding values. That is, each polygonal line of $p - 1$ segments in the parallel coordinates represents a point in d-dimensional space. Wegman was the pioneer in giving parallel coordinates meaningful statistical interpretations (Wegman, 1990). Parallel coordinates provides a means to visualize higher-order geometries in an easily recognizable 2D representation, which in addition helps find the patterns, trends, and correlations in the data set (Wegman and Luo, 1997).

The purpose of using parallel coordinates is to find certain features in the data set through visualization. Consider a series of points on a straight line in Cartesian coordinates: $y = mx + b$. If we display these points in a parallel coordinates, the points on a line in Cartesian coordinates become line segments in parallel coordinates. These line segments intersect at a point. This point in the parallel coordinates is called the dual of the line in the Cartesian coordinates. The point–line duality extends to conic sections. An ellipse in Cartesian coordinates maps into a hyperbola in parallel coordinates and vice versa. Rotations in Cartesian coordinates become translations in parallel coordinates and vice versa. More importantly, clustering is easily isolated and visualized in parallel coordinates. An individual parallel coordinate axis represents a 1D projection of the data set. Thus, separation between or among sets of data on one axis represents a view of the data of isolated clusters.

The brushing technique is to interactively separate a cluster of data by painting it with a unique color. The brushed color becomes an attribute of the cluster. Different clusters can be brushed with different colors and relations among clusters can then be visually detected. Heavily plotted areas can be blended with color mixes and transparencies. Animation of the colored clusters through time allows visualization of the data evolution history.

Fig. 1. Parallel coordinates (courtesy of E.J. Wegman and Q. Luo, 1997).

The tour method is to search for patterns by looking at the high dimensional data from all different angles (Buja and Asimov, 1985, Wegman and Luo, 1997). That is, to project the data into all possible d-planes through generalized rotations. The purpose of the grand tour animation is to look for unusual configurations of the data that may reflect some structure from a specific angle. The rotation, projection, and animation methods vary depending on specific assumptions and implementations. Currently, there are multidimensional data visualization tools that include parallel coordinates and grand tours available completely or partially for free, such as such as

ExplorN (ftp://www.galaxy.gmu.edu/pub/software/ExplorN_v1.tar),
CrystalVision (ftp://www.galaxy.gmu.edu/pub/software/CrystalVisionDemo.exe), and
XGobi (http://www.research.att.com/areas/stat/xgobi/).

The parallel coordinates method, which handles multiple dimensional data, is limited to data sets that have a few dimensions more than 3D. For data sets that have more or less hundreds of dimensions, parallel coordinates will be crowded and visualization will be difficult. Also, parallel coordinates treat every coordinate equally. There is no emphasis of certain scalar or vector values. It may be better to emphasis certain scalar values separately in different forms.

4.9. Linked micromap plots

Linked Micromap plots (LM plots) constitute a new template for the display of spatially indexed statistical summaries (Carr and Olsen, 1998). This template has four key features: displaying at least three parallel sequences of panels (micromap, label, and statistical summary) that are linked by position, sorting the units of study, partitioning the study units into panels to focus attention on a few units at a time, and linking the highlighted study units across corresponding panels of the sequences. Displaying LM plots on the Internet introduces a new and effective way of visualizing various statistical summaries. A reader can easily access and view the statistical data in LM plots everywhere around the World. A demo system with all the features is available on the website at http://graphics.gmu.edu/~jchen/cancer/. The following functions are implemented:

Display panels and study units. The LM plots for displaying the cancer summary statistics have four parallel sequences of display panels (columns), which are US/State micromap, State/County name, and two cancer statistical summaries. Figure 2 is a snapshot showing the basic layouts of the cancer statistical LM plots. In a row of a linked study unit, the geographic location, the leading dot before the name, and the statistics are all in the same color. On each statistics panel, the corresponding statistical value is shown as a dot with the confidence interval (CI) or bound displayed as line segments on both sides of the dot.

Sorting the study units. When sorting on a statistical variable, the displayed dot curve can show the relative relationships among the study units clearly. A sorting button with a triangle icon is located in front of the corresponding panel title. An up-triangle icon on the button represents an ascending sorting, and a down-triangle icon shows a descending sorting. Figure 2 shows that the study units (states) are sorted by one statistical summary (cancer mortality rate) in the descending order.

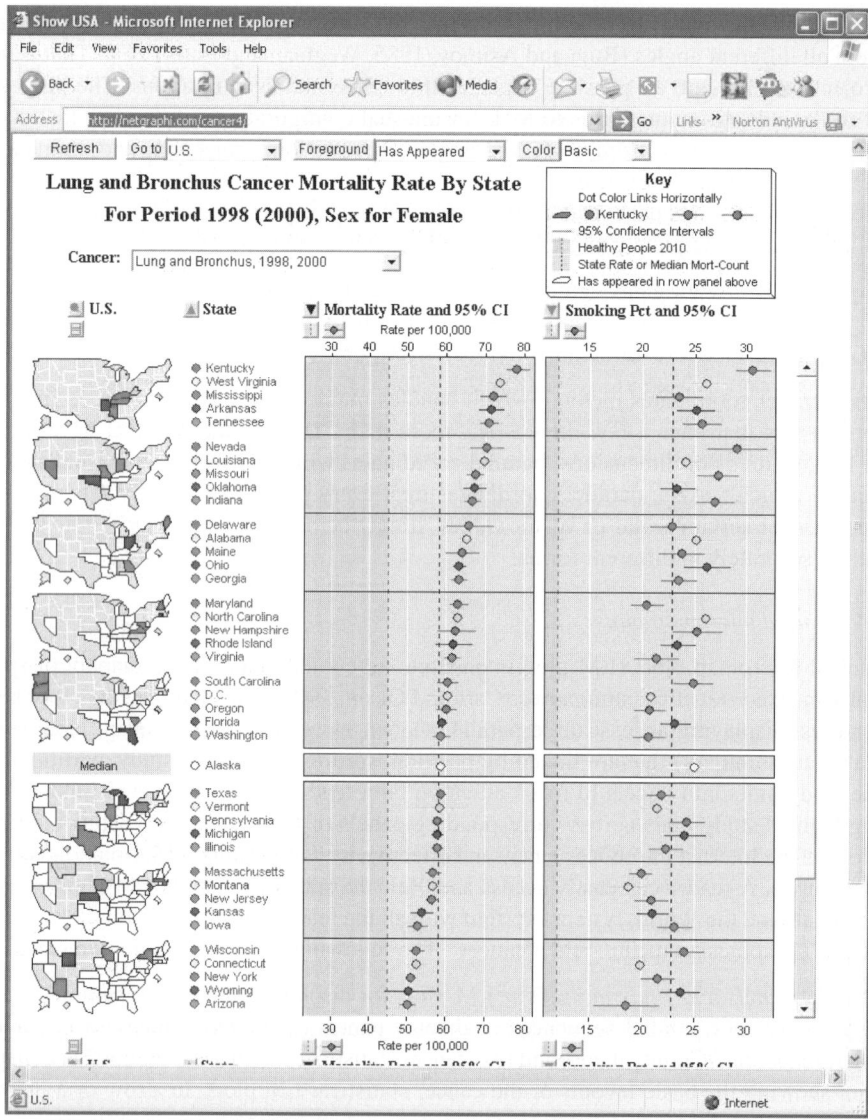

Fig. 2. A snapshot of the LM plots of the US cancer statistics. For a color reproduction of this figure see the color figures section, page 607.

Linking the related elements of a study unit. All the elements across the corresponding panels of the sequences are represented by the same color. When the user moves the mouse cursor onto any one of the related elements (micromap, unit name, or colored dots), all the linked elements in a study unit will blink. Meanwhile, the corresponding statistical summaries are displayed in text at the bottom of the browser window (in the browser's status line).

Grouping the study units. The study units are partitioned into a number of sub-panels vertically. This helps to focus attention on a few units at a time. In each group, a coloring scheme of five different colors is assigned to the five study units, respectively. This can show the linked elements of one study unit within a group more clearly.

Micromap magnification. Through the micromaps, a reader can find the locations of the study units intuitively. Some states may include many counties (study units). For example, Texas has over 200 counties. Since each micromap occupies a small fixed area, the area of a study unit in a micromap may be too small for a reader to see its shape. Magnifying the micromap helps the reader to view all study units clearly.

Drill-down and navigation. Drill-down is an operation that zooms in from a high-level LM plot to the corresponding low-level LM plot to view the detailed statistical data and the relationships among different hierarchical levels of detail. When using LM plots to visualize the national cancer statistical summaries, the US LM plot provides a starting place for subsequent drill-down to any other state LM plot. In the US LM plot, the state names and all the micromap regions are the active links to drill down to the corresponding state LM plots.

Overall look of the statistical summaries. When the number of the study units is bigger than the maximum that the display panel can hold, some study units will not be in the display panel. At this time, the reader cannot see the whole statistical curves formed by the dots shown in the statistical panels. Therefore, the overall look of the whole pattern presents very useful information. We have added a pop-up window to display the pattern in a scaled down fashion.

Displaying different statistical data sets. Interactively displaying different statistical data allows a reader to view and compare the relationships among different statistical results. In our system, we provide a pull-down menu for a reader to choose and view several different statistical cancer data sets. Once a reader selects a new cancer type, the data will be downloaded from the web-server and displayed in the display panel with a new calculated scale.

Statistical data retrieval. In a statistical visualization application, efficient retrieval of the statistical data is very important. In a statistical visualization system, the statistical data may be saved in different formats, places or modes to expedite retrieval. In our application, the statistical data is saved in the file mode and on the web-server. We do not need to write any code for the web-server. Instead, the Java applet directly retrieves the cancer data from the files.

The above set of web-based interactive LM plots provides a statistical data visualization system that integrates geographical data manipulation, visualization, interactive statistical graphics, and web-based Java technologies. The system is effective in presenting the complex and large-volume sample data on the national cancer statistics. With some modifications, the web-based interactive LM plots can be easily applied to visualize many other spatially indexed statistical datasets over the Internet.

4.10. Genetic algorithm data visualization

Genetic algorithms are search algorithms of better individuals based on the mechanics of natural selection and natural genetics (De Jong, 1990). The struggle for existence and better fitness that we see in nature is inherent in the algorithms. Some individuals have a higher probability of being recombined into a new generation and survive, while others have a higher probability of being eliminated from the population. To determine how good or bad a particular individual is, the genetic algorithm evaluates its *fitness* through the use of appropriate metrics on different fields. Each individual is an array of fields and a fitness value. Given an individual, its fitness is defined by the values of its fields. Many genetic algorithms produce data (individuals) of hundreds of fields with large population sizes over a long time (many generations). People are interested in visualizing the evolution of the population clusters and searching for the individuals that survive better, so to find good solutions to the target problem. The parallel coordinates method is limited to data of relatively ($p < 100$) low number of dimensions. We have to find some other methods to effectively visualize the very high-dimensional multivariate evolution data set. Also, the fitness is an important attribute in the visualization, which should be displayed in a separate scale. In the following, we introduce a dimension reduction method that transforms an arbitrary high-dimensional data into 2D with the third dimension representing the fitness.

Each individual encoding is a possible solution in a given problem space. This space, referred to as the search space, comprises all possible dimension field values and solutions to the problem at hand. We introduce a method called Quad-Tree Mapping (QTM) that uses the quad-tree subdivision method to map an ND data with a fitness value into a 3D scatter data at multi-levels of detail. Using animation and zooming techniques, the search space and the evolution process can be visualized. QTM is discussed in detail as follows.

Given an ND data, we use an m-bit binary value to represent each field value. At this point of time, we only consider the most significant bit in each field. Therefore, the N-dimension data is an N-bit binary. We take the first 2 bits in the data as the first level quad code, which has four possible values (00, 01, 10, and 11) placing the data into one of the four quadrants on a 2D rectangular plane (as shown in Figure 3). Then, we divide each quadrant into 4 sub-quadrants, and apply the next 2 bits in the data as the second level quad code placing the data into one of the four sub-quadrants on the 2D plane. For example, if the first two bits are 10, then the data will be in one of the upper-left sub-quadrants. This subdivision process continues recursively until all the bits are evaluated. If N is an odd number, we cut the last sub-quadrant into half instead of four pieces to make a decision. For each N-bit binary, there is a unique point on the 2D plane. Given

1010	1011	11
1000	1001	
00		01

Fig. 3. QTM subdivision to map an N-dimensional data into a 2D space.

a point on a plane, we can calculate and find its corresponding N-bit binary data as well.

We can draw the 2D quad plane on the x–z plane and raise each data along the y-axis direction according to its fitness value. This way, we have a scatter plot of the whole gene population, search space, as well as the fitness values.

Because we only used one bit in the m-bit data value, there may be many overlapping points. This problem can be solved by the following methods. First, we can zoom-in to a specific point (area), which will invoke displaying the second bit of the original data set that have the same first N-bit array. Again, we can use the same method to display the second N-bit data. Therefore, we can zoom in and out m different levels corresponding to m bits of data. Second, after finishing the QTM subdivision for the first N-bit array, we can continue the subdivision with the second N-bit array. The process continues until we finish the mth N-bit array. Third, we can consider N m-bit data in a row. That is, we use the m-bit first field as an input to the QTM, and the second field, and so on. There are pros and cons for each of the methods. They can be interchanged or combined in a visualization tool. Of course, the plane can be divided into 6, 10 or other numbers of rectangles as well, instead of a fixed 4 (quadrants). Our CRT display usually has confined number of pixels (1280 × 1024). Therefore, if we subdivide an area so many times that the distance between the points are shorter than a pixel, the display will not able to do any better. Therefore, depending on the area used, the number of fields N, and the number of bits m of the given data set, we have to decide the specific criteria of subdivision or levels of detail.

After the QTM transformation, we have 3D scatter points of the individuals. At a low-level of detail where no overlapping points exit, we can use space curve fitting to find a fitness surface that smoothly goes through all the individual points. The surface represents the search space with all solutions. The peaks and ridges represent those

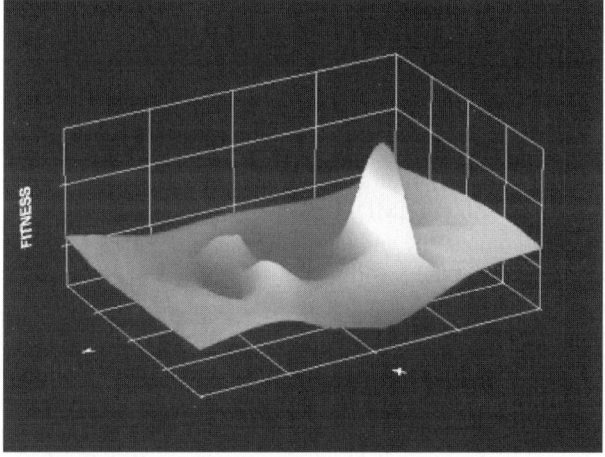

Fig. 4. An example of dimension reduction (QTM). For a color reproduction of this figure see the color figures section, page 608.

individual values (and possible solutions to the target problem) that are good. The bottoms and valleys represent the individuals that are likely to terminate. Figure 4 is an example of data with 226 fields, population size = 3, number of generation = 100, gene pool data is storied in an array of [300, 226].

5. Virtual reality

Virtual Reality (VR) extends 3D graphics world to include stereoscopic, acoustic, haptic, tactile, and other feedbacks to create a sense of immersion. A 3D image is like an ordinary picture we see, but a stereo image gives a strong sense of depth in 3D. It is generated by providing two slightly different views (images) of the same object to our two eyes separately. The head-mounted device (HMD), the ImmersaDesk/CAVE, and the VREX stereo projectors are different kinds of display devices for stereo images. A HMD has two separate display channels/screens to cover our two eyes. An ImmersaDesk or CAVE has only one channel like an ordinary display screen, except that it displays two different images alternatively for our two eyes. Viewers wear lightweight liquid crystal shutter glasses. These glasses activate each eye in succession. The glasses are kept synchronized with the two images through an infrared emitter. CAVE is the predecessor of ImmersaDesk, which is more expensive and has multiple display screens surrounding the viewers. An ImmersaDesk can be considered to be a one-wall CAVE. VREX's stereo projectors generate two images at the same time that can be viewed through lightweight, inexpensive polarized glasses. Wegman and his colleagues developed a mini-CAVE, which is a PC-based multiple stereo display system that saves both cost and space (Wegman, 2000, Wegman and Symanzik, 2002). Their papers (Wegman and Carr, 1993, Wegman, 2000, Wegman and Symanzik, 2002, Wegman and Luo, 2002) include valuable information concerning stereoscopic data visualization methods. They also include advantages and disadvantages of 1-wall versus 4-wall CAVES and an extensive history of stereoscopic visualization.

5.1. Hardware and software

The key hardware technologies in achieving VR are real-time graphics, stereo displays/views, tracking sensors, sound machines, and haptic devices. Real-time graphics (computer) and stereo displays (HMD, ImmersaDesk, CAVE, or VREX projectors) allow us to view stereoscopic scene and animation, and provide us a sense of immersion. Tracking sensors, which get the position and orientation of the viewer's head, hands, body parts, or other inputs, will enable us manipulate models and navigate in the virtual environment. Sound machines provide a sense of locations and orientations of certain objects and activities in the environment. Like sound machines, haptic devices vibrate and touch user's body, generating another feedback from the virtual environment in addition to stereoscopic view and 3D sound, enhancing the sense of immersion.

Some VR software tools are available that recognize well-defined commercial tracking sensors, sound machines, and haptic devices, in addition to functions in developing 3D virtual environment. Sense8's WorldToolKit and World_Up are cross-platform software development system for building real-time integrated 3D applications. World-ToolKit also supports network-based distributed simulations, CAVE-like immersive display options, and many interface devices, such as HMDs, trackers, and navigation controllers. Lincom's VrTool is an OpenInventor-based toolkit to provide a rapid prototyping capability to enable VR users to quickly get their application running with the minimum amount of effort. MultiGen-Paradigm's Vega is a real-time visual and audio simulation software tool that includes stereo imaging. MR (Minimal Reality) Toolkit by the graphics group at University of Alberta is a set of software tools for the production of virtual reality systems and other forms of three-dimensional user interfaces.

5.2. Non-immersive systems

Often non-immersive 3D graphics systems are also called VR systems by some people. Users can change the viewpoint and navigate in the virtual world through input devices interactively. VRML (Virtual Reality Modeling Language) is a web-based 3D modeling and animation language – a subset of OpenInventor. Java3D, similar to VRML, is also a web-based graphics tool to assemble and manipulate predefined geometric models. DIVE (Distributed Interactive Virtual Environment) is an internet-based multi-user VR system where participants navigate in 3D space and see, meet and interact with other users and applications. Alice is a scripting and prototyping environment for 3D object behavior. By writing simple scripts, Alice users can control object appearance and behavior, and while the scripts are executing, objects respond to user input via mouse and keyboard.

5.3. Basic VR system properties

In an immersive VR system, users wear head-mounted devices (HMD) or special glasses to view stereoscopic images. The viewpoint usually follows the viewer's head movement in real time. In a non-immersive VR, which is usually a lot cheaper, users usually do not wear any device, and the viewpoint does not follow the user's head movement. Users navigate in the virtual world through input devices interactively and the image is usually a first-person view. In a VR system, navigation allows a user to move around and to view virtual objects and places, and interaction provides an active way for a user to control the appearance and behavior of objects. 3D navigation, probably with interaction, stereoscopes, and visualization, is the main property of a VR system, immersive or not.

Simulation is another property of a VR system. Simulations integrate scientific results and rules to control, display, and animate virtual objects, behaviors, and environments. Without simulation, the virtual world will not be able to describe and represent real world phenomena correctly. Different VR applications may simulate different objects, phenomena, behaviors, and environments, mostly in real time. These properties make the VR technology able to be applied in various areas such as data visualization, training, surgery, scientific studying, science learning, and game playing.

5.4. VR tools

A VR system often simulates certain real-world activities in various areas, such as training, education, and entertainment. A VR system always repeats the following processing steps:

(a) Handle user inputs from various devices – keyboard, mouse, VR trackers, sensors, voice recognition systems, etc.
(b) Calculate the new state of the objects and the environment according to the simulation models.
(c) Pre-process 3D objects including collision detection, levels of detail, clipping/culling, etc.
(d) Render the virtual world.

In order to achieve the above process, the software in the VR system has to be able to create a virtual world, handle various events from input devices, control the appearances and behaviors of the 3D objects, render the virtual world and display it on the display devices. In step (b), different VR applications may use different simulation models. No matter what application a VR system implements, the software to handle the other three steps, a high-level graphics library called a VR tool (or VR toolkit), is always needed. Therefore, VR tools, which are built on a low-level graphics library, are usually independent of the applications.

5.4.1. VR simulation tools

A VR system is usually a VR application implemented on top of a VR tool, which provides an API for the VR application to manipulate the objects according to the simulation models. VR tools are likely to be device dependent, built on low-level basic graphics libraries with interfaces to sensory devices. Some VR tools, such as MR Toolkit, OpenInventor, and WorldToolkit, only provide APIs embedded in certain programming languages for VR developers. It requires more knowledge and programming skills to employ these toolkits, but they provide more flexibility in application implementations. Others, such as Alice and WorldUp (often called VR simulation tools), provide graphical user interfaces (GUIs) for the developers to build applications. Developers achieve virtual worlds and simulations by typing, clicking, and dragging through GUIs. Sometimes simple script languages are used to construct simulation processes. VR simulation tools allow developing a VR system quicker and easier, but the application developed is an independent fixed module that cannot be modified or integrated in a user-developed program. A VR simulation tool, which is part of VR tools, is generally developed on top of another VR tool, so it is one level higher than the basic VR tools in software levels.

5.4.2. A list of VR tools

Some companies and academic departments have provided various VR tools for different computer platforms. Table 1 lists a few VR tools that are available and well-known. When we want to build a VR system, we do not have to develop everything by ourselves. Instead, we can use an existing VR tool and the necessary hardware, and add very limited application codes.

Table 1
A list of VR tools

Names	Platforms	Functions	References
ActiveWorlds	Windows	Network VR	http://www.activeworlds.com
Alice	Windows	VR simul. tool	http://www.alice.org
AVRIL	Windows	VR toolkit	http://sunee.uwaterloo.ca/~broehl/avril.html
DIVE	Windows	Network VR	http://www.sics.se/dive/dive.html
DIVERSE	Irix	VR toolkit	http://diverse.sourceforge.net
EON Studio	Multiple	VR toolkit	http://www.eonreality.com
GHOST	PC	VR toolkit	http://www.sensable.com
Java3D	Multiple	VR toolkit	http://www.java3d.org
MEME	Windows	Network VR	http://www.immersive.com
Summit3D	Windows	Web VR	http://www.summit3d.com
VREK	Windows	VR toolkit	http://www.themekit.com
Vega	Multiple	VR toolkit	http://www.multigen.com
VRSG	Multiple	VR toolkit	http://www.metavr.com
VrTool	SGI	VR toolkit	http://www.lincom-asg.com
X3D/VRML	Multiple	VR toolkit	http://www.web3d.org
WorldToolKit	Windows	VR toolkit	http://www.sense8.com/
WorldUp	Windows	VR simul. tool	http://www.sense8.com/

5.4.3. Basic functions in VR tool

In addition to a simulation loop and basic graphics functions, a VR tool usually provides the following functions as well:

Import that loads 3D objects or worlds from files on the hard disk into computer internal memory as data structures (called scene graphs) for manipulation and rendering. The 3D virtual world is usually generated with a 3D modeling tool.

Stereo display that allows two different projections of the VR environment to appear in our two eyes. For different display devices, such as HMD, CAVE, and Workbench, the display channels and operating mechanisms are different. A VR tool should support different display devices as well.

Event handling that accepts and processes user interaction and control. Various input from users and external devices are generated in the form of events. The event handling must be fast enough to guarantee the system to run in real time.

Audio and haptic output that generates sounds through the computer speaker or headphone and signals to drive the haptic devices.

Collision detection that prevents two objects to collide with each other and to touch or pick up virtual objects. Collision detection is a time-consuming operation, so most VR tools provide collision Enable/Disable switching functions for VR applications to turn it on/off if necessary.

Level of detail (LOD) that optimizes the rendering detail for faster display and animation. To provide LOD, a VR tool should save multiple different models for one object. VR tool will choose a corresponding model to render according to the distance between the viewpoint and the object.

User interface that accepts user inputs for data and status managements.

5.5. Characteristics of VR

We have briefly introduced VR. What does a high-end VR offer data visualization that conventional technologies do not? While a number of items could be cited, here is a list of those that are important:

Immersion, which implies realism, multi-sensory coverage, and freedom from distractions. Immersion is more an ultimate goal than a complete virtue due to the hardware limitations. For data visualization, immersion should provide a user with an increased ability to identify patterns, anomalies, and trends in data that is visualized.

Multisensory, which allows user input and system feedback to users in different sensory channels in addition to traditional hand (mouse/keyboard) input and visual (screen display) feedback. For data visualization, multisensory allows multimodal manipulation and perception of abstract information in data.

Presence, which is more subjective – a feel of being in the environment, probably with other realistic, sociable, and interactive objects and people. Presence can contribute to the "naturalness" of the environment in which a user works and the ease with which the user interacts with that environment. Clearly, the "quality" of the virtual reality – as measured by display fidelity, sensory richness, and real-time behavior – is critical to a sense of presence.

Navigation, which provides users to move around and investigate virtual objects and places not only by 3D traversal, but also through multisensory interactions and presence. Navigation motivates users to "visualize" and investigate data in multiple perspectives that goes beyond traditional 3D graphics.

Multi-modal displays, which "displays" the VR contents through auditory, haptic, vestibular, olfactory, and gustatory senses in addition to the visual sense. The mapping of information onto more than one sensory modality may well increase the "human bandwidth" for understanding complex, multivariate data. Lacking a theory of multi-sensory perception and processing of information, the critical issue is determining what data "best" maps onto what sensory input channel. Virtual reality offers the opportunity to explore this interesting frontier to find a means of enabling users to effectively work with more and more complex information.

6. Some examples of visualization using VR

Today, more complex data is generated and collected than ever before in human history. This avalanche creates opportunities and information as well as difficulties and challenges. Many people and places are dedicated to data acquisition, computing, and visualization, and there is a great need for sharing information and approaches. Here we list of some different research and applications in visualizing data using VR, so as to foster more insight among this fast-growing community. Visualization is a tool that many use to explore data in a large number of domains. Hundreds of publications address the application of virtual reality in visualization in the related conferences (IEEE VR, IEEE Visualization, ACM SIGGRAPH, etc.), journals (PRESENCE, IEEE CG&A, CiSE, etc.), and books (Burdea and Coiffet, 2003, Chen, 1999,

Göbel, 1996, Durlach and Mavor, 1995, Nielson et al., 1997, Rosenblum et al., 1994, Stanney, 2002, etc.). Below are specific projects that demonstrate the breath of applicability of virtual reality to visualization. The following examples are just demonstrative at random:

- archeology (Acevedo et al., 2001),
- architectural design (Leigh et al., 1996),
- battlespace simulation (Hix et al., 1999, Durbin et al., 1998),
- cosmology (Song and Norman, 1993),
- genome visualization (Kano et al., 2002),
- geosciences (Loftin et al., 1999),
- meteorology (Ziegler et al., 2001),
- materials simulations (Nakano et al., 1998, Sharma et al., 2002),
- oceanography (Gaither et al., 1997),
- protein structures (Akkiraju et al., 1996),
- software systems (Amari et al., 1993),
- scientific data (Hasse et al., 1997),
- statistical data (Arns et al., 1999, Wegman, 2000),
- vector fields (Kuester et al., 2001),
- vehicle design (Kuschfeldt et al., 1997),
- virtual wind tunnel (Bryson and Levit, 1992, Bryson, 1993, Bryson, 1994, Bryson et al., 1997, Severance et al., 2001).

References

Acevedo, D., Vote, E., Laidlaw, D.H., Joukowsky, M.S. (2001). Archaeological data visualization in VR: Analysis of Lamp Finds at the Great Temple of Petra, a Case Study. In: *Proceedings of the 2001 Conference on Virtual Reality, Archeology, and Cultural Heritage*. Glyfada, North Athens, Greece, November 28–30, pp. 493–496.

Akkiraju, N., Edelsbrunner, H., Fu, P., Qian, J. (1996). Viewing geometric protein structures from inside a CAVE. *IEEE Comput. Graph. Appl.* **16** (4), 58–61.

Amari, H., Nagumo, T., Okada, M., Hirose, M., Ishii, T. (1993). A virtual reality application for software visualization. In: *Proceedings of the 1993 Virtual Reality Annual International Symposium*. Seattle, WA, September 18–22, pp. 1–6.

Arns, L., Cook, D., Cruz-Neira, C. (1999). The benefits of statistical visualization in an immersive environment. In: *Proceedings of the 1999 IEEE Virtual Reality Conference*. Houston, TX, March 13–17, IEEE Press, Los Alamitos, CA, pp. 88–95.

Burdea, G.C., Coiffet, P. (2003). *Virtual Reality Technology*, second ed. Wiley.

Bryson, S., Levit, C. (1992). The virtual wind tunnel. *IEEE Comput. Graph. Appl.* **12** (4), 25–34.

Bryson, S. (1993). The Virtual Windtunnel: a high-performance virtual reality application. In: *Proceedings of the 1993 Virtual Reality Annual International Symposium*. Seattle, WA, September 18–22, pp. 20–26.

Bryson, S. (1994). Real-time exploratory scientific visualization and virtual reality. In: Rosenblum, L., Earnshaw, R.A., Encarnação, J., Hagen, H., Kaufman, A., Klimenko, S., Nielson, G., Post, F., Thalman, D. (Eds.), *Scientific Visualization: Advances and Challenges*. Academic Press, London, pp. 65–85.

Bryson, S., Johan, S., Schlecht, L. (1997). An extensible interactive visualization framework for the Virtual Windtunnel. In: *Proceedings of the 1997 Virtual Reality Annual International Symposium*. Albuquerque, NM, March 1–5, pp. 106–113.

Buja, A., Asimov, D. (1985). Grand tour methods: an outline. In: Allen, D. (Ed.), *Computer Science and Statistics: Proceedings of the Seventeenth Symposium on the Interface*. North-Holland, Amsterdam, pp. 63–67.

Carr, D.B., Olsen, A.R., et al. (1998). Linked micromap plots: named and described. *Statist. Comput. Statist. Graph. Newsletter* **9** (1), 24–31.

Chen, C. (1999). *Information Visualization and Virtual Environments*. Springer-Verlag, Berlin.

Chen, J.X., Yang, Y. (2000). 3D graphics formats and conversions. *IEEE Comput. Sci. Engrg.* **2** (5), 82–87.

Chen, J.X., Wang, S. (2001). Data visualization: parallel coordinates and dimension reduction. *IEEE Comput. Sci. Engrg.* **3** (5), 110–113.

Chen, J.X. (2002). *Guide to Graphics Software Tools*. Springer-Verlag.

De Jong, K. (1990). Genetic algorithm-based learning. In: Kodratoff, Y., Michalski, R.S. (Eds.), *Machine Learning: An Artificial Intelligence Approach, vol. 3*. Morgan Kaufmann, San Mateo, CA, pp. 611–638.

Durbin, J., Swan, J.E. II, Colbert, B., Crowe, J., King, R., King, T., Scannell, C., Wartell, Z., Welsh, T. (1998). Battlefield visualization on the responsive workbench. In: *Proceedings of the 1998 IEEE Visualization Conference*. Research Triangle Park, NC, October 18–23, IEEE Press, Los Alamitos, CA, pp. 463–466.

Durlach, N., Mavor, A. (Eds.) (1995). *Virtual Reality: Scientific and Technological Challenges*. National Academy Press, Washington, DC.

Foley, J.D., van Dam, A., Feiner, S.K., Hughes, J.F. (1996). *Computer Graphics: Principles and Practice in C*, second ed. Addison–Wesley, Reading, MA.

Gaither, K., Moorhead, R., Nations, S., Fox, D. (1997). Visualizing ocean circulation models through virtual environments. *IEEE Comput. Graph. Appl.* **17** (1), 16–19.

Göbel, M. (Ed.) (1996). *Virtual Environments and Scientific Visualization 96*. Proceedings of the Eurographics Workshops in Monte Carlo, Monaco, February 19–20, 1996 and in Prague, Czech Republic, April 23–23, 1996, Springer-Verlag, Berlin.

Hasse, H., Dai, F., Strassner, J., Göbel, M. (1997). Immersive investigation of scientific data. In: Nielson, G.M., Hagen, H., Müller, H. (Eds.), *Scientific Visualization: Overviews, Methods, and Techniques*. IEEE Press, Los Alamitos, CA, pp. 35–58.

Hix, D., Swan, J.E. II, Gabbard, J.L., McGree, M., Durbin, J., King, T. (1999). User-centered design and evaluation of a real-time battlefield visualization virtual environment. In: *Proceedings of the 1999 IEEE Virtual Reality Conference*. Houston, TX March 13–17, IEEE Press, Los Alamitos, CA, pp. 96–103.

Kano, M., Tsutsumi, S., Nishimura, K. (2002). Visualization for genome function analysis using immersive projection technology. In: *Proceedings of the 2002 IEEE Virtual Reality Conference*. Orlando, FL, October 24–28, IEEE Press, Los Alamitos, CA, pp. 224–231.

Kuester, F., Bruckschen, R., Hamann, B., Joy, K.I. (2001). Visualization of particle traces in virtual environments. In: *Proceedings of the 2001 ACM Symposium on Virtual Reality Software and Technology*. November, ACM Press.

Kuschfeldt, S., Schultz, M., Ertl, T., Reuding, T., Holzner, M. (1997). The use of a virtual environment for FE analysis of vehicle crash worthiness. In: *Proceedings of the 1997 Virtual Reality Annual International Symposium*. Albuquerque, NM, March 1–5, p. 209.

Leigh, J., Johnson, A.E., Vasilakis, C.A., DeFanti, T.A. (1996). Multi-perspective collaborative design in persistent networked virtual environments. In: *Proceedings of the 1996 Virtual Reality Annual International Symposium*. Santa Clara, CA, March 30–April 3, pp. 253–260.

Loftin, R.B., Harding, C., Chen, D., Lin, C., Chuter, C., Acosta, M., Ugray, A., Gordon, P., Nesbitt, K. (1999). Advanced visualization techniques in the geosciences. In: *Proceedings of the Nineteenth Annual Research Conference of the Gulf Coast Section Society of Economic Paleontologists and Mineralogists Foundation*. Houston, Texas, December 5–8.

Nakano, O., Bachlechner, M.E., Campbell, T.J., Kalia, R.K., Omeltchenko, A., Tsuruta, K., Vashishta, P., Ogata, S., Ebbsjo, I., Madhukar, A. (1998). Atomistic simulation of nanostructured materials. *IEEE Comput. Sci. Engrg.* **5** (4), 68–78.

Nielson, G.M., Hagen, H., Müller, H. (1997). *Scientific Visualization: Overviews, Methods, and Techniques*. IEEE Press, Los Alamitos, CA.

Rosenblum, L., Earnshaw, R.A., Encarnação, J., Hagen, H., Kaufman, A., Klimenko, S., Nielson, G., Post, F., Thalman, D. (Eds.) (1994). *Scientific Visualization: Advances and Challenges*. Academic Press, London.

Severance, K., Brewster, P., Lazos, B., Keefe, D. (2001). Wind tunnel data fusion and immersive visualization. In: *Proceedings of the 2001 IEEE Visualization Conference*. October, IEEE Press, Los Alamitos, CA.

Sharma, A., Liu, X., Miller, P., Nakano, N., et al. (2002). Immersive and interactive exploration of billion-atom systems. In: *IEEE Virtual Reality Conference 2002*. Orlando, FL, IEEE Press, Los Alamitos, CA, pp. 217–225.

Song, D., Norman, M.L. (1993). Cosmic explorer: a virtual reality environment for exploring cosmic data. In: *Proceedings of the 1993 IEEE Symposium on Research Frontiers in Virtual Reality*. San Jose, CA, October 25–26, IEEE Press, Los Alamitos, CA, pp. 75–79.

Stanney, K.M. (Ed.) (2002). *Handbook of Virtual Environments: Design, Implementation, and Applications*. Lawrence Erlbaum Associates, Mahwah, NJ.

Vince, J. (1995). *Virtual Reality Systems*. Addison–Wesley, Reading, MA.

Vince, J. (1998). *Essential Virtual Reality Fast*. Springer-Verlag.

Wang, X., Chen, J.X., Carr, D.B., et al. (2002). Geographic statistics visualization: web-based linked micromap plots. *IEEE/AIP Comput. Sci. Engrg.* **4** (3), 90–94.

Watt, A.H., Watt, M. (1992). *Advanced Animation and Rendering Techniques: Theory and Practice*. Addison–Wesley, Reading, MA.

Wegman, E.J. (1990). Hyperdimensional data analysis using parallel coordinates. *J. Amer. Statist. Assoc.* **85**, 664–675.

Wegman, E.J., Carr, D.J. (1993). Statistical graphics and visualization. In: Rao, C.R. (Ed.), *Computational Statistics*. In: *Handbook of Statistics*, vol. 9. North-Holland, Amsterdam, pp. 857–958.

Wegman, E.J., Luo, Q. (1997). High dimensional clustering using parallel coordinates and the grand tour. *Comput. Sci. Statist.* **28**, 352–360.

Wegman, E.J. (2000). Affordable environments for 3D collaborative data visualization. *Comput. Sci. Engrg.* **2** (6), 68–72, 74.

Wegman, E.J., Symanzik, J. (2002). Immersive projection technology for visual data mining. *J. Comput. Graph. Statist.* **11** (1), 163–188.

Wegman, E.J., Luo, Q. (2002). On methods of computer graphics for visualizing densities. *J. Comput. Graph. Statist.* **11** (1), 137–162.

Ziegler, S., Moorhead, R.J., Croft, P.J., Lu, D. (2001). The MetVR case study: meteorological visualization in an immersive virtual environment. In: *Proceedings of the 2001 IEEE Visualization Conference*. October, IEEE Press, Los Alamitos, CA.

Colour figures

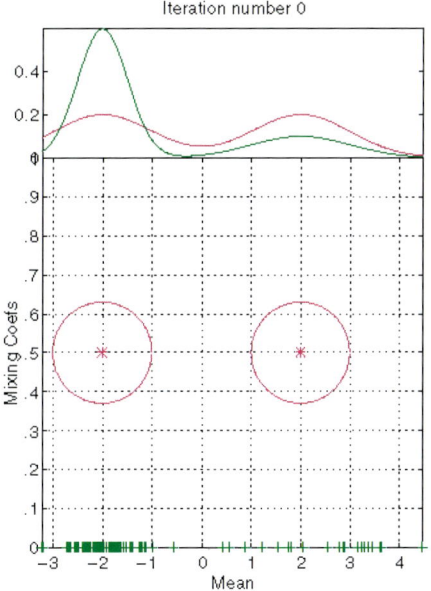

Chapter 1, Fig. 7, p. 23. Visualization of normal mixture model parameters in a one-dimensional setting.

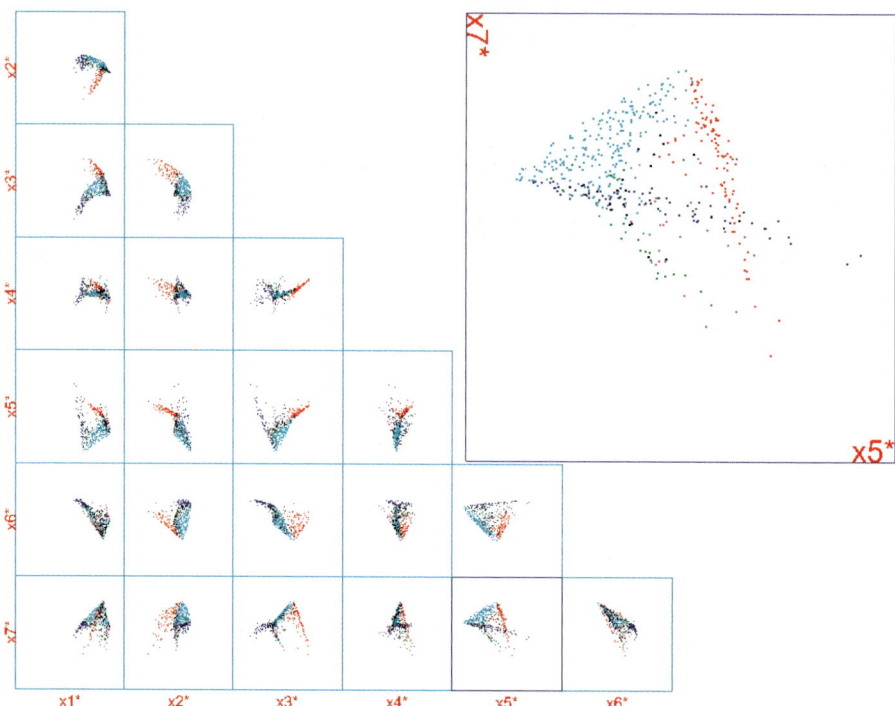

Chapter 1, Fig. 11, p. 33. Scatterplot matrix of the PRIM 7 data after GRAND-TOUR rotation with features highlighted in different colors.

Chapter 1, Fig. 12, p. 33. A scatterplot of the PRIM 7 data after GRAND-TOUR illustrating the three triangular features in the data, again highlighted in different colors.

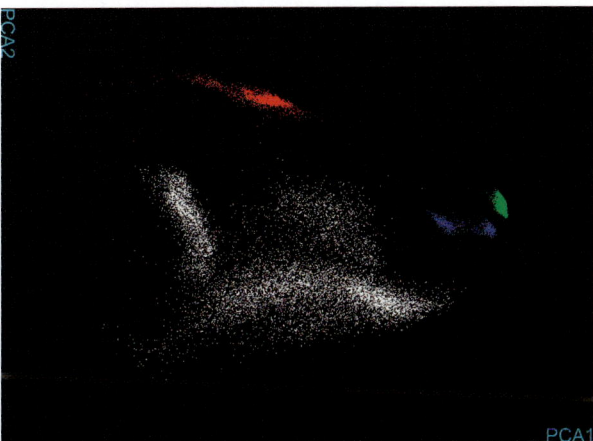

Chapter 1, Fig. 15, p. 35. The first two principal components of the hyperspectral imagery. There are 7 classes of pixels including runway, water, swamp, grass, scrub, pine, and unknown (really oaks). The water and runway are isolated in this figure.

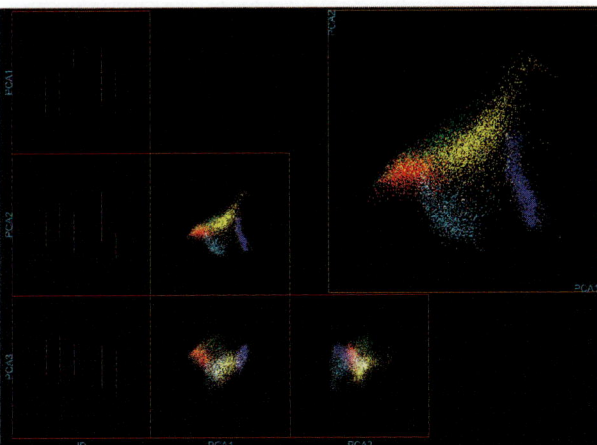

Chapter 1, Fig. 16, p. 36. The recomputed principal components after denoising by removing water and runway pixels. The swamp and grass pixels are colored are colored respectively by cyan and blue. These may be removed and the principal components once again computed.

568 *Colour figures*

Chapter 1, Fig. 17, p. 36. The penultimate denoised image. The scrub pixels are shown in blue. They are removed and one final computation of the principal components is completed.

Chapter 1, Fig. 18, p. 37. The final denoised image show heavy overlap between the pine and unknown (oak) pixels. Based on this analysis, we classify the unknowns as being closest to pines. In fact, both are trees and are actually intermingled when the hyperspectral imagery was ground truthed.

Colour figures

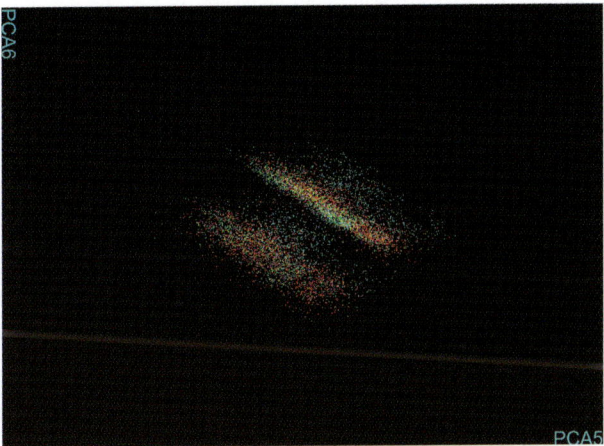

Chapter 1, Fig. 19, p. 37. The same hyperspectral data but based on 10 principal components instead of just 3. The plot of PC 5 versus PC 6 shows that there is additional structure in this dataset not captured by the seven classes originally conjectured.

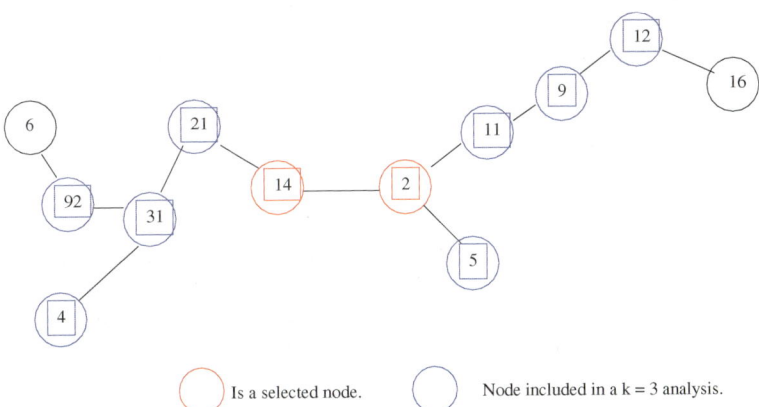

Chapter 5, Fig. 1, p. 135. User specification of drill down focus region. Nodes 14 and 2 have been selected.

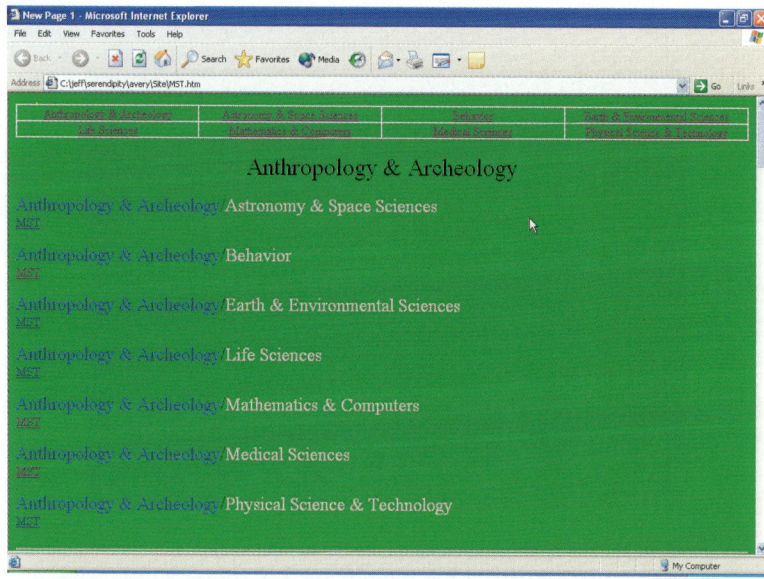

Chapter 5, Fig. 2, p. 136. Opening screen of the automated serendipity search engine.

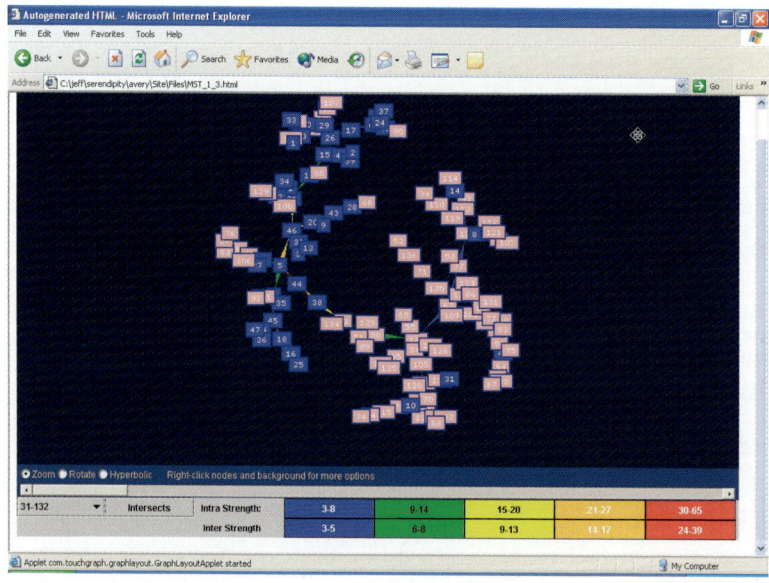

Chapter 5, Fig. 3, p. 137. An example of the MST layout screen.

Colour figures 571

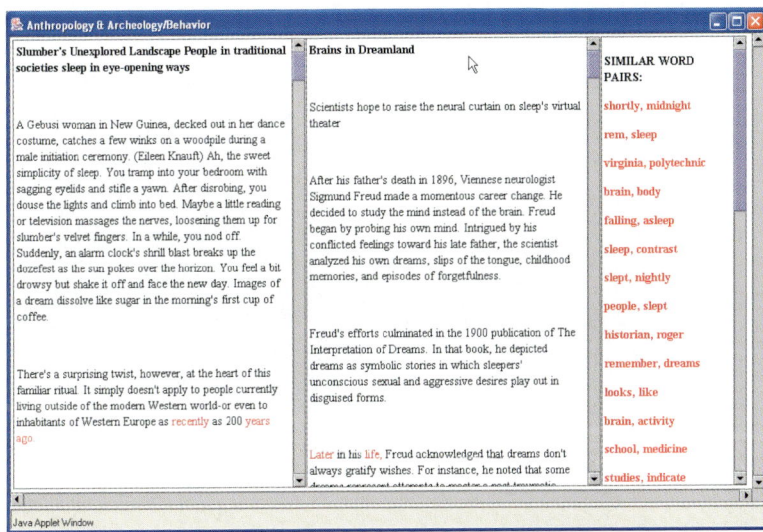

Chapter 5, Fig. 4, p. 138. A sample comparison file.

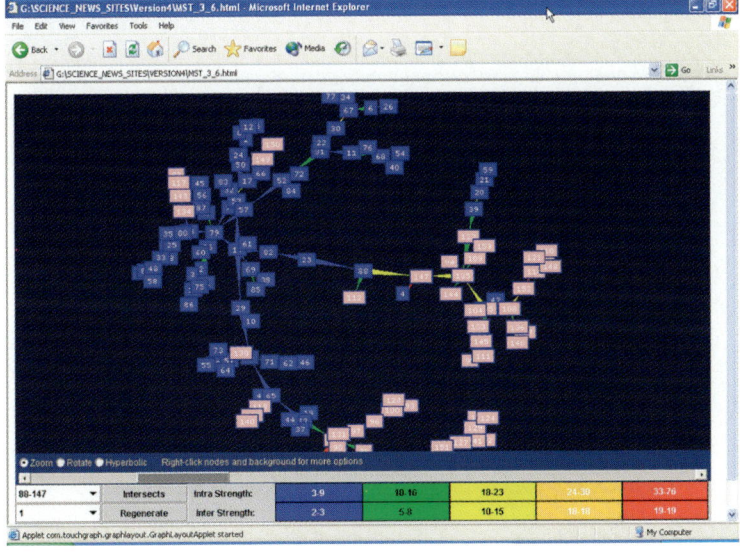

Chapter 5, Fig. 5, p. 138. An example screen for the automated serendipity system with included point picking.

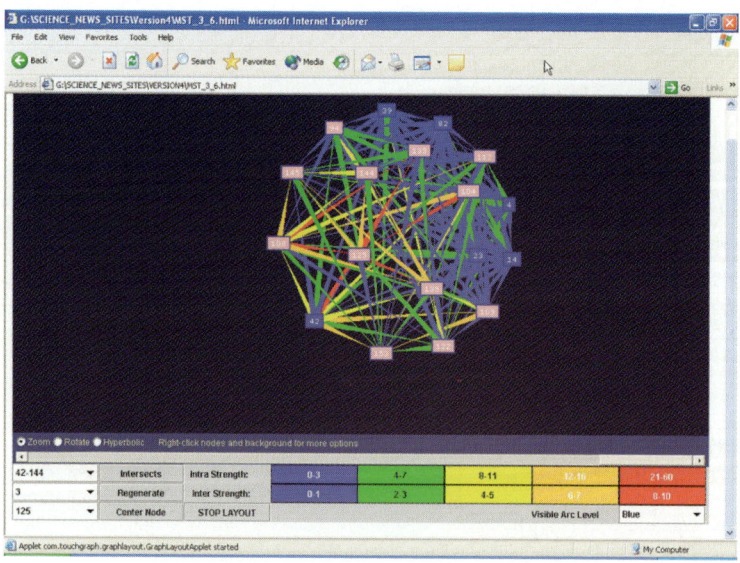

Chapter 5, Fig. 6, p. 139. An example screen for the automated serendipity system with point picking where the user has chosen to limit the focus of attention based on $k = 3$.

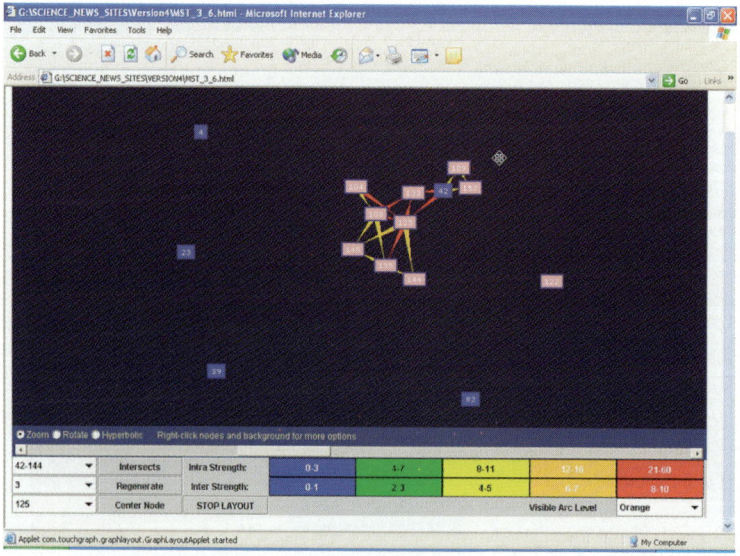

Chapter 5, Fig. 7, p. 139. An example screen for the automated serendipity system with included point picking with point picking where the user has chosen to limit the focus of attention based on $k = 3$ after the user has deleted the edges with a strength less than orange.

Colour figures 573

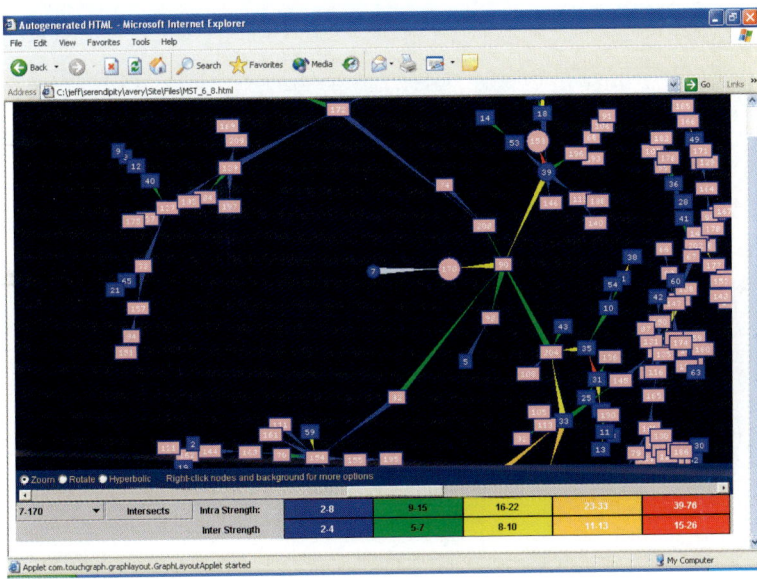

Chapter 5, Fig. 8, p. 141. The MST for the Mathematics and Computer Sciences category as compared to the Physical Sciences and Technology category. The edge indicating the strongest association has been highlighted.

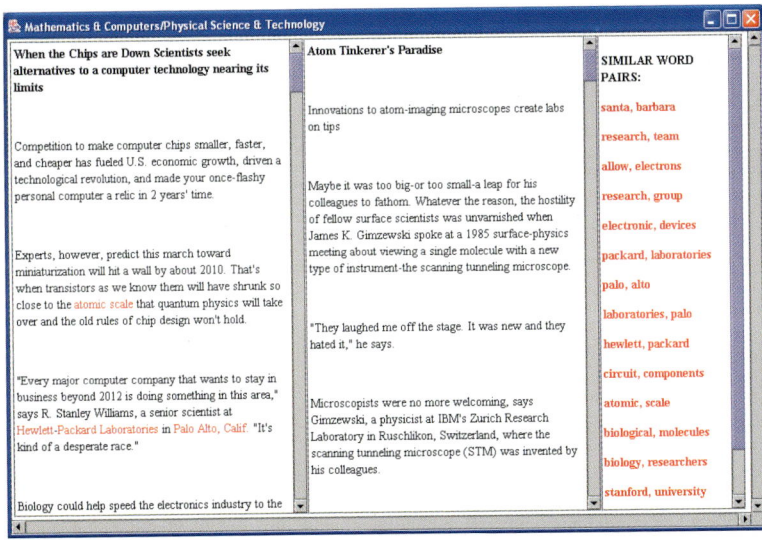

Chapter 5, Fig. 9, p. 142. The BPM comparison file for the strongest association between the Mathematics and Computer Sciences category and the Physical Sciences and Technology category.

574   Colour figures

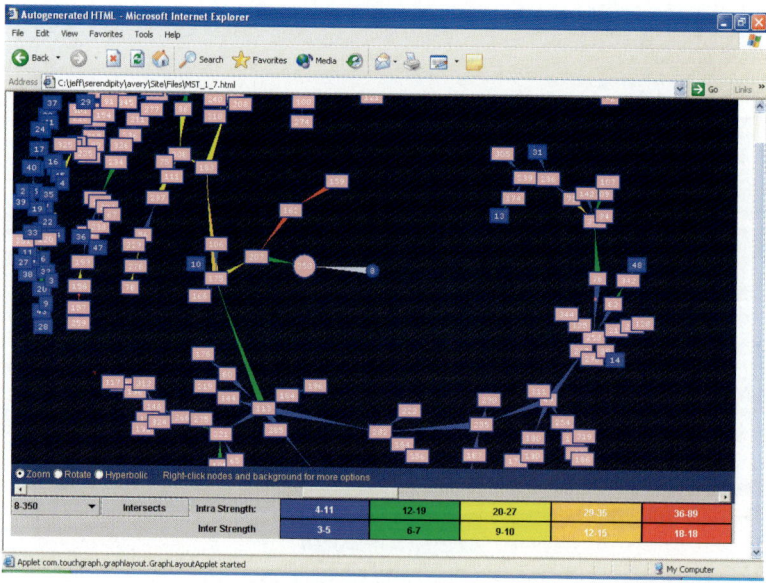

Chapter 5, Fig. 10, p. 142. The MST and highlighted strongest edge between the Anthropology and Archeology discipline area and the Medical Sciences discipline area.

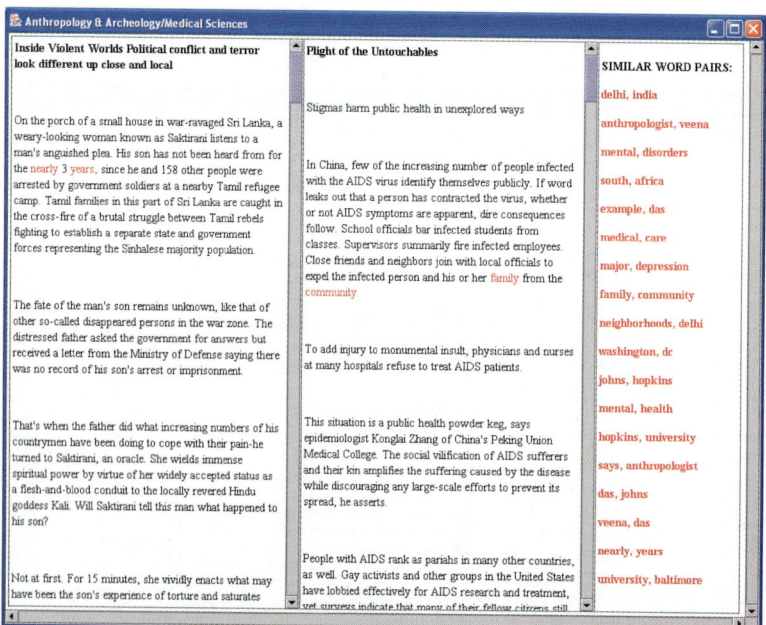

Chapter 5, Fig. 11, p. 143. The BPM comparison file for the strongest association between the Anthropology and Archeology category and the Medical Sciences category.

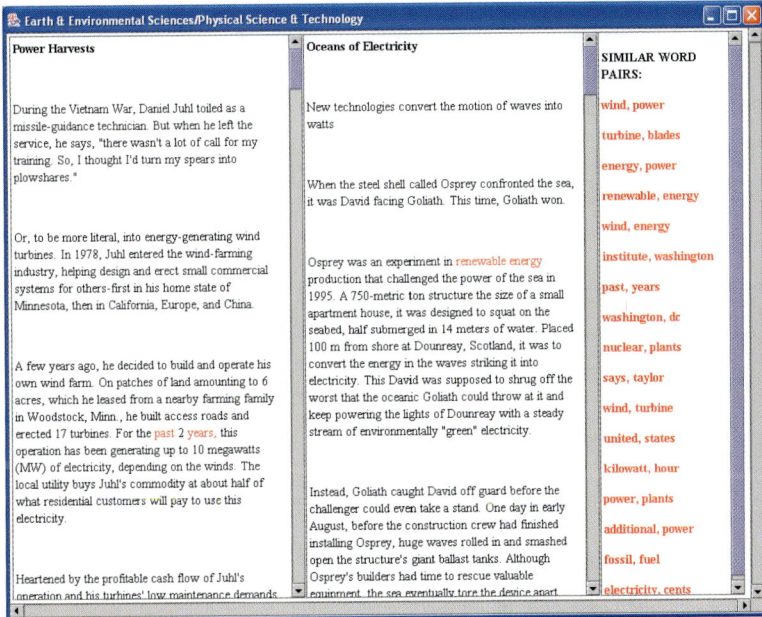

Chapter 5, Fig. 12, p. 144. The BPM comparison file for the strongest association between the Earth and Environmental Sciences category and the Physical Sciences and Technology category.

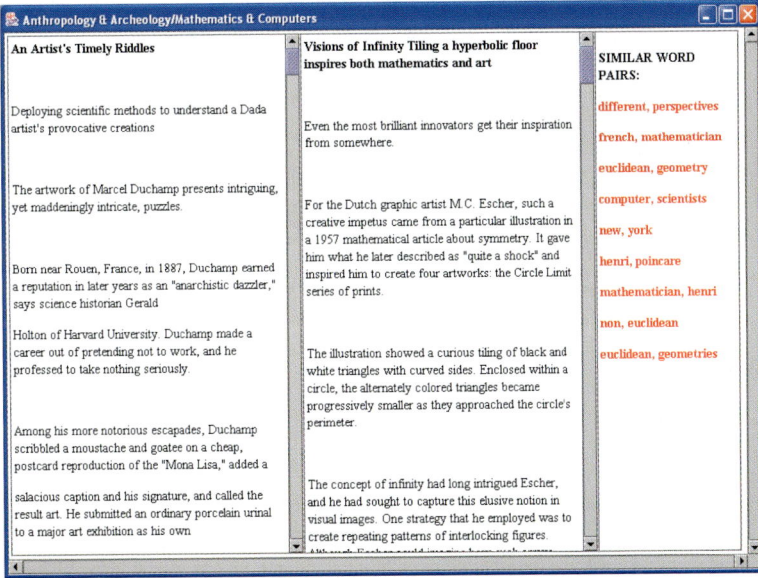

Chapter 5, Fig. 13, p. 145. The BPM comparison file for an interesting association between the Anthropology and Archeology and Mathematics and Computer Sciences categories.

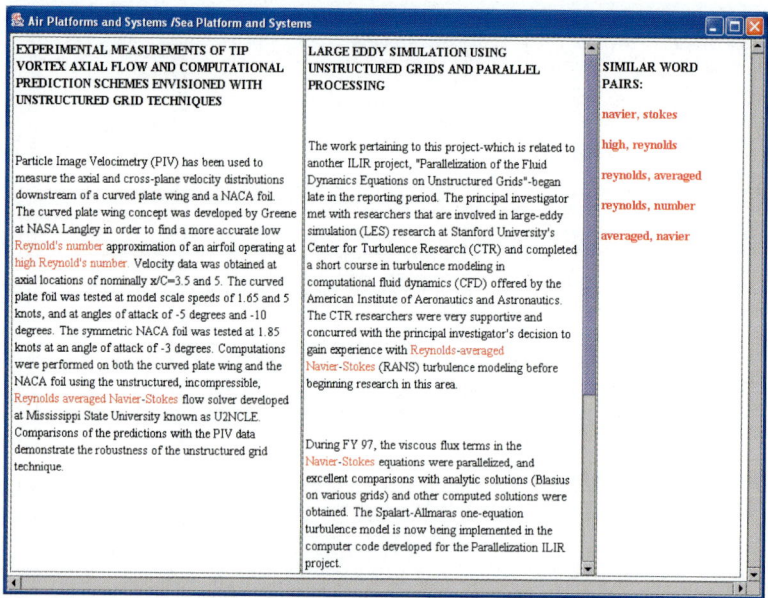

Chapter 5, Fig. 14, p. 146. The BPM comparison file for a computational fluid dynamics association between the Air Platform and Systems and the Sea Platform and Systems classes.

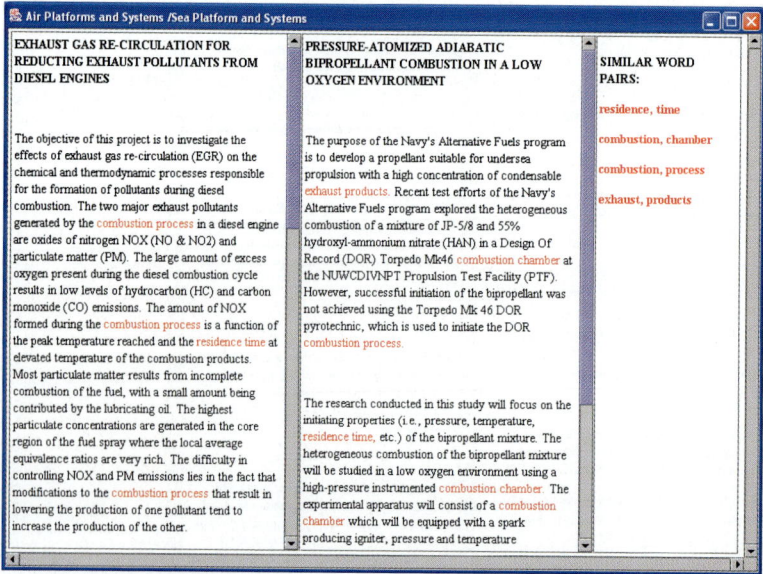

Chapter 5, Fig. 15, p. 146. The BPM comparison file for a combustion physics association between the Air Platform and Systems and the Sea Platform and Systems classes.

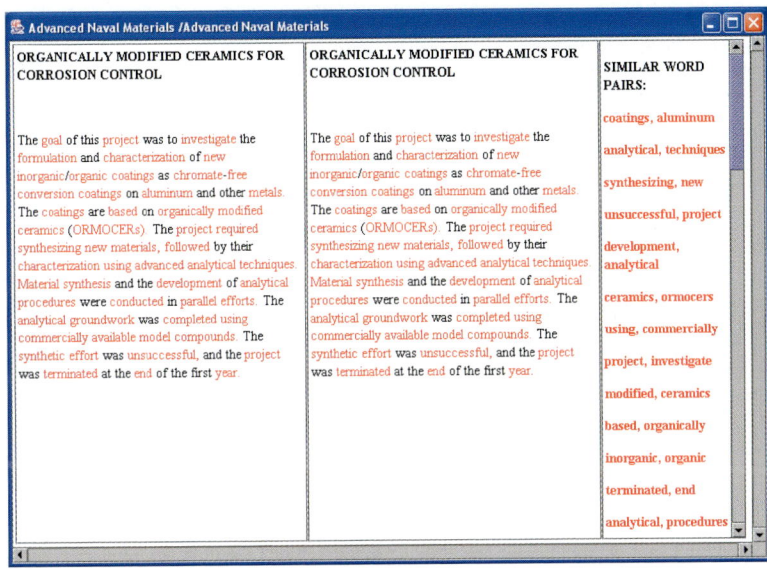

Chapter 5, Fig. 16, p. 147. An apparent duplication entry in the Advanced Naval Material category.

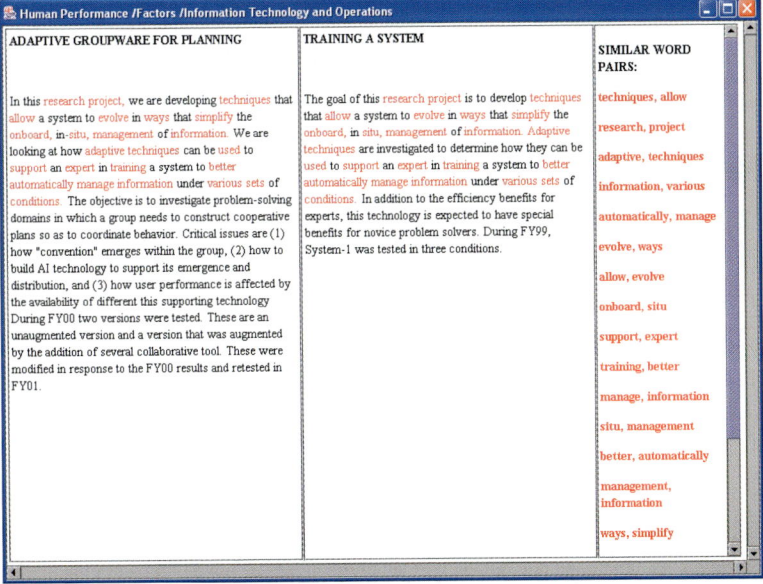

Chapter 5, Fig. 17, p. 148. An unusual association across the Human Performance Factors and the Information Technology and Operation focus areas.

578   Colour figures

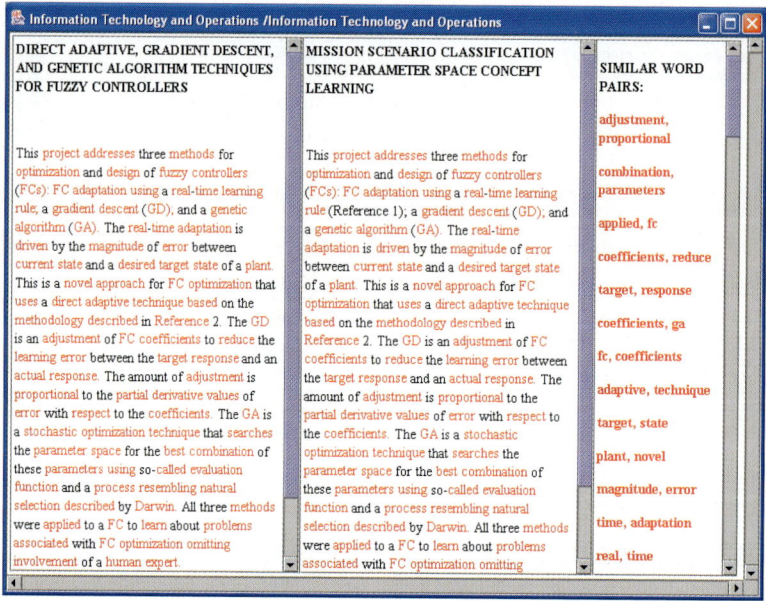

Chapter 5, Fig. 18, p. 148. A highly unusual association between two articles in the Information Technology and Operations focus areas.

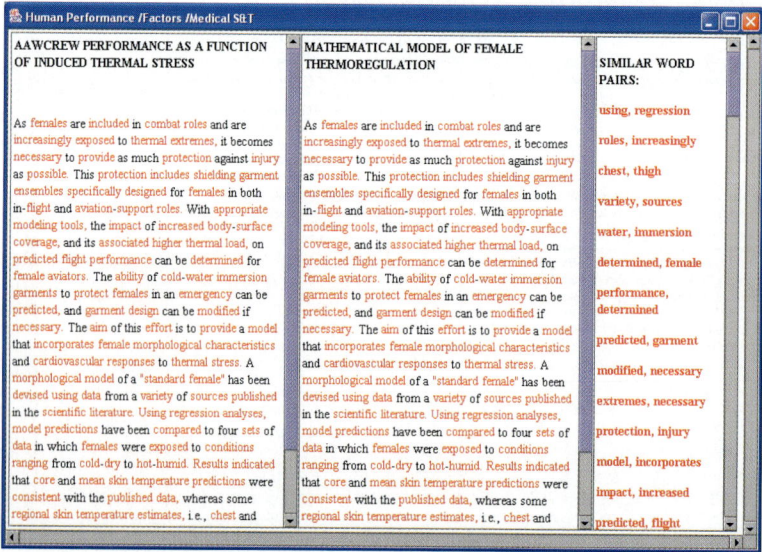

Chapter 5, Fig. 19, p. 149. A highly unusual association between two articles in the Human Performance Factors and the Medical Sciences and Technology focus areas.

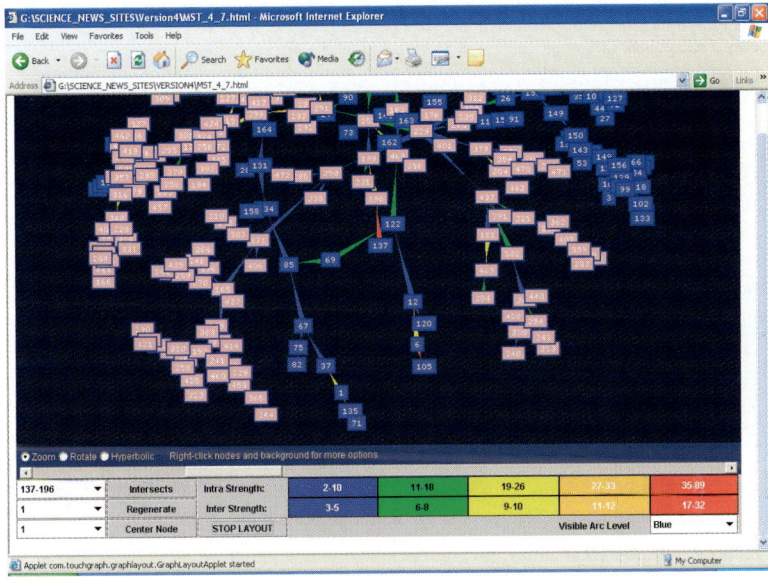

Chapter 5, Fig. 20, p. 150. The user has selected nodes 196 and 137 as the "focus of attention".

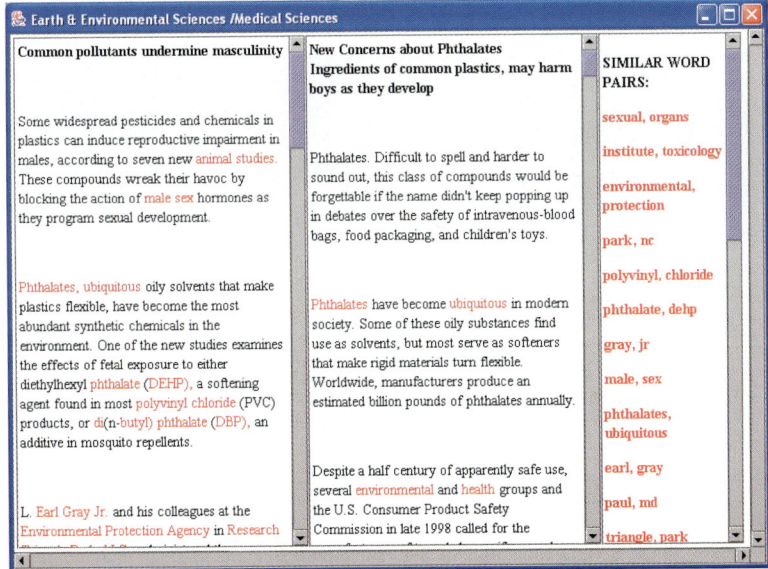

Chapter 5, Fig. 21, p. 150. A comparison for the two articles selected for the "focus of attention".

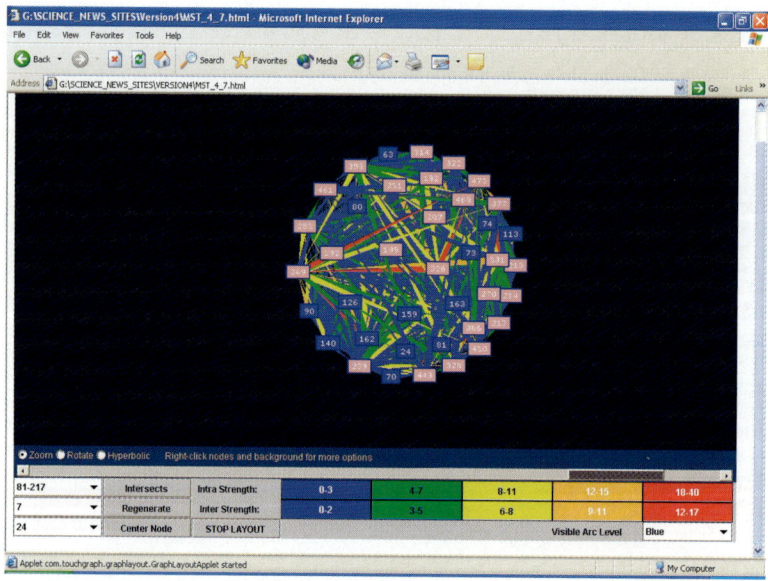

Chapter 5, Fig. 22, p. 151. The complete graph based on those articles within 7 links of articles 196 and 197.

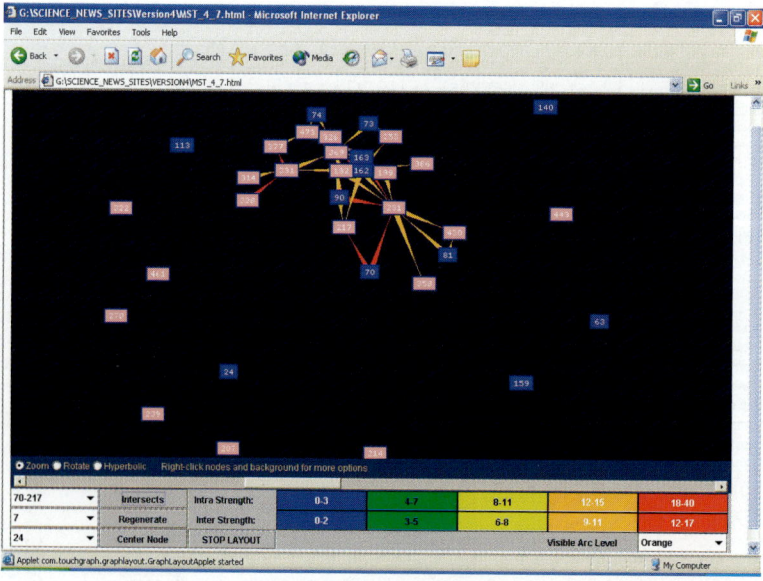

Chapter 5, Fig. 23, p. 152. The complete graph based on those articles within 7 links of articles 196 and 137 with the edges of strength less than orange removed. The graph has been centered on the edge extending between nodes 70 and 217.

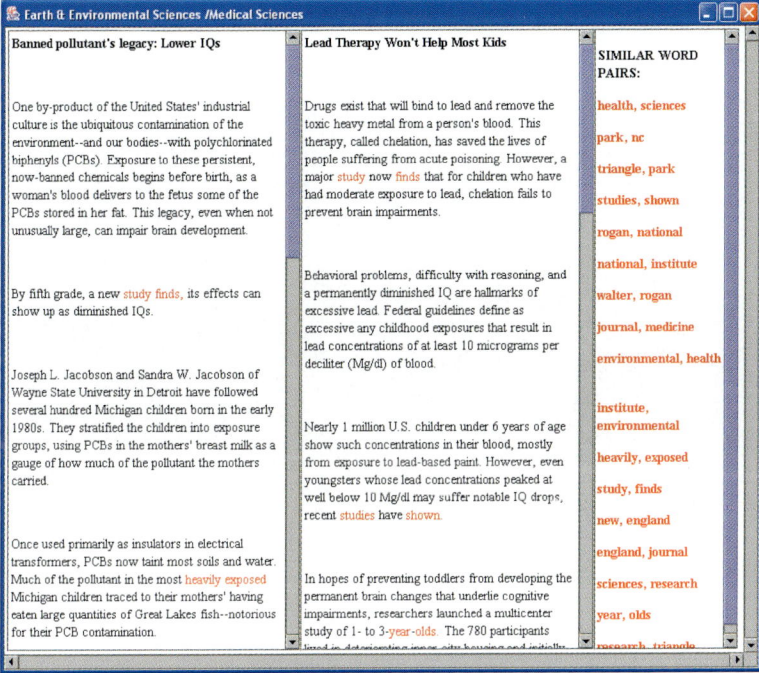

Chapter 5, Fig. 24, p. 152. A closer look at the associations between nodes 70 and 217.

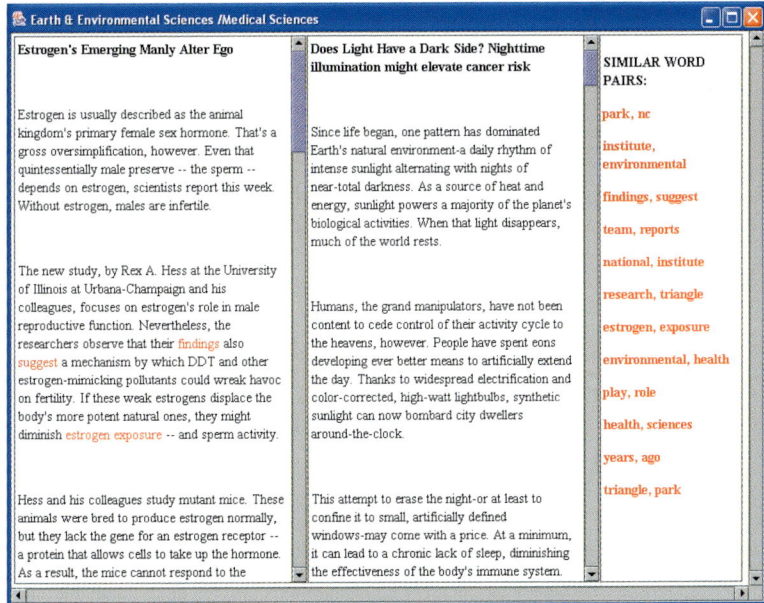

Chapter 5, Fig. 25, p. 153. A closer look at the associations between nodes 162 and 369.

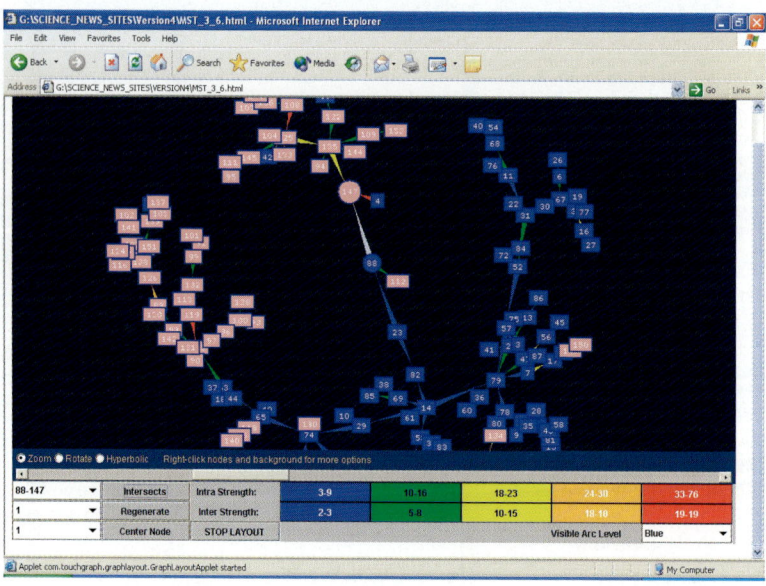

Chapter 5, Fig. 26, p. 154. The MST and nodes 88 and 147 of the Behavior and Computer Science and Mathematics categories.

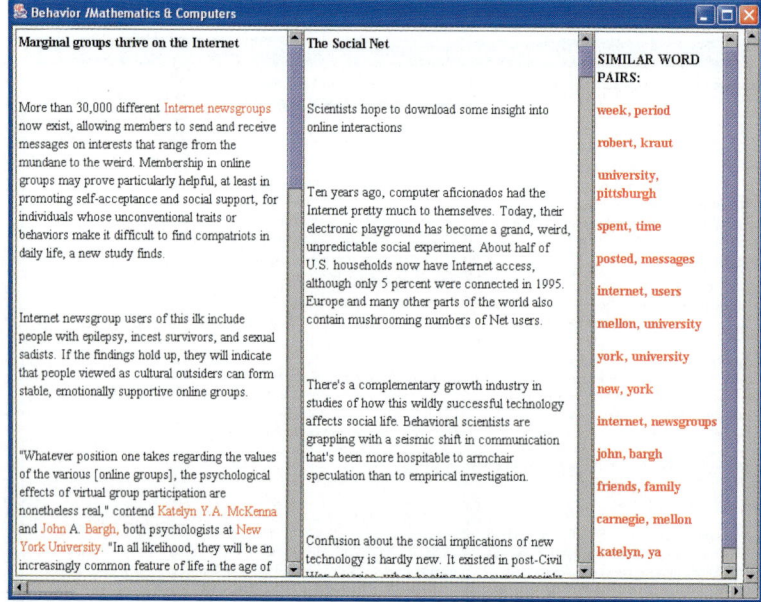

Chapter 5, Fig. 27, p. 154. The BPM comparison file nodes 88 and 147 of the Behavior and Computer Science and Mathematics categories.

Colour figures 583

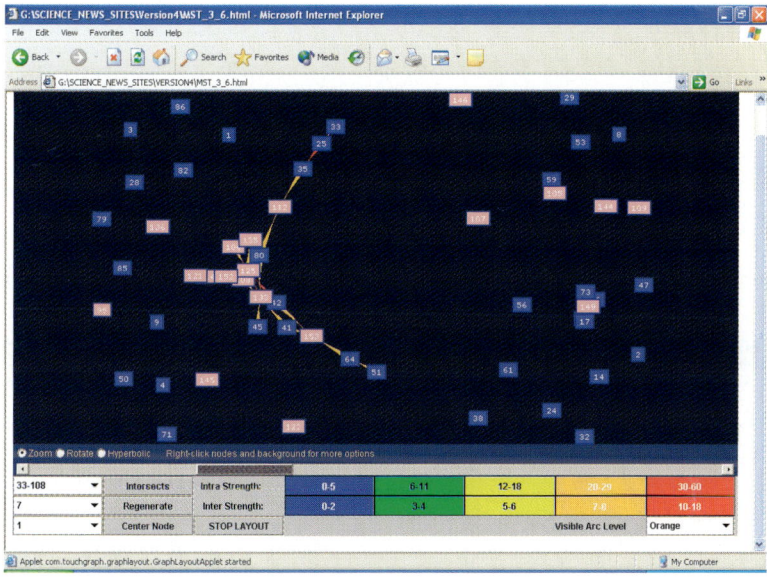

Chapter 5, Fig. 28, p. 155. The articles within 7 links of 88 and 147 with a strength of at least orange.

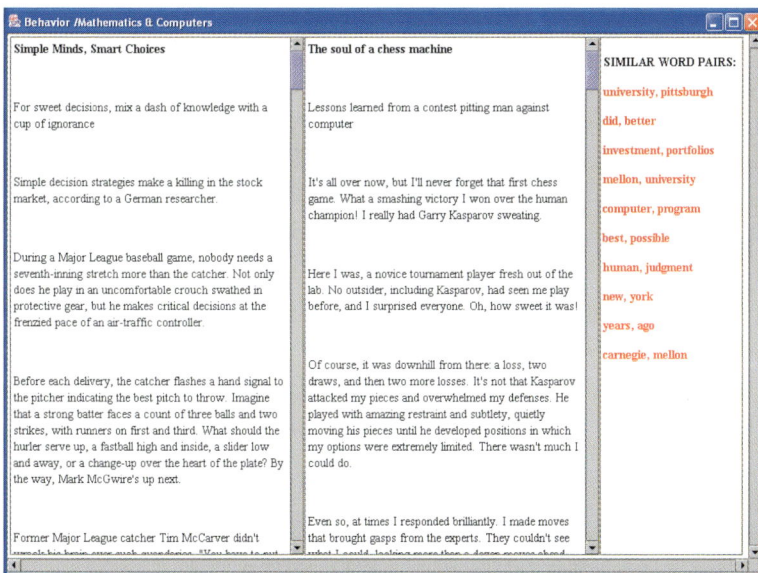

Chapter 5, Fig. 29, p. 156. A closer look at the between-article associations between Behavior article 80 and Mathematics and Computer Science article 108.

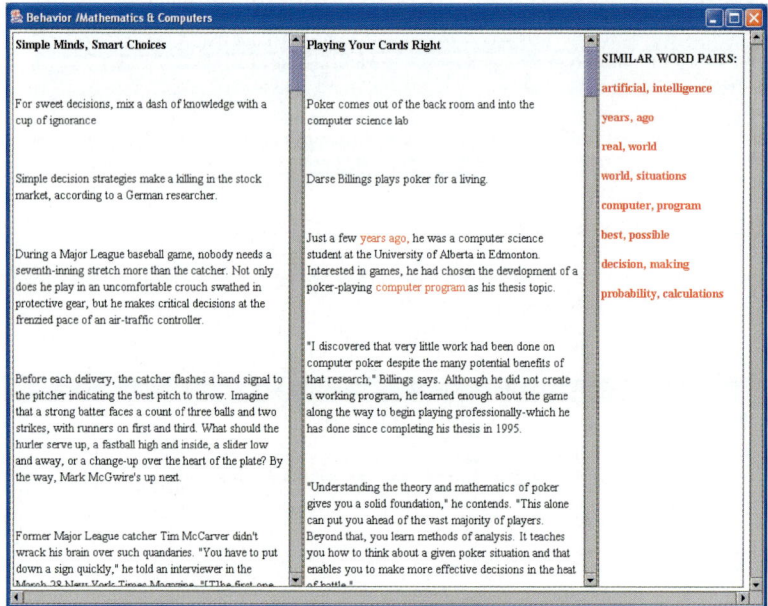

Chapter 5, Fig. 30, p. 157. A closer look at the between article associations between article 80 a Behavior article and article 133 a Mathematics and Computer Science article.

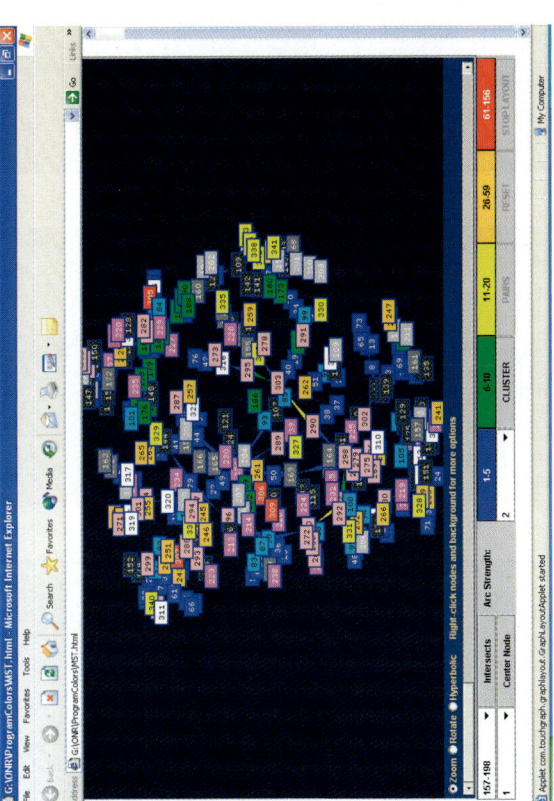

Chapter 5, Fig. 31, p. 158. The opening screen for the clustering program when loaded with the ILIR data and associated color key.

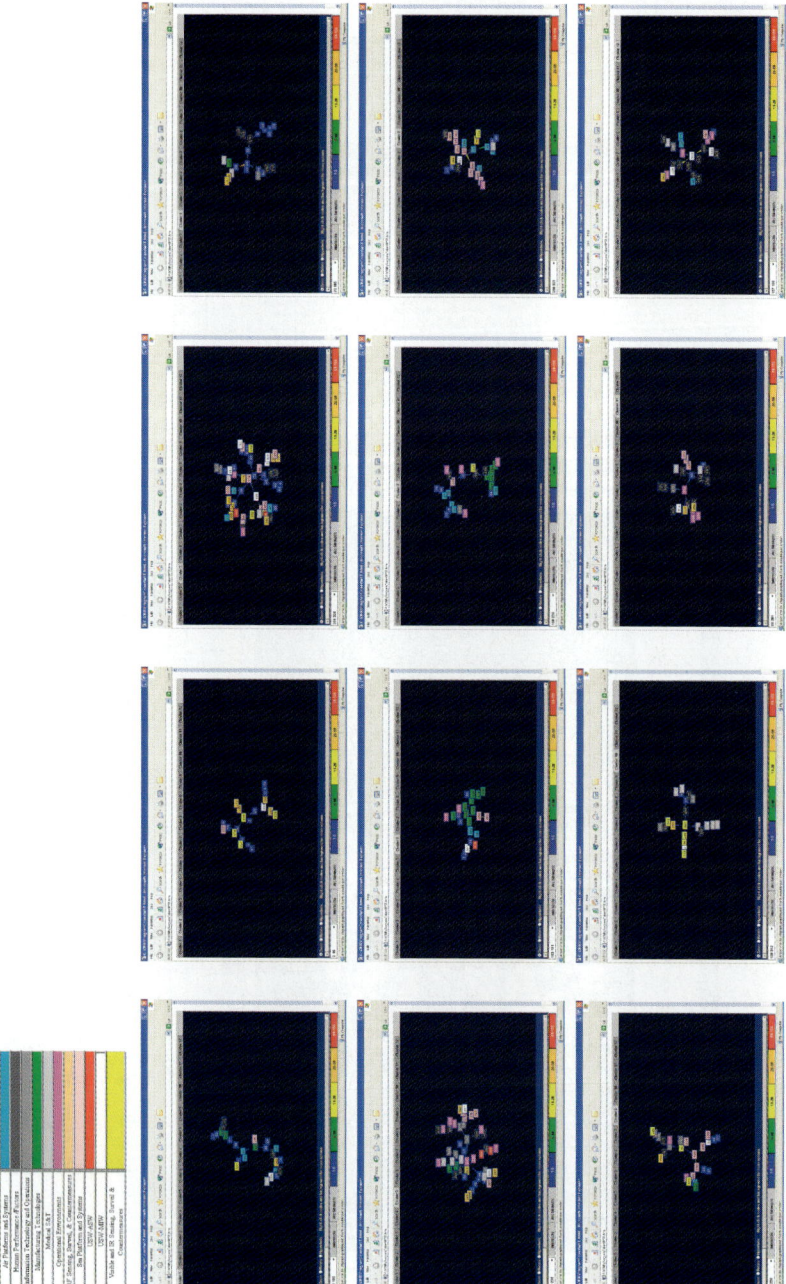

Chapter 5, Fig. 32, p. 159. Overview of the ILIR cluster structure.

Colour figures 587

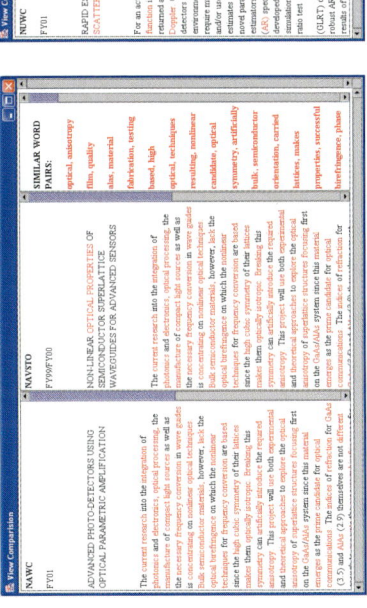

Chapter 5, Fig. 33, p. 161. Two interesting article pairs from cluster 2 of the ILIR dataset.

Chapter 5, Fig. 34, p. 161. Two interesting article pairs from cluster 3 of the ILIR dataset.

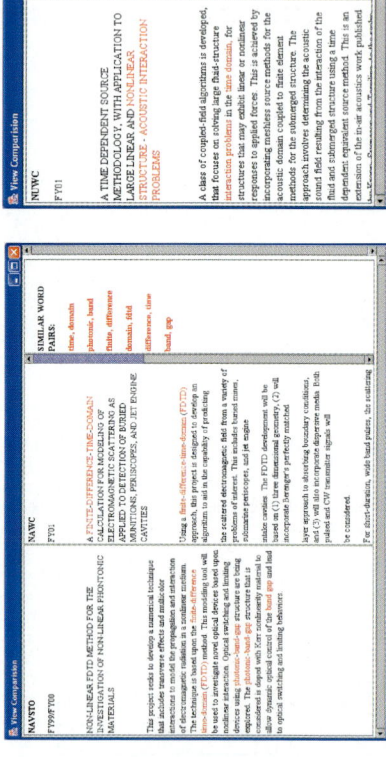

Chapter 5, Fig. 35, p. 162. Two interesting article pairs from cluster 3 of the ILIR dataset.

Chapter 5, Fig. 36, p. 162. Some interesting article pairs as revealed in cluster 5 of the ILIR dataset.

Colour figures

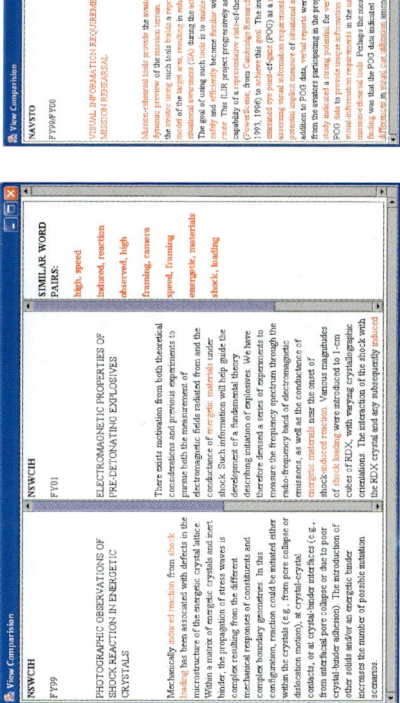

Chapter 5, Fig. 37, p. 163. Some interesting article pairs as revealed in clusters 7, left figure, and 12, right figure, of the ILR dataset.

590 *Colour figures*

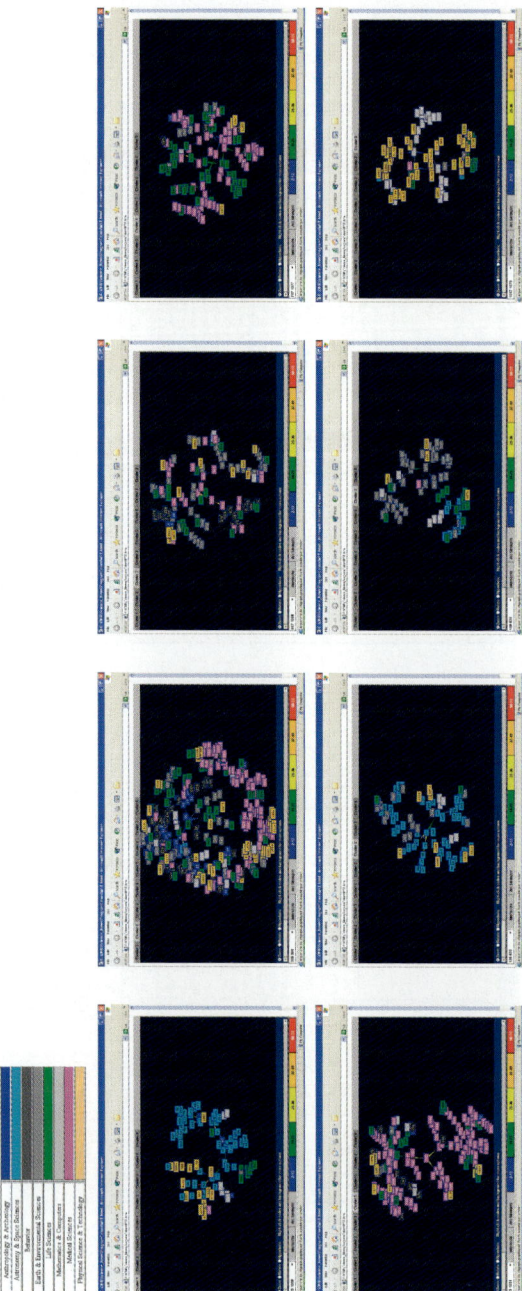

Chapter 5, Fig. 38, p. 164. Overview of the Science News cluster structure.

Colour figures 591

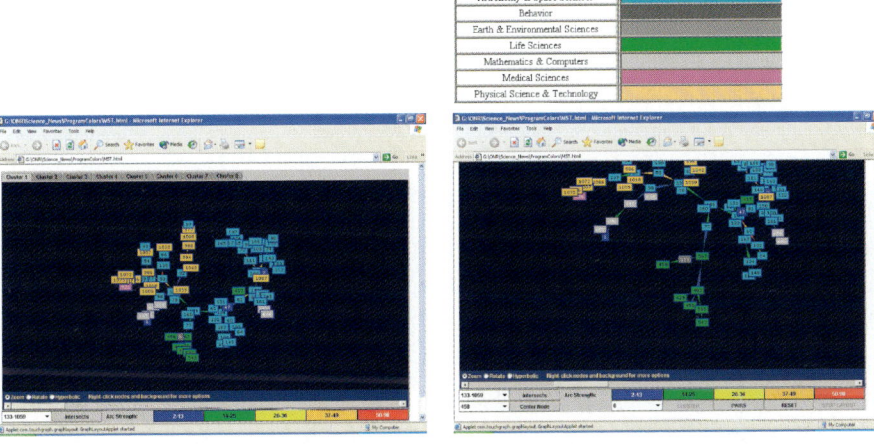

Chapter 5, Fig. 39, p. 165. A subcluster of Science News cluster 1 that is focused on animal behavior and sexuality.

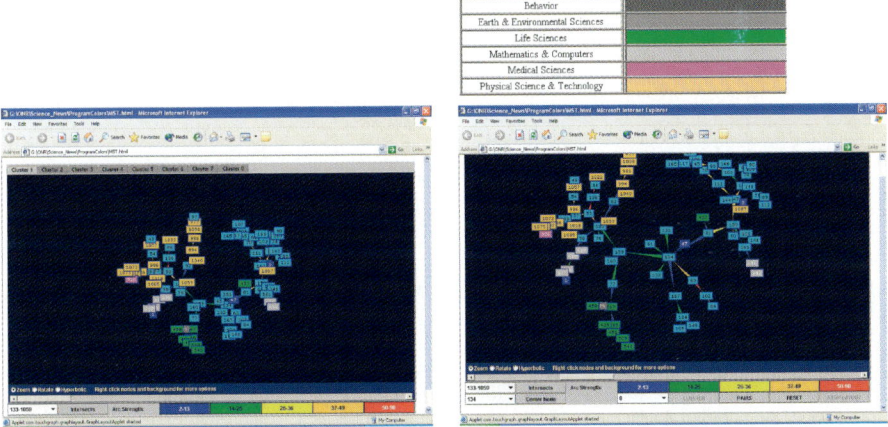

Chapter 5, Fig. 40, p. 166. A subcluster of Science News cluster 1 that is focused on an infrared camera and its applications.

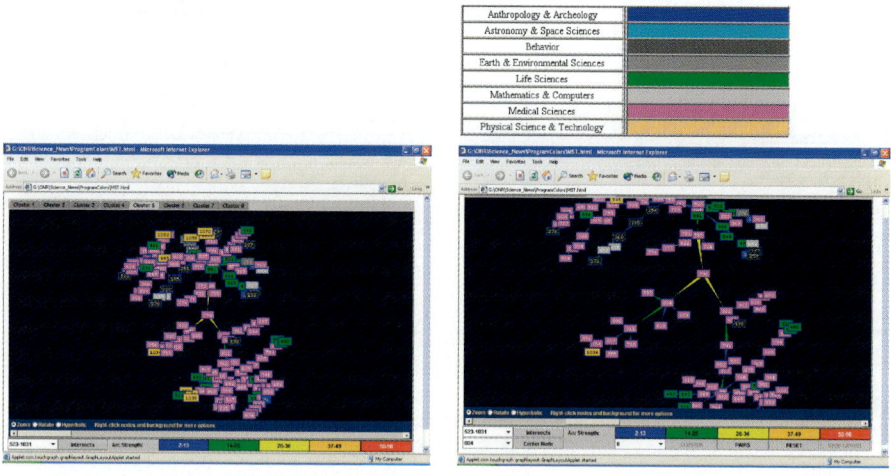

Chapter 5, Fig. 41, p. 166. A subcluster of Science News cluster 5 that is associated with AIDs.

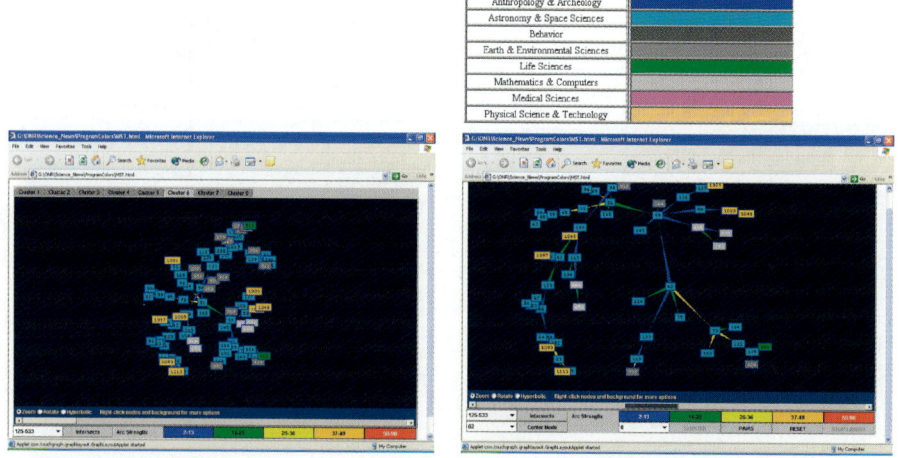

Chapter 5, Fig. 42, p. 167. A subcluster of Science News cluster 6 that is associated with solar activity.

Colour figures 593

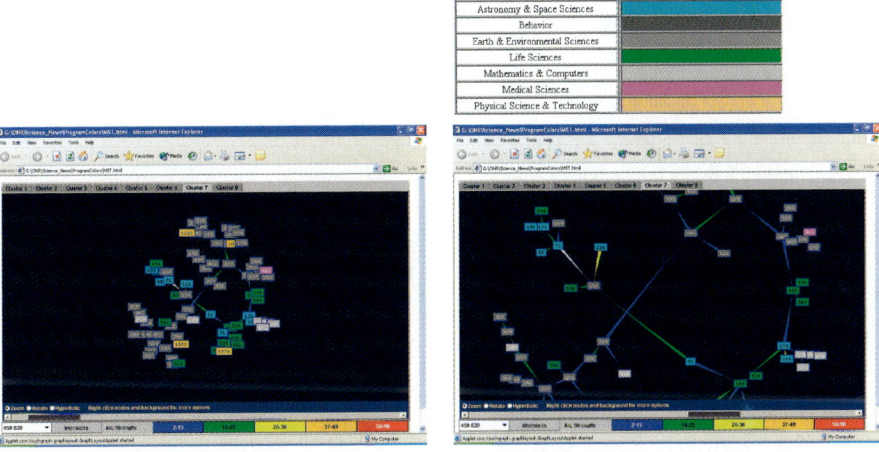

Chapter 5, Fig. 43, p. 167. A subcluster of Science News cluster 7 that is associated with the origins of life.

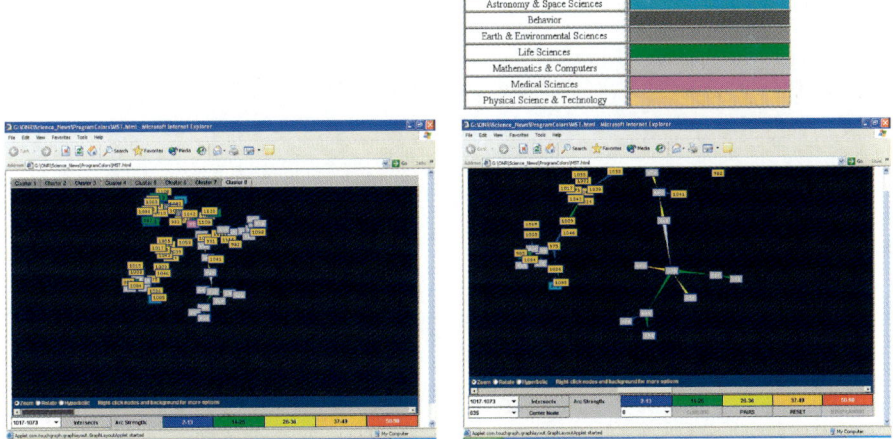

Chapter 5, Fig. 44, p. 168. A subcluster of Science News cluster 8 that is associated with artificial intelligence.

594 Colour figures

Chapter 6, Fig. 1, p. 176. Uncompressed color images (upper row) were saved as JPEG images at 80% quality (lower row). The JPEG images were inserted into the LSBs of the uncompressed images. Differences between uncompressed original images cannot be observed by a naked eye.

Chapter 6, Fig. 3, p. 181. Two natural images: a falcon (top) and a barley field (bottom). Image sizes are 450 × 292 (falcon) and 480 × 320 (barley).

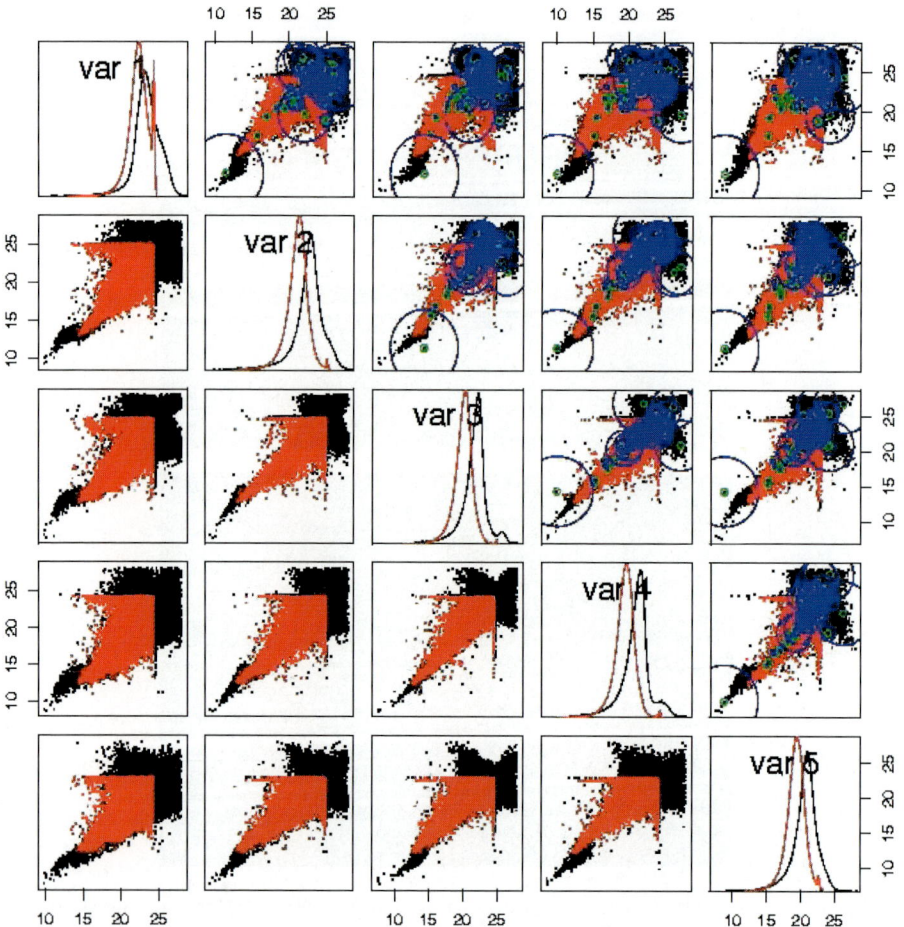

Chapter 12, Fig. 19, p. 351. Pairs plot of Sloan data. Circles are overlaid in blue in upper triangle, with centers indicated in green. Kernel densities of the two classes are plotted on diagonal.

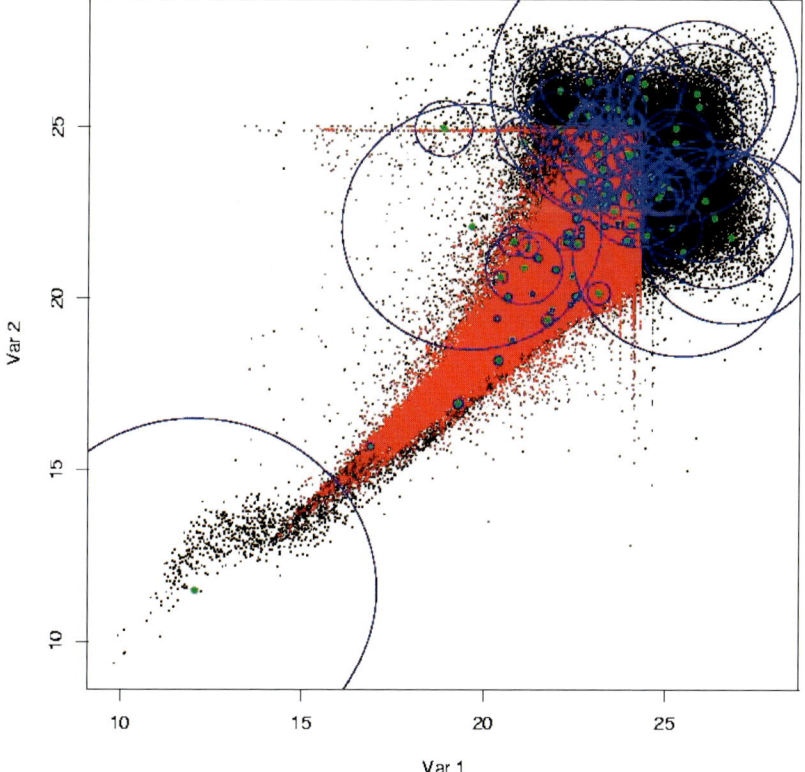

Chapter 12, Fig. 20, p. 352. Plot of Sloan data for variables 1 and 2. Circles are overlaid in blue, with their centers indicated in green.

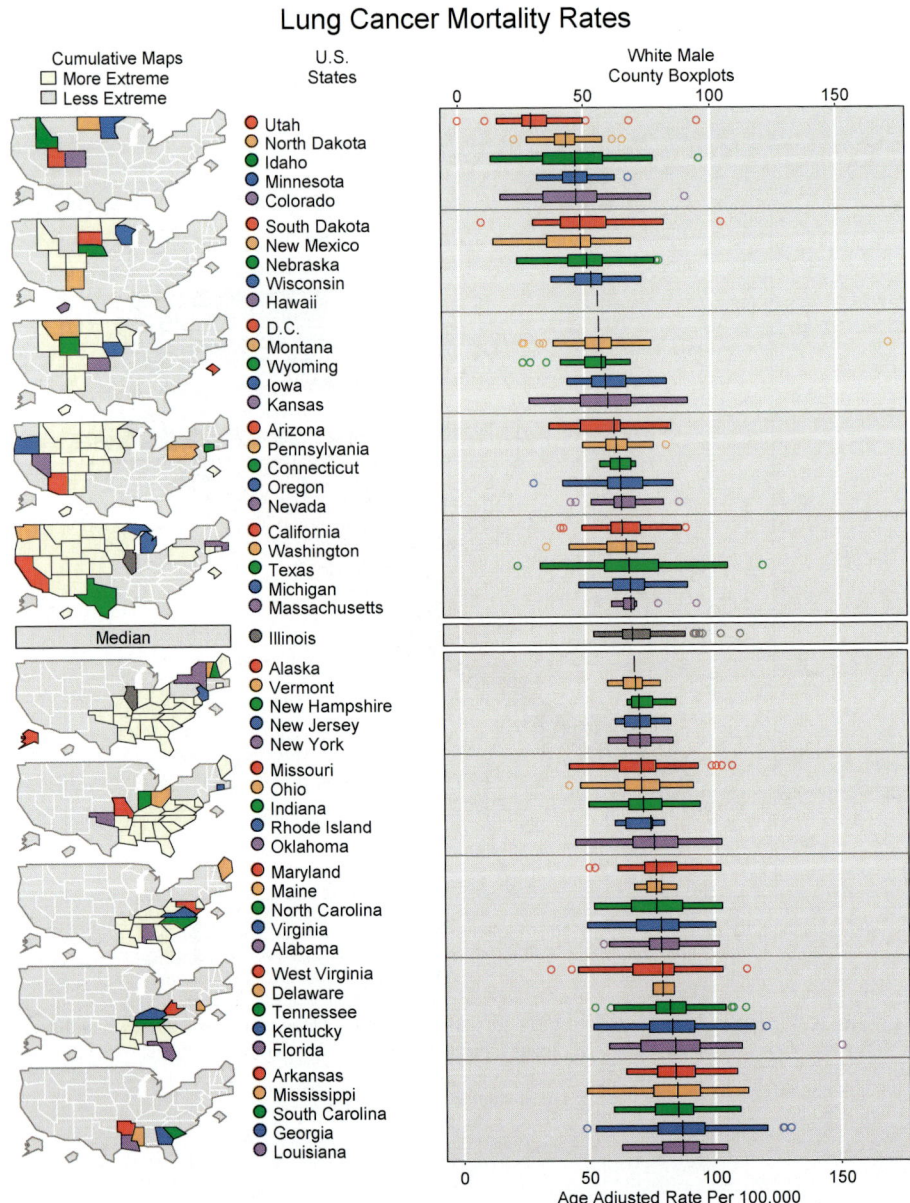

Chapter 15, Fig. 2, p. 421. An LM plot with a box plot schema.

Colour figures 599

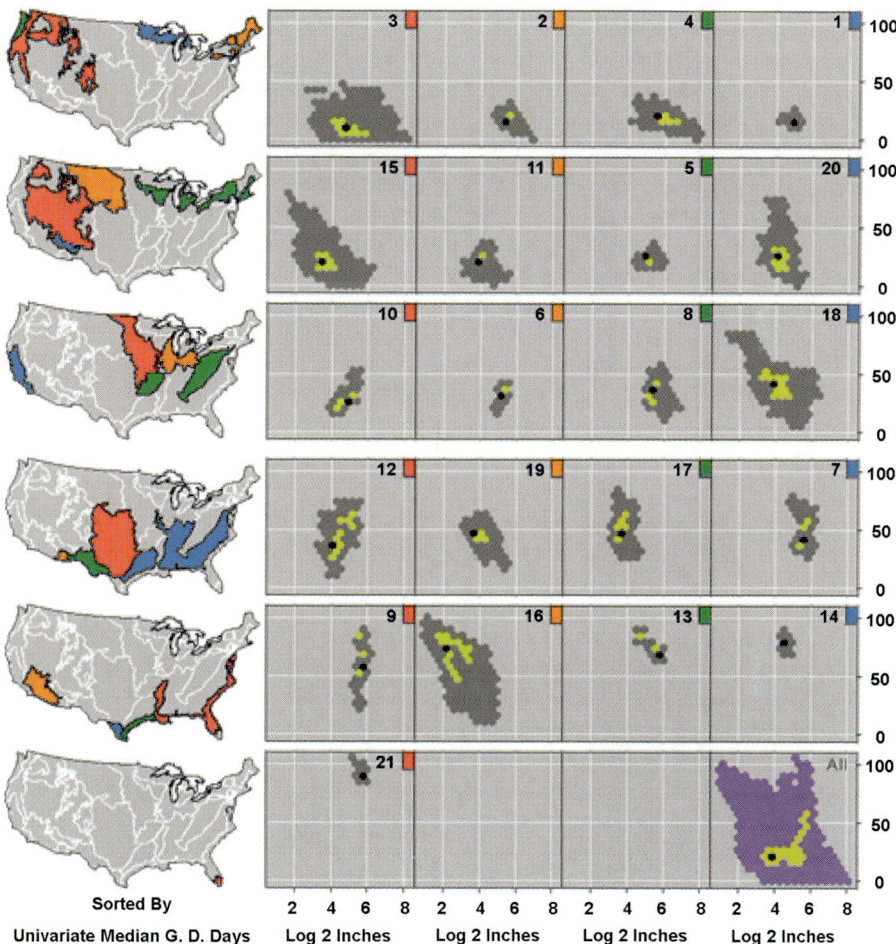

Chapter 15, Fig. 3, p. 423. LM bivariate boxplots. 1961–1990 precipitation (x) versus growing degree days/100 (y).

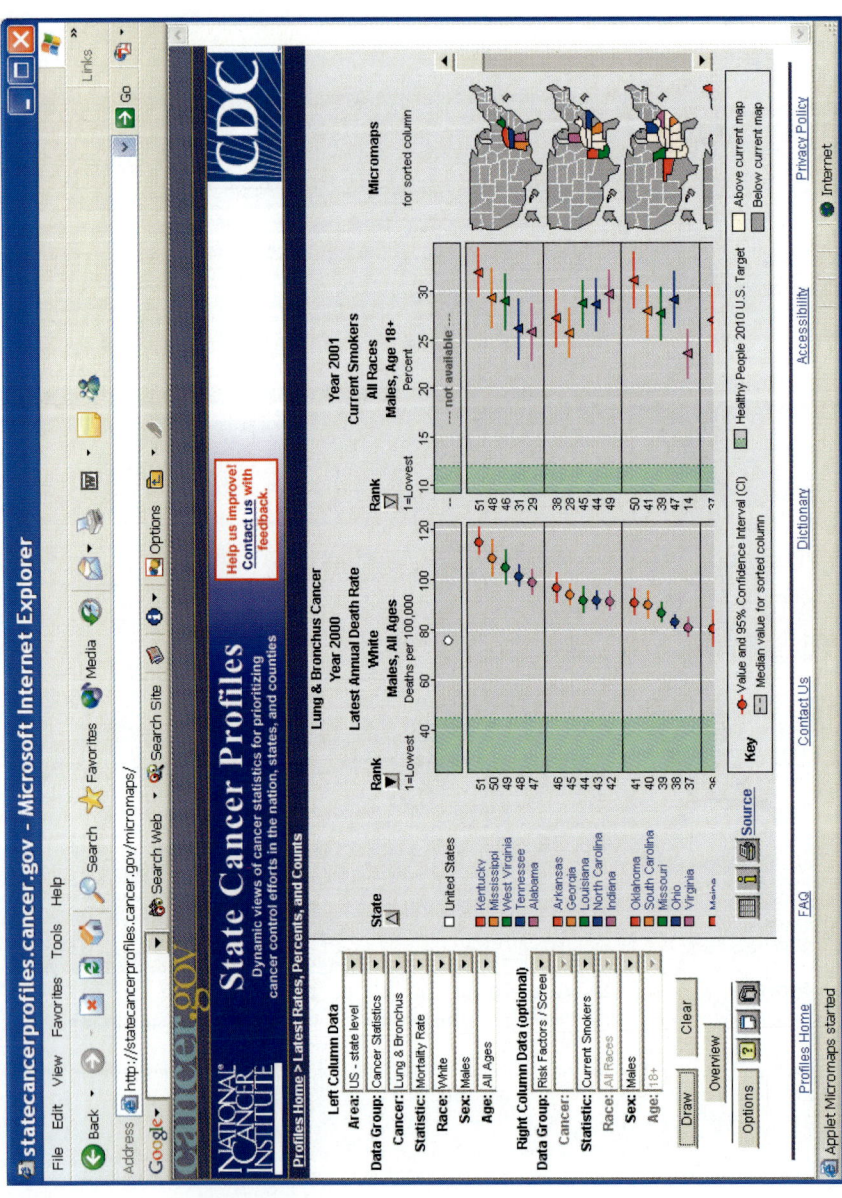

Chapter 15, Fig. 4, p. 425. Web-based LM plots with many interactive and dynamic features.

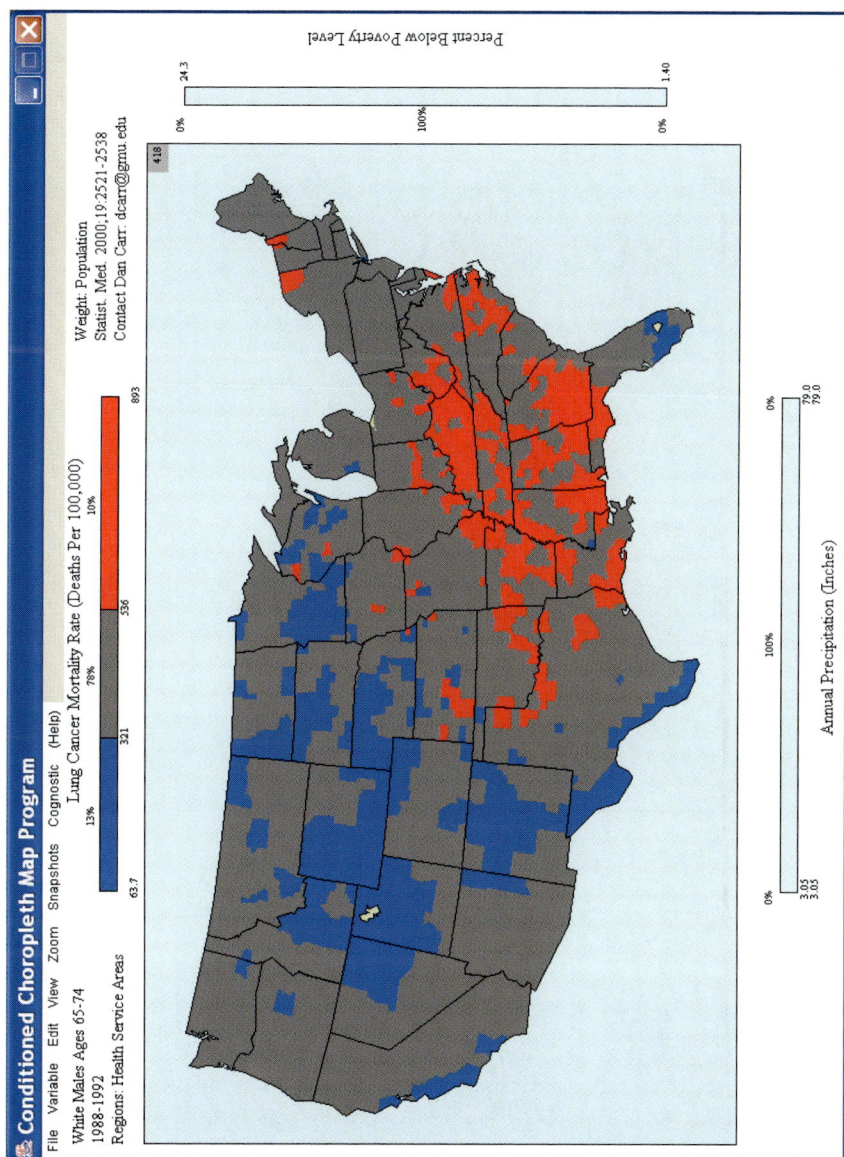

Chapter 15, Fig. 5, p. 429. CCmaps: an enlarged view of one conditioned panel with the conditioning constraints removed.

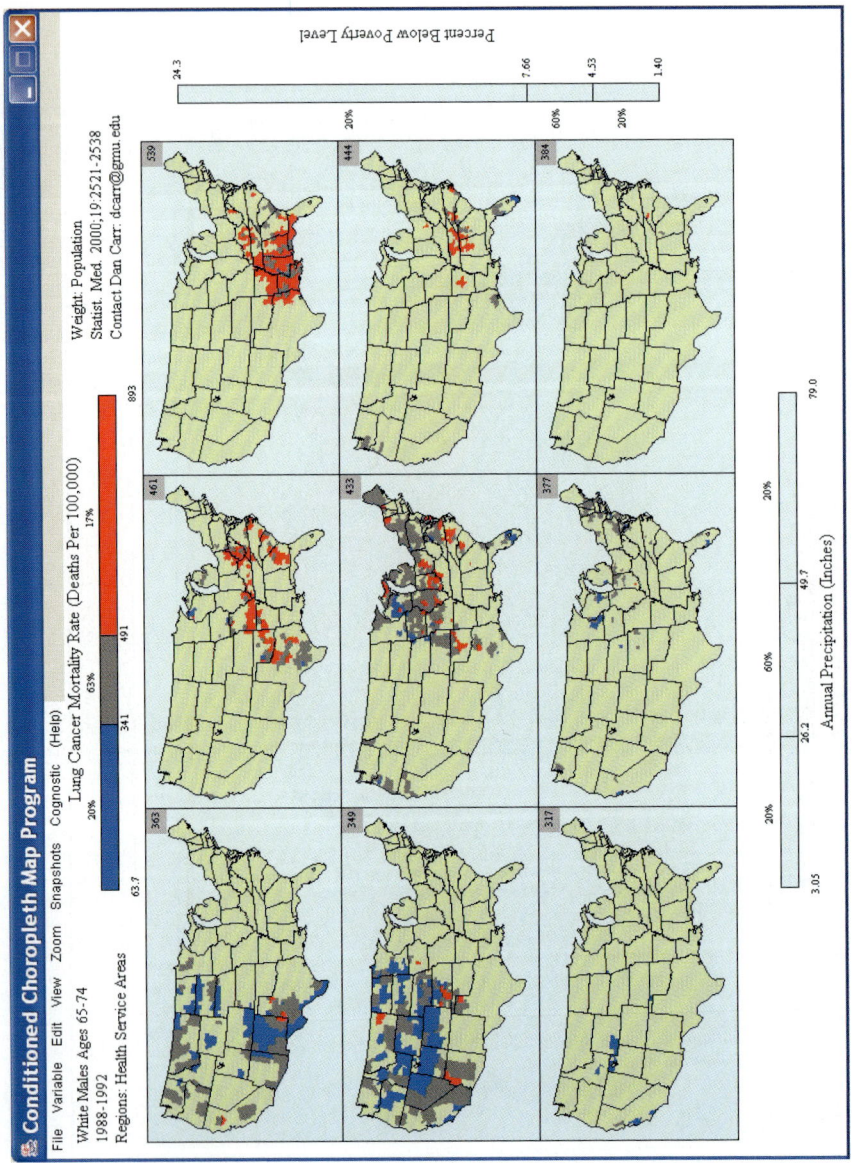

Chapter 15, Fig. 6, p. 430. CCmaps: a dynamic two-way partitioning of regions.

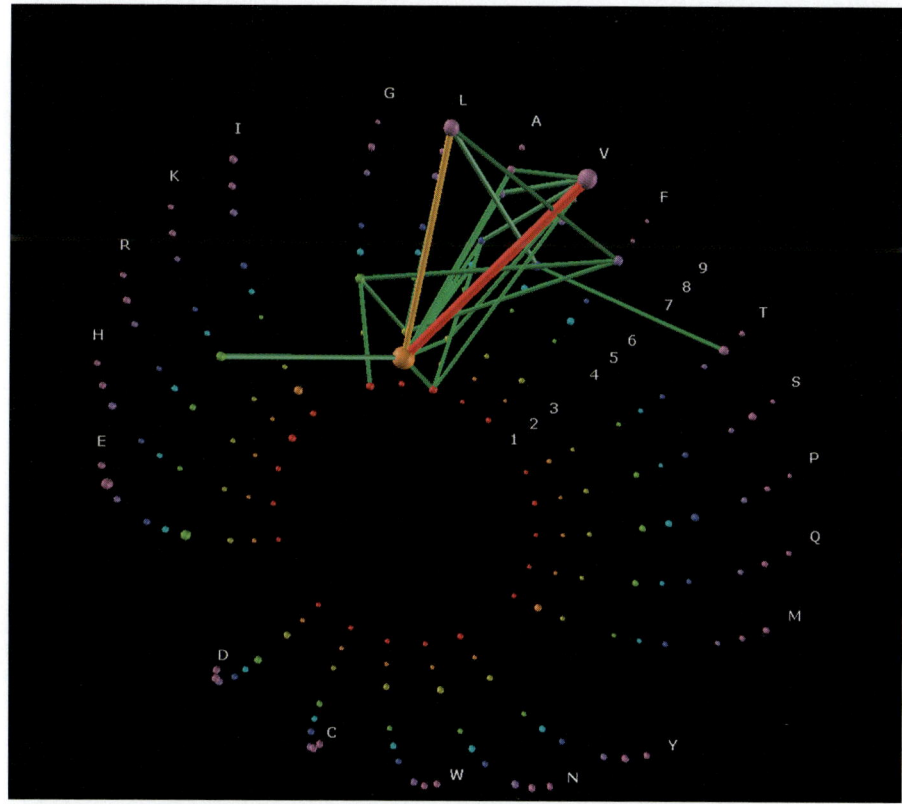

Chapter 15, Fig. 7, p. 432. GLISTEN: capless hemisphere coordinates with perceptual grouping alterations. Path shows immune system molecule docking statistics. The large count (thick path) is for Leucine in position 2 and Valine in position 9.

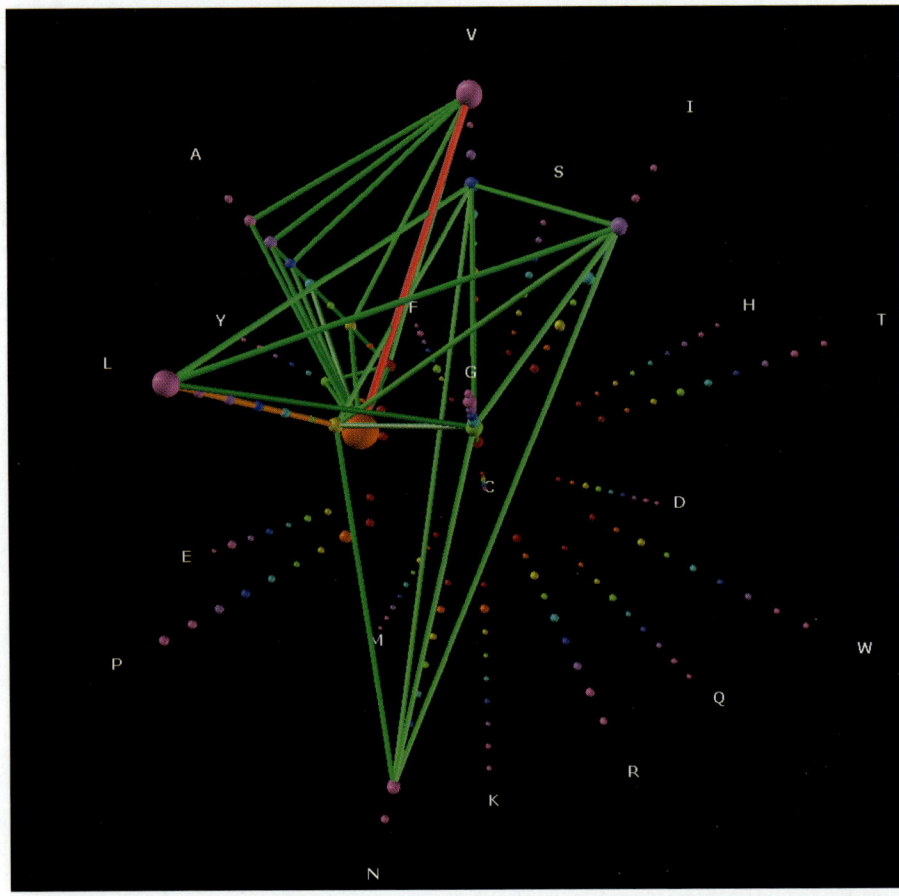

Chapter 15, Fig. 8, p. 433. GLISTEN: icosahedron face-center axes. High two-way table counts for different positions along amino acid 9-mers.

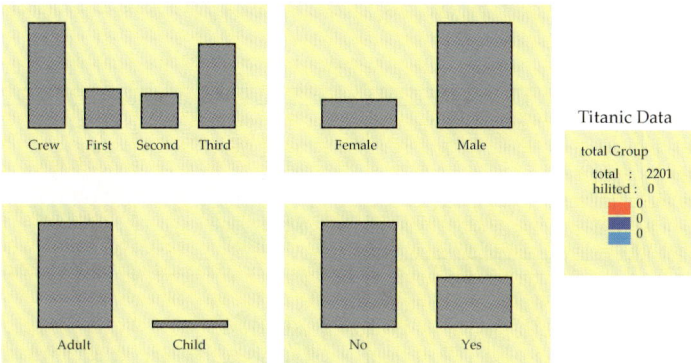

Chapter 16, Fig. 1, p. 451. Bar charts showing the marginal distributions of all variables in the Titanic data set. As can be seen in the group window on the right there are 2201 individuals in the data set. The data are stored in a case by case format which can be seen from the fact that there is no weighting variable indicated in the plot (cf. Figure 2).

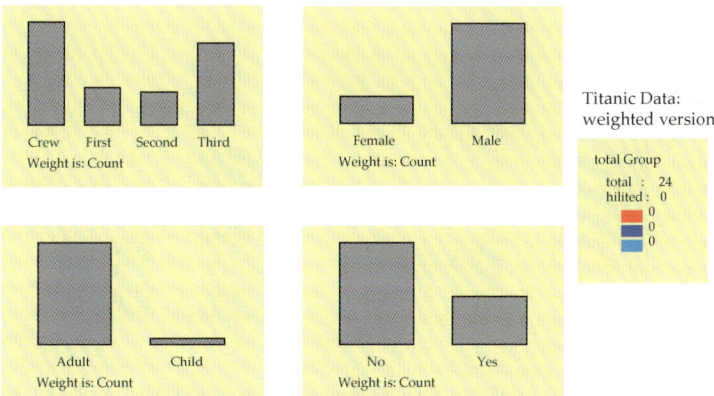

Chapter 16, Fig. 2, p. 451. Weighted bar charts showing the marginal distributions according to class, age, sex and survival of those on board the Titanic. As can be seen in the group window on the right only 24 items are indicated here that correspond to the non-empty cells in the contingency table.

606 *Colour figures*

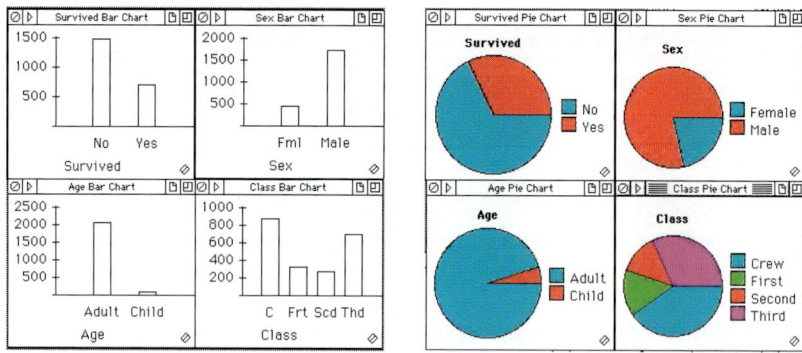

Chapter 16, Fig. 8, p. 461. Bar charts and pie charts for the four variables in the Titanic data set.

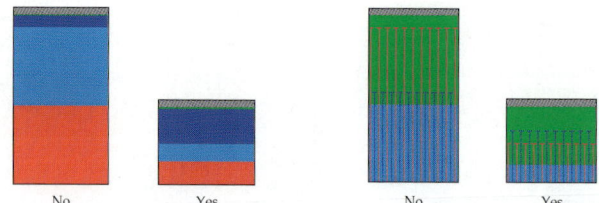

Chapter 16, Fig. 35, p. 508. Two different ways of coloring subgroups (and their intersection) in a bar chart. On the left-hand side the colors are *stacked*, on the right-hand side, colored bar charts are *overlayed*.

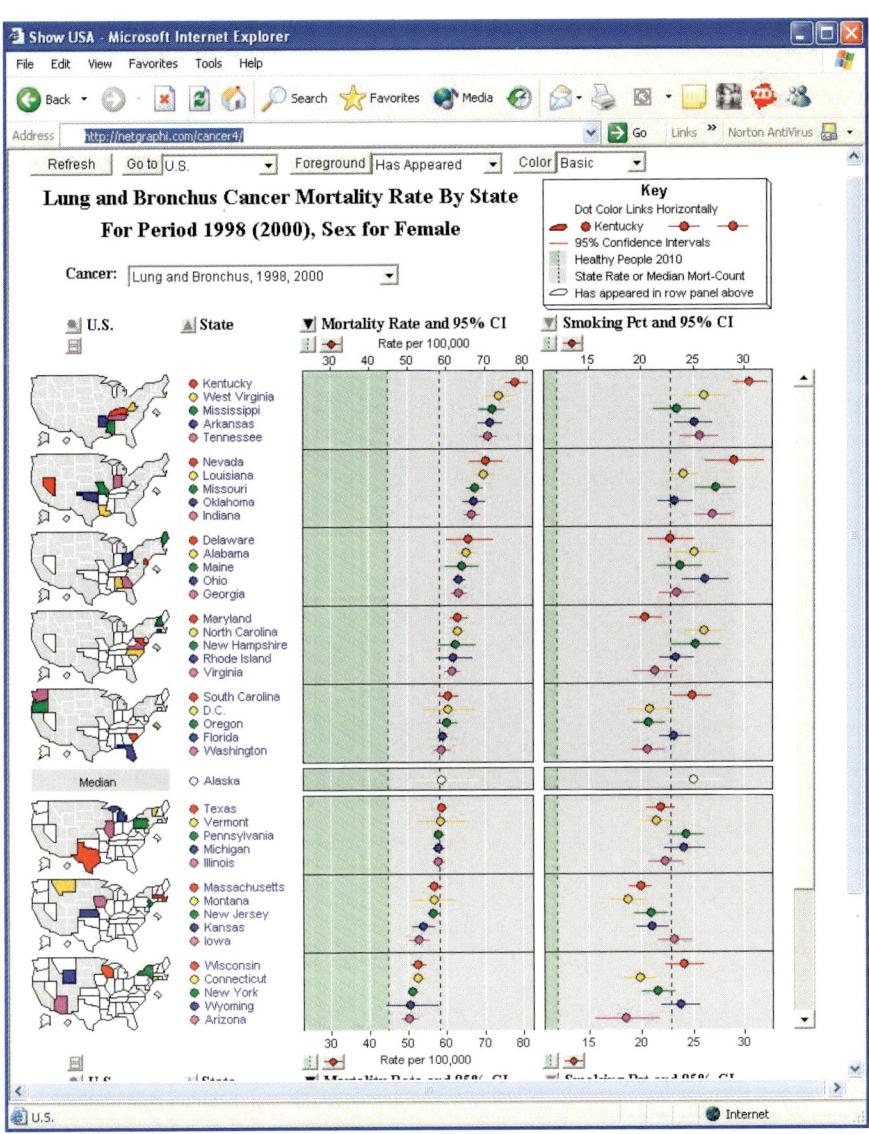

Chapter 17, Fig. 2, p. 552. A snapshot of the LM plots of the US cancer statistics.

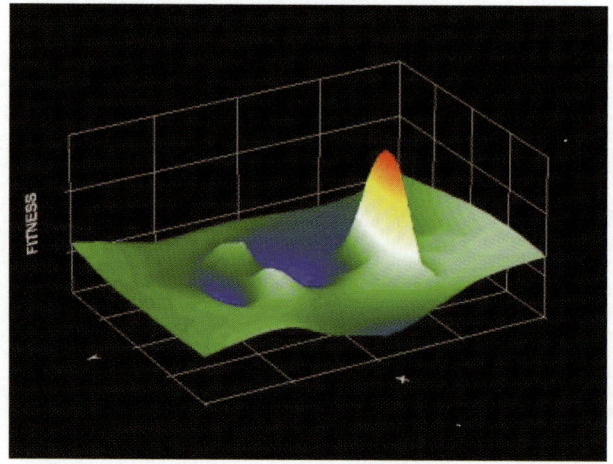

Chapter 17, Fig. 4, p. 555. An example of dimension reduction (QTM).

Subject Index

3D rotating plots, 437, 464
3D stereoscopic plots, 30
3DStudioMax, 545
3DVIEWNIX, 543
3DWin, 547

ABACUS, 56
ACK packet, 82
acknowledgment number, 82
activation function, 28
AdaBoost, 324
AdaBoost algorithm, 305
adaptive mixtures, 22, 251
addition operator, 386
advanced crossover techniques, 382
advanced mutation, 385
Advizor, 442
affine equivariant, 269, 279
agglomerative clustering algorithm, 25
AIC, Akaike's information criterion, 250
AID, automatic interaction detection, 304
algebraic linking, 440
algorithmic complexity, 1, 2
Alias|Wavefront's Studio, 545
Alice, 557
Amapi3D, 547
ANALYZE, 543
Andrew curves, 438
Andrews plots, 190
angular resolution of the human eye, 8
angular separation metric, 25
animation, 545
anonymity, 171
API, application programmer's interface, 543
application domain, 10
approximate MCD solution, 266
AQ learning, 60
AQ learning methodology, 50
arcing algorithms, 305
artificial neural networks, 27
artificial neural systems, 1
ASA Data Exposition dataset, 521

ASH, averaged shifted histogram, 229, 239, 241, 254, 257, 485
ASH software, 254
aspect ratio, 482
association rules, 11, 47
asymptotic properties, 21
attributes, 14
attributional calculus, 51
attributional descriptions, 51
attributional rules, 51
attributive highlighting, 505
AutoCAD, 544
automated serendipity system, 139
average shifted histogram, 229, 239, 241, 254, 257, 485
AVHRR, 424
AVS/Express, 543
axes based selection, 474
axon, 27

back propagation, 29, 224
backscatter, 87
BACON, 56
bagging, 226, 303, 305, 318
balloon estimator, 243
bandwidth matrices, 247
bar charts, 460, 547
base learner, 324
Bayes classifier, 321
Bayes error rate, 216
BIC, Bayesian information criterion, 251
Bayesian misclassification rate, 257
Bayesian nets, 47
Bayesian networks, 14
Bertin's semiology, 439
bi-directional linking, 509
bigram proximity matrix, 109, 122, 124, 133
binary alphabet, 371
binary tree, 308
binned bivariate scatterplots, 422
biplot, 189, 438, 464
bit parity, 177

block-recursive plots, 42
BMDP, 441
BMP, 172
BMRT, 544
Boolean splits, 310
boosting, 226, 303, 305, 323
bootstrap aggregation, 226, 305, 318
bounded rationality, 155
boxplot, 418, 463
BPM, bigram proximity matrix, 109, 122, 124, 133
breakdown value, 275, 279
BRUSH-TOUR, 31
brushing, 441
Building Block Hypothesis, 372, 379
building blocks, 379

CAD, 544
CAG1, 63
calculus-based schemes, 362
Canberra metric, 25
canonical coordinates, 190
canonical variate analysis, 189
CART, classification and regression trees, 21, 47, 303, 304
cases, 14
categorical data analysis, 21
categorize operator, 456
CBC, clustering by committee, 118
CC, conditioned choropleth, 415, 416
CC map sliders, 418
CC maps, 416, 418, 427
CCmaps, 428
CDC, Centers for Disease Control and Prevention, 426
change rate, 175
changing frame, 475
characteristic descriptions, 52
characteristics, 214
Chernoff faces, 190
CHAID, chi-squared automatic interaction detector, 304
choropleth maps, 479
chromosome, 365
city block metric, 25
CCCD, class cover catch digraph, 331, 332
classical density estimators, 230
classification, 14, 21, 282, 331, 519
classification and regression trees, 21, 47, 303, 304
classification complexity, 135
classification rule, 215
classification tree, 306
classifier, 51

cluster analysis, 24, 284
cluster catch digraph, 338
clustering, 1, 14, 21, 519
coding, 365
cold linking, 497
color histogram, 16
Committee on Applied and Theoretical Statistics, 38
complete linkage, 25
computational complexity, 18, 331
computational feasibility analysis, 4
computational linguistics, 109
computational steering, 549
computationally feasible, 6
computer intrusion detection, 47
computer security, 77
computer security system, 77
concept association graphs, 62
conceptual clustering, 53
conceptual cohesiveness, 54
conditional plots, 42, 465
conditional probabilities, 511
confidence, 12
confirmatory analysis, 1, 21
CONSEQUENT, 50
conservation of energy and momentum, 32
constructive induction, 55
contingency table, 449
continuous animation, 392
continuous response models, 524
contours, 229
COPER, 56
copyright marking, 171
correlation and regression, 21
correspondence analysis, 21, 190, 201
cost–complexity measure, 313
covariates, 214
covert channels, 171
CPCA, 290
CPU performance, 4
CQDA, classical quadratic discriminant analysis, 282
create rule, 23
cross-corpora relationships, 135
cross-validation, 125, 229, 235, 312
cross-validation estimate, 312
crossover, 360
crossover operator, 365, 367
Crossroads, 547
CrystalVision, 396, 551
CT, computed tomography, 548
current solutions, 10
curse of dimensionality, 18, 22, 229, 257

Subject Index

d-dimensional simplex, 18
data cleaning/data quality assurance, 15
data cube, 20
Data Desk, 440
data displays, 446
Data Explorer, 543
data image, 16
data inconsistencies, 15
data management operators, 67
data mining, 1, 9, 109, 439, 532
data preparation, 11, 14
data queries, 475
data rotations, 391
data visualization, 391, 539
data warehousing, 532
data-driven choice, 394
databases, 19
datacube, 20
DataViewer, 395
decision surface, 215
decision tree, 51
decomposition model, 58
deletion operator, 386
DEM, digital elevation model, 424
dendrites, 27
dendrogram, 26
denial of service attacks, 84, 86, 88–90, 97
density curves, 477
density plots, 31
design set, 216
destination port, 81
diagnostic paradigm, 214
diagnostic plot, 289
diagrammatic visualization, 62
digital watermarking, 181
Digitized Palomar Sky Survey, 270
dimensionality reduction, 1, 15, 194
Direct3D, 542
directed graph, 332
discounting older data, 39
discrete cosine transform, 173
discriminant analysis, 14, 21, 263, 284, 303
discriminant descriptions, 52
discriminant function, 216
display device, 539
dissimilarity matrix, 189
distributed grid type architecture, 5
DIVE, distributed interactive virtual environment, 557
document retrieval, 119
dominating set, 333
dotplots, 439, 458
drag box, 471
drill down, 20, 553

drill through, 20
drill up, 20
duplicate data removal, 15
dynamic graphics, 30, 445

Earth Simulator Center, 5
EDA, 1
elitism, 381
EM, expectation-maximization, 249
EM algorithm, 22, 117, 249, 250
EM iterations, 252
embedding efficiency, 175
embedding rate, 175
encryption, 369
ensemble classifiers, 225
EnSight, 544
enumerative-based optimization schemes, 363
error rate, 214
error-correcting procedure, 220
estimating bandwidth matrices, 247
Euclidean distance, 25
evolution strategies, 359
evolutionary algorithms, 359
evolutionary computation, 359
evolutionary graphics, 30, 40
evolutionary programming, 359
evolutionary systems, 360
EXCEPTION, 50
excitatory, 27
Explor4, 443
exploratory data analysis, 1, 229, 437
ExplorN, 396, 443, 551
exponential smoother, 39
external linking structure, 496

FAHRENHEIT, 56
FAST, 544
fast algorithms, 331, 340
FAST-LTS algorithm, 275, 297, 299
FAST-MCD algorithm, 264, 266, 268, 271, 297
feature aggregation, 10
feature construction, 10
feature discretization, 10
feature extraction, 140
feature selection, 226
feature subset selection, 10
features, 214
FERRET, 544
FFT, Fast Fourier Transform, 241
FIELDVIEW, 544
file transfer performance, 7
fingerprinting, 83
finite-sample breakdown value, 265

firewall, 92
Fisher linear discriminant analysis, 253
Fisher's Iris data, 449
Fisher's linear discriminant, 282
fitness technique, 381
flags, 80, 82
flow chart for the data mining process, 10
fluid visualization, 548
focus of attention, 149
fragment offset, 80
frame, 447
frame buffer, 540
frame color, 476
frame interpolation, 406
frame regions, 469
fraud detection, 47
frequency histogram, 231
frequency polygons, 229, 230, 237, 240
Frobenius norm, 400
full-dimensional tours, 395
fuzzy logic controller, 376

gaining access, 91
Gaussian kernel, 241
gene, 365
gene regulation nucleotides, 431
generalization error, 311
generalized penalty methods, 372
genetic algorithm fitness, 380
genetic algorithm parameters, 385
genetic algorithms, 359, 548, 554
genetic operators, 359
geodesic, 403
Geographic Information Systems, 440, 543
geometric structure, 520
geometric transformation, 540
geometry-based tessellation, 17
GGobi, 396
GIF, 172, 173
Gini index of diversity, 310
GIS, geographic information systems, 440, 543
Givens paths, 408
Givens rotation, 399
GLISTEN, geometric letter-indexed statistical
    table encoding, 415, 432–434
GLUI, 547
Gnuplot, 544
golden rectangle, 482
gradient boosting, 326
Gram–Schmidt, 405
grammar induction, 114
grand tour, 31, 392, 437
graph-theoretical classifier, 331
graphical elements, 469

graphical simulation, 546
GUI, graphical user interface, 419
graphics library, 540, 542
graphics templates, 415
greedy algorithm, 317, 333, 341
grouping, 487
grouping categories, 484
guided tours, 392

handling of missing data points, 15
header, 172
Hellinger distance, 204, 233
hidden layers, 28
hierarchical clustering, 25
hierarchical linking, 504
high-dimensional data spaces, 391
high-dimensional rotations, 391
high-interaction graphic, 445
histograms, 229, 462, 477, 547
HMD, head-mounted device, 556
HMM, hidden Markov models, 106, 109, 111
Householder paths, 410
Huffman coding, 174
Huffman decoding, 174
Huffman encoded, 174
human perception and cognition, 416
hybridization, 380
hypercube, 18
hyperspectral imagery data, 34

ICMP, internet control message protocol, 81
identity linking, 503
identity model, 453
illumination, 542
image tour, 395
ImmersaDesk/CAVE, 556
immersion, 560
impurity function, 310
In-house Laboratory Independent Research
    (ILIR) Program, 140
incremental learning, 58
independence Bayes model, 219
independence models, 530
INDUCE, 51
inductive databases, 49
inductive learning, 48
information hiding, 171
information radius measure, 124
information visualization, 548
inhibitory, 27
INLEN system, 66
integrated absolute error, 233
integrated squared error, 233

interactive feasibility, 6
interactive graphics, 30, 445
interactive GUI, 426
interactive statistical graphics, 437, 439
interactive visualization, 549
InterChange, 547
Interface Foundation of North America, 438
internal linking, 488, 489
Internet, 77
interpoint distance, 341
interpoint distance matrix, 16
interpoint distance measure, 133
interrogation, 446, 495
inventory resources, 10
inversion operator, 386
IP, internet protocol, 78, 81
IP address, 85, 92
– of source, 80, 86, 88, 90
– of destination, 80, 89
IP header, 80, 81
IP layer, 79, 80
IP scans, 85
IP version, 80
IRad, 124
IRIS Explorer, 543
IRIS Performer, 546
Irix, 96

Jaccard coefficient, 124
Java AWT, 547
Java3D, 557
JMP, 442
JND, just noticeable difference, 183
JPEG, 172, 173
juxtaposition, 507

k-dimensional tours, 395
k-nearest neighbour, 222
kd-trees, 343
KDD, 2
kernel density estimation, 222
kernel density estimator, 39
kernel density smoother, 22
kernel estimators, 239, 240
kernel-based methods, 303, 327
keystroke timings, 102
Khoros, 543
knowledge discovery in databases, 2, 9
knowledge generation operators, 49, 66, 67
knowledge management operators, 67
knowledge mining, 47
Kruskal's algorithm, 136
Kullback–Liebler distance, 233

$L_2 E$ method, 250
L_2 theory of histograms, 233
large datasets, 548
lasso, 470
latent semantic indexing, 109, 122, 130
Lawrence Livermore National Laboratory, 5
LDA, linear discriminant analysis, 253
learning sample, 308
least significant bits, 172
leverage points, 273, 279, 289
lighting, 542
LightScape, 544
LightWave, 545
linear combination split, 309
LDA, linear discriminant analysis, 217
linear models, 456
linear normalization technique, 381
linear probability, 382
link layer, 79
linked low-dimensional views, 509
linked micromap plots, 417, 420, 422, 551
linked micromaps, 415, 416, 482
linked scatterplots, 440
linked views, 31, 437
linking, 446
– 1-to-1, 491
– 1-to-n, 493
linking frames, 498
linking observations, 500
linking sample populations, 503
linking scales, 501
linking types, 499
Linux, 96
Lisp-Stat, 397
LispStat, 440
LM, linked micromap, 415, 416, 482
LM bivariate boxplots, 423
LM plot applet, 424
LM plots, 417, 420, 422, 551
local nonparametric methods, 222
locally adaptive estimators, 243
log files, 77, 91
log-sqrt transform, 42
logical data analysis, 49, 59
logical zooming, 486
logistic curve, 28
logistic discrimination, 218
low-dimensional projections, 391
low-dimensional views, 439
LSBs, 172
LSI, latent semantic indexing, 109, 122, 130
LTS, least trimmed squares, 264, 274
LTS estimate, 274

LTS location estimator, 276
LTS objective function, 275
LTS regression, 275, 292

machine learning, 47
MacOS, 96
magic pen, 539
Mahalanobis distance, 23, 190, 263
Mallows' C_p, 318
Manet, 440, 443
manual choice, 394
manual tours, 392
market-based analysis, 11
MARS, multivariate adaptive regression splines, 316
masking effect, 265
massive datasets, 2, 435
Materialize3D, 547
Mathematica, 546
Matlab, 304, 543
Matlab library LIBRA, 299
matrix encoding, 177
maximum likelihood, 232
Maya, 545
MCD, minimum covariance determinant, 264
MCD breakdown value, 266
MCD estimate
– of location, 265, 269, 288
– of scatter, 265, 269
MCD estimator, 264
MCD method, 265
MCD objective function, 266
MCD regression, 279, 280
MCD regression estimates, 279
measures of semantic similarity, 123
membership function, 377
method of moments, 231
micromap magnification, 553
Microsoft Excel, 452
Microsoft Visual C++ MFC/Studio, 547
MicroStation, 544
mini-CAVE, 556
minimal spanning tree, 27, 133, 135, 422
minimal subspace restriction, 398
minimum cost–complexity pruning, 313
minimum dominating set, 333
Minitab, 441
misclassification costs, 308
misclassification rate, 214, 283, 311
missing value imputation, 15
missing values, 15, 317, 466
mixture density estimation, 248
mixture models, 229
mixture of normal densities, 22

MLTS, multivariate least trimmed squares estimator, 280
model, 447
model linking, 492
model operations, 452
model-based clustering, 109, 125
modeling, 544
models with missing values, 457
MOLAP, 20
mosaic plots, 437, 462, 515
Motif, 547
MPEG3, 172
MRI, magnetic resonance imaging, 548
MST, 135
multi-corpora document sets, 140
multi-modal displays, 560
multi-objective optimization, 375
multidimensional density estimation, 229
multidimensional scaling, 190
MultigenPro, 544
multilayer perceptrons, 223
multiple regression, 272
multiple views, 441
multisensory, 560
multivariate contingency tables, 512
multivariate outlier detection, 263
multivariate regression, 263, 278
mushroom dataset, 63
mutation operator, 365, 368
mutual information, 118

n-gram, 106
naive Bayes model, 219
naive greedy algorithm, 334
National LambdaRail, 7
natural language processing, 109, 110
natural selection, 359
navigation, 553, 560
NCAR, 544
NCI, National Cancer Institute, 424
nearest neighbors, 331
nearest neighbors methods, 303
nearest-neighbor classification, 257
neighborhood linking, 504
network data, 77
network monitoring, 92
network sensor, 92
neural networks, 223
neuron, 27
neurotransmitters, 27
Neyman–Pearson criterion, 215
nonparametric density estimation, 21, 229
nonparametric kernel methods, 229

nonparametric regression, 21
normal mixture model, 23
number of groups problem, 26

Ochiai measure, 124
Ochiai similarity measure, 134
OLAP, 20
OLS, ordinary least squares, 303
OLS regression, 316, 320
Omernik ecoregions, 422
online analytical processing, 20
OpenGL, 542
OpenInventor, 557
optimization mechanism, 360
order of magnitude, 2
orthogonal outliers, 289
orthogonal series density estimators, 248
orthonormal frames, 397
out-of-bag estimates, 320
outlier detection, 21, 263
outliers, 15
overfit predictor, 327
overlaying, 505

P-splines, 248
painting, 441
pair operator, 454
pairs of values, 178
parallel coordinate plots, 30, 395, 398 432, 437, 438, 464
parallel coordinates, 16, 31, 96, 190, 550
parallel profiles, 190
parallel projection, 541
PRISM, Parameter-elevation Regression and Independent Slopes Model, 424
Pareto-optimal optimization, 375
parsing, 110
part-of-speech tagging, 110
partial least squares regression, 296
partially mixed crossover, 383
particle-physics scattering experiment, 32
passive fingerprinting, 83
pattern recognition, 213
PCA, 197, 283, 464
perceptron, 220
perceptron criterion function, 221
perceptual accuracy, 417
performance assessment, 226
periodic model, 58
perspective projection, 541
perturb and combine methods, 303
PGM, 172
phenotype, 365
PHIGS, 542

pi-mesons, 32
pie charts, 460, 547
planar rotations, 397, 399
PLANET, 426
plausibility values, 325
plausible knowledge, 49
PLSR, 296
point-of-sale transactions, 12
pointer, 470
PolyForm, 547
polygon map, 463, 477
PolyTran, 547
posterior probability, 22
potentially subsampling, 15
POVRay, 544
PoVs, 178
Precomputed choice, 394
PREMISE, 50
preprojection, 398
presence, 560
PRIM 7 data, 32
PRIM-9, 444
principal component analysis, 197, 283, 464
principal component regression, 292
principal components, 37, 263
prior domain knowledge, 10
probabilistic context-free grammars, 109, 112, 114
probes, 85
problem objectives, 10
program profiling, 104
projection operator, 456
projection pursuit, 286, 394, 437
projection views, 446
PROLOG, 51
PROMISE, 61
proportional highlighting, 506
proportionate grouping, 366
prot scans, 85
protocol layer, 79
pruning, 313
pruning and cropping, 31
PSH, 83
PV-WAVE, 543
pV3, 544

QQplot, 271, 431
QDA, quadratic discriminant analysis, 218
QTM, quad-tree mapping, 554
Quail, 443
qualitative prediction, 57
quantitative graphics, 439
quantization, 15, 17

querying, 494
QUEST, quick unbiased efficient statistical tree, 304

R, 441
Radiance, 544
Random choice, 394
random digraph, 332
random grand tours, 395
rank grouping, 366
ranking technique, 366
Rao distance, 210
raster data, 172
rasterization, 544
raw image format, 172
ray-tracing, 542
Rayshade, 544
real-coded crossover operator, 373
receptor site, 27
ReClus, 126
recombination, 360
records, 14
rectangular brush, 470
recursive binary partitioning, 307
recursive formulation, 39
recursive methodologies, 2
regression tree, 315
relational databases, 9, 19
rendering, 544
RenderMan, 544
renormalization, 256
reproduction operator, 365
resizing frame, 476
resubstitution estimate, 311
reweighted estimate, 269
reweighted MCD estimates, 279, 285, 288
reweighted regression estimates, 279
RIPPER, 61
risk, 215
RMSECV$_k$, 292
ROBPCA, 290
robust estimation, 263
robust LTS-subspace estimator, 289
robust PCA, 285
RQDA, robust quadratic discriminant analysis, 282
robustness, 263
rotation-invariant Riemannian metrics, 400
RSIMPLS, 296
ruleset, 51
ruleset family, 51
ruleset visualization, 62

sample population, 447

SAS, 298, 441
SAS Insight, 442
saturation brushing, 31
SC, self-similar coordinate, 415
SC plots, 416, 432
scalability, 2
scaled fitness, 380
scaling, 445
scan-conversion, 540, 544
scans, 85
scatterplot, 459
scatterplot brushing, 439
scatterplot matrices, 30, 395, 465
scatterplot smoother view, 431
schema, 370, 371
schema theorem, 378
SciAn, 544
Science News, 140
Scott's rule, 235
selection, 446, 469
selection memory, 472
selection of axes, 469
selection operation, 472
selection tools, 472
self-consistent, 17
self-organizing maps, 109, 129
semantic content, 133
shading, 542
sigmoid curve, 28
signatures, 101
Simpson's paradox, 517
SIMUL8, 546
simulation, 545
single linkage (nearest neighbor), 25
singular value decomposition, 191, 404
SiZer, 242
skyline plots, 43
slicer, 470, 471
Sloan Digital Sky Survey, 331, 351
smoothing models, 457
sniffer, 92
SoftImage, 545
Solaris, 96
source port, 81
space-filling curve, 392
spatial data, 440
spine plots, 462, 514
spline estimators, 248
split operator, 455
S-Plus, 254, 297, 304, 441
spoofing, 86
SPSS, 441
SQL, 9, 19

Subject Index

squashing the dataset, 17
squishing the dataset, 17
standard plot types, 457
standardization, 15
StarCD, 544
static graphic, 30
static graphics, 30
Statistica, 441
statistical data mining, 1
statistical grammar learning, 114
statistical optimality, 21
statistical pattern recognition, 1, 21
statistical summaries, 415, 553
steady-state technique, 382
steganalysis, 171, 179
steganography, 171
stereoscopic visualization, 556
Stiefel manifold, 400, 409
Stoke's law, 56
stop splitting rule, 312
stopper words, 140
streaming data, 2, 37, 344
streaming data sets, 435
structured query language, 9, 19
Sturges' rule, 231, 526
sub-dominating set, 354
substring-swapping operator, 371
success criteria, 10
supervised classification, 213
supervised disambiguation, 115
supervised learning, 14
support, 12
support vector machines, 221, 224, 327
surrogate splits, 317
survival of the fittest, 359
SVD, singular value decomposition, 191, 404
symbolic reasoning, 48
SYN flood, 90
SYN packet, 82
SYN/ACK packet, 82
synapse, 27
Systat, 441

tables, 19
TCP, transmission control protocol, 78, 81
TCP header, 81, 99
TCP packets, 86, 92
– malformed, 97
– unsolicited, 88
TCP session, 82, 97
TCP/IP, 78
telnet packet, 99
term weights, 121
terrorism prevention, 47

test sample, 311
test set, 216
text classification, 14
text data mining, 109, 133
text list, 465
text processing, 355
text retrieval, 109, 119
text understanding, 109
texture mapping, 542
Theta AID, THAID, 304
TIFF format, 173
time to live, 80
time-series methods, 21
torus method, 395
TouchGraph, 136
tour method, 551
TOUR-PRUNE, 31, 34
tournament grouping, 366
trace plot, 459, 509
training data, 331
training error, 311
training set, 216
transfer function, 28
transient geographic mapping, 40
transmission coefficients, 542
transparency, 542
tree classifiers, 221
trivariate ASH, 255, 256
Trojan programs, 85, 99
TrueSpace, 545
truncated octahedra, 18
tunneling, 97
two-point crossover, 383
two-way contingency tables, 201
two-way tables, 431
type, 447
type of service, 80

UDP, user datagram protocol, 81
UDP packet, 101
unbiased cross-validation, 236
uniform crossover, 383
uniform order-based crossover, 384
uniform order-based mutation, 384
unsupervised disambiguation, 116
unsupervised learning, 14
unsupervised pattern recognition, 213
update rule, 23
user profiling, 102

validation set, 216
variable list, 465
variable transformations, 453

variable-mesh histogram, 236
variables, 14, 214
Vector Data, 548
vector space model, 119
vertex random graphs, 332
vertical outliers, 279
video controller, 540
VIEW3D, 547
viewing, 541
viewing transformation, 541
VINLEN, 66
virtual environment, 539
virtual reality, 470, 539, 556
virtual world, 539
viruses, 77
Vis5D, 544
Visible Human Project, 543
visual analytics, 415
visual resolution, 8
Visual3, 544
visualization, 21, 229, 543
volume rendering, 543
Volumizer, 543
VolVis, 543
Voronoi tessellations, 27
voxel, 547
VREX stereo projectors, 556
VRML, virtual reality modeling language, 543, 557

VR simulation tools, 558
VR tools, 558
Vtk, 543

waterfall diagrams, 40
watermarking, 171
wavelet bases, 248
weak learners, 323
weakest-link cutting, 313
WEBSOM, 129
weight operator, 455
weighted resubstitution error rate, 324
weighting, 484
winding algorithms, 408
windowing, 380
Windows, 96
word sense disambiguation, 109, 115
worms, 77
WYSIWYG, 495

XForms, 547
XGobi, 393, 395, 396, 442, 551
Xgvis, 442
Xlisp-Stat, 442
XmdvTool, 443, 474
XploRe, 442

zooming, 478

Handbook of Statistics
Contents of Previous Volumes

Volume 1. Analysis of Variance
Edited by P.R. Krishnaiah
1980 xviii + 1002 pp.

1. Estimation of Variance Components by C.R. Rao and J. Kleffe
2. Multivariate Analysis of Variance of Repeated Measurements by N.H. Timm
3. Growth Curve Analysis by S. Geisser
4. Bayesian Inference in MANOVA by S.J. Press
5. Graphical Methods for Internal Comparisons in ANOVA and MANOVA by R. Gnanadesikan
6. Monotonicity and Unbiasedness Properties of ANOVA and MANOVA Tests by S. Das Gupta
7. Robustness of ANOVA and MANOVA Test Procedures by P.K. Ito
8. Analysis of Variance and Problems under Time Series Models by D.R. Brillinger
9. Tests of Univariate and Multivariate Normality by K.V. Mardia
10. Transformations to Normality by G. Kaskey, B. Kolman, P.R. Krishnaiah and L. Steinberg
11. ANOVA and MANOVA: Models for Categorical Data by V.P. Bhapkar
12. Inference and the Structural Model for ANOVA and MANOVA by D.A.S. Fraser
13. Inference Based on Conditionally Specified ANOVA Models Incorporating Preliminary Testing by T.A. Bancroft and C.-P. Han
14. Quadratic Forms in Normal Variables by C.G. Khatri
15. Generalized Inverse of Matrices and Applications to Linear Models by S.K. Mitra
16. Likelihood Ratio Tests for Mean Vectors and Covariance Matrices by P.R. Krishnaiah and J.C. Lee
17. Assessing Dimensionality in Multivariate Regression by A.J. Izenman
18. Parameter Estimation in Nonlinear Regression Models by H. Bunke
19. Early History of Multiple Comparison Tests by H.L. Harter
20. Representations of Simultaneous Pairwise Comparisons by A.R. Sampson
21. Simultaneous Test Procedures for Mean Vectors and Covariance Matrices by P.R. Krishnaiah, G.S. Mudholkar and P. Subbaiah
22. Nonparametric Simultaneous Inference for Some MANOVA Models by P.K. Sen

23. Comparison of Some Computer Programs for Univariate and Multivariate Analysis of Variance by R.D. Bock and D. Brandt
24. Computations of Some Multivariate Distributions by P.R. Krishnaiah
25. Inference on the Structure of Interaction Two-Way Classification Model by P.R. Krishnaiah and M. Yochmowitz

Volume 2. Classification, Pattern Recognition and Reduction of Dimensionality
Edited by P.R. Krishnaiah and L.N. Kanal
1982 xxii + 903 pp.

1. Discriminant Analysis for Time Series by R.H. Shumway
2. Optimum Rules for Classification into Two Multivariate Normal Populations with the Same Covariance Matrix by S. Das Gupta
3. Large Sample Approximations and Asymptotic Expansions of Classification Statistics by M. Siotani
4. Bayesian Discrimination by S. Geisser
5. Classification of Growth Curves by J.C. Lee
6. Nonparametric Classification by J.D. Broffitt
7. Logistic Discrimination by J.A. Anderson
8. Nearest Neighbor Methods in Discrimination by L. Devroye and T.J. Wagner
9. The Classification and Mixture Maximum Likelihood Approaches to Cluster Analysis by G.J. McLachlan
10. Graphical Techniques for Multivariate Data and for Clustering by J.M. Chambers and B. Kleiner
11. Cluster Analysis Software by R.K. Blashfield, M.S. Aldenderfer and L.C. Morey
12. Single-link Clustering Algorithms by F.J. Rohlf
13. Theory of Multidimensional Scaling by J. de Leeuw and W. Heiser
14. Multidimensional Scaling and its Application by M. Wish and J.D. Carroll
15. Intrinsic Dimensionality Extraction by K. Fukunaga
16. Structural Methods in Image Analysis and Recognition by L.N. Kanal, B.A. Lambird and D. Lavine
17. Image Models by N. Ahuja and A. Rosenfield
18. Image Texture Survey by R.M. Haralick
19. Applications of Stochastic Languages by K.S. Fu
20. A Unifying Viewpoint on Pattern Recognition by J.C. Simon, E. Backer and J. Sallentin
21. Logical Functions in the Problems of Empirical Prediction by G.S. Lbov
22. Inference and Data Tables and Missing Values by N.G. Zagoruiko and V.N. Yolkina
23. Recognition of Electrocardiographic Patterns by J.H. van Bemmel
24. Waveform Parsing Systems by G.C. Stockman

14. Measuring Attenuation by M.A. Cameron and P.J. Thomson
15. Speech Recognition Using LPC Distance Measures by P.J. Thomson and P. de Souza
16. Varying Coefficient Regression by D.F. Nicholls and A.R. Pagan
17. Small Samples and Large Equations Systems by H. Theil and D.G. Fiebig

Volume 6. Sampling
Edited by P.R. Krishnaiah and C.R. Rao
1988 xvi + 594 pp.

1. A Brief History of Random Sampling Methods by D.R. Bellhouse
2. A First Course in Survey Sampling by T. Dalenius
3. Optimality of Sampling Strategies by A. Chaudhuri
4. Simple Random Sampling by P.K. Pathak
5. On Single Stage Unequal Probability Sampling by V.P. Godambe and M.E. Thompson
6. Systematic Sampling by D.R. Bellhouse
7. Systematic Sampling with Illustrative Examples by M.N. Murthy and T.J. Rao
8. Sampling in Time by D.A. Binder and M.A. Hidiroglou
9. Bayesian Inference in Finite Populations by W.A. Ericson
10. Inference Based on Data from Complex Sample Designs by G. Nathan
11. Inference for Finite Population Quantiles by J. Sedransk and P.J. Smith
12. Asymptotics in Finite Population Sampling by P.K. Sen
13. The Technique of Replicated or Interpenetrating Samples by J.C. Koop
14. On the Use of Models in Sampling from Finite Populations by I. Thomsen and D. Tesfu
15. The Prediction Approach to Sampling Theory by R.M. Royall
16. Sample Survey Analysis: Analysis of Variance and Contingency Tables by D.H. Freeman Jr
17. Variance Estimation in Sample Surveys by J.N.K. Rao
18. Ratio and Regression Estimators by P.S.R.S. Rao
19. Role and Use of Composite Sampling and Capture-Recapture Sampling in Ecological Studies by M.T. Boswell, K.P. Burnham and G.P. Patil
20. Data-based Sampling and Model-based Estimation for Environmental Resources by G.P. Patil, G.J. Babu, R.C. Hennemuth, W.L. Meyers, M.B. Rajarshi and C. Taillie
21. On Transect Sampling to Assess Wildlife Populations and Marine Resources by F.L. Ramsey, C.E. Gates, G.P. Patil and C. Taillie
22. A Review of Current Survey Sampling Methods in Marketing Research (Telephone, Mall Intercept and Panel Surveys) by R. Velu and G.M. Naidu
23. Observational Errors in Behavioural Traits of Man and their Implications for Genetics by P.V. Sukhatme
24. Designs in Survey Sampling Avoiding Contiguous Units by A.S. Hedayat, C.R. Rao and J. Stufken

21. M-, L- and R-estimators by J. Jurečková
22. Nonparametric Sequential Estimation by P.K. Sen
23. Stochastic Approximation by V. Dupač
24. Density Estimation by P. Révész
25. Censored Data by A.P. Basu
26. Tests for Exponentiality by K.A. Doksum and B.S. Yandell
27. Nonparametric Concepts and Methods in Reliability by M. Hollander and F. Proschan
28. Sequential Nonparametric Tests by U. Müller-Funk
29. Nonparametric Procedures for some Miscellaneous Problems by P.K. Sen
30. Minimum Distance Procedures by R. Beran
31. Nonparametric Methods in Directional Data Analysis by S.R. Jammalamadaka
32. Application of Nonparametric Statistics to Cancer Data by H.S. Wieand
33. Nonparametric Frequentist Proposals for Monitoring Comparative Survival Studies by M. Gail
34. Meteorological Applications of Permutation Techniques Based on Distance Functions by P.W. Mielke Jr
35. Categorical Data Problems Using Information Theoretic Approach by S. Kullback and J.C. Keegel
36. Tables for Order Statistics by P.R. Krishnaiah and P.K. Sen
37. Selected Tables for Nonparametric Statistics by P.K. Sen and P.R. Krishnaiah

Volume 5. Time Series in the Time Domain
Edited by E.J. Hannan, P.R. Krishnaiah and M.M. Rao
1985 xiv + 490 pp.

1. Nonstationary Autoregressive Time Series by W.A. Fuller
2. Non-Linear Time Series Models and Dynamical Systems by T. Ozaki
3. Autoregressive Moving Average Models, Intervention Problems and Outlier Detection in Time Series by G.C. Tiao
4. Robustness in Time Series and Estimating ARMA Models by R.D. Martin and V.J. Yohai
5. Time Series Analysis with Unequally Spaced Data by R.H. Jones
6. Various Model Selection Techniques in Time Series Analysis by R. Shibata
7. Estimation of Parameters in Dynamical Systems by L. Ljung
8. Recursive Identification, Estimation and Control by P. Young
9. General Structure and Parametrization of ARMA and State-Space Systems and its Relation to Statistical Problems by M. Deistler
10. Harmonizable, Cramér, and Karhunen Classes of Processes by M.M. Rao
11. On Non-Stationary Time Series by C.S.K. Bhagavan
12. Harmonizable Filtering and Sampling of Time Series by D.K. Chang
13. Sampling Designs for Time Series by S. Cambanis

12. Threshold Autoregression and Some Frequency-Domain Characteristics by J. Pemberton and H. Tong
13. The Frequency-Domain Approach to the Analysis of Closed-Loop Systems by M.B. Priestley
14. The Bispectral Analysis of Nonlinear Stationary Time Series with Reference to Bilinear Time-Series Models by T. Subba Rao
15. Frequency-Domain Analysis of Multidimensional Time-Series Data by E.A. Robinson
16. Review of Various Approaches to Power Spectrum Estimation by P.M. Robinson
17. Cumulants and Cumulant Spectra by M. Rosenblatt
18. Replicated Time-Series Regression: An Approach to Signal Estimation and Detection by R.H. Shumway
19. Computer Programming of Spectrum Estimation by T. Thrall
20. Likelihood Ratio Tests on Covariance Matrices and Mean Vectors of Complex Multivariate Normal Populations and their Applications in Time Series by P.R. Krishnaiah, J.C. Lee and T.C. Chang

Volume 4. Nonparametric Methods
Edited by P.R. Krishnaiah and P.K. Sen
1984 xx + 968 pp.

1. Randomization Procedures by C.B. Bell and P.K. Sen
2. Univariate and Multivariate Multisample Location and Scale Tests by V.P. Bhapkar
3. Hypothesis of Symmetry by M. Hušková
4. Measures of Dependence by K. Joag-Dev
5. Tests of Randomness against Trend or Serial Correlations by G.K. Bhattacharyya
6. Combination of Independent Tests by J.L. Folks
7. Combinatorics by L. Takács
8. Rank Statistics and Limit Theorems by M. Ghosh
9. Asymptotic Comparison of Tests – A Review by K. Singh
10. Nonparametric Methods in Two-Way Layouts by D. Quade
11. Rank Tests in Linear Models by J.N. Adichie
12. On the Use of Rank Tests and Estimates in the Linear Model by J.C. Aubuchon and T.P. Hettmansperger
13. Nonparametric Preliminary Test Inference by A.K.Md.E. Saleh and P.K. Sen
14. Paired Comparisons: Some Basic Procedures and Examples by R.A. Bradley
15. Restricted Alternatives by S.K. Chatterjee
16. Adaptive Methods by M. Hušková
17. Order Statistics by J. Galambos
18. Induced Order Statistics: Theory and Applications by P.K. Bhattacharya
19. Empirical Distribution Function by F. Csáki
20. Invariance Principles for Empirical Processes by M. Csörgő

25. Continuous Speech Recognition: Statistical Methods by F. Jelinek, R.L. Mercer and L.R. Bahl
26. Applications of Pattern Recognition in Radar by A.A. Grometstein and W.H. Schoendorf
27. White Blood Cell Recognition by F.S. Gelsema and G.H. Landweerd
28. Pattern Recognition Techniques for Remote Sensing Applications by P.H. Swain
29. Optical Character Recognition – Theory and Practice by G. Nagy
30. Computer and Statistical Considerations for Oil Spill Identification by Y.T. Chien and T.J. Killeen
31. Pattern Recognition in Chemistry by B.R. Kowalski and S. Wold
32. Covariance Matrix Representation and Object-Predicate Symmetry by T. Kaminuma, S. Tomita and S. Watanabe
33. Multivariate Morphometrics by R.A. Reyment
34. Multivariate Analysis with Latent Variables by P.M. Bentler and D.G. Weeks
35. Use of Distance Measures, Information Measures and Error Bounds in Feature Evaluation by M. Ben-Bassat
36. Topics in Measurement Selection by J.M. Van Campenhout
37. Selection of Variables Under Univariate Regression Models by P.R. Krishnaiah
38. On the Selection of Variables Under Regression Models Using Krishnaiah's Finite Intersection Tests by J.L. Schmidhammer
39. Dimensionality and Sample Size Considerations in Pattern Recognition Practice by A.K. Jain and B. Chandrasekaran
40. Selecting Variables in Discriminant Analysis for Improving upon Classical Procedures by W. Schaafsma
41. Selection of Variables in Discriminant Analysis by P.R. Krishnaiah

Volume 3. Time Series in the Frequency Domain
Edited by D.R. Brillinger and P.R. Krishnaiah
1983 xiv + 485 pp.

1. Wiener Filtering (with emphasis on frequency-domain approaches) by R.J. Bhansali and D. Karavellas
2. The Finite Fourier Transform of a Stationary Process by D.R. Brillinger
3. Seasonal and Calendar Adjustment by W.S. Cleveland
4. Optimal Inference in the Frequency Domain by R.B. Davies
5. Applications of Spectral Analysis in Econometrics by C.W.J. Granger and R. Engle
6. Signal Estimation by E.J. Hannan
7. Complex Demodulation: Some Theory and Applications by T. Hasan
8. Estimating the Gain of a Linear Filter from Noisy Data by M.J. Hinich
9. A Spectral Analysis Primer by L.H. Koopmans
10. Robust-Resistant Spectral Analysis by R.D. Martin
11. Autoregressive Spectral Estimation by E. Parzen

Volume 7. Quality Control and Reliability
Edited by P.R. Krishnaiah and C.R. Rao
1988 xiv + 503 pp.

1. Transformation of Western Style of Management by W. Edwards Deming
2. Software Reliability by F.B. Bastani and C.V. Ramamoorthy
3. Stress–Strength Models for Reliability by R.A. Johnson
4. Approximate Computation of Power Generating System Reliability Indexes by M. Mazumdar
5. Software Reliability Models by T.A. Mazzuchi and N.D. Singpurwalla
6. Dependence Notions in Reliability Theory by N.R. Chaganty and K. Joagdev
7. Application of Goodness-of-Fit Tests in Reliability by B.W. Woodruff and A.H. Moore
8. Multivariate Nonparametric Classes in Reliability by H.W. Block and T.H. Savits
9. Selection and Ranking Procedures in Reliability Models by S.S. Gupta and S. Panchapakesan
10. The Impact of Reliability Theory on Some Branches of Mathematics and Statistics by P.J. Boland and F. Proschan
11. Reliability Ideas and Applications in Economics and Social Sciences by M.C. Bhattacharjee
12. Mean Residual Life: Theory and Applications by F. Guess and F. Proschan
13. Life Distribution Models and Incomplete Data by R.E. Barlow and F. Proschan
14. Piecewise Geometric Estimation of a Survival Function by G.M. Mimmack and F. Proschan
15. Applications of Pattern Recognition in Failure Diagnosis and Quality Control by L.F. Pau
16. Nonparametric Estimation of Density and Hazard Rate Functions when Samples are Censored by W.J. Padgett
17. Multivariate Process Control by F.B. Alt and N.D. Smith
18. QMP/USP – A Modern Approach to Statistical Quality Auditing by B. Hoadley
19. Review About Estimation of Change Points by P.R. Krishnaiah and B.Q. Miao
20. Nonparametric Methods for Changepoint Problems by M. Csörgő and L. Horváth
21. Optimal Allocation of Multistate Components by E. El-Neweihi, F. Proschan and J. Sethuraman
22. Weibull, Log-Weibull and Gamma Order Statistics by H.L. Herter
23. Multivariate Exponential Distributions and their Applications in Reliability by A.P. Basu
24. Recent Developments in the Inverse Gaussian Distribution by S. Iyengar and G. Patwardhan

Volume 8. Statistical Methods in Biological and Medical Sciences
Edited by C.R. Rao and R. Chakraborty
1991 xvi + 554 pp.

1. Methods for the Inheritance of Qualitative Traits by J. Rice, R. Neuman and S.O. Moldin
2. Ascertainment Biases and their Resolution in Biological Surveys by W.J. Ewens
3. Statistical Considerations in Applications of Path Analytical in Genetic Epidemiology by D.C. Rao
4. Statistical Methods for Linkage Analysis by G.M. Lathrop and J.M. Lalouel
5. Statistical Design and Analysis of Epidemiologic Studies: Some Directions of Current Research by N. Breslow
6. Robust Classification Procedures and their Applications to Anthropometry by N. Balakrishnan and R.S. Ambagaspitiya
7. Analysis of Population Structure: A Comparative Analysis of Different Estimators of Wright's Fixation Indices by R. Chakraborty and H. Danker-Hopfe
8. Estimation of Relationships from Genetic Data by E.A. Thompson
9. Measurement of Genetic Variation for Evolutionary Studies by R. Chakraborty and C.R. Rao
10. Statistical Methods for Phylogenetic Tree Reconstruction by N. Saitou
11. Statistical Models for Sex-Ratio Evolution by S. Lessard
12. Stochastic Models of Carcinogenesis by S.H. Moolgavkar
13. An Application of Score Methodology: Confidence Intervals and Tests of Fit for One-Hit-Curves by J.J. Gart
14. Kidney-Survival Analysis of IgA Nephropathy Patients: A Case Study by O.J.W.F. Kardaun
15. Confidence Bands and the Relation with Decision Analysis: Theory by O.J.W.F. Kardaun
16. Sample Size Determination in Clinical Research by J. Bock and H. Toutenburg

Volume 9. Computational Statistics
Edited by C.R. Rao
1993 xix + 1045 pp.

1. Algorithms by B. Kalyanasundaram
2. Steady State Analysis of Stochastic Systems by K. Kant
3. Parallel Computer Architectures by R. Krishnamurti and B. Narahari
4. Database Systems by S. Lanka and S. Pal
5. Programming Languages and Systems by S. Purushothaman and J. Seaman
6. Algorithms and Complexity for Markov Processes by R. Varadarajan
7. Mathematical Programming: A Computational Perspective by W.W. Hager, R. Horst and P.M. Pardalos

8. Integer Programming by P.M. Pardalos and Y. Li
9. Numerical Aspects of Solving Linear Least Squares Problems by J.L. Barlow
10. The Total Least Squares Problem by S. van Huffel and H. Zha
11. Construction of Reliable Maximum-Likelihood-Algorithms with Applications to Logistic and Cox Regression by D. Böhning
12. Nonparametric Function Estimation by T. Gasser, J. Engel and B. Seifert
13. Computation Using the OR Decomposition by C.R. Goodall
14. The EM Algorithm by N. Laird
15. Analysis of Ordered Categorial Data through Appropriate Scaling by C.R. Rao and P.M. Caligiuri
16. Statistical Applications of Artificial Intelligence by W.A. Gale, D.J. Hand and A.E. Kelly
17. Some Aspects of Natural Language Processes by A.K. Joshi
18. Gibbs Sampling by S.F. Arnold
19. Bootstrap Methodology by G.J. Babu and C.R. Rao
20. The Art of Computer Generation of Random Variables by M.T. Boswell, S.D. Gore, G.P. Patil and C. Taillie
21. Jackknife Variance Estimation and Bias Reduction by S. Das Peddada
22. Designing Effective Statistical Graphs by D.A. Burn
23. Graphical Methods for Linear Models by A.S. Hadi
24. Graphics for Time Series Analysis by H.J. Newton
25. Graphics as Visual Language by T. Selkar and A. Appel
26. Statistical Graphics and Visualization by E.J. Wegman and D.B. Carr
27. Multivariate Statistical Visualization by F.W. Young, R.A. Faldowski and M.M. McFarlane
28. Graphical Methods for Process Control by T.L. Ziemer

Volume 10. Signal Processing and its Applications
Edited by N.K. Bose and C.R. Rao
1993 xvii + 992 pp.

1. Signal Processing for Linear Instrumental Systems with Noise: A General Theory with Illustrations from Optical Imaging and Light Scattering Problems by M. Bertero and E.R. Pike
2. Boundary Implication Results in Parameter Space by N.K. Bose
3. Sampling of Bandlimited Signals: Fundamental Results and Some Extensions by J.L. Brown Jr
4. Localization of Sources in a Sector: Algorithms and Statistical Analysis by K. Buckley and X.-L. Xu
5. The Signal Subspace Direction-of-Arrival Algorithm by J.A. Cadzow
6. Digital Differentiators by S.C. Dutta Roy and B. Kumar
7. Orthogonal Decompositions of 2D Random Fields and their Applications for 2D Spectral Estimation by J.M. Francos

8. VLSI in Signal Processing by A. Ghouse
9. Constrained Beamforming and Adaptive Algorithms by L.C. Godara
10. Bispectral Speckle Interferometry to Reconstruct Extended Objects from Turbulence-Degraded Telescope Images by D.M. Goodman, T.W. Lawrence, E.M. Johansson and J.P. Fitch
11. Multi-Dimensional Signal Processing by K. Hirano and T. Nomura
12. On the Assessment of Visual Communication by F.O. Huck, C.L. Fales, R. Alter-Gartenberg and Z. Rahman
13. VLSI Implementations of Number Theoretic Concepts with Applications in Signal Processing by G.A. Jullien, N.M. Wigley and J. Reilly
14. Decision-level Neural Net Sensor Fusion by R.Y. Levine and T.S. Khuon
15. Statistical Algorithms for Noncausal Gauss Markov Fields by J.M.F. Moura and N. Balram
16. Subspace Methods for Directions-of-Arrival Estimation by A. Paulraj, B. Ottersten, R. Roy, A. Swindlehurst, G. Xu and T. Kailath
17. Closed Form Solution to the Estimates of Directions of Arrival Using Data from an Array of Sensors by C.R. Rao and B. Zhou
18. High-Resolution Direction Finding by S.V. Schell and W.A. Gardner
19. Multiscale Signal Processing Techniques: A Review by A.H. Tewfik, M. Kim and M. Deriche
20. Sampling Theorems and Wavelets by G.G. Walter
21. Image and Video Coding Research by J.W. Woods
22. Fast Algorithms for Structured Matrices in Signal Processing by A.E. Yagle

Volume 11. Econometrics
Edited by G.S. Maddala, C.R. Rao and H.D. Vinod
1993 xx + 783 pp.

1. Estimation from Endogenously Stratified Samples by S.R. Cosslett
2. Semiparametric and Nonparametric Estimation of Quantal Response Models by J.L. Horowitz
3. The Selection Problem in Econometrics and Statistics by C.F. Manski
4. General Nonparametric Regression Estimation and Testing in Econometrics by A. Ullah and H.D. Vinod
5. Simultaneous Microeconometric Models with Censored or Qualitative Dependent Variables by R. Blundell and R.J. Smith
6. Multivariate Tobit Models in Econometrics by L.-F. Lee
7. Estimation of Limited Dependent Variable Models under Rational Expectations by G.S. Maddala
8. Nonlinear Time Series and Macroeconometrics by W.A. Brock and S.M. Potter
9. Estimation, Inference and Forecasting of Time Series Subject to Changes in Time by J.D. Hamilton

10. Structural Time Series Models by A.C. Harvey and N. Shephard
11. Bayesian Testing and Testing Bayesians by J.-P. Florens and M. Mouchart
12. Pseudo-Likelihood Methods by C. Gourieroux and A. Monfort
13. Rao's Score Test: Recent Asymptotic Results by R. Mukerjee
14. On the Strong Consistency of M-Estimates in Linear Models under a General Discrepancy Function by Z.D. Bai, Z.J. Liu and C.R. Rao
15. Some Aspects of Generalized Method of Moments Estimation by A. Hall
16. Efficient Estimation of Models with Conditional Moment Restrictions by W.K. Newey
17. Generalized Method of Moments: Econometric Applications by M. Ogaki
18. Testing for Heteroscedasticity by A.R. Pagan and Y. Pak
19. Simulation Estimation Methods for Limited Dependent Variable Models by V.A. Hajivassiliou
20. Simulation Estimation for Panel Data Models with Limited Dependent Variable by M.P. Keane
21. A Perspective Application of Bootstrap Methods in Econometrics by J. Jeong and G.S. Maddala
22. Stochastic Simulations for Inference in Nonlinear Errors-in-Variables Models by R.S. Mariano and B.W. Brown
23. Bootstrap Methods: Applications in Econometrics by H.D. Vinod
24. Identifying Outliers and Influential Observations in Econometric Models by S.G. Donald and G.S. Maddala
25. Statistical Aspects of Calibration in Macroeconomics by A.W. Gregory and G.W. Smith
26. Panel Data Models with Rational Expectations by K. Lahiri
27. Continuous Time Financial Models: Statistical Applications of Stochastic Processes by K.R. Sawyer

Volume 12. Environmental Statistics
Edited by G.P. Patil and C.R. Rao
1994 xix + 927 pp.

1. Environmetrics: An Emerging Science by J.S. Hunter
2. A National Center for Statistical Ecology and Environmental Statistics: A Center Without Walls by G.P. Patil
3. Replicate Measurements for Data Quality and Environmental Modeling by W. Liggett
4. Design and Analysis of Composite Sampling Procedures: A Review by G. Lovison, S.D. Gore and G.P. Patil
5. Ranked Set Sampling by G.P. Patil, A.K. Sinha and C. Taillie
6. Environmental Adaptive Sampling by G.A.F. Seber and S.K. Thompson
7. Statistical Analysis of Censored Environmental Data by M. Akritas, T. Ruscitti and G.P. Patil

8. Biological Monitoring: Statistical Issues and Models by E.P. Smith
9. Environmental Sampling and Monitoring by S.V. Stehman and W. Scott Overton
10. Ecological Statistics by B.F.J. Manly
11. Forest Biometrics by H.E. Burkhart and T.G. Gregoire
12. Ecological Diversity and Forest Management by J.H. Gove, G.P. Patil, B.F. Swindel and C. Taillie
13. Ornithological Statistics by P.M. North
14. Statistical Methods in Developmental Toxicology by P.J. Catalano and L.M. Ryan
15. Environmental Biometry: Assessing Impacts of Environmental Stimuli Via Animal and Microbial Laboratory Studies by W.W. Piegorsch
16. Stochasticity in Deterministic Models by J.J.M. Bedaux and S.A.L.M. Kooijman
17. Compartmental Models of Ecological and Environmental Systems by J.H. Matis and T.E. Wehrly
18. Environmental Remote Sensing and Geographic Information Systems-Based Modeling by W.L. Myers
19. Regression Analysis of Spatially Correlated Data: The Kanawha County Health Study by C.A. Donnelly, J.H. Ware and N.M. Laird
20. Methods for Estimating Heterogeneous Spatial Covariance Functions with Environmental Applications by P. Guttorp and P.D. Sampson
21. Meta-analysis in Environmental Statistics by V. Hasselblad
22. Statistical Methods in Atmospheric Science by A.R. Solow
23. Statistics with Agricultural Pests and Environmental Impacts by L.J. Young and J.H. Young
24. A Crystal Cube for Coastal and Estuarine Degradation: Selection of Endpoints and Development of Indices for Use in Decision Making by M.T. Boswell, J.S. O'Connor and G.P. Patil
25. How Does Scientific Information in General and Statistical Information in Particular Input to the Environmental Regulatory Process? by C.R. Cothern
26. Environmental Regulatory Statistics by C.B. Davis
27. An Overview of Statistical Issues Related to Environmental Cleanup by R. Gilbert
28. Environmental Risk Estimation and Policy Decisions by H. Lacayo Jr

Volume 13. Design and Analysis of Experiments
Edited by S. Ghosh and C.R. Rao
1996 xviii + 1230 pp.

1. The Design and Analysis of Clinical Trials by P. Armitage
2. Clinical Trials in Drug Development: Some Statistical Issues by H.I. Patel
3. Optimal Crossover Designs by J. Stufken
4. Design and Analysis of Experiments: Nonparametric Methods with Applications to Clinical Trials by P.K. Sen

5. Adaptive Designs for Parametric Models by S. Zacks
6. Observational Studies and Nonrandomized Experiments by P.R. Rosenbaum
7. Robust Design: Experiments for Improving Quality by D.M. Steinberg
8. Analysis of Location and Dispersion Effects from Factorial Experiments with a Circular Response by C.M. Anderson
9. Computer Experiments by J.R. Koehler and A.B. Owen
10. A Critique of Some Aspects of Experimental Design by J.N. Srivastava
11. Response Surface Designs by N.R. Draper and D.K.J. Lin
12. Multiresponse Surface Methodology by A.I. Khuri
13. Sequential Assembly of Fractions in Factorial Experiments by S. Ghosh
14. Designs for Nonlinear and Generalized Linear Models by A.C. Atkinson and L.M. Haines
15. Spatial Experimental Design by R.J. Martin
16. Design of Spatial Experiments: Model Fitting and Prediction by V.V. Fedorov
17. Design of Experiments with Selection and Ranking Goals by S.S. Gupta and S. Panchapakesan
18. Multiple Comparisons by A.C. Tamhane
19. Nonparametric Methods in Design and Analysis of Experiments by E. Brunner and M.L. Puri
20. Nonparametric Analysis of Experiments by A.M. Dean and D.A. Wolfe
21. Block and Other Designs in Agriculture by D.J. Street
22. Block Designs: Their Combinatorial and Statistical Properties by T. Calinski and S. Kageyama
23. Developments in Incomplete Block Designs for Parallel Line Bioassays by S. Gupta and R. Mukerjee
24. Row-Column Designs by K.R. Shah and B.K. Sinha
25. Nested Designs by J.P. Morgan
26. Optimal Design: Exact Theory by C.S. Cheng
27. Optimal and Efficient Treatment – Control Designs by D. Majumdar
28. Model Robust Designs by Y.-J. Chang and W.I. Notz
29. Review of Optimal Bayes Designs by A. DasGupta
30. Approximate Designs for Polynomial Regression: Invariance, Admissibility, and Optimality by N. Gaffke and B. Heiligers

Volume 14. Statistical Methods in Finance
Edited by G.S. Maddala and C.R. Rao
1996 xvi + 733 pp.

1. Econometric Evaluation of Asset Pricing Models by W.E. Person and R. Jegannathan
2. Instrumental Variables Estimation of Conditional Beta Pricing Models by C.R. Harvey and C.M. Kirby

3. Semiparametric Methods for Asset Pricing Models by B.N. Lehmann
4. Modeling the Term Structure by A.R. Pagan, A.D. Hall and V. Martin
5. Stochastic Volatility by E. Ghysels, A.C. Harvey and E. Renault
6. Stock Price Volatility by S.F. LeRoy
7. GARCH Models of Volatility by F.C. Palm
8. Forecast Evaluation and Combination by F.X. Diebold and J.A. Lopez
9. Predictable Components in Stock Returns by G. Kaul
10. Interset Rate Spreads as Predictors of Business Cycles by K. Lahiri and J.G. Wang
11. Nonlinear Time Series, Complexity Theory, and Finance by W.A. Brock and P.J.F. deLima
12. Count Data Models for Financial Data by A.C. Cameron and P.K. Trivedi
13. Financial Applications of Stable Distributions by J.H. McCulloch
14. Probability Distributions for Financial Models by J.B. McDonald
15. Bootstrap Based Tests in Financial Models by G.S. Maddala and H. Li
16. Principal Component and Factor Analyses by C.R. Rao
17. Errors in Variables Problems in Finance by G.S. Maddala and M. Nimalendran
18. Financial Applications of Artificial Neural Networks by M. Qi
19. Applications of Limited Dependent Variable Models in Finance by G.S. Maddala
20. Testing Option Pricing Models by D.S. Bates
21. Peso Problems: Their Theoretical and Empirical Implications by M.D.D. Evans
22. Modeling Market Microstructure Time Series by J. Hasbrouck
23. Statistical Methods in Tests of Portfolio Efficiency: A Synthesis by J. Shanken

Volume 15. Robust Inference
Edited by G.S. Maddala and C.R. Rao
1997 xviii + 698 pp.

1. Robust Inference in Multivariate Linear Regression Using Difference of Two Convex Functions as the Discrepancy Measure by Z.D. Bai, C.R. Rao and Y.H. Wu
2. Minimum Distance Estimation: The Approach Using Density-Based Distances by A. Basu, I.R. Harris and S. Basu
3. Robust Inference: The Approach Based on Influence Functions by M. Markatou and E. Ronchetti
4. Practical Applications of Bounded-Influence Tests by S. Heritier and M.-P. Victoria-Feser
5. Introduction to Positive-Breakdown Methods by P.J. Rousseeuw
6. Outlier Identification and Robust Methods by U. Gather and C. Becker
7. Rank-Based Analysis of Linear Models by T.P. Hettmansperger, J.W. McKean and S.J. Sheather
8. Rank Tests for Linear Models by R. Koenker
9. Some Extensions in the Robust Estimation of Parameters of Exponential and Double Exponential Distributions in the Presence of Multiple Outliers by A. Childs and N. Balakrishnan

10. Outliers, Unit Roots and Robust Estimation of Nonstationary Time Series by G.S. Maddala and Y. Yin
11. Autocorrelation-Robust Inference by P.M. Robinson and C. Velasco
12. A Practitioner's Guide to Robust Covariance Matrix Estimation by W.J. den Haan and A. Levin
13. Approaches to the Robust Estimation of Mixed Models by A.H. Welsh and A.M. Richardson
14. Nonparametric Maximum Likelihood Methods by S.R. Cosslett
15. A Guide to Censored Quantile Regressions by B. Fitzenberger
16. What Can Be Learned About Population Parameters When the Data Are Contaminated by J.L. Horowitz and C.F. Manski
17. Asymptotic Representations and Interrelations of Robust Estimators and Their Applications by J. Jurečková and P.K. Sen
18. Small Sample Asymptotics: Applications in Robustness by C.A. Field and M.A. Tingley
19. On the Fundamentals of Data Robustness by G. Maguluri and K. Singh
20. Statistical Analysis With Incomplete Data: A Selective Review by M.G. Akritas and M.P. La Valley
21. On Contamination Level and Sensitivity of Robust Tests by J.Á. Visšek
22. Finite Sample Robustness of Tests: An Overview by T. Kariya and P. Kim
23. Future Directions by G.S. Maddala and C.R. Rao

Volume 16. Order Statistics – Theory and Methods
Edited by N. Balakrishnan and C.R. Rao
1997 xix + 688 pp.

1. Order Statistics: An Introduction by N. Balakrishnan and C.R. Rao
2. Order Statistics: A Historical Perspective by H. Leon Harter and N. Balakrishnan
3. Computer Simulation of Order Statistics by Pandu R. Tadikamalla and N. Balakrishnan
4. Lorenz Ordering of Order Statistics and Record Values by Barry C. Arnold and Jose A. Villasenor
5. Stochastic Ordering of Order Statistics by Philip J. Boland, Moshe Shaked and J. George Shanthikumar
6. Bounds for Expectations of L-Estimates by T. Rychlik
7. Recurrence Relations and Identities for Moments of Order Statistics by N. Balakrishnan and K.S. Sultan
8. Recent Approaches to Characterizations Based on Order Statistics and Record Values by C.R. Rao and D.N. Shanbhag
9. Characterizations of Distributions via Identically Distributed Functions of Order Statistics by Ursula Gather, Udo Kamps and Nicole Schweitzer
10. Characterizations of Distributions by Recurrence Relations and Identities for Moments of Order Statistics by Udo Kamps

11. Univariate Extreme Value Theory and Applications by Janos Galambos
12. Order Statistics: Asymptotics in Applications by Pranab Kumar Sen
13. Zero-One Laws for Large Order Statistics by R.J. Tomkins and Hong Wang
14. Some Exact Properties of Cook's D_I by D.R. Jensen and D.E. Ramirez
15. Generalized Recurrence Relations for Moments of Order Statistics from Non-Identical Pareto and Truncated Pareto Random Variables with Applications to Robustness by Aaron Childs and N. Balakrishnan
16. A Semiparametric Bootstrap for Simulating Extreme Order Statistics by Robert L. Strawderman and Daniel Zelterman
17. Approximations to Distributions of Sample Quantiles by Chunsheng Ma and John Robinson
18. Concomitants of Order Statistics by H.A. David and H.N. Nagaraja
19. A Record of Records by Valery B. Nevzorov and N. Balakrishnan
20. Weighted Sequential Empirical Type Processes with Applications to Change-Point Problems by Barbara Szyszkowicz
21. Sequential Quantile and Bahadur–Kiefer Processes by Miklós Csörgő and Barbara Szyszkowicz

Volume 17. Order Statistics: Applications
Edited by N. Balakrishnan and C.R. Rao
1998 xviii + 712 pp.

1. Order Statistics in Exponential Distribution by Asit P. Basu and Bahadur Singh
2. Higher Order Moments of Order Statistics from Exponential and Right-truncated Exponential Distributions and Applications to Life-testing Problems by N. Balakrishnan and Shanti S. Gupta
3. Log-gamma Order Statistics and Linear Estimation of Parameters by N. Balakrishnan and P.S. Chan
4. Recurrence Relations for Single and Product Moments of Order Statistics from a Generalized Logistic Distribution with Applications to Inference and Generalizations to Double Truncation by N. Balakrishnan and Rita Aggarwala
5. Order Statistics from the Type III Generalized Logistic Distribution and Applications by N. Balakrishnan and S.K. Lee
6. Estimation of Scale Parameter Based on a Fixed Set of Order Statistics by Sanat K. Sarkar and Wenjin Wang
7. Optimal Linear Inference Using Selected Order Statistics in Location-Scale Models by M. Masoom Ali and Dale Umbach
8. L-Estimation by J.R.M. Hosking
9. On Some L-estimation in Linear Regression Models by Soroush Alimoradi and A.K.Md. Ehsanes Saleh
10. The Role of Order Statistics in Estimating Threshold Parameters by A. Clifford Cohen

11. Parameter Estimation under Multiply Type-II Censoring by Fanhui Kong
12. On Some Aspects of Ranked Set Sampling in Parametric Estimation by Nora Ni Chuiv and Bimal K. Sinha
13. Some Uses of Order Statistics in Bayesian Analysis by Seymour Geisser
14. Inverse Sampling Procedures to Test for Homogeneity in a Multinomial Distribution by S. Panchapakesan, Aaron Childs, B.H. Humphrey and N. Balakrishnan
15. Prediction of Order Statistics by Kenneth S. Kaminsky and Paul I. Nelson
16. The Probability Plot: Tests of Fit Based on the Correlation Coefficient by R.A. Lockhart and M.A. Stephens
17. Distribution Assessment by Samuel Shapiro
18. Application of Order Statistics to Sampling Plans for Inspection by Variables by Helmut Schneider and Frances Barbera
19. Linear Combinations of Ordered Symmetric Observations with Applications to Visual Acuity by Marios Viana
20. Order-Statistic Filtering and Smoothing of Time-Series: Part I by Gonzalo R. Arce, Yeong-Taeg Kim and Kenneth E. Barner
21. Order-Statistic Filtering and Smoothing of Time-Series: Part II by Kenneth E. Barner and Gonzalo R. Arce
22. Order Statistics in Image Processing by Scott T. Acton and Alan C. Bovik
23. Order Statistics Application to CFAR Radar Target Detection by R. Viswanathan

Volume 18. Bioenvironmental and Public Health Statistics
Edited by P.K. Sen and C.R. Rao
2000 xxiv + 1105 pp.

1. Bioenvironment and Public Health: Statistical Perspectives by Pranab K. Sen
2. Some Examples of Random Process Environmental Data Analysis by David R. Brillinger
3. Modeling Infectious Diseases – Aids by L. Billard
4. On Some Multiplicity Problems and Multiple Comparison Procedures in Biostatistics by Yosef Hochberg and Peter H. Westfall
5. Analysis of Longitudinal Data by Julio M. Singer and Dalton F. Andrade
6. Regression Models for Survival Data by Richard A. Johnson and John P. Klein
7. Generalised Linear Models for Independent and Dependent Responses by Bahjat F. Qaqish and John S. Preisser
8. Hierarchial and Empirical Bayes Methods for Environmental Risk Assessment by Gauri Datta, Malay Ghosh and Lance A. Waller
9. Non-parametrics in Bioenvironmental and Public Health Statistics by Pranab Kumar Sen
10. Estimation and Comparison of Growth and Dose-Response Curves in the Presence of Purposeful Censoring by Paul W. Stewart

11. Spatial Statistical Methods for Environmental Epidemiology by Andrew B. Lawson and Noel Cressie
12. Evaluating Diagnostic Tests in Public Health by Margaret Pepe, Wendy Leisenring and Carolyn Rutter
13. Statistical Issues in Inhalation Toxicology by E. Weller, L. Ryan and D. Dockery
14. Quantitative Potency Estimation to Measure Risk with Bioenvironmental Hazards by A. John Bailer and Walter W. Piegorsch
15. The Analysis of Case-Control Data: Epidemiologic Studies of Familial Aggregation by Nan M. Laird, Garrett M. Fitzmaurice and Ann G. Schwartz
16. Cochran–Mantel–Haenszel Techniques: Applications Involving Epidemiologic Survey Data by Daniel B. Hall, Robert F. Woolson, William R. Clarke and Martha F. Jones
17. Measurement Error Models for Environmental and Occupational Health Applications by Robert H. Lyles and Lawrence L. Kupper
18. Statistical Perspectives in Clinical Epidemiology by Shrikant I. Bangdiwala and Sergio R. Muñoz
19. ANOVA and ANOCOVA for Two-Period Crossover Trial Data: New vs. Standard by Subir Ghosh and Lisa D. Fairchild
20. Statistical Methods for Crossover Designs in Bioenvironmental and Public Health Studies by Gail E. Tudor, Gary G. Koch and Diane Catellier
21. Statistical Models for Human Reproduction by C.M. Suchindran and Helen P. Koo
22. Statistical Methods for Reproductive Risk Assessment by Sati Mazumdar, Yikang Xu, Donald R. Mattison, Nancy B. Sussman and Vincent C. Arena
23. Selection Biases of Samples and their Resolutions by Ranajit Chakraborty and C. Radhakrishna Rao
24. Genomic Sequences and Quasi-Multivariate CATANOVA by Hildete Prisco Pinheiro, Françoise Seillier-Moiseiwitsch, Pranab Kumar Sen and Joseph Eron Jr
25. Statistical Methods for Multivariate Failure Time Data and Competing Risks by Ralph A. DeMasi
26. Bounds on Joint Survival Probabilities with Positively Dependent Competing Risks by Sanat K. Sarkar and Kalyan Ghosh
27. Modeling Multivariate Failure Time Data by Limin X. Clegg, Jianwen Cai and Pranab K. Sen
28. The Cost–Effectiveness Ratio in the Analysis of Health Care Programs by Joseph C. Gardiner, Cathy J. Bradley and Marianne Huebner
29. Quality-of-Life: Statistical Validation and Analysis An Example from a Clinical Trial by Balakrishna Hosmane, Clement Maurath and Richard Manski
30. Carcinogenic Potency: Statistical Perspectives by Anup Dewanji
31. Statistical Applications in Cardiovascular Disease by Elizabeth R. DeLong and David M. DeLong
32. Medical Informatics and Health Care Systems: Biostatistical and Epidemiologic Perspectives by J. Zvárová
33. Methods of Establishing In Vitro–In Vivo Relationships for Modified Release Drug Products by David T. Mauger and Vernon M. Chinchilli

34. Statistics in Psychiatric Research by Sati Mazumdar, Patricia R. Houck and Charles F. Reynolds III
35. Bridging the Biostatistics–Epidemiology Gap by Lloyd J. Edwards
36. Biodiversity – Measurement and Analysis by S.P. Mukherjee

Volume 19. Stochastic Processes: Theory and Methods
Edited by D.N. Shanbhag and C.R. Rao
2001 xiv + 967 pp.

1. Pareto Processes by Barry C. Arnold
2. Branching Processes by K.B. Athreya and A.N. Vidyashankar
3. Inference in Stochastic Processes by I.V. Basawa
4. Topics in Poisson Approximation by A.D. Barbour
5. Some Elements on Lévy Processes by Jean Bertoin
6. Iterated Random Maps and Some Classes of Markov Processes by Rabi Bhattacharya and Edward C. Waymire
7. Random Walk and Fluctuation Theory by N.H. Bingham
8. A Semigroup Representation and Asymptotic Behavior of Certain Statistics of the Fisher–Wright–Moran Coalescent by Adam Bobrowski, Marek Kimmel, Ovide Arino and Ranajit Chakraborty
9. Continuous-Time ARMA Processes by P.J. Brockwell
10. Record Sequences and their Applications by John Bunge and Charles M. Goldie
11. Stochastic Networks with Product Form Equilibrium by Hans Daduna
12. Stochastic Processes in Insurance and Finance by Paul Embrechts, Rüdiger Frey and Hansjörg Furrer
13. Renewal Theory by D.R. Grey
14. The Kolmogorov Isomorphism Theorem and Extensions to some Nonstationary Processes by Yûichirô Kakihara
15. Stochastic Processes in Reliability by Masaaki Kijima, Haijun Li and Moshe Shaked
16. On the supports of Stochastic Processes of Multiplicity One by A. Kłopotowski and M.G. Nadkarni
17. Gaussian Processes: Inequalities, Small Ball Probabilities and Applications by W.V. Li and Q.-M. Shao
18. Point Processes and Some Related Processes by Robin K. Milne
19. Characterization and Identifiability for Stochastic Processes by B.L.S. Prakasa Rao
20. Associated Sequences and Related Inference Problems by B.L.S. Prakasa Rao and Isha Dewan
21. Exchangeability, Functional Equations, and Characterizations by C.R. Rao and D.N. Shanbhag
22. Martingales and Some Applications by M.M. Rao
23. Markov Chains: Structure and Applications by R.L. Tweedie

24. Diffusion Processes by S.R.S. Varadhan
25. Itô's Stochastic Calculus and Its Applications by S. Watanabe

Volume 20. Advances in Reliability
Edited by N. Balakrishnan and C.R. Rao
2001 xxii + 860 pp.

1. Basic Probabilistic Models in Reliability by N. Balakrishnan, N. Limnios and C. Papadopoulos
2. The Weibull Nonhomogeneous Poisson Process by A.P. Basu and S.E. Rigdon
3. Bathtub-Shaped Failure Rate Life Distributions by C.D. Lai, M. Xie and D.N.P. Murthy
4. Equilibrium Distribution – its Role in Reliability Theory by A. Chatterjee and S.P. Mukherjee
5. Reliability and Hazard Based on Finite Mixture Models by E.K. Al-Hussaini and K.S. Sultan
6. Mixtures and Monotonicity of Failure Rate Functions by M. Shaked and F. Spizzichino
7. Hazard Measure and Mean Residual Life Orderings: A Unified Approach by M. Asadi and D.N. Shanbhag
8. Some Comparison Results of the Reliability Functions of Some Coherent Systems by J. Mi
9. On the Reliability of Hierarchical Structures by L.B. Klebanov and G.J. Szekely
10. Consecutive k-out-of-n Systems by N.A. Mokhlis
11. Exact Reliability and Lifetime of Consecutive Systems by S. Aki
12. Sequential k-out-of-n Systems by E. Cramer and U. Kamps
13. Progressive Censoring: A Review by R. Aggarwala
14. Point and Interval Estimation for Parameters of the Logistic Distribution Based on Progressively Type-II Censored Samples by N. Balakrishnan and N. Kannan
15. Progressively Censored Variables-Sampling Plans for Life Testing by U. Balasooriya
16. Graphical Techniques for Analysis of Data From Repairable Systems by P.A. Akersten, B. Klefsjö and B. Bergman
17. A Bayes Approach to the Problem of Making Repairs by G.C. McDonald
18. Statistical Analysis for Masked Data by B.J. Flehinger[†], B. Reiser and E. Yashchin
19. Analysis of Masked Failure Data under Competing Risks by A. Sen, S. Basu and M. Banerjee
20. Warranty and Reliability by D.N.P. Murthy and W.R. Blischke
21. Statistical Analysis of Reliability Warranty Data by K. Suzuki, Md. Rezaul Karim and L. Wang
22. Prediction of Field Reliability of Units, Each under Differing Dynamic Stresses, from Accelerated Test Data by W. Nelson

23. Step Stress Accelerated Life Test by E. Gouno and N. Balakrishnan
24. Estimation of Correlation under Destructive Testing by R. Johnson and W. Lu
25. System-Based Component Test Plans for Reliability Demonstration: A Review and Survey of the State-of-the-Art by J. Rajgopal and M. Mazumdar
26. Life-Test Planning for Preliminary Screening of Materials: A Case Study by J. Stein and N. Doganaksoy
27. Analysis of Reliability Data from In-House Audit Laboratory Testing by R. Agrawal and N. Doganaksoy
28. Software Reliability Modeling, Estimation and Analysis by M. Xie and G.Y. Hong
29. Bayesian Analysis for Software Reliability Data by J.A. Achcar
30. Direct Graphical Estimation for the Parameters in a Three-Parameter Weibull Distribution by P.R. Nelson and K.B. Kulasekera
31. Bayesian and Frequentist Methods in Change-Point Problems by N. Ebrahimi and S.K. Ghosh
32. The Operating Characteristics of Sequential Procedures in Reliability by S. Zacks
33. Simultaneous Selection of Extreme Populations from a Set of Two-Parameter Exponential Populations by K. Hussein and S. Panchapakesan

Volume 21. Stochastic Processes: Modelling and Simulation
Edited by D.N. Shanbhag and C.R. Rao
2003 xxviii + 1002 pp.

1. Modelling and Numerical Methods in Manufacturing System Using Control Theory by E.K. Boukas and Z.K. Liu
2. Models of Random Graphs and their Applications by C. Cannings and D.B. Penman
3. Locally Self-Similar Processes and their Wavelet Analysis by J.E. Cavanaugh, Y. Wang and J.W. Davis
4. Stochastic Models for DNA Replication by R. Cowan
5. An Empirical Process with Applications to Testing the Exponential and Geometric Models by J.A. Ferreira
6. Patterns in Sequences of Random Events by J. Gani
7. Stochastic Models in Telecommunications for Optimal Design, Control and Performance Evaluation by N. Gautam
8. Stochastic Processes in Epidemic Modelling and Simulation by D. Greenhalgh
9. Empirical Estimators Based on MCMC Data by P.E. Greenwood and W. Wefelmeyer
10. Fractals and the Modelling of Self-Similarity by B.M. Hambly
11. Numerical Methods in Queueing Theory by D. Heyman
12. Applications of Markov Chains to the Distribution Theory of Runs and Patterns by M.V. Koutras
13. Modelling Image Analysis Problems Using Markov Random Fields by S.Z. Li
14. An Introduction to Semi-Markov Processes with Application to Reliability by N. Limnios and G. Oprişan

15. Departures and Related Characteristics in Queueing Models by M. Manoharan, M.H. Alamatsaz and D.N. Shanbhag
16. Discrete Variate Time Series by E. McKenzie
17. Extreme Value Theory, Models and Simulation by S. Nadarajah
18. Biological Applications of Branching Processes by A.G. Pakes
19. Markov Chain Approaches to Damage Models by C.R. Rao, M. Albassam, M.B. Rao and D.N. Shanbhag
20. Point Processes in Astronomy: Exciting Events in the Universe by J.D. Scargle and G.J. Babu
21. On the Theory of Discrete and Continuous Bilinear Time Series Models by T. Subba Rao and Gy. Terdik
22. Nonlinear and Non-Gaussian State-Space Modeling with Monte Carlo Techniques: A Survey and Comparative Study by H. Tanizaki
23. Markov Modelling of Burst Behaviour in Ion Channels by G.F. Yeo, R. K. Milne, B.W. Madsen, Y. Li and R.O. Edeson

Volume 22. Statistics in Industry
Edited by R. Khattree and C.R. Rao
2003 xxi + 1150 pp.

1. Guidelines for Selecting Factors and Factor Levels for an Industrial Designed Experiment by V. Czitrom
2. Industrial Experimentation for Screening by D.K.J. Lin
3. The Planning and Analysis of Industrial Selection and Screening Experiments by G. Pan, T.J. Santner and D.M. Goldsman
4. Uniform Experimental Designs and their Applications in Industry by K.-T. Fang and D.K.J. Lin
5. Mixed Models and Repeated Measures: Some Illustrative Industrial Examples by G.A. Milliken
6. Current Modeling and Design Issues in Response Surface Methodology: GLMs and Models with Block Effects by A.I. Khuri
7. A Review of Design and Modeling in Computer Experiments by V.C.P. Chen, K.-L. Tsui, R.R. Barton and J.K. Allen
8. Quality Improvement and Robustness via Design of Experiments by B.E. Ankenman and A.M. Dean
9. Software to Support Manufacturing Experiments by J.E. Reece
10. Statistics in the Semiconductor Industry by V. Czitrom
11. PREDICT: A New Approach to Product Development and Lifetime Assessment Using Information Integration Technology by J.M. Booker, T.R. Bement, M.A. Meyer and W.J. Kerscher III
12. The Promise and Challenge of Mining Web Transaction Data by S.R. Dalal, D. Egan, Y. Ho and M. Rosenstein

13. Control Chart Schemes for Monitoring the Mean and Variance of Processes Subject to Sustained Shifts and Drifts by Z.G. Stoumbos, M.R. Reynolds Jr and W.H. Woodall
14. Multivariate Control Charts: Hotelling T^2, Data Depth and Beyond by R.Y. Liu
15. Effective Sample Sizes for T^2 Control Charts by R.L. Mason, Y.-M. Chou and J.C. Young
16. Multidimensional Scaling in Process Control by T.F. Cox
17. Quantifying the Capability of Industrial Processes by A.M. Polansky and S.N.U.A. Kirmani
18. Taguchi's Approach to On-line Control Procedure by M.S. Srivastava and Y. Wu
19. Dead-Band Adjustment Schemes for On-line Feedback Quality Control by A. Luceño
20. Statistical Calibration and Measurements by H. Iyer
21. Subsampling Designs in Industry: Statistical Inference for Variance Components by R. Khattree
22. Repeatability, Reproducibility and Interlaboratory Studies by R. Khattree
23. Tolerancing – Approaches and Related Issues in Industry by T.S. Arthanari
24. Goodness-of-fit Tests for Univariate and Multivariate Normal Models by D.K. Srivastava and G.S. Mudholkar
25. Normal Theory Methods and their Simple Robust Analogs for Univariate and Multivariate Linear Models by D.K. Srivastava and G.S. Mudholkar
26. Diagnostic Methods for Univariate and Multivariate Normal Data by D.N. Naik
27. Dimension Reduction Methods Used in Industry by G. Merola and B. Abraham
28. Growth and Wear Curves by A.M. Kshirsagar
29. Time Series in Industry and Business by B. Abraham and N. Balakrishna
30. Stochastic Process Models for Reliability in Dynamic Environments by N.D. Singpurwalla, T.A. Mazzuchi, S. Özekici and R. Soyer
31. Bayesian Inference for the Number of Undetected Errors by S. Basu

Volume 23. Advances in Survival Analysis
Edited by N. Balakrishnan and C.R. Rao
2003 xxv + 795 pp.

1. Evaluation of the Performance of Survival Analysis Models: Discrimination and Calibration Measures by R.B. D'Agostino and B.-H. Nam
2. Discretizing a Continuous Covariate in Survival Studies by J.P. Klein and J.-T. Wu
3. On Comparison of Two Classification Methods with Survival Endpoints by Y. Lu, H. Jin and J. Mi
4. Time-Varying Effects in Survival Analysis by T.H. Scheike
5. Kaplan–Meier Integrals by W. Stute
6. Statistical Analysis of Doubly Interval-Censored Failure Time Data by J. Sun
7. The Missing Censoring-Indicator Model of Random Censorship by S. Subramanian
8. Estimation of the Bivariate Survival Function with Generalized Bivariate Right Censored Data Structures by S. Keleş, M.J. van der Laan and J.M. Robins

9. Estimation of Semi-Markov Models with Right-Censored Data by O. Pons
10. Nonparametric Bivariate Estimation with Randomly Truncated Observations by Ü. Gürler
11. Lower Bounds for Estimating a Hazard by C. Huber and B. MacGibbon
12. Non-Parametric Hazard Rate Estimation under Progressive Type-II Censoring by N. Balakrishnan and L. Bordes
13. Statistical Tests of the Equality of Survival Curves: Reconsidering the Options by G.P. Suciu, S. Lemeshow and M. Moeschberger
14. Testing Equality of Survival Functions with Bivariate Censored Data: A Review by P.V. Rao
15. Statistical Methods for the Comparison of Crossing Survival Curves by C.T. Le
16. Inference for Competing Risks by J.P. Klein and R. Bajorunaite
17. Analysis of Cause-Specific Events in Competing Risks Survival Data by J. Dignam, J. Bryant and H.S. Wieand
18. Analysis of Progressively Censored Competing Risks Data by D. Kundu, N. Kannan and N. Balakrishnan
19. Marginal Analysis of Point Processes with Competing Risks by R.J. Cook, B. Chen and P. Major
20. Categorical Auxiliary Data in the Discrete Time Proportional Hazards Model by P. Slasor and N. Laird
21. Hosmer and Lemeshow type Goodness-of-Fit Statistics for the Cox Proportional Hazards Model by S. May and D.W. Hosmer
22. The Effects of Misspecifying Cox's Regression Model on Randomized Treatment Group Comparisons by A.G. DiRienzo and S.W. Lagakos
23. Statistical Modeling in Survival Analysis and Its Influence on the Duration Analysis by V. Bagdonavičius and M. Nikulin
24. Accelerated Hazards Model: Method, Theory and Applications by Y.Q. Chen, N.P. Jewell and J. Yang
25. Diagnostics for the Accelerated Life Time Model of Survival Data by D. Zelterman and H. Lin
26. Cumulative Damage Approaches Leading to Inverse Gaussian Accelerated Test Models by A. Onar and W.J. Padgett
27. On Estimating the Gamma Accelerated Failure-Time Models by K.M. Koti
28. Frailty Model and its Application to Seizure Data by N. Ebrahimi, X. Zhang, A. Berg and S. Shinnar
29. State Space Models for Survival Analysis by W.Y. Tan and W. Ke
30. First Hitting Time Models for Lifetime Data by M.-L.T. Lee and G.A. Whitmore
31. An Increasing Hazard Cure Model by Y. Peng and K.B.G. Dear
32. Marginal Analyses of Multistage Data by G.A. Satten and S. Datta
33. The Matrix-Valued Counting Process Model with Proportional Hazards for Sequential Survival Data by K.L. Kesler and P.K. Sen
34. Analysis of Recurrent Event Data by J. Cai and D.E. Schaubel
35. Current Status Data: Review, Recent Developments and Open Problems by N.P. Jewell and M. van der Laan

36. Appraisal of Models for the Study of Disease Progression in Psoriatic Arthritis by R. Aguirre-Hernández and V.T. Farewell
37. Survival Analysis with Gene Expression Arrays by D.K. Pauler, J. Hardin, J.R. Faulkner, M. LeBlanc and J.J. Crowley
38. Joint Analysis of Longitudinal Quality of Life and Survival Processes by M. Mesbah, J.-F. Dupuy, N. Heutte and L. Awad
39. Modelling Survival Data using Flowgraph Models by A.V. Huzurbazar
40. Nonparametric Methods for Repair Models by M. Hollander and J. Sethuraman